AF794862

Nuclear Reactor Physics and Engineering

# Nuclear Reactor Physics and Engineering

*John C. Lee*
University of Michigan
Ann Arbor, Michigan

*Second Edition*

WILEY

Copyright © 2025 by John Wiley & Sons, Inc. All rights reserved, including rights for text and data mining and training of artificial technologies or similar technologies.

Published by John Wiley & Sons, Inc., Hoboken, New Jersey.

Published simultaneously in Canada.

No part of this publication may be reproduced, stored in a retrieval system, or transmitted in any form or by any means, electronic, mechanical, photocopying, recording, scanning, or otherwise, except as permitted under Section 107 or 108 of the 1976 United States Copyright Act, without either the prior written permission of the Publisher, or authorization through payment of the appropriate per-copy fee to the Copyright Clearance Center, Inc., 222 Rosewood Drive, Danvers, MA 01923, (978) 750-8400, fax (978) 750-4470, or on the web at www.copyright.com. Requests to the Publisher for permission should be addressed to the Permissions Department, John Wiley & Sons, Inc., 111 River Street, Hoboken, NJ 07030, (201) 748-6011, fax (201) 748-6008, or online at http://www.wiley.com/go/permission.

Trademarks: Wiley and the Wiley logo are trademarks or registered trademarks of John Wiley & Sons, Inc. and/or its affiliates in the United States and other countries and may not be used without written permission. All other trademarks are the property of their respective owners. John Wiley & Sons, Inc. is not associated with any product or vendor mentioned in this book.

Limit of Liability/Disclaimer of Warranty: While the publisher and author have used their best efforts in preparing this book, they make no representations or warranties with respect to the accuracy or completeness of the contents of this book and specifically disclaim any implied warranties of merchantability or fitness for a particular purpose. No warranty may be created or extended by sales representatives or written sales materials. The advice and strategies contained herein may not be suitable for your situation. You should consult with a professional where appropriate. Further, readers should be aware that websites listed in this work may have changed or disappeared between when this work was written and when it is read. Neither the publisher nor authors shall be liable for any loss of profit or any other commercial damages, including but not limited to special, incidental, consequential, or other damages.

For general information on our other products and services or for technical support, please contact our Customer Care Department within the United States at (800) 762-2974, outside the United States at (317) 572-3993 or fax (317) 572-4002.

Wiley also publishes its books in a variety of electronic formats. Some content that appears in print may not be available in electronic formats. For more information about Wiley products, visit our web site at www.wiley.com.

***Library of Congress Cataloging-in-Publication Data Applied for:***

Hardback ISBN: 9781394283552

Cover Design: Wiley
Cover Image: © Aitor Diago/Getty Images

Set in 9.5/12.5pt STIXTwoText by Straive, Chennai, India

# Contents

# List of Tables

# List of Figures

# Preface

The book has been developed to introduce undergraduate and graduate students in nuclear engineering, as well as practicing engineers, to basic concepts of nuclear reactor physics and applications of the concepts to the analysis, design, control, and operation of nuclear reactors. The basic concepts are discussed and the associated mathematical formulations presented with the understanding that the reader has solid background in differential equations and linear algebra. A focus is placed on the use of neutron diffusion theory, with a minimum use of the neutron transport equation, for the development of techniques for lattice physics and global reactor system studies. When the neutron transport equation is used, effort is made to stay with one-dimensional forms of the Boltzmann equation and Legendre polynomials, without invoking the full-blown three-dimensional Boltzmann equation and spherical harmonics. Recent developments in numerical algorithms, including the Krylov subspace method, and the MATLAB software, including the Simulink toolbox, are discussed for efficient studies of steady-state and transient reactor configurations. In addition, nuclear fuel cycle and associated economics analysis are presented together with the application of modern control theory to reactor operation. A self-contained derivation of fluid conservation equations is presented, together with relevant examples, so that the material could be used in a sequence of courses in nuclear reactor physics and engineering to cover thermal-hydraulic analysis of nuclear systems.

The overall structure of the book allows the coverage of fundamental concepts and tools necessary for nuclear reactor physics studies with the first half of the book up to Chapter 10, as it is usually done in a one-semester senior nuclear engineering course at the University of Michigan. Some of the remaining chapters of the book could be covered in a follow-up semester in the undergraduate curriculum or in graduate courses; Chapters 16 and 17 could likely be covered in separate courses dealing with nuclear reactor kinetics and reactor design analysis, respectively, The author sincerely hopes that the book will augment and update several excellent textbooks that have been used in the nuclear science and engineering curriculum in the United States and elsewhere for the first half century of nuclear energy development.

The material for the book has originated from several reactor physics and engineering courses that the author has taught over the past 45 years at the University of Michigan and also on a part-time basis at the Korea Advanced Institute of Science and Technology and Pohang University of Science and Technology. Some of the material also reflects industrial experience he gained through his early employments at Westinghouse Electric Corporation and General Electric Company. Selection of the topics and presentation of the material have greatly benefited from discussions with the students in and outside the classroom and the author wishes to express appreciation to a generation of bright, young students at all of the three institutions.

In addition, the author acknowledges help and support from a number of current and former students including Matthew Krupcale and Junjie Guo. He offers thanks for help and encouragement from his mentors, the late Professors Thomas Pigford, John King, and William Kerr, as well as his colleagues including William Martin, Ziya Akcasu, James Duderstadt, the late Professor Glenn Knoll, Thomas Downar, Won Sik Yang, Volkan Seker, Frederick Buckman, and David Wehe. Special appreciation is expressed to the late Professor Byung Ho Lee for introducing the author to the beauty and challenges of nuclear reactor physics and to Professor Hans M. Mark for providing opportunities for graduate study at Berkeley during the exciting days of the free speech movement. Finally, he offers thanks to his wife Theresa and daughter Nina for their loving care and support.

September 2019

*John C. Lee*
*Ann Arbor, Michigan*

# Preface to the Second Edition

This second edition of the book includes updated coverages on Generation IV and modular nuclear power plants, used nuclear fuel management, and nuclear energy economics, reflecting renewed interest world-wide in developing clean and safe nuclear energy for the 21st century.

June 2024

*John C. Lee*
*Ann Arbor, Michigan*

## Permissions and Copyrights

Many figures and tables in this book have been reproduced from copyrighted sources. Permission from the publishers and authors for the use of the material is gratefully acknowledged. Some of the sources are directly identified in captions and footnotes, while many others are cited by alphanumeric references. Citations for these sources are listed below:

*Convective Boiling and Condensation*, J.G. Collier.
Copyright © 1972 by McGraw-Hill. Figure 13.17.

*Nuclear Reactor Kinetics*, 2nd edition, M. Ash.
Copyright © 1979 by McGraw-Hill. Figure 8.16.

*Mathematical Methods for Physicist: A Comprehensive Guide*, 7th ed., G.B. Arfken, J.J. Weber, and F.E. Harris.
Copyright © 2013 by Academic Press. Figures C.1, C.2, C.3, and C.4.

*Nuclear Science and Engineering.*
Copyright ©1956, 1957, 1980, 1982, 1989, 1993, 2015, by Taylor & Francis Group. Figures 9.8, 10.5, 16.5, 16.6, 16.14, 16.15, 16.16, 16.19, 16.20, 16.23.

*Nuclear Technology.*
Copyright ©1978 by Taylor & Francis Group. Figures 10.3, 10.4, and 12.6.

*Science.*
Copyright ©2022 by American Association for the Advancement of Science, 1200 New York Avenue NW, Washington, DC 20005. Figure 12.30.

*SKB.*
Copyright ©1997 by Svensk Kärnbränslehantering AB, Evenemangsgatan 13, Box 3091, SE-169 03 Solna, Sweden. Figure 12.22.

A number of figures were also obtained from publications of various US government agencies and laboratories and other open-source publications: Figures 1.1, 1.2, 1.3, 1.4, 1.6, 1.7, 1.8, 1.9, 1.10, 1.11, 1.12, 1.13, 1.14, 1.15, 1.16, 1.17,1.18, 1.19, 1.20, 11.8, 12.16, 12.17, 12.18, 12.19, 12.20, 12.23, 12.24, 12.25, 12.26, 12.27, 12.28, 12.29, 12.31, 13.10, 13.26, 14.4, and 16.11.

# About the Companion Website

This book is accompanied by a companion website:

www.wiley.com/go/LeeNRPE2

The website includes:

## Repository for Instructor Resources

Solution manual and collection of figures and tables are available to instructors who adopt this book for their course. Please register in the Wiley repository www.wiley.com/go/LeeNRPE2 to access the files.

# 1

# Nuclear Power Plants

As of April 2024, 94 nuclear power plants provide an installed electrical generating capacity of 99 GWe and account for about 20% of electricity generated in the United States, while 436 nuclear power plants provide an installed capacity of 393 GWe worldwide. All of the nuclear power plants in the United States and 90% worldwide utilize light-water cooled reactors (LWRs), which may be grouped into pressurized water reactors (PWRs) and boiling water reactors (BWRs). About 70 $\sim$ 80% of LWRs operating in the United States and around the world are PWRs. We begin this introductory chapter with Section 1.1 covering a brief history and the current status of nuclear power plants (NPPs) in the United States and elsewhere. This is followed by Sections 1.2–1.4, providing an introduction to the basic operating features of the reactor core and nuclear steam supply system (NSSS) that produce heat and steam through the self-sustaining fission process. The focus is primarily on LWRs, which are expected to serve as the key reactor type for the foreseeable future. Advanced reactor concepts including small and modular reactor (SMR) designs under development are discussed in Section 1.5.

## 1.1 History and Current Status of Nuclear Power Plants

The development and deployment of full-scale NPPs began essentially with the announcement in 1964 that the 625-MWe Oyster Creek Power Plant would be built in Forked River, New Jersey, with an expected capital cost of ~$100/kWe. The capital cost would make the plant competitive with coal-fired power plants. The plant went into commercial operation in 1969. Although there were a few smaller NPPs that began operation earlier, the prospect of economically competitive LWR plants made many utility companies rush to order NPPs during the next decade until the 1978 accident at the Three Mile Island (TMI) Unit 2 Plant. Approximately 300 NPP orders were canceled during the several years

*Nuclear Reactor Physics and Engineering*, Second Edition. John C. Lee.
© 2025 John Wiley & Sons, Inc. Published 2025 by John Wiley & Sons, Inc.
Companion website: www.wiley.com/go/LeeNRPE2

following the TMI accident due to public concerns over the safety of NPPs and the difficulty encountered in completing construction of the plants on schedule and within the initial budget estimates.

Approximately 100 NPPs were constructed by the 1980s, and after that, no new NPPs went into operation for two decades until 2016, when the Watts Bar Unit 2 (WB2) Plant began commercial operation. The WB2 project in fact started in 1972 and was suspended in 1988 when the growth in power demand began to decline for the Tennessee Valley Authority. Its sister unit Watts Bar Unit 1 began operation in 1996 and was the last nuclear plant to do so in the United States until the WB2 Plant. During the decades following the TMI Unit 2 accident, through improved operator training and by installing back-fit safety features, the fleet of 100 NPPs provided approximately 20% of electricity in the United States at a competitive generation cost. Beginning in the late 1970s, France adopted the PWR technology and constructed a fleet of economically competitive NPPs over a period of two decades, with 58 plants providing ~75% of electricity for the country in 2018. Several other countries, including Japan, Korea, China, and Russia, also currently operate fleets of nuclear plants with power ratings in the range of 500 ~ 1200 MWe. The current fleets of NPPs operating in the United States and elsewhere are generally known as Generation II plants, while much smaller units, including the 60-MWe Shippingport PWR plant, 200-MWe Dresden Unit 1 BWR plant, and 61-MWe Fermi Unit 1 sodium-cooled fast reactor (SFR) plant are known as Generation I plants. Figure 1.1 displays the evolution of the NPP generations.

Illustrated in Figure 1.1 are advanced LWRs including the Advanced BWR (ABWR), System 80+, and AP600 designs classified as Generation III plants, together with the evolutionary Generation III+ plants that offer improved safety

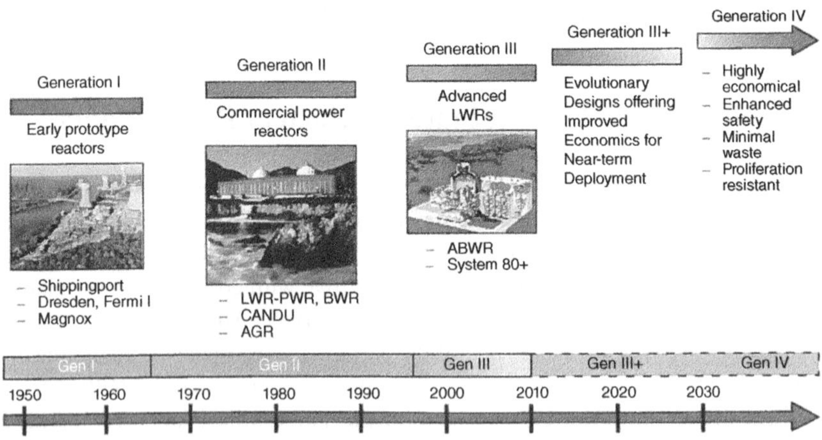


**Figure 1.1** Nuclear power plant evolution. *Source*: US Department of Energy [DOE02].

features and economics. Primary examples of Generation III+ plants are the 1.1-GWe AP1000 PWR,[1] 1.4-GWe APR-1400, 1.5-GWe ESBWR, and 1.6-GWe EPR plants, some of which began operation during the last several years. Beginning in the early 2000s, effort was initiated under the aegis of the US Department of Energy (DOE) to develop advanced reactor designs, labeled as Generation IV plants, which could provide enhanced safety and economics of power generation. Beginning with late 2010s, significant effort has been underway to develop SMRs that could allow timely construction and reduced capital outlay per unit, while accomplishing key objectives of Generation IV plants for efficient fuel utilization and used fuel management.

## 1.2 Basic Features of Nuclear Power Plants

In the bulk of NPPs, energy released in the fission process is deposited as heat energy initially in fuel pins enclosed in metallic tubes. This energy is eventually transmitted through heat conduction and convection to fluid circulating through the reactor core, which is located within a steel pressure vessel, with wall thickness of $0.17 \sim 0.2$ m. In the case of LWRs, water is used as the circulating fluid known as the *reactor coolant*. In gas-cooled reactors, pressurized gases, e.g. helium or carbon dioxide, may serve the role of reactor coolant, while circulating liquid metal, e.g. sodium or lead, picks up the heat in liquid-metal cooled reactors. The CANDU (Canadian Deuterium Uranium) reactor may be cooled either with heavy or light water.

Once the fission energy is picked up by the reactor coolant in the PWR, the coolant circulates through a heat exchanger, where the heat is transferred from the primary loop to the secondary loop, as illustrated schematically in Figure 1.2. The heat exchanger in the PWR is known as a *steam generator*, since the circulating fluid in the secondary heat-transfer loop is allowed to boil and the resulting steam is separated from the liquid. The steam is used to turn the steam turbines and electrical generators, thereby producing electricity. Included in the schematics in Figure 1.2 are a pressurizer, which is essentially an extension of the primary loop to regulate the pressure of the primary system, and a reactor coolant pump, which circulates the reactor coolant. The circulating fluid in the secondary heat-transfer loop is known as *feedwater* and the steam that exits from the turbines is condensed into feedwater in the condenser and associated machinery. The feedwater system reheats the condensed steam and regulates

---

1 AP1000 is a trademark or registered trademark of Westinghouse Electric Company LLC, its affiliates and/or its subsidiaries in the United States of America and may be registered in other countries throughout the world. All rights reserved. Unauthorized use is strictly prohibited.

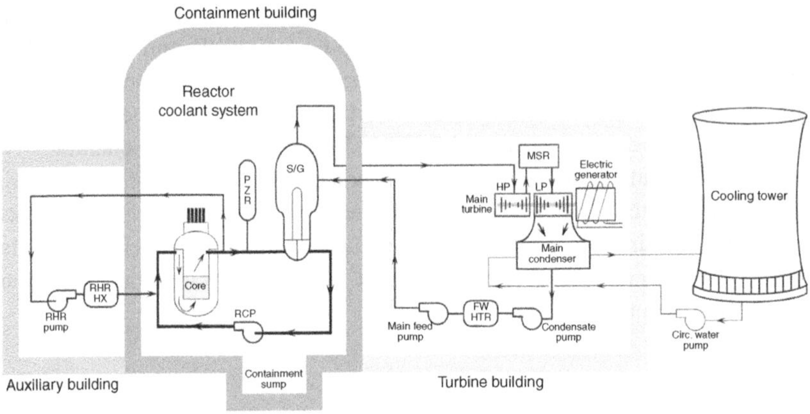


**Figure 1.2** Overall layout of a PWR plant. *Source*: US Nuclear Regulatory Commission [NRC08]/John Wiley & Sons.

the temperature of the feedwater before it recirculates into the secondary side of the steam generator. The heat transferred from the steam into the condenser is eventually rejected to the atmosphere through a cooling pond or cooling tower in a tertiary loop, which is the final heat-transfer loop shown in Figure 1.2.

In a PWR plant, the reactor pressure vessel (RPV), coolant pump, steam generator, and pressurizer are enclosed in a concrete containment structure, built with an inner steel liner. The plant components located within the containment building are collectively known as the nuclear steam supply system (NSSS), while those located outside the containment are generally known as the balance of plant (BOP). Particular attention is given to the reliability and integrity of NSSS components, which are subject to specific regulations and oversight by the US Nuclear Regulatory Commission. In modern BWRs employing a direct cycle, coolant water circulating in the primary loop is allowed to boil inside the reactor vessel. Steam is separated from liquid water in the reactor vessel and is used to turn the turbo-generators in much the same way steam extracted from the steam generators in PWRs is used to generate electricity. Incorporation of a direct cycle in BWRs eliminates a heat-transfer loop and allows for simplifications in the plant system design. Production of a significant amount of steam within the reactor core, however, requires a number of special considerations for the design and analysis of reactor core and fuel elements in BWRs.

## 1.3 Pressurized Water Reactor Systems

Figure 1.3, borrowed from a PWR training manual [NRC08], presents the overall PWR system with a focus on the engineered safety features (ESFs) [Lee17] provided to handle operational transients and accident scenarios. For the primary

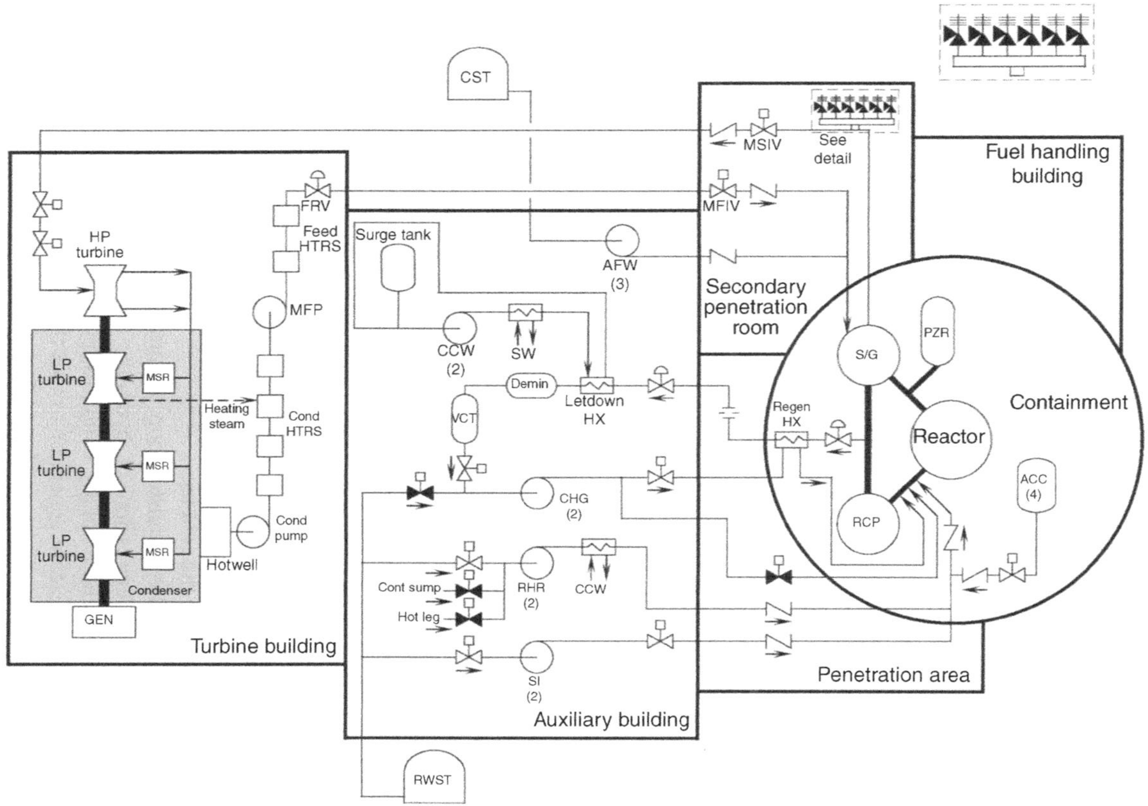


**Figure 1.3** Schematic diagram of a PWR plant. *Source*: US Nuclear Regulatory Commission [NRC08]/John Wiley & Sons.

loop, the charging and letdown lines connected to the cold and hot legs, respectively, of the reactor coolant system (RCS), and the safety injection (SI) pump, reactor coolant pump (RCP), and accumulator connected to the cold leg are indicated. The diagram also illustrates that the accumulator discharge line has a check valve with an arrow pointing in the flow direction and a motor-operated valve (MOV) in a normally open position. Note also that the discharge from the accumulator is aided by nitrogen gas pressure. Not shown in Figure 1.3 is the boron injection tank (BIT), through which the charging pump could be routed for the injection of boric acid to the cold leg. Together with the pressurizer discussed earlier, Figure 1.3 illustrates that the residual heat removal (RHR) system, delivering coolant to the cold leg, may take suction from the refueling water storage tank (RWST) or containment sump as well as from the RCS hot leg. After the reactor is shut down, the primary system is cooled by the RHR system, which removes the heat produced through the decay of fission products. The RHR heat exchanger in turn dissipates heat through the component cooling water (CCW) heat exchanger. Note also that the CCW heat exchanger itself is cooled by the service water system. Thus, similar to the three-level heat-transfer loop structure of the plant in the normal operating mode, the CCW and service water systems serve as the secondary and tertiary heat-transfer loops, respectively, for the RHR system. The two MOVs connecting the RHR charging line to the containment sump and RCS hot leg are blackened, indicating that they are normally in a closed position.

We note also in Figure 1.3 that, as part of the primary loop, the SI pump takes suction from the RWST via an MOV. The valve is shown in an open position, which is not illustrative of a normal operating mode. The charging pump takes suction normally from the chemical and volume control (CVC) tank (labeled VCT) but may switch to the RWST as necessary. The regenerative and letdown heat exchangers coupled to the demineralizer provide the means to cool down the primary coolant that is discharged from the RCS hot leg and returned to the cold leg via the charging line. The demineralizer and CVC system also serve to filter out unwanted contaminants in the coolant water and maintain the desired soluble boron concentration in the primary loop. If it becomes necessary to increase the soluble boron concentration in an accident situation, the charging flow is switched through the BIT before it is returned to the cold leg.

For the secondary heat-transfer loop, Figure 1.3 shows that the main feedwater line, with a succession of hydraulic or air-operated valves (AOVs) through the turbine and auxiliary buildings, provides feedwater to the shell of a tube-and-shell-type steam generator so that the feedwater picks up heat from a cluster of U-shaped tubes through which the primary coolant circulates. The main steam line delivers hot steam from the secondary or shell side of the steam generator to a series of high-pressure (HP) and low-pressure (LP) turbines in the turbine building. The exhaust steam discharged from the final LP turbine is sent to the hotwell of the steam condenser, from which the condensate and feedwater pumps deliver the condensed water through a series of components in the condensate and feedwater systems to the steam generator. Finally, the auxiliary feedwater (AFW) pump takes suction from the condensate storage tank (CST). Note also a series of main steam isolation valves (MSIVs) outside the containment but upstream of the pipe tunnel in the auxiliary building.

Internal structures of the RPV are illustrated in Figure 1.4, including (a) fuel assemblies, (b) inlet and outlet coolant nozzles, (c) clustered control rod

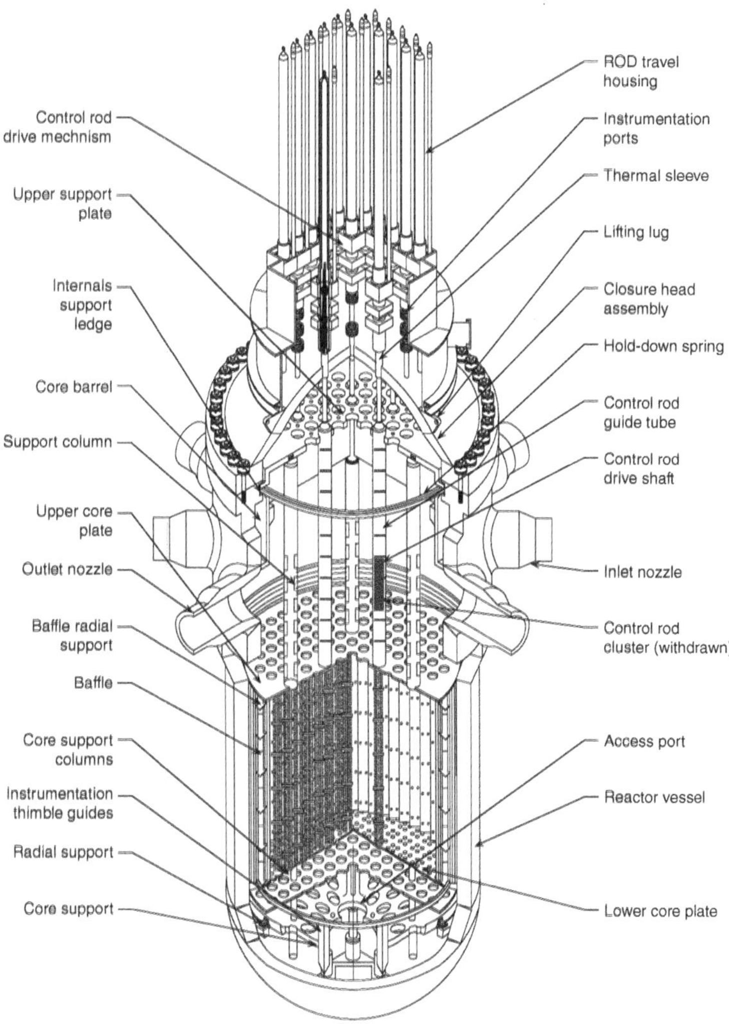


**Figure 1.4** Cutaway view of a PWR pressure vessel illustrating key components including fuel elements and supporting structures. *Source*: US Nuclear Regulatory Commission [NRC08]/John Wiley & Sons.

assemblies, (d) lower and upper core plates, (e) instrumentation thimble guides, and (e) core barrel and baffle. The number of inlet and outlet coolant nozzles depends naturally on the number of heat-transfer loops, each comprising a reactor coolant pump and steam generator. Located inside the RPV of a PWR plant, illustrated schematically in Figure 1.5a, is a reactor core comprising 150 ~ 200 fuel assemblies, surrounded by steel plates that form the flow baffle. A cylindrical barrel separates the upward flow of coolant through the core from the inlet coolant flowing downward in the annulus formed by the barrel and pressure vessel. The core baffle provides structural support to the fuel elements and channels the coolant water to flow primarily through the heat-producing fuel elements. Neutron shield panels are located in the lower portion of the vessel to attenuate high-energy gamma rays and neutrons leaking out of the core, thereby reducing the radiation-induced embrittlement of the vessel. Specimens to monitor radiation exposure of the vessel are also indicated in Figure 1.5a. Figure 1.5b illustrates a typical PWR fuel assembly with an array of approximately 250 fuel rods, each consisting of a stack of $UO_2$ pellets loaded in zirconium-alloy tubes with a diameter of 9 ~ 12 mm and an effective fuel length of 3.6 ~ 4.3 m. Other prominent structures for the fuel assembly include the spacer grids and clustered control absorbers, usually known as the rod cluster control (RCC) elements, inserted into the top of the fuel assembly.

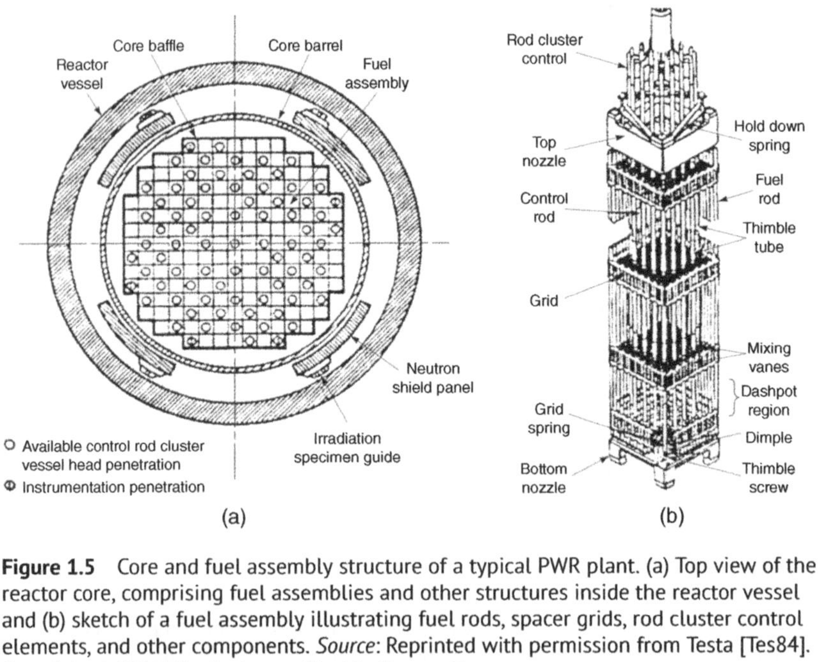


**Figure 1.5** Core and fuel assembly structure of a typical PWR plant. (a) Top view of the reactor core, comprising fuel assemblies and other structures inside the reactor vessel and (b) sketch of a fuel assembly illustrating fuel rods, spacer grids, rod cluster control elements, and other components. *Source*: Reprinted with permission from Testa [Tes84]. Copyright ©1984 Westinghouse Electric Corporation.

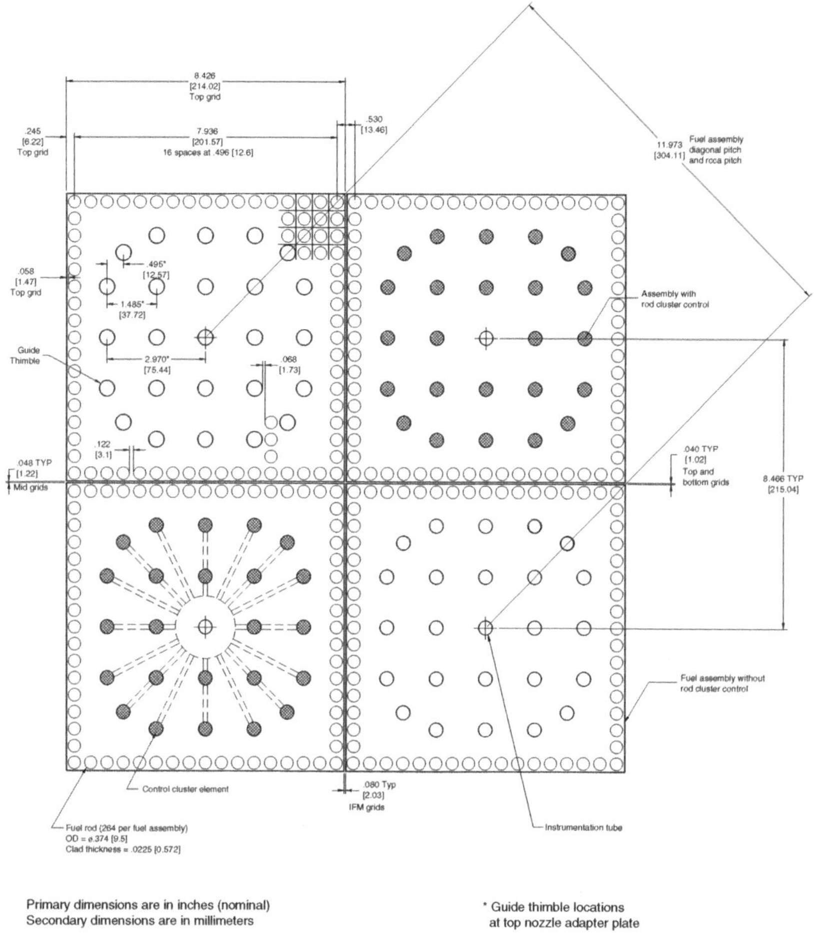


**Figure 1.6** Cross-section view of PWR fuel assemblies for the AP1000 design. *Source*: Adapted from Hone et al. [Hon12].

A cross-section view of a set of four fuel assemblies for the AP1000 design [Hon12, Lee17] is illustrated in Figure 1.6. The top-left and bottom-right assemblies indicate fuel elements with guide tubes or thimbles, while the top-right and bottom-left assemblies indicate RCC assemblies with the control elements withdrawn and inserted, respectively. The guide tubes are used to accommodate burnable absorber elements to control both the spatial power distribution and neutron reaction rates. The fuel element design features 254 fuel rods per assembly, with a rod diameter of 9.5 mm. At the center of each assembly lies an instrumentation tube, which accommodates either fixed or movable incore detectors. It should be noted that PWR fuel assemblies are structurally supported by a set of spacer grids distributed over the length of fuel rods and bottom and top nozzles, allowing for coolant flow freely moving around the fuel rods as it flows from the bottom to the top of the core. Any reactor physics analysis of a PWR core requires a detailed representation of the heterogeneous structure of each fuel assembly as well as the RCC and burnable absorber elements loaded into the guide thimbles. Methods to represent various structures in the reactor core will be a major focus of discussion in the subsequent chapters of the book.

The AP1000 design, as a prime example of the Generation III+ plants being deployed in the United States and elsewhere, features a number of enhanced safety features [Hon12, Lee17] as well as advanced fuel assembly designs accommodating load-follow maneuver capability and improved fuel cycle characteristics. No pumps, fans, diesel generators, chillers, or other rotating machinery are required for the safety systems in normal operating conditions and postulated accidents. Enhanced passive safety features include a large-volume pressurizer, obviating the need for a power-operated relief valve (PORV), and an in-containment refueling water storage tank (IRWST) providing gravity-driven coolant water for 72 hr via a squib-actuated automatic depressurization system (ADS). Length of the fuel elements is increased from traditional 3.66 m (12 ft) to 4.27 m (14 ft) with an advanced mechanical shim (MSHIM) control system allowing for an efficient control of power distribution and power output of the core, as discussed in Chapter 16. As an enhanced two-loop plant, the AP1000 core layout otherwise maintains most of the traditional PWR features.

Four PWR plants featuring the AP1000 design started full-power operation in 2018 and 2019 in China. Two AP1000 units at the Vogtle Plant, after considerable delays in the construction schedule and associated increases in the construction cost, have finally been connected to the power grid in 2024. Likewise, the 1.6-GWe EPR unit at the Olkiluoto Plant also started producing electricity in 2023. Utility companies have been able to expedite the process of combined construction and operating license applications with the certified design. The AP1000 design certification process required more than two decades of development, starting with its predecessor AP600, and a cumulative expenditure of more than $400 million by Westinghouse Electric Company.

## 1.4 Boiling Water Reactor Systems

The schematic diagram in Figure 1.7 presents the overall BWR plant layout [NRC08, Lee17] starting with the reactor vessel on the far left of the figure. The main difference between the BWR layout and that of the PWR system in Figure 1.3 is the obvious lack of the steam-generator and presence of the steam-separation equipment located in the upper region of the reactor vessel.

Primary coolant pumps, which are called *recirculation pumps* in BWR plants, are illustrated, together with the control rod drives located at the bottom of the vessel. The control rods, in the shape of cruciform blades, are inserted through the bottom head of the vessel to accommodate the presence of the steam separation equipment in the upper region of the vessel. An equally important reason for the bottom-entry control blades is to control the axial power distribution, which has to be shaped and controlled allowing for sharp variations in the coolant density due to boiling in the fuel region. The coolant water cleanup system, featuring a filtration and demineralization system, cleanup pumps, and heat exchangers, is coupled to the recirculation pumps.

The BOP structure for BWR plants is fairly similar to that of PWR plants, with one obvious difference due to the use of a direct steam cycle, which does not require steam generators. This implies that the steam is radioactive, and hence access to the turbine room has to be limited during operation. The connections between multiple stages of HP and LP turbines are indicated in Figure 1.7. We also note the MSIVs and safety relief valves in the steam line, inside, and outside the primary containment boundary. A number of AOVs as well as MOVs are noted in various flow paths. The RHR system serves as a normal shutdown cooling system and is cooled by the service water system in the RHR heat exchanger. As an alternate feedwater delivery system, reactor core isolation cooling (RCIC) pumps take suction from the CST and deliver feedwater in case of core isolation transients, involving the loss of feedwater coupled with the closure of main steam-isolation valves. The RHR system also provides the vessel head spray to the steam dome in the upper region of the reactor vessel above the steam-separation equipment.

The BWR containment structure is illustrated in Figure 1.8 for the Mark I system employed in Units 1 through 5 of the ill-fated Fukushima Daiichi NPPs. Noteworthy in the figure is the primary containment building in the shape of an inverted light bulb, known as the *drywell*, which houses the steel RPV with a wall thickness of $0.15 \sim 0.18$ m. The RPV is connected through relief valves to the pressure-suppression pool or *wetwell* in the shape of a torus as it is often called. The drywell consists of a 50-mm thick steel shell surrounded by $0.6 \sim 1.8$ m of reinforced concrete. When it becomes necessary to control the steam pressure within the reactor vessel, steam is discharged from the RPV to the wetwell, where steam is condensed, as part of the ADS. The drywell, wetwell, and other NSSS components are housed in a concrete structure serving as the secondary containment.

A cutaway view of BWR reactor vessel internal structures is presented in Figure 1.9, where two stages of steam-separation equipment above the core are clearly illustrated, together with the core spray and sparger lines. The core shroud, which provides the same function as the PWR core barrel, separates the downward flow of coolant in the downcomer from the upward coolant water

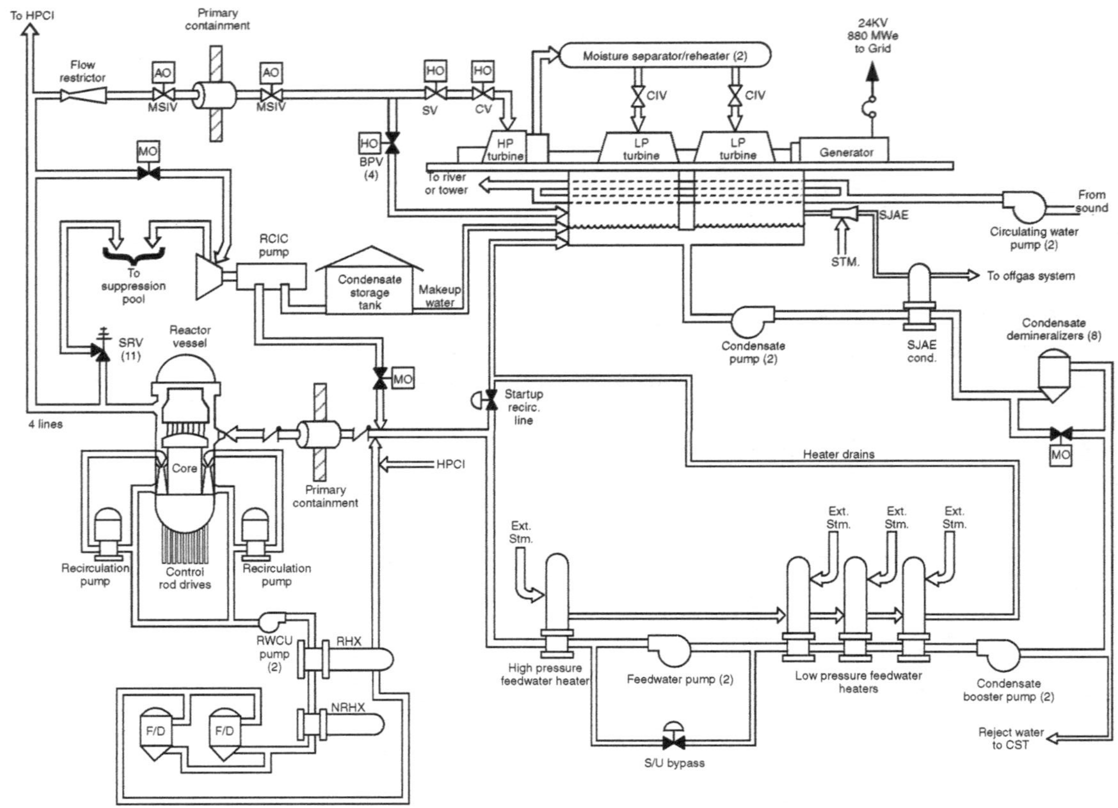


**Figure 1.7** Schematic diagram of a BWR plant. Abbreviations: BPV = bypass valve, CV = control valve, CBP = condensate booster pump. CP = condensate pump, F/D = filter demineralizer, HTX = heat exchanger, SRV = safety relief valve, SV = stop valve. *Source*: US Nuclear Regulatory Commission [NRC08].

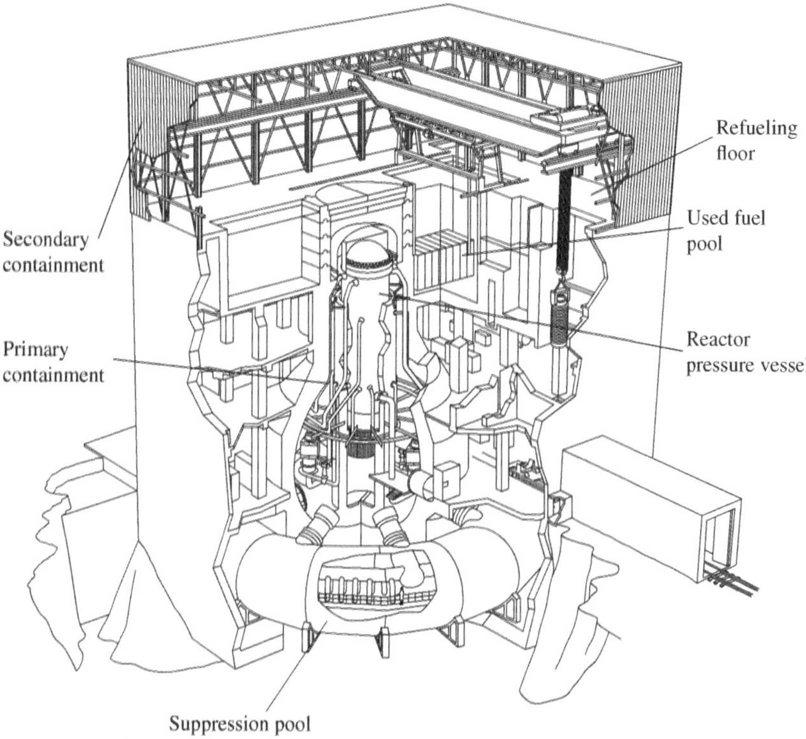


**Figure 1.8** Cutaway view of Mark I BWR containment structure. *Source*: Adapted from US Nuclear Regulatory Commission [NRC08].

flow through the core. In the direct-cycle BWR plant illustrated, feedwater is delivered directly to the feedwater sparger located above the core, mixed with recirculating water through jet pumps and pumped to the fuel region of the core via the recirculation pumps located within the drywell. Steam is separated from liquid in the upper region of the reactor vessel and delivered to steam turbines. Exhaust steam is condensed in the condenser and returned via the feedwater system to the core, closing the feedwater-steam loop for the BWR plant. The funnel-shaped jet pumps pick up the downward flow of liquid, separated from steam in the steam separator and dryer assemblies, and mixed with the feedwater delivered through the feedwater sparger and recirculating flow. The recirculation pumps located outside the RPV deliver the mixed flow of coolant through the downcomer and eventually upward through the core. The control rod drive and incore flux monitoring mechanisms are located under the reactor vessel.

In BWR plants, reactivity control is primarily achieved via control blades in the form of a cross, often known as the cruciform control blades, with the wings loaded

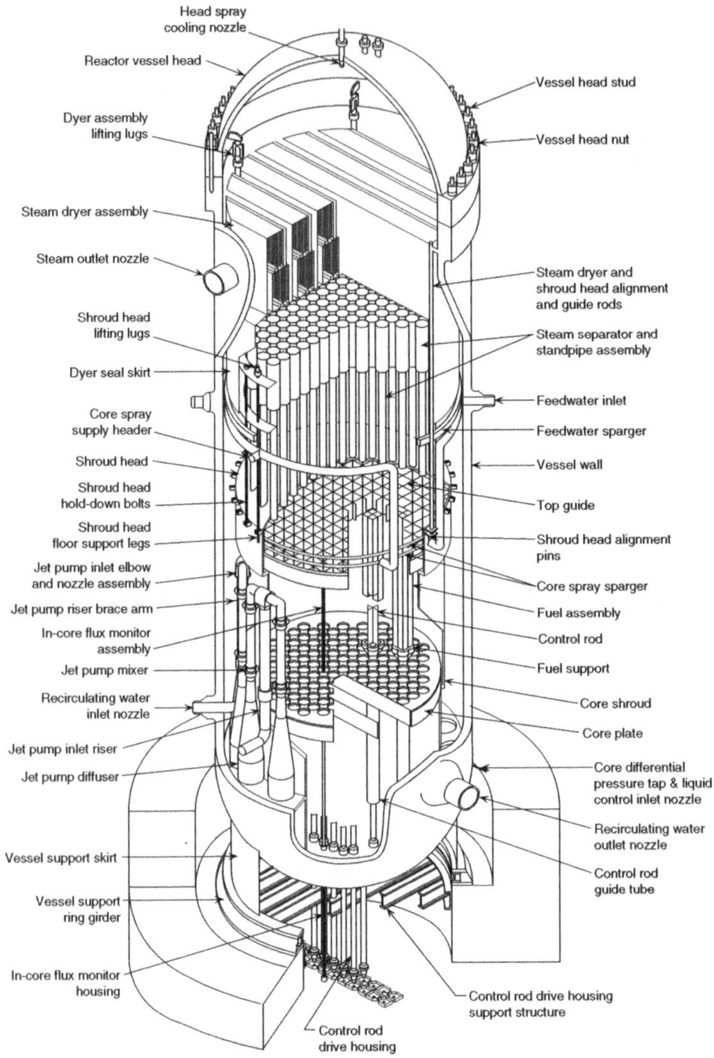


**Figure 1.9** Cutaway view of a BWR pressure vessel illustrating detailed coolant flow and core spray arrangement. *Source*: US Nuclear Regulatory Commission [NRC08].

with tubes of neutron absorbers. A typical arrangement for the cruciform blade inserted in the wide-wide (W-W) gap of a cluster of 2×2 BWR bundles, as they are often called, is presented in Figure 1.10. The diagonal opposite of the W-W corner is the narrow-narrow (N-N) corner, where an incore instrumentation tube marked *R* is located. Illustrated also are the Zircaloy channel box and tie rods

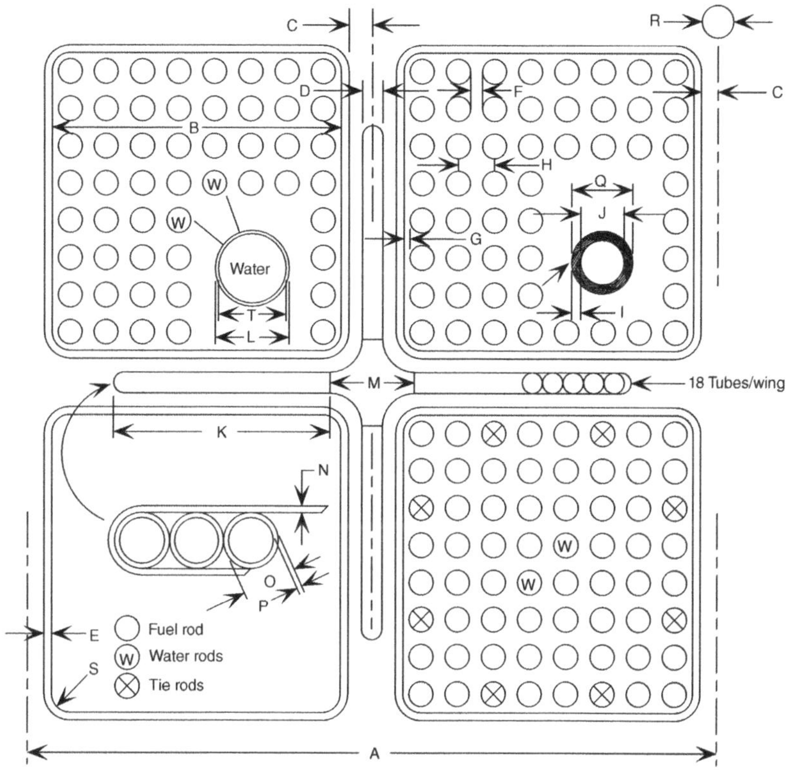


| DIM. IDENT. | A | B | C | D | E | F | G | H | I | J |
|---|---|---|---|---|---|---|---|---|---|---|
| DIM. INCHES | 12.0 | 5.215 | 0.2725 | 0.323 | 0.120 | 0.153 | 0.140 | 0.636 | 0.032 | 0.410 |

| DIM. IDENT. | K | L | M | N | O | P | Q | R | S | T |
|---|---|---|---|---|---|---|---|---|---|---|
| DIM. INCHES | 4.094 | 0.591 | 1.562 | 0.046 | 0.220 | 0.027 | 0.483 | 1.062 | 0.380 | 0.531 |

**Figure 1.10** BWR fuel bundle cluster illustrating the W-W and N-N gaps. *Source:* US Nuclear Regulatory Commission [NRC08].

introduced for structural support and water rods introduced for power distribution and reactivity control purposes. Coolant boiling takes place within the channel box in a BWR core, and the flow outside the box in the N-N and W-W gaps is essentially single-phase liquid. With a combination of two-phase flow within the channel box and single-phase flow outside the box, the average coolant density in a BWR core is maintained similar to that in a PWR core.

Boiling water reactor designs evolved over the years from the plant layout illustrated in Figure 1.7 to the ABWR design where the recirculation pumps are located inside the reactor vessel, thereby reducing the likelihood of loss of coolant accidents (LOCAs). The ABWR subsequently evolved into the 600-MWe Simplified BWR (SBWR) design that eliminated the recirculation pumps altogether, relying entirely on natural circulation cooling for normal and emergency operations. The Economic SBWR (ESBWR) [GEH14], with key safety features illustrated in Figure 1.11, increases the power rating to 1.55 GWe (4.5 GWt), with natural circulation cooling of the core achieved through the installation of a tall chimney and associated increase in the RPV height, combined with a decrease in the active fuel length from the conventional 3.67 m (12 ft) to 3.0 m. The increase in power output is obtained by increasing the number of fuel assemblies from 800 to 872 for the BWR/6 and SBWR designs, respectively, to 1,132 for the ESBWR. A large inventory of water and steam in the RPV, combined with passive safety features, eliminates safety-grade pumps and AC power for managing postulated accidents. The ESBWR safety-grade system [GEH14, Lee17] consists of the emergency core cooling system (ECCS) and passive containment cooling system (PCCS). The ECCS comprises the ADS and gravity-driven cooling system (GDCS), while the PCCS relies on isolation

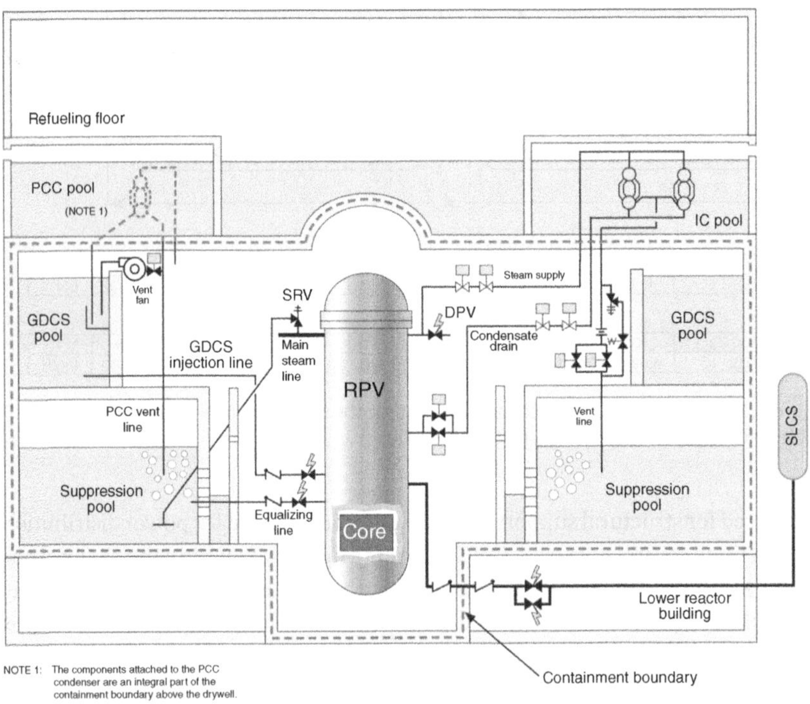


**Figure 1.11** Schematic diagram of the ESBWR safety system. Abbreviations: DPV = depressurization valve, IC = isolation condenser, SRV = safety relief valve, GDCS = gravity-driven cooling system, PCC = passive containment cooling. *Source*: GE Hitachi Nuclear Energy [GEH14].

and passive containment cooling condensers. Rejection of a full load subject to a turbine trip is allowed without the need to shut down the reactor, allowing for quick power recovery from secondary-system malfunctions. A noteworthy feature of the ESBWR is the basemat internal melt arrest and coolability (BiMAC) core catcher installed below the RPV to protect the plant in accidents resulting in containment failures. The electro-hydraulic fine-motion control rod drive (FMCRD) system also improves the controllability and reliability of the ESBWR plant. The ESBWR design was docketed in December 2005 for review by the NRC and received the design certification in 2014.

## 1.5 Advanced Reactor Designs

In spite of the excellent safety records of LWR plants, both PWR and BWR, the Fukushima accidents in 2011 indicated the need to develop new reactor and plant designs that reflect lessons learned from the current generation of power reactors. These advanced reactor designs cover a number of different features that may be classified as evolutionary in nature as well as those representing more radical changes and providing enhanced passive safety characteristics receiving additional attention after the Fukushima accidents. Several power plants featuring evolutionary LWR designs include the OPR-1000/APR-1400 and General Electric ABWR that have been operating in Korea and Japan, respectively, for a number of years, together with the AP1000 and ESBWR designs discussed in Sections 1.3 and 1.4, respectively.

### 1.5.1 Key Features of Advanced Reactors

Together with the evolutionary Generation III+ designs, the Generation IV initiative promotes innovative designs that will facilitate improved safety and high sustainability, including (a) increased economic competitiveness, (b) enhanced safety and reliability, (c) minimizing radioactive waste generation, and (d) increasing nuclear proliferation resistance. Under the leadership of the US DOE, a multinational study was performed to develop the Generation IV roadmap [DOE02] and to select the six most promising systems for detailed design and development. The DOE initially selected [GIF14, TRP14] to focus on the very high-temperature gas-cooled reactor (VHTR), SFR, and molten-salt-cooled reactor (MSR) designs for development in the United States among the six designs included in the roadmap.

(1) The SFR, operating with neutron energies around 0.1 MeV, offers the best potential for transmuting the entire transuranic elements, not just plutonium, from the used LWR fuel inventory. The SFR design evolved from the 19-MWe Experimental Breeder Reactor Unit 2 that operated for 30 yr (1964–1994)

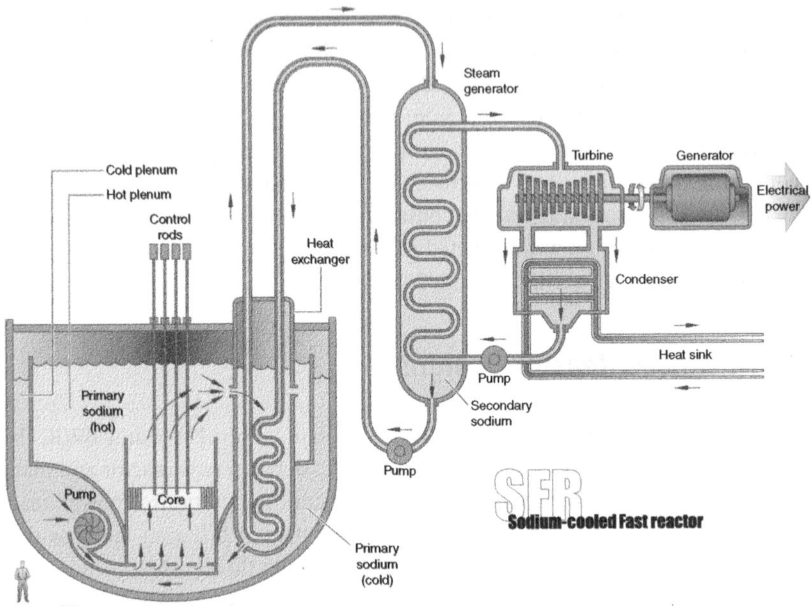


**Figure 1.12** Sodium-cooled fast reactor plant. *Source*: US Department of Energy [DOE02].

at the Argonne National Laboratory in Idaho. Two SFRs, 0.25-GWe Phénix and 1.2-GWe Superphénix, operated in France for 1973–2009 and 1986–1998, respectively. Several modular SFR plant designs, including the S-PRISM design [Boa99], were also proposed. The pool-type SFR plant illustrated in Figure 1.12 features a flat core immersed in a large pool of molten sodium, serving as the primary coolant, and an intermediate heat exchanger. The intermediate heat exchanger would provide a barrier between the radioactive sodium pool and the steam cycle employed for power generation. Fuel elements as illustrated in Figure 1.13 are structured in hexagonal arrays to allow tight coupling with a small sodium coolant volume fraction to retain a fast neutron flux spectrum with an average neutron energy of ~0.1 MeV. The design includes two driver regions surrounded by the reflector and shield regions, together with the primary control and secondary shutdown systems. The core is configured in a flat pancake structure to enhance the axial neutron leakage. Gas expansion modules are located near the periphery of the core to promote neutron leakage and help avoid the potential for a positive sodium void coefficient, discussed further in Chapter 14.

(2) Figure 1.14 illustrates a dual-purpose VHTR or high-temperature gas-cooled reactor (HTGR) design with hexagonal fuel blocks, with a He-water

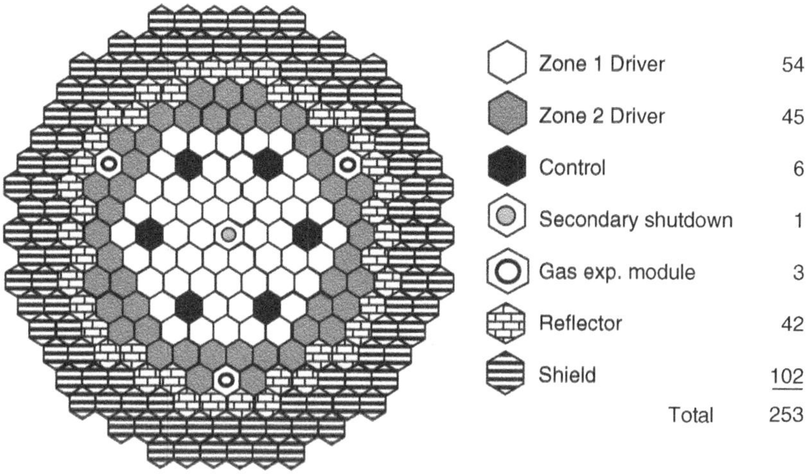


**Figure 1.13** Top view of the reactor core of a SFR plant. *Source*: Petti et al. [Pet16]/U.S. Department of Energy.

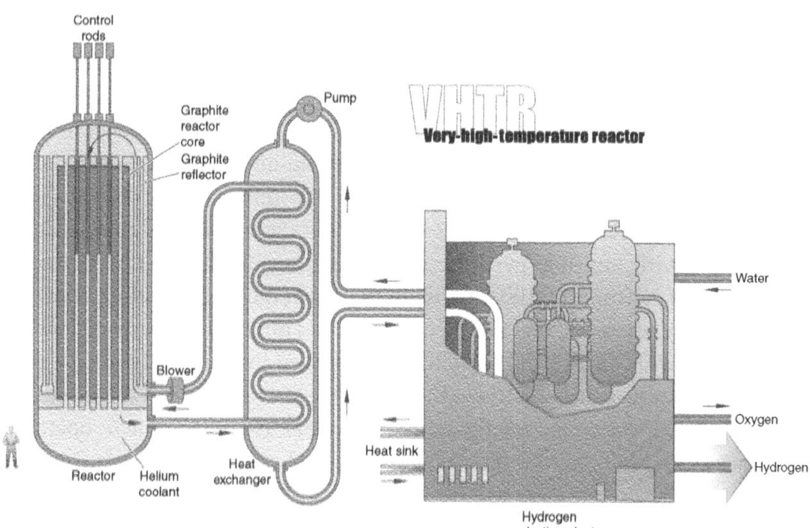


**Figure 1.14** Very high-temperature reactor plant. *Source*: US Department of Energy [DOE02].

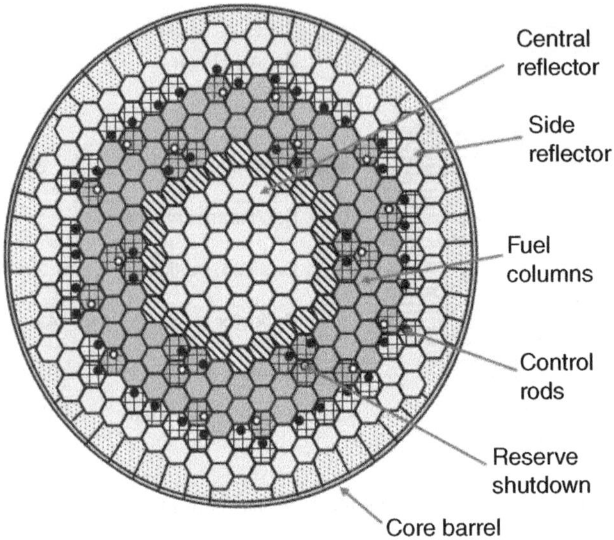


**Figure 1.15** Top view inside the reactor vessel of a VHTR plant. *Source*: Petti et al. [Pet16]/U.S. Department of Energy.

steam generator providing high-temperature steam and hydrogen in a co-generation plant. The VHTR design has the capacity to heat the He coolant to temperatures in excess of 1100 K, suitable for the generation of hydrogen via dissociation of water. The graphite-moderated gas-cooled core layout in Figure 1.15 features central and side reflectors surrounding the hexagonal fuel blocks. The VHTR design offers additional safety measures associated with multiple pyrolytic carbon coatings in the 1.0-mm diameter tristructural-isotropic (TRISO) particles that form the basic building block for the core. The prismatic fuel block comprises fuel pin cells or compacts, which are packed with TRISO fuel particles as illustrated in Figure 1.16. The VHTR was demonstrated successfully in the 0.33-GWe Fort St. Vrain Plant, featuring a prestressed concrete reactor pressure vessel (PCRV), which operated during 1979–1989. In the alternate pebble bed reactor (PBR) design, the

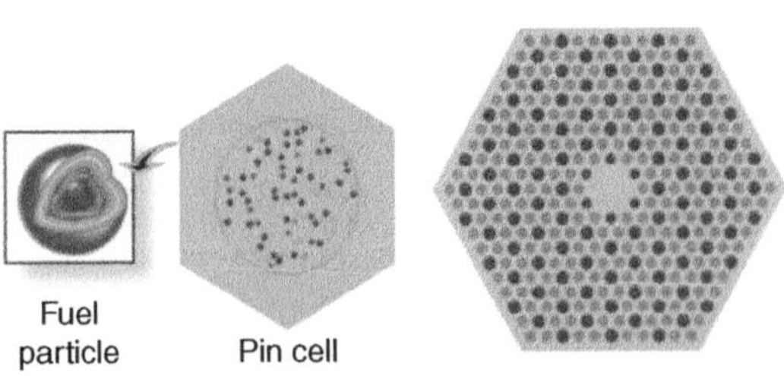


**Figure 1.16** TRISO particle, pin cell, and prismatic fuel assembly for the VHTR plant. *Source*: US Department of Energy [DOE02].

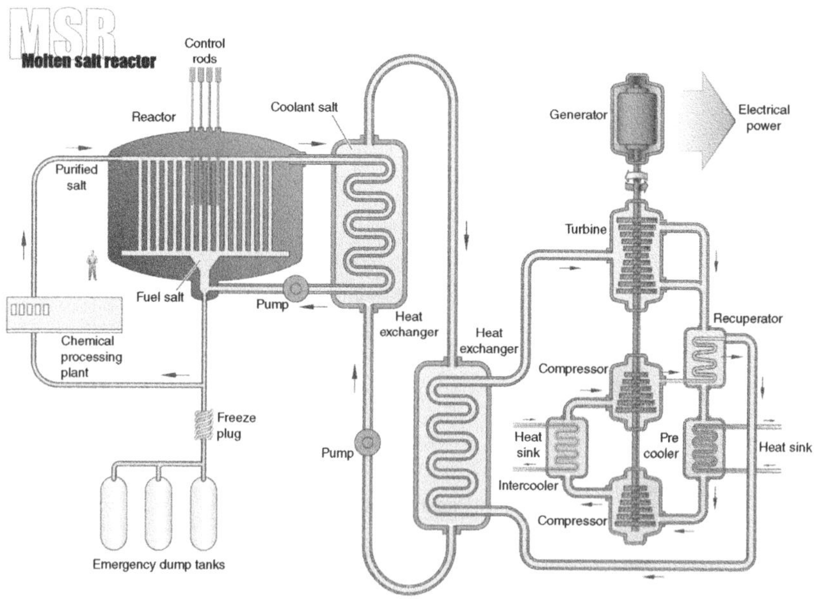


**Figure 1.17** Molten-salt reactor plant. *Source*: US Department of Energy [DOE02].

TRISO particles are packed into graphite spheres with a diameter of 60 mm, which are then loaded and circulated in the reactor vessel. The concept was demonstrated in the 15-MWe Arbeitsgemeinschaft Versuchsreaktor (AVR) during 1967–1988 in Germany.

(3) The MSR design illustrated in Figure 1.17 offers the potential use of flexible fuel and coolant compositions, with a high thermal efficiency. A popular fluoride coolant considered for the MSR design is FLiBe (LiF-$BeF_2$), with several alternate fuel designs [Pet16] typically featuring a graphite core, illustrated in Figure 1.18. Fuel dissolved in the molten salt as well as TRISO particles discussed for the VHTR design have been considered for the MSR design. A direct reactor auxiliary cooling system (DRACS) is considered as part of the passive reactor cooling system and is included in the design illustrated in Figure 1.18. The concept benefits from the design and operating experience of the 7.4-MWt Molten Salt Reactor Experiment (MSRE) [Hau70]] that operated for five years (1964–1969) at the Oak Ridge National Laboratory. Additional demonstration and testing facilities are under construction for the Kairo project are discussed further in Section 1.5.2

The key design features of the three promising Generation IV NPP designs are compared in Table 1.1 to augment the graphical illustrations in Figures 1.12–1.18. A systematic comparison of reactor physics, thermal hydraulics, and safety

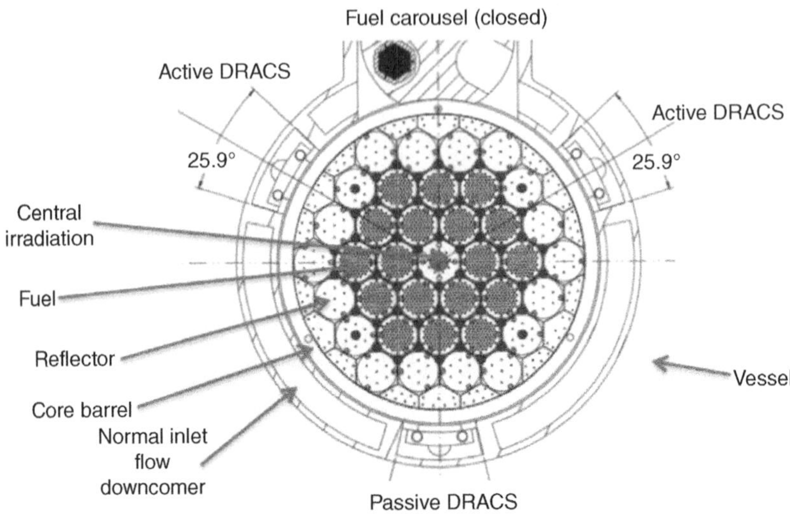


**Figure 1.18** Top view inside a MSR reactor vessel. *Source*: Petti et al. [Pet16]/U.S. Department of Energy.

**Table 1.1** Key features of three Generation IV plants.

| SFR | VHTR | MSR |
|---|---|---|
| Fast spectrum, closed fuel cycle, TRU transmutation | High He outlet temperature facilitating $H_2$ production | Low pressure, high thermal efficiency with fluoride salt |
| Enhanced resource utilization | Improved overall economics | Generation of electricity and $H_2$ |
| Pyrometallurgical processing | Pebble bed or prismatic design | Closed fuel cycle possible |
| Need to reduce cost for intermediate heat exchanger | Need to develop heat-resistant fuel and materials for 1273 K | Need to develop improved corrosion-resistant materials |

characteristics of three major reactor types, LWR, SFR, and VHTR, is presented in Appendix B.

With the realization that the AP1000 plants at the Vogtle and Summer Plants in the United States have incurred large capital investments, with an estimate of $30B for the two Vogtle units that have finally begun to produce power in 2024 [Cho23b], significant effort is underway to develop SMRs that would require much smaller initial capital outlays per unit, with the 77-MWe NuScale design [Doy16, Cho19] as a primary example. Each 77-MWe module illustrated in Figure 1.19 features PWR fuel elements 2.0 m in length, compact helical-coil steam generators, and a reactor

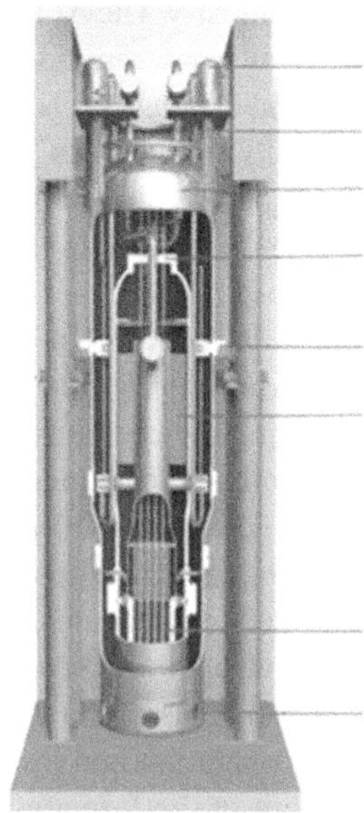


**Figure 1.19** Schematic illustration of the NuScale module. *Source*: U.S.NRC [NRC17]/ US Nuclear Regulatory Commission/Public domain.

vessel inside an integrated containment vessel and relies on natural circulation cooling for normal operation and passive shutdown. The NuScale containment vessel is designed to be 25 m tall, compared with the 82-m AP1000 containment structure [Cho19]. Incorporating advanced passive safety features in new innovative designs, the SMRs may offer meaningful alternatives to the full-size AP1000 and ESBWR designs, especially in regions with smaller energy markets.

Realization of the goals enunciated for the Generation IV initiatives will continue to present new challenges to nuclear engineers. Power plant designs, including fuel, coolant, and engineered safety systems, should be optimized systematically with energy-generation cost under consideration. Despite much hope generated for the successful launch of the NuScale project for which the construction permit from the US Nuclear Regulatory Commission had been granted in 2020, the construction funding arrangement fell apart in 2023 and effort is underway for alternative arrangement [Cho23b]. Issues associated with capital costs for nuclear plants and methods for determining nuclear electricity generation costs are discussed in Chapter 15.

Despite difficulties experienced both with the Vogtle AP1000 project and Olkiluoto EPR project, the successful deployment of four APR-1400 units at the Barakah Plant providing 5.0-GWe to the United Arab Emirates within the budget of $20B and construction period of 14 years for the entire four units raises some hope for the nuclear power industry. The APR-1400 success that was achieved in the recent UAE project, in many ways, reminds us of the significant success that the French nuclear program and Framatome achieved with 57 PWR plants constructed over a period of essentially two decades. The successful French PWR program started with the first PWR plant, Bugey Unit 2, which began operation after six years of construction in 1978. The recent UAE APR-1400 project has benefited significantly [Ste22] from sustained reliance on Generation II and Generation III PWR designs in Korea, starting with the Kori Unit 1 project featuring a two-loop Westinghouse PWR in 1978, evolving through the Combustion Engineering System 80 design, the OPR-1000 design, and currently the 1.4-GWe APR-1400 design for a total of 27 nuclear plants constructed. Under a joint construction management of Korea Hydro and Nuclear Power (KHNP) and Doosan Enerbility, the success of the UAE project clearly emphasizes the importance of the robust, sustained infrastructure and consistent project management [Ste22]. The APR-1400 design received the Design Certification from the US Nuclear Regulatory Commission in 2019.

Nuclear power plants operating in the United States have achieved an impressive record of safe operation and low electricity generation cost in recent years, especially as a result of the formation of large operating companies comprising multiple nuclear plants each. This does not, however, guarantee continued operation of nuclear plants in the current deregulated merchant-fleet structure, as exemplified by recent decisions to decommission several nuclear plants with operating licenses remaining, e.g. the Kewaunee and Vermont Yankee Plants. It appears imperative to reevaluate the rate structure for the electric market in the United States, in light of the increasing need for carbon-free, base-load energy sources around the world.

### 1.5.2 Recent Developments in Advanced Reactors

Together with the Generation IV designs supported through the Generation IV International Forum [GIF14], the International Atomic Energy Agency maintains the Advanced Reactors Information System (ARIS) [IAE24], as well as periodic publications summarizing key technology developments [IAE22]. There are 78 designs featured in the ARIS database as of March 2024, with several others under development that are yet to be included in the database. As part of the 2023 US National Academy of Sciences study [NAS23], developers of 18 advanced reactor designs made presentations to the committee and the key features of the designs are summarized in the published report. The key features of the 18 advanced reactor designs together with the AP1000, as a fiducial design, summarized in Table 1.2 will serve as the representative concepts for our discussion and analysis here.

**Table 1.2** Key attributes of advanced nuclear reactors under development.

| Project description | Power | Features | Fuel cycle | Additional comments |
|---|---|---|---|---|
| *Base case: Light water reactor* | | | | |
| AP1000 Reactor<br>Westinghouse Electric | 3.4 GWt<br>1.09 GWe | Pressurized water coolant<br>15.5 MPa | LEU once-through<br>49 MWd/kgHM | Operating |
| *Modular light water reactor* | | | | |
| NuScale Small<br>Modular Reactor | 250 MWt<br>77 MWe | Natural circulation cooling<br>12 PWR modules | LEU once-through<br>41 ~ 60 MWd/kgHM | Construction permit approved<br>Project suspended |
| *Fast spectrum reactor* | | | | |
| Versatile Test Reactor<br>U.S. Dept. of Energy | 300 MWt<br>0.0 MWe | Na cooled<br>Na-bonded fuel | U-20Pu-10Zr metallic fuel<br>5.0 wt% $^{235}U$ | Activity suspended |
| Natrium<br>TerraPower | 840 MWt<br>339 MWe | Na cooled<br>1.0-m fuel column | U-10Zr HALEU 16.7 wt%<br>150 MWd/kgHM | Construction permit submitted<br>for demonstration project |
| ARC-100<br>Advanced Rx Concepts | 286 MWt<br>100 MWe | Na cooled, superheated<br>steam for balance of plant | HALEU metallic alloy<br>13.1 wt%, 77 MWd/kgHM | 20-year refueling cycle<br>Load follow capability |
| SEALER-55<br>LeadCold | 140 MWt<br>55 MWe | Lead cooled, Hf and<br>enriched $^{15}N$ in fuel | UN HALEU 12.0 wt%<br>60 MWd/kgHM | 27-year core life, SFR flux<br>spectrum, Sweden project |
| MCFR<br>TerraPower | 0.5 ~<br>1.2 GWe | Molten chloride fuel online<br>refueling | U-Pu HALEU 12.0 wt%<br>183 MWd/kgHM | HEU-fueled MCRE reactor<br>experiment planed, 150 kWt |
| eVinci Microreactor<br>Westinghouse | 15 MWt<br>5 MWe | Heat pipe, C core block fuel<br>pin design | TRISO UCO HALEU<br>19.8 wt%, 45 MWd/kgHM | Passive heat removal, rotating<br>drum reactivity control |
| Aurora Powerhouse<br>Oklo | 4.0 MWt<br>1.5 MWe | Heat pipe metallic annular<br>fuel | U-10Zr HALEU 12-19.8 wt%<br>20 MWd/kgHM | Combined construction and<br>operating license submitted |
| Lead Fast Reactor<br>Westinghouse | 950 MWt<br>450 MWe | Lead cooled, secondary<br>supercritical water | $UO_2$ HALEU, 13.8 wt%<br>80 MWd/kgHM | Energy storage system proposed |

*(Continued)*

**Table 1.2** (Continued)

| Project description | Power | Features | Fuel cycle | Additional comments |
|---|---|---|---|---|
| *Gas cooled reactor* | | | | |
| Xe-100<br>X-Energy | 200 MWt<br>80 MWe | He cooled pebble bed reactor | TRISO UCO HALEU<br>15.5 wt% 160 MWd/kgHM | No containment, TRISO fuel production planned |
| KP-X FHR<br>Kairos Power | 320 MWt<br>140 MWe | 2LiF-$BeF_2$ molten salt pebble bed reactor | TRISO 19.5 wt% 40-mm pebble, 194 MWd/kgHM | Double-walled piping<br>Engineering Test completed |
| SC-HTGR<br>Framatome | 625 MWt<br>272 MWe | He cooled prismatic HTGR | TRISO UC HALEU 14.5 wt%<br>160 MWd/kgHM | 10 × 0.8-m core block<br>Process heat production |
| BANR HTGR<br>BWXT Technologies | 50 MWt<br>17 MWe | He cooled prismatic HTGR | TRISO UCO 19.8 wt%<br>Alternate: UN HALEU | Down blend HEU to HALEU<br>Started TRISO production |
| Energy Multiplier EM2<br>General Atomics | 500 MWt<br>250 MWe | Pressurized He coolant | UC pellet, SiC cladding<br>14.0 wt%, 143 MWd/kgHM | Convert-burn design<br>Containment system proposed |
| *Molten salt reactor* | | | | |
| IMSR-400<br>Terrestrial Energy | 400 MWt<br>195 MWe | C-moderated molten salt fuel | $UF_4$ LEU 4.95 wt%<br>29 MWd/kgHM | Double-walled piping, seven-year cycle, no online salt processing |
| ThorCon | 557 MWt<br>250 MWe | Fluoride molten salt<br>5% fuel salt, 95% C | 85% $ThF_4$, 15% $UF_4$<br>19.7 wt%, 509 MWd/kgHM | Continuous fuel feed<br>MSRE concept |
| LFTR<br>Flibe Energy | 600 MWt<br>250 MWe | LiF-$BeF_2$ coolant salt<br>LiF-$ThF_4$ blanket salt | LiF-$BeF_2$- $^{233}UF_4$, 19.6 wt%<br>194 MWd/kgHM | Double-walled piping, inter-mediate heat exchanger |
| Moltex SSR-W<br>Moltex Clean Energy | 750 MWt<br>300 MWe | KCl coolant in cylindrical fuel pins | CANDU spent fuel as feed<br>150 MWd/kgHM | BWR-type hexagonal fuel channel, online refueling |

Fuel cycle characteristics in a rector core are represented primarily in terms of the fuel composition and coolant that play a dominant role in determining the neutron flux spectrum in the core. The flux spectrum, together with the fuel composition, in turn characterizes the depletion and isotopic evolution of nuclear fuel materials. This observation suggests classifying advanced reactor designs as (a) fast spectrum reactor, (b) gas cooled reactor, and (c) molten salt reactor, together with (d) modular light water reactor, as summarized in Table 1.2. The basic advanced reactor concepts have been already discussed as key Generation IV features through Figures 1.12–1.18 and Table 1.1, together with the modular NuScale design illustrated in Figure 1.19. One could, however, regroup some of the 18 designs, especially the Kairo design, which features pyrolytic graphite fuel elements, together with molten salt coolant, not typical He coolant. As discussed further in connections with fuel cycle characteristics in detail in Chapter 12, it was decided, however, to include the Kairo design in the VHTR group, rather than in the MSR group, and not to consider a sub-group. The tabulations present both thermal and electric power levels proposed, together with key core characteristics and fuel composition, fissile enrichment, fuel discharge burnup, whenever such data are available either from the NAS presentations or the ARIS database. Additional comments on other design features or the current state of project development effective May 2024 are also included for each project.

In addition to the power rating, key features of the system, and fuel type and enrichment, the fuel cycle data include discharge fuel burnup in units of [MWd/kgHM = GWD/MTHM]. The discharge burnup data play an important role in estimating the inventory of used nuclear fuel (UNF) from the plant, since the inverse of the discharge fuel burnup determines the mass of discharged fuel per unit energy generated in units of MgHM/GWe-yr, as discussed in detail in Chapter 12. In fact, this was an issue that required serious discussions for the recent US National Academy of Sciences report [NAS23], because a paper [Kra22] published before the release of the report suggested, based on preliminary data and a simple analysis, that small modular reactors would have a lower power density. This would then imply a low discharge fuel burnup per unit fuel mass, thereby generating a larger UNF inventory for a unit of energy generated in units of MgHM/GWe-yr. With updated NuScale design data provided by the NuScale design team [Rey22], Table 1.2 indicates that the discharge fuel burnup of 41 ~ 60 MWd/kgHM for the 77-MWe modular core featuring 13 PWR fuel assemblies could be comparable to the discharge burnup of 49 MWd/kgHM for the 157-assembly 1.09-GWe PWR core. This indicates that a SMR core could be designed to produce a UNF inventory per unit energy generated, certainly comparable to that of a GWe-scale power plant. It is possible, at the same time, that a smaller reactor core could increase the neutron leakage into the ex-core regions, including the neutron shields and reflectors, resulting in increased production of low-level wastes. This issue is discussed further in Chapter 12.

Several of the advanced reactor designs summarized in Table 1.2 are supported by the US Department of Energy (DOE) via the Advanced Reactor Development Program (ARDP) and Risk Reduction Pathway Program (RRPP) initiated in 2020. The ARDP selected the TerraPower Natrium project and the X-Energy Xe-100 project, with the initial funding of $160M in 2020, as two mature projects that could possibly become operational by 2027. The RRPP awards include five projects: (a) KP-X FHR (Kairos Power), (b) eVinci Microreactor (Westinghouse Electric), (c) BANR HTGR (BWRX Technologies), (d) SMR 160 (Holtec), and (e) MCFR (TerraPower) with the plan to provide $600M over a period of seven years. All of the seven projects are progressing as of May 2024, with additional private funding.

A number of other advanced designs require some discussion at this point in the emerging era of nuclear energy. The economics of nuclear power plants was not a key feature of the recent NAS Committee charge and only passing comments are included on economics in the report [NAS23]. The Russian invasion of Ukraine in February 2022, with no sign of a near-term resolution of the conflict, clearly made the final stage of the committee deliberations shifting significantly from the initial discussions on the clean energy resources and economics. During the past two years, many nations of the world have come to realize that each country has to be much more self-reliant on energy resources and that nuclear energy has to play a major role in the future. Thus, the nuclear energy communities around the world have begun to search for various approaches to add clean nuclear power plants as expeditiously as possible. This entails the need to demonstrate the feasibility and applicability of SMRs and revisit some of the earlier concepts that have been put in the reserve in favor of full-size LWRs of $1.0 \sim 1.6$-GWe capacity.

The 600-MWe designs, AP600 and SBWR, were developed during the late 1980s, in the aftermath of the 1986 Chernobyl accident and yet encouraged by the successful passive safety demonstration tests at the Experimental Breeder Reactor Unit 2 (EBR-II) SFR facility [Lee17]. The AP600 and SBWR designs introduced passive safety features to augment the first-generation LWRs and received design certifications. The Generation III+ designs were, however, not seriously considered during the Generation IV Roadmap discussions [DOE02] with the overriding desire to achieve economic parity with the 1.1-GWe AP1000 plant serving as the fiducial design. The economy of scale expected with the AP1000 design, however, has not been realized, certainly not in the United States with the Vogtle and Summer projects, and there has been increased interest in the 300-MWe SMRs, AP300 and BWRX-300, together with micro reactors with power outputs in the range of $1 \sim 5$ MWe illustrated in Figure 1.20 for the micro or modular reactor projects underway at the Idaho National Laboratory [INL24]. One example of the micro

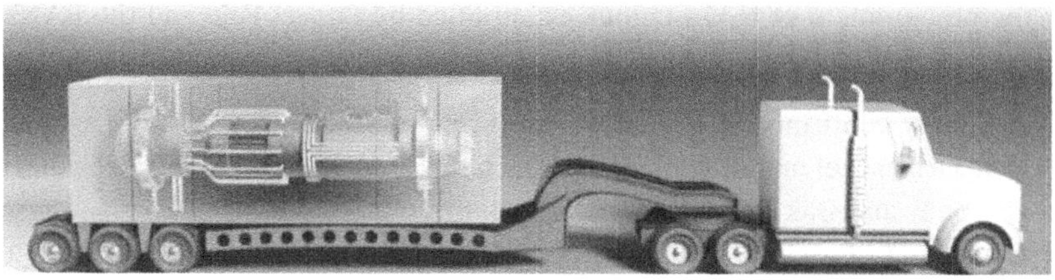

**Figure 1.20** Micro modular reactor. *Source*: Idaho National Laboratory [INL24].

reactor is Project Pele [Wak22], which features transportability and initial military implementation with power output in the range of $1 \sim 5$ MWe. The Pele design utilizes the basic VHTR concept with SiC-based TRISO particles and is able to bypass US Nuclear Regulatory Commission (NRC) regulations, although it will seek NRC approval for the reactor module transportation package. The BWXT Company with DOE support has developed a commercial-scale TRISO manufacturing facility and is moving ahead expeditiously with the project with the hope to get the first Pele unit operational by 2025 or 2026. There is considerable hope and excitement that an operating micro reactor could spur further investments in SMRs and various advanced reactors around the world.

Most of the advanced reactor designs summarized in Table 1.2 feature the high-assay low enrichment uranium (HALEU) fuel, with the $^{235}$U enrichment in the range of $5 \sim 19.5$ wt%, in contrast to the fissile enrichment of $3 \sim 5$ wt% for all LWR plants currently operating around the world. The suggestion for HALEU reflects the advantages resulting from (a) a smaller fuel inventory and associated reduction in the size of the core and reactor and (b) a reduced UNF inventory resulting in reduced requirements for long-term management and disposal of the UNF. These advantages should, of course, be contrasted with the increased requirements and costs associated with higher separative work unit (swu) in the enrichment process. The potential fuel cycle advantages of the HALEU composition in advanced reactors are illustrated via a few representative fuel arrangements in Chapter 12, with their effects on fuel cycle costs due to higher swu discussed in Chapter 15. Another issue associated with the HALEU introduction into the commercial nuclear energy market is the nuclear proliferation and security concerns for Category II special nuclear material (SNM) designation in effect for uranium with the fissile enrichment greater than 10.0 wt% in the United States; it would be easier to produce weapons-grade high-enrichment uranium (HEU) from the HALEU fuel than from the low enrichment uranium (LEU) fuel, with Category III designation, currently used in LWR plants.

At this point, however, in the active development of SMRs and Generation IV plants, the nuclear energy community is moving ahead with various arrangements for the procurement of HALEU fuel, exemplified by Project Pele and the BANR project with BWXT. Other HALEU sources include:

(a) X-Energy confirmed in the 2024 American Nuclear Society forum [ANS24] that TRISO fuel manufacturing had been initiated at the TRISO-X subsidiary company and Oak Ridge National Laboratory.
(b) Centrus Energy Corporation has been developing 16 advanced centrifuges to produce 600 kg HALEU with $170M DOE support [NAS23].
(c) About 374 Mg of HEU and reprocessed U at the Savannah River Site are available [NAS23] for down-blending, if necessary.
(d) Orano USA, with the enrichment capability estimated as 7.5 M swu/year, 15,000 MgHM/year, and a revenue of $1.0 B/year, could produce HALEU, rather than relying on Russian sources in the near future [Vex21]. Some of the issues associated with the introduction of HALEU designs are discussed further in the 2023 NAS report [NAS23].

Additional encouraging developments for advanced nuclear reactors include:

(a) The progress made with the Kairos KP-X FHR project, where the non-nuclear Engineering Test (ET) Unit 1 has been successfully completed and the plan for the ET Unit 2 and Hermes Model is moving ahead with the DOE support as one of the five RRPP awards and private funding.
(b) Auroa Powerhouse also suggests increasing the power output for the Oklo project to 15 MWt or larger [ANS24].
(c) For supplying metallic fuel for the Natrium project, one of the two ARDP awards, effort has been underway to develop and test SFR fuel without Na bonding, very likely in the form of annular fuel rod. This is mostly in recognition of the difficulty associated with geological disposal of Na-bonded used fuel elements [NAS23]. Together with private funding of $1.0B, TerraPower has also submitted to the US Nuclear Regulatory Commission the construction permit for the Natrium demonstration project, and preparations are underway for the preliminary safety analysis report.
(d) With a RRPP award, TerraPower is also progressing with the HEU-fueled Molten Chloride Reactor Experiment (MCRE) project [Cho23a], with a power rating of 150 kWt and plans for Pu-burner prototype designs [Tou24].

## References

**[ANS24]** American Nuclear Society, "Advanced Reactor Roundup," (2024). https://www.ans.org/webinars/view-advreac2024/ (accessed 28 March 2024).

**[Boa99]** C.E. Boardman, A. Fanning, D. Carroll et al., "A Description of the S-PRISM Plant," *Proc. 8th Int. Conf. Nucl. Eng. (ICONE-8)* (1999).

**[Cho19]** A. Cho, "The Little Reactors That Could," *Science* **363**, 806 (2019).

**[Cho23a]** A. Cho, "U.S. Planning Test Reactor to Run on Weapons-Grade Uranium," *Science* **380**, 783 (2023).

**[Cho23b]** A. Cho, "Deal to Build Pint-size Nuclear Reactors Canceled," *Science* **382**, 749 (2023).

**[DOE02]** US Department of Energy, *A Technology Roadmap for the Generation IV Nuclear Energy Systems*, https://www.gen-4.org/gif/jcms/c40473/a-technology-roadmap-for-generation-iv-nuclear-energy-systems (2002).

**[Doy16]** J. Doyle, B. Haley, C. Fachiol et al., "Highly Reliable Nuclear Power for Mission-Critical Applications," *Proc. Int. Cong. Adv. Nucl. Power Plants (ICAPP)* (2016).

**[GEH14]** GE Hitachi Nuclear Energy, "ESBWR Design Control Document, Tier 2," 26A6642AD, rev. 10 (2014).

**[GIF14]** Generation IV International Forum, *Technology Roadmap Update for Generation IV Nuclear Energy Systems*, https://www.gen-4.org/gif/upload/docs/application/pdf/2014-03/gif-tru2014.pdf (2014).

**[Hau70]** P.N. Haubenreich and J.R. Engel, "Experience with the Molten-Salt Reactor Experiment," *Nucl. Appl. Technol.* **4**, 118 (1970).

**[Hon12]** M. Hone, S. Skidmore, M. Misvel, and E. Resch, "AP1000 Core Reference Report," WCAP-17524-NP, Westinghouse Electric Company (2012).

**[IAE22]** International Atomic Energy Agency, *Advances in Small Modular Reactor Technology Developments* (2022).

**[IAE24]** International Atomic Energy Agency, "Advanced Reactors Information System (ARIS)," (2024) https://aris.iaea.org/sites/overview.html (accessed 1 April 2024).

**[INL24]** Idaho National Laboratory, "Micro Modular Reactor," (2024) INL.gov/trending-topics/microreactors (accessed 24 April 2024).

**[Kra22]** L.M. Krall, A.M. Macfarlane, and R.C. Ewing, "Nuclear Waste From Small Modular Reactors," *Proc. Natl. Acad. Sci. U.S.A.* **119**, 2111833119 (2022).

**[Lee17]** J.C. Lee and N.J. McCormick, *Risk and Safety Analysis of Nuclear Systems*, Wiley (2011/2017).

**[NAS23]** J.D. Lee, P.A. Baisden, R.C. Ewing et al., "Merits and Viability of Different Nuclear Fuel Cycles and Technology Options and the Waste Aspects of Advanced Nuclear Reactors," US National Academy of Sciences, https://doi.org/10.17226/26500 (2023).

**[NRC08]** US Nuclear Regulatory Commission, "NRC Reactor Systems Training Manual," (2008).

**[NRC17]** U.S.NRC, "Design Certification Application - NuScale," www.nrc.gov/reactors/new-reactors/design-cert/nuscale.html (2017).

**[Pet16]** D. Petti, R. Hill, and J. Gehin, "Advanced Demonstration and Test Reactor Options Study," INL/EXT-16-37867, rev. 1, Idaho National Laboratory (2016).

**[Rey22]** J. Reyes, "Overview of Design Optimization Process for a Commercial Nuclear Core," transmittal to the US National Academy of Sciences committee, 26 August 2022, https://doi.org/10.17226/26500 (2023).

**[Ste22]** W.R. Stewart and K. Shirvan, "Capital Cost Estimation for Advanced Nuclear Power Plant," *Renewable Sustainable Energy Rev.* **155**, 111880 (2022).

**[Tes84]** D. Testa, ed., *The Westinghouse Pressurized Water Reactor Nuclear Power Plant*, Westinghouse Electric Corporation (1984).

**[Tou24]** N. Touran, TerraPower, private communication, 11 April 2024.

**[TRP14]** US Department of Energy, "Advanced Reactor Concepts Technical Review Panel Public Report," (2014).

**[Vex21]** A. Vexler, "HALEU," presentation to the US National Academy of Sciences Committee, 15 September 2021, https://doi.org/10.17226/26500 (2023).

**[Wak22]** J. Waksman, "Project Pele Overview, Mobile Nuclear Power for Future DoD Needs," (2022) www.nrc.gov/docs/ML2212/ML22126A059 (Accessed 1 April 2024).

## Problems

**1.1** Study the Generation IV Roadmap and prepare a comprehensive comparison of the key features of the six concept groups selected. Discuss how the concepts satisfy the five objectives and goals of Generation IV designs.

**1.2** Tabulate and compare the key safety features that should be enhanced for the six Generation IV concept groups.

**1.3** Discuss the key technology developments that are required for successful development and deployment of the Generation IV groups selected.

# 2

# Neutron–Nucleus Reaction and Neutron Cross Section

The design and analysis of a nuclear reactor core require the determination of nuclear fuel element configurations and mechanical and control devices that can provide self-sustaining chain reactions and produce power safely and economically. This in turn requires an accurate representation of interactions of nuclear radiation with matter. For the nuclear reactor physics analysis that we focus on in this book, the mechanisms of neutrons undergoing collisions with nuclei of core materials are of primary interest. Thus, interactions of $\gamma$-rays with various materials, resulting in the deposition of heat in the core, will not be explicitly considered.

We begin in Section 2.1 with a discussion of how the probability of neutrons interacting with the surrounding nuclei is represented in terms of the microscopic cross section, together with the physical meaning of the macroscopic cross section. We present a brief review of the mechanisms of neutron interactions with matter in Section 2.2, focusing on the reactions resulting in the formation of compound nuclei. The concepts of mass defect and binding energy will be discussed together with various reactions that are of significance in our subsequent effort to obtain relationships governing the transport of neutrons in space, time, and energy in a reactor core. Various parameters and data related to the fission process are presented in Section 2.3. To accurately represent neutrons undergoing scattering collisions in the core, Section 2.4 discusses two-body collision mechanics and transformation from the laboratory to the center-of-mass (CM) coordinate system. Section 2.5 introduces a particular type of cross section that represents a resonance phenomenon in neutron–nucleus interactions. Scattering collisions discussed in Section 2.4 will allow us to relate the change in the energy of a neutron to the change in the direction of the neutron motion, as it undergoes an elastic scattering collision with a nucleus, leading to the concept of scattering kernel in Section 2.6. We conclude the chapter with general remarks on key features of neutron cross sections of importance in Section 2.7.

*Nuclear Reactor Physics and Engineering*, Second Edition. John C. Lee.
© 2025 John Wiley & Sons, Inc. Published 2025 by John Wiley & Sons, Inc.
Companion website: www.wiley.com/go/LeeNRPE2

## 2.1 Neutron–Nucleus Reaction Probability and Neutron Cross Section

To introduce the concept of neutron cross section, consider a simple experiment where a collimated beam of neutrons of intensity $I$ [neutron $\cdot$ cm$^{-2}$ s$^{-1}$] is incident uniformly on a slab of thickness $a$ [cm]. The slab consists of a certain nuclide with number density $N$ [nucleus $\cdot$ cm$^{-3}$], some of which are illustrated as spheres in Figure 2.1. The number of neutrons $-dI$ suffering collisions in a thin layer $dx$ of the slab per cm$^2$ per s will then be proportional to beam intensity $I$ and the number $Ndx$ of nuclei in unit cm$^2$ of the layer exposed to the beam. With a proportionality constant for the interaction selected as $\sigma$, we obtain

$$-dI = \sigma INdx, \tag{2.1}$$

or alternately,

$$-\frac{dI}{I} = \sigma\left(\frac{\text{cm}^2}{\text{nucleus}}\right)Ndx\left(\frac{\text{number of nuclei}}{\text{cm}^2}\right). \tag{2.2}$$

Thus, the ratio $(-dI)/I$ yields the probability of neutron interactions in distance $dx$ and also represents the fraction of the nominal slab cross-sectional area that serves as the effective target area. This leads to the definition of the reaction probability $\sigma$ as the *microscopic cross section*, expressed in units of [barn $= 10^{-24}$ cm$^2$]. Defining the *macroscopic cross section* $\Sigma = N\sigma$, in units of [cm$^{-1}$], recasts Eq. (2.2) as

$$\frac{dI}{dx} = -\Sigma I. \tag{2.3}$$

The microscopic cross section $\sigma$ introduced in Eq. (2.2) suggests that we may interpret $\sigma$ as an effective surface area of the nucleus that is presented to the neutrons on their paths. Since the neutron–nucleus reactions are characterized by quantum mechanical phenomena, the microscopic cross section is characteristic of each nuclide and is a function of reaction type, e.g. scattering, capture, and fission, and depends heavily on the relative speed between the neutron and nucleus. The microscopic cross section for each nuclide and each reaction type has to be measured in the laboratory, although there are theoretical models that provide physical understanding of general cross section behavior. One such measurement could indeed involve a simple neutron penetration experiment as visualized in Figure 2.1.

We may readily integrate Eq. (2.3), representing the neutron beam intensity $I$ decreasing as the beam penetrates the slab, to determine the intensity $I(a)$ of the beam of neutrons that penetrate, without suffering any collision, the entire slab of thickness $a$

$$I(a) = I(0)\exp(-\Sigma a), \tag{2.4}$$

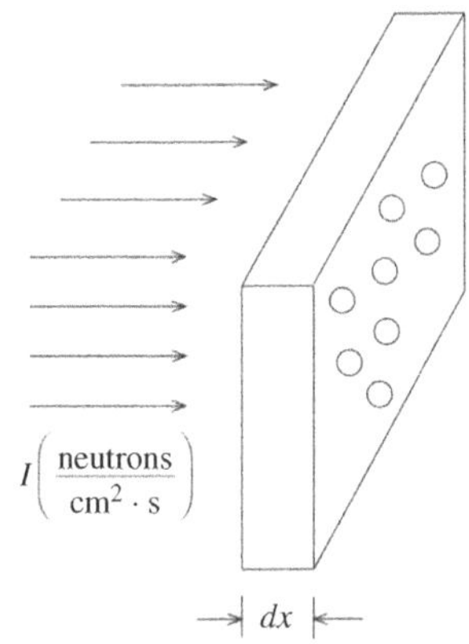


**Figure 2.1** A collimated beam of neutron incident on a slab.

where $I(0)$ is the intensity of the beam incident on the left-hand side (LHS) of the slab. Equation (2.2), written in terms of the macroscopic cross section $\Sigma$, suggests that $\Sigma$ represents the probability of neutron–nucleus reactions per unit distance of neutron travel. In an infinite medium, this observation also leads to the interpretation that $1/\Sigma$ represents the average distance a neutron travels between collisions, or the *mean free path* (mfp). This interpretation of $1/\Sigma$ as the mfp is analogous to the interpretation of $1/\lambda$ as the mean life of a radioactive species with decay constant $\lambda$, where $\lambda$ represents the probability of radioactive decay per unit time.

## 2.2 Mechanisms of Neutron–Nucleus Interaction

We now present a brief review of mechanisms of neutron interaction with matter, recognizing again that the interactions are characterized by quantum mechanical phenomena. We discuss interactions that result in the formation of a compound nucleus (CN) [Gla52] as well as those involving elastic scattering and direct interactions.

**1. Compound nucleus formation**

We consider in general particle $a$ incident on target nucleus $X$ forming a CN, which eventually decays into residual nucleus $Y$ and ejects particle $b$

$$X + a \rightarrow CN^{\star} \rightarrow Y + b. \tag{2.5}$$

In the reaction depicted in Eq. (2.5), we indicate the *CN* with an asterisk to highlight the point that it is left in an excited state. The reaction would typically be written in standard notation: $X(a, b)Y$. For neutron reactions of our primary interest, the incident particle $a$ is a neutron, and the ejected particle could be another neutron, a photon, or another particle, e.g. a proton.

That a neutron–nucleus reaction results in the formation of another nucleus, i.e. a CN, rather than a direct interaction yielding a residual nucleus and an ejected particle may first be indicated [Gla52] by a relatively long reaction time on the order of $10^{-14}$ s. This reaction time may be compared with the time required for a slow neutron of speed $v = 10^3$ m $\cdot$ s$^{-1}$ to travel a distance of $10^{-14}$ m, which is equal to a typical diameter of a nucleus. Thus, the reaction time of $10^{-14}$ s is three orders of magnitude longer than the transit time of $10^{-17}$ s, suggesting the formation of an intermediate nucleus, i.e. a CN.

An important feature of the CN resulting from a neutron interaction is that mass $m(CN)$ of the CN often is less than the sum of mass $m(X)$ of target nucleus $X$ and mass $m(n)$ of the incident neutron, with the result that the mass defect appears as the binding energy $E_b$ of the CN

$$E_b = [m(X) + m(n) - m(CN)]c^2, \tag{2.6}$$

where $c$ is the speed of light. The binding energy is often referred to as the *Q-value* of the reaction and is the energy that has to be supplied to break up the CN back to its constituents: nucleus $X$ and a neutron. The CN would be left in an excited state, with the excitation energy $E_{ex}$ including in general the *kinetic energy* $E_k$ of the neutron

$$E_{ex} = E_b + E_k. \tag{2.7}$$

The energy levels of a CN clarifying Eq. (2.7) are illustrated in Figure 2.2, where we indicate that for neutron interactions with many heavy nuclides of interest in reactor physics, $E_b$ is on the order of 8 MeV. Determination of the excited level of a CN through Eq. (2.7) is strictly valid in the CM framework, but only approximate in the laboratory. This reflects the fact that the CN itself will acquire some kinetic energy of its own, although the target nucleus is at rest, as discussed further in Section 2.4.

When the excited state of the CN lands in the proximity of one of quantum levels of the nucleus, then the probability of the CN formation increases markedly.

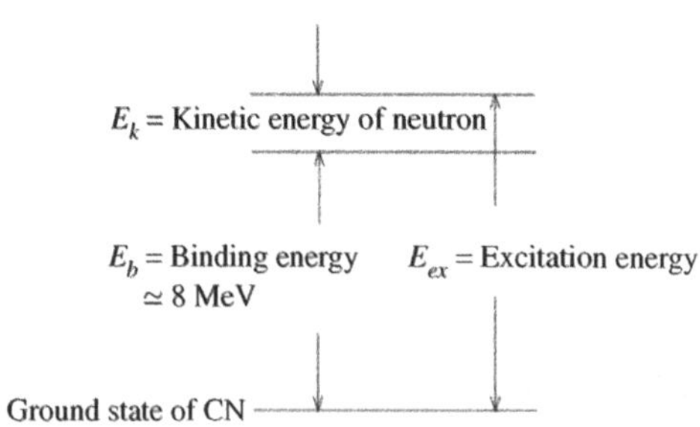


**Figure 2.2** Energy levels for a compound nucleus.

This phenomenon is known as the *resonance reaction* and involves both the resonance capture and resonance scattering of neutrons. A CN may decay in a number of different modes or channels:

(1) *Radiative capture* refers to the case when the CN decays from the excited state to ground state with the emission of a photon. In this case, the cross section is written as $\sigma(n, \gamma)$, $\sigma_\gamma$, or $\sigma_c$.
(2) When the CN is rather unstable, it may split into two lighter nuclei together with the emission of $2 \sim 3$ neutrons. This is the *fission* process, with cross section $\sigma_f$ significant for heavy nuclides, and it is of fundamental importance to any nuclear reactor. The fission process is discussed further in Section 2.3.
(3) The CN may also decay with the emission of a neutron, typically through a resonance reaction. This is known as *resonance elastic scattering*, with cross section $\sigma_s(n, n)$ or $\sigma_{se,res}$.
(4) Neutrons may experience *inelastic scattering*, whereby a neutron is ejected with energy $E < E_k$, while the CN is left in an excited state, which subsequently decays with the emission of a photon. The inelastic scattering cross section $\sigma_{in}$ is usually significant for $E_k > 0.1$ MeV and hence is important for the slowing down of neutrons for sodium-cooled fast reactors, where the average energy of neutrons may lie close to $0.1 \sim 0.2$ MeV. Note that in an inelastic scattering, momentum is conserved but kinetic energy is not, while both momentum and kinetic energy are conserved in an elastic scattering.

**2. Potential scattering**

In addition to various neutron–nucleus reactions involving the formation of a CN, as just discussed, neutrons also undergo ordinary elastic scattering collisions with the surrounding nuclei in a reactor core. This reaction may be visualized as the scattering of two billiard balls and characterized by a short reaction time compared with that associated with CN formation. Such scattering collisions are known as *potential scattering*, where the cross section is represented simply as $\sigma_p = 4\pi R^2$ in terms of nuclear radius $R$. Note here that $\sigma_p$ represents the surface area, not the projected area $\pi R^2$, of the nucleus visualized as a sphere of radius $R$.

**3. Direct interactions**

For high-energy neutrons with $E_k > 10$ MeV, there is also another mode of neutron–nucleus reactions that is of importance. The reaction in this case is a knockout process, where a neutron literally knocks a chunk of nucleons off the nucleus, resulting in $(n, p)$, $(n, 2n)$, and similar reactions.

As a summary of various neutron–nucleus reactions discussed in this section, we introduce notations for cross sections of relevance to reactor physics. We begin with the total cross section written as a sum of the cross sections representing radiative capture, fission, elastic and inelastic scattering, and other reactions including the $(n, 2n)$ reaction

$$\sigma_t = (\sigma_\gamma + \sigma_f) + (\sigma_{se} + \sigma_{in}) + \sigma(n, 2n) + \cdots, \tag{2.8}$$

which may be routinely grouped into

$$\sigma_t = \sigma_a + \sigma_s, \tag{2.9}$$

if we neglect the $(n, 2n)$ and other minor reactions. The absorption cross section may be rewritten in terms of the smooth background cross section and the resonance component

$$\sigma_a = \sigma_{a,smooth} + \sigma_{a,res}, \tag{2.10}$$

while the elastic scattering cross section may be split into the potential and resonance components

$$\sigma_{se} = \sigma_p + \sigma_{se,res}. \tag{2.11}$$

We discuss one particular model for representing the resonance cross section in Section 2.5, together with further elaboration on the angular and energy distributions of neutrons emerging from elastic scattering.

## 2.3 Nuclear Fission Process

The neutron–nucleus reaction of particular interest to nuclear reactor physics is naturally the fission process. The fission reaction, as part of the general CN reaction process, occurs primarily with heavy nuclides, often referred to as *actinides* with the atomic number $Z \geq 90$. Splitting of the CN may occur with the absorption of both low-energy neutrons, known as *thermal neutrons*, and high-energy neutrons and may be visualized through a liquid drop model. The CN with excess energy may undergo oscillations and deform into a shape of liquid drop, which may eventually split and release one or more neutrons. The probability of fission reaction increases for a target nucleus with an even number of protons and an odd number of neutrons such that even with a negligible kinetic energy of incident neutrons, the resulting CN may undergo a fission process and release neutrons. The primary examples of nuclides with the particular combination of proton and neutron numbers are $^{233}$U, $^{235}$U, $^{239}$Pu, and $^{241}$Pu, known as *fissile nuclides*. On the other hand, a target nucleus with an even number each of both protons and neutrons usually requires neutrons with kinetic energy $E \geq 0.1$ MeV to undergo fission. Nuclides of this type are referred to as *fertile nuclides*, e.g. $^{234}$U, $^{238}$U, $^{240}$Pu, and $^{242}$Pu. A fertile nucleus upon neutron capture typically undergoes a radiative capture process resulting in a fissile nucleus.

**Table 2.1** Fission energy breakdown for $^{233}U$, $^{235}U$, $^{238}U$, $^{239}Pu$, and $^{241}Pu$.

| **Component** | $^{233}U$ | $^{235}U$ | $^{238}U$ | $^{239}Pu$ | $^{241}Pu$ |
|---|---|---|---|---|---|
| Fission product | 168.34 | 169.13 | 169.54 | 175.66 | $175.36 \pm 0.68$ |
| Prompt neutron | 4.904 | 4.828 | 4.631 | 6.071 | $5.99 \pm 0.13$ |
| Delayed neutron | 0.003 | 0.008 | 0.023 | 0.003 | $0.005 \pm 0.001$ |
| Prompt gamma | 7.736 | 7.281 | 5.610 | 6.369 | $7.64 \pm 0.69$ |
| Delayed gamma | 5.235 | $6.33 \pm 0.05$ | $8.25 \pm 0.07$ | $5.17 \pm 0.06$ | $6.40 \pm 0.09$ |
| Delayed beta | 5.020 | $6.50 \pm 0.05$ | $8.48 \pm 0.08$ | $5.31 \pm 0.06$ | $6.58 \pm 0.09$ |
| Neutrino | 6.875 | $8.75 \pm 0.07$ | $11.39 \pm 0.11$ | $7.14 \pm 0.09$ | $8.85 \pm 0.12$ |
| Total | 198.116 | 202.827 | 207.925 | 205.719 | $210.83 \pm 0.25$ |

*Source*: ENDF/B-VIII.0 International Atomic Energy Agency [NDS18].

The fission process typically results in two light nuclei known as *fission fragments* or *fission products*. In addition to the binary fission process, there exists a small, finite probability of a ternary fission process releasing three fission products.

Together with the fission products, the fission process also releases a significant amount of energy and a number of neutrons, with the energy breakdown tabulated for four major fissile nuclides $^{233}U$, $^{235}U$, $^{239}Pu$, and $^{241}Pu$ and the major fertile nuclide $^{238}U$ in Table 2.1. Approximately 200 MeV of energy is released together with ~2.45 neutrons from the thermal fission of $^{235}U$. The fission energy is carried off by fission products, neutrons, photons, and $\beta$-particles. Note that the fission products decay through $\beta$- and $\gamma$-decay processes and generate heat continually even after the fission process is terminated in a reactor core. The *fission product decay heat* accounts for $6 \sim 7\%$ of the total fission energy release of ~200 MeV and decreases as a function of the decay constants of the fission products. Most of the fission products are rich in neutrons and decay with the release of $\beta$-particles. In the decay process, energetic photons are also emitted. Thus, the kinetic energy of electrons and photons are deposited in the fuel, coolant, and structural materials as part of the radioactive decay process of the fission products and has to be duly considered in the safety assessment of any nuclear reactor core.

The yield of fission products is plotted in Figure 2.3 as a function of the fission product mass number $A$ for the major fissile nuclide $^{235}U$ for thermal neutron fission events at 0.1 eV. The double-humped plots are peaked around atomic number $Z = 35$ and $Z = 60$ and are fairly similar for other fissile and fertile nuclides. Vertical bars denoting approximate estimates of 1.0 standard deviation for the data points are used in the plot. Fast fission events produce a profile that is a bit more peaked than indicated in Figure 2.3.

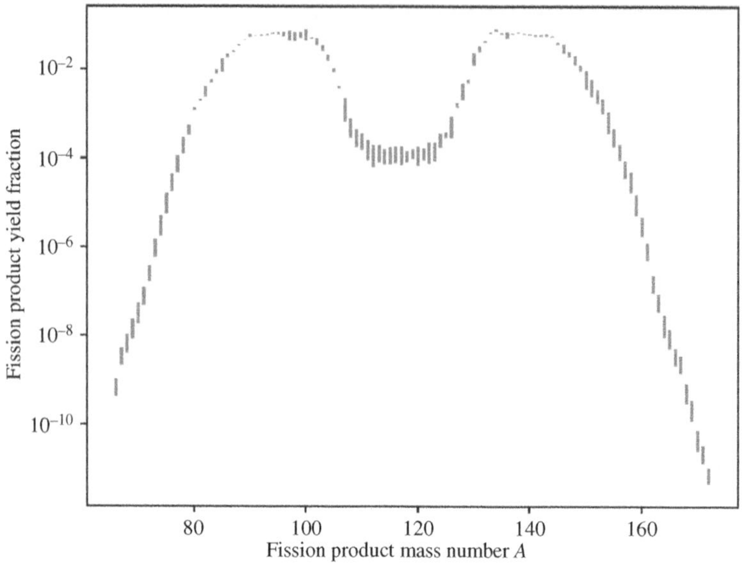


**Figure 2.3** Fraction of fission product released vs. mass number for thermal fission of $^{235}$U for incident neutron energy of 0.1 eV, ENDF/B-VIII.0 MT = 454, MF = 8. *Source*: ENDF/B-VIII.0 International Atomic Energy Agency [NDS18].

Most of the neutrons released from the fission process are released immediately upon the decay of the CN and are known as *prompt neutrons*, while a small fraction of neutrons released through the decay of fission products are known as *delayed neutrons*. The fission products that produce delayed neutrons are known as the *delayed neutron precursors* and they are usually collected into six groups for convenient handling in reactor kinetics studies as detailed in Chapter 8. The total fraction $\beta$ of delayed neutrons, decay constants, and half-lives together with fractional yields, $a_i = \beta_i/\beta, i = 1, \ldots, 6$, in a six-group structure are summarized in Table 2.2. It is worth noting that the total delayed neutron fraction $\beta$ is significantly different among the major fissile and fertile nuclides of interest.

We also note from the distribution of fission energy in Table 2.1 that the energy associated with both prompt and delayed neutrons varies visibly from nuclide to nuclide. This implies that the number of neutrons released from the fission process also depends on the fissioning nuclide, as illustrated in Figures 2.4 and 2.5. Figure 2.4 indicates that the average total number $\overline{\nu}(E)$ of neutrons released from the fission process is generally larger for Pu fission than U fission, as illustrated for two key nuclides, supporting the tabulated data of Table 2.1. It should be noted that an *overbar* symbol is used for $\overline{\nu}(E)$ to indicate that some fission events may release more or less than two neutrons as discussed earlier.

**Table 2.2** Delayed neutron data for $U^{233}$, $U^{235}$, $Pu^{239}$, $Pu^{241}$, and $U^{238}$ for incident neutron energy of 0.1 eV, ENDF/B-VIII.0 MF = 1, 5, 7, 9, 11, MT = 455.

| | **$^{233}$U: $\beta = 0.002696$** | | | **$^{235}$U: $\beta = 0.006523$** | | |
|---|---|---|---|---|---|---|
| **Precursor group** | **Decay constant $\lambda_i$ (s$^{-1}$)** | **Half-life $t_{1/2}$ (s)** | **Fractional yield $a_i = \beta_i/\beta$** | **Decay constant $\lambda_i$ (s$^{-1}$)** | **Half-life $t_{1/2}$ (s)** | **Fractional yield $a_i = \beta_i/\beta$** |
| 1 | 0.0129 | 53.7323 | 0.0859 | 0.0133 | 51.9756 | 0.0350 |
| 2 | 0.0347 | 19.9754 | 0.2292 | 0.0327 | 21.1719 | 0.1807 |
| 3 | 0.1193 | 5.8101 | 0.1781 | 0.1208 | 5.7389 | 0.1725 |
| 4 | 0.2862 | 2.4219 | 0.3516 | 0.3028 | 2.2893 | 0.3868 |
| 5 | 0.7877 | 0.8800 | 0.1142 | 0.8495 | 0.8160 | 0.1586 |
| 6 | 2.4417 | 0.2839 | 0.0409 | 2.8530 | 0.2430 | 0.0664 |

| | **$^{239}$Pu: $\beta = 0.002251$** | | | **$^{241}$Pu: $\beta = 0.005500$** | | |
|---|---|---|---|---|---|---|
| **Precursor group** | **Decay constant $\lambda_i$ (s$^{-1}$)** | **Half-life $t_{1/2}$ (s)** | **Fractional yield $a_i = \beta_i/\beta$** | **Decay constant $\lambda_i$ (s$^{-1}$)** | **Half-life $t_{1/2}$ (s)** | **Fractional yield $a_i = \beta_i/\beta$** |
| 1 | 0.0133 | 52.2302 | 0.0363 | 0.0136 | 50.9705 | 0.0180 |
| 2 | 0.0309 | 22.4457 | 0.2364 | 0.0300 | 23.1311 | 0.2243 |
| 3 | 0.1134 | 6.1140 | 0.1789 | 0.1167 | 5.9380 | 0.1426 |
| 4 | 0.2925 | 2.3697 | 0.3267 | 0.3069 | 2.2585 | 0.3493 |
| 5 | 0.8575 | 0.8083 | 0.1702 | 0.8701 | 0.7966 | 0.1976 |
| 6 | 2.7297 | 0.2539 | 0.0515 | 3.0028 | 0.2308 | 0.0682 |

| | **$^{238}$U: $\beta = 0.018010$** | | |
|---|---|---|---|
| **Precursor group** | **Decay constant $\lambda_i$ (s$^{-1}$)** | **Half-life $t_{1/2}$ (s)** | **Fractional yield $a_i = \beta_i/\beta$** |
| 1 | 0.0136 | 50.8545 | 0.0139 |
| 2 | 0.0313 | 22.1212 | 0.1128 |
| 3 | 0.1233 | 5.6198 | 0.1310 |
| 4 | 0.3237 | 2.1411 | 0.3851 |
| 5 | 0.9060 | 0.7651 | 0.2540 |
| 6 | 3.0487 | 0.2274 | 0.1031 |

*Source*: ENDF/B-VIII.0 International Atomic Energy Agency [NDS18].

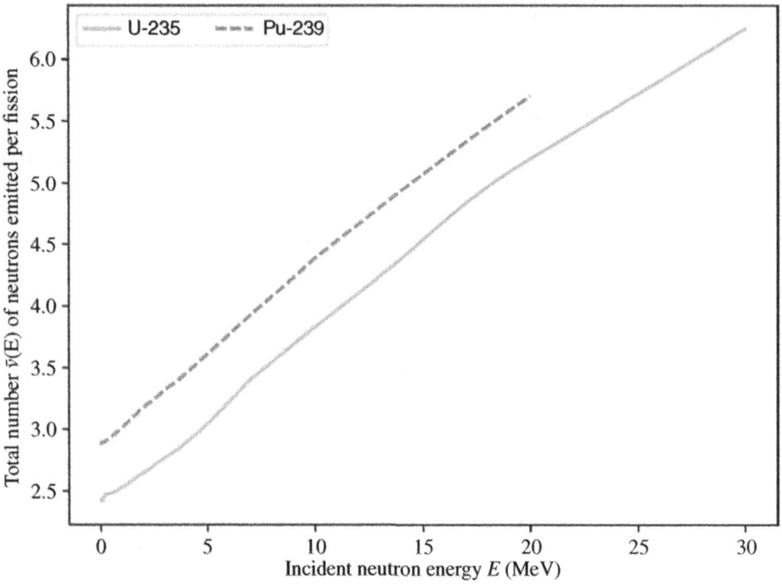


**Figure 2.4** Number of total fission neutrons emitted per fission for $^{235}U$ and $^{239}Pu$, ENDF/B-VIII.0 MF = 1, MT = 452. *Source*: International Atomic Energy Agency [NDS18]/International Atomic Energy Agency/Public domain.

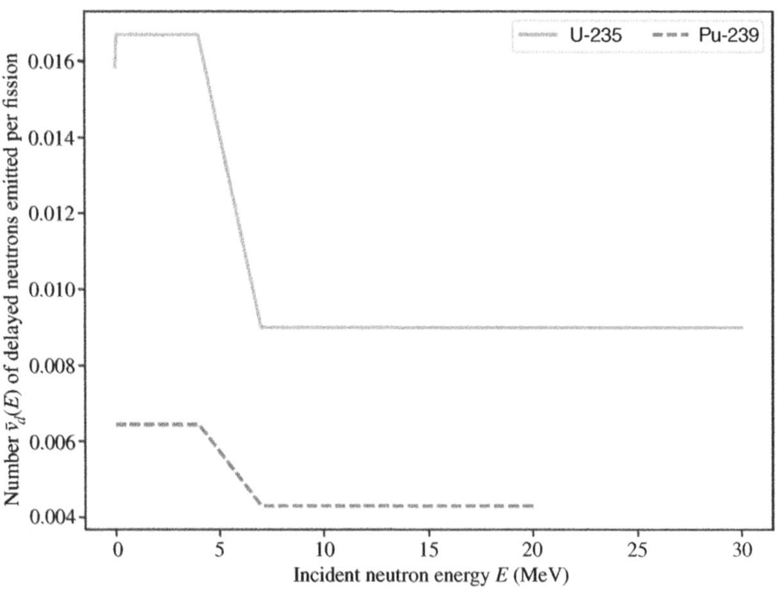


**Figure 2.5** Number of delayed neutrons released as a function of incident neutron energy for $^{235}U$ and $^{239}Pu$, ENDF/B-VIII.0 MF = 1, MT = 455. *Source*: International Atomic Energy Agency [NDS18]/International Atomic Energy Agency/Public domain.

Note also that the average number $\overline{\nu}(E)$ of neutrons emitted in the fission process depends heavily on the neutron energy $E$ that induces nuclear fission. We verify that the average number of neutrons released from thermal fission of neutrons for $^{235}$U with the incident energy $E = 0.025 \sim 0.1$ eV is equal to 2.45, as generally assumed. For other U and Pu nuclides, the general trends illustrated in Figure 2.4 for $^{235}$U and $^{239}$Pu, respectively, hold. Associated also with the dependence of $\overline{\nu}(E)$ on the incident neutron energy $E$ is the energy dependence of the average number $\overline{\nu}_d(E) = \beta \cdot \overline{\nu}(E)$ of delayed neutrons released from the fission process, as illustrated for $^{235}$U and $^{239}$Pu fission in Figure 2.5. The energy dependence for $\overline{\nu}_d(E)$ noted is, however, of little significance for nuclear reactor physics considerations because practically all fission events take place with energy far below 5 MeV.

To conclude the discussion on characteristic features of nuclear fission, we compare in Figure 2.6 the emission energy spectra $\chi(E' \to E)$ for prompt and delayed neutron emissions as a function of the emerging neutron energy $E$ for $^{235}$U fission. Both spectra follow approximately a Maxwell–Boltzmann distribution but with the peak energy significantly lower for the delayed neutron emission spectrum. The difference in emitted energy spectra plays a role in determining the *effective* delayed neutron fraction for proper representation of delayed neutron effects in reactor kinetics discussed in Chapter 8.

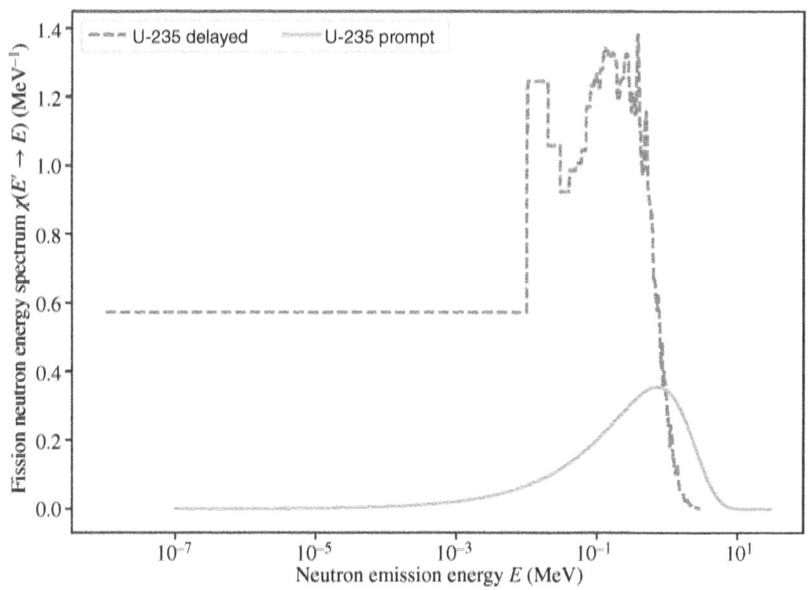


**Figure 2.6** Energy spectrum of fission neutrons emitted for $^{235}$U, ENDF/B-VIII.0 MF = 5, MT = 18 and 455. *Source*: International Atomic Energy Agency [NDS18]/International Atomic Energy Agency/Public domain.

**Table 2.3** Contents of evaluated nuclear data file.

| File number | Section and material | Contents |
|---|---|---|
| MF = 1 | | General information |
| MF = 1 | MT = 451, … , 458 | Number of neutrons and energy released in fission |
| MF = 2 | MT = 151 | Resonance parameters, resolved and unresolved |
| MF = 3 | MT = 1, … , 107 | Reaction cross sections |
| MF = 4 | | Angular distributions of emitted particles |
| MF = 5 | | Energy distributions of emitted particles |
| MF = 7 | | Thermal neutron scattering |
| MF = 8, … , 10 | | Fission product and radioactivity data |
| MF = 12,13,15 | | Photon production and energy spectra |
| MF = 31, … , 40 | | Covariance data |
| | MAT = 1, … , 9999 | Nuclides, elements, compounds |

*Source*: Adapted from Chadwick et al. [Cha06].

We also note that various experimental data summarized in this section for fission reaction physics are extracted from the Evaluated Nuclear Data File (ENDF), Part B, Version VIII.0 [Bro18]. The contents and notational convention for the ENDF library are summarized in Table 2.3. The file structure is carried over from earlier versions of the ENDF databases and is documented in the Version VII release report [Cha06]. Part B of the data libraries refers to a fully evaluated set of data, while Part A contains data that have yet to be closely evaluated and documented. The ENDF database is developed and maintained by the Cross Section Evaluation Working Group (CSEWG) at the National Nuclear Data Center (NNDC) of the Brookhaven National Laboratory. The database was initiated with the Brookhaven report series BNL-325, often referred to as the *Barn Book*, and subsequently evolved into the ENDF series with the final hard copies [Mug81, Mug84, McL88] published for ENDF/B-V. Other cross section libraries widely used in the nuclear community include the Joint Evaluated Fission and Fusion File (JEFF) maintained by the Nuclear Energy Agency (NEA) of the Organization of Economic Cooperation and Development (OECD), with the latest version 3.3 released in November 2017[JEF17].

## 2.4 Two-Body Collision Mechanics and Center-of-Mass System

We derive in this section the relationships governing the elastic scattering [Gol80] of a neutron of mass $m$ and a nucleus of mass $M$. Of particular interest are

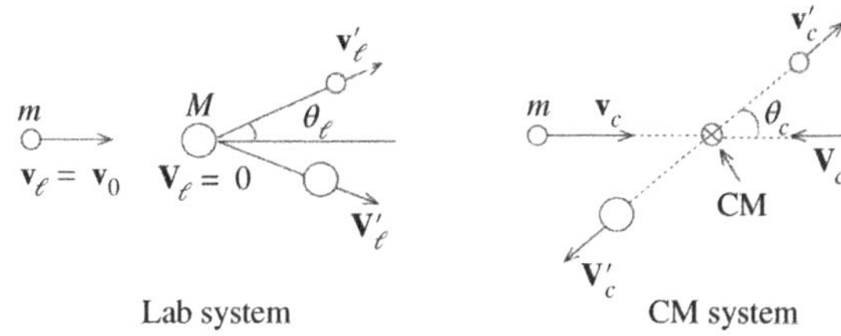


**Figure 2.7** Velocities before and after the collision in Lab and CM systems.

the angular distribution and energy of the neutron emerging from the elastic scattering process in the laboratory (Lab) system. This requires the transformation of the two-body collision mechanics from the CM to Lab coordinate.

To simplify our mathematical treatment, we consider somewhat of an idealized problem where the neutron moves with speed $v_0$ toward the nucleus at rest in the Lab, as illustrated in Figure 2.7. Here, the velocities after the collision are indicated with primes. In the elastic collision under consideration, both linear momentum and kinetic energy of the colliding particles are preserved in both coordinate systems. In the CM system, the scattering particles move directly toward each other before the collision and move away from each other scattered through an angle $\theta_c$. Furthermore, for an observer located at the CM, the total linear momentum of the scattering particles is zero throughout the entire collision process. This follows from the definition of a center of mass.

**1. Basic relationships between the CM and Lab frames**
Represent the conservation of momentum and kinetic energy in the CM system using the velocities defined in Figure 2.7

$$m\mathbf{v}_c + M\mathbf{V}_c = 0 = m\mathbf{v}'_c + M\mathbf{V}'_c, \tag{2.12}$$

$$\frac{1}{2}mv_c^2 + \frac{1}{2}MV_c^2 = \frac{1}{2}m(v'_c)^2 + \frac{1}{2}M(V'_c)^2. \tag{2.13}$$

From these two conservation equations, it may be readily shown that both particles move away in the CM frame, each with the same speeds after the collision as those when they enter the collision

$$v'_c = |\mathbf{v}'_c| = v_c, \tag{2.14a}$$

$$V'_c = |\mathbf{V}'_c| = V_c. \tag{2.14b}$$

With the understanding that neutron velocity in the Lab frame can be written as a vector sum of its velocity in the CM frame and velocity $\mathbf{v}_{CM}$ of the CM itself

$$\mathbf{v}'_\ell = \mathbf{v}'_c + \mathbf{v}_{CM}, \tag{2.15}$$

Figure 2.8 clarifies the relationship between the velocities and scattering angles in the Lab and CM frames before and after the collision.

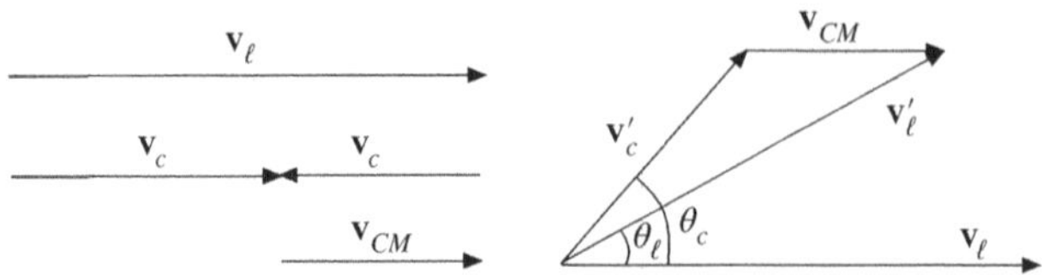


(a) Before collision (b) After collision

**Figure 2.8** Relationship between velocities and scattering angles (a) before and (b) after the collision.

**2. Linear momentum in the CM and Lab frames**

Since the CM itself moves with velocity $\mathbf{v}_{CM}$ with a combined mass of two scattering particles and the linear momentum of the CM is equal to the total linear momentum before the collision in the Lab system, we obtain

$$(m+M)\mathbf{v}_{CM} = m\mathbf{v}_\ell + M\mathbf{V}_\ell = m\mathbf{v}_\ell, \mathbf{V}_\ell = 0. \tag{2.16}$$

Thus, we obtain an expression for the speed of the CM

$$v_{CM} = \frac{m}{m+M}v_0 = \frac{\mu}{M}v_0 = \frac{1}{A+1}v_0, \tag{2.17}$$

which remains constant throughout the collision process due to the conservation of linear momentum. In Eq. (2.17), we have introduced the *mass number* $A$ of the nucleus and *reduced mass* $\mu$

$$A = \frac{M}{m},\ \mu = \frac{mM}{m+M} \quad \text{or} \quad \frac{1}{\mu} = \frac{1}{m} + \frac{1}{M}. \tag{2.18}$$

Equations (2.14a) and (2.17) then yield expressions for the CM velocities for the scattering particles:

$$v'_c = v_c = v_0 - v_{CM} = \frac{\mu}{m}v_0 = Av_{CM}, \tag{2.19a}$$

$$V'_c = |\mathbf{V}_c| = |\mathbf{V}_\ell - \mathbf{v}_{CM}| = \frac{\mu}{M}v_0. \tag{2.19b}$$

In Eqs. (2.19), we also invoked the relationship between the Lab and CM frames equivalent to Eq. (2.15).

**3. Kinetic energy in the CM and Lab frames**

Determine first an expression for the total kinetic energy $E_c$ of the scattering particles in the CM, written in terms of reduced mass $\mu$

$$E_c = \frac{1}{2}mv_c^2 + \frac{1}{2}MV_c^2 = \frac{1}{2}\mu v_0^2, \tag{2.20}$$

noting that the initial Lab speed $v_0$ of the neutron is simply the relative speed between the two particles in the Lab system. With Eq. (2.19), note also that $v_0$ is the relative speed between the particles in the CM frame. This observation allows us to interpret the kinetic energy $E_c$ as that of one combined particle with mass $\mu$ moving with speed $v_0$. This then suggests that the two-body collision mechanics in the Lab frame is converted into one-body mechanics in the CM frame, which provides a powerful technique to obtain complex equations in a significantly simplified manner [Gol80]. One such example is the Breit–Wigner formula discussed in Section 2.5, which was derived [Eng66, Bre36] in terms of the CM energy $E_c$, rather than in terms of the kinetic energy of the incident neutron in the Lab system

$$E_\ell = \frac{1}{2}mv_0^2 = E_0. \tag{2.21}$$

Without too much effort, we may rewrite Eq. (2.20)

$$E_c = E_\ell \frac{M}{m+M} = E_\ell - \frac{1}{2}(m+M)v_{CM}^2 < E_\ell, \tag{2.22}$$

noting clearly that $E_c < E_\ell$, with the difference equal to the kinetic energy of the CM movement.

**4. Scattering angle and neutron kinetic energy after collision**

With the relationship between the scattering angles and velocities after the scattering illustrated in Figure 2.8, we now turn our attention to the task of obtaining equations connecting the Lab neutron scattering angle to the CM scattering angle:

$$(v'_\ell)^2 = (v'_c)^2 + v_{CM}^2 + 2v'_c v_{CM} \cos\theta_c, \tag{2.23a}$$

$$v'_\ell \cos\theta_\ell = v'_c \cos\theta_c + v_{CM}. \tag{2.23b}$$

Combining Eqs. (2.23) and using Eqs. (2.17) and (2.19a) yields

$$\cos\theta_\ell = \frac{v'_c \cos\theta_c + v_{CM}}{v'_\ell} = \frac{1 + A\cos\theta_c}{\sqrt{1 + A^2 + 2A\cos\theta_c}}, \quad A = \frac{M}{m}. \tag{2.24}$$

From Figure 2.8, note $v'_\ell \sin\theta_\ell = v'_c \sin\theta_c$, which yields, together with Eqs. (2.19a) and (2.23b),

$$\tan\theta_\ell = \frac{\sin\theta_c}{\cos\theta_c + \dfrac{1}{A}}. \tag{2.25}$$

Finally, combining Eqs. (2.23a), (2.17), and (2.19a), yields a ratio for kinetic energy $E$ of the neutron emerging from the collision to energy $E_0$ before the collision in the Lab frame

$$\frac{E}{E_0} = \frac{(v'_\ell)^2}{v_0^2} = \frac{1 + A^2 + 2A\cos\theta_c}{(A+1)^2} = \frac{(1+\alpha) + (1-\alpha)\cos\theta_c}{2}, \tag{2.26}$$

with the definition

$$\alpha = \frac{(A-1)^2}{(A+1)^2}. \tag{2.27}$$

Equation (2.26) yields the maximum and minimum values of the Lab kinetic energy of the scattered neutron as a function of the CM scattering angle:

$$\begin{cases} E_{max} = E_0 & \text{for} \quad \theta_c = 0\text{: glancing collision,} \\ E_{min} = \alpha E_0 & \text{for} \quad \theta_c = \pi\text{: head-on collision.} \end{cases} \tag{2.28}$$

The expression for $E_{min}$ provides a physical interpretation for the parameter $\alpha$ of Eq. (2.27) as the *minimum fraction* of the neutron energy attainable after collision.

## 2.5 Single-Level Breit–Wigner Formula for Resonance Reaction

Breit and Wigner [Bre36, Eng66], using the concept of virtual potential in partial-wave expansions for the Schrödinger equation [Kra88, Ser89], derived a theoretical model to represent resonance reactions resulting in a CN formation. As discussed in Section 2.2, the probability of the CN formation increases sharply when the excited energy level of the reaction lies in the vicinity of a quantum state of the CN. For neutron reactions with moderate and high mass number with $E_b \simeq 8.0\,\text{MeV}$, the separation of energy levels is $1 \sim 10\,\text{eV}$ so that the resonance reaction involving neutrons occurs typically for neutron energies of this order of magnitude. We will not attempt to derive [Eng66] the *single-level Breit–Wigner formula* in this section, but merely make use of it, with some physical interpretation of the formula provided.

For a neutron undergoing low-energy *s*-wave scattering with a nucleus, the reaction probability or the reaction cross section associated with a CN formation is written as a sum of the radiative capture cross section $\sigma_\gamma = \sigma(n,\gamma)$ and the scattering cross section $\sigma_n = \sigma(n,n)$

$$\sigma_\gamma = \frac{\pi}{k^2}\frac{\Gamma_n\Gamma_\gamma}{(E_c - E_0)^2 + (\Gamma/2)^2}, \quad k = \frac{2\pi}{\lambda}, \tag{2.29a}$$

$$\sigma_n = \frac{\pi}{k^2}\left|\frac{\Gamma_n}{(E_c - E_0) + i\Gamma/2} + 2kR\right|^2, \tag{2.29b}$$

as a function of kinetic energy $E_c$ of the neutron and nucleus in the CM system and the CM energy $E_0$ corresponding to the peak of the resonance, together with the total level width $\Gamma$, partial level widths $\Gamma_n$ and $\Gamma_\gamma$, wave number $k$ and wavelength $\lambda$ of the de Broglie wave [Kra88, Ser89] associated with the CN formation, and the nucleus radius $R$. Explicitly writing out the magnitude of the complex expression for $\sigma_n$ yields

$$\sigma_n = \frac{\pi}{k^2}\frac{\Gamma_n^2}{(E_c - E_0)^2 + (\Gamma/2)^2} + \frac{4\pi R}{k}\frac{\Gamma_n(E_c - E_0)}{(E_c - E_0)^2 + (\Gamma/2)^2} + 4\pi R^2, \tag{2.30}$$

where the three terms represent (i) resonance scattering, (ii) interference, and (iii) potential scattering cross sections, respectively. Add the statistical factor

$$g = \begin{cases} \dfrac{2J+1}{2(2I+1)}, & I \neq 0, \\ 1, & I = 0, \end{cases} \tag{2.31}$$

with $I$ representing the spin of the target nucleus and $J = I \pm 1/2$ representing the spin of the CN, and define the total CN cross section:

$$\sigma_0 = \frac{4\pi}{k^2}\frac{\Gamma_n}{\Gamma}g\bigg|_{k=k_0,E_c=E_0} = \frac{4\pi}{k_0^2}\frac{\Gamma_n(E_0)}{\Gamma}g. \tag{2.32}$$

For the $s$-wave scattering associated with low-energy neutron interactions, $I = 0$ and $g = 1$, corresponding to the quantum number $\ell = 0$ for the angular momentum of the relative motion between the neutron and nucleus. Note also that the de Broglie wave number $k$ and the scattering partial width $\Gamma_n$ are both proportional to $\sqrt{E_c}$, while the total width $\Gamma$ is nearly independent of $E_c$ and $\Gamma \gg \Gamma_n$. These observations allow recasting Eq. (2.29) as

$$\sigma_\gamma(E_c) = \sigma_0 \frac{k_0^2}{k^2}\frac{\Gamma_n(E_c)}{\Gamma_n(E_0)}\frac{\Gamma_\gamma}{\Gamma}\frac{1}{1+x^2} = \sigma_0\sqrt{\frac{E_0}{E_c}}\frac{\Gamma_\gamma}{\Gamma}\frac{1}{1+x^2}, \tag{2.33a}$$

$$\sigma_n(E_c) = \sigma_0\sqrt{\frac{E_0}{E_c}}\frac{\Gamma_n(E_c)}{\Gamma}\frac{1}{1+x^2} + \frac{4\pi R\sigma_0}{\lambda_0}\frac{x}{1+x^2} + 4\pi R^2, \tag{2.33b}$$

with

$$x = \frac{E_c - E_0}{\Gamma/2}, \lambda_0 = \frac{h}{\sqrt{2\mu E_0}} = \frac{h}{\sqrt{2mE_{\ell,0}}}\frac{A+1}{A},$$

the CM kinetic energy $E_0 = \frac{1}{2}\mu v_0^2$ of interacting particles defined in Eq. (2.20), and the Planck constant $h = 6.626 \times 10^{-34}$ J·s. Combining the resonance portion of $\sigma_n$ with $\sigma_\gamma$ finally yields the *total resonance cross section* as a function of neutron energy $E_c$ in the CM system

$$\sigma_t(E_c) = \sigma_0\sqrt{\frac{E_0}{E_c}}\frac{1}{1+x^2}, \tag{2.34}$$

where $E_0$ is the resonance energy corresponding to the peak of the resonance and is equal to a quantum level of the CN formed. The parameter $\sigma_0$ is the *peak total cross section* of the resonance, while the *level width* $\Gamma$ is the full width at half maximum (FWHM) of the resonance, as illustrated in Figure 2.9.

According to Heisenberg's uncertainty principle [Kra88, Ser89], the product of the uncertainty $\Delta E$ in energy measurement and the uncertainty $\Delta t$ in time

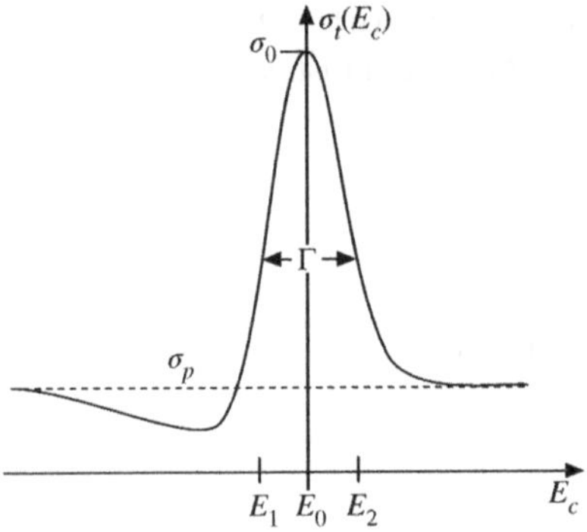


**Figure 2.9** Resonance cross section as a function of neutron energy.

measurement is constrained by $\hbar = h/2\pi$

$$\Delta E \Delta t \geq \frac{\hbar}{2}. \tag{2.35}$$

Thus, we may interpret [Gla52] the level width $\Gamma$ as the uncertainty or indefiniteness associated with the determination of the resonance energy $E_0$, which then suggests

$$\Gamma \propto \frac{1}{\tau}, \tag{2.36}$$

where $\tau$ is the mean life of the particular quantum level of the CN. Since the inverse of the mean life is the decay constant associated with a particular event, we may finally interpret that $\Gamma$ represents the decay probability of the resonance. Thus, $\Gamma$ may be constructed as a sum of *partial widths* associated with various decay channels, radiative capture, scattering, and fission

$$\Gamma = \Gamma_\gamma + \Gamma_n + \Gamma_f. \tag{2.37}$$

We then note that the cross section associated with a particular decay mode is proportional to the partial width of that particular mode, with the peak total cross section written as a sum of three partial cross sections

$$\sigma_0 = \sigma_t(E_0) = \sigma_\gamma(E_0) + \sigma_n(E_0) + \sigma_f(E_0). \tag{2.38}$$

The peak resonance cross section $\sigma_0$ of Eq. (2.32) may now be explicitly written as

$$\sigma_0\ [\text{b}] = 2.604 \times 10^6 \frac{1}{E_{\ell,0}} \left(\frac{A+1}{A}\right)^2 \frac{\Gamma_n}{\Gamma} g, \tag{2.39}$$

where $A$ is the mass number of the nuclide and g is the statistical factor given in Eq. (2.31). Note that, in Eq. (2.39), $E_{\ell,0}$ is the total kinetic energy in units of eV of the neutron and nucleus in the laboratory system corresponding to the CM resonance energy $E_c$, and is the energy used in all cross section databases and libraries, e.g. BNL-325 (barn book) [McL88] and the Evaluated Nuclear Data File, part B, version VIII (ENDF/B-VIII) [Bro18].

To conclude our discussion of the Breit–Wigner formula, we offer a few remarks:

(1) Note that for a sharp resonance, with the approximation that $E_c \simeq E_0$ in Eq. (2.34), it may readily be shown that $E_2 - E_1 = \Gamma$, verifying the statement made earlier that $\Gamma$ is equal to the FWHM of the resonance.

(2) The actual range of the influence of a resonance is better represented by the *practical width*

$$\Gamma_p = \Gamma\sqrt{\frac{\sigma_0}{\sigma_p}}, \tag{2.40}$$

which can be shown to be approximately equal to the energy interval over which the sum of resonance scattering and absorption cross sections is equal to or larger than the potential scattering cross section $\sigma_p$ illustrated in Figure 2.9.

(3) For two limiting cases [Gla52] when (a) $E_c \ll E_0$ and (b) $\Gamma \gg E_c - E_0$, obtain an approximate relationship $\sigma_t \propto 1/v$, where $v$ is the neutron speed. Case (a) represents the behavior of many materials well below the first resonance energy, where the neutron cross section often exhibits a "$1/v$-behavior." The broad resonance approximation of case (b) may explain the behavior of $^{10}B$, which is one of the most common neutron-absorbing materials used in a variety of applications.

## 2.6 Differential Scattering Cross Section and Scattering Kernel

Through the simple neutron penetration experiment considered in Section 2.1, we determined in Eq. (2.1) an expression for the number $-dI$ of neutrons suffering collisions in the interval $dx$ per unit cm$^2$ per s. We now wish to extend the simple experiment to obtain an expression for the number $-dI(\mathbf{\Omega})$ of neutrons scattered into solid angle $d\mathbf{\Omega}$ around direction $\mathbf{\Omega}$ in the interval $dx$ per unit cm$^2$ per s. As illustrated in Figure 2.10, we now explicitly assign direction $\mathbf{\Omega}_0$ to the neutrons incident on the slab, again in a collimated beam. Recall that the solid angle $d\mathbf{\Omega}$ is defined as

$$d\mathbf{\Omega} = \frac{dA}{r^2}, \tag{2.41}$$

where $dA$ is the cross-sectional area of a differential cone of length $r$, or the differential cross-sectional area subtended at distance $r$. The solid angle is dimensionless, usually expressed in units of steradian, and should not be considered a differential of unit directional vector $\mathbf{\Omega}$.

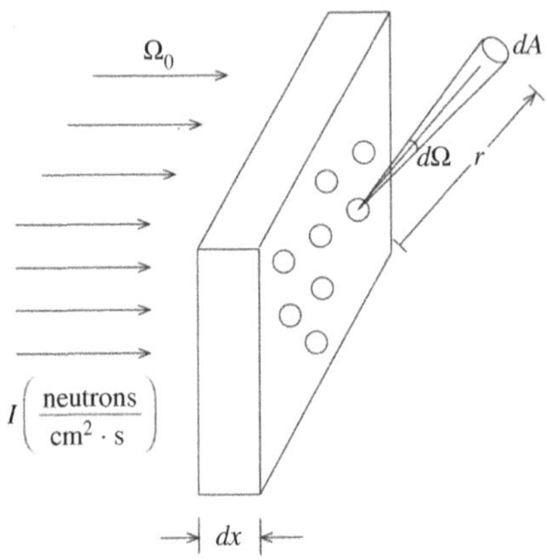


**Figure 2.10** Scattering of collimated beam into solid angle $d\boldsymbol{\Omega}$ around $\boldsymbol{\Omega}$ in distance $dx$.

### 2.6.1 Differential Microscopic Scattering Cross Section

Similar to the procedure we followed in Section 2.1 to determine the total number of neutrons suffering collision in $dx$, we introduce the scattering probability and obtain an expression for the number of neutrons scattered in the interval $dx$ per unit $\text{cm}^2$ per s into solid angle $d\boldsymbol{\Omega}$ around direction $\boldsymbol{\Omega}$

$$-dI(\boldsymbol{\Omega}) = I\sigma(\boldsymbol{\Omega})d\boldsymbol{\Omega}Ndx = -d^2I(\boldsymbol{\Omega}), \tag{2.42}$$

where we indicate that it is a doubly differential quantity associated with both $dx$ and $d\boldsymbol{\Omega}$. Equation (2.42) is recast to determine the fraction of incident neutrons scattered in $dx$ into $d\boldsymbol{\Omega}$

$$-\frac{dI(\boldsymbol{\Omega})}{I} = \sigma(\boldsymbol{\Omega})d\boldsymbol{\Omega}\left(\frac{\text{cm}^2}{\text{nucleus}}\right) \cdot Ndx\left(\frac{\text{number of nuclei}}{\text{cm}^2}\right). \tag{2.43}$$

Equation (2.43) suggests a more precise interpretation for the *differential microscopic scattering cross section* as the effective cross section area of the nucleus for scattering into unit solid angle (steradian), which may be written more formally as:

$$\sigma(\boldsymbol{\Omega}) = \frac{d\sigma(\boldsymbol{\Omega}_0 \rightarrow \boldsymbol{\Omega})}{d\boldsymbol{\Omega}} = \sigma(\boldsymbol{\Omega}_0 \rightarrow \boldsymbol{\Omega}). \tag{2.44}$$

The last two notations clearly indicate that the differential cross section is associated with scattering from direction $\boldsymbol{\Omega}_0$ into unit solid angle around $\boldsymbol{\Omega}$.

Integrating the differential scattering cross section over all possible scattering angles yields the total or integral microscopic cross section

$$\sigma = \int_{4\pi} \sigma(\boldsymbol{\Omega})d\boldsymbol{\Omega}, \tag{2.45}$$

which is the microscopic cross section $\sigma$ introduced in Eqs. (2.1) and (2.2) if the slab under consideration comprises a purely scattering medium. In this case, integrating Eq. (2.43) over $\mathbf{\Omega}$ yields Eq. (2.2) representing the total fraction of neutrons suffering collisions in $dx$, regardless of the scattering angle. Since neutrons undergoing scattering collisions with surrounding nuclei will in general undergo changes in their energy as well as their direction of motion, we may generalize Eq. (2.44)

$$\sigma(E \to E', \mathbf{\Omega} \to \mathbf{\Omega}') = \frac{d^2\sigma(E \to E', \mathbf{\Omega} \to \mathbf{\Omega}')}{dE' d\mathbf{\Omega}'}, \tag{2.46}$$

so that

$$\int_{4\pi} \sigma(E \to E', \mathbf{\Omega} \to \mathbf{\Omega}') d\mathbf{\Omega}' = \sigma(E \to E') = \frac{d\sigma(E \to E')}{dE'}. \tag{2.47}$$

Similarly, integration of the energy-dependent scattering cross section over all energies that the neutron may acquire following the scattering collision yields the total microscopic cross section of Eq. (2.45)

$$\sigma(E) = \int_0^\infty \sigma(E \to E') dE', \tag{2.48}$$

where we emphasize the point that the microscopic cross section is in general a function of the incident neutron energy $E$. Note also that $\sigma(E)$ is a cross section for neutrons of energy $E$ expressed in units of $cm^2$, or more commonly, in units of barn.

### 2.6.2 Scattering Kernel for Isotropic Scattering in CM Frame

For elastic scattering of low-energy neutrons of interest in reactor physics, often referred to as the *s-wave* scattering associated with low-level quantum levels excited in the process, we may assume to a good approximation that the neutrons are scattered *isotropically* in the CM system. Furthermore, recall from Eq. (2.26) that the neutron energy $E$ after the collision is uniquely determined by the CM scattering angle $\theta_c$. Thus, the differential scattering cross section of Eq. (2.46) may be rewritten in a somewhat simplified, and perhaps more intuitive, manner to represent the probability of neutrons suffering scattering collisions at energy $E_0$ and emerging in differential solid angle $d\mathbf{\Omega}_c$ around $\mathbf{\Omega}_c$

$$\sigma_s(E_0, \mathbf{\Omega}_c) d\mathbf{\Omega}_c = \frac{\sigma_s(E_0)}{4\pi} 2\pi \sin\theta_c d\theta_c, \tag{2.49}$$

where we clearly highlight the scattering cross section with the subscript $s$ and indicate the incident energy $E_0$. The solid angle $d\mathbf{\Omega}_c$ is illustrated explicitly in Figure 2.11, with azimuthal symmetry inherent in the CM scattering process

$$d\mathbf{\Omega}_c = \frac{dA}{r^2} = \frac{2\pi r \sin\theta_c \cdot r d\theta_c}{r^2} = 2\pi \sin\theta_c d\theta_c. \tag{2.50}$$

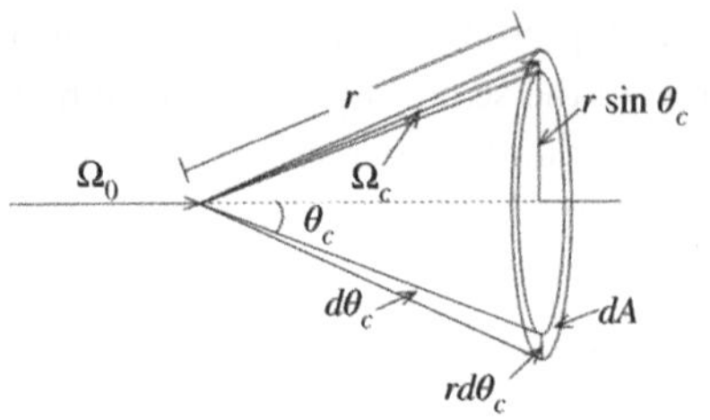


**Figure 2.11** Solid angle with azimuthal symmetry.

Equation (2.49) may be rewritten in terms of the outgoing neutron energy $E$

$$\sigma_s(E_0, \mathbf{\Omega}_c)d\mathbf{\Omega}_c = \frac{\sigma_s(E_0)}{2}\sin\theta_c d\theta_c = -\sigma_s(E_0 \to E)dE, \tag{2.51}$$

where a negative sign is introduced in the last expression to render the microscopic scattering cross section $\sigma_s(E_0 \to E)$ positive with the realization that an elastic scattering generally entails a decrease in the neutron energy, i.e. $-dE > 0$. Solve for $\sigma_s(E_0 \to E)$ from Eq. (2.51)

$$\sigma_s(E_0 \to E) = -\frac{\sigma_s(E_0)}{2}\sin\theta_c\frac{d\theta_c}{dE},$$

and, recalling Eq. (2.26) finally obtain the *elastic scattering kernel*:

$$\sigma_s(E_0 \to E) = \sigma_s(E_0)p(E_0 \to E) = \begin{cases} \dfrac{\sigma_s(E_0)}{E_0(1-\alpha)}, & \alpha E_0 \le E \le E_0, \\ 0, & \text{otherwise.} \end{cases} \tag{2.52}$$

The microscopic scattering cross section $\sigma_s(E_0)$ is the probability of scattering for neutrons of energy $E_0$ in units of barn, and the term $p(E_0 \to E)$ represents the *conditional probability* that neutrons, having suffered scattering at energy $E_0$, will emerge in a unit energy interval around $E$. The scattering kernel is graphically displayed in Figure 2.12. Integrating Eq. (2.52) over the entire neutron energy below

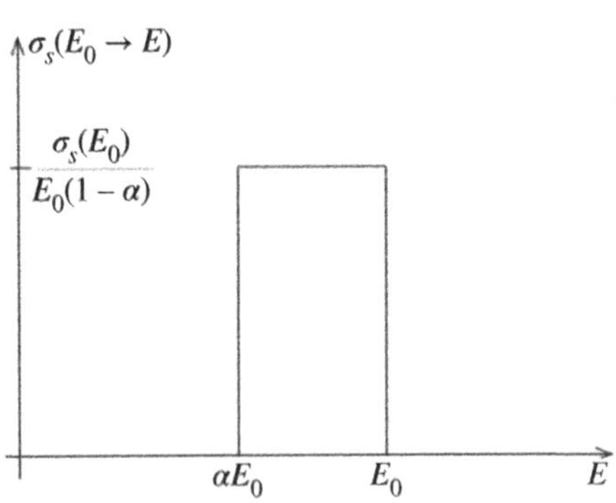


**Figure 2.12** Elastic scattering kernel as a function of outgoing neutron energy.

the incident energy $E_0$ yields

$$\int_0^{E_0} \sigma_s(E_0 \rightarrow E)dE = \int_{\alpha E0}^{E_0} \sigma_s(E_0 \rightarrow E)dE = \sigma_s(E_0). \tag{2.53}$$

This is an intuitive result, since the integral of the scattering kernel over all possible energies of the scattered neutron should simply yield the probability of scattering at energy $E_0$. This is another way of stating that the integral of the conditional probability $p(E_0 \rightarrow E)$ over all outgoing energies should be 1.0.

## 2.7 Further Remarks on Neutron Cross Section

Having discussed key features of neutron–nucleus reactions, followed by the derivation of fundamental relationships governing neutron cross sections, in particular, the scattering cross section and the Breit–Wigner formula for resonance reactions, we now conclude with a few remarks on general behavior [Lam66] of the neutron cross section.

**1. Total cross section $\sigma_t$**

(1) **Light nuclides**: The total cross section indicates a region of nearly constant value, which may sometimes be coupled to a $1/v$-region

$$\sigma_t = C_1 + \frac{C_2}{v}, \tag{2.54}$$

where the constant value $C_1$ may be associated with the elastic scattering cross section and the $1/v$-region associated with the $(n, \gamma)$ radiative capture process with some constant $C_2$. Resonances appear mostly in the keV ~ MeV region, but the cross section values are rather small. This may be qualitatively understood [Gla52] if we recall the concept of de Broglie wavelength [Kra88, Ser89]

$$\lambda = \frac{h}{p} = \frac{h}{mv} = \frac{h}{\sqrt{2mE}} \propto \frac{1}{\sqrt{E}}. \tag{2.55}$$

and visualize the wavelength $\lambda$ as a measure of the wave packet, and hence the probability, associated with the neutron–nucleus interaction. Thus, we can expect that the resonance cross sections will exhibit relatively small values for light nuclides, since the resonances occur at much higher energies for light nuclides than for heavy nuclides.

Figures 2.13 and 2.14 present a few cross section plots for $^{10}B$ and $^{12}C$, respectively, from the latest edition of ENDF/B library, ENDF/B-VIII. For $^{10}B$, the total cross section, which is mostly associated with the absorption cross

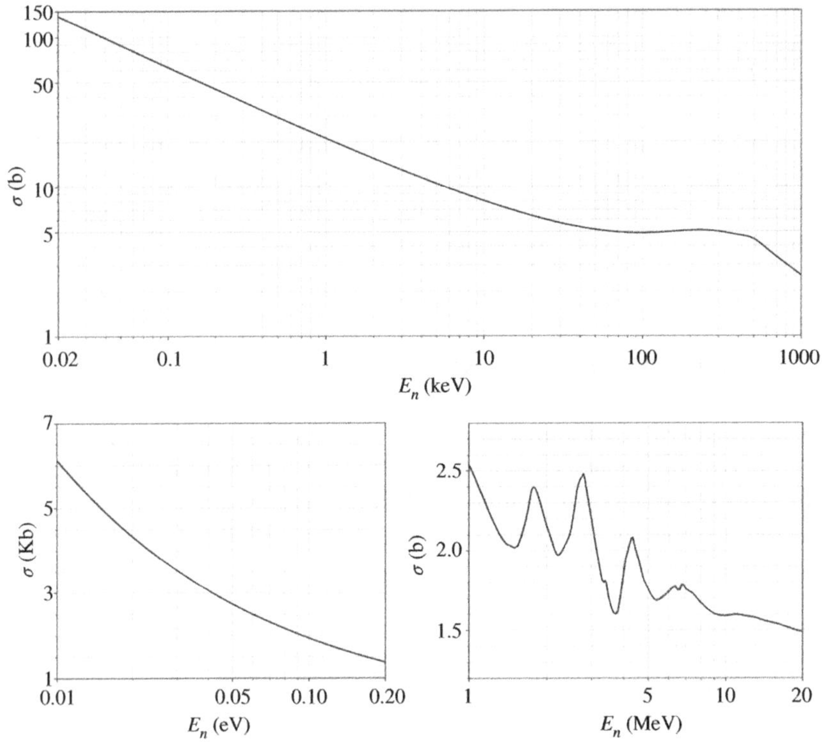


**Figure 2.13** Total cross section $\sigma_t$ of $^{10}B$, ENDF/B-VIII.

section, shows a $1/v$-behavior for a large range of energy, 10 meV ~ 50 keV, as suggested in Eq. (2.54). This point was also discussed as a limiting case of the Breit–Wigner resonance cross section in Section 2.5 and is the reason why $^{10}B$ is the favorite choice as a neutron absorber. In contrast, $^{12}C$ exhibits a nearly constant total cross section, as suggested by $C_1$ of Eq. (2.54), for a broad energy interval all the way up to 50 keV, with some resonances observed at high energy reaching into the MeV range. In passing, we should note that the tabulated cross section data are subject to significant experimental uncertainties in general, with the uncertainty quantification of various cross sections currently in active study with covariance matrices documented in the ENDF/B-VIII databases.

(2) **Heavy nuclides**: The total cross section often shows a $1/v$-behavior for low energies, followed by a multitude of large resonances in the eV ~ keV range. As discussed in connection with Eq. (2.55), since the resonances occur at much lower energies than those for light nuclides, the resonance absorption

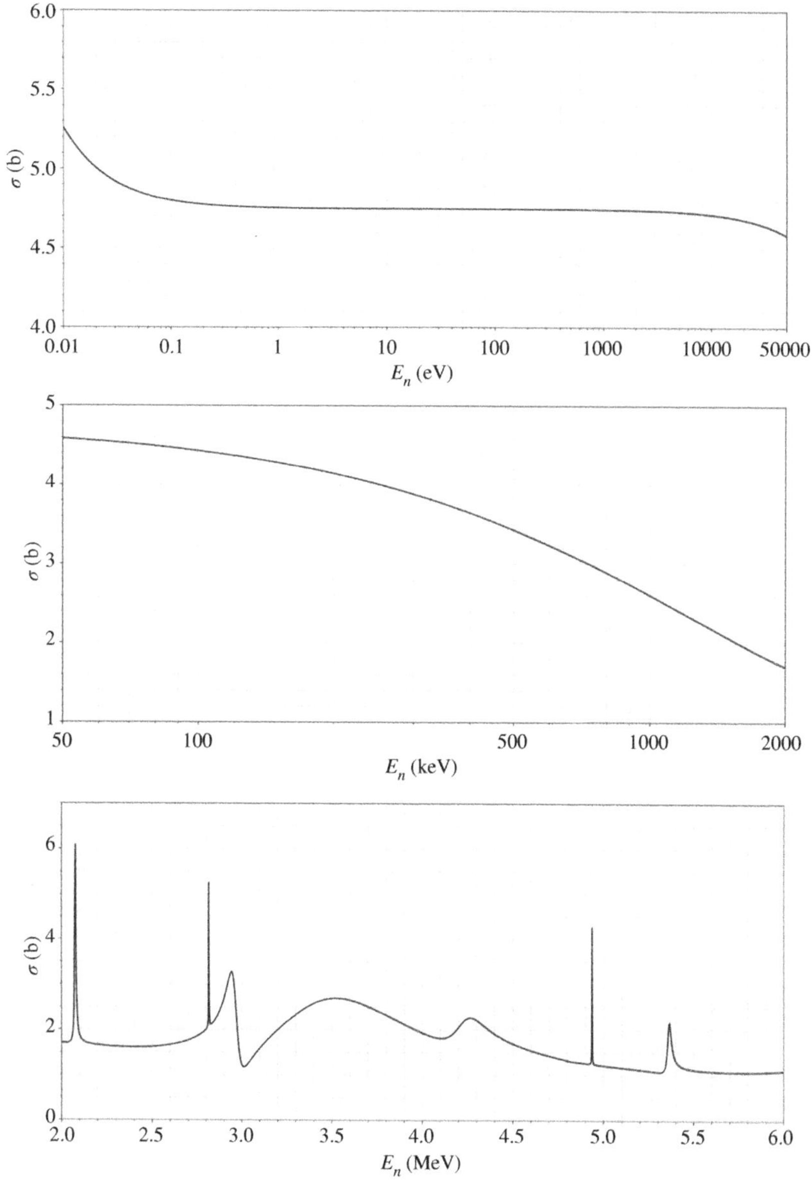


**Figure 2.14** Total cross section $\sigma_t$ of $^{12}C$, ENDF/B-VIII.

cross sections are large and play a significant role in determining reaction rates of neutrons in this energy range. This point is highlighted by low-lying resonances, in the eV range, with peak cross sections of several kb for both $^{238}U$ and $^{239}Pu$ plotted in Figures 2.15 and 2.16, respectively. A number of

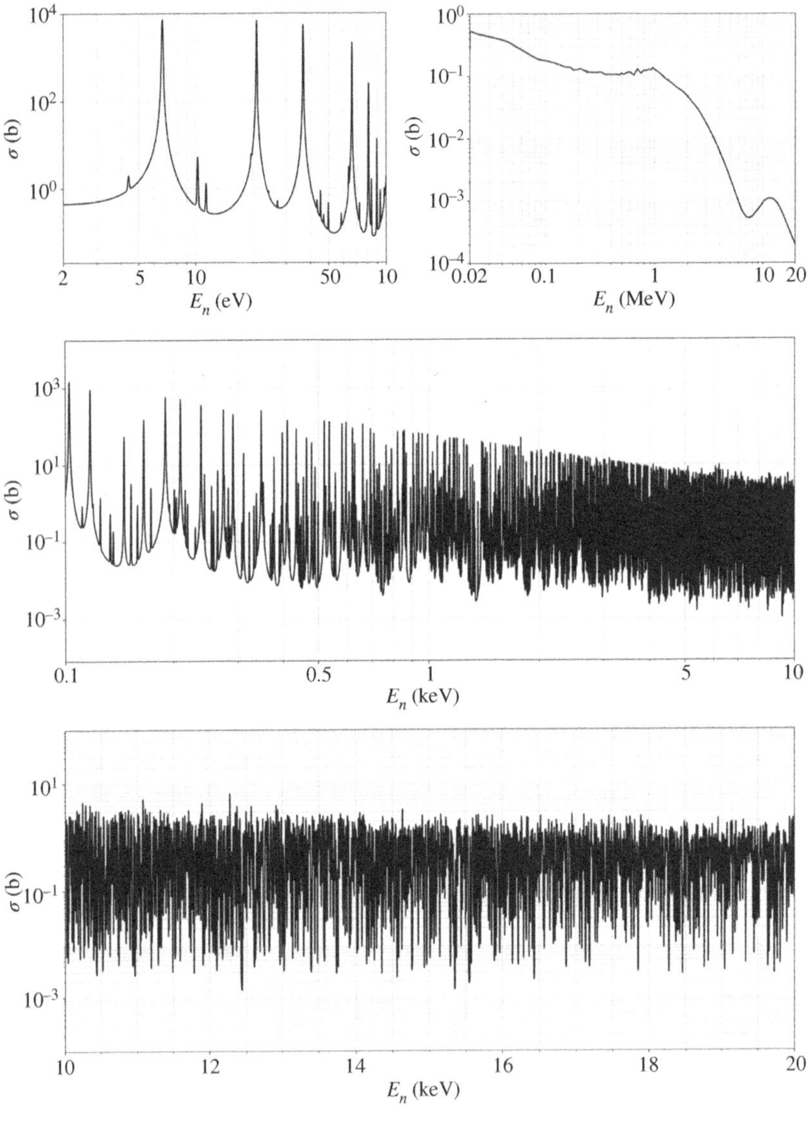


**Figure 2.15** Radiative capture cross section $\sigma_\gamma$ for $^{238}U$, ENDF/B-VIII.

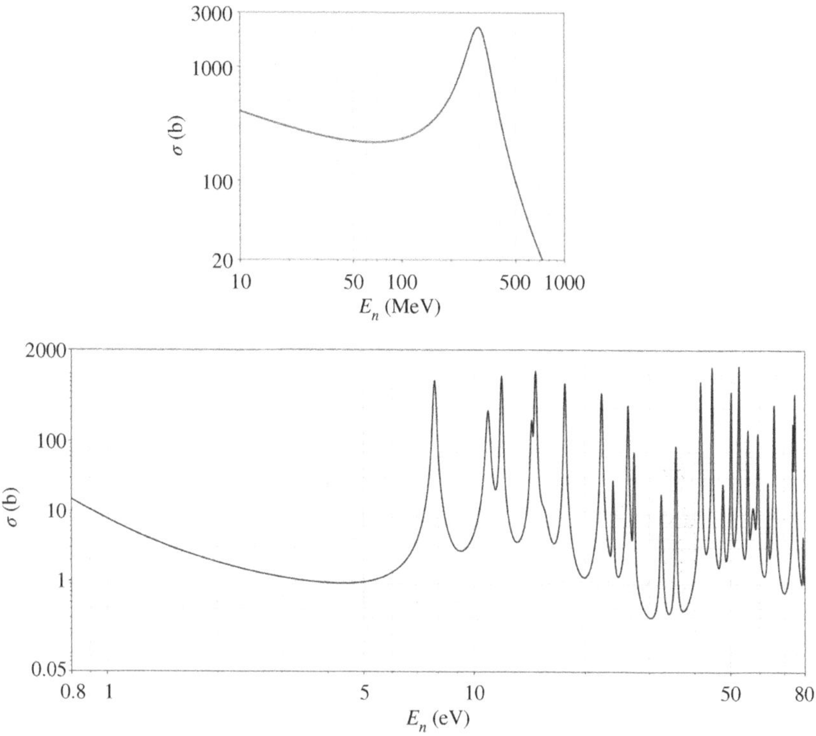


**Figure 2.16** Radiative capture cross section $\sigma_\gamma$ for $^{239}Pu$, ENDF/B-VIII.

nearly overlapping resonances are also noted. These *unresolved resonances* require statistical rather than individual representation [Heb10].

**2. Elastic scattering cross section**

As discussed in Section 2.2, in particular with Eq. (2.11) as part of the general discussion on neutron–nucleus interactions, the elastic scattering cross section $\sigma_{se}$ comprises both the potential scattering cross section $\sigma_p$ and resonance scattering cross section $\sigma_{se,res}$. The combined $\sigma_{se}$ is thus typically nearly constant, characterized by the hardball-type potential cross section, except near localized resonances. In addition, note that for scattering cross sections, crystalline or molecular effects of the particular atom or compound also play an important role, especially for low-energy neutron scattering events.

An example in this regard is a relationship between bound- and free-atom scattering cross sections, which can be studied [Bel70] with Eq. (2.24). For a tightly bound atom in a molecule, the effective mass number $A \gg 1.0$ so that the scattering in the laboratory system may be considered isotropic as in the CM System, with $\cos\theta_c = \cos\theta_\ell = \mu_\ell$,

$$\sigma_{bound}(\mu_\ell) = \frac{1}{2}\sigma_{bound}.$$

Furthermore, for a glancing collision with $\mu_c = \mu_\ell = 1.0$, the nucleus does not recoil and hence the mass of the nucleus is immaterial, resulting in

$$\begin{aligned}\sigma_{bound}(\mu_\ell = 1.0) = \sigma_{free}(\mu_\ell = 1.0) = \sigma_{free}(\mu_c)\left.\frac{d\mu_c}{d\mu_\ell}\right|_{\mu_c=\mu_\ell=1} \\ = \frac{1}{2}\sigma_{free}\left(\frac{A+1}{A}\right)^2,\end{aligned} \tag{2.56}$$

which yields

$$\sigma_{bound} = \left(\frac{A+1}{A}\right)^2\sigma_{free}. \tag{2.57}$$

This indicates that, for a proton with mass number $A = 1$, the bound-atom scattering cross section for a molecule of water may be four times as large as that of a free, unbound proton. Effects of thermal motion of neutrons and nuclei on scattering cross sections for moderating materials, e.g. light water, heavy water, and graphite, are often represented in Monte Carlo calculations by a tabulated function $S(\alpha, \beta)$, where the gain in the linear momentum and kinetic energy of neutrons due to thermal motion of nuclei are represented by dimensionless parameters $\alpha$ and $\beta$, respectively [Heb10].

## References


**[Bel70]** G.I. Bell and S. Glasstone, *Nuclear Reactor Theory*, Van Nostrand Reinhold (1970).

**[Bre36]** G. Breit and E. Wigner, "Capture of Slow Neutrons," *Phys. Rev.* **49**, 519 (1936).

**[Bro18]** D.A. Brown et al., "ENDF/B-VIII.0: The 8th Major Release of the Nuclear Reaction Data Library with CIELO-project Cross Section, New Standards and Thermal Scattering Data," *Nucl. Data Sheets* **148**, 1 (2018).

**[Cha06]** M.B. Chadwick et al., "ENDF/B-VII.0: Next Generation Evaluated Nuclear Data Library for Nuclear Science and Technology," *Nucl. Data Sheets* **107**, 2931 (2006).

**[Eng66]** H.A. Enge, *Introduction to Nuclear Physics*, Addison-Wesley (1966).

**[Gla52]** S. Glasstone and M.C. Edlund, *The Elements of Nuclear Reactor Theory*, D. Van Nostrand (1952).

**[Gol80]** H. Goldstein, *Classical Mechanics*, Addison-Wesley (1980).

**[Heb10]** A. Hébert, *Applied Reactor Physics*, Presses Internationales Polytechnique (2010).

**[JEF17]** Nuclear Energy Agency, http://www.oecd-nea.org/dbdata/jeff/jeff33 (2017).

**[Kra88]** K.S. Krane, *Introductory Nuclear Physics*, Wiley (1988).

**[Lam66]** J.R. Lamarsh, *Introduction to Nuclear Reactor Theory*, Addison-Wesley (1966).

**[McL88]** V. McLane, D.L. Dunford, and P.F. Rose, *Neutron Cross Sections, Neutron Cross Section Curves*, vol. **2**, Academic Press (1988).

**[Mug81]** S.F. Mughabghab, M. Divadeenam, and N.E. Holden, *Neutron Cross Section, Neutron Resonance Parameters and Thermal Cross Sections, Part A, Z = 1 – 60*, vol. **1**, Academic Press (1981).

**[Mug84]** S.F. Mughabghab, *Neutron Cross Section, Neutron Resonance Parameters and Thermal Cross Sections, Part B, Z = 61 – 100*, vol. **1**, Academic Press (1984).

**[NDS18]** International Atomic Energy Agency, https://www-nds.iaea.org/public/download-endf/ENDF-B-VIII.0/000-NSUB-index.htm (2018).

**[Ser89]** R.A. Serway, C.J. Moses, and C.A. Moyer, *Modern Physics*, Saunders College Publishing (1989).

## Problems

**2.1** Prove that for a sharp resonance, the FWHM of a resonance is equal to its level width $\Gamma$.

**2.2** Californium-252 is widely used as a compact neutron source. ${}^{252}_{98}\text{Cf}$ decays by either spontaneous fission or emission of $\alpha$-particles. About 3.09% of the total decay events are spontaneous fission with a half-life of 85.60 yr, and the remaining 96.91% are $\alpha$-decay with a half-life of 2.729 yr. For each spontaneous fission of ${}^{252}\text{Cf}$, 3.75 neutrons are released on the average. (a) What is the neutron generation rate in units of [neutron $\cdot$ s$^{-1}$ mg$^{-1}$] of ${}^{252}_{98}\text{Cf}$? (b) What is the neutron generation rate in units of [neutron $\cdot$ s$^{-1}$ Bq$^{-1}$]? (c) What is the half-life of ${}^{252}\text{Cf}$, accounting for both spontaneous fission and $\alpha$-decay?

**2.3** An alpha particle reacts with a ${}^{23}\text{Na}$ nucleus and emits a proton. (a) What are the compound and residual nuclei in this $(\alpha, p)$ reaction? (b) If the kinetic energy of the alpha particle is 2.0 MeV in the laboratory, what is the excitation energy of the compound nucleus in the center of mass?

**2.4** Starting from Eq. (2.34), derive the expression for the practical width Eq. (2.40).

**2.5** Extend the two-body elastic collision model of Section 2.4 to the case when a nucleus of mass $M$ and a neutron of mass $m$ move toward each other with speed $V_\ell$ and speed $v_\ell$, respectively, and derive an expression for the ratio of the energy $E$ after the collision to $E_0$ before the collision. Determine the condition under which the ratio may be greater than 1.0.

**2.6** For the $^{235}$U fission spectrum $\chi(E) = 0.77\sqrt{E}\exp(-0.775E), E$ in MeV, (a) determine the fraction of fission neutrons that can cause fission in $^{238}$U with the fission threshold of 1.0 MeV and (b) the average energy of fission neutrons.

**2.7** Prove that the kinetic energy of two interacting particles in the CM can be represented by the kinetic energy of a single particle of reduced mass $\mu$ and relative speed between the particles.

**2.8** In the boron neutron capture therapy, thermal neutrons are absorbed in $^{10}$B nuclei via an $(n, \alpha)$ reaction. The kinetic energy of the alpha particles and residual nuclei is then deposited in tumor cells. (a) What are the compound and residual nuclei in this $(n, \alpha)$ reaction? (b) If the neutron energy is 0.025 eV in the laboratory, what is the excitation energy of the compound nucleus in the center of mass? (c) If the kinetic energy of the $\alpha$-particle is 1.47 MeV and a 0.48-MeV $\gamma$-ray is emitted in the reaction, what is the kinetic energy of the residual nucleus obtained in part (a)?

**2.9** A compact neutron source consists of 13 g of $^{238}_{94}$Pu mixed with 7 g of $^{9}_{4}$Be. In the neutron source, 5.5-MeV $\alpha$-particles, emitted from the decay of $^{238}$Pu with a half-life of 87.7 yr, react with $^{9}$Be and produce neutrons. (a) Determine the radioactivity of $^{238}_{94}$Pu in the neutron source in units of Ci. (b) What are the residual nuclei produced through the interaction of $\alpha$-particles with $^{9}$Be nuclei? (c) Determine the maximum energy of neutrons produced and the neutron production rate, with the observation that 30 neutrons are produced typically from $10^6$ $\alpha$-particles.

**2.10** A neutron of speed $v_0$ undergoes an elastic scattering collision with a proton, which is initially stationary in the laboratory. In the CM system, the neutron is scattered off by $\theta_c = \pi/2$. (a) Sketch the velocity of the particles in the CM system after the collision. (b) What is the speed $v_{CM}$ of the center of mass itself? (c) What is the neutron speed $v'_c$ after collision in the

CM system? (d) Based on the results of parts (b) and (c), determine the scattering angle $\theta_\ell$ of the neutron in the laboratory system.

**2.11** In a Compton scattering [Kra88] between a photon and an electron, it may be assumed that the electron of mass $m$ is initially at rest. The photon with incident energy $E$ and wavelength $\lambda$ emerges from the collision at an angle $\theta$ from the initial direction, while the electron moves away from the collision with speed $v$. (a) Set up the energy and momentum balance statements for the photon-electron scattering process, accounting for the relativistic energy of the scattered electron. (b) Using the results of part (a), obtain an expression for the wavelength $\lambda'$ after the collision in terms of $\lambda$ and $\theta$.

**2.12** A photon of energy $E$ emerges at an angle $\theta$ from the initial direction with energy $E'$ following a Compton scattering with an electron. (a) Obtain an expression relating $E'$ to $E$ in units of the rest mass of an electron. (b) What is the maximum kinetic energy that the electron could gain for $E = 2.044$ MeV?

**2.13** A neutron of energy $E_0$ suffers an elastic scattering collision with a nucleus of mass number $A$ initially at rest in the Lab system. (a) What is the probability that the neutron will have energy below $E_0/2$ after scattering? (b) What is the average energy $\overline{E}$ that the neutron is expected to have after scattering? (c) How many scattering collisions with deuterons would a neutron of energy 1.0 MeV undergo, on average, to attain an energy of 1.0 eV?

**2.14** One method of producing neutrons in the laboratory is to bombard alpha particles on a $^7$Li target. (a) What is the residual nucleus in this $(\alpha, n)$ reaction? (b) What is the minimum kinetic energy of the alpha particles in the CM frame to produce neutrons? (c) What is the minimum energy of the alpha particles in the laboratory system?

**2.15** Prove that the interacting particles undergoing elastic collisions move away in the CM system, each with the same speeds as those entering the collision, as indicated in Eqs. (2.14).

**2.16** In a binary fission event in $^{235}$U induced by a thermal neutron capture, one of the fission products is $^{141}$Cs. (a) What is the other fission product if no prompt neutrons are released? (b) If both fission products undergo $\beta$-decays until they become stable, what will be the final products? (c) Estimate the energy released in the fission process. (d) Repeat parts

(a) and (b) with the emission of two prompt neutrons. For atomic mass data, consult https://www-nds.iaea.org/amdc/.

**2.17** The radiative capture cross section of $^{113}$Cd for thermal neutrons is 25 kb. Estimate the thickness of Cd metal foil that would be required to attenuate a thermal neutron beam by a factor of 100.

**2.18** Neutrons of mass $m$ and kinetic energy $E_0$ are incident on a target of stationary nuclei of mass $M$. The neutrons are inelastically scattered from the target nuclei through head-on collisions, i.e. the scattering angle $\theta_c = \pi$ in the CM system and the neutrons have zero kinetic energy after collision in the laboratory system. (a) Determine the speed $v_c'$ of neutrons in the CM system after the head-on collision. (b) Obtain the energy level $Q$ for the state to which the target nuclei are excited in this inelastic collision process.

# 3

# Neutron Flux, Reaction Rate, and Effective Cross Section

In many problems of importance in nuclear reactor physics, we need to determine the rate at which neutrons interact with the surrounding medium and the rate at which they leak out of a given volume. The neutron transport and diffusion equations, for example, represent the balance of neutrons in a reactor core or in general for any medium of interest. To set up a neutron balance equation, we need to write the neutron leakage and reaction rates in terms of the neutron flux and introduce the concept of effective neutron cross section. Power density in a reactor core is directly proportional to the fission reaction rate, and the determination of the critical size of a chain reacting system depends on the neutron leakage and reaction rates as well. In this chapter, we introduce the definition of neutron flux and derive general expressions for neutron reaction rates, which then provide us with the concept of effective neutron cross section.

We begin with the definition of the angular number density and angular neutron flux in Section 3.1, together with physical interpretations for the angular flux. In terms of the angular flux, we then define the scalar flux and net current. In Section 3.2, a general expression for the neutron–nucleus reaction rate is presented in terms of the angular number density of neutrons and that of nuclei, with which the neutrons interact. Preserving the neutron–nucleus reaction rate, while accounting for thermal motion of target nuclei, introduces the concept of an effective reaction cross section. We then consider in Section 3.3 the case of neutrons in thermal equilibrium with the surrounding nuclei and illustrate the concept of an effective thermal cross section. Section 3.4 derives an expression for the effective cross section of a $1/v$-absorber, which plays an important role in determining the thermal neutron reaction rates of various materials.

*Nuclear Reactor Physics and Engineering*, Second Edition. John C. Lee.
© 2025 John Wiley & Sons, Inc. Published 2025 by John Wiley & Sons, Inc.
Companion website: www.wiley.com/go/LeeNRPE2

## 3.1 Neutron Flux and Current

Since the probability of neutrons interacting with nuclei of the surrounding medium depends on the relative speeds between the neutrons and nuclei, we need to introduce the definitions related to the neutron population in terms of the neutron velocity $\mathbf{v}$, i.e. speed $v$ and direction of motion $\mathbf{\Omega}$. We are, however, dealing with a large population of neutrons and nuclei in a given volume, and we limit ourselves to statistically averaged or statistically expected quantities in a steady-state configuration for our definition of neutron and nuclear number densities. With this perspective in mind, we consider a differential volume $d\mathbf{r}$ in space located at position $\mathbf{r}$ and a differential volume $d\mathbf{v}$ in velocity space located at velocity $\mathbf{v}$, illustrated in Figure 3.1.

**1. Angular neutron number density** $n(\mathbf{r}, \mathbf{v})$
We define the angular number density

$$\begin{aligned} n(\mathbf{r}, \mathbf{v}) = &\text{ expected number of neutrons located in unit volume at } \mathbf{r} \\ &\textit{and} \text{ in unit velocity volume at } \mathbf{v}, \end{aligned} \tag{3.1}$$

so that the number of neutrons located simultaneously in the differential volumes $d\mathbf{r}$ and $d\mathbf{v}$ in Figure 3.1 may be written as $n(\mathbf{r}, \mathbf{v})d\mathbf{r}d\mathbf{v}$. The product $d\mathbf{r}d\mathbf{v}$ is often called the *differential phase volume* and may be considered a shorthand notation for the product $(dxdydz) \times (dv_x dv_y dv_z)$ in Cartesian coordinates; $d\mathbf{r}$ and $d\mathbf{v}$ should not be interpreted as differential vectors. We should also recognize that $n(\mathbf{r}, \mathbf{v})$ is a density function, given in units of $[\text{neutron} \cdot \text{cm}^{-3} (\text{cm} \cdot \text{s}^{-1})^{-3}]$, so that only when it is multiplied by a volume element, in this case, the phase volume $d\mathbf{r}d\mathbf{v}$, we may obtain the actual number of neutrons that can be counted. The term $n(\mathbf{r}, \mathbf{v})d\mathbf{r}d\mathbf{v}$ represents the number of neutrons, on a statistically averaged basis,

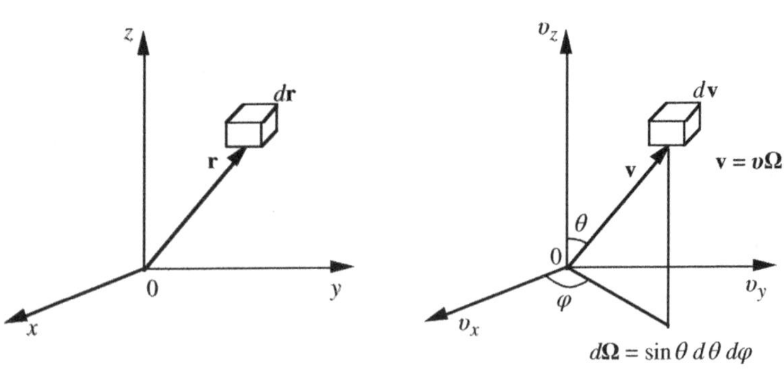


**Figure 3.1** Differential volume elements in physical and velocity spaces.

that are located in the physical volume $\mathbf{r} \sim \mathbf{r} + d\mathbf{r}$ and at the same time traveling with velocity in the interval $\mathbf{v} \sim \mathbf{v} + d\mathbf{v}$.

Remembering that the angular number density is a density function, we may establish the relationships connecting three different expressions for the quantity written in different units:

$$n(\mathbf{r}, \mathbf{v})d\mathbf{r}d\mathbf{v} = n(\mathbf{r}, \mathbf{v})d\mathbf{r}\upsilon^2 d\upsilon d\mathbf{\Omega} = n(\mathbf{r}, \upsilon, \mathbf{\Omega})d\mathbf{r}d\upsilon d\mathbf{\Omega} = n(\mathbf{r}, E, \mathbf{\Omega})d\mathbf{r}dEd\mathbf{\Omega}. \tag{3.2}$$

We write $\mathbf{v} = \upsilon\mathbf{\Omega}$ in terms of speed $\upsilon$ and unit directional vector $\mathbf{\Omega}$ and consider the velocity volume element in spherical polar coordinates $d\mathbf{v} = \upsilon^2 d\upsilon d\mathbf{\Omega} = \upsilon^2 d\upsilon \sin\theta d\theta d\varphi$, where $d\mathbf{\Omega}$ is the differential solid angle in the velocity space, not a differential of the directional vector $\mathbf{\Omega}$. The concept of solid angle was discussed in connection with the differential scattering cross section in Eqs. (2.41) and (2.50). Similarly, one example of a probability density function is the conditional probability $p(E_0 \to E)$ introduced in Eq. (2.52). Since the angular number density is a density function, when the velocity $\mathbf{v}$ is written separately in terms of speed $\upsilon$ and direction $\mathbf{\Omega}$ in Eq. (3.2), the number density $n(\mathbf{r}, \upsilon, \mathbf{\Omega})$ has to be multiplied by the proper phase volume $d\mathbf{r}d\upsilon d\mathbf{\Omega}$ to yield the number of neutrons that equals that represented by $n(\mathbf{r}, \mathbf{v})d\mathbf{r}d\mathbf{v}$. The need to consider the appropriate phase volumes in Eq. (3.2) becomes more evident in taking the coordinate transformation from speed $\upsilon$ to energy $E$. Thus, Eq. (3.2) merely represents a certain number of neutrons in phase volume $d\mathbf{r}d\mathbf{v}$, which is expressed, however, in alternate but equivalent units.

**2. Total neutron number density** $n(\mathbf{r})$

In terms of the angular number density, define the *total number density* of neutrons:

$$n(\mathbf{r}) = \int_{\mathbf{v}} n(\mathbf{r}, \mathbf{v})d\mathbf{v}. \tag{3.3}$$

Thus, $n(\mathbf{r})$ is given in units of $[\text{neutron} \cdot \text{cm}^{-3}]$ and simply represents the number of neutrons in unit volume around position $\mathbf{r}$, regardless of their velocity, i.e. speed or direction of motion. The quantity $n(\mathbf{r})$ is usually known simply as the *neutron density*.

**3. Angular neutron flux** $\psi(\mathbf{r}, \mathbf{v})$

Define the *angular neutron flux* also in terms of the angular number density

$$\psi(\mathbf{r}, \mathbf{v}) = \upsilon n(\mathbf{r}, \mathbf{v}) \tag{3.4}$$

= total track length traveled in unit time by neutrons in unit volume at $\mathbf{r}$ and in unit velocity volume at $\mathbf{v}$.

This interpretation of $\psi(\mathbf{r},\mathbf{v})$ follows simply from the recognition that the speed $v$ equals the distance traveled in unit time by each of the neutrons with their velocity in the interval $\mathbf{v} \sim \mathbf{v} + d\mathbf{v}$, and hence $\psi(\mathbf{r},\mathbf{v})$ should be expressed in units of $[\text{neutron} \cdot \text{cm}^{-2}\,\text{s}^{-1}\,(\text{cm} \cdot \text{s}^{-1})^{-3}]$. This interpretation of $\psi(\mathbf{r},\mathbf{v})$ allows us to determine the rate of neutrons interacting with target nuclei as $\Sigma\psi(\mathbf{r},\mathbf{v})d\mathbf{v}$ $[\text{neutron} \cdot \text{cm}^{-3}\,\text{s}^{-1}]$, remembering that the macroscopic cross section $\Sigma$ represents the interacting probability of a neutron per unit distance of its travel. This is the primary reason why we define $\psi(\mathbf{r},\mathbf{v})$ in the particular way we have done in Eq. (3.4).

With a cylinder of cross-sectional area $dA$ and height $vdt$ positioned at $\mathbf{r}$, as illustrated in Figure 3.2, consider neutrons located within the cylinder and with velocity in the interval $\mathbf{v} \sim \mathbf{v} + d\mathbf{v}$ that will pass through the top surface of the cylinder in time $dt$. Since the volume of the differential cylinder is $dA \cdot vdt$, the number of neutrons with velocity in $d\mathbf{v}$ around $\mathbf{v}$ passing through the cross-sectional area $dA$ in $dt$ is $n(\mathbf{r},\mathbf{v})d\mathbf{v} \cdot dAvdt = \psi(\mathbf{r},\mathbf{v})d\mathbf{v}dAdt$. This suggests that the angular flux may also be interpreted as

$$\psi(\mathbf{r},\mathbf{v}) = \text{number of neutrons in unit velocity around } \mathbf{v} \text{ crossing in unit time a unit cross-sectional area perpendicular to } \mathbf{v}. \tag{3.5}$$

This implies that $\psi(\mathbf{r},\mathbf{v})$ represents the current of neutrons with respect to a surface area that is perpendicular to $\mathbf{v}$ or $\boldsymbol{\Omega}$. This second interpretation of angular flux is useful in determining the neutron leakage rate: $\psi(\mathbf{r},\mathbf{v})d\mathbf{v}$ represents the rate of neutrons with velocity in the interval $\mathbf{v} \sim \mathbf{v} + d\mathbf{v}$ leaking through a unit cross-sectional area perpendicular to $\mathbf{v}$, given in units of $[\text{neutron} \cdot \text{cm}^{-2}\,\text{s}^{-1}]$. This interpretation will be also used to define the net neutron current as shown later.

**4. Scalar neutron flux $\phi(\mathbf{r})$, $\phi(\mathbf{r}, v)$**

Integrating the angular flux over the entire possible directions $4\pi$ of neutron motion, define the *speed-dependent scalar flux*:

$$\phi(\mathbf{r}, v) = \int_{4\pi} \psi(\mathbf{r}, v, \boldsymbol{\Omega})d\boldsymbol{\Omega}. \tag{3.6}$$

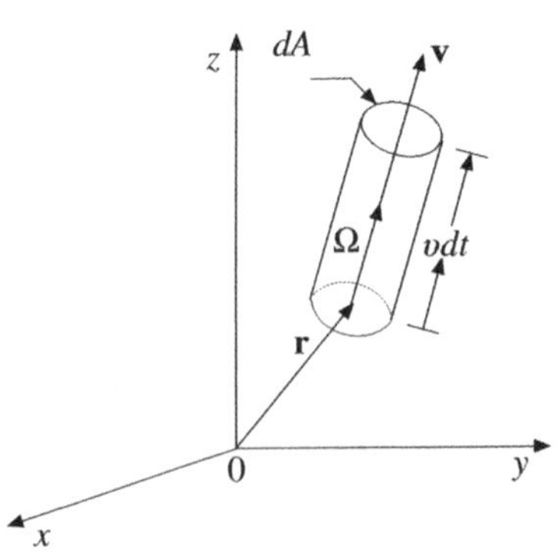


**Figure 3.2** Differential cylinder for visualizing angular flux as neutron current.

Based on the interpretation of the angular flux in Eq. (3.4) given in terms of total track length, Eq. (3.6) offers a simple interpretation for the scalar flux as the *total track length traveled in unit time by neutrons located in unit volume around* **r** *and in unit speed interval around* $v$, *regardless of their direction of motion*. Similarly, the speed-independent or *total scalar flux* is defined as

$$\phi(\mathbf{r}) = \int_0^\infty \phi(\mathbf{r}, v)dv = \int_{\mathbf{v}} \psi(\mathbf{r}, \mathbf{v})d\mathbf{v}, \tag{3.7}$$

which suggests that $\phi(\mathbf{r})$ represents the *total track length traveled in unit time by neutrons in unit volume around* **r**, *regardless of their speed or direction of motion*. Thus, the scalar flux $\phi(\mathbf{r}, v)$ is given in units of [neutron $\cdot$ cm$^{-2}$ s$^{-1}$ (cm $\cdot$ s$^{-1}$)$^{-1}$], with $\phi(\mathbf{r})$ in units of [neutron $\cdot$ cm$^{-2}$ s$^{-1}$]. Scalar neutron flux is usually referred to as the *neutron flux*, unless otherwise specified. With the interpretation of scalar flux in Eq. (3.6) or (3.7), given in terms of the neutron track length, we may readily obtain neutron reaction rates as $\Sigma\phi(\mathbf{r}, v)dv$ or $\Sigma\phi(\mathbf{r})$, both in units of [neutron $\cdot$ cm$^{-3}$ s$^{-1}$]. Equation (3.7) may also be written as

$$\phi(\mathbf{r}) = \int_{\mathbf{v}} vn(\mathbf{r}, \mathbf{v})d\mathbf{v} = n(\mathbf{r})\overline{v}, \tag{3.8}$$

where $n(\mathbf{r})$ is given by Eq. (3.3) and the average neutron speed $\overline{v}$ is defined as

$$\overline{v} = \frac{\int_{\mathbf{v}} vn(\mathbf{r}, \mathbf{v})d\mathbf{v}}{\int_{\mathbf{v}} n(\mathbf{r}, \mathbf{v})d\mathbf{v}} = \frac{\int_{\mathbf{v}} vn(\mathbf{r}, \mathbf{v})d\mathbf{v}}{n(\mathbf{r})}. \tag{3.9}$$

Thus, $\phi(\mathbf{r})$ may also be interpreted as the total track length traveled in unit time by monoenergetic neutrons of speed $\overline{v}$ located in unit volume around **r**, regardless of their direction of motion.

To offer another *restricted interpretation* of the scalar flux, consider a collimated beam of neutrons traveling in direction $\boldsymbol{\Omega}$ and with intensity $I(\mathbf{r}, v)$ [neutron $\cdot$ cm$^{-2}$ s$^{-1}$ (cm $\cdot$ s$^{-1}$)$^{-1}$], incident normally on a cylinder of cross-sectional area $dA$ and height $vdt$, as illustrated in Figure 3.3. Quite similar to our interpretation of angular flux $\psi(\mathbf{r}, \mathbf{v})$ in Figure 3.2, neutrons located in the cylinder with speed in the interval $v \sim v + dv$ will pass through the right-hand surface of the cylinder in $dt$. That is, the number of neutrons with speed in $dv$ around $v$ passing through $dA$ in time $dt$ will be

$$I(\mathbf{r}, v)dtdAdv = n(\mathbf{r}, v)vdtdAdv = \phi(\mathbf{r}, v)dtdAdv. \tag{3.10}$$

Thus, $\phi(\mathbf{r}, v)$ is simply equal to the beam intensity $I(\mathbf{r}, v)$ and represents the number of collimated neutrons in unit speed around $v$ crossing in unit time a unit cross-sectional area normal to the neutron direction $\boldsymbol{\Omega}$. We should emphasize that the interpretation of the scalar neutron flux offered here as the neutron current is strictly valid only for the special case of a collimated beam. For all other general cases, the scalar flux should be considered in terms of the neutron track length, not in terms of the neutron current.

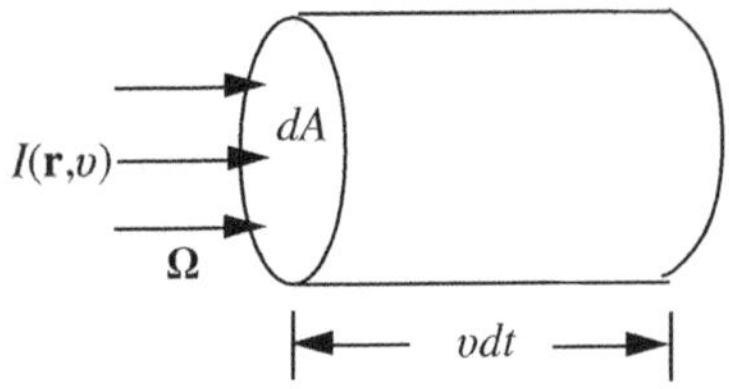


**Figure 3.3** Neutron flux for a collimated neutron beam.

**5. Net current** $J(\mathrm{r})$

Given the interpretation of angular flux $\psi(\mathbf{r},\mathbf{v})$ as the current of neutrons with respect to a surface area perpendicular to the direction of neutron motion $\mathbf{\Omega}$, as illustrated in Figure 3.2, now turn to the task of determining the current of neutrons relative to an area fixed in space. For this purpose, consider a unit cross-sectional area located at $\mathbf{r}$ with an outward normal vector $\mathbf{n}$, and determine the area whose outward normal vector is $\mathbf{\Omega}$, onto which the unit cross-sectional area is projected. As illustrated in Figure 3.4, the projected area is $\mathbf{n}\cdot\mathbf{\Omega} = \cos\theta = \mu$ [cm$^2$] and the number of neutrons with velocity in unit velocity around $\mathbf{v}$ crossing the unit cross-sectional area in unit time should equal the angular flux multiplied by the projected area:

$$\boldsymbol{n}\cdot\mathbf{\Omega}\psi(\mathbf{r},\mathbf{v}) = \mu\psi(\mathbf{r},\mathbf{v}). \tag{3.11}$$

Thus, the *net current* $J(\mathbf{r})$ [neutron $\cdot$ cm$^{-2}$ s$^{-1}$], representing the number of neutrons crossing in unit time a unit cross-sectional area whose outward normal vector is $\mathbf{n}$, is given as the integral of Eq. (3.11) over all possible neutron velocities $\mathbf{v}$

$$J(\mathbf{r}) = \int_{\mathbf{v}} \mathbf{n}\cdot\mathbf{\Omega}\psi(\mathbf{r},\mathbf{v})d\mathbf{v} = \mathbf{n}\cdot\int_{\mathbf{v}} \mathbf{\Omega}\psi(\mathbf{r},\mathbf{v})d\mathbf{v} = \mathbf{n}\cdot\mathbf{J}(\mathbf{r}), \tag{3.12}$$

where we introduce the *vector current*

$$\mathbf{J}(\mathbf{r}) = \int_{\mathbf{v}} \mathbf{\Omega}\psi(\mathbf{r},\mathbf{v})d\mathbf{v}. \tag{3.13}$$

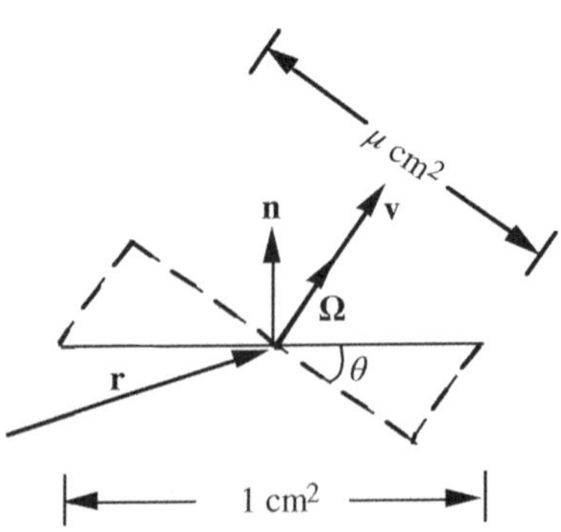


**Figure 3.4** Projection of a unit cross-sectional area.

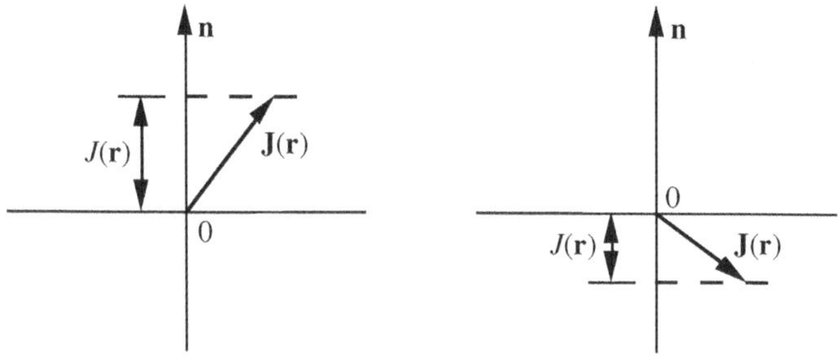


**Figure 3.5** Relationship between vector current **J** and net current $J$.

Similarly, an expression for the speed-dependent net current of neutrons may be defined

$$J(\mathbf{r}, v) = \int_{4\pi} \mathbf{n} \cdot \mathbf{\Omega} \psi(\mathbf{r}, v, \mathbf{\Omega}) d\mathbf{\Omega}, \tag{3.14}$$

in analogy to Eq. (3.6). When the velocity integral is written explicitly in polar coordinates with the integration for the polar angle $\theta$ converted to that in terms of $\mu = \cos\theta$, Eq. (3.12) becomes

$$\begin{aligned} J(\mathbf{r}) &= \int_0^\infty dv v^2 \int_{-1}^{1} d\mu\mu \int_0^{2\pi} d\varphi \psi(\mathbf{r}, \mathbf{v}) \\ &= \int_0^\infty dv \int_{-1}^{1} d\mu\mu \int_0^{2\pi} d\varphi \psi(\mathbf{r}, v, \mu, \varphi). \end{aligned} \tag{3.15}$$

The relationship between vector current $\mathbf{J}(\mathbf{r})$ and net current $J(\mathbf{r})$ is illustrated in Figure 3.5 for cases where $J(\mathbf{r}) > 0$ and $J(\mathbf{r}) < 0$.

**6. Partial current $J_\pm(\mathbf{r})$**

Similar to the definition of the net current of neutrons in Eq. (3.12), define the components of $J(\mathbf{r})$ associated with neutrons whose velocity or direction of motion is in the positive and negative directions, respectively. Thus, the positive partial current $J_+(\mathbf{r})$ is defined as

$$J_+(\mathbf{r}) = \int_{\mathbf{n}\cdot\mathbf{\Omega}>0} \mathbf{n} \cdot \mathbf{\Omega} \psi(\mathbf{r}, \mathbf{v}) d\mathbf{v}, \tag{3.16}$$

with a similar definition for $J_-(\mathbf{r})$. We can also write the partial currents more explicitly:

$$J_\pm(\mathbf{r}) = \int_0^\infty dv \int_0^{\pm 1} d\mu\mu \int_0^{2\pi} d\varphi \psi(\mathbf{r}, v, \mu, \varphi). \tag{3.17}$$

Comparing Eq. (3.17) with Eq. (3.15) shows that

$$J(\mathbf{r}) = J_+(\mathbf{r}) - J_-(\mathbf{r}), \tag{3.18}$$

which simply restates the definition of the net current $J(\mathbf{r})$ as the current resulting from the difference between the positive flow and negative flow of neutrons. Furthermore, it can be shown that application of the $P_1$ approximation or diffusion theory approximation to the angular flux in Eq. (3.17) yields

$$J_{\pm}(\mathbf{r}) = \frac{\phi(\mathbf{r})}{4} \pm \frac{J(\mathbf{r})}{2}, \tag{3.19}$$

which will be derived later in Chapter 4.

## 3.2 Rate of Neutron–Nucleus Interaction

When the interpretation of neutron flux was introduced in terms of total track length in Section 3.1, we noted that a product of the reaction cross section $\Sigma$ and flux yields the rate of neutrons interacting with surrounding nuclei. We now wish to be a bit more precise in representing the neutron–nucleus reaction rate, explicitly accounting for the relative motion between the neutron and nucleus. This will allow us to indicate clearly how the macroscopic cross section $\Sigma$ should be determined. For this purpose, we need to introduce a definition for the angular number density of nuclei $N(\mathbf{r}, \mathbf{V})$ in terms of the velocity $\mathbf{V}$ of the nuclei that neutrons would interact with:

$$\begin{aligned} N(\mathbf{r}, \mathbf{V}) = {} & \text{expected number of nuclei located in unit volume at } \mathbf{r} \textit{ and} \\ & \text{in unit velocity volume at } \mathbf{V}. \end{aligned} \tag{3.20}$$

This definition for $N(\mathbf{r}, \mathbf{V})$ [nucleus $\cdot$ cm$^{-3}$ (cm $\cdot$ s$^{-1}$)$^{-3}$] is, of course, identical to that of the angular neutron number density $n(\mathbf{r}, \mathbf{v})$ introduced in Eq. (3.1), except that $N(\mathbf{r}, \mathbf{V})$ is defined in the velocity space for the nuclei, not the neutrons. Note also that the interaction probability $\sigma$ between a neutron of velocity $\mathbf{v}$ and a nucleus of velocity $\mathbf{V}$ is in general a function of the relative speed $|\mathbf{v} - \mathbf{V}|$ between the interacting particles.

With these observations, it also becomes clear that we need to rewrite the neutron track length, used in the interpretation of the angular flux of Eq. (3.4), in terms of the relative distance between the interacting particles. Thus, the total track length traveled in unit time by neutrons located in unit volume around $\mathbf{r}$ and in unit velocity at $\mathbf{v}$, relative to the target nuclei of velocity $\mathbf{V}$, has to be written as $|\mathbf{v} - \mathbf{V}|n(\mathbf{r}, \mathbf{v})$. The macroscopic cross section also has to be written explicitly as a product of the microscopic cross section $\sigma(|\mathbf{v} - \mathbf{V}|)$ and the nuclear density $N(\mathbf{r}, \mathbf{V})$. We may then introduce an expression for the neutron–nucleus reaction rate $R(\mathbf{r}, \mathbf{v})$ [neutron $\cdot$ cm$^{-3}$ s$^{-1}$ (cm $\cdot$ s$^{-1}$)$^{-3}$] as the *number of neutrons interacting with the surrounding nuclei of any velocity per unit time, per unit volume at* $\mathbf{r}$ *and per neutron velocity volume at* $\mathbf{v}$:

$$R(\mathbf{r}, \mathbf{v}) = \int_{\mathbf{V}} d\mathbf{V}\sigma(|\mathbf{v} - \mathbf{V}|)N(\mathbf{r}, \mathbf{V})|\mathbf{v} - \mathbf{V}|n(\mathbf{r}, \mathbf{v}). \tag{3.21}$$

Here, $\sigma(|\mathbf{v} - \mathbf{V}|)$ is the *true* microscopic cross section, which is a function of the relative speed $|\mathbf{v} - \mathbf{V}|$ and hence may be determined theoretically in the center-of-mass system.

For notational convenience, we now drop the explicit dependence on the position vector $\mathbf{r}$ and rewrite Eq. (3.21):

$$R(\mathbf{v}) = n(\mathbf{v}) \int_{\mathbf{V}} d\mathbf{V}|\mathbf{v} - \mathbf{V}|\sigma(|\mathbf{v} - \mathbf{V}|)N(\mathbf{V}). \tag{3.22}$$

Integrating Eq. (3.22) over all possible neutron velocities, we may readily obtain an expression for the *total neutron reaction rate* $R = R(\mathbf{r})$ [neutron $\cdot$ cm$^{-3}$ s$^{-1}$], i.e. the *total number of neutrons interacting with the surrounding nuclei per unit time and per unit volume at* $\mathbf{r}$:

$$R \equiv R(\mathbf{r}) = \int_{\mathbf{v}} d\mathbf{v}n(\mathbf{v}) \int_{\mathbf{V}} d\mathbf{V}|\mathbf{v} - \mathbf{V}|\sigma(|\mathbf{v} - \mathbf{V}|)N(\mathbf{V}). \tag{3.23}$$

It should be noted that the integral in Eq. (3.22) represents the probability $P(\mathbf{v})$ that a neutron of velocity $\mathbf{v}$ will interact with nuclei of any velocity per unit time:

$$P(\mathbf{v}) = \int_{\mathbf{V}} d\mathbf{V}|\mathbf{v} - \mathbf{V}|\sigma(|\mathbf{v} - \mathbf{V}|)N(\mathbf{V}). \tag{3.24}$$

In a typical experiment measuring the neutron cross section, a beam of mono-energetic neutrons is incident upon a sample of target nuclei, and the reaction rate $R(\mathbf{v})$ or $P(\mathbf{v})$ is measured. In such an experiment, the target nuclei are in constant thermal motion at some finite temperature $T$, resulting in the velocity distribution $N(\mathbf{V})$, and the true cross section $\sigma(|\mathbf{v} - \mathbf{V}|)$ cannot be measured in the laboratory. One would in fact be measuring, through $P(\mathbf{v})$, an effective cross section $\overline{\sigma}(v)$ averaged over the thermal motion of the target nuclei

$$P(\mathbf{v}) = P(v) = v\overline{\sigma}(v)N, \tag{3.25}$$

where

$$N = \int_{\mathbf{V}} d\mathbf{V}N(\mathbf{V}) \equiv N(\mathbf{r})$$

is the total target nuclear density given in units of [nucleus $\cdot$ cm$^{-3}$]. [*Note*: we denote here and subsequently a density function, with an argument missing, to indicate that it represents the same function but with the integration carried out over the missing argument, as is the case here for $N = N(\mathbf{r})$, without the argument $\mathbf{V}$, in contrast to $N(\mathbf{r}, \mathbf{V})$. We will adopt this notational convention for the rest of the chapter whenever it is feasible without risking confusion.] It is worth noting that $\overline{\sigma}(v)$ defined in Eq. (3.25)

$$\overline{\sigma}(v) = \frac{\displaystyle\int_{\mathbf{V}} d\mathbf{V}|\mathbf{v} - \mathbf{V}|\sigma(|\mathbf{v} - \mathbf{V}|)N(\mathbf{V})}{\displaystyle v\int_{\mathbf{V}} d\mathbf{V}N(\mathbf{V})}, \tag{3.26}$$

represents an average cross section that preserves the reaction rate $P(v)$. Written in terms of $\overline{\sigma}(v)$, Eq. (3.25) shows the intuitive result that $P(\mathbf{v}) = P(v)$, i.e. that the reaction rate depends simply on the neutron speed alone, not on the direction of motion. It should be also emphasized that $\overline{\sigma}(v)$ is the cross section that is tabulated in cross section libraries, e.g. the ENDF/B-VIII library [Bro18], and that it is the cross section that should be used in reactor physics from now on. Substituting Eq. (3.25) into Eq. (3.23) yields a familiar expression for the total reaction rate $R$ in units of $[\text{neutron} \cdot \text{cm}^{-3}\,\text{s}^{-1}]$

$$R \equiv R(\mathbf{r}) = \int_{\mathbf{v}} d\mathbf{v}N\overline{\sigma}(v)vn(\mathbf{v}) = \int_{\mathbf{v}} d\mathbf{v}\Sigma(v)\psi(\mathbf{v}) = \int_0^{\infty} dv\Sigma(v)\phi(v), \tag{3.27}$$

where $\psi(\mathbf{v}) \equiv \psi(\mathbf{r}, \mathbf{v})$ is the angular flux, $\phi(v) \equiv \phi(\mathbf{r}, v)$ is the scalar flux, and the macroscopic cross section $\Sigma(v) = N\overline{\sigma}(v) = N(\mathbf{r})\overline{\sigma}(v) \equiv \Sigma(\mathbf{r}, v)$ is a function of neutron speed $v$, not of direction of neutron motion $\mathbf{\Omega}$.

One important example where the thermal motion of nuclei has a major impact on the neutron reaction rate is Doppler broadening of resonance cross sections. As the temperature $T$ at which the nuclei are in thermal equilibrium increases, thermal motion of the nuclei increases, resulting in spreading or smearing of the relative speeds between the neutrons and nuclei. This causes the sharp resonance line shapes, represented by the Breit–Wigner formula from Eq. (2.34), to broaden, since the resonance cross sections depend on the center-of-mass speed or the neutron–nucleus relative speed. This is known as *Doppler broadening of resonances*, in analogy to the acoustic Doppler phenomena. Around an absorption resonance, the neutron flux $\phi(E)$ is depressed due to the increased neutron absorption, resulting in a decrease in the absorption rate per absorber atom compared with the case without the flux depression. This phenomenon is known as *energy self-shielding*, because the absorption rate of the resonance absorber decreases as the result of its own presence. As $T$ increases, due to Doppler broadening of resonances, the flux depression and hence the energy self-shielding decrease, and the energy interval over which the resonance is felt increases. This in turn causes the absorption rate around the resonance to increase, thereby decreasing the overall probability of neutrons slowing down without absorption, especially for light water reactors (LWRs) with low fissile enrichment. Thus, if the fuel elements are overheated in an accident, Doppler broadening of resonances guarantees a rapid insertion of negative reactivity, which is an inherent safety mechanism of considerable importance in LWRs. The concepts of energy self-shielding and Doppler broadening of resonances are discussed further in Chapter 9.

## 3.3 Neutron Energy Distribution and Effective Thermal Cross Section

The neutron reaction rate in Eq. (3.27) is written in terms of $\overline{\sigma}(v)$, which explicitly accounts for thermal motion of nuclei represented by $N(\mathbf{V})$ or effectively $N(\mathbf{V}, T)$. Continuing the averaging process introduced in Eq. (3.26), we may rewrite the total reaction rate $R$ [cm$^{-3}$ s$^{-1}$] from Eq. (3.27) in terms of the scalar flux $\phi(\mathbf{r}, v)$, integrated over the entire energy interval, and introduce another effective cross section $\sigma_{eff}$:

$$R = R(\mathbf{r}) = N\sigma_{eff}\phi_0 = N\int_0^{\infty} dv\overline{\sigma}(v)\phi(v). \tag{3.28}$$

Invoking Eq. (3.7) for the total scalar flux $\phi_0 \equiv \phi(\mathbf{r})$, we recognize that $\sigma_{eff}$ is given by

$$\sigma_{eff} = \frac{\int_0^{\infty} dv\overline{\sigma}(v)\phi(v)}{\int_0^{\infty} dv\phi(v)} \equiv \frac{\int_0^{\infty} dv\overline{\sigma}(v)\phi(v)}{\phi_0}, \tag{3.29}$$

which simply indicates that the effective cross section is obtained as a flux-weighted average cross section. In much the same way that $\overline{\sigma}(v)$ is introduced in Eq. (3.26), the averaging process of Eq. (3.29) preserves the neutron reaction rate. That we should always endeavor to preserve the reaction rate is a point we have to keep in mind whenever we generate an effective or average cross section. Because we have gone through two stages of averaging to obtain $\sigma_{eff}$, we should also recognize that $\sigma_{eff}$ accounts for the velocity distributions of both neutrons and target nuclei.

If neutrons are in thermal equilibrium with target nuclei of a medium at a certain temperature $T$, then the neutron energy spectrum can be represented by the Maxwell–Boltzmann distribution [Eng72, Ser89] at $T$, and the average cross section $\sigma_{eff}$ determined through Eq. (3.29) will be the proper effective cross section for the thermal neutrons. The energy of neutrons in a nuclear reactor core actually covers a broad range, all the way from the fission energy down to the sub-eV range, attained when the neutrons are brought into thermal equilibrium through repeated collisions with the nuclei. Because of this broad range of neutron energy, it is often not very accurate to talk in terms of a single effective cross section covering the entire energy interval. In thermal reactors, including LWR cores, however, where the bulk of fissions take place with energy below 1.0 eV, we may effectively consider the flux spectrum approximated by the Maxwell–Boltzmann distribution, with the integrals in Eqs. (3.28) and (3.29) covering the entire energy range. In this sense, the effective cross section $\sigma_{eff}$ may be considered the average cross section for thermal neutrons, and we may write the effective thermal cross section $\Sigma_{th} = N\sigma_{eff}, N \equiv N(\mathbf{r})$.

For neutrons in thermal equilibrium with target nuclei at $T$, the angular number density will follow the Boltzmann factor $\exp(-E/kT)$ or, with the proper normalization, the Maxwell–Boltzmann distribution

$$n(\mathbf{v}) = n_0 \left(\frac{m}{2\pi kT}\right)^{3/2} \exp\left(-\frac{mv^2}{2kT}\right), \tag{3.30}$$

where $k = 1.381 \times 10^{-23}\ \mathrm{J \cdot K^{-1}}$ is the Boltzmann constant and $m$ is the neutron mass. In Eq. (3.30), the normalization is chosen such that the total number density $n_0$ is given by Eq. (3.3)

$$n_0 = \int_{\mathbf{v}} d\mathbf{v} n(\mathbf{v}) \equiv n(\mathbf{r}).$$

[Recall here the notational convention introduced in connection with Eq. (3.25).] The number of neutrons per unit volume and per unit speed about $v$ can now be readily obtained:

$$n(v) \equiv n(\mathbf{r}, v) = \int_{4\pi} d\Omega\, n(\mathbf{v})v^2 = 4\pi n(\mathbf{v})v^2$$

or

$$n(v) = n_0 \left(\frac{2}{\pi}\right)^{1/2} \left(\frac{m}{kT}\right)^{3/2} v^2 \exp\left(-\frac{mv^2}{2kT}\right). \tag{3.31}$$

Similarly, remembering that $n(E)dE = n(v)dv$, convert the speed distribution function of Eq. (3.31) into the energy distribution function

$$n(E) = n(v)\frac{dv}{dE} = \frac{n(v)}{\sqrt{2mE}}, \tag{3.32}$$

or

$$n(E) = n_0 \frac{2}{\sqrt{\pi}} \left(\frac{1}{kT}\right)^{3/2} E^{1/2} \exp\left(-\frac{E}{kT}\right). \tag{3.33}$$

From Eq. (3.32), obtain the energy-dependent scalar flux:

$$\phi(E) = \phi(v)\frac{dv}{dE} = vn(v)\frac{dv}{dE} = \frac{n(v)}{m}. \tag{3.34}$$

Equation (3.7) then yields the total scalar flux for neutrons obeying the Maxwell–Boltzmann distribution

$$\phi_0 \equiv \phi(\mathbf{r}) = \int_0^\infty dv\phi(v) = \int_0^\infty dvn(v)v = n_0\sqrt{\frac{8kT}{\pi m}}, \tag{3.35}$$

and the speed-dependent scalar flux may be written as

$$\phi(v) = \frac{\phi_0}{2}\left(\frac{m}{kT}\right)^2 v^3 \exp\left(-\frac{mv^2}{2kT}\right). \tag{3.36}$$

Similarly, Eq. (3.34) with Eq. (3.36) yields

$$\phi(E) = \phi_0\frac{E}{(kT)^2}\exp\left(-\frac{E}{kT}\right). \tag{3.37}$$

The distribution functions represented by Eqs. (3.31), (3.33), (3.36), and (3.37) are obviously equivalent to one another, and any one of them may be used to evaluate speeds and energies with certain attributes. Some of them are:

(1) The *most probable speed* $v_0$ is the speed for which Eq. (3.31) reaches the maximum

$$v_0 = \sqrt{\frac{2kT}{m}}. \tag{3.38}$$

The energy corresponding to $v_0$ is

$$E_0 = kT. \tag{3.39}$$

(2) The *average speed* is obtained as

$$\overline{v} = \frac{1}{n_0}\int_0^\infty dvn(v)v = \sqrt{\frac{8kT}{\pi m}} = \frac{2}{\sqrt{\pi}}v_0. \tag{3.40}$$

It should be noted that Eq. (3.35) may be written as $\phi_0 = n_0\overline{v}$, in agreement with Eq. (3.8).

(3) The *most probable energy* is the energy for which Eq. (3.33) reaches the maximum

$$E_{mp} = \frac{kT}{2}. \tag{3.41}$$

(4) The *average energy* is given by

$$\overline{E} = \frac{1}{n_0}\int_0^\infty dEn(E)E = \frac{3kT}{2}. \tag{3.42}$$

(5) The energy at which $\phi(E)$ of Eq. (3.37) is maximized is also $E_0 = kT$ of Eq. (3.39), the energy corresponding to the most probable speed $v_0$. It has become conventional to refer to $E_0 = kT$ as the *neutron energy* for neutrons in

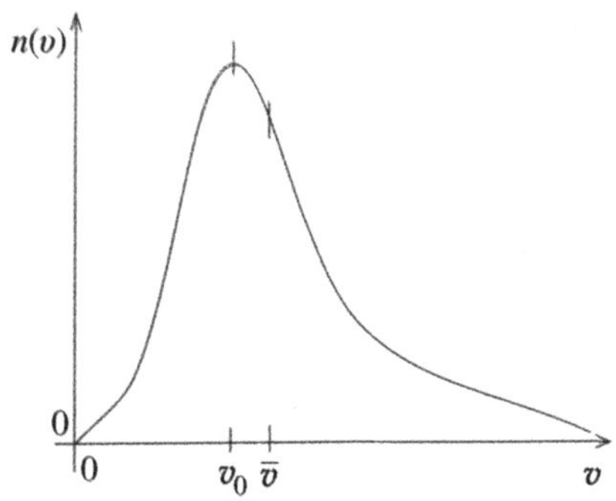


**Figure 3.6** Maxwell–Boltzmann distribution as a function of speed.

thermal equilibrium at $T$, while the temperature $T$ itself is often referred to as the *neutron temperature*. In particular, for neutrons in thermal equilibrium at room temperature, i.e. $T = 300$ K, $E_0 = kT = 0.025\,\text{eV}$, corresponding to $v_0 = 2{,}200\,\text{m}\cdot\text{s}^{-1}$.

The Maxwell–Boltzmann distribution $n(v)$ of Eq. (3.31) is illustrated in Figure 3.6, where the relationship between the most probable speed $v_0$ and average speed $\overline{v}$ is highlighted.

The four equivalent density functions, Eqs. (3.31), (3.33), (3.36), and (3.37), for neutrons obeying the Maxwell–Boltzmann distribution may now be written in simple dimensionless form [Ben81] in terms of the most probable speed $v_0$ and the neutron energy $E_0$. For example, the speed distribution $n(v)$ may be written in terms of the normalized speed variable $v/v_0$

$$n(v)dv = n\left(\frac{v}{v_0}\right)\frac{dv}{v_0}, \tag{3.43}$$

or

$$\frac{n(v/v_0)}{n_0} = \frac{4}{\sqrt{\pi}}\left(\frac{v}{v_0}\right)^2 \exp\left(-\frac{v^2}{v_0^2}\right), \tag{3.44a}$$

where $n_0$ is the total number density introduced in Eq. (3.30). Similarly, we obtain:

$$\frac{n\left(E/E_0\right)}{n_0} = \frac{2}{\sqrt{\pi}}\left(\frac{E}{E_0}\right)^{1/2} \exp\left(-\frac{E}{E_0}\right), \tag{3.44b}$$

$$\frac{\phi(v/v_0)}{\phi_0} = 2\left(\frac{v}{v_0}\right)^3 \exp\left(-\frac{v^2}{v_0^2}\right), \text{ and} \tag{3.44c}$$

$$\frac{\phi\left(E/E_0\right)}{\phi_0} = \frac{E}{E_0} \exp\left(-\frac{E}{E_0}\right). \tag{3.44d}$$

Equation (3.43) clearly indicates that both speed distributions, $n(v)$ and $n(v/v_0)$, are density functions and hence have to be multiplied by the differential speeds

$dv$ and $d(v/v_0)$, respectively, before they are equated. This follows the discussion presented in connection with Eq. (3.2) and applies to all three of the remaining normalized distribution functions above. Note that Eq. (3.44d) is in a particularly simple form suitable for many applications, since the scalar flux of neutrons at energy $E$ obeying the Maxwell–Boltzmann distribution may be obtained by merely remembering that, for $T = 300$ K, $E_0 = kT = 0.025$ eV. This is illustrated in Section 3.4.

## 3.4 Application to a 1/V-Absorber

For many reactions involving neutron absorption, e.g. $(n,\gamma), (n,p), (n,\alpha)$, and $(n,f)$, cross sections for many nuclides of interest in reactor physics are approximately inversely proportional to the neutron speed, as discussed in Section 2.7. For such a $1/v$-absorber, we may write the energy-dependent cross section as

$$\overline{\sigma}(E) = \overline{\sigma}(v) = \overline{\sigma}(v_0)\frac{v_0}{v} = \overline{\sigma}(E_0)\sqrt{\frac{E_0}{E}}. \tag{3.45}$$

If we assume that Eq. (3.45) holds for all energies and that neutrons are in thermal equilibrium at temperature $T$, the effective thermal neutron cross section $\sigma_{eff} = \sigma_{eff}(T) = \sigma_{eff}(E_0)$ can be calculated from Eqs. (3.29) and (3.44d):

$$\begin{aligned}\sigma_{eff}(T) &= \frac{1}{\phi_0}\int_0^\infty dE\phi(E)\overline{\sigma}(E) = \frac{1}{\phi_0}\int_0^\infty \frac{dE}{E_0}\phi\left(\frac{E}{E_0}\right)\overline{\sigma}(E)\\ &= \overline{\sigma}(E_0)\int_0^\infty \frac{dE}{E_0}\sqrt{\frac{E_0}{E}}\frac{E}{E_0}\exp\left(-\frac{E}{E_0}\right).\end{aligned}$$

which may be evaluated through the gamma function [Arf13] [Appendix C] to yield

$$\sigma_{eff}(T) = \overline{\sigma}(E_0)\frac{\sqrt{\pi}}{2}. \tag{3.46}$$

The gamma function is a generalization of the factorial $n!$ such that $\Gamma(z+1) = z\Gamma(z)$ when $z$ is in general a complex number. Thus, when $z$ is an integer, $\Gamma(z+1) = z!$ We note one simple representation

$$\Gamma(z) = \int_0^\infty e^{-t}t^{z-1}dt, \tag{3.47}$$

which yields a useful result

$$\Gamma\left(\frac{1}{2}\right) = \sqrt{\pi}. \tag{3.48}$$

With Eq. (3.46) substituted into Eq. (3.28) for the total reaction rate $R$, together with Eq. (3.35), we obtain

$$R = N\sigma_{eff}\phi_0 = N\overline{\sigma}(E_0)\frac{\sqrt{\pi}}{2} \cdot n_0 \mathrm{v}_0 \frac{2}{\sqrt{\pi}} = N\overline{\sigma}(E_0)n_0\mathrm{v}_0. \tag{3.49}$$

This is an important conclusion that the reaction rate $R/N$ per absorber atom is independent of the neutron temperature $T$ for a $1/v$-absorber when the neutrons are in thermal equilibrium at $T$. When it becomes necessary to represent the actual non-$1/v$ behavior of various nuclides, introduce a correction factor into Eq. (3.46)

$$\sigma_{eff}(T) = \overline{\sigma}(E_0)\frac{\sqrt{\pi}}{2}g(T_n), \tag{3.50}$$

where $g(T_n)$ is known as Westcott's $g$-factor [Mag82], often tabulated as a function of the effective neutron temperature $T_n$. To the first order of accuracy, $T_n$ is simply set equal to the neutron temperature $T$. If enhanced accuracy is desired, Eq. (3.50) may be used with a corrected or adjusted $T_n$ for the $g$-factor, together with $E_0 = kT_n$.

## References

**[Arf13]** G.B. Arfken, H.J. Weber, and F.E. Harris, *Mathematical Methods for Physicist: A Comprehensive Guide*, 7th ed., Academic Press (2013).

**[Ben81]** M. Benedict, T.H. Pigford, and H.W. Levi, *Nuclear Chemical Engineering*, 2nd ed., McGraw-Hill (1981).

**[Bro18]** D.A. Brown, "ENDF/B-VIII.0: The 8th Major Release of the Nuclear Reaction Data Library with CIELO-project Cross Section, New Standards and Thermal Scattering Data," *Nucl. Data Sheets* **148**, 1 (2018).

**[Eng72]** H.A. Enge, M.R. Wehr, and J.A. Richards, *Introduction to Atomic Physics*, Addison-Wesley (1972).

**[Mag82]** B.A. Magurno, R. Kinsey, and F.M. Scheffel, "Guidebook for the ENDF.B-V Nuclear Data Files," EPRI NP-2510, BNL-NCS-31451, ENDF-328, Electric Power Research Institute (1982).

**[Ser89]** R.A. Serway, C.J. Moses, and C.A. Moyer, *Modern Physics*, Saunders College Publishing (1989).

## Problems

**3.1** The scalar flux in a spherical reactor of radius $R = 0.5$ m is described by

$$\Phi(\mathbf{r}) = \phi_0 \frac{\sin(\pi r/R)}{\pi r/R}.$$

(a) If the constant $\phi_0$ is $3 \times 10^{13}$ neutron $\cdot$ cm$^{-2}$ s$^{-1}$, calculate the number of neutrons contained in the core volume. Assume the flux corresponds to an average neutron speed $\overline{v} = 3{,}000$ m $\cdot$ s$^{-1}$. (b) What is the value of the flux at the core center? (c) With the fission cross section $\Sigma_f = 0.05$ cm$^{-1}$ given for the core material, determine the total power of the core in units of MW. (d) The angular flux for the core is represented as

$$\psi(\mathbf{r}, E, \mathbf{\Omega}) = \frac{E}{4\pi E_0^2} \exp\left(-\frac{E}{E_0}\right) \Phi(\mathbf{r}),$$

with neutron energy $E_0$ corresponds to $\overline{v}$. Determine the corresponding expressions for angular flux $\psi(\mathbf{r}, v, \mathbf{\Omega})$ and $\psi(\mathbf{r}, \mathbf{v})$ and scalar flux $\phi(\mathbf{r}, v)$ and $\phi(\mathbf{r})$.

**3.2** The core configuration of a research reactor is represented by a $(6 \times 7)$ array of fuel elements, each of which has dimensions $(76.2 \times 76.2 \times 609.6)$ mm. The scalar flux distribution at position $(x, y, z)$ in the core is given as $\phi(x, y, z) = A\cos(\pi x/X)\cos(\pi y/Y)\cos(\pi z/Z)$, where the distances $x, y,$ and $z$ are measured from the center of the core with sides $X, Y,$ and $Z$, respectively. If the constant $A$ is determined experimentally to be $3 \times 10^{13}$ neutron $\cdot$ cm$^{-2}$ s$^{-1}$, calculate the number of neutrons contained in the core volume. Assume that neutrons in the core are in thermal equilibrium with an average neutron speed $v_0 = 3{,}000$ m $\cdot$ s$^{-1}$.

**3.3** The microscopic absorption cross section of stationary $^{10}$B nuclei for neutrons of speed $v$ is $\sigma(v)$ [b]. A small sample of $^{10}$B is shot with speed $V_0$ at angle $\theta$ through a uniform, collimated beam of neutrons of speed $v_0$. Assume $^{10}$B is a pure absorber, and neglect thermal motion of $^{10}$B nuclei in the sample. Obtain an expression for the number of absorptions per s per $^{10}$B nucleus while the sample is in the neutron beam of intensity $I_0$ [neutron $\cdot$ cm$^{-2}$ s$^{-1}$].

**3.4** A collimated beam of neutrons is incident normally on a slab of thickness $H$. The beam intensity is $I_0(E)$ [neutron $\cdot$ cm$^{-2}$ s$^{-1}$ eV$^{-1}$], and the slab material is a $1/E$-absorber with absorption cross section $\overline{\sigma}(E_0) = \sigma_0$. The scattering cross section of the slab material is negligibly small. (a) Determine an expression for the total intensity $I(E)$ of neutrons emerging from the slab. (b) Assuming that the incident beam intensity $I_0(E)$ may be represented by the Maxwell–Boltzmann distribution, perform the integral of part (a) over the energy interval [0.1 meV, 6 eV]. Contributions to the integral from neutrons of energy outside the interval may be neglected. The slab consists of gadolinium at $T = 696$ K, with $\overline{\sigma}(E_0) = 24$ kb at $E_0 = 60$ meV. Given also are the slab thickness $H = 1.0$ mm and the total incident beam intensity of

$10^8$ neutron · $cm^{-2}\,s^{-1}$. Write a computer program in C/C++, Fortran, or MATLAB to perform the numerical integration using Simpson's formula, with 100 divisions or meshes.

**3.5** (a) For the gadolinium nuclei in thermal equilibrium at $T = 696$ K in Problem 3.4, determine the effective absorption cross section $\sigma_{eff}(T)$. (b) If $\sigma_{eff}(T) = \sigma^*$ for $1/v$-absorber nuclei interacting with neutrons in thermal equilibrium at $T = 300$ K, determine $\sigma_{eff}(T_1)$ with $T_1 = 450$ K.

**3.6** Mono energetic neutrons are distributed isotropically at position $\mathbf{r}$ in a reactor core with the scalar flux $\phi(\mathbf{r}) = 3 \times 10^{13}$ neutron · $cm^{-2}\,s^{-1}$. Obtain an expression for (a) angular flux $\psi(\mathbf{r}, \boldsymbol{\Omega})$, (b) negative partial current $J\_(\mathbf{r})$, and (c) net current $J(\mathbf{r})$.

**3.7** An infinite homogeneous medium consists of nuclei of mass number $A$ at temperature $T_0 = 300$ K, where a steady-state distribution of neutrons is established. The *true* absorption cross section $\sigma(|\mathbf{v} - \mathbf{V}|)$ of the nuclide is inversely proportional to the relative speed $|\mathbf{v} - \mathbf{V}|$ between neutrons of velocity $\mathbf{v}$ and nuclei of velocity $\mathbf{V}$. (a) Show that the effective neutron absorption cross section $\overline{\sigma}(v)$ for neutrons of speed $v$, averaged over thermal motions of the target nuclei, is inversely proportional to $v$. (b) Accounting for the absorption of neutrons in the medium, approximate the neutron velocity distribution in the medium as a Maxwellian, with an effective neutron temperature $T_1 = 350$ K. Given the absorption cross section $\sigma(v_0) = 2.0$ kb for $v_0 = 2,200$ m · $s^{-1}$, obtain an effective neutron absorption cross section $\langle\sigma\rangle$ averaged over the neutron flux spectrum $\phi(E)$.

# 4

# Derivation of the Neutron Diffusion Equation

The neutron diffusion equation is a basic balance equation that describes the transport of neutrons in space, energy, and time. This equation plays a central role in reactor physics, because solution of the diffusion equation provides the neutron flux, which is required to represent the rate at which neutrons interact with the surrounding medium and the rate at which they leak out of a given volume. For example, the power density in a reactor core is directly proportional to the fission reaction rate, and the determination of the critical size of a chain-reacting system depends on the neutron leakage and reaction rates as well.

To be proper, the neutron diffusion equation should be derived from the neutron transport equation, which describes the balance of neutrons in space, velocity, and time. The derivation would involve integrating out the directional dependence of neutron motion explicitly represented in the transport equation. To avoid a bit of extra mathematics involved in this approach, we first take an alternate route to set up a balance equation that represents the distribution of neutrons in space, energy, and time only. This requires an approximate treatment of neutron migration in space as a diffusion process, i.e. neutrons are considered diffusing from a region of high density to that of low density.

Apart from this approximation regarding the neutron migration process, we derive the neutron balance equation from first principles with a minimum number of approximations. We discuss in Section 4.1 the simplifying assumptions that allow us to write the neutron balance equation in a tractable form. The establishment of the balance equation in Section 4.2 makes a direct use of the concepts of neutron flux and current as well as effective cross section introduced in Chapter 3. The source term appearing in the diffusion equation is discussed further in Section 4.3, followed by the derivation of Fick's law of diffusion for the neutron current in Section 4.4. To gain better understanding of the diffusion approximation as compared with the rigorous transport theory, in Section 4.5, we use the energy- and time-independent one-dimensional (1-D) transport equation and the $P_1$ approximation for the angular flux to derive the

*Nuclear Reactor Physics and Engineering*, Second Edition. John C. Lee.
© 2025 John Wiley & Sons, Inc. Published 2025 by John Wiley & Sons, Inc.
Companion website: www.wiley.com/go/LeeNRPE2

steady-state 1-D diffusion equation. In Section 4.6, we provide an interpretation of the diffusion coefficient for the current and in Section 4.7, we discuss limitations and applicability of diffusion theory. A simplified, energy-independent form of the diffusion equation is derived in Section 4.8 and is used in the bulk of reactor physics analysis in the subsequent chapters. Section 4.9 presents some concluding remarks regarding diffusion theory.

## 4.1 Basic Assumptions for Neutron Balance Statement

We are interested in setting up a balance equation that describes the distribution of neutrons in space, energy, and time, as they undergo collisions with nuclei of the surrounding medium and travel from one collision site to another or leak out of the medium. To cast the balance equation in a manageable form, we note characteristic features of the nuclear reactions involved and introduce a few simplifying assumptions:

(1) Since we are dealing with a large population of neutrons and nuclei, consider only a statistically averaged distribution of neutrons in unit volume both in physical and velocity space, instead of representing the state of individual neutrons. Thus, the diffusion equation we derive cannot be directly used to describe the statistical fluctuations in the neutron population at very low power levels in a nuclear reactor core.

(2) The neutron number density is on the order of $10^9$ neutron $\cdot$ cm$^{-3}$ or lower in a typical reactor core, as compared with the nuclear number density, which is on the order of $10^{22}$ nucleus $\cdot$ cm$^{-3}$. Hence, neutron–neutron interactions may be neglected, and particle interactions will be limited to nucleus–neutron binary collisions. This simplifies the form of the diffusion equation considerably and also allows for the representation of the neutron population due to multiple sources through a linear superposition of individual contributions.

(3) The reaction time for nucleus–neutron interactions is usually on the order of $10^{-14}$ s or less and is several orders of magnitude shorter than the time interval between successive collisions, which is typically on the order of μs ∼ ms. Thus, it is an excellent approximation to assume that neutron–nucleus collisions take place instantaneously and that neutrons do not suffer any displacement in space due to and during collisions with the target nuclei.

(4) The natural radioactive decay of neutrons, with the half-life $t_{1/2} = 610$ s [Gre16], is a slow process compared with nuclear reactions taking place in a reactor core, and hence will be neglected. For special applications, however, radioactive decay of neutrons may readily be added to the diffusion equation.

(5) Body forces, e.g. gravitational and electromagnetic forces, are negligibly small compared with nuclear forces involved in the neutron–nucleus interactions. Thus, we limit ourselves to classical, unpolarized neutrons in our neutron balance setup.

## 4.2 Neutron Balance Equation

For the purpose of including time dependence in the diffusion equation that we derive, formally extend the definitions for the neutron flux and current introduced in Chapter 3 to add the time variable as a parameter. In particular, rewrite the scalar neutron flux of Eq. (3.6) as $\phi(\mathbf{r}, v, t)$ to represent the total track length traveled in unit time at $t$ by neutrons located in unit volume around $\mathbf{r}$ and in unit speed interval around $v$, regardless of their direction of motion. Recall that the neutron flux is defined in terms of track length of neutrons so that the product of $\phi(\mathbf{r}, v, t)dv$ and reaction cross section $\Sigma(\mathbf{r}, v)$ yields the number of neutrons, with speed in the interval $v \sim v + dv$, interacting with target nuclei in unit volume around $\mathbf{r}$ and in unit time around $t$. This interpretation of the speed-dependent neutron flux can be readily extended to that of the energy-dependent scalar flux $\phi(\mathbf{r}, E, t)$, together with the relationship $\phi(\mathbf{r}, E, t)dE = \phi(\mathbf{r}, v, t)dv$. We also recall that the macroscopic cross section $\Sigma(\mathbf{r}, v)$ should be written as the product of the nuclear number density $N(\mathbf{r}, t)$ and the effective cross section $\overline{\sigma}(v)$, which is an average cross section accounting for thermal motion of target nuclei.

In terms of the scalar flux $\phi(\mathbf{r}, E, t)$ and effective cross section $\overline{\sigma}(v)$, we are now ready to set up a general balance equation for neutrons whose population density may vary from point to point in space and in time. In addition, we desire to account fully for the variation in the energy of the neutrons as they undergo scattering and absorption collisions and move around in a medium. We also need to represent the leakage of neutrons from the medium and the production of neutrons from external sources or through the fission process in a multiplying medium.

Consistent with assumption (1) from Section 4.1, consider a population of neutrons at time $t$ in a phase volume, consisting of a unit volume around $\mathbf{r}$ and a unit energy interval around $E$, and write a general balance statement:

$$\begin{aligned}&(\text{Time rate of change of neutron population})\\&= (\text{Rate of production of neutrons}) - (\text{Rate of loss of neutrons}).\end{aligned} \tag{4.1}$$

Accounting for the neutron source, leakage and collisions, break up each of the production and loss terms into two components:

$$\begin{aligned}&(\text{Time rate of change of neutron population}, I)\\&\quad = (\text{Rate of generation of neutrons from fission and external sources}, I_1)\end{aligned}$$

+ (Rate of neutrons scattered into the phase volume, $I_2$)
– (Rate of neutrons absorbed in and scattered out of the phase volume, $I_3$)
– (Rate of leakage of neutrons from the phase volume, $I_4$). (4.2)

Write the left-hand side (LHS) of the balance equation simply as the rate of change of the number density of neutrons:

$$I = \frac{\partial}{\partial t} n(\mathbf{r}, E, t) = \frac{1}{v}\frac{\partial}{\partial t}\phi(\mathbf{r}, E, t). \tag{4.3}$$

For now, also write the source term $I_1$ simply as the number $S(\mathbf{r}, E, t)$ of neutrons produced in unit physical volume around $\mathbf{r}$ and in unit energy interval around $E$ per unit time at $t$:

$$I_1 = S(\mathbf{r}, E, t). \tag{4.4}$$

After deriving the desired balance equation, we will develop a full expression for $S(\mathbf{r}, E, t)$ representing both the fission and external sources.

The terms $I_2$ and $I_3$ both are related to neutron collision rates. We begin with the loss term $I_3$ and build upon it to derive the in-scattering term $I_2$. The total collision rate $I_3$ accounts for absorption and scattering collisions, both of which result in the removal of neutrons from the phase volume of our interest around $\{\mathbf{r}, E\}$ at time $t$. Thus, the collision rate $I_3$ may be written as

$$I_3 = \Sigma_t(\mathbf{r}, E, t)\phi(\mathbf{r}, E, t) = N(\mathbf{r}, t)\overline{\sigma}_t(E)\phi(\mathbf{r}, E, t), \tag{4.5}$$

where the effective cross section $\overline{\sigma}_t(E)$ is the sum of absorption and scattering cross sections

$$\overline{\sigma}_t(E) = \overline{\sigma}_t(v) = \frac{\int_{\mathbf{V}} d\mathbf{V}\sigma_t(|\mathbf{v}-\mathbf{V}|)N(\mathbf{r}, \mathbf{V}, t)|\mathbf{v}-\mathbf{V}|}{N(\mathbf{r}, t)v}, \tag{4.6}$$

with the target nuclear number density

$$N(\mathbf{r}, t) = \int_{\mathbf{V}} d\mathbf{V}\, N(\mathbf{r}, \mathbf{V}, t). \tag{4.7}$$

Thus, the collision rate $I_3$ fully accounts for the thermal motion of target nuclei. Note also that, based on assumption (2) from Section 4.1, we have neglected neutron–neutron collisions and limited ourselves to the neutron–nucleus binary collisions only.

To derive the in-scattering term $I_2$, consider explicitly the rate of neutrons undergoing scattering collisions in unit energy interval around $E'$ and in unit physical volume around $\mathbf{r}$ at time $t$:

$$\Sigma_s(\mathbf{r}, E', t)\phi(\mathbf{r}, E', t) = N(\mathbf{r}, t)\overline{\sigma}_s(E')\phi(\mathbf{r}, E', t). \tag{4.8}$$

Equation (4.8) represents the out-scattering rate of neutrons from the phase volume around $\{\mathbf{r}, E'\}$. We now define a conditional probability $p(E' \rightarrow E)$ such that

$$p(E' \rightarrow E)dE = \text{probability that a neutron with energy } E'\text{, when scattered, emerges from the collision with energy distributed in the interval } E \sim E + dE. \tag{4.9}$$

Since by definition, the probability that a neutron of energy $E'$, following a scattering event, emerges with any energy is unity, the conditional probability is properly normalized:

$$\int_0^\infty p(E' \rightarrow E)dE = 1. \tag{4.10}$$

Recall that $p(E' \rightarrow E)$ of Eq. (4.9) is equal to the conditional probability $p(E_0 \rightarrow E)$ introduced in Eq. (2.52) with $E_0 = E'$. We may now obtain the rate $I_2$ of neutrons scattered into unit energy interval around $E$ and in unit physical volume around $\mathbf{r}$ at time $t$ by multiplying the scattering rate of Eq. (4.8) by the conditional probability of Eq. (4.9) and integrating the product over all possible energies $E'$ of neutrons that undergo scattering:

$$\begin{aligned} I_2 &= \int_0^\infty dE' \overline{\sigma}_s(E') p(E' \rightarrow E) N(\mathbf{r}, t) \phi(\mathbf{r}, E', t) \\ &= \int_0^\infty dE' \Sigma_s(\mathbf{r}, E' \rightarrow E, t) \phi(\mathbf{r}, E', t). \end{aligned} \tag{4.11}$$

The scattering cross section $\Sigma_s(\mathbf{r}, E' \rightarrow E, t)$ represents the probability that neutrons of energy $E'$ suffer scattering collisions and emerge in unit energy interval around $E$ at position $\mathbf{r}$ and time $t$. The cross section is also called the *scattering kernel*, introduced in Eq. (2.52) and may be written more explicitly as

$$\begin{aligned} \Sigma_s(\mathbf{r}, E' \rightarrow E, t) &= N(\mathbf{r}, t) \overline{\sigma}_s(E') p(E' \rightarrow E) \\ &= \Sigma_s(\mathbf{r}, E', t) p(E' \rightarrow E) = N(\mathbf{r}, t) \overline{\sigma}_s(E' \rightarrow E). \end{aligned} \tag{4.12}$$

Having obtained the expression for the in-scattering term $I_2$, we should note that the conditional probability of Eq. (4.9) is defined completely in terms of the incident and ejected neutron energies. This is possible because, invoking assumption (3) from Section 4.1, we neglect any displacement in the physical space during the short time interval over which the scattering reaction takes place. Without this assumption, the conditional probability should also explicitly include the reaction time and physical displacement, as is the case with the scattering of charged particles in an electromagnetic field. Note also that Eq. (4.10) implies

$$\int_0^\infty dE\, \Sigma_s(\mathbf{r}, E' \rightarrow E, t) = \Sigma_s(\mathbf{r}, E', t) = N(\mathbf{r}, t) \overline{\sigma}_s(E'). \tag{4.13}$$

Remembering that the conditional probability $p(E' \rightarrow E)$ is a density function, defined in a manner similar to the scalar flux $\phi(\mathbf{r}, E, t)$, we may establish the relationship

$$\overline{\sigma}_s(v' \rightarrow v)dv = \overline{\sigma}_s(E' \rightarrow E)dE. \tag{4.14}$$

Equation (4.14) merely represents the probability that a neutron of speed $v'$, corresponding to energy $E'$, suffers a scattering collision and emerges with speed in the interval $dv$ around $v$, and hence is equal to the probability that such a neutron will emerge from the collision with energy in the interval $dE$ around $E$.

Now that we have derived the two expressions, $I_2$ and $I_3$, associated with neutron reactions, we turn our attention to the leakage term $I_4$. For this purpose, recall the definition of the net current $J(\mathbf{r}, v)$ from Eq. (3.14), representing the current of neutrons relative to a surface area whose outward normal vector is $\mathbf{n}$. Thus, converting the speed variable $v$ to energy $E$, we obtain an expression for the number of neutrons in unit energy interval about $E$, streaming, per unit time at $t$, out of a small volume element at position $\mathbf{r}$, with volume $\Delta V = \Delta x \Delta y \Delta z$, surface area $\Delta A$, and outward normal vector $\mathbf{n}$

$$\begin{aligned} I_4 \Delta V &= J(\mathbf{r}, E, t)\Delta A = \mathbf{n} \cdot \mathbf{J}(\mathbf{r}, E, t)\Delta A = \int_{\Delta A} \mathbf{n} \cdot \mathbf{J}(\mathbf{r}, E, t) dA \\ &= \int_{\Delta V} \nabla \cdot \mathbf{J}(\mathbf{r}, E, t) d\mathbf{r} = \nabla \cdot \mathbf{J}(\mathbf{r}, E, t)\Delta V, \end{aligned} \tag{4.15}$$

where we have applied the Gauss divergence theorem [Arf13] to convert a surface integral to a volume integral. To be proper, we need to set up the leakage term, as well as the rest of the terms of the balance equation, over a finite volume $\Delta V$, but we have taken a shortcut to consider $\Delta V$ and $\Delta A$ in the limit as they approach differential elements. This allows us to equate the surface integral to a product of the integrand and surface area $\Delta A$, and likewise for the volume integral. With Eq. (4.15), obtain the rate of leakage of neutrons per unit physical volume around $\mathbf{r}$ and per unit energy interval around $E$ at time $t$:

$$I_4 = \nabla \cdot \mathbf{J}(\mathbf{r}, E, t). \tag{4.16}$$

Note also that, consistent with assumption (5) from Section 4.1, we neglect body forces and hence any leakage in the velocity or energy space.

Finally, collecting terms, set $I = I_1 + I_2 - I_3 - I_4$ to obtain the *energy-dependent neutron diffusion equation*:

$$\begin{aligned} \frac{1}{v}\frac{\partial \phi(\mathbf{r}, E, t)}{\partial t} &= S(\mathbf{r}, E, t) + \int_0^\infty dE' \, \Sigma_s(E' \rightarrow E)\phi(\mathbf{r}, E', t) \\ &\quad - \Sigma_t(E)\phi(\mathbf{r}, E, t) - \nabla \cdot \mathbf{J}(\mathbf{r}, E, t). \end{aligned} \tag{4.17}$$

For notational convenience, we have introduced in Eq. (4.17)

*Assumption* (6): $N(\mathbf{r},t) = N =$ constant.

Through this simplifying assumption, we may suppress the spatial and temporal dependence in the nuclear number density, essentially ignoring the variations in the densities of various materials due to thermal-hydraulic feedback, fuel depletion, and transient effects as well as material heterogeneities inherent in any reactor core. The diffusion equation may also be written in terms of the speed variable $v$, instead of $E$:

$$\begin{aligned}\frac{1}{v}\frac{\partial\phi(\mathbf{r},v,t)}{\partial t} &= S(\mathbf{r},v,t) + \int_0^\infty dv'\,\Sigma_s(v' \to v)\phi(\mathbf{r},v',t)\\ &\quad - \Sigma_t(v)\phi(\mathbf{r},v,t) - \nabla\cdot\mathbf{J}(\mathbf{r},v,t).\end{aligned} \tag{4.18}$$

## 4.3 Neutron Source Term

As indicated in Section 4.2, the neutron source term $S(\mathbf{r},E,t)$ in Eq. (4.17) comprises, in general, the fission neutron source $S_f(\mathbf{r},E,t)$ and the external source $Q(\mathbf{r},E,t)$:

$$S(\mathbf{r},E,t) = S_f(\mathbf{r},E,t) + Q(\mathbf{r},E,t). \tag{4.19}$$

The fission neutron source $S_f(\mathbf{r},E,t)$ may be written in a form similar to the in-scattering term $I_2$ from Eq. (4.11), with the fission spectrum $\chi(E' \to E)$ replacing the conditional scattering probability $p(E' \to E)$ and the addition of the *average number* $\nu$ of neutrons released per fission:

$$S_f(\mathbf{r},E,t) = \int_0^\infty dE'\,\nu\Sigma_f(E')\phi(\mathbf{r},E',t)\chi(E' \to E). \tag{4.20}$$

The term $\chi(E' \to E)dE$ represents the conditional probability that neutrons of energy $E'$ undergoing fission will produce neutrons with energy distributed in the interval $E \sim E + dE$. The combination of different source distributions in Eq. (4.19) reflects assumption (2) from Section 4.1, which allows for the determination of the neutron population due to multiple sources through a linear superposition of individual contributions. Once again, primarily for notational convenience, we have introduced in Eq. (4.20)

*Assumption* (7): All fission neutrons are released instantaneously, thereby ignoring the presence of delayed neutrons.

Furthermore, introduce one last approximation.

*Assumption* (8): Fission neutrons are emitted isotropically in the laboratory system and energy $E$ of neutrons emitted in the fission process is independent of energy $E'$ of the incident neutron.

Assumption (7) is clearly an approximation and will be removed in Chapter 8, when we discuss nuclear reactor kinetics. In contrast, assumption (8) represents the actual physics of the nuclear fission process to a degree of accuracy that is routinely acceptable in reactor physics.

Under these assumptions, the fission spectrum may be written as

$$\chi(E' \rightarrow E) = \chi(E), \tag{4.21}$$

with the normalization

$$\int_0^\infty dE\chi(E' \rightarrow E) = \int_0^\infty dE\chi(E) = 1.0. \tag{4.22}$$

The fission source term, Eq. (4.20), is then written as

$$S_f(\mathbf{r}, E, t) = \chi(E) \int_0^\infty dE' \nu\Sigma_f(E')\phi(\mathbf{r}, E', t), \tag{4.23}$$

which yields, together with the external source term, an expression for the total neutron source:

$$S(\mathbf{r}, E, t) = \chi(E) \int_0^\infty dE' \nu\Sigma_f(E')\phi(\mathbf{r}, E', t) + Q(\mathbf{r}, E, t). \tag{4.24}$$

## 4.4 Fick's Law of Neutron Current

In order to solve the neutron diffusion equation (4.17) for the scalar flux $\phi(\mathbf{r}, E, t)$, we still need to express the leakage term $\nabla \cdot \mathbf{J}(\mathbf{r}, E, t)$ in terms of $\phi(\mathbf{r}, E, t)$. Starting from the neutron transport equation, which describes the neutron balance in terms of the angular neutron flux $\psi(\mathbf{r}, \mathbf{v}, t)$ or $\psi(\mathbf{r}, E, \boldsymbol{\Omega}, t)$, we may derive an expression for the leakage term systematically in terms of scalar flux $\phi(\mathbf{r}, E, t)$. This basically entails the assumption that the angular flux is linearly anisotropic. This assumption is known as the $P_1$ approximation and corresponds, in one-dimensional slab geometry, to truncating an expansion of the angular $\psi(z, E, \mu, t)$ in Legendre polynomials $P_n(\mu)$ at $n = 1$, where $\mu = \cos\theta$. Here $\theta$ represents the angle between the direction of neutron motion and the spatial coordinate $z$-axis for the slab. This approach is discussed further in Section 4.5.

We derive in this section an expression for the vector current $\mathbf{J}$ based on a classical approach taken by Glasstone and Edlund [Gla52]. The derivation assumes that the medium is purely scattering and that neutrons undergoing scattering are released isotropically in the laboratory. After we derive Fick's law connecting the gradient of the scalar flux to neutron current, we make provisions to allow for anisotropic scattering in general. Furthermore, for notational convenience in the derivation, we limit ourselves to mono-energetic neutrons in a steady-state configuration.

Consider a unit cross-sectional area, located at the origin and lying in the ($x$-$y$) plane, as illustrated in Figure 4.1, and calculate the number $j_-(0)$ of neutrons

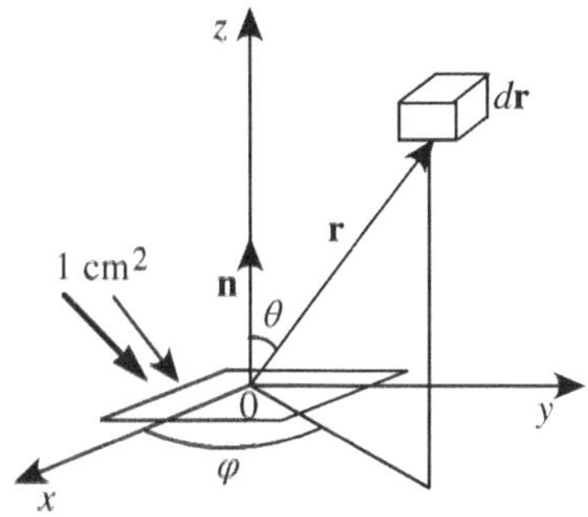


**Figure 4.1** Unit cross-sectional area for the negative partial current.

suffering collisions in a differential volume $d\mathbf{r}$ and reaching the area in unit time. We begin with the term $\Sigma_s\phi(\mathbf{r})d\mathbf{r}$ representing the rate of neutrons suffering collisions in $d\mathbf{r}$. Next, calculate the probability $p(\mathbf{r} \to 0)$ that neutrons scattered isotropically out of the differential volume $d\mathbf{r}$ at $\mathbf{r}$ will reach, without making another collision, a unit cross-sectional area perpendicular to $\mathbf{r}$ at the origin:

$$p(\mathbf{r} \to 0) = \frac{\exp(-\Sigma_s r)}{4\pi r^2}. \tag{4.25}$$

The exponential term represents the probability that neutrons travel a distance $r$ without suffering any collision in a purely scattering medium, while the factor $1/(4\pi r^2)$ accounts for the geometrical distribution of neutrons isotropically arriving at the surface of a sphere of radius $r$ anchored at position $\mathbf{r}$.

Note that the unit cross-sectional area of our interest in Figure 4.1 has an outward normal vector $\mathbf{n}$ and makes an angle $\theta$ to the vector $\mathbf{r}$ along which neutrons travel from the volume $d\mathbf{r}$ to the surface at the origin, and hence that the effective area the neutrons see, traveling along $\mathbf{r}$, is reduced by a factor of $\cos\theta$. Thus, we evaluate the rate $j_-(0)$ as the product of the scattering rate and $p(\mathbf{r} \to 0)$ multiplied by $\mu = \cos\theta$:

$$j_-(0) = \cos\theta \frac{\exp(-\Sigma_s r)}{4\pi r^2}\Sigma_s\phi(\mathbf{r})d\mathbf{r} = \frac{\exp(-\Sigma_s r)}{4\pi}\Sigma_s\phi(\mathbf{r})\mu d\mu d\varphi dr. \tag{4.26}$$

Hence, the negative *partial current* $J_-(0)$ of neutrons at the unit cross-sectional area located at the origin, representing the current of neutrons moving downward, is obtained by integrating $j_-(0)$ over the upper half space:

$$J_-(0) = \frac{\Sigma_s}{4\pi}\int_0^{2\pi} d\varphi \int_0^1 d\mu\mu \int_0^\infty dr\, e^{-\Sigma_s r}\phi(\mathbf{r}). \tag{4.27}$$

We now make the assumption that the neutron flux $\phi(\mathbf{r})$ varies slowly in space, which is conceptually equivalent to the linearly anisotropic behavior of the angular flux $\psi(z, \mu, t)$ assumed in the $P_1$ approximation. This reflects the viewpoint that if $\phi(\mathbf{r})$ varies rapidly in space, there has to be a significant gradient in the neutron population and this in turn implies that the neutrons must be traveling preferentially in some direction, i.e. the angular flux $\psi(z, \mu, t)$ has a significant dependence on the neutron direction of motion. With this perspective in mind, represent $\phi(\mathbf{r})$ in terms of a Taylor's expansion around the origin:

$$\phi(\mathbf{r}) = \phi(0) + x\left(\frac{\partial \phi}{\partial x}\right)_0 + y\left(\frac{\partial \phi}{\partial y}\right)_0 + z\left(\frac{\partial \phi}{\partial z}\right)_0. \tag{4.28}$$

Substituting Eq. (4.28) into Eq. (4.27) and remembering

$$\begin{aligned} x &= r \sin\theta \cos\varphi, \\ y &= r \sin\theta \sin\varphi, \\ z &= r \cos\theta, \end{aligned} \tag{4.29}$$

obtain

$$J_-(0) = \frac{\Sigma_s}{2}\int_0^1 d\mu\mu \int_0^\infty dr e^{-\Sigma_s r}\left[\phi(0) + r\mu\left(\frac{\partial \phi}{\partial z}\right)_0\right]. \tag{4.30}$$

The terms involving $x$ and $y$ in Eq. (4.28) are dropped because

$$\int_0^{2\pi} d\varphi \cos\varphi = \int_0^{2\pi} d\varphi \sin\varphi = 0. \tag{4.31}$$

Carrying on the integration over $\mu$ reduces Eq. (4.30) to

$$J_-(0) = \Sigma_s \int_0^\infty dr e^{-\Sigma_s r}\left[\frac{\phi(0)}{4} + \frac{r}{6}\left(\frac{\partial \phi}{\partial z}\right)_0\right],$$

which yields

$$J_-(0) = \frac{\phi(0)}{4} + \frac{1}{6\Sigma_s}\left(\frac{\partial \phi}{\partial z}\right)_0. \tag{4.32}$$

A similar calculation yields for the positive partial current of neutrons at the origin

$$J_+(0) = \frac{\phi(0)}{4} - \frac{1}{6\Sigma_s}\left(\frac{\partial \phi}{\partial z}\right)_0, \tag{4.33}$$

rendering the desired expression for the net current

$$J(0) = J_+(0) - J_-(0) = -\frac{1}{3\Sigma_s}\left(\frac{\partial \phi}{\partial z}\right)_0. \tag{4.34}$$

Recognize that Eq. (4.34) provides the net current with respective to a surface area whose outward normal vector $\mathbf{n}$ is parallel to the $z$-axis, i.e.

$$J(0) = \mathbf{n} \cdot \mathbf{J}(0) = -\frac{1}{3\Sigma_s}\left(\frac{\partial \phi}{\partial z}\right)_0,$$

which suggests that

$$\mathbf{J}(0) = -\frac{1}{3\Sigma_s}\nabla\phi(0). \tag{4.35}$$

With a definition for the *diffusion coefficient* $D = 1/3\Sigma_s$, we simply generalize Eq. (4.35) to obtain *Fick's law of diffusion* for neutron current:

$$\mathbf{J}(\mathbf{r}) = -D\nabla\phi(\mathbf{r}). \tag{4.36}$$

Equation (4.36) physically represents an intuitive result that neutrons move from a region of high concentration to that of low concentration with a judicious proportionality constant $D$. The derivation of Eq. (4.36), however, has provided us with an explicit expression for $D$ and also allows us to evaluate its applicability in Section 4.7. Furthermore, we may use Eqs. (4.32) and (4.33) to obtain a general expression for partial currents

$$J_{\pm}(\mathbf{r}) = \frac{\phi(\mathbf{r})}{4} \pm \frac{\mathbf{n}\cdot\mathbf{J}(\mathbf{r})}{2}, \tag{4.37}$$

which was suggested in Eq. (3.19). We now assume that Fick's law is valid for time- and energy-dependent diffusion of neutrons, although further approximations have to be introduced to arrive at the result. Thus, substitute

$$\mathbf{J}(\mathbf{r}, E, t) = -D(E)\nabla\phi(\mathbf{r}, E, t) \tag{4.38}$$

together with Eq. (4.24) into Eq. (4.17) to arrive at the final form of the *neutron diffusion equation*:

$$\begin{aligned}\frac{1}{v}\frac{\partial\phi(\mathbf{r}, E, t)}{\partial t} &= \chi(E)\int_0^\infty dE'\nu\Sigma_f(E')\phi(\mathbf{r}, E', t) + Q(\mathbf{r}, E, t)\\ &\quad + \int_0^\infty dE'\,\Sigma_s(E' \rightarrow E)\phi(\mathbf{r}, E', t) - \Sigma_t(E)\phi(\mathbf{r}, E, t)\\ &\quad + \nabla\cdot D(E)\nabla\phi(\mathbf{r}, E, t).\end{aligned} \tag{4.39}$$

In Eq. (4.39), for notational convenience, the spatial and temporal dependencies are suppressed in all of the cross sections and the diffusion coefficient, with assumption (6) introduced in connection with Eq. (4.17).

## 4.5 Neutron Transport Equation and $P_1$ Approximation

To gain a better understanding of the approximations and hence the limitations of diffusion theory, we take a detour now and briefly discuss the neutron transport equation, to sketch out the path to the diffusion equation from the transport equation. Without retracing the derivation of the diffusion equation (4.17), write down the corresponding transport equation in terms of $\psi(\mathbf{r}, \mathbf{v}, t)$ by incorporating the direction $\mathbf{\Omega}$ into the neutron balance equation (4.18). This requires converting speed $v$ into velocity vector $\mathbf{v}$, and rewriting the neutron leakage rate in terms of the gradient of angular flux:

$$\frac{1}{v}\frac{\partial\psi(\mathbf{r},\mathbf{v},t)}{\partial t} = S(\mathbf{r},\mathbf{v},t) + \int_{\mathbf{v}'} d\mathbf{v}'\,\Sigma_s(\mathbf{v}'\rightarrow\mathbf{v})\psi(\mathbf{r},\mathbf{v}',t) - \Sigma_t(v)\psi(\mathbf{r},\mathbf{v},t) - \mathbf{\Omega}\cdot\nabla\psi(\mathbf{r},\mathbf{v},t)\,. \tag{4.40}$$

The leakage rate is derived in a manner similar to the steps taken for Eqs. (4.15) and (4.16), except that we make use of the interpretation for the angular flux given in Eq. (3.5) as the current with respect to an area perpendicular to $\mathbf{\Omega}$:

$$\begin{aligned} I_4\Delta V &= \mathbf{n}\cdot\mathbf{\Omega}\psi(\mathbf{r},\mathbf{v},t)\Delta A = \int_{\Delta A}\mathbf{n}\cdot\mathbf{\Omega}\psi(\mathbf{r},\mathbf{v},t)dA \\ &= \int_{\Delta V}\nabla\cdot\mathbf{\Omega}\psi(\mathbf{r},\mathbf{v},t)d\mathbf{r} = \nabla\cdot\mathbf{\Omega}\psi(\mathbf{r},\mathbf{v},t)\Delta V = \mathbf{\Omega}\cdot\nabla\psi(\mathbf{r},\mathbf{v},t)\Delta V. \end{aligned} \tag{4.41}$$

In the last step, note that $\mathbf{\Omega}$ is a unit directional vector in the velocity space and is therefore independent of the divergence operation in the physical space. In fact, we could have started by deriving Eq. (4.41) instead of Eq. (4.15) or (4.16), and then integrated Eq. (4.41) over solid angle $\mathbf{\Omega}$, recalling the definition for the vector current $\mathbf{J}(\mathbf{r}, E, t)$ in Eq. (3.13), to arrive at Eq. (4.16).

Derivation of Fick's law of neutron diffusion, Eq. (4.36), from Eq. (4.41) requires expanding the angular flux in terms of spherical harmonics [Mar76, Arf13]

$$\psi(\mathbf{r},E,\Omega,t) \simeq \frac{1}{4\pi}\left[\phi(\mathbf{r},E,t) + 3\mathbf{\Omega}\cdot\mathbf{J}(\mathbf{r},E,t)\right]. \tag{4.42}$$

To understand the angular expansion with a simple notation, consider the angular flux for steady-state, 1-D slab geometry, with the energy dependence suppressed

$$\psi(z,\mu) = \int_0^{2\pi}\psi(z,\mu,\varphi)d\varphi, \tag{4.43}$$

with $\mu = \cos\theta$ as illustrated in Figure 4.1. The azimuthal symmetry inherent in the one-dimensional symmetry is represented through integrating over the azimuthal angle $\varphi$, as illustrated in Figure 4.2.

The transport equation (4.40) may then be simplified to

$$\Sigma_t\psi(z,\mu) + \mu\frac{\partial\psi(z,\mu)}{\partial z} = S(z,\mu) + \rho(z,\mu). \tag{4.44}$$

with the in-scattering rate $\rho(z,\mu)$ represented as

$$\rho(z,\mu) = \int_{-1}^{1} d\mu'\Sigma_s(\mu'\rightarrow\mu)\psi(z,\mu'). \tag{4.45}$$

For an *isotropic medium*, the differential scattering cross section is independent of the incoming neutron direction $\mathbf{\Omega}'$ and outgoing neutron direction $\mathbf{\Omega}$ separately and may be written in terms of the scattering angle $\theta_0$ in the laboratory or

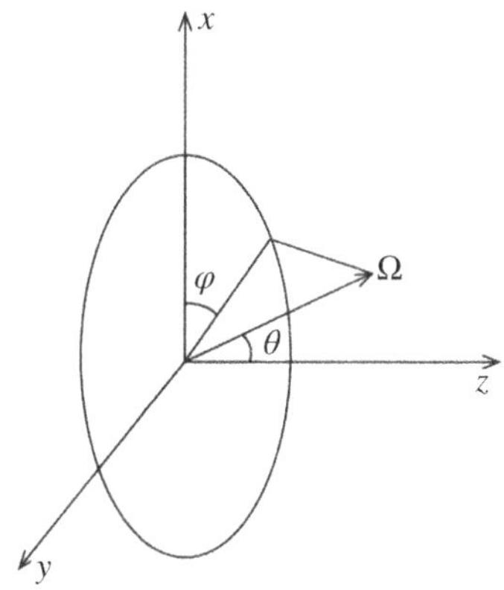


**Figure 4.2** Directional vector Ω projected via polar and azimuthal angles.

$\mu_0 = \cos\theta_0 = \boldsymbol{\Omega}' \cdot \boldsymbol{\Omega}$:

$$\Sigma_s(\boldsymbol{\Omega}' \to \boldsymbol{\Omega}) = \Sigma_s(\mu' \to \mu) = \Sigma_s(\boldsymbol{\Omega}' \cdot \boldsymbol{\Omega}) = \Sigma_s(\mu_0). \tag{4.46}$$

We may then expand the one-dimensional angular flux in terms of Legendre polynomials [Mar76, Arf13] [Appendix C]

$$\psi(z,\mu) = \sum_{\ell=0}^{\infty} \left(\frac{2\ell+1}{2}\right) \phi_\ell(z) P_\ell(\mu), \tag{4.47}$$

where Legendre polynomials have the following expressions:

$$P_0(\mu) = 1,\ P_1(u) = \mu, P_2(u) = \frac{3\mu^2-1}{2},\ P_3(u) = \frac{5\mu^3-3\mu}{2}, \ldots. \tag{4.48}$$

In the $P_1$ approximation, the summation is truncated after the first two terms, yielding a linear dependence on $\mu$, in terms of the two lowest-order expansion coefficients $\phi_0(z)$ and $\phi_1(z)$:

$$\psi(z,\mu) \simeq \frac{1}{2}\phi_0(z) + \frac{3\mu}{2}\phi_1(z). \tag{4.49}$$

Recalling the definition of the scalar flux in Eqs. (3.6) and (3.7) and that of the net current in Eq. (3.15), we may show via direct integrations

$$\phi_0(z) = \phi(z), \tag{4.50a}$$

$$\phi_1(z) = J(z). \tag{4.50b}$$

Invoking the orthonormality property of Legendre polynomials

$$\int_{-1}^{1} P_\ell(\mu) P_m(\mu) d\mu = \frac{2}{2\ell+1}\delta_{\ell m}, \tag{4.51}$$

with the Kronecker delta

$$\delta_{\ell m} = \begin{cases} 1, & \ell = m, \\ 0, & \ell \neq m, \end{cases}$$

also obtain Eqs. (4.50) as the two lowest-order expansion coefficients for the angular flux in Eq. (4.47):

$$\phi_n(z) = \int_{-1}^{1} d\mu P_n(\mu)\psi(z,\mu)$$
$$= \int_{-1}^{1} d\mu P_n(\mu)\sum_{\ell=0}^{\infty}\left(\frac{2\ell+1}{2}\right)\phi_\ell(z)P_\ell(\mu), n = 0, 1. \tag{4.52}$$

Substituting the Legendre polynomial expansion from Eq. (4.49) into the in-scattering integral of Eq. (4.45) yields

$$\rho(z,\mu) = \sum_{\ell=0}^{1}\left(\frac{2\ell+1}{2}\right)\phi_\ell(z)\int_{-1}^{1} d\mu'\Sigma_s(\mu_0)P_\ell(\mu'). \tag{4.53}$$

With another Legendre polynomial expansion for the scattering kernel

$$\Sigma_s(\mu_0) = \sum_{n=0}^{1}\left(\frac{2n+1}{2}\right)\Sigma_{sn}P_n(\mu_0) \tag{4.54}$$

and the addition theorem [Mar76, Arf13] [Appendix C] for the Legendre polynomials with azimuthal symmetry

$$P_n(\mu_0) = P_n(\mathbf{\Omega}' \cdot \mathbf{\Omega}) = P_n(\mu')P_n(\mu) \tag{4.55}$$

substituted into Eq. (4.53), we obtain

$$\rho(z,\mu) = \sum_{\ell=0}^{1}\left(\frac{2\ell+1}{2}\right)\phi_\ell(z)\Sigma_{s\ell}P_\ell(\mu), \tag{4.56}$$

where

$$\Sigma_{s\ell} = \int_{-1}^{1} d\mu_0\Sigma_s(\mu_0)P_\ell(\mu_0), \ell = 0, 1. \tag{4.57}$$

It should be noted that the addition theorem from Eq. (4.55) for $n = 1$ represents a simple geometrical relationship that the cosine of the angle between two directions $\mathbf{\Omega}$ and $\mathbf{\Omega}'$ may be represented by the product of cosines of the polar angles for $\mathbf{\Omega}$ and $\mathbf{\Omega}'$. The two lowest-order components of the scattering kernel in Eq. (4.57) may explicitly be written as

$$\Sigma_{s0} = \int_{-1}^{1} \Sigma_s(\mu_0)d\mu_0 = \Sigma_s, \tag{4.58a}$$

$$\Sigma_{s1} = \int_{-1}^{1} \Sigma_s(\mu_0)\mu_0 d\mu_0 = \overline{\mu}_0\Sigma_s, \tag{4.58b}$$

where $\Sigma_s$ is simply the total scattering cross section and $\overline{\mu}_0$ is the weighted-average cosine of the scattering angle. The expansion for the in-scattering integral obtained in Eq. (4.56) may be written conveniently as

$$\rho(z,\mu) = \sum_{\ell=0}^{1} \left(\frac{2\ell+1}{2}\right) \rho_\ell(z) P_\ell(\mu) \tag{4.59}$$

in terms of consolidated expansion coefficients

$$\rho_0(z) = \Sigma_s \phi_0(z), \tag{4.60a}$$

$$\rho_1(z) = \overline{\mu}_0 \Sigma_s \phi_1(z). \tag{4.60b}$$

Finally, the source term may be expanded in terms of Legendre polynomials:

$$S(z,\mu) = \sum_{\ell=0}^{1} \left(\frac{2\ell+1}{2}\right) S_\ell(z) P_\ell(\mu). \tag{4.61}$$

Substitute Eqs. (4.49) and (4.59) through (4.61) into the 1-D transport equation (4.44), multiply by the Legendre polynomials $P_\ell(\mu)$, $\ell = 0, 1$, and integrate over $\mu \in [-1, +1]$ to pick up the two lowest moments of the transport equation:

$$\Sigma_t \phi_0(z) + \frac{d\phi_1(z)}{dz} = S_0(z) + \Sigma_s \phi_0(z), \tag{4.62a}$$

$$\Sigma_t \phi_1(z) + \frac{1}{3}\frac{d\phi_0(z)}{dz} = S_1(z) + \overline{\mu}_0 \Sigma_s \phi_1(z). \tag{4.62b}$$

Equation (4.62b), with the source term $S_1(z)$ dropped, yields the energy-independent, 1-D form of the net current $\phi_1(z) = J(z)$ given in Eq. (4.36), with the diffusion coefficient $D$ generalized to

$$D = \frac{1}{3\Sigma_{tr}} = \frac{1}{3(\Sigma_t - \overline{\mu}_0 \Sigma_s)} = \frac{\lambda_{tr}}{3}. \tag{4.63}$$

The *transport cross section* $\Sigma_{tr}$ and *transport mean free path* $\lambda_{tr}$ introduced in Eq. (4.63) are not physical parameters that may be actually measured in the laboratory; rather, they are convenient parameters introduced in the $P_1$ approximation or diffusion theory to represent the net current in the form of Fick's law of diffusion. Section 4.6 presents a physical interpretation of the transport cross section.

Continuing a step further, use Eq. (4.49), together with Eqs. (4.50), to determine the positive and negative partial currents defined in Eq. (3.17):

$$J_\pm(z) = \int_0^{\pm 1} \psi(z,\mu)\mu d\mu = \frac{\phi(z)}{4} \pm \frac{J(z)}{2}. \tag{4.64}$$

This is obviously the 1-D form of Eqs. (3.19) and (4.37). Derivation of the full-blown three-dimensional (3-D) form of the $P_1$ equations requires a spherical harmonics expansion of Eq. (4.42), instead of the Legendre polynomial expansion from Eq. (4.47), thereby explicitly accounting for both the azimuthal and polar angle dependencies in the angular flux.

## 4.6 Remarks on Diffusion Coefficient

The diffusion coefficient $D$ in Eqs. (4.35) and (4.36) has been obtained for a purely scattering medium with isotropic scattering in the laboratory frame. For mono-energetic neutrons, the $P_1$ approximation to the neutron transport equation yields a more general expression for $D$ in Eq. (4.63), which accounts for anisotropic scattering in the laboratory in terms of the average $\overline{\mu}_0$ of the cosine of the scattering angle in the laboratory system. We return now to Eq. (4.58b) to obtain an explicit expression

$$\overline{\mu}_0 = \frac{\int_{-1}^{1} d\mu_0 \mu_0 \Sigma_s(\mu_0)}{\int_{-1}^{1} d\mu_0 \Sigma_s(\mu_0)}, \tag{4.65}$$

where the scattering cross section $\Sigma_s(\mu_0)$ is chosen naturally as the weighting factor in the averaging process. For a medium that is isotropic, we may place, without loss of generality, the incident neutron direction $\mathbf{\Omega}'$ along the $z$-axis. Then, with $\mu_0 = \mu = \cos\theta$, Eq. (4.65) is simply rewritten as

$$\overline{\mu}_0 = \frac{\int_{-1}^{1} d\mu \mu \Sigma_s(\mu)}{\int_{-1}^{1} d\mu \Sigma_s(\mu)}. \tag{4.66}$$

Now assume that elastic neutron scattering is isotropic in the center-of-mass (CM) system, which is an excellent approximation for low-energy neutrons. Thus, similar to Eq. (2.49), we write the scattering cross section in terms of the direction $\mathbf{\Omega}_c$ of the neutron motion after the collision in the CM system

$$\Sigma_s(\mathbf{\Omega}_c) = \frac{\Sigma_s}{4\pi}, \tag{4.67}$$

with the incident neutron direction along the $z$-axis. We then integrate Eq. (4.67) over the azimuthal angle $\varphi$ over $2\pi$, and obtain the probability of neutrons scattered into $-d\mu = \sin\theta d\theta$ in the laboratory, which is equal to that associated with the scattering into $-d\mu_c = \sin\theta_c d\theta_c$ in the CM system:

$$\Sigma_s(\mu)d\mu = \Sigma_s(\mu_c)d\mu_c = \frac{\Sigma_s}{2}d\mu_c. \tag{4.68}$$

One of the key results obtained in Chapter 2, when we studied two-body collision mechanics, involving a neutron suffering an elastic scattering with a nucleus of mass number $A$ initially stationary in the laboratory system, is a relationship connecting the scattering angles in the two coordinate systems, Eq. (2.24). This may now be rewritten as

$$\mu = \frac{1 + A\mu_c}{\sqrt{A^2 + 2A\mu_c + 1}}. \tag{4.69}$$

Substituting Eqs. (4.68) and (4.69) into Eq. (4.66) finally yields

$$\overline{\mu}_0 = \frac{1}{2}\int_{-1}^{1} \mu d\mu_c = \frac{1}{2}\int_{-1}^{1} \frac{1 + A\mu_c}{\sqrt{A^2 + 2A\mu_c + 1}} d\mu_c = \frac{2}{3A}. \tag{4.70}$$

A physical interpretation [Pig65, Lam66] of the diffusion coefficient defined in Eq. (4.63) is possible for a purely scattering medium, with

$$D = \frac{\lambda_{tr}}{3} = \frac{\lambda_s}{1 - \overline{\mu}_0}. \tag{4.71}$$

To obtain an interpretation of the transport mean free path (mfp) $\lambda_{tr}$, follow schematically in Figure 4.3 the path of a neutron undergoing a sequence of collisions with a scattering mfp $\lambda_s$. Projecting the distances traveled by the neutron onto the initial direction of travel and summing up the projected distances, obtain

$$\lambda_s(1 + \overline{\mu}_0 + \overline{\mu}_0^2 + \overline{\mu}_0^3 + \cdots) = \frac{\lambda_s}{1 - \overline{\mu}_0} = \lambda_{tr}. \tag{4.72}$$

We may thus interpret $\lambda_{tr}$ as the net distance traveled on average by a neutron after an infinite number of collisions in a purely scattering medium.

Note that, if the scattering is isotropic in the laboratory, $\overline{\mu}_0 = 0$, and $\lambda_{tr}$ becomes equal to the scattering mfp $\lambda_s = 1/\Sigma_s$. Hence, we may now interpret $\lambda_{tr}$ as an effective neutron mfp that accounts for anisotropy in scattering collisions in the laboratory. For the general case with $\Sigma_a \neq 0$, we may also interpret that the transport mfp $\lambda_{tr} = 1/(\Sigma_t - \overline{\mu}_0\Sigma_s)$ physically represents the total collision mfp with the scattering anisotropy taken into account and that it is the proper expression to use in describing the neutron diffusion process. It is indeed a generally accepted empiricism that neutron transport calculations with an isotropic scattering model yield a first-order accuracy for an anisotropic representation, provided the total cross

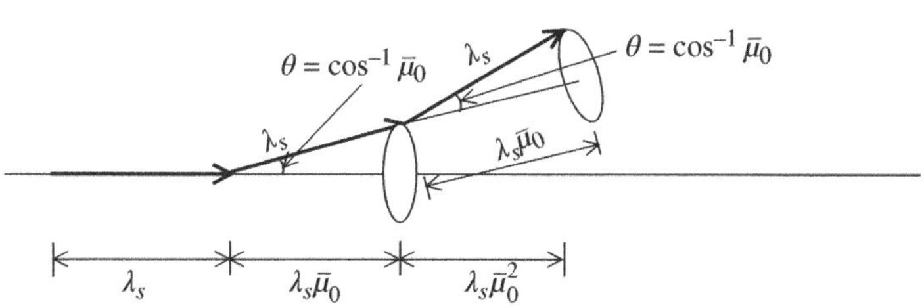


**Figure 4.3** Physical interpretation of transport mean free path $\lambda_{tr}$.

section $\Sigma_t$ appearing in the isotropic model is replaced by an equivalent transport cross section $\Sigma_{tr}$ from Eq. (4.63).

## 4.7 Limitations of Neutron Diffusion Theory

We note here that the neutron diffusion equation (4.17) or (4.18) accurately represents the space-, energy-, and time-dependent transport of neutrons in any system undergoing fission chain reactions, subject essentially to the five assumptions introduced in Section 4.1. The three additional assumptions introduced in Sections 4.2 and 4.3 are primarily for notational convenience during the early stage of our reactor physics analysis and do not entail inherent limitations of diffusion theory. The final form of the neutron balance equation (4.39) was derived, however, with an approximate relationship for the neutron current, i.e. Fick's law of diffusion. For this reason, the diffusion equation may not be as accurate as the transport equation in many problems of practical interest.

Since Fick's law, Eqs. (4.34), was obtained for a purely scattering medium and even with Eq. (4.63) in the $P_1$ approximation, we expect that diffusion theory will not be accurate for a heavily absorbing medium subject to large anisotropy in the angular flux or in a medium where the flux varies rapidly in space. If the spatial variation of the scalar flux is significant, there is a significant transport of neutrons from one region to another, which also implies that the angular neutron flux will be substantially anisotropic. Thus, the linear anisotropy assumed in the $P_1$ approximation is valid only up to a certain degree of accuracy.

Diffusion theory is, therefore, inaccurate for a medium or region with heavy absorption, large leakage, or highly anisotropic neutron flux. This observation may be translated into the recognition that diffusion theory is inaccurate or invalid

(1) Near or in vacuum, and near material interfaces,
(2) Near or within strong absorbers,
(3) Near a localized source, and
(4) In a thin region, which may be considered a special example of case (1).

## 4.8 One-Group Neutron Diffusion Equation

Having derived the energy-dependent neutron diffusion equation (4.17) or (4.18) and discussed the limitations of diffusion theory, we now turn our attention finally to a simplified form of Eq. (4.17). The simplification entails suppressing the energy dependence in the neutron balance equation, as we did in Section 4.5, so that we may obtain both analytic and numerical solutions of the resulting neutron balance equation with minimal effort.

Instead of merely dropping the energy dependence in Eq. (4.17), however, introduce a more rigorous use of the definition of the *total scalar flux*

$$\phi(\mathbf{r},t) = \int_0^\infty \phi(\mathbf{r},E,t)dE, \tag{4.73}$$

which then requires integrating Eq. (4.17) over neutron energy $E$. Begin with the fission source term

$$\begin{aligned} S_f(\mathbf{r},t) &= \int_0^\infty dE\,\chi(E)\int_0^\infty dE'\nu\Sigma_f(E')\phi(\mathbf{r},E',t) \\ &= \int_0^\infty dE'\nu\Sigma_f(E')\phi(\mathbf{r},E',t) \simeq \left\langle \nu\Sigma_f\right\rangle_E \phi(\mathbf{r},t) \equiv \nu\Sigma_f\phi(\mathbf{r},t), \end{aligned} \tag{4.74}$$

where we invoke the normalization of the fission spectrum in Eq. (4.22) and introduce a simplifying notation that the energy-average cross section $\left\langle \nu\Sigma_f\right\rangle_E$ is written as the energy-independent cross section $\nu\Sigma_f$. Likewise, interchanging the order of integration and invoking Eq. (4.13), simplify the in-scattering term:

$$\int_0^\infty dE' \int_0^\infty dE\,\Sigma_s(E'\to E)\phi(\mathbf{r},E',t) = \int_0^\infty dE'\Sigma_s(E')\phi(\mathbf{r},E',t) = \Sigma_s\phi(\mathbf{r},t). \tag{4.75}$$

With the corresponding simplifications of other terms and the recognition that

$$\Sigma_a = \Sigma_t - \Sigma_s \tag{4.76}$$

in the diffusion theory formulation, obtain the desired energy-independent form of the diffusion equation:

$$\begin{aligned} \frac{1}{v}\frac{\partial\phi(\mathbf{r},t)}{\partial t} &= S(\mathbf{r},t) - \Sigma_a\phi(\mathbf{r},t) - \nabla\cdot\mathbf{J}(\mathbf{r},t) \\ &= \nu\Sigma_f\phi(\mathbf{r},t) + Q(\mathbf{r},t) - \Sigma_a\phi(\mathbf{r},t) + \nabla\cdot D\nabla\phi(\mathbf{r},t). \end{aligned} \tag{4.77}$$

Equation (4.77) is usually known as the *one-group, one-speed*, or *mono-energetic diffusion equation*. The terminology *one-group* originates as a limiting case of a multigroup representation of Eq. (4.17) in Chapter 7, where the energy interval for the diffusion equation is subdivided in multiple energy intervals or groups.

The one-group neutron diffusion equation (4.77) is the simplest balance statement for neutron transport in space and time, with the energy dependence of neutrons integrated out. Thus, applications of Eq. (4.77) are generally limited to survey calculations involving the space- and time-dependent scalar flux and to the analysis of well-thermalized reactor problems. To the first order, however, the energy dependence of the scalar flux may be assumed separable from its space and time dependence in many practical problems, and hence Eq. (4.77) plays an important role in nuclear reactor physics. In fact, we will concentrate on analytical and numerical solutions of Eq. (4.77) in Chapters 5 and 6, respectively.

## 4.9 Summary Discussion of Diffusion Equation

In this chapter, we have presented a careful derivation of the neutron diffusion equation. For this purpose, we introduced simplifying assumptions for the neutron balance statement, which is cast in terms of the neutron flux and current defined in Chapter 3. We also made use of the concept of effective cross section accounting for thermal motion of the target nuclei. The derivation of the balance equation clarifies the reason behind the linear nature of the diffusion equation and explains why the scattering cross section is only a function of the incident and emerging neutron speeds. The diffusion equation (4.17) or (4.18) represents an accurate balance statement, apart from the fact that the direction of neutron motion is not explicitly represented. In fact, Eq. (4.17) or (4.18) may be derived directly from the corresponding transport equation by merely integrating out the explicit directional dependence of the neutron flux, as indicated in Section 4.5. The introduction of Fick's law of neutron diffusion, Eq. (4.38), to represent the neutron current, however, renders the final diffusion equation (4.39) only approximate. Through an explicit derivation of Fick's law of neutron diffusion, however, we are able to recognize the limitations of the diffusion equation. The use of the $P_1$ approximation for the angular flux to derive Fick's law in Section 4.5 similarly indicates that the diffusion equation is expected to be valid when the angular flux is not highly anisotropic.

Although neutron diffusion theory is approximate because of its use of Fick's law, the diffusion equation, especially the one-group equation (4.77), is the simplest form of the neutron balance statement and is used extensively in reactor physics. In fact, in many applications where diffusion theory is not expected to be accurate, it still provides useful, and often surprisingly accurate, results. Nonetheless, we should always keep in mind the inherent limitations of the neutron diffusion equation. It is important to evaluate, whenever possible, the physical validity or consistency of any solution to the neutron diffusion equation. In addition, the accuracy of the solution should be ascertained through comparison with applicable experimental data or with a relevant solution to the more rigorous neutron transport equation.

## References

**[ANL63]** Argonne National Laboratory, *Reactor Physics Constants*, ANL-5800, 2nd ed. (1963).

**[Arf13]** G.B. Arfken, H.J. Weber, and F.E. Harris, *Mathematical Methods for Physicist: A Comprehensive Guide*, 7th ed., Academic Press (2013).

**[Gla52]** S. Glasstone and M.C. Edlund, *The Elements of Nuclear Reactor Theory*, Van Nostrand (1952).

**[Gre16]** G.L. Greene and P. Geltenbort, "The Neutron Enigma," *Sci. Am.* **314**, 37 (2016).

**[Lam66]** J.R. Lamarsh, *Introduction to Nuclear Reactor Theory*, Addison-Wesley (1966).

**[Mar76]** H. Margenau and G.M. Murphy, *The Mathematics of Physics and Chemistry*, Krieger Publishing (1976).

**[Pig65]** T.H. Pigford, "Random Walk Problem with No Absorption – The Transport Mean Free Path," unpublished lecture notes, Department of Nuclear Engineering, University of California, Berkeley, CA, USA (1965).

## Problems

**4.1** An isotropic point source of strength $S_0$ [neutron $\cdot$ s$^{-1}$] is located at the left-hand surface of a semi-infinite, purely absorbing slab of thickness $H$ surrounded by vacuum. A small foil with surface area $A$ is located at the right-hand surface of the slab at a distance $H$ measured from the point opposite the point source. Determine the number of source neutrons arriving at the foil per s without suffering collisions.

**4.2** For a purely scattering medium, the scattering cross section is linearly anisotropic in the laboratory and given by

$$\Sigma_s(\mu_0) = \frac{\Sigma_{s0}}{2}\left(1 + \frac{\mu_0}{2}\right), \mu_0 = \mathbf{\Omega}' \cdot \mathbf{\Omega}.$$

Using the $P_1$ approximation to the one-group 1-D transport equation, derive expressions for (a) net current $J(z)$ and (b) partial currents $J_\pm(z)$ in terms of net current $J(z)$ and scalar flux $\phi(z)$.

**4.3** A collimated beam of mono-energetic neutrons is incident at an angle $\theta$ on a half space described by $\Sigma_a$ and $\Sigma_s$. (a) Derive an expression for the probability $p(\theta \to \theta')d\theta'$ that the neutrons will emerge from the half space in $d\theta'$ around $\theta'$. (b) Determine the probability $p(\theta)$ that the neutron incident at $\theta$ is re-emitted after one collision in any direction. Show that your answer is reasonable for $\theta = 0$ and $\theta = \pi/2$.

**4.4** Extend the analysis of Problem 4.3 to determine the probability $p(\theta')d\theta'$ that the neutrons will emerge from the half space in $d\theta'$ around $\theta'$ after two scattering collisions in the half space.

**4.5** Applying the steps taken to obtain partial current $j_{-}(0)$ via Eqs. (4.25) and (4.26), show that the radiation dose rate at 1.0 m from a 1.0-Curie $^{60}$Co source is approximately 1.0 rem · hr$^{-1}$. You may use the energy-absorption mass attenuation coefficient [ANL63] for 1.25-MeV $\gamma$-rays, $\mu_e/\rho = 2.7 \times 10^{-3}$ m$^2$ · kg$^{-1}$ for air and $\mu_e/\rho = 2.9 \times 10^{-3}$ m$^2$ · kg$^{-1}$ for tissue.

**4.6** A plane source of mono-energetic neutrons in an infinite medium emits $S_0$ [neutron·cm$^{-2}$ s$^{-1}$] isotropically on each side of the plane. The nuclei of the surrounding medium are purely absorbing with total cross section $\Sigma$. Derive an expression for the number of neutrons absorbed per cm$^3$ per s at a distance $x$ from the plane in terms of an exponential integral [Appendix C].

**4.7** Repeat Problem 4.6 with the source distribution altered to $S(\theta)d\theta = S_0 \sin\theta \cos\theta d\theta$.

# 5

# Applications of the One-Group Neutron Diffusion Equation

We discuss in this chapter how the one-group neutron diffusion equation may be utilized in the analysis of nuclear reactor systems by obtaining some representative solutions to the equation both for multiplying and non-multiplying media. We will study some of the analytical techniques that can be used for solution of the diffusion equation and also present certain basic concepts that can be understood in terms of the equation. Although essentially the same mathematical approaches and boundary conditions may be applied to the solution of the one- and multi-group diffusion equations, we intentionally limit our discussion in this chapter to the simplest form of the diffusion equation so that we may clearly understand key concepts in reactor physics. In particular, we will discuss the boundary conditions applicable to the diffusion equation, and through the solution of the time-dependent one-group diffusion equation, derive the criticality condition and introduce the concept of buckling.

With these objectives in mind, we will not be exhaustive in our discussion of both the solution techniques and the geometries for which the solution is obtained. For notational convenience, spatial dependence of the cross sections and diffusion coefficient is suppressed in the diffusion equation, allowing us to consider the diffusion equation written in terms of the energy-independent scalar flux $\phi(\mathbf{r}, t)$:

$$\frac{1}{v}\frac{\partial \phi(\mathbf{r}, t)}{\partial t} = D\nabla^2\phi(\mathbf{r}, t) - \Sigma_a\phi(\mathbf{r}, t) + S(\mathbf{r}, t). \tag{5.1}$$

This is the one-group diffusion equation (4.77), with the space-independent diffusion coefficient $D$ taken outside the divergence operator. As it becomes necessary, however, we will allow the cross sections and diffusion coefficient to be distinct from region to region in multi-region geometries.

Section 5.1 begins with a discussion of the boundary conditions necessary for the solution of Eq. (5.1); we move into representative solutions of the

*Nuclear Reactor Physics and Engineering*, Second Edition. John C. Lee.
© 2025 John Wiley & Sons, Inc. Published 2025 by John Wiley & Sons, Inc.
Companion website: www.wiley.com/go/LeeNRPE2

steady-state diffusion equation in Section 5.2. This is followed by a solution of the time-dependent diffusion equation in Section 5.3, which introduces the concept of material and geometric bucklings together with the one-group form of the effective multiplication factor $k_{eff}$. A sample criticality calculation is also presented in Section 5.3 for a typical pressurized water reactor (PWR) configuration. To clarify the concept of $k_{eff}$, Section 5.4 presents the well-known four-factor and six-factor formulas for the multiplication factor, which account for the process of neutron slowing down and qualitatively represent the evolution of the neutron population over successive generations. The chapter concludes with brief remarks in Section 5.5.

## 5.1 Boundary Conditions for Diffusion Equation

Since the diffusion equation (5.1) is a partial differential equation involving both space and time, we need both boundary conditions applicable to the space variable and initial conditions representing the time dependence. In particular, since the time-independent form of the equation is a second-order differential equation, we have to provide two boundary conditions for each region and in each dimension for which a distinct form of the solution is desired. With this understanding, consider the following conditions:

(1) Neutron flux should be non-negative everywhere. It should also be finite everywhere except perhaps at the location of localized sources, which are usually introduced as a mathematical idealization of the actual distributed sources and should thus be treated as points of singularity. The finiteness condition also implies that the neutron flux, due to localized finite sources, should vanish at infinity.

(2) At an interface between two different materials, partial currents $J_+$ and $J_-$, net current $J$, and scalar flux $\phi$ should all be continuous. All of these conditions represent the simple requirement that the number of neutrons has to be conserved at such a material interface. Depending on the particular interface, we may want to impose, at our convenience, the continuity condition in terms of $J$, $\phi$, or $J_+$ and $J_-$.

(3) At a free surface, e.g. a non-reentrant surface in contact with vacuum and a surface surrounded by a black absorber, we assume the return current, or negative partial current $J_-$, should vanish. We refer to the limiting case of a thick absorber region with a large absorption cross section as a perfect or *black absorber*. By linearly extrapolating the flux near the free surface, as illustrated in Figure 5.1, we also note that the free surface condition may be approximated

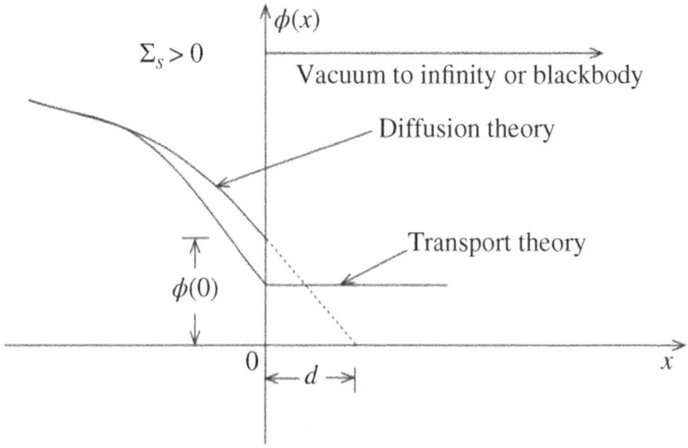


**Figure 5.1** Linear extrapolation of flux at a free surface.

by the condition that the flux $\phi$ vanishes at a distance $d$ beyond the physical boundary. For a slab geometry, the extrapolation distance $d$ can be obtained from the zero return current condition applied at the boundary $x = 0$:

$$J\_(0) = \frac{\phi(0)}{4} + \frac{\lambda_{tr}}{6}\frac{d\phi(0)}{dx} = 0. \tag{5.2}$$

Assuming $\phi(x)$ extrapolates linearly beyond the free surface into vacuum or a black absorber region converts Eq. (5.2) to

$$\frac{d\phi(0)}{dx} = -\frac{6\phi(0)}{4\lambda_{tr}} = -\frac{\phi(0)}{d}, \tag{5.3}$$

which then yields

$$\phi(d) = 0, \quad \text{with} \quad d = \frac{2}{3}\lambda_{tr}. \tag{5.4}$$

More accurate transport theory calculations [Bel70] show that $d = 0.7104$ $\lambda_{tr}$. Since the extrapolated endpoint boundary condition, Eq. (5.4), requires a linear extrapolation on top of the rigorous boundary condition of Eq. (5.2), Eq. (5.4) is an approximation of Eq. (5.2). Lacking any collisions in vacuum, the neutrons actually maintain a constant flux in vacuum, as illustrated by the exact transport theory solution in Figure 5.1. Equation (5.4) is, however, often used obviously for mathematical convenience. We may show that Eq. (5.4) becomes equivalent to Eq. (5.2) for a large, weakly absorbing medium.

One exception to the vacuum boundary or free surface boundary condition arises in a configuration with two finite media separated by vacuum. For example, we consider in Figure 5.2 two semi-infinite slabs separated by vacuum, where the continuity condition (2) applies. In addition, because no

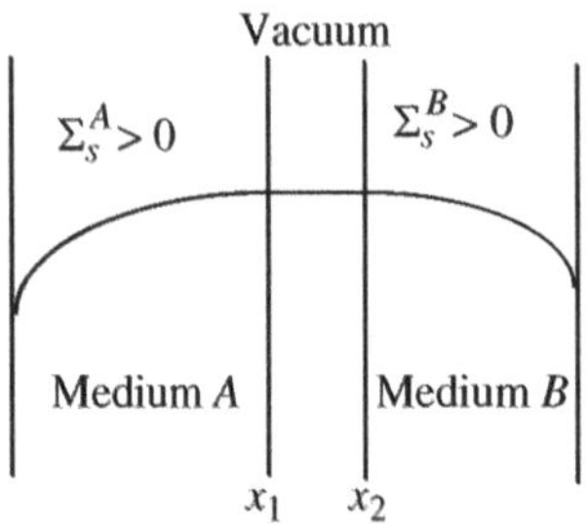


**Figure 5.2** Two material regions separated by vacuum.

reactions take place in the vacuum region surrounded by the media, we also require:

$$J(x_1) = J(x_2), \tag{5.5a}$$

$$\phi(x_1) = \phi(x_2). \tag{5.5b}$$

Equations (5.5) are natural requirements for this configuration because the intervening vacuum region cannot affect the net current or flux of neutrons.

(4) Source conditions are introduced to idealize, for mathematical convenience, localized neutron source distributions. We consider three representative source configurations:

(a) Plane isotropic source of strength $S$ [neutron $\cdot$ cm$^{-2}$ $\cdot$ s$^{-1}$]. In this case, as illustrated in Figure 5.3 for slab geometry, the conservation of neutrons across the source plane at $x_0$ demands

$$J_+^B(x_0^+) = \frac{S}{2} + J_+^A(x_0^-), \tag{5.6a}$$

$$J_-^A(x_0^-) = \frac{S}{2} + J_-^B(x_0^+). \tag{5.6b}$$

where $x_0^\pm = x_0 \pm \varepsilon$, for a small positive number $\varepsilon$, represents the surface of a small capsule or pillbox built around the source plane. Equations (5.6) explicitly account for the fact that the source is isotropic, i.e. one half of the source neutrons would be traveling to the right of the source plane, while the other half would be traveling to the left. The source condition of Eqs. (5.6) also represents the fact that, in the limit as $\varepsilon \to 0$, the absorption of neutrons in the pillbox vanishes so that the differences in the partial currents at the left and right surfaces of the pillbox are only due to the source of neutrons enclosed in the pillbox. Adding Eqs. (5.6a) and (5.6b) first and then subtracting Eq. (5.6b) from Eq. (5.6a) provide two equivalent source conditions:

$$J^B(x_0^+) - J^A(x_0^-) = S, \tag{5.7a}$$

$$\phi(x_0^+) = \phi(x_0^-). \tag{5.7b}$$

**Figure 5.3** Planar source in slab geometry.

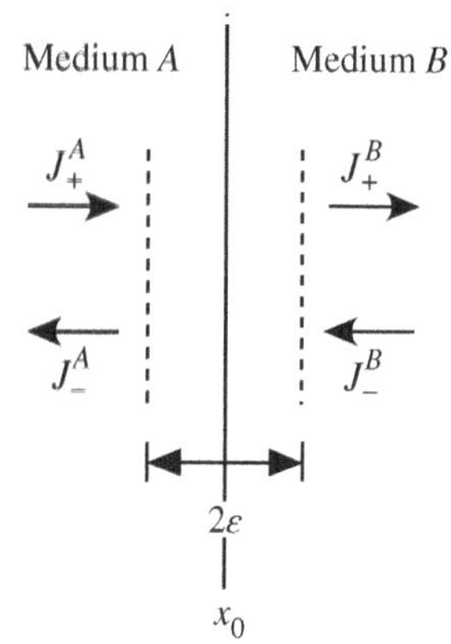


While Eq. (5.7b) represents the general interface condition requiring the continuity of flux, Eq. (5.7a) accounts for the presence of the source plane in terms of the net currents at the right- and left-hand sides of the source. In the special case where regions $A$ and $B$ are identical in material composition and geometrically symmetric, i.e. if

$$J_+^B(x_0^+) = J_-^A(x_0^-),$$

$$J_-^B(x_0^+) = J_+^A(x_0^-),$$

then Eq. (5.7a) degenerates to a simpler and more intuitive condition:

$$J^B(x_0^+) = -J^A(x_0^-) = \frac{S}{2}. \tag{5.8}$$

We emphasize here that Eqs. (5.6) and (5.7) are equivalent and represent source conditions applicable to general planar geometry, while Eq. (5.8) is more restrictive and applicable only to a symmetric geometry. Note here that $J^A < 0$ while $J^B > 0$.

(b) Point source of strength $S$ [neutron $\cdot$ s$^{-1}$] at the origin $r = 0$. The number of neutrons passing through a small sphere of radius $r$ surrounding the source should equal the number of neutrons produced at the point source, in the limit of infinitesimally small $r$ when the number of neutrons absorbed in the sphere becomes negligibly small. This provides the applicable source condition

$$S = \lim_{r\to 0}\,[4\pi r^2\, J(r)] = \lim_{r\to 0}\left[4\pi r^2\left\{-D\frac{d\phi(r)}{dr}\right\}\right]. \tag{5.9}$$

(c) Line source of strength $S$ [neutron $\cdot$ cm$^{-1}$ $\cdot$ s$^{-1}$] at the axis of an infinitely long cylinder. By constructing a cylindrical tube of radius $r$ surrounding the line source, conservation of the number of neutrons across the tube wall requires

$$S = \lim_{r \to 0} [2\pi r J(r)] = \lim_{r \to 0} \left[ 2\pi r \left\{ -D \frac{d\phi(r)}{dr} \right\} \right]. \tag{5.10}$$

(5) Finally, to handle time-dependent problems, an initial condition has to be specified:

$$\phi(\mathbf{r}, 0) = \phi_0(\mathbf{r}). \tag{5.11}$$

## 5.2 Solution of Steady-State Diffusion Equation

We study in this section simple solutions to the neutron diffusion equation (5.1) in non-multiplying media to illustrate the basic approach we take for the solution including the application of boundary conditions. Section 5.2.1 begins with solutions in simple geometries, subject to localized source distributions, which are idealized as singularities to simplify the necessary mathematical steps. We then discuss in Section 5.2.2 how the elementary solutions obtained in Section 5.2.1 may be used as a kernel or Green's function to yield solutions for more realistic source distributions.

### 5.2.1 Flux in Non-Multiplying Media with Localized Sources

When the time dependence is dropped, Eq. (5.1) takes the form of a Helmholtz or wave equation:

$$D\nabla^2\phi(\mathbf{r}) - \Sigma_a\phi(\mathbf{r}) + S(\mathbf{r}) = 0. \tag{5.12}$$

We begin with the solution to Eq. (5.12) for slab geometries with a plane isotropic source and then obtain the solution to a point isotropic source in an infinite medium.

1. **Isotropic plane source of strength $S$ [neutron $\cdot$ cm$^{-2}$ $\cdot$ s$^{-1}$] in an infinite medium**

   For this symmetric one-dimensional slab geometry problem, obtain first an expression for the scalar flux satisfying the source-free form of Eq. (5.12):

   $$\frac{d^2\phi(x)}{dx^2} - \kappa^2\phi(x) = 0 \quad \text{for} \quad x \neq 0, \tag{5.13}$$

   with

   $$\kappa^2 = \frac{\Sigma_a}{D} = \frac{1}{L^2}. \tag{5.14}$$

   The quantity $L^2$ can be shown to be 1/6 of the mean square distance that a neutron travels before capture in an infinite medium, and $L$ is known as the *diffusion length*. Once the solution to the homogeneous, source-free diffusion equation (5.13) is obtained, the source condition, Eq. (5.8), is applied. A general solution to Eq. (5.13) is readily obtained:

$$\phi(x) = Ae^{-\kappa x} + Be^{\kappa x}, \ x > 0. \tag{5.15}$$

Note that Eq. (5.15) contains two unknown constants $A$ and $B$, as it should for a solution to any second-order differential equation. Applying condition (1) from Section 5.1 stipulating that flux remain non-negative and finite everywhere yields $B = 0$. The other constant of integration should reflect the plane source condition, Eq. (5.8), which for this symmetric problem may be written as

$$\lim_{x\to 0} J(x) = \frac{S}{2} = -D\frac{d\phi(0)}{dx} = \kappa DA \lim_{x\to 0} e^{-\kappa x}, \ x > 0, \tag{5.16}$$

yielding $A = S/2\kappa D$. Hence, the solution to the steady-state one-group neutron diffusion equation, subject to an isotropic plane source at $x = 0$, is obtained as

$$\phi(x) = \frac{S}{2\kappa D}e^{-\kappa|x|}. \tag{5.17}$$

To gain a better understanding of the source condition, Eq. (5.16), consider an alternate approach in applying the source condition, by rewriting Eq. (5.13) more in the form of Eq. (5.12), with an explicit representation of the source term:

$$-D\frac{d^2\phi(x)}{dx^2} + \Sigma_a\phi(x) = S\delta(x). \tag{5.18}$$

We integrate Eq. (5.18) over the interval $[-\varepsilon, \varepsilon]$, as visualized in Figure 5.3, to obtain

$$-D\int_{-\varepsilon}^{\varepsilon}\frac{d^2\phi(x)}{dx^2}dx = S = -D\frac{d\phi(\varepsilon)}{dx} + D\frac{d\phi(-\varepsilon)}{dx} = J(\varepsilon) - J(-\varepsilon), \tag{5.19}$$

where the integral involving $\Sigma_a\phi(x)$ vanishes in the limit as $\varepsilon \to 0$ because the flux has to remain finite even across the source plane. Remembering the symmetry of the problem, we recognize that Eq. (5.19) yields the source condition of Eq. (5.8), as it should.

2. **Alternate solution for the plane source problem**
   For the slab geometry problem studied previously, we introduce an alternate approach to gain some understanding of a general technique for solving the diffusion equation. Instead of solving the source-free diffusion equation together with the appropriate boundary conditions, use an integral transform method [Appendix D], which includes the boundary conditions automatically as an integral part of the solution technique. Rewrite Eq. (5.18) in the form of Eq. (5.13), but with the source term retained:

$$\frac{d^2\phi(x)}{dx^2} - \kappa^2\phi(x) = -\frac{S\delta(x)}{D}. \tag{5.20}$$

Defining the Fourier transform of the scalar flux $\phi(x)$ as

$$\overline{\phi}(k) = \frac{1}{\sqrt{2\pi}} \int_{-\infty}^{\infty} e^{-ikx}\phi(x)dx, \tag{5.21}$$

take the Fourier transform of Eq. (5.20) to obtain

$$-k^2\overline{\phi}(k) - \kappa^2\overline{\phi}(k) = -\frac{1}{\sqrt{2\pi}}\frac{S}{D}, \tag{5.22}$$

with the finiteness condition, $\phi(x) < \infty$ for all $x$. Solving the algebraic equation (5.22) for the transform and taking the inverse Fourier transform

$$\phi(x) = \frac{1}{\sqrt{2\pi}} \int_{-\infty}^{\infty} \overline{\phi}(k)e^{ikx}dk, \tag{5.23}$$

provide a solution for the flux:

$$\phi(x) = \frac{S}{2\pi D} \int_{-\infty}^{\infty} \frac{e^{ikx}}{k^2 + \kappa^2}dk. \tag{5.24}$$

The complex integral in Eq. (5.24) can be evaluated by extending the integral path along a semicircle $\Gamma$ of infinite radius and using the residue theorem [Arf13] for the simple pole at $k = i\kappa$ illustrated in Figure 5.4. Note, in passing, that the other simple pole at $k = -i\kappa$ lies outside the closed contour and hence does not have to be considered. The integral in Eq. (5.24) becomes

$$\int_{\Gamma} \frac{e^{ikx}dk}{k^2 + \kappa^2} + \int_{-\infty}^{\infty} \frac{e^{ikx}dk}{k^2 + \kappa^2} = 2\pi i \frac{e^{-\kappa x}}{2i\kappa} = \frac{\pi e^{-\kappa x}}{\kappa}. \tag{5.25}$$

By Jordan's lemma [Arf13] [Appendix D], the first integral over the semicircle $\Gamma$ vanishes, and we obtain the solution to Eq. (5.20):

$$\phi(x) = \frac{S}{2\pi D}\frac{\pi e^{-\kappa x}}{\kappa} = \frac{Se^{-\kappa x}}{2\kappa D}, x > 0. \tag{5.26}$$

Of course, Eq. (5.26) has been obtained much more easily by a direct solution in Eqs. (5.15) through (5.17). The application of Fourier transform techniques,

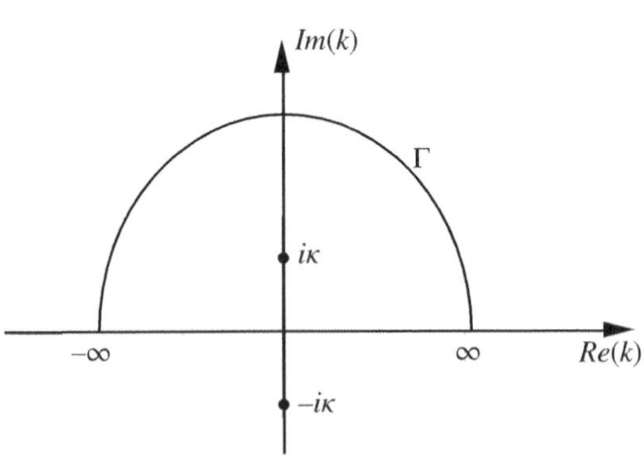


**Figure 5.4** Closed contour in the Fourier domain $k$-space.

however, accounts for all the necessary boundary conditions and the presence of the plane source, as an integral part of the transform process. For this simple problem, the Fourier transform approach is taken to illustrate the power and elegance of the integral transform methods in general.

3. **Point source of strength $S$ [neutron $\cdot$ s$^{-1}$] in an infinite medium, $S(\mathbf{r}) = S\delta(r)/4\pi r^2$**
For this problem with spherical geometry, take a direct approach of Eqs. (5.15) through (5.17). Solve the source-free diffusion equation, which is valid at every point away from the source at the origin $r = 0$, and then apply the source condition of Eq. (5.9). For this spherically symmetric problem, Eq. (5.12) is written as
$$\frac{1}{r^2}\frac{d}{dr}\left[r^2\frac{d\phi(r)}{dr}\right] - \kappa^2\phi(r) = 0, \quad r > 0. \tag{5.27}$$
Substituting
$$u(r) = r\phi(r)$$
into Eq. (5.27) yields a simple second-order ordinary differential equation for $u(r)$
$$\frac{d^2u(r)}{dr^2} - \kappa^2 u(r) = 0, \tag{5.28}$$
which is in the same mathematical form as the slab-geometry equation (5.13). Thus, a general solution to Eq. (5.27) follows
$$\phi(r) = A\frac{e^{-\kappa r}}{r}, \tag{5.29}$$
where the other solution containing $\exp(\kappa r)$ is eliminated to satisfy condition (1) from Section 5.1, $\phi(r) < \infty$.
The remaining constant of integration $A$ has to reflect the point source at $r = 0$, i.e. Eq. (5.9)
$$S = \lim_{r\to 0} 4\pi r^2\left[-DA\left(\frac{-\kappa r - 1}{r^2}\right)e^{-\kappa r}\right], \tag{5.30}$$
which yields $A = S/4\pi D$. Hence, the solution to the steady-state one-group neutron diffusion equation, subject to a point source at $r = 0$, is written as
$$\phi(r) = \frac{Se^{-\kappa r}}{4\pi D r}. \tag{5.31}$$

4. **Plane source in a semi-infinite slab of extrapolated thickness $2a$**
We assume the same source strength $S$ [neutron $\cdot$ cm$^{-2}$ $\cdot$ s$^{-1}$] as in the infinite medium case from Eq. (5.13) but need to account for the finite slab thickness explicitly. For mathematical convenience, consider the extrapolated endpoint boundary condition from Eq. (5.4), as illustrated in Figure 5.5, and begin with a general solution to Eq. (5.13) without the source
$$\phi(x) = A\sinh\kappa x + B\cosh\kappa x, \tag{5.32}$$

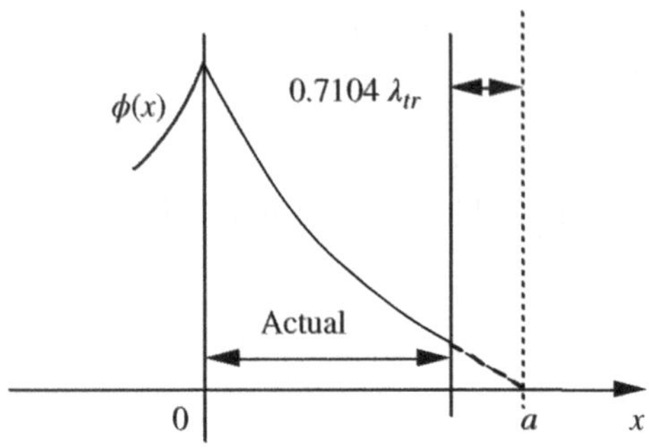


**Figure 5.5** Extrapolated boundary for slab geometry.

with two constants $A$ and $B$ yet to be determined. For mathematical convenience, write the general solution in terms of the hyperbolic sine and cosine functions, rather than the exponential functions. We should, however, note that the hyperbolic and exponential functions are completely equivalent to one another and that Eq. (5.32) may be written equivalently in terms of $\exp(\pm\kappa x)$, if we prefer.

The vacuum boundary condition, Eq. (5.4), requires $\phi(a) = 0$, which yields the constant $B$ in terms of the constant $A$:

$$B = -A \tanh \kappa a. \tag{5.33}$$

Substituting Eq. (5.33) into Eq. (5.32) yields

$$\phi(x) = \frac{A \sinh \kappa(x-a)}{\cosh \kappa a}. \tag{5.34}$$

The source condition, Eq. (5.8), requires

$$J(0+) = \frac{S}{2} = -D\frac{d\phi(0+)}{dx} = -A\kappa D,$$

which finally generates the desired solution for the scalar flux

$$\phi(x) = \frac{S \sinh \kappa(a-x)}{2\kappa D \cosh \kappa a}, \ x > 0. \tag{5.35}$$

We may also verify that Eq. (5.35) can be obtained by starting with the solution written in terms of the exponential functions, rather than Eq. (5.32). We have to contend, however, with some extra amount of algebra in this approach, since the vacuum boundary condition cannot be applied in the succinct manner that we have seen in Eq. (5.33). In general, the hyperbolic functions yield efficient solutions for diffusion theory problems dealing with finite media, and that is precisely why we began with the general solution in the form of Eq. (5.32).

5. **Plane source with two slabs of finite thickness**

Consider an extension of the slab geometry problem just analyzed by adding another slab region on each side of the original slab such that the half thickness

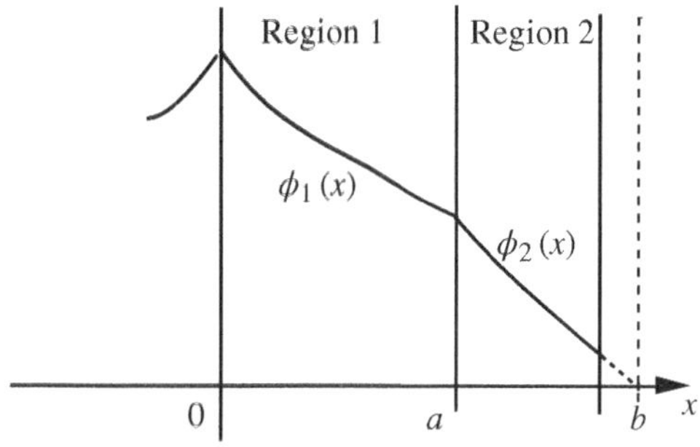


**Figure 5.6** Two-region slab with plane source.

of the first region is $a$ and the extrapolated half thickness of the entire slab is $b$, as illustrated in Figure 5.6. Subject to the same plane source at $x = 0$, write general solutions $\phi_1(x)$ and $\phi_2(x)$ for regions 1 and 2, respectively, in terms of $\kappa_1^2 = \Sigma_{a1}/D_1$, $\kappa_2^2 = \Sigma_{a2}/D_2$, and four unknown constants $A_1, B_1, A_2$, and $B_2$

$$\phi_1(x) = A_1 \sinh \kappa_1 x + B_1 \cosh \kappa_1 x, \tag{5.36a}$$

$$\phi_2(x) = A_2 \sinh \kappa_2 x + B_2 \cosh \kappa_2 x. \tag{5.36b}$$

The source condition, Eq. (5.8), yields

$$A_1 = -\frac{S}{2\kappa_1 D_1},$$

so that we may rewrite Eq. (5.36a) as

$$\phi_1(x) = -\frac{S}{2\kappa_1 D_1} \sinh \kappa_1 x + B_1 \cosh \kappa_1 x. \tag{5.37}$$

The vacuum boundary condition, Eq. (5.4), reduces Eq. (5.36b) to

$$\phi_2(x) = A_2' \sinh \kappa_2 (b - x), \tag{5.38}$$

where $A_2'$ is another constant yet to be determined. Equation (5.38) is obviously equivalent to Eq. (5.34). Continuity of the net current $J(x)$ and flux $\phi(x)$ at the interface $x = a$ provides two equations for the two remaining constants $B_1$ and $A_2'$ in Eqs. (5.37) and (5.38), respectively, and the solutions $\phi_1(x)$ and $\phi_2(x)$ both can be fully obtained.

Instead of pursuing this straightforward path, we take an alternative approach, which provides a useful means to obtain the flux $\phi_1(x)$ for the first region based on the properties of the second region. Define the *albedo* $\beta$ as the ratio of the number of neutrons crossing the interface from region 2 into region 1 to that of neutrons crossing the interface into region 2:

$$\beta = \left.\frac{J_{out}}{J_{in}}\right|_{region\ 2} = \frac{J_-(a)}{J_+(a)}. \tag{5.39}$$

Thus, $\beta$ represents the probability that neutrons leaving region 1 return to the region after undergoing collisions in region 2, i.e. the reflection probability of region 2. The partial currents $J_-(a)$ and $J_+(a)$ are continuous at the interface and may be evaluated in terms of the material properties and the flux of either region 1 or 2. Since we are primarily interested in obtaining the flux $\phi_1(x)$ for region 1, we evaluate the partial currents and $\beta$ for region 2:

$$\beta = \frac{\left[\frac{\phi_2}{4} + \frac{D_2}{2}\frac{d\phi_2}{dx}\right]_{x=a}}{\left[\frac{\phi_2}{4} - \frac{D_2}{2}\frac{d\phi_2}{dx}\right]_{x=a}} = \frac{1 - 2\kappa_2 D_2 \coth \kappa_2(b-a)}{1 + 2\kappa_2 D_2 \coth \kappa_2(b-a)}. \tag{5.40}$$

Expressing $\beta$ in terms of the properties of region 1 yields likewise

$$\beta = \frac{\left[1 + 2D_1 \frac{d\ln\phi_1}{dx}\right]_{x=a}}{\left[1 - 2D_1 \frac{d\ln\phi_1}{dx}\right]_{x=a}}$$

or

$$D_1 \frac{d\ln\phi_1}{dx}\bigg|_{x=a} = -\frac{1}{2}\left(\frac{1-\beta}{1+\beta}\right). \tag{5.41}$$

Hence, the unknown constant $B_1$ in Eq. (5.37) may be evaluated by substituting Eq. (5.37) for $\phi_1(x)$ into Eq. (5.41), where the albedo $\beta$ may be obtained from Eq. (5.40) or from appropriate measurements. Equivalently, substituting Eq. (5.40) into Eq. (5.41) yields

$$D_1 \frac{d\ln\phi_1}{dx}\bigg|_{x=a} = -\kappa_2 D_2 \coth \kappa_2(b-a). \tag{5.42}$$

The actual evaluation of the constant $B_1$ in terms of Eq. (5.41) or (5.42) will be left as an exercise. It should be emphasized, however, that through the introduction of the albedo $\beta$, the flux $\phi_1(x)$ for region 1 can be obtained without explicit evaluation of the flux for region 2, provided that we determine $\beta$ itself through separate calculations or measurements. In particular, at the interface, the combined or mixed boundary condition is given by Eq. (5.42), which obviously is not a function of the unknown constant $A_2'$ for region 2 so that $B_1$ may be determined without having to evaluate $A_2'$. This technique of essentially replacing a region by a mixed boundary condition involving the albedo is often used when region 1 is a multiplying medium and region 2 is a non-multiplying, diffusing medium. The albedo approach is particularly useful when region 2 consists of materials that have a small neutron absorption cross section and large scattering cross section. Such materials often yield large values of albedo and hence are often used as *neutron moderators* or reflectors. It should be noted that although

**Table 5.1** Thermal diffusion properties of moderators.

| Material | Density ($g \cdot cm^{-3}$) | $\Sigma_a$ ($cm^{-1}$) | $D$ (cm) | $L$ (cm) | $\beta(\infty)$ |
|---|---|---|---|---|---|
| $H_2O$ | 1.0 | $1.7 \times 10^{-2}$ | 0.142 | 2.88 | 0.821 |
| $D_2O$ | 1.1 | $8.0 \times 10^{-5}$ | 0.80 | 100.0 | 0.968 |
| C | 1.62 | $3.6 \times 10^{-4}$ | 0.903 | 50.0 | 0.930 |
| Be | 1.84 | $1.3 \times 10^{-3}$ | 0.70 | 23.6 | 0.889 |

*Source:* [Gla52].

the concept of albedo is introduced in terms of a non-multiplying medium with a plane source in region 1, the albedo defined in Eq. (5.39) reflects the continuity of partial currents at a material interface and hence is also valid when region 1 comprises a multiplying medium surrounded by a reflector.

Thermal diffusion properties of a few moderating materials are given in Table 5.1, together with the albedo $\beta(\infty)$ for the limiting case of an infinitely thick reflector

$$\beta(\infty) = \frac{1 - 2\kappa_2 D_2}{1 + 2\kappa_2 D_2}. \tag{5.43}$$

When the thickness of the reflector is at least twice the diffusion length of the material, i.e. when $(b - a) \geq 2L_2 = 2/\kappa_2$ in Eq. (5.42), the reflector acts for all practical purposes as though it is infinitely thick. Although the basic mathematical relationship given in Eq. (5.41) is valid for actual operating reactors, the thermal diffusion length tabulated in Table 5.1 has to be extended to account for non-thermal or epithermal neutrons present in the core, resulting in a much larger effective diffusion length or migration length. This will be discussed later in connection with multi-group diffusion theory in Chapter 7.

### 5.2.2 Flux in Non-Multiplying Media with Distributed Sources

Solution of the diffusion equation (5.12) with distributed sources may in principle be obtained in a manner similar to those for localized sources considered in Section 5.2.1. Thus, the solution may be written as a sum of the solution for the homogeneous equation, i.e. for the source-free equation, and a particular solution corresponding to the distributed source. In many cases, however, the search for a particular solution may take some effort even for a relatively simple form of the distributed source $S(\mathbf{r})$.

We take an alternative approach based on the kernel or Green's function method to obtain the solution to the diffusion equation (5.12) for an arbitrary form of the source distribution. Since neutron–neutron collisions may be neglected in all of our solutions, the diffusion equation is linear in flux $\phi(\mathbf{r})$, and hence the linear superposition of the fluxes due to more than one source of neutrons is mathematically possible.

To understand the Green's function approach, we begin with a one-dimensional diffusion equation similar to Eq. (5.18), except that we now have a distributed source $S(x)$, not a localized source, in an infinite medium:

$$-D\frac{d^2\phi(x)}{dx^2} + \Sigma_a\phi(x) = S(x). \tag{5.44}$$

To build a full solution to Eq. (5.44), obtain first an elementary solution to Eq. (5.44) where a unit plane source is located at $x = x'$, i.e. $S(x) = \delta(x - x')$. Recalling that Eq. (5.17) is the solution to Eq. (5.18), write the solution for flux at $x$ due to a unit plane source at $x'$

$$\phi(x' \to x) = \frac{e^{-\kappa|x-x'|}}{2\kappa D} = \phi(x, x'), \tag{5.45}$$

where we recognize that the term $|x - x'|$ merely represents the distance between the source plane and the observation point, equivalent to the distance between the source plane at the origin and the observation point at $x$ in Eq. (5.17), with the source strength reduced to unity. The solution to a unit plane source problem given in Eq. (5.45) is called the *infinite-medium plane diffusion kernel*, since with Eq. (5.45) serving as a kernel or elementary solution, we may build the solution to Eq. (5.44) for an arbitrary source distribution $S(x)$:

$$\phi(x) = \int_{-\infty}^{\infty} dx'\, S(x')\, \phi(x' \to x). \tag{5.46}$$

The volumetric source $S(x')$ in units of [neutron·cm$^{-3}$ · s$^{-1}$] may be interpreted as the number of neutrons produced per [cm$^2$ · s] per cm of distance perpendicular to the source plane. This then allows us to interpret the product $S(x')dx'$ as the number of neutrons produced per [cm$^2$ · s] in thickness $dx'$, i.e. the total plane source strength contained in thickness $dx'$ of the medium. When $S(x')dx'$ is multiplied by the elementary solution, Eq. (5.45), the product yields the contribution to the flux $\phi(x)$ from the sum total of plane sources in $dx'$, which is then summed over or integrated over all possible contributions from the distributed sources potentially covering the entire space $[-\infty, \infty]$ to obtain the scalar flux $\phi(x)$ desired.

We may look at Eq. (5.46) a bit more mathematically by rewriting Eq. (5.44) as an inhomogeneous differential equation

$$L\phi(x) = S(x), \tag{5.47}$$

with the *diffusion operator* $L$ defined as

$$L = -D\frac{d^2}{dx^2} + \Sigma_a. \tag{5.48}$$

The operator notation is introduced primarily for notational convenience so that the diffusion kernel, Eq. (5.45), may be explicitly written as a solution to

$$LG(x, x') = \delta(x - x'). \tag{5.49}$$

The function $G(x, x')$ is usually called the *Green's function* [Fri56] of the differential operator $L$ of Eq. (5.48) and is simply equal to the plane diffusion kernel of Eq. (5.45). In terms of the Green's function $G(x, x')$, the solution to the inhomogeneous differential equation (5.47) is written as

$$\phi(x) = \int_{-\infty}^{\infty} dx'\, S(x')\, G(x, x'). \tag{5.50}$$

We may formally provide a proof of Eq. (5.50) by operating on both sides of the equation by the diffusion operator $L$ of Eq. (5.48)

$$L\phi(x) = \int_{-\infty}^{\infty} dx'\, S(x')\, LG(x, x') = \int_{-\infty}^{\infty} dx'\, S(x')\, \delta(x - x') = S(x), \tag{5.51}$$

where the definition of the Green's function, Eq. (5.49), is used.

Thus, the scalar flux due to a distributed source can be readily constructed from a superposition of the fluxes due to point sources making up the distributed source. For general three-dimensional geometry, we may rewrite Eq. (5.31) to obtain the neutron flux at position $\mathbf{r}$ due to a unit source at $\mathbf{r}'$

$$\phi(\mathbf{r}' \to \mathbf{r}) = \frac{e^{-\kappa|\mathbf{r}-\mathbf{r}'|}}{4\pi D|\mathbf{r} - \mathbf{r}'|} = G_{pt}(\mathbf{r}, \mathbf{r}'), \tag{5.52}$$

which may be considered a kernel or a "point" Green's function connecting flux at $\mathbf{r}$ due to a source at $\mathbf{r}'$. Thus, considering $S(\mathbf{r}')d\mathbf{r}'$ as the strength of a collection of point sources in $d\mathbf{r}'$ at $\mathbf{r}'$, the total flux $\phi(\mathbf{r})$ due to a distributed source $S(\mathbf{r})$ within a volume $V$ may be evaluated conveniently by "summing up" the product of the source $S(\mathbf{r}')d\mathbf{r}'$ and the kernel $\phi(\mathbf{r}' \to \mathbf{r})$ over $V$:

$$\phi(\mathbf{r}) = \int_V d\mathbf{r}' S(\mathbf{r}')\phi(\mathbf{r}' \to \mathbf{r}). \tag{5.53}$$

The quantity $\phi(\mathbf{r}' \to \mathbf{r})$ defined in Eq. (5.52) is known as the *infinite-medium point diffusion kernel*. Other diffusion kernels can be obtained corresponding to the particular geometries making up the source distributions. The kernels have to be derived in general with due accounting given for the boundary conditions of the problem at hand. Thus, for a finite slab, Eq. (5.35) has to be converted into a plane diffusion kernel, while Eq. (5.45) suffices for infinite slab geometry.

**Example 5.1** Apply Eq. (5.53) to solve for the scalar flux $\phi(x)$ due to a plane isotropic source of strength $S_0$ at the origin using the infinite-medium point diffusion kernel from Eq. (5.52). Of course, we have already obtained a direct solution to this problem in Eq. (5.17).

We represent here the plane source consisting of annular ring sources as illustrated in Figure 5.7, where the variable $r$ represents the distance $|\mathbf{r} - \mathbf{r}'|$

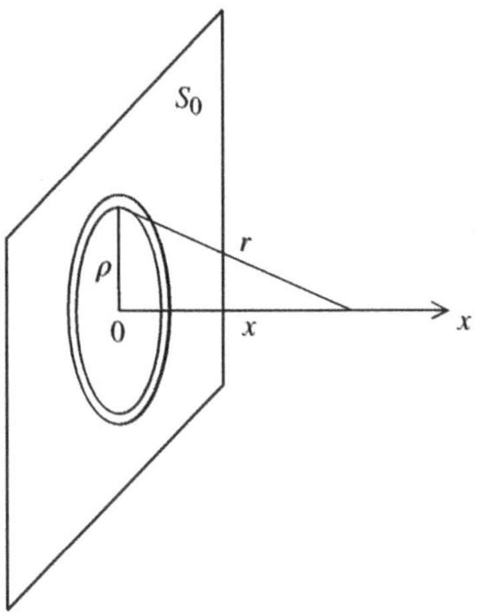


**Figure 5.7** Construction of a plane source from annular ring sources.

between the source and the observation point, and $2\pi\rho d\rho$ is the area of the annular ring of thickness $d\rho$. The flux $\phi(x)$ at position $x$ for the plane source may be constructed via Eqs. (5.52) and (5.53):

$$\phi(x) = \int_0^\infty S_0 \frac{2\pi\rho d\rho}{4\pi Dr} e^{-\kappa r}. \tag{5.54}$$

From the geometrical relationship

$$r^2 = \rho^2 + x^2 \text{ and } rdr = \rho d\rho, \tag{5.55}$$

for a given point $x$, we may evaluate Eq. (5.54)

$$\phi(x) = \frac{S_0}{2D}\int_x^\infty e^{-\kappa r}dr = \frac{S_0 e^{-\kappa x}}{2kD}, x > 0, \tag{5.56}$$

in agreement with Eq. (5.17).

For more complex geometries, most of the techniques presented so far may not yield a tractable solution. In such cases, one may have to turn to alternative techniques including the eigenfunction or modal expansion methods as well as direct numerical techniques, e.g. the finite difference, finite element, and nodal methods. A finite difference formulation of the diffusion equation, together with a nodal expansion method, are discussed in Chapter 6. Modal expansion techniques are presented in Chapter 10.

## 5.3 Neutron Flux in Multiplying Medium and Criticality Condition

In this section, we study solutions to the diffusion equation with an explicit representation of fission events taking place in a multiplying medium. Through this study, comparing steady-state and time-dependent solutions, we will establish criteria for the criticality of a chain-reacting system.

### 5.3.1 Criticality and Buckling

With the general source term $S(\mathbf{r}, t)$ replaced by the fission source term, the time-dependent diffusion equation (5.1) takes the form

$$\frac{1}{v}\frac{\partial \phi(\mathbf{r}, t)}{\partial t} = D\nabla^2\phi(\mathbf{r}, t) + (\nu\Sigma_f - \Sigma_a)\phi(\mathbf{r}, t). \tag{5.57}$$

For a critical reactor in steady-state operation, $\partial\phi/\partial t = 0$, and hence we would have

$$\nabla^2\phi(\mathbf{r}) + \frac{\nu\Sigma_f - \Sigma_a}{D}\phi(\mathbf{r}) = 0,$$

which may also be written as

$$\nabla^2\phi(\mathbf{r}) + B^2\phi(\mathbf{r}) = 0, \tag{5.58}$$

with the definition

$$B^2 = \frac{\nu\Sigma_f - \Sigma_a}{D} = \frac{\Sigma_a}{D}\left(\frac{\nu\Sigma_f}{\Sigma_a} - 1\right) = \frac{k_\infty - 1}{L^2}. \tag{5.59}$$

In Eq. (5.59), $k_\infty = \nu\Sigma_f/\Sigma_a$ represents the ratio of the number of neutrons produced from the fission process to that absorbed, thereby representing the self-sustaining potential of the system without leakage and is called the *infinite multiplication factor*. The parameter $B^2$ is called the *buckling* of the system, representing a measure of the curvature in the flux distribution.

Assuming the time and spatial dependencies in $\phi(\mathbf{r}, t)$ are separable, substitute

$$\phi(\mathbf{r}, t) = \psi(\mathbf{r})T(t),$$

into Eq. (5.57) and divide through both sides of the equation by $\phi(\mathbf{r}, t)$ to obtain

$$\frac{1}{vD}\frac{1}{T}\frac{dT}{dt} = \frac{1}{\psi}(\nabla^2\psi + B^2\psi) = -\alpha, \tag{5.60}$$

where the time- and space-dependent terms are clearly separated and hence the parameter $\alpha$ is introduced as a constant. The temporal portion of Eq. (5.60) may be simply integrated to yield

$$T(t) = T(0)e^{-\alpha vDt}, \tag{5.61}$$

with $T(0)$ to be obtained from a judicious initial condition. On the other hand, the spatial component of the solution has to be obtained from

$$\nabla^2\psi(\mathbf{r}) + (B^2 + \alpha)\psi(\mathbf{r}) = 0,$$

subject to the proper boundary conditions at the physical boundary. Hence, the spatial component $\psi(\mathbf{r})$ satisfies an eigenvalue problem such that the solution exists only for certain values of the eigenvalue $B_n^2$ in the equation of the form

$$\nabla^2\psi_n(\mathbf{r}) + B_n^2\psi_n(\mathbf{r}) = 0, \tag{5.62}$$

where

$$B_n^2 = B^2 + \alpha_n. \tag{5.63}$$

Equation (5.62) is known as the *Helmholtz* or *wave equation*. Hence, Eq. (5.61) should properly be rewritten as

$$T_n(t) = T_n(0)e^{-\alpha_n vDt}, \tag{5.64}$$

and the complete solution to Eq. (5.57) is given as

$$\phi(\mathbf{r}, t) = \sum_{n=0}^{\infty}\psi_n(\mathbf{r})T_n(t). \tag{5.65}$$

Arrange the eigenvalues in an ascending order such that

$$B_0^2 < B_1^2 < B_2^2 \cdots < B_n^2 < \cdots.$$

For a reactor to be critical, i.e. for $\phi(\mathbf{r}, t)$ to be constant in time, we require that $\alpha_0 = 0$ or $B_0^2 = B^2$ and that all higher harmonics with $\alpha_n > 0, n = 1, 2, \ldots,$ should vanish in a short time. Here, $B_0^2$ is the lowest eigenvalue of Eq. (5.62), corresponding to the fundamental mode $\psi_0(\mathbf{r})$, and is determined entirely by the geometry of the system, and hence is known as the *geometrical buckling*. The buckling $B^2$ defined in Eq. (5.59) is a function only of the material properties of the system and is called the *material buckling*, which is rewritten now as $B_m^2$. Hence, a chain-reacting system becomes critical and the neutron population remains at a constant level, when the geometrical buckling of the system is equal to the material buckling:

$$B_g^2 = B_m^2. \tag{5.66}$$

It is also clear that if $B_g^2 > B_m^2$, i.e. $\alpha_0 > 0$, the system is subcritical and the neutron flux will die away in due time. Likewise, if $B_g^2 < B_m^2$, i.e. $\alpha_0 < 0$, the system becomes supercritical, resulting in an uncontrolled growth of the neutron population.

### 5.3.2 Effective Multiplication Factor

For a critical system, Eq. (5.59) now yields

$$\frac{\nu\Sigma_f - \Sigma_a}{D} = B_g^2,$$

or

$$\frac{\nu\Sigma_f}{\Sigma_a + DB_g^2} = 1.0 = \frac{k_\infty\Sigma_a}{\Sigma_a + DB_g^2} = k_\infty P_{NL}, \tag{5.67}$$

where $DB_g^2$ is the leakage rate relative to the absorption rate represented by $\Sigma_a$, and hence $P_{NL}$ represents the non leakage probability of neutrons for the reactor. Before proceeding to further discussion on criticality, we will pause here to clarify how the neutron leakage rate is represented by the term $DB_g^2$. With the basic neutron balance of Eq. (5.1), the leakage rate $R_L$ out of the reactor volume $V$ and surface area $A$ may be obtained as

$$\begin{aligned} R_L &= -\int_V D\nabla^2\phi(\mathbf{r})d\mathbf{r} = -\int_V \nabla \cdot D\nabla\phi(\mathbf{r})d\mathbf{r} = -\int_A \mathbf{n} \cdot D\nabla\phi(\mathbf{r})dA \\ &= \int_A \mathbf{n} \cdot \mathbf{J}(\mathbf{r})dA = \int_A J(\mathbf{r})dA = DB_g^2 \int_V \phi(\mathbf{r})d\mathbf{r}. \end{aligned} \tag{5.68}$$

Several equivalent expressions for the leakage rate are intentionally listed, with the last expression obtained from Eq. (5.58) with buckling $B^2$ now set equal to the geometrical buckling $B_g^2$. Performing a similar integration for the absorption rate of Eq. (5.1), we simply recognize that $DB_g^2$ represents the net leakage rate relative to the absorption rate represented by $\Sigma_a$ in Eq. (5.67). With this perspective, confirm also that the non leakage probability $P_{NL}$ is properly defined as the ratio of the absorption rate to the total loss rate of neutrons comprising both the absorption and leakage rates.

Even when the system is not exactly critical, i.e. when $B_g^2 \neq B_m^2$, it is desirable to obtain an expression for the flux $\phi(\mathbf{r})$ as a solution to the eigenvalue equation

$$\nabla^2\phi(\mathbf{r}) + (B_m^2 + \alpha_0)\phi(\mathbf{r}) = 0, \tag{5.69}$$

by introducing the value $\alpha_0$, often known as the *dynamic eigenvalue* of the system, and by setting the flux $\phi(\mathbf{r})$ equal to the fundamental mode $\psi_0(\mathbf{r})$. When Eq. (5.69) yields a nontrivial solution for some $\alpha_0 \neq 0$, such a solution implies that the material composition and/or arrangement of the reactor should be adjusted until $\alpha_0 = 0$ or $B_g^2 = B_m^2$. Alternatively, we may recast Eq. (5.69) in terms of yet another eigenvalue $\lambda$:

$$D\nabla^2\phi(\mathbf{r}) + \left(\frac{\nu\Sigma_f}{\lambda} - \Sigma_a\right)\phi(\mathbf{r}) = 0. \tag{5.70}$$

Equation (5.70) is equivalent to Eq. (5.69), in the sense that there is an adjustable parameter introduced in either equation. The parameter $\lambda$ is known as the *static eigenvalue*, or very often simply as the *eigenvalue* of the system. In recent years, it is also referred to as the *k-eigenvalue*. Representing the leakage rate in terms of the buckling in Eq. (5.70) yields the one-group definition for the *effective multiplication factor* $k_{eff}$:

$$\lambda = k_{eff} = \frac{\nu\Sigma_f}{\Sigma_a + DB_g^2}. \tag{5.71}$$

Comparison with Eq. (5.67) simply shows that $k_{eff} = 1$ for a critical system, which is the most direct statement of neutron balance in a multiplying medium expressed in terms of one-group neutron diffusion theory. It is obvious that $k_{eff} = k_\infty$ for an infinitely large system, from which no neutrons may leak out.

For a noncritical system, i.e. when $\lambda \neq 1.0$, Eq. (5.70) again implies that the system configuration has to be adjusted until a critical state is attained. Equation (5.70) is, however, quite useful for noncritical systems, because we are able to obtain solution $\phi(\mathbf{r})$, albeit with $\lambda \neq 1.0$, and gain understanding of the degree of adjustments required to arrive at a critical configuration. Furthermore, even without obtaining a precisely critical configuration, we may determine the relative changes in the eigenvalue of the system due to perturbations in core parameters. Such perturbation calculations would not be possible without the introduction of the eigenvalue $\lambda$ in Eq. (5.70), because it is an eigenvalue equation and renders a nontrivial solution only if the criticality condition $B_g^2 = B_m^2$ is satisfied.

Together with $k_\infty$ and $k_{eff}$, defined in this section, the proximity to criticality may also be represented in terms of the *reactivity* defined as the fractional difference in $k_{eff}$ from unity:

$$\rho\left[\%\frac{\Delta k}{k}\right] = \frac{k_{eff} - 1}{k_{eff}}. \tag{5.72}$$

Although $\rho$ is obviously dimensionless and so is $\Delta k/k$, it is customary to add $\Delta k/k$ in the reactivity unit to clearly indicate that we are referring to reactivity. To handle small reactivity values, we also refer to $10^{-3}\ \%\Delta k/k$ as percent mille or *pcm*. Another reactivity unit in common use is the reactivity normalized by the delayed neutron fraction $\beta$

$$K\,[\$] = \frac{k_{eff} - 1}{k_{eff}\beta}, \tag{5.73}$$

expressed in units of dollar. For a thermal reactor core fueled with $^{235}$U, $\beta \simeq 0.0065$, as indicated in Table 2.2.

### 5.3.3 Eigenfunctions of Diffusion Equation and Buckling

To clarify the concept of eigenfunctions and associated eigenvalues for Eq. (5.62), consider one-dimensional slab geometry with core height $H$, where the extrapolation length is negligibly small compared with $H$ and may be dropped. Thus, the eigenfunction $\psi_n(x)$ should satisfy the eigenvalue equation

$$\frac{d^2\psi_n(x)}{dx^2} + B_n^2\psi_n(x) = 0, \tag{5.74}$$

subject to the boundary conditions that the eigenfunctions, representing the spatial components of the time-dependent scalar flux of Eq. (5.65), vanish at the core boundaries. This is equivalent to stipulating that the steady-state scalar flux vanishes at the extrapolated boundary, Eq. (5.4). For the slab geometry under consideration, the boundary conditions are simply

$$\text{B.C.: (i) } \psi_n(+H/2) = 0, \quad \text{(ii) } \psi_n(-H/2) = 0. \tag{5.75}$$

The solutions to Eq. (5.74) satisfying the boundary conditions of Eq. (5.75) are

$$\psi_n(x) = \begin{cases} \cos B_n x = \cos \dfrac{(n+1)\pi}{H} x, \; n = 0, 2, 4, \ldots \\ \sin B_n x = \sin \dfrac{(n+1)\pi}{H} x, \; n = 1, 3, 5, \ldots \end{cases} \tag{5.76}$$

with the eigenvalues

$$B_n^2 = \left[ \frac{(n+1)}{H} \pi \right]^2, \; n = 0, 1, 2, 3, \ldots . \tag{5.77}$$

The three lowest-order eigenfunctions of Eq. (5.76) are plotted in Figure 5.8.

The fundamental mode shape for a bare slab reactor is given simply by Eq. (5.76) for $n = 0$ with $B_g^2 = (\pi/H)^2$. For a right circular cylinder with radius $R$ and core height $H$, consider the Helmholtz equation

$$\nabla^2 \phi(r, z) + B^2 \phi(r, z) = 0, \tag{5.78}$$

into which we substitute and divide through by $\phi(r, z) = \theta(r)Z(z)$ to obtain

$$\frac{\nabla_r^2 \theta(r)}{\theta(r)} + \frac{\nabla_z^2 Z(z)}{Z(z)} = -B^2.$$

Standard separation-of-variable techniques allow setting

$$\nabla_r^2 \theta(r) + \alpha^2 \theta(r) = 0 \text{ and } \nabla_z^2 Z(z) + \lambda^2 Z(z) = 0, B^2 = \alpha^2 + \lambda^2. \tag{5.79}$$

The radial equation is written as

$$r^2 \frac{d^2\theta(r)}{dr^2} + r \frac{d\theta(r)}{dr} + \alpha^2 r^2 \theta(r) = 0, \tag{5.80}$$

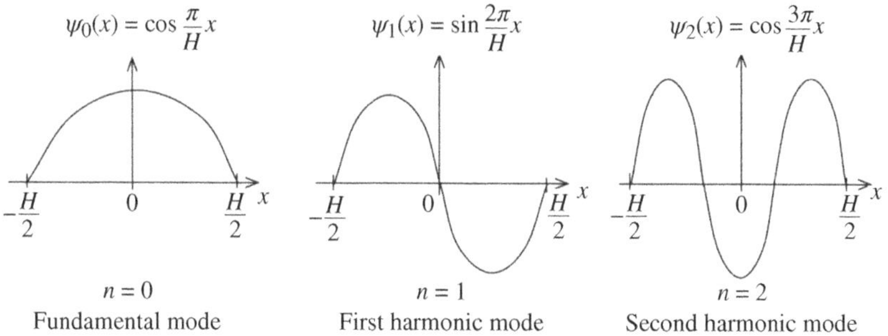


**Figure 5.8** Three lowest-order flux modes for slab reactor of height $H$.

with the solution obtained as

$$\theta(r) = A_1 J_0(\alpha r) + C_1 Y_0(\alpha r),$$

where $J_0$ and $Y_0$ are the Bessel functions [Appendix C] of the first and second kinds, respectively, of the zeroth order, together with two constants $A_1$ and $C_1$. Boundary condition (i) $\theta(r) < \infty$ requires setting $C_1 = 0$, since $\lim_{x\to 0} Y_0(x) = -\infty$. Boundary condition (ii) $\theta(R) = 0$ introduces the first zero of $J_0$, i.e. $J_0(\nu_0) = 0$ for $\nu_0 = 2.405$, which yields the eigenfunction $\theta(r) = A_1 J_0(\nu_0 r/R)$. The axial component $Z(z)$ of the eigenfunction $\theta(r, z)$ is simply that of the bare slab reactor, providing the combined solution

$$\phi(r,z) = AJ_0\left(\frac{2.405r}{R}\right)\cos\frac{\pi z}{H}, \tag{5.81}$$

with the geometrical buckling

$$B_g^2 = \left(\frac{2.405}{R}\right)^2 + \left(\frac{\pi}{H}\right)^2. \tag{5.82}$$

Results of the fundamental mode solution, i.e. $n = 0$ for Eq. (5.62), for representative geometries are presented in Table 5.2. In general, the solutions both for $\phi(\mathbf{r}) = \psi_0(\mathbf{r})$ and $B_g^2$ should be represented in terms of extrapolated core dimensions. For example, the core height $H$ for a slab reactor should be replaced by an extrapolated height $H' = H + 2d$, where $d$ is the extrapolation distance into vacuum. For slab geometry, the flux shape with an extrapolated height or thickness translates into a "chopped cosine" power shape, since the actual power distribution vanishes outside the physical boundary of the core, i.e. outside the core height of $H$. For a reflected reactor, the core-reflector interfaces may conveniently be replaced by albedo boundary conditions of the type given in Eq. (5.41). In particular, the critical core height $H$ for a reflected slab reactor may be obtained directly from Eq. (5.41) in terms of the albedo $\beta$ of the reflector.

**Example 5.2** Develop a one-group model of a bare, homogeneous, cylindrical reactor given number densities of various nuclides in Table 5.3 corresponding to

**Table 5.2** Flux and geometric buckling for bare critical reactors.

| Geometry | Flux | Geometric buckling |
|---|---|---|
| Infinite slab of thickness $H$ | $\cos\frac{\pi x}{H}$ | $\left(\frac{\pi}{H}\right)^2$ |
| Cylinder of radius $R$ and height $H$ | $J_0\left(\frac{2.405r}{R}\right)\cos\frac{\pi z}{H}$ | $\left(\frac{2.405}{R}\right)^2 + \left(\frac{\pi}{H}\right)^2$ |
| Sphere of radius $R$ | $\frac{1}{r}\sin\frac{\pi r}{R}$ | $\left(\frac{\pi}{R}\right)^2$ |

a PWR core operating at a hot full power (HFP) condition and fueled with $UO_2$ containing 2.5 wt% of $^{235}U$, similar to the AP600 mid-enrichment zone. Burnable absorber rods and boric acid dissolved in the coolant are represented by $^{10}B$ at a concentration of 1970 ppm by weight of natural boron in the water. The microscopic cross sections given in Table 5.3 are the spectrum-average cross sections for the core.

Based on the microscopic cross sections and number densities presented in Table 5.3, macroscopic cross sections are calculated for each nuclide and for the total core composition in Table 5.4. Also included in Table 5.4 are fractional absorption rates, which serve as a measure of overall neutron balance for the core. With microscopic cross sections obtained separately via Serpent Monte Carlo calculations [Fri11] for reaction rate edits, the macroscopic cross sections $\Sigma_a$ and $\nu\Sigma_f$ obtained in Table 5.4 are slightly different from those obtained from another

**Table 5.3** Number densities and microscopic cross sections for a PWR core.

| Material | $N$ (b$^{-1}$ cm$^{-1}$) | $\sigma_a$ (b) | $\nu$ | $\sigma_f$ (b) |
|---|---|---|---|---|
| $^{235}U$ | $1.900 \times 10^{-4}$ | $5.373 \times 10^{1}$ | 2.44 | $4.381 \times 10^{1}$ |
| $^{238}U$ | $7.315 \times 10^{-3}$ | $9.879 \times 10^{-1}$ | 2.79 | $9.467 \times 10^{-2}$ |
| $^{1}H$ | $2.611 \times 10^{-2}$ | $4.718 \times 10^{-2}$ | – | – |
| $^{16}O$ | $2.806 \times 10^{-2}$ | $2.354 \times 10^{-3}$ | – | – |
| $^{10}B$ | $9.219 \times 10^{-6}$ | $3.412 \times 10^{2}$ | – | – |
| Zr | $4.664 \times 10^{-3}$ | $4.386 \times 10^{-2}$ | – | – |
| Trace elements | $2.861 \times 10^{-4}$ | $3.340 \times 10^{-1}$ | – | – |

**Table 5.4** Macroscopic cross sections and absorption rates.

| Material | $N$ (b$^{-1}$ cm$^{-1}$) | $\Sigma_a$ (cm$^{-1}$) | $\nu\Sigma_f$ (cm$^{-1}$) | Absorption fraction |
|---|---|---|---|---|
| $^{235}U$ | $1.900 \times 10^{-4}$ | $1.021 \times 10^{-2}$ | $2.030 \times 10^{-2}$ | 0.460 |
| $^{238}U$ | $7.315 \times 10^{-3}$ | $7.226 \times 10^{-3}$ | $1.932 \times 10^{-3}$ | 0.326 |
| $^{1}H$ | $2.611 \times 10^{-2}$ | $1.232 \times 10^{-3}$ | – | 0.056 |
| $^{16}O$ | $2.806 \times 10^{-2}$ | $6.605 \times 10^{-5}$ | – | 0.003 |
| $^{10}B$ | $9.219 \times 10^{-6}$ | $3.146 \times 10^{-3}$ | – | 0.142 |
| Zr | $4.664 \times 10^{-3}$ | $2.045 \times 10^{-4}$ | – | 0.009 |
| Trace elements | $2.861 \times 10^{-4}$ | $9.560 \times 10^{-5}$ | – | 0.004 |
| Core total | $6.663 \times 10^{-2}$ | $2.218 \times 10^{-2}$ | $2.223 \times 10^{-2}$ | 1.000 |

two-group Serpent criticality calculation. With a standard deviation of 0.1% for $k_\infty = 1.0232$ and much larger standard deviations for individual reaction rates, the discrepancies between the two different edits are likely statistical in nature. We select for our PWR analysis $\Sigma_a = 2.168 \times 10^{-2}\,\text{cm}^{-1}$ and $\nu\Sigma_f = 2.215 \times 10^{-2}\,\text{cm}^{-1}$, together with $D = 1.0442\,\text{cm}$, obtained from the two-group criticality calculation discussed in Example 7.1.

- Infinite multiplication factor $k_\infty = \nu\Sigma_f/\Sigma_a = 1.022$,
- Material buckling $B_m^2 = (\nu\Sigma_f - \Sigma_a)/D = 4.501 \times 10^{-4}\,\text{cm}^{-2}$,
- Extrapolation distance $d = 0.7104\ \lambda_{tr} = 2.225\,\text{cm}$,
- Leakage probability for a critical core $P_L = 1 - P_{NL} = 0.022$.

For a critical core with height $H = 3.658$ m (12 ft), calculate the axial geometric buckling $B_z^2$

$$B_z^2 = (\pi/370.2)^2 = 7.201 \times 10^{-5}\ \text{cm}^{-2},$$

and the radial buckling $B_r^2$

$$B_r^2 = B_m^2 - B_z^2 = [2.405/(R+d)]^2 = 3.781 \times 10^{-4}\ \text{cm}^{-2}.$$

The radius of the critical PWR core with a height of 3.658 m is estimated to be $R =$ 1.21 m, which is somewhat smaller than the AP600 design calculation of 1.46 m.

**Example 5.3** Determine the reduction in the thickness $H$ of a slab reactor surrounded by a reflector of thickness $T = 25\,\text{cm}$ on each side of the core. Use the one-group constants of Example 5.2 for the core and diffusion coefficient $D_r =$ 1.443 cm and absorption cross section $\Sigma_{ar} = 9.06 \times 10^{-3}\,\text{cm}^{-1}$ for the reflector.

The flux distribution $\phi_c(x)$ for the core obeys the eigenvalue equation (5.74):

$$\frac{d^2\phi(x)}{dx^2} + B_c^2\phi(x) = 0,\ \ B_c^2 = \left.\frac{\nu\Sigma_f - \Sigma_a}{D}\right|_{core}.$$

The albedo boundary condition from Eq. (5.41) provides the criticality condition

$$D_c\left.\frac{d\ln\phi_c(x)}{dx}\right|_{H/2} = -\frac{J(H/2)}{\phi(H/2)} = -\frac{1}{2}\left(\frac{1-\beta}{1+\beta}\right) = -\kappa_r D_r \coth\kappa_r T,$$

which yields

$$D_c B_c \tan B_c\frac{H}{2} = \kappa_r D_r \coth\kappa_r T.$$

Instead of solving for the thickness $H$, determine the *reflector savings* $\delta$ representing the reduction in the core thickness due to the reflector:

$$\delta = \frac{1}{2}(H_0 - H) = \frac{\pi}{2B_c} - \frac{H}{2}.$$

Substituting $H/2$ in terms of $\delta$ recasts the criticality condition

$$D_c B_c \tan B_c \left( \frac{\pi}{2B_c} - \delta \right) = D_c B_c \cot B_c \delta = \kappa_r D_r \coth \kappa_r T,$$

which yields the desired expression for the reflector savings

$$\delta = \frac{1}{B_c} \tan^{-1} \left( \frac{D_c B_c}{D_r \kappa_r} \tanh \kappa_r T \right)$$
$$= \frac{1}{0.0212} \tan^{-1} \left[ \frac{1.044 \times 0.0212}{1.443 \times 0.0792} \tanh(0.0792 \times 25) \right] = 8.72 \text{ cm}.$$

The reflector savings is significant compared with the critical thickness $H_0 = \pi/B_c = 148.2$ cm for an unreflected reactor.

## 5.4 Four- and Six-Factor Formulas for Multiplication Factor

Before concluding this chapter on applications of one-group diffusion theory, with sample solutions and the definition of infinite and effective multiplication factors, it would be instructive to consider the well-known four-factor and six-factor formulas for the multiplication factors. Although these formulas in their original forms are seldom used in the actual core physics analysis, they provide useful physical insights to $k_\infty$ and $k_{eff}$ defined in terms of the one-group diffusion equation and prepare us for more accurate and practical definitions for the multiplication factors that two-group diffusion theory offers in Chapter 7.

The life cycle of a generation of neutrons, accounting for fission, slowing down, leakage, and absorption as thermal neutrons, as well as during slowing down, is traced in Figure 5.9, leading to the four- and six-factor formulas. We begin with one thermal neutron absorbed in the fuel in a multiplying system and end with the number of thermal neutrons absorbed again in the fuel after a full generation of neutron life cycle. The number of thermal neutron absorptions in the fuel at the end of one generation then yields the *effective multiplication factor* $k_{eff}$, which is often written simply also as $k$:

$$k_{eff} = \varepsilon \eta p f P_{NLF} P_{NLT} = k_\infty P_{NL} = k. \tag{5.83}$$

The *six-factor formula* from Eq. (5.83) is equivalent to Eq. (5.71), with the infinite multiplication factor $k_\infty$ written in terms of the *four-factor formula*

$$k_\infty = \varepsilon \eta p f, \tag{5.84}$$

and the nonleakage probability $P_{NL}$ is broken up into the fast and thermal non leakage probabilities $P_{NLF}$ and $P_{NLT}$, respectively. The four factors appearing in

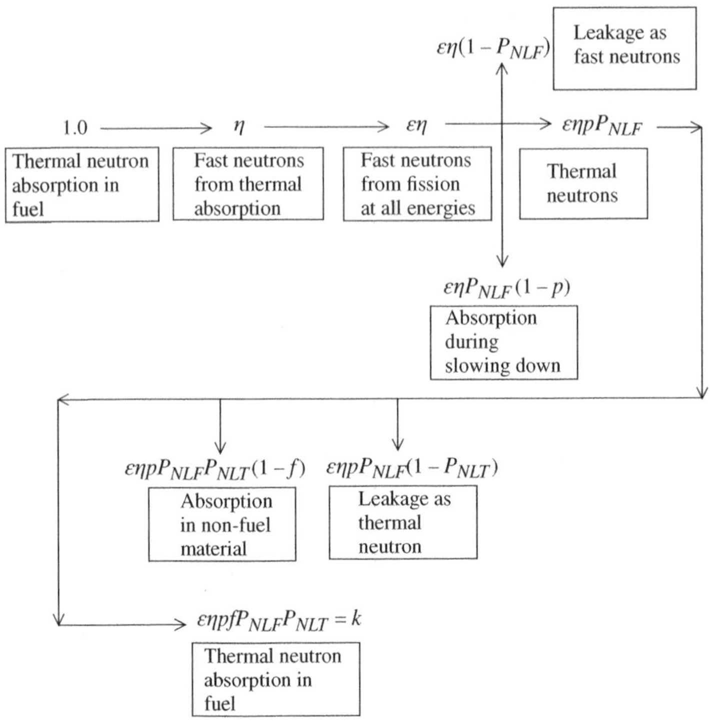


**Figure 5.9** Life cycle of neutrons illustrating the six-factor formula.

Eq. (5.84) are defined as

$$\varepsilon = \text{fast fission factor}$$

$$= \frac{\text{number of fast neutrons produced from fission at all energies}}{\text{number of fast neutrons produced from thermal fission}},$$

$$\eta = \frac{\text{number of fast neutrons produced from thermal fission}}{\text{number of thermal neutrons absorbed in fuel}} = \frac{\nu\Sigma_f}{\Sigma_a^F},$$

$$p = \text{resonance escape probability}$$

$$= \text{probability of neutrons escaping absorption during slowing down},$$

$$f = \text{thermal utilization}$$

$$= \frac{\text{number of thermal neutrons absorbed in fuel}}{\text{total number of thermal neutrons absorbed}} = \frac{\Sigma_a^F}{\Sigma_a^F + \Sigma_a^{NF}}.$$

For the parameters $\eta$ and $f$, we break up the thermal neutron absorption cross section $\Sigma_a$ into the fuel and non-fuel components, $\Sigma_a^F$ and $\Sigma_a^{NF}$, respectively. The product $\eta f$ represents the number of fast neutrons released from the fission caused by a thermal neutron absorbed in fuel and is equivalent to $k_\infty$ of Eq. (5.67). The fast fission factor $\varepsilon$ is then introduced to account for non-thermal neutrons causing fission, while the resonance escape probability $p$ accounts for the probability of neutrons not getting absorbed during slowing down from the fission energy to thermal energy. The four factors defined in Eq. (5.84) will be rederived quantitatively via two-group neutron diffusion theory in Chapter 7.

## 5.5 Concluding Remarks

In this chapter, we have discussed techniques for solving the one-group neutron diffusion equation and presented a few representative solutions. An emphasis was placed on the delineation of boundary conditions required for the solution in general. Through a direct solution of the time-dependent diffusion equation for multiplying media, a criticality condition is obtained in terms of the geometric and material bucklings, and an expression for $k_{eff}$ is obtained. It should be noted that we may readily generalize to a multi-group structure many of the basic parameters, including $k_{eff}$ and the albedo introduced here in the one-group framework.

It is obvious that the simple one-group form of the neutron diffusion equation plays a critical role in reactor physics, providing valuable insights and often yielding sufficiently accurate estimates of the flux and reaction rates in a multiplying or non-multiplying medium. We should, however, keep in mind general limitations or inaccuracies of neutron diffusion theory, as compared with transport theory. The numerical examples presented for a PWR core indicate further limitations or approximations inherent in the simple one-group form of the neutron balance equation.

## References

**[Arf13]** G.B. Arfken, H.J. Weber, and F.E. Harris, *Mathematical Methods for Physicist: A Comprehensive Guide*, 7th ed., Academic Press (2013).

**[Bel70]** G.I. Bell and S. Glasstone, *Nuclear Reactor Theory*, Van Nostrand Reinhold (1970).

**[Fri56]** B. Friedman, *Principles and Techniques of Applied Mathematics*, Wiley (1956).

**[Fri11]** E. Fridman and J. Leppänen, "On the Use of the Serpent Monte Carlo Code for Few-group Cross Section Generation," *Ann. Nucl. Energy* **38**, 1399 (2011).

**[Gla52]** S. Glasstone and M.C. Edlund, *The Elements of Nuclear Reactor Theory*, Van Nostrand (1952).

## Problems

**5.1** An isotropic point source of monoenergetic neutrons of intensity $S_0$ [neutron $\cdot$ s$^{-1}$] is located at the origin of an infinite diffusing medium with absorption cross section $\Sigma_a$ and diffusion coefficient $D$. (a) Obtain an expression for the rate of neutrons that are absorbed in a spherical shell of thick $dr$ at radius $r$. (b) Using the result of part (a), obtain an expression for the mean square distance the source neutrons travel between their birth and capture in terms of $L^2$.

**5.2** Obtain an interpretation for $L^2$ equivalent to that of Problem 5.1 for an infinite slab with an isotropic plane source.

**5.3** A collimated beam of monoenergetic neutrons is incident normally on a half space consisting of purely absorbing material with absorption cross section $\Sigma$. The half space is surrounded by vacuum. (a) Obtain an expression for the fraction of the incident beam that is absorbed in an 1.0-cm interval at distance $x$ into the half space. (b) Derive an expression for the mean square distance the neutrons travel before they are captured in the half space.

**5.4** Consider a semi-infinite slab of thickness $2H$ containing an isotropic plane source emitting $S$ [neutron $\cdot$ cm$^{-2} \cdot$ s$^{-1}$] at its midplane, $z = 0$. The slab is surrounded by vacuum. Using the one-group neutron diffusion equation: (a) Obtain the flux distribution $\phi(z)$ within the slab with the boundary condition that no neutrons return from the surrounding vacuum, i.e. $J_{-}(H) = J_{+}(-H) = 0$. (b) Obtain $\phi(z)$ within the slab using the boundary condition that $\phi(z) = 0$ at $|z| = H + 2D$. (c) Compare the two expressions for $\phi(z)$ obtained in parts (a) and (b), and show that the two expressions are equal for a large, weakly absorbing medium, i.e. when $\Sigma_a \ll \Sigma_s$, or equivalently $\Sigma_a \ll \Sigma_{tr}$, and $H \gg \lambda_{tr}$.

**5.5** A semi-infinite slab of thickness $2H$ is surrounded by vacuum. The slab consists of non-multiplying diffusing material with one-group constants $D$ and $\Sigma_a$. Neutrons are produced uniformly and isotropically at a rate of $Q$ [neutron$\cdot$cm$^{-3} \cdot$ s$^{-1}$] throughout the slab. Neutron extrapolation distance is negligible compared with $H$. Determine the scalar flux distribution $\phi(x)$ within the slab at distance $x$ from the midplane of the slab and the fraction of source neutrons leaking out of the slab into vacuum.

**5.6** Thermal neutrons are produced uniformly and isotropically at the rate of $Q$ [neutron$\cdot$cm$^{-3} \cdot$ s$^{-1}$] in an infinite medium, with one-group constants

$D$ and $\Sigma_a$. (a) Obtain an expression for the thermal neutron flux in the medium. (b) A large indium foil with thickness $a = 25\ \mu m$, which may be considered infinite in extent in two dimensions, is placed in the medium. Calculate the depression $F$ of the neutron flux and the rate of activation $A$ of the foil per $cm^2$ if the medium is water with $D = 0.14\,cm$ and $\Sigma_a = 0.015\,cm^{-1}$. Assume the indium foil is a pure absorber with cross section $\Sigma_a^{In} = 7.39\,cm^{-1}$. Express the flux depression $F$ as the ratio of the neutron flux at the foil position to that at the same position when the foil is not present. The foil is thin enough so that we assume spatial variation of the neutron flux within the foil can be neglected.

**5.7** Repeat the solution for Problem 5.6 using the infinite-medium plane diffusion kernel and treating the absorber foil as a localized plane source of negative strength.

**5.8** Using the steady-state, one-group neutron diffusion equation: (a) Determine the neutron flux in an infinite medium with a spherical cavity of radius $R$ containing an isotropic point source of neutrons of strength $S$ $[neutron \cdot s^{-1}]$ at the center. (b) Compare the result with that for a point source in an infinite medium with no cavity, and explain the difference between the two cases physically for the region outside the cavity. For numerical comparison, assume that the medium consists of graphite, with $\Sigma_a = 3.33 \times 10^{-3}\,cm^{-1}$ and $D = 0.90\,cm$, and use $R = 0.1\,m$.

**5.9** A collimated beam of neutrons of speed $v_0$ is incident normally on a purely absorbing slab of thickness $H$. The absorption cross section $\sigma$ of the slab material is independent of the relative speed between neutrons and nuclei. Assume thermal motion of the target nuclei is negligible. If the slab is moving toward the neutron beam with speed $v_0$, parallel to the beam direction, determine the fraction of the incident neutrons that are absorbed in the slab.

**5.10** A collimated beam of neutrons of intensity $I_0$ $[neutron \cdot cm^{-2} \cdot s^{-1}]$ and speed $v_0$ is normally incident on the left surface of a semi-infinite slab of thickness $H$. The one-group cross sections for neutrons of speed $v_0$ are $\Sigma_{a0}$, $\Sigma_{s0}$, and $D$. (a) If a detector with a cross-sectional area of $1.0\,cm^2$ is located at the right surface of the slab and counts only neutrons transmitted through the slab with no change in the direction, what is the count rate of the detector? (b) The slab now moves normal to the direction of the beam with speed $V$. The absorption cross section of the slab has the $1/v$ variation, while the scattering cross section is independent of neutron energy. Determine the detector count rate.

**5.11** A semi-infinite slab of diffusing material of thickness $H$ is surrounded by vacuum. The diffusing material has absorption cross section $\Sigma_a$ and diffusion coefficient $D$. The slab has an isotropic plane source of strength $S_0$ [neutron $\cdot$ cm$^{-2}$ $\cdot$ s$^{-1}$] on the left face of the slab at $x = 0$. Assume that the neutron extrapolation distance $d$ is negligibly small compared with $H$. (a) What are the boundary conditions that may be used to solve the one-group neutron diffusion equation? (b) Obtain the scalar flux $\phi(x)$ at position $x$ within the slab. (c) Obtain an expression for the number of neutrons leaking out of the slab at $x = H/\text{cm}^2/\text{s}$.

**5.12** A bare homogeneous slab reactor of thickness $2H$ is critical with $k_\infty$, $D$, and $\Sigma_a$. The thickness of the core is reduced to $H$, without changing the material composition of the core, and the reduced core is surrounded on one side by a reflector of infinite thickness. The reflector has the same $D$ and $\Sigma_a$ as the core material but contains no fuel. A volumetric source of neutrons of strength $S_0$ [neutron $\cdot$ cm$^{-3}$ $\cdot$ s$^{-1}$] is distributed uniformly in the reflector region. Obtain the flux $\phi_c(x)$ in the core.

**5.13** A half space consisting of diffusing material with absorption cross section $\Sigma_a$ and diffusion constant $D$ is surrounded by vacuum. A plane isotropic source of mono-energetic neutrons of strength $S_0$ [neutron $\cdot$ cm$^{-2}$ $\cdot$ s$^{-1}$] is located at the vacuum interface. Obtain an expression for the total number of neutrons streaming into vacuum per cm$^2$ of the source plane per s.

**5.14** An infinite, non-multiplying medium is characterized by one-group constants $D$ and $\Sigma_a$. An isotropic source of mono-energetic neutrons of strength $Q$ [neutron $\cdot$ cm$^{-3}$ $\cdot$ s$^{-1}$] is distributed over a region of thickness $2H$. Using the one-group neutron diffusion equation and setting up the boundary conditions between the regions, obtain the neutron flux distribution $\phi(x)$ throughout the infinite medium.

**5.15** Repeat the solution for Problem 5.14 using the infinite-medium slab diffusion kernel from Eq. (5.45).

**5.16** A semi-infinite slab of diffusing material of thickness $H$ is surrounded by vacuum. The diffusing material has absorption cross section $\Sigma_a$ and diffusion coefficient $D$. A plane source of strength $S_0$ [neutron $\cdot$ cm$^{-2}$ $\cdot$ s$^{-1}$] is placed on the left face at $x = 0$ of the slab, with one-third of the source neutrons released in the positive direction and two-thirds released in the negative direction. The neutron extrapolation distance $d$ is negligibly small compared with $H$. (a) Obtain the scalar flux $\phi(x)$ at position $x$ within the slab. (b) Obtain an expression for the number of neutrons leaking out of the slab at $x = 0$ into vacuum per $cm^2 \cdot s^{-1}$.

**5.17** The microscopic cross sections given in Table 5.3 correspond to a PWR core with soluble boron concentration $C_B = 1970$ ppm by weight of natural boron in water. Using the PWR configuration considered in Section 5.3, obtain the reactivity $\rho$ of the core in units of $[\%\Delta k/k]$, if $C_B$ is reduced to 1870 ppm. What is the *differential boron worth* defined as $\Delta\rho/\Delta C_B$ [pcm/ppm of boron]? Assume that the microscopic cross sections are not affected by the change in boron concentration.

**5.18** A bare slab core of extrapolated thickness $H_0 = 50$ cm is just critical. The core material is described by one-group constants $D = 5$ cm and $L^2 = 50\,cm^2$. If a central 5-cm region of the slab is replaced by vacuum, sketch and obtain an expression for the flux distribution, and calculate effective multiplication factor $k_{eff}$ for the perturbed system.

**5.19** A bare slab core of extrapolated thickness $H = 100$ cm is just critical. The core material is described by one-group constants $D = 5$ cm and $\Sigma_a = 0.1\,cm^{-1}$. A central 10-cm region of the slab is replaced by a highly absorbing material so that the neutron flux may be assumed to vanish identically throughout the absorber. Determine the effective multiplication factor $k_{eff}$ for the perturbed system.

**5.20** A *thin* fuel plate of thickness $2a$ is surrounded by a reflector of thickness $b$ on each side of the plate. The fuel has one-group cross sections $\nu\Sigma_f$ and $\Sigma_{aF}$, while the reflector material is described by $D_M$ and $\Sigma_{aM}$. Set up the current-to-flux ratio $J(a)/\phi(a)$, and derive a criticality condition for the reflected fuel plate.

**5.21** A bare spherical reactor with cross sections $\Sigma_a$, $\nu\Sigma_f$, and $D$ and radius $R$ contains a uniformly distributed source $Q$ [neutron $\cdot\, cm^{-3} \cdot s^{-1}$]. (a) Derive the scalar flux $\phi(r)$ within the reactor with $\Sigma_a < \nu\Sigma_f$. (b) Determine the radius $R_c$ for which the reactor is critical and $\phi(0)$.

**5.22** A neutron randomly enters a sphere of radius $R$ described by $\Sigma_a$ and $D$. Determine the probability that the neutron will eventually leak out of the sphere.

**5.23** For the $^{252}Cf$ source considered in Problem 2.2, a radiative capture cross section $\sigma_c = 20$ b is estimated for thermal neutrons, with negligibly small

$\sigma_c$ for fast or epithermal neutrons. If a startup source consisting of 1.0-mg $^{252}$Cf is used in a nuclear reactor core with thermal neutron flux of $5 \times 10^{13}$ neutron $\cdot$ cm$^{-2}$ $\cdot$ s$^{-1}$, what is the effective half-life of the $^{252}$Cf source in the core?

**5.24** A semi-infinite slab of diffusing material with thickness $H$ [cm] is surrounded by vacuum. The diffusing material has absorption cross section $\Sigma_a$ and diffusion coefficient $D$. A collimated beam of neutrons of intensity $I_0$ [neutron $\cdot$ cm$^{-2}$ $\cdot$ s$^{-1}$] is incident on the left face of the slab at $x = 0$. Assume that the neutron extrapolation distance $d$ is negligibly small compared with $H$. (a) Derive an expression for the scalar flux $\phi(x)$ at position $x$ within the slab. (b) Obtain an expression for the number of neutrons leaking out of the slab at $x = 0/\text{cm}^2/\text{s}$.

**5.25** Using the one-group neutron diffusion equation, (a) obtain the criticality condition for an unreflected spherical reactor with inner radius $R_1$, outer radius $R_0$, and material buckling $B^2$. (b) Show that the expression reduces to that of a sphere without a hollow center. (c) If a small amount of core material is removed from the center of a critical spherical reactor, how much must be added to its outer surface in a uniform layer to restore criticality?

**5.26** A thin absorbing foil is inserted into a slab reactor of thickness $H$ at its midplane so that only a fraction $\gamma$ of neutrons incident on one side of the foil passes through the foil without absorption. Obtain the criticality condition for the slab reactor with material buckling $B^2$ and diffusion coefficient $D$.

**5.27** A slab reactor core of thickness $H$ with one-group constants $D$, $\Sigma_a$, and $\nu\Sigma_f$ is surrounded by vacuum on one side and by a reflector of infinite thickness on the other side. The reflector material has the same $D$ and $\Sigma_a$ as the core material but contains no fuel. Calculate the core thickness $H$ that would make the system critical.

# 6

# Numerical Solution of the Neutron Diffusion Equation

Having completed in Chapter 5 a study of analytical solutions to the neutron diffusion equation, we now turn to the task of obtaining numerical solutions of the steady-state neutron diffusion equation primarily via finite difference (FD) techniques, concentrating on one-dimensional geometries. We begin with the steady-state one-group diffusion equation in the form

$$\nabla \cdot \mathbf{J}(\mathbf{r}) + \Sigma(\mathbf{r})\phi(\mathbf{r}) = S(\mathbf{r}). \tag{6.1}$$

This is the same as Eq. (5.12), considered for analytical solutions of the diffusion equation in Chapter 5, with the recognition that, in the one-group model, $\Sigma = \Sigma_a$ and the source term $S(\mathbf{r}) = \nu\Sigma_f\phi(\mathbf{r})/\lambda$, where $\lambda$ is the eigenvalue of the system. In a multigroup structure, the neutron balance equation for each group may be put in the form of Eq. (6.1), with some modifications. Hence, the finite-difference formulation we develop in this chapter can be applied equally well to the solution of the multigroup neutron diffusion equation, although we limit ourselves formally to the solution of a one-group model given in Eq. (6.1).

Discretization of the basic diffusion equation (6.1) into a three-point FD equation is presented in Section 6.1, followed by a discussion of the numerical solution of the FD equation in Section 6.2. Handling standard boundary conditions imposed on the diffusion equation is discussed in Section 6.3, together with the treatment of the fission source in Eq. (6.1) through an iterative technique in Section 6.4. The concept of a normalized or relative power distribution is discussed in Section 6.5, followed by a flux synthesis approach in Section 6.6 to determine the two-dimensional power distributions with one-dimensional diffusion theory codes. Section 6.7 presents the extension of the 1-D techniques to the FD solution of multigroup and multidimensional diffusion equations. Section 6.8 presents follow-up discussions on coarse-mesh diffusion solver algorithms, with Section 6.9 covering numerical techniques for the solution of diffusion equations with more recent techniques that could be efficiently applied to the solution of large sparse matrices.

*Nuclear Reactor Physics and Engineering*, Second Edition. John C. Lee.
© 2025 John Wiley & Sons, Inc. Published 2025 by John Wiley & Sons, Inc.
Companion website: www.wiley.com/go/LeeNRPE2

## 6.1 Finite Difference Form of Diffusion Equation

A FD form of Eq. (6.1) for a medium with $N$ mesh intervals is derived in this section with the assumption that the source distribution $S$ is known. Once we have discussed the FD formulation of the fixed source problem, we investigate in Section 6.4 how the eigenvalue problem with $S(\mathbf{r}) = \nu\Sigma_f\phi(\mathbf{r})/\lambda$ can be solved through an iterative technique. To facilitate the integration of Eq. (6.1) over a grid of mesh intervals and thereby obtain a discretized representation, Eq. (6.1) is explicitly written as

$$-\frac{1}{x^p}\frac{d}{dx}\left[D(x)x^p\frac{d\phi(x)}{dx}\right] + \Sigma(x)\phi(x) = S(x), \tag{6.2}$$

where the parameter $p = 0$ for slab geometry and $p = 1$ for cylindrical geometry. A sufficiently fine mesh description is chosen for either the slab or cylindrical geometry so that the material composition and cross sections are uniform within each mesh interval of width $h_n, n = 1, 2, \ldots, N$. The neutron flux is determined at each mesh boundary [Bar67, Cad67, Lee74] and the derivative of flux is evaluated through a simple linear difference scheme.

The discretization schemes follows with

$$D(x) = D_n \text{ and } \Sigma(x) = \Sigma_n \text{ for } x \in [x_n, x_{n+1}] \tag{6.3}$$

and

$$\phi(x) = \phi_n \text{ for } x \in [x_{n-1}^0, x_n^0], \tag{6.4}$$

where

$$x_n^0 = x_n + \frac{h_n}{2} \text{ with } h_n = x_{n+1} - x_n. \tag{6.5}$$

In addition, determine the flux derivative:

$$\frac{d\phi(x)}{dx} = \frac{\phi_{n+1} - \phi_n}{h_n} \text{ for } x \in [x_n, x_{n+1}]. \tag{6.6}$$

Equations (6.4) and (6.6) imply that the flux $\phi(x)$ is continuous across each mesh boundary but that a discontinuity is allowed for the flux derivative $d\phi(x)/dx$ at the mesh boundary. The designation of mesh intervals, cross sections, fluxes, and flux derivatives is illustrated in Figure 6.1.

A FD form of Eq. (6.2) may now be obtained by multiplying the equation by $x^p$ and integrating the resulting the equation over the interval $[x_{n-1}^0, x_n^0]$, with the flux $\phi_n$ constant throughout the interval. This is equivalent to integrating over the volume associated with the mesh interval for both the slab and cylindrical

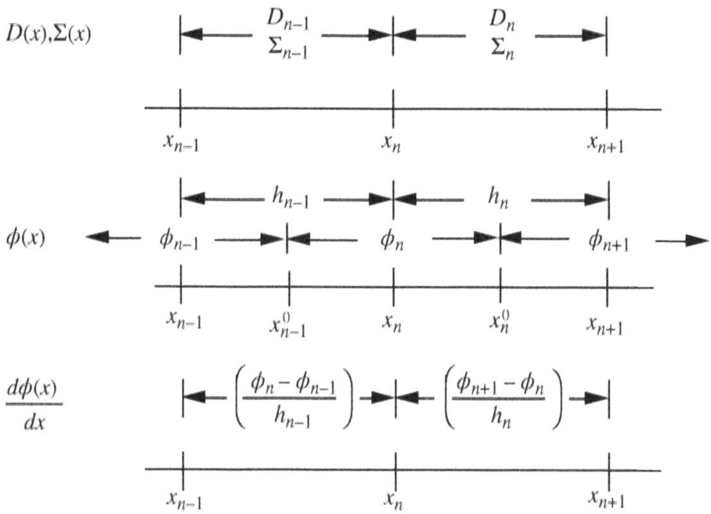


**Figure 6.1** Discretization scheme for the flux and flux derivative.

geometries. We begin with the leakage term for mesh $n$, which takes the form

$$I_1 = -\int_{x^0_{n-1}}^{x^0_n} \frac{d}{dx}\left[Dx^p \frac{d\phi}{dx}\right] dx = -Dx^p \left.\frac{d\phi}{dx}\right|_{x^0_{n-1}}^{x^0_n}$$

or with Eqs. (6.3) and (6.6)

$$I_1 = -D_n\left(x^0_n\right)^p \left(\frac{\phi_{n+1}-\phi_n}{h_n}\right) + D_{n-1}\left(x^0_{n-1}\right)^p \left(\frac{\phi_n-\phi_{n-1}}{h_{n-1}}\right)$$
$$= -\langle D_n\rangle\left(\phi_{n+1}-\phi_n\right) + \langle D_{n-1}\rangle\left(\phi_n-\phi_{n-1}\right),$$

where

$$\langle D_n\rangle = \frac{D_n\left(x^0_n\right)^p}{h_n} = \frac{D_n}{h_n}\left(x_n + \frac{h_n}{2}\right)^p. \tag{6.7}$$

For slab geometry with $p = 0$, Eq. (6.7) simply yields a dimensionless quantity $\langle D_n\rangle = D_n/h_n$, and the leakage term can be written in a three-point difference form:

$$I_1 = -\langle D_{n-1}\rangle\phi_{n-1} + \left(\langle D_{n-1}\rangle + \langle D_n\rangle\right)\phi_n - \langle D_n\rangle\phi_{n+1}. \tag{6.8}$$

Note that Eq. (6.8) applies strictly for internal mesh points, i.e. for $n = 2, \ldots, N$, and that some adjustments have to be made to Eq. (6.8) for boundary points, $n = 1$

and $n = N+1$: $\langle D_0 \rangle = \langle D_{N+1} \rangle = 0$. This simply states that, with the diffusion coefficient $D$ for regions outside the volume of our interest represented, Eq. (6.8) applies to all of the mesh points, $n = 1, \dots, N+1$.

With Eqs. (6.3) and (6.4), the FD form of the removal term for mesh $n$ is obtained as

$$I_2 = \int_{x_{n-1}^0}^{x_n^0} \Sigma\phi x^p dx = \phi_n \int_{x_{n-1}^0}^{x_n^0} \Sigma x^p dx$$
$$= \phi_n \left[ \Sigma_{n-1} \int_{x_{n-1}^0}^{x_n} x^p dx + \Sigma_n \int_{x_n}^{x_n^0} x^p dx \right].$$

Carrying on the integrations yields

$$I_2 = \frac{\phi_n}{p+1} \left[ \Sigma_{n-1} \left\{ x_n^{p+1} - \left(x_{n-1}^0\right)^{p+1} \right\} + \Sigma_n \left\{ \left(x_n^0\right)^{p+1} - x_n^{p+1} \right\} \right]$$

or

$$I_2 = \langle \Sigma_n \rangle \phi_n = \left( \langle \Sigma_n^- \rangle + \langle \Sigma_n^+ \rangle \right) \phi_n, \tag{6.9}$$

with

$$\langle \Sigma_n^- \rangle = \frac{\Sigma_{n-1}}{p+1} \left[ x_n^{p+1} - \left( x_{n-1} + \frac{h_{n-1}}{2} \right)^{p+1} \right] = \frac{\Sigma_{n-1} h_{n-1}}{2} \left( x_n - \frac{h_{n-1}}{4} \right)^p, \tag{6.10}$$

$$\langle \Sigma_n^+ \rangle = \frac{\Sigma_n}{p+1} \left[ \left( x_n + \frac{h_n}{2} \right)^{p+1} - x_n^{p+1} \right] = \frac{\Sigma_n h_n}{2} \left( x_n + \frac{h_n}{4} \right)^p. \tag{6.11}$$

Note again that for slab geometry with $p = 0$, Eq. (6.9) simplifies to $\langle \Sigma_n \rangle = (\Sigma_n h_n + \Sigma_{n-1} h_{n-1})/2$, which is a mesh-spacing weighted cross section corresponding to the flux $\phi_n$ over the interval $[x_{n-1}^0, x_n^0]$ straddling the mesh boundary $x_n$. This coefficient is also dimensionless. Equation (6.9) holds for boundary cells as well as internal cells if we remember that

$$\langle \Sigma_1^- \rangle = \langle \Sigma_{N+1}^+ \rangle = 0. \tag{6.12}$$

Similarly, discretize the source term

$$I_3 = \int_{x_{n-1}^0}^{x_n^0} S x^p dx = \int_{x_{n-1}^0}^{x_n^0} \frac{\nu\Sigma_f}{\lambda} \phi x^p dx,$$

or

$$I_3 = \langle S_n \rangle = \frac{1}{\lambda} \langle \nu\Sigma_{fn} \rangle \phi_n = \frac{1}{\lambda} \left( \langle \nu\Sigma_{fn}^- \rangle + \langle \nu\Sigma_{fn}^+ \rangle \right) \phi_n, \tag{6.13}$$

with

$$\langle \nu\Sigma_{fn}^{-} \rangle = \frac{\nu\Sigma_{f,n-1}h_{n-1}}{2}\left(x_n - \frac{h_{n-1}}{4}\right)^p, \tag{6.14}$$

$$\langle \nu\Sigma_{fn}^{+} \rangle = \frac{\nu\Sigma_{fn}h_n}{2}\left(x_n + \frac{h_n}{4}\right)^p. \tag{6.15}$$

For slab geometry with $p = 0$, note as in Eq. (6.9) that the mesh-spacing weighted fission cross section simplifies to $\langle \nu\Sigma_{fn} \rangle = (\nu\Sigma_{fn}h_n + \nu\Sigma_{f,n-1}h_{n-1})/2$. Equation (6.15) is also valid for boundary as well as internal cells, with

$$\langle \nu\Sigma_{f1}^{-} \rangle = \langle \nu\Sigma_{f,N+1}^{+} \rangle = 0. \tag{6.16}$$

Summing up Eqs. (6.8), (6.9), and (6.13), we obtain the FD form of Eq. (6.2)

$$-\langle D_{n-1} \rangle \phi_{n-1} + (\langle D_{n-1} \rangle + \langle D_n \rangle + \langle \Sigma_n \rangle)\phi_n - \langle D_n \rangle \phi_{n+1} = \langle S_n \rangle,$$

or

$$a_n\phi_{n-1} + b_n\phi_n + c_n\phi_{n+1} = S_n, n = 1, \ldots, N+1, \tag{6.17}$$

where

$$a_n = -\langle D_{n-1} \rangle, \tag{6.18a}$$

$$b_n = \langle D_{n-1} \rangle + \langle D_n \rangle + \langle \Sigma_n \rangle = \langle D_{n-1} \rangle + \langle D_n \rangle + \langle \Sigma_n^{-} \rangle + \langle \Sigma_n^{+} \rangle, \tag{6.18b}$$

$$c_n = -\langle D_n \rangle, \tag{6.18c}$$

$$S_n = \langle S_n \rangle. \tag{6.18d}$$

In Eqs. (6.17) and (6.18), note $a_1 = c_{N+1} = 0$, which corresponds to $\langle D_0 \rangle = \langle D_{N+1} \rangle = 0$. With this understanding, it becomes clear that Eq. (6.17) is a discretized form of Eq. (6.2) providing a set of discrete flux values $\{\phi_1, \phi_2, \ldots, \phi_{N+1}\}$, as a histogram, for the continuum distribution of scalar flux $\phi(x)$ subject to the proper boundary conditions.

Particular expressions for $b_1$ and $b_{N+1}$, at the system boundaries, will be derived in Section 6.3 for representative boundary conditions on the neutron flux and current. Note that Eq. (6.17) takes the form of a three-point difference equation because the leakage term of Eq. (6.8) couples two adjacent cells to mesh $n$. Although Eqs. (6.18a) and (6.18c) indicate that $a_n = c_{n-1}$, we maintain, for clarity of presentation, distinct notations for these two coefficients appearing in Eq. (6.17). For the simple case of uniform mesh width $h$ and uniform cross sections in slab geometry, Eqs. (6.18) simplify to

$$a_n = -D/h = c_{n-1}, \tag{6.19a}$$

$$b_n = 2D/h + \Sigma h, \tag{6.19b}$$

$$S_n = \frac{1}{\lambda}\nu\Sigma_f\phi_n h. \tag{6.19c}$$

## 6.2 Flux Solution Algorithm: Inner Iteration

The set of linear algebraic equations (6.17) may be solved by various techniques including iterative methods. For one-dimensional geometry, the solution can be obtained efficiently by using the Gaussian elimination technique [Bar67, Dah74, Der84], consisting of the forward elimination and backward substitution steps. To clarify the Gaussian elimination algorithm, cast the three-point difference equation (6.17) into a matrix equation

$$A\mathbf{\Phi} = \mathbf{S}. \tag{6.20}$$

by defining

$$A = \begin{bmatrix} b_1 & c_1 & 0 & \cdot & \cdot & 0 \\ a_2 & b_2 & c_2 & \cdot & \cdot & 0 \\ 0 & a_3 & b_3 & c_3 & \cdot & 0 \\ 0 & \cdot & \cdot & \cdot & \cdot & \cdot \\ 0 & 0 & \cdot & a_N & b_N & c_N \\ 0 & 0 & 0 & \cdot & a_{N+1} & b_{N+1} \end{bmatrix}, \mathbf{\Phi} = \begin{bmatrix} \phi_1 \\ \phi_2 \\ \phi_3 \\ \cdot \\ \phi_N \\ \phi_{N+1} \end{bmatrix}, \boldsymbol{S} = \begin{bmatrix} S_1 \\ S_2 \\ S_3 \\ \cdot \\ S_N \\ S_{N+1} \end{bmatrix}. \tag{6.21}$$

Equation (6.20) may also be rewritten in terms of the source vector $\mathbf{Q}$ and the diagonal matrix $F$ representing the spatial distribution of $\nu\Sigma_f$

$$A\mathbf{\Phi} = \frac{1}{\lambda}\mathbf{Q} = \frac{1}{\lambda}F\mathbf{\Phi}, \tag{6.22}$$

with

$$\mathbf{Q} = \begin{bmatrix} \langle \nu\Sigma_{f1}\rangle\phi_1 \\ \langle \nu\Sigma_{f2}\rangle\phi_2 \\ \langle \nu\Sigma_{f3}\rangle\phi_3 \\ \cdot \\ \langle \nu\Sigma_{fN}\rangle\phi_N \\ \langle \nu\Sigma_{f,N+1}\rangle\phi_{N+1} \end{bmatrix}, F = \begin{bmatrix} \langle \nu\Sigma_{f1}\rangle & 0 & 0 & \cdot & \cdot & 0 \\ 0 & \langle \nu\Sigma_{f2}\rangle & 0 & \cdot & \cdot & 0 \\ 0 & 0 & \langle \nu\Sigma_{f3}\rangle & 0 & \cdot & 0 \\ 0 & \cdot & \cdot & \cdot & \cdot & \cdot \\ 0 & 0 & \cdot & 0 & \langle \nu\Sigma_{fN}\rangle & 0 \\ 0 & 0 & 0 & \cdot & 0 & \langle \nu\Sigma_{f,N+1}\rangle \end{bmatrix}. \tag{6.23}$$

We note that the matrix $A$ is a tridiagonal matrix, i.e. the nonzero elements of the matrix lie along the major diagonal and two adjacent diagonals only, with $a_1 = c_{N+1} = 0$.

The Gaussian elimination technique consists of breaking up the tridiagonal matrix $A$ into a product of a lower triangular matrix $L$ and an upper triangular matrix $U$ and successively inverting the product matrix $A = LU$

$$L = \begin{bmatrix} \alpha_1 & 0 & 0 & \cdot & \cdot & 0 \\ a_2 & \alpha_2 & 0 & \cdot & \cdot & 0 \\ 0 & a_3 & \alpha_3 & 0 & \cdot & 0 \\ 0 & \cdot & \cdot & \cdot & \cdot & \cdot \\ 0 & 0 & \cdot & a_N & \alpha_N & 0 \\ 0 & 0 & 0 & \cdot & a_{N+1} & \alpha_{N+1} \end{bmatrix}, \; U = \begin{bmatrix} 1 & q_1 & 0 & \cdot & \cdot & 0 \\ 0 & 1 & q_2 & \cdot & \cdot & 0 \\ 0 & 0 & 1 & q_3 & \cdot & 0 \\ 0 & \cdot & \cdot & \cdot & \cdot & \cdot \\ 0 & 0 & \cdot & 0 & 1 & q_N \\ 0 & 0 & 0 & \cdot & 0 & 1 \end{bmatrix}. \tag{6.24}$$

The diagonal elements of $L$ and upper diagonal elements of $U$ are given by

$$\alpha_n = b_n - a_n q_{n-1}, q_n = \frac{c_n}{\alpha_n}, n = 1, \ldots, N+1, \tag{6.25}$$

with $a_1 = c_{N+1} = 0$. Invert first the lower triangular matrix $L$

$$L^{-1} = \begin{bmatrix} \frac{1}{\alpha_1} & 0 & 0 & \cdot & \cdot & 0 \\ -\frac{1}{\alpha_1}\frac{a_2}{\alpha_2} & \frac{1}{\alpha_2} & 0 & \cdot & \cdot & 0 \\ \frac{a_2}{\alpha_1\alpha_2}\frac{a_3}{\alpha_3} & -\frac{1}{\alpha_2}\frac{a_3}{\alpha_3} & \frac{1}{\alpha_3} & 0 & \cdot & 0 \\ \cdot & \cdot & \cdot & \cdot & \cdot & \cdot \\ \cdot & \cdot & \cdot & \cdot & \cdot & 0 \\ \cdot & \cdot & \cdot & \cdot & \cdot & \frac{1}{\alpha_{N+1}} \end{bmatrix}, \tag{6.26}$$

and obtain

$$U\mathbf{\Phi} = L^{-1}\mathbf{S} = \mathbf{M}, \tag{6.27}$$

where $\mathbf{M}$ is an intermediate column vector with elements $m_n$ given by

$$m_n = \frac{S_n - a_n m_{n-1}}{\alpha_n} = \frac{S_n - a_n m_{n-1}}{b_n - a_n q_{n-1}}, n = 1, \ldots, N+1. \tag{6.28}$$

The upper triangular matrix $U$ is then inverted

$$U^{-1} = \begin{bmatrix} 1 & -q_1 & q_1 q_2 & -q_1 q_2 q_3 & \cdot & \cdot \\ 0 & 1 & -q_2 & q_2 q_3 & \cdot & \cdot \\ 0 & 0 & 1 & -q_3 & \cdot & \cdot \\ 0 & \cdot & \cdot & \cdot & \cdot & \cdot \\ 0 & 0 & \cdot & 0 & 1 & -q_N \\ 0 & 0 & 0 & \cdot & 0 & 1 \end{bmatrix}, \tag{6.29}$$

finally providing the solution for the flux vector

$$\mathbf{\Phi} = U^{-1}\mathbf{M}. \tag{6.30}$$

We may show readily that the matrix operation of Eq. (6.27) represents the *forward elimination step* of the Gaussian elimination algorithm with Eqs. (6.25) and (6.28), while the matrix inversion of Eq. (6.30) represents the *backward substitution step*

$$\phi_n = m_n - q_n\phi_{n+1}, n = 1, \ldots, N. \tag{6.31}$$

The last linear relationship of Eq. (6.31) for the boundary flux is simply $\phi_{N+1} = m_{N+1}$, which follows naturally from the boundary condition at $x_{N+1}$. This is clarified in the next section.

The Gaussian elimination algorithm may be initiated, without invoking the formality of matrix algebra, by substituting the backward substitution equation (6.31) for $\phi_{n-1}$ into Eq. (6.17) and eliminating $\phi_n$ by using Eq. (6.31) once more to obtain

$$(c_n - b_n q_n + a_n q_{n-1} q_n)\phi_{n+1} + (a_n m_{n-1} + b_n m_n - a_n m_n q_{n-1} - S_n) = 0. \tag{6.32}$$

For Eq. (6.32) to hold in general, it is necessary that the sum of terms in each pair of parentheses vanishes, which yields a pair of recursion relationships of Eqs. (6.25) and (6.28) for $q_n$ and $m_n$, respectively. We recall that $a_1 = 0$, which yields

$$\alpha_1 = b_1, q_1 = \frac{c_1}{b_1}, m_1 = \frac{S_1}{b_1}. \tag{6.33}$$

Thus, the flux solution algorithm is initiated by evaluating the set $\{q_n, m_n, n = 2, \ldots, N\}$ through Eqs. (6.25) and (6.28) beginning with the boundary values $(q_1, m_1)$ in the forward elimination step. The backward substitution equation (6.31), beginning with $\phi_{N+1}$, without the need for $q_{N+1}$ and $m_{N+1}$, provides the desired discretized solution $\{\phi_n, n = 1, \ldots, N\}$ of the flux distribution $\phi(x)$.

Although the inversion of Eq. (6.20) for one-dimensional geometry does not involve any iterations, the corresponding algorithm for two- and three-dimensional geometries usually involves iterations. Hence, the flux solution algorithm presented in this section is often referred to as an *inner iteration*, as compared with an outer iteration algorithm as discussed in Section 6.4.

## 6.3 Boundary Conditions for Difference Equation

As discussed in Section 6.2, the initiation of the forward elimination algorithm given in Eqs. (6.25) and (6.28) requires the determination of $m_1$ and $q_1$, which are supplied by the boundary condition at $x_1$. Similarly, the boundary

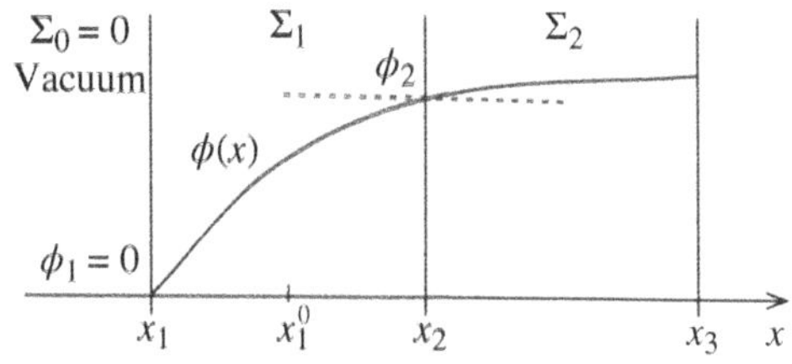


**Figure 6.2** Discretized flux distribution at the vacuum boundary.

condition at $x_{N+1}$ is required to determine $\phi_{N+1}$ so that the backward substitution algorithm of Eq. (6.31) may be carried out. In this section, common boundary conditions representing the zero flux and zero current at the boundaries are considered. Other boundary conditions, e.g. albedo boundary conditions, may be implemented with a little additional effort.

1. **Zero flux at $x = x_1$, $\phi(x_1) = \phi_1 = 0$, at a vacuum boundary**
   For this boundary condition, Figure 6.2 illustrates the discretized flux distribution $\phi(x)$ near the vacuum boundary, where the flux vanishes at $x = x_1$, so that Eq. (6.31) yields

   $$m_1 - q_1\phi_2 = \phi_1 = 0,$$

   which, together with $a_1 = 0$ and $S_1 = m_1 = 0$, yields

   $$q_1 = 0. \tag{6.34}$$

2. **Zero current at $x = x_1$, $d\phi(x_1)/dx = 0$**
   With the flux distribution near a reflecting boundary $x_1$ illustrated in Figure 6.3, rederive the leakage term of Eq. (6.8):

   $$l_1 = -\int_{x_1}^{x_1^0} \frac{d}{dx}\left[Dx^p \frac{d\phi}{dx}\right] dx = -D_1\left(x_1^0\right)^p \left(\frac{\phi_2 - \phi_1}{h_1}\right) = \langle D_1\rangle\phi_1 - \langle D_1\rangle\phi_2.$$

   Hence, $\langle D_0\rangle = \langle \Sigma_1^-\rangle = 0$ and $a_1 = 0$, as noted earlier, and $b_1$ in Eq. (6.19b) is now reduced to

   $$b_1 = \langle D_1\rangle + \langle \Sigma_1^+\rangle, \tag{6.35}$$

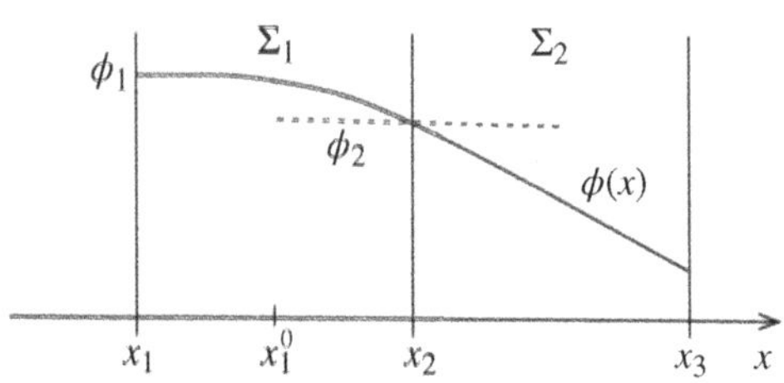


**Figure 6.3** Flux distribution near a reflecting boundary at $x_1$.

and Eq. (6.17) becomes

$$b_1\phi_1 + c_1\phi_2 = S_1.$$

This yields, when compared with Eq. (6.31),

$$q_1 = \frac{c_1}{b_1}, m_1 = \frac{S_1}{b_1}, \tag{6.36}$$

which reproduces the general relationship of Eq. (6.33) but with a reduced expression for $b_1$ in Eq. (6.35).

3. **Zero flux at $x = x_{N+1}$, $\phi(x_{N+1}) = \phi_{N+1} = 0$, at a vacuum boundary**
For this boundary condition, Eq. (6.31) immediately yields

$$\phi_N = m_N, \tag{6.37}$$

and the backward recursion proceeds directly.

4. **Zero current at $x = x_{N+1}$, $d\phi(x_{N+1})/dx = 0$**
Similar to Eq. (6.35), recall that $\langle D_{N+1}\rangle = \langle \Sigma^+_{N+1}\rangle = 0$ and $c_{N+1} = 0$, but $b_{N+1}$ in Eq. (6.18) now reduces to

$$b_{N+1} = \langle D_N\rangle + \langle \Sigma^-_{N+1}\rangle. \tag{6.38}$$

Thus, Eq. (6.17) now yields for the boundary cell

$$a_{N+1}\phi_N + b_{N+1}\phi_{N+1} = S_{N+1}$$

or, with Eq. (6.31) to eliminate $\phi_N$, we obtain

$$\phi_{N+1} = \frac{S_{N+1} - a_{N+1}m_N}{b_{N+1} - a_{N+1}q_N}, \tag{6.39}$$

which serves as the starting point for the backward substitution algorithm of Eq. (6.31). Note that Eq. (6.39) is in the same form as Eq. (6.28) for $m_{N+1}$, thereby satisfying the matrix multiplication of Eq. (6.30) for the last element of vector **M**. This was discussed in connection with Eq. (6.30) in Section 6.2.

## 6.4 Source or Outer Iteration

In Section 6.1, a three-point FD formulation was obtained for the one-group neutron diffusion equation, which may be solved through the Gaussian elimination algorithm described in Section 6.2. The boundary conditions required to initiate the inversion of the tridiagonal matrix $A$ from Eq. (6.21) were discussed in Section 6.3. The numerical algorithm discussed so far may thus yield a discretized flux distribution, provided the source distribution $S(x)$ is known in Eq. (6.2). In an eigenvalue problem, however, the source distribution $S(x) = \nu\Sigma_f\phi(x)/\lambda$ itself contains the unknown flux $\phi(x)$ with the eigenvalue $\lambda$ yet to be determined. Hence, the solution to an eigenvalue problem can be obtained only through an iterative scheme.

In the one-dimensional geometries considered so far, the flux distribution $\phi(x)$ for a given source distribution $S(x)$ can be obtained by a direct inversion of matrix $A$ through the Gaussian elimination algorithm, thus without any iteration. In contrast, in multidimensional geometries, the flux solution for a given source distribution usually requires some iterative techniques. Such iteration for a flux solution is known as the *inner iteration*, discussed in Section 6.2, while the iteration required to solve the eigenvalue problem through an iterative updating of the source distribution itself is called the *outer iteration* or *source iteration*.

The most common technique used for source iteration is the *power iteration method* discussed here with Eq. (6.22). Start the iteration with an arbitrary guess for $\mathbf{Q}^0$ and the eigenvalue $\lambda^0$. The first estimate of the flux distribution $\mathbf{\Phi}^1$ may then be obtained by solving

$$A\mathbf{\Phi}^1 = \frac{1}{\lambda^0}\mathbf{Q}^0$$

through the Gaussian elimination algorithm of Section 6.2, or an inner iteration scheme in general. The next estimate of the source distribution $\mathbf{Q}^1$ is then determined as $Q_n^1 = \langle \nu\Sigma_{fn}\rangle \phi_n^1, n = 1, \ldots, N+1$, via Eq. (6.13), and the eigenvalue $\lambda^1$ is calculated so that the eigenvalue equation itself is satisfied:

$$A\mathbf{\Phi}^1 = \frac{1}{\lambda^1}\mathbf{Q}^1.$$

The outer iteration proceeds so that, in general, at the $i$th iteration, we solve

$$A\mathbf{\Phi}^{i+1} = \frac{1}{\lambda^i}\mathbf{Q}^i. \tag{6.40}$$

for $\mathbf{\Phi}^{i+1}$, which yields

$$Q_n^{i+1} = \langle \nu\Sigma_{fn}\rangle \phi_n^{i+1}, n = 1, \ldots .N+1, \tag{6.41}$$

and an update for the eigenvalue so that the eigenvalue equation

$$A\mathbf{\Phi}^{i+1} = \frac{1}{\lambda^{i+1}}\mathbf{Q}^{i+1} \tag{6.42}$$

is satisfied. Since Eq. (6.42) actually represents, however, $N+1$ linear equations relating $\{\phi_1^{i+1}, \phi_2^{i+1}, \ldots, \phi_{N+1}^{i+1}\}$ to $\{Q_1^{i+1}, Q_2^{i+1}, \ldots, Q_{N+1}^{i+1}\}$, it becomes necessary to find a value of $\lambda^{i+1}$ that satisfies the entire set of linear equations. One obvious approach is to find $\lambda^{i+1}$ such that Eq. (6.42) is satisfied in an integral sense, i.e. when the $N+1$ elements of Eq. (6.42) are summed up:

$$\lambda^{i+1} = \frac{\sum_{n=1}^{N+1} Q_n^{i+1}}{\sum_{n=1}^{N+1} \{A\mathbf{\Phi}^{i+1}\}_n} = \frac{\sum_{n=1}^{N+1} Q_n^{i+1}}{\frac{1}{\lambda^i}\sum_{n=1}^{N+1} Q_n^i}. \tag{6.43}$$

Note that the denominator in the second ratio of Eq. (6.43) follows from Eq. (6.40). If the initial normalization on the source distribution is chosen such that

$$\lambda^0 = \sum_{n=1}^{N+1} Q_n^0,$$

then, by mathematical induction, we obtain

$$\lambda^{i+1} = \sum_{n=1}^{N+1} Q_n^{i+1} = \sum_{n=1}^{N+1} \langle \nu\Sigma_{fn} \rangle \phi_n^{i+1}, \tag{6.44}$$

which is computationally preferable to Eq. (6.43). The iteration represented by Eqs. (6.40) and (6.44) may be terminated when the solution has converged sufficiently to yield

$$\left| \frac{\lambda^{i+1} - \lambda^i}{\lambda^{i+1}} \right| < \varepsilon_\lambda, \tag{6.45}$$

$$\max_{1 \le n \le N+1} \left| \frac{Q_n^{i+1} - Q_n^i}{Q_n^{i+1}} \right| < \varepsilon_p, \tag{6.46}$$

for some desired convergence criteria $\varepsilon_\lambda$ and $\varepsilon_p$ for the eigenvalue and pointwise source distribution, respectively. In order to accelerate the convergence of the source iteration, an extrapolation factor $\omega$ is often introduced so that

$$\mathbf{S}^{i+1} = \frac{1}{\lambda^i}\mathbf{Q}^i + \omega \left[ \frac{1}{\lambda^{i+1}}\mathbf{Q}^{i+1} - \frac{1}{\lambda^i}\mathbf{Q}^i \right], \tag{6.47}$$

where $\mathbf{Q}^i$ and $\lambda^i$ are retained from the previous iteration and $\mathbf{Q}^{i+1}$ and $\lambda^{i+1}$ are evaluated from the current iteration through Eqs. (6.41) and (6.44). The extrapolation factor $\omega$ usually lies between 1.0 and 2.0, with a common recommendation of $\omega = 1.8$ for one-dimensional problems. In Eq. (6.47), the difference in the source vector $\mathbf{S}$ over two successive iterations is amplified by $\omega > 1.0$ and added to the previous iterate $\mathbf{S}^i$ to arrive at the updated estimate $\mathbf{S}^{i+1}$ for the current iteration, thereby accelerating the convergence. The iteration scheme is usually called an *over-relaxation scheme*, with an *over-relaxation parameter* $\omega$.

## 6.5 Relative Power Distribution and Overall Flow Chart

Once a converged flux distribution is obtained, the power distribution can be determined as

$$P(\mathbf{r}) = E_f \Sigma_f(\mathbf{r}) \phi(\mathbf{r}) \tag{6.48}$$

in terms of the energy $E_f$ released per fission. Depending on the unit chosen for $E_f$, the power distribution may be written in units of $[\mathrm{W} \cdot \mathrm{cm}^{-3}]$ or $[\mathrm{MW} \cdot \mathrm{m}^{-3}]$.

For a variety of applications, however, it is often convenient if we determine a normalized power distribution, often referred to as the *relative power distribution*, such that the average of the distribution over the core volume $V$ is unity. One obvious use of the relative power distribution is that we may multiply it by the *average core power density* in a unit we desire and quickly obtain the core power distribution in that particular unit.

From Eq. (6.44), confirm the normalization on the converged neutron flux and fission source rate such that

$$\int_V S(\mathbf{r})d\mathbf{r} = \frac{1}{\lambda}\int_V Q(\mathbf{r})d\mathbf{r} = \frac{1}{\lambda}\int_V \nu\Sigma_f\phi(\mathbf{r})d\mathbf{r} = 1.0, \tag{6.49}$$

which implies that one neutron is produced in the entire core volume $V$ per second. This then suggests that the relative power distribution may be obtained by

$$P_{rel}(\mathbf{r}) = V\frac{1}{\lambda}\nu\Sigma_f\phi(\mathbf{r}), \tag{6.50}$$

since the volume average of $P_{rel}(\mathbf{r})$ would simply be unity due to the normalization of Eq. (6.49). For a bare slab reactor of height $H$, with a uniform fuel composition and negligible extrapolation distance, Eq. (6.50) may be written explicitly as

$$P_{rel}(z) = S(z)H = H\frac{Q(z)}{\lambda} = H\frac{1}{\lambda}\nu\Sigma_f\phi(z) = \frac{\pi}{2}\cos\left(\frac{\pi z}{H}\right), \tag{6.51}$$

recalling the fundamental-mode flux profile $\psi_0(z)$ of Eq. (5.76), as illustrated in Figure 5.8. This establishes the normalization on flux $\phi(z)$ such that one neutron is produced per second out of a unit cross-sectional area of the entire slab reactor and clarifies the point that the magnitude of the flux and power distribution cannot be determined by the solution of the eigenvalue equation (5.62) or (5.74). This implies then that the power output of a critical reactor can be set arbitrarily, subject to operational constraints, and not by the criticality consideration. This is strictly true, however, up to the power level where the thermal-hydraulic feedback in an operating reactor is negligible.

The axial relative power distribution of Eq. (6.51), averaged over the core height $H$, yields unity again, recalling the normalization on $S(z)$. We may also prove this relationship by using the property of the cosine function. We can also establish that the maximum value of $P_{rel}(z)$ is equal to the peak-to-average power ratio for the axial power distribution given in any unit. This can be also verified directly using the cosine distribution given in Eq. (6.51). The peak-to-average power ratio is known as the *axial power peaking factor* $F_z$ and plays an important role in thermal-hydraulic and safety analyses of nuclear reactors. Extending this concept to the general power distribution of Eq. (6.50), we may determine the overall peak-to-average power ratio or the *overall power peaking factor* $F_q$ in a three-dimensional reactor core by simply evaluating the maximum value of the relative power distribution $P_{rel}(\mathbf{r})$. This is yet another reason why it is often important to determine the relative or normalized power distribution in a numerical solution of the neutron diffusion equation.

With this perspective in mind, by integrating the relative power distribution over the $n$th cell, in a manner similar to the discretization scheme of Section 6.1, we may derive a discretized form of the relative power distribution using an *average flux* for the cell

$$P_n = \frac{\nu\Sigma_{fn}}{\lambda}\left(\frac{x_n^+\phi_n + x_{n+1}^-\phi_{n+1}}{x_n^+ + x_{n+1}^-}\right)V,\ n = 1, \ldots, N, \tag{6.52}$$

where

$$x_n^+ = \left(x_n + \frac{h_n}{4}\right)^p \text{ and } x_{n+1}^- = \left(x_{n+1} - \frac{h_n}{4}\right)^p. \tag{6.53}$$

For a slab reactor with core height $H$, the average flux for mesh $n$ is obtained as an arithmetic average of two mesh-boundary fluxes, simplifying Eq. (6.52) to

$$P_n = \frac{\nu\Sigma_{fn}}{\lambda}\left(\frac{\phi_n + \phi_{n+1}}{2}\right)H, \quad n = 1, \ldots, N. \tag{6.54}$$

The discretized axial relative power distribution is normalized such that

$$\frac{1}{N}\sum_{n=1}^{N} P_n = 1.0 \tag{6.55}$$

and the maximum value of $P_n$ yields the axial power peaking factor $F_z$.

The steps involved in the solution of the 1-D diffusion equation are summarized as a flow chart in Figure 6.4. We illustrate how a combination of inner and outer iterations is used to solve eigenvalue problems in general, although the inner iteration involves a non-iterative inversion of a tridiagonal matrix for the 1-D geometries considered here.

The three-point difference formulation for a discrete solution of the 1-D diffusion equation assumes that the flux distribution is continuous across the mesh boundary and hence automatically guarantees the continuity of neutron flux across the cell boundaries. This *face-centered differencing scheme* in the PANDA code [Bar67] and the PDQ-7 code [Cad67] as well as the ONED code [Lee74], however, fails to satisfy the continuity of neutron current at the cell boundaries. An alternative scheme where the flux is *body-centered* can easily be constructed, where the continuity of current is satisfied at the price of violating the continuity of flux. To satisfy the continuity of current and flux simultaneously, we need to go to higher-order discretization schemes, e.g. nodal methods [Law86], or define volume-weighted diffusion coefficients across the cell boundaries [Lit68, Der84] illustrated for the finite-difference formulation of the two-dimensional diffusion equation in Section 6.7.

**Figure 6.4** Inner and outer iterations for solution of the diffusion equation.

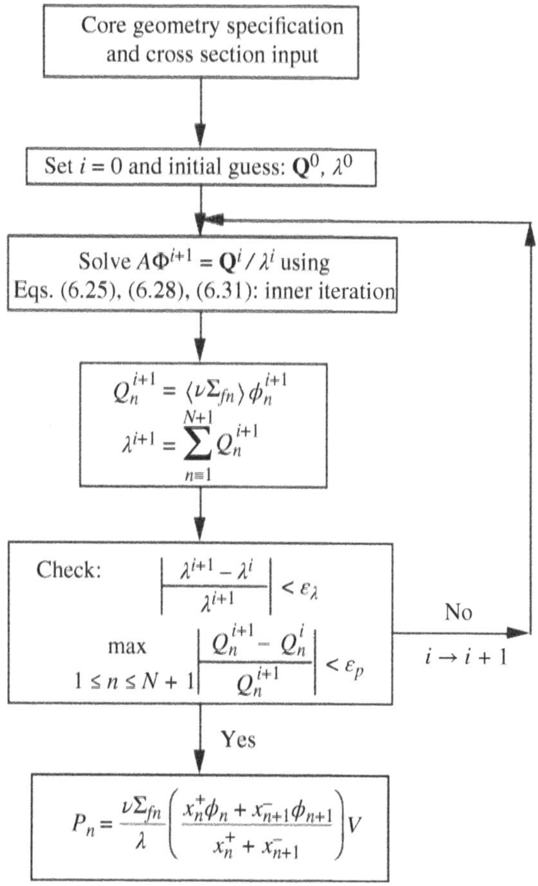


## 6.6 Single-Channel Flux Synthesis

Having developed numerical algorithms for determining the axial and radial flux distributions for slab and cylindrical geometries, respectively, we now discuss how the FD flux solutions may be combined or synthesized to yield the two-dimensional flux distribution $\phi(r, z)$ in a cylindrical reactor. This is the simplest form of the flux synthesis approaches [Sta01] and is known as the *single-channel flux synthesis.*

Assume that $\phi(r, z)$ may be separated into the radial and axial flux distributions

$$\phi(r, z) = \theta(r)Z(z) \tag{6.56}$$

and obtain the solutions for the radial flux distribution $\theta(r)$ and axial flux distribution $Z(z)$ separately, while accounting for the neutron leakage in the missing or transverse direction. This is accomplished by substituting Eq. (6.56) into the one-group form of Eq. (6.1), with constant diffusion coefficient $D$, and writing the cylindrical Laplacian operator as a sum of the radial and axial Laplacian operators:

$$D\left(\nabla_r^2 + \nabla_z^2\right)\theta(r)Z(z) + \left(\frac{\nu\Sigma_f}{\lambda} - \Sigma_a\right)\theta(r)Z(z) = 0. \tag{6.57}$$

The axial Laplacian operator is simply

$$\nabla_z^2 = \frac{d^2}{dz^2}, \tag{6.58}$$

while the radial Laplacian operator represents

$$\nabla_r^2 = \frac{1}{r}\frac{d}{dr}\left(r\frac{d}{dr}\right). \tag{6.59}$$

Dividing Eq. (6.57) by $D$ and carrying on the radial and axial Laplacian operations yields

$$\left[Z(z)\nabla_r^2\theta(r) + \theta(r)\nabla_z^2 Z(z)\right] + \frac{1}{D}\left(\frac{\nu\Sigma_f}{\lambda} - \Sigma_a\right)\theta(r)Z(z) = 0. \tag{6.60}$$

Using the expression for the effective multiplication factor $k_{eff} = \lambda$ from Eq. (5.71) and recalling the concept of material and geometrical bucklings of Eq. (5.66), recognize that

$$\frac{1}{D}\left(\frac{\nu\Sigma_f}{\lambda} - \Sigma_a\right) = B_m^2 = B_g^2. \tag{6.61}$$

We showed in Eq. (5.68) that the term $DB_g^2$ represents the total leakage rate relative to the absorption rate represented by $\Sigma_a$, and the total buckling from Eq. (6.61) may thus be broken up into the radial buckling $B_r^2$ and the axial buckling $B_z^2$, as in the sample PWR criticality calculation of Section 5.3.3. Substitute the radial component of the eigenvalue equation (5.70)

$$\nabla_r^2\theta(r) + B_r^2\theta(r) = 0 \tag{6.62}$$

into the two-dimensional diffusion equation (6.60) to obtain

$$D\nabla_z^2 Z(z) + \left[\frac{\nu\Sigma_f}{\lambda} - \left(\Sigma_a + DB_r^2\right)\right] Z(z) = 0. \tag{6.63}$$

Comparing this axial component of Eq. (6.57) with the general eigenvalue equation (5.70) suggests that Eq. (6.63) is simply equal to the axial form of Eq. (5.70), provided we replace the regular absorption cross section with an effective absorption cross section that includes the radial leakage rate

$$\Sigma_{a,z}^{eff} = \Sigma_a + DB_r^2. \tag{6.64}$$

Equation (6.63) may then be numerically solved for $Z(z)$ and the eigenvalue $\lambda$ through the FD formulation for axial slab geometry developed in this chapter. The solution $Z(z)$ then represents the axial flux distribution in a cylindrical reactor, duly accounting for the *transverse leakage in the radial direction*. It should be noted that the eigenvalue $\lambda$ obtained from the FD solution of Eq. (6.63) accounts for the radial leakage through the term $DB_r^2$, while the axial leakage is directly represented via the axial FD solution, thus yielding the effective multiplication factor $\lambda = k_{eff}$ for the two-dimensional cylindrical reactor.

In a similar fashion, to determine the radial flux distribution $\theta(r)$ in a cylindrical reactor, we may substitute the axial eigenvalue equation

$$\nabla_z^2 Z(z) + B_z^2 Z(z) = \frac{d^2 Z(z)}{dz^2} + B_z^2 Z(z) = 0 \tag{6.65}$$

into Eq. (6.60) and obtain the radial component of the diffusion equation

$$D\nabla_r^2 \theta(r) + \left[\frac{\nu\Sigma_f}{\lambda} - (\Sigma_a + DB_z^2)\right]\theta(r) = 0. \tag{6.66}$$

Equation (6.66) may then be solved again using the cylindrical FD formulation developed in this chapter for the radial component $\theta(r)$ of the flux $\phi(r, z)$, in terms of an effective absorption cross section accounting for the axial leakage:

$$\Sigma_{a,r}^{eff} = \Sigma_a + DB_z^2. \tag{6.67}$$

Note again that the eigenvalue $\lambda$ obtained from the solution of Eq. (6.66) represents the axial leakage through the axial buckling and hence is the proper $k_{eff}$ for the entire cylindrical reactor. Through the single-channel synthesis developed in this section, we are able to calculate both the axial and radial flux distributions in a cylindrical reactor using a simple one-dimensional FD diffusion theory code. One such code is the ONED code [Lee74] that provides FD solutions of the one- and two-group neutron diffusion equations in both axial slab and radial cylindrical geometries.

**Example 6.1** Set up a FD formulation of the one-group slab-geometry diffusion equation with four uniform intervals, following the face-centered structure of Section 6.1, to illustrate and benchmark the FD algorithm developed in Sections 6.1 through 6.6. Solve the resulting matrix equation directly to obtain the eigenvalue and eigenfunction representing the scalar flux. The mesh structure is illustrated in Figure 6.5 with mesh size $h = 73.315$ cm; one-group constants $D = 4.61$ cm, $\Sigma_a = 0.157$ cm$^{-1}$, $\nu\Sigma_f = 0.161$ cm$^{-1}$, and radial buckling $B_r^2 = 3.39 \times 10^{-4}$ cm$^{-2}$.

The face-centered flux requires a half-mesh interval at both ends of the slab and, with zero-flux boundary conditions at both boundaries, we need to evaluate three

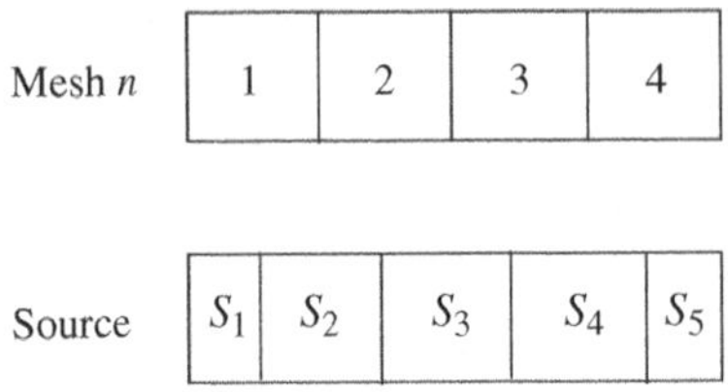


**Figure 6.5** Finite-difference mesh structure for $N = 4$.

flux values $\mathbf{\Phi} = [\phi_2, \phi_3, \phi_4]^T$. Thus, we set up a $(3 \times 3)$ matrix $A$ for the loss term of the diffusion equation with the eigenvalue modified to $\lambda = \nu\Sigma_f h/k$

$$A\mathbf{\Phi} = \begin{bmatrix} b_2 & c_2 & 0 \\ a_3 & b_3 & c_3 \\ 0 & a_4 & b_4 \end{bmatrix} \begin{bmatrix} \phi_2 \\ \phi_3 \\ \phi_4 \end{bmatrix} = \lambda \begin{bmatrix} \phi_2 \\ \phi_3 \\ \phi_4 \end{bmatrix}, \tag{6.68}$$

with

$$b_2 = b_3 = b_4 = 11.7508, c_2 = a_3 = c_3 = a_4 = -0.0628794, \nu\Sigma_f h = 11.80371.$$

Solving the matrix equation (6.68) through the *eig* function of *MATLAB* yields the fundamental mode eigenvalue $\lambda = 11.6619$ and eigenfunction

$$\mathbf{\Phi} = [0.5000, 0.7071, 0.5000]^T.$$

The matrix solution corresponds to $k$-eigenvalue = 1.0122 and the normalized flux, with boundary flux values included,

$$\mathbf{\Phi} = [0.0, 0.02513, 0.03550, 0.02513, 0.0]^T,$$

in agreement with the power-iteration solution via the ONED code. The $k$-eigenvalue result agrees favorably with the analytical solution $k_{eff} = \nu\Sigma_f/(\Sigma_a + DB_g^2) = 1.0120$ for a bare slab reactor.

## 6.7 Multidimensional Finite Difference Formulation

Recognizing that the flux synthesis technique can only provide approximate numerical solutions to the diffusion equation in cylindrical geometry with non-uniform cross sections, we now turn to the task of directly discretizing the 2-D diffusion equation. This is followed by a brief discussion on extending the task to the solution of 3-D diffusion equation and general properties of the convergence of the iterative matrix equations.

### 6.7.1 Two-Dimensional Matrix Formulation

Integrate the one-group diffusion equation (6.1) over a cell with volume $V_0$ and cell-average flux $\phi_0$, absorption cross section $\Sigma_0$, and source $S_0$, surrounded by

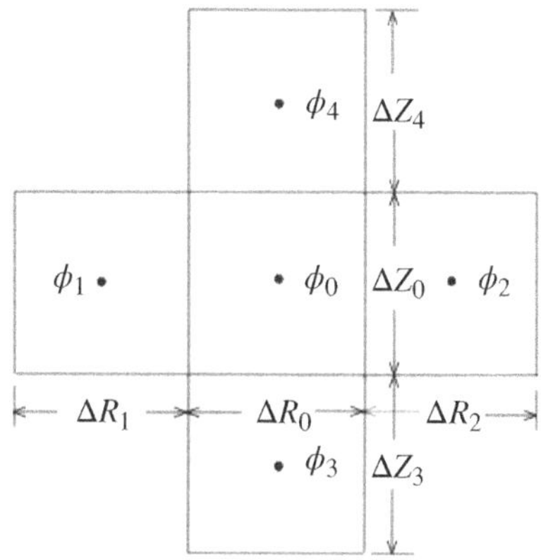


**Figure 6.6** 2-D discretization scheme.

four cells in the ($r$–$z$) geometry shown in Figure 6.6. By defining the net current $J_k = \mathbf{n} \cdot \mathbf{J}$ at surfaces $A_k$, $k = 1, \ldots, 4$, making up $V_0$, where $\mathbf{n}$ is the outward normal vector to each surface, the diffusion equation may be written in a discretized form [Lit68, Der84]:

$$\sum_{k=1}^{4} J_k A_k + \Sigma_0 \phi_0 V_0 = S_0 V_0. \tag{6.69}$$

The net current $J_1$ for surface $A_1$ may be explicitly written with the diffusion coefficient $D_0$ for $V_0$ and $D_1$ for the cell left of $V_0$, and the corresponding flux gradients in terms of the flux $\phi_{1/2}$ at the interface $A_1$:

$$J_1 = D_0 \frac{\phi_0 - \phi_{1/2}}{\Delta R_0/2} = D_1 \frac{\phi_{1/2} - \phi_1}{\Delta R_1/2}. \tag{6.70}$$

Note that the flux gradients are represented properly in terms of the widths $\Delta R_0$ and $\Delta R_1$ of the two adjacent cells in the $r$-direction, and the continuity of the current at the interface $A_1$ is naturally satisfied. Solve for the interface flux $\phi_{1/2}$ in terms of the cell-average fluxes $\phi_0$ and $\phi_1$

$$\phi_{1/2} = \frac{D_1 \phi_1 \Delta R_0 + D_0 \phi_0 \Delta R_1}{D_1 \Delta R_0 + D_0 \Delta R_1}, \tag{6.71}$$

which may be substituted back into Eq. (6.70) for the interface current in terms of the cell-average fluxes:

$$J_1 = \frac{D_0 D_1 (\phi_0 - \phi_1)}{D_1 \Delta R_0 + D_0 \Delta R_1} \cdot \frac{\Delta R_0 + \Delta R_1}{(\Delta R_0 + \Delta R_1)/2}. \tag{6.72}$$

Equation (6.71) is then rewritten and generalized for the four interface currents of control volume $V_0$

$$J_k = \frac{\langle D_k \rangle (\phi_0 - \phi_k)}{\ell_k}, \quad k = 1, \ldots, 4, \tag{6.73}$$

where $\ell_k$ is the distance between two cell centers

$$\ell_k = \begin{cases} (\Delta R_0 + \Delta R_k)/2, & k = 1, 2, \\ (\Delta Z_0 + \Delta Z_k)/2, & k = 3, 4, \end{cases} \tag{6.74}$$

and $\langle D_k \rangle$ is the weighted-average diffusion coefficient

$$\langle D_k \rangle = \begin{cases} \dfrac{D_0 D_k (\Delta R_0 + \Delta R_k)}{D_0 \Delta R_k + D_k \Delta R_0}, & k = 1, 2, \\[2ex] \dfrac{D_0 D_k (\Delta Z_0 + \Delta Z_k)}{D_0 \Delta Z_k + D_k \Delta Z_0}, & k = 3, 4. \end{cases} \tag{6.75}$$

Rewriting $\langle D_k \rangle$ for $k = 1$ or 2

$$\frac{1}{\langle D_k \rangle} = \frac{\Delta R_0 / D_0 + \Delta R_k / D_k}{\Delta R_0 + \Delta R_k} \tag{6.76}$$

reveals that the weighted diffusion coefficient is based on a weighted average of the transport cross sections across two adjacent cells.

The difference equation (6.69) may be cast in the form of the FD equation (6.17) or matrix equation (6.20) for the 1-D diffusion equation by introducing $c_k = \langle D_k \rangle A_k / \ell_k$:

$$\sum_{k=1}^{4} c_k (\phi_0 - \phi_k) + \Sigma_0 \phi_0 V_0 = S_0 V_0. \tag{6.77}$$

We may finally generalize the difference equation by setting the control cell in a formal 2-D mesh structure

$$a_{ij}\phi_{i-1,j} + b_{ij}\phi_{ij} + c_{ij}\phi_{i+1,j} + d_{ij}\phi_{i,j-1} + e_{ij}\phi_{i,j+1} = S_{ij}, \\ i = 1, \ldots, I,\ j = 1, \ldots, J, \tag{6.78}$$

by defining the matrix elements

$$a_{ij} = -c_1,\ b_{ij} = \sum_{k=1}^{4} c_k + \Sigma_0 V_0,\ c_{ij} = -c_2,\ d_{ij} = -c_3,\ e_{ij} = -c_4, \\ \phi_{ij} = \phi_0, \phi_{i-1,j} = \phi_1, \phi_{i+1,j} = \phi_2, \phi_{i,j-1} = \phi_3, \phi_{i,j+1} = \phi_4, S_{ij} = S_0 V_0. \tag{6.79}$$

The five-band matrix structure of Eq. (6.78) may be explicitly written in the form of Eq. (6.20)

$$A\mathbf{\Phi} = \mathbf{S} \tag{6.80}$$

and illustrated in Figure 6.7.

As noted with Eq. (6.20) for the 1-D diffusion equation, where the tridiagonal matrix structure originates from the leakage term of Eq. (6.8) connecting three adjacent cells, so does the five-band matrix structure from the leakage term

$$A = \left[\begin{array}{cccccc|cccccc|cccccc|cccccc}
b_{11} & c_{11} & 0 & \cdot & \cdot & 0 & e_{11} & \cdot & \cdot & \cdot & \cdot & \cdot & & & & & & & & & & & & \\
a_{21} & b_{21} & c_{21} & \cdot & \cdot & 0 & \cdot & e_{21} & \cdot & \cdot & \cdot & \cdot & & & & & & & & & & & & \\
0 & a_{31} & b_{31} & c_{31} & \cdot & 0 & \cdot & \cdot & \cdot & \cdot & \cdot & \cdot & & & & & & & & & & & & \\
0 & \cdot & \cdot & \cdot & \cdot & \cdot & 0 & \cdot & \cdot & \cdot & \cdot & \cdot & & & & & & & & & & & & \\
0 & 0 & \cdot & \cdot & \cdot & c_{I-1,1} & \cdot & \cdot & \cdot & \cdot & \cdot & \cdot & & & & & & & & & & & & \\
0 & 0 & 0 & \cdot & a_{I1} & b_{I1} & 0 & 0 & 0 & 0 & 0 & e_{I1} & & & & & & & & & & & & \\
\hline
d_{12} & \cdot & \cdot & \cdot & \cdot & \cdot & b_{12} & c_{12} & \cdot & \cdot & \cdot & \cdot & e_{12} & \cdot & \cdot & \cdot & \cdot & \cdot & & & & & & \\
\cdot & d_{22} & \cdot & \cdot & \cdot & \cdot & a_{22} & b_{22} & c_{22} & \cdot & \cdot & \cdot & \cdot & e_{22} & \cdot & \cdot & \cdot & \cdot & & & & & & \\
\cdot & \cdot & \cdot & \cdot & \cdot & \cdot & \cdot & \cdot & \cdot & \cdot & \cdot & \cdot & \cdot & \cdot & \cdot & \cdot & \cdot & \cdot & & & & & & \\
\cdot & \cdot & \cdot & \cdot & \cdot & \cdot & \cdot & \cdot & \cdot & \cdot & \cdot & \cdot & \cdot & \cdot & \cdot & \cdot & \cdot & \cdot & & & & & & \\
\cdot & \cdot & \cdot & \cdot & \cdot & \cdot & \cdot & \cdot & \cdot & \cdot & \cdot & \cdot & \cdot & \cdot & \cdot & \cdot & e_{I-1,2} & \cdot & & & & & & \\
\cdot & \cdot & \cdot & \cdot & \cdot & d_{I2} & 0 & 0 & 0 & 0 & a_{I2} & b_{I2} & \cdot & \cdot & \cdot & \cdot & \cdot & e_{I2} & & & & & & \\
\hline
 & & & & & & \cdot & \cdot & \cdot & \cdot & \cdot & \cdot & \cdot & \cdot & \cdot & \cdot & \cdot & \cdot & e_{1,J-1} & \cdot & \cdot & \cdot & \cdot & \cdot \\
 & & & & & & \cdot & \cdot & \cdot & \cdot & \cdot & \cdot & \cdot & \cdot & \cdot & \cdot & \cdot & \cdot & \cdot & e_{2,J-1} & \cdot & \cdot & \cdot & \cdot \\
 & & & & & & \cdot & \cdot & \cdot & \cdot & \cdot & \cdot & \cdot & \cdot & \cdot & \cdot & \cdot & \cdot & \cdot & \cdot & \cdot & \cdot & \cdot & \cdot \\
 & & & & & & \cdot & \cdot & \cdot & \cdot & \cdot & \cdot & \cdot & \cdot & \cdot & \cdot & \cdot & \cdot & \cdot & \cdot & \cdot & \cdot & \cdot & \cdot \\
 & & & & & & \cdot & \cdot & \cdot & \cdot & \cdot & \cdot & \cdot & \cdot & \cdot & \cdot & \cdot & \cdot & \cdot & \cdot & \cdot & \cdot & e_{I-1,J-1} & \cdot \\
 & & & & & & \cdot & \cdot & \cdot & \cdot & \cdot & \cdot & \cdot & \cdot & \cdot & \cdot & a_{I,J-1} & b_{I,J-1} & \cdot & \cdot & \cdot & \cdot & \cdot & e_{I,J-1} \\
\hline
 & & & & & & & & & & & & d_{1J} & \cdot & \cdot & \cdot & \cdot & \cdot & b_{1J} & c_{1J} & \cdot & \cdot & \cdot & \cdot \\
 & & & & & & & & & & & & \cdot & d_{2J} & \cdot & \cdot & \cdot & \cdot & a_{2J} & b_{2J} & c_{2J} & \cdot & \cdot & \cdot \\
 & & & & & & & & & & & & & \cdot & \cdot & \cdot & \cdot & \cdot & \cdot & \cdot & \cdot & \cdot & \cdot & \cdot \\
 & & & & & & & & & & & & \cdot & \cdot & \cdot & \cdot & \cdot & \cdot & \cdot & \cdot & \cdot & \cdot & \cdot & \cdot \\
 & & & & & & & & & & & & \cdot & \cdot & \cdot & \cdot & d_{I-1,J} & \cdot & \cdot & \cdot & \cdot & a_{I-1,J} & b_{I-1,J} & c_{I-1,J} \\
 & & & & & & & & & & \cdot & \cdot & \cdot & \cdot & \cdot & \cdot & \cdot & d_{IJ} & \cdot & \cdot & \cdot & \cdot & a_{IJ} & b_{IJ}
\end{array}\right]$$

$$\mathbf{\Phi} = (\phi_{11}, \phi_{21}, \ldots, \phi_{I1}, \phi_{12}, \ldots, \phi_{I2}, \ldots, \phi_{1J}, \ldots, \phi_{IJ})^T,\ \mathbf{S} = (S_{11}, S_{21}, \ldots, S_{I1}, S_{12}, \ldots, S_{I2}, \ldots, S_{1J}, \ldots, S_{IJ})^T$$

**Figure 6.7** Two-dimensional finite-difference structure.

of Eq. (6.69) or (6.77) involving five adjacent cells. The difference structure of Eq. (6.69) leading eventually to Eq. (6.78) also could easily be recast for the 2-D ($x$-$y$) geometry by changing the mesh structure from ($\Delta R, \Delta Z$) to ($\Delta X, \Delta Y$). For an ($x$-$y$) representation of the 2-D diffusion equation, note that the neutron leakage in the missing $z$-direction should be represented by the axial buckling, as in Eqs. (6.66) and (6.67). Once the solution of the one-group diffusion equation is achieved through a combination of inner and outer iterations, it is straightforward to extend the formulation to the multigroup diffusion equation. All that is required in practice is to include the down- and up-scattering terms as part of the source term $S$ and iteratively update it as an integral part of the outer iteration, as discussed further in Chapter 7.

### 6.7.2 Three-Dimensional Formulation

As expected from the 2-D FD formulation, 3-D FD formulations of the diffusion equation result in a seven-band matrix, involving six adjacent neighboring cells. The multiband matrices invariably have to be solved through various inner iteration algorithms. Some of the more popular iteration algorithms make use of the Gaussian elimination algorithm as part of the overall inner iteration technique. One such technique [Lit68, Der84] involves recasting the matrix equation (6.78) in the form of a block-tridiagonal matrix equation:

$$A = \begin{bmatrix} B_1 & C_1 & 0 & \cdot & \cdot & 0 \\ A_2 & B_2 & C_2 & \cdot & \cdot & 0 \\ 0 & A_3 & B_3 & C_3 & \cdot & 0 \\ 0 & 0 & \cdot & \cdot & \cdot & \cdot \\ 0 & 0 & \cdot & \cdot & \cdot & \cdot \\ 0 & 0 & 0 & \cdot & A_J & B_J \end{bmatrix}, \boldsymbol{\Phi} = \begin{bmatrix} \boldsymbol{\Phi}_1 \\ \boldsymbol{\Phi}_2 \\ \boldsymbol{\Phi}_3 \\ \cdot \\ \cdot \\ \boldsymbol{\Phi}_J \end{bmatrix}, \mathbf{S} = \begin{bmatrix} \mathbf{S}_1 \\ \mathbf{S}_2 \\ \mathbf{S}_3 \\ \cdot \\ \cdot \\ \cdot \\ \mathbf{S}_J \end{bmatrix}. \tag{6.81}$$

or equivalently

$$A_j \boldsymbol{\Phi}_{j-1} + B_j \boldsymbol{\Phi}_j + C_j \boldsymbol{\Phi}_{j+1} = \mathbf{S}_j, \quad j = 1, \dots, J. \tag{6.82}$$

Recognizing that $B_j$ is a tridiagonal matrix

$$B_j = \begin{bmatrix} b_{1j} & c_{1j} & 0 & 0 & 0 \\ a_{2j} & b_{2j} & c_{2j} & 0 & 0 \\ 0 & \cdot & \cdot & \cdot & 0 \\ 0 & 0 & \cdot & \cdot & c_{I-1,j} \\ 0 & 0 & 0 & a_{Ij} & b_{Ij} \end{bmatrix}, \tag{6.83}$$

and $A_j$ and $C_j$ are diagonal matrices with elements $d_{ij}$ and $e_{ij}$, respectively, we may recast, for iteration $p+1$, the block tridiagonal matrix $A$ in the form

$$B_j \boldsymbol{\Phi}_j^{\star p+1} = -A_j \boldsymbol{\Phi}_{j-1}^{p+1} - C_j \boldsymbol{\Phi}_{j+1}^{p} + \mathbf{S}_j = \mathbf{Q}_j, \quad j = 1, \dots, J, \tag{6.84}$$

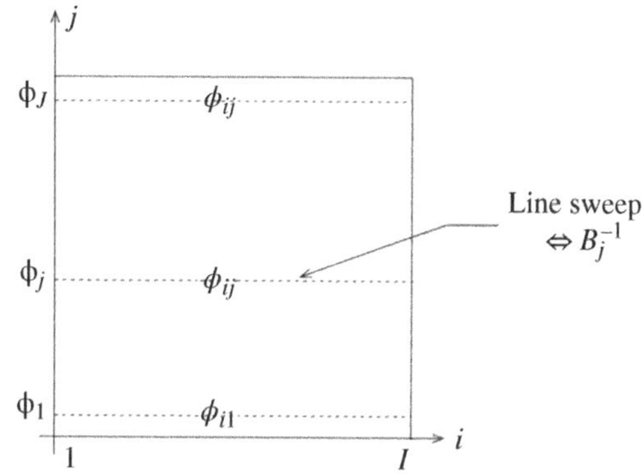


**Figure 6.8** 2-D line relaxation scheme.

with a column vector $\mathbf{Q}_j$. The flux along line $j$ is then obtained by inverting matrix $B_j$ for $\mathbf{\Phi}_j^{\star p+1}$ via the standard Gaussian elimination algorithm of Section 6.2. Then a *successive over-relaxation* (SOR) is performed at the block matrix level to yield

$$\mathbf{\Phi}_j^{p+1} = \mathbf{\Phi}_j^p + \omega(\mathbf{\Phi}_j^{\star p+1} - \mathbf{\Phi}_j^p), \tag{6.85}$$

where $\omega$ is the over-relaxation or extrapolation parameter. The inversion of tridiagonal matrix $B_j$ in Eq. (6.84) corresponds to a line sweep of the two-dimensional mesh, illustrated in Figure 6.8. The *successive relaxation* (SR) or *Gauss–Seidel algorithm* [Dah74, Saa03] is illustrated in Figure 6.9 for the case of inverting the simple matrix equation (6.22), although the *successive line over-relaxation* (SLOR) algorithm [Lit68, Der84] combining Eqs. (6.84) and (6.85) employ the SR algorithm at the block matrix level. Thus, Eq. (6.22) may now be written as

$$(L + D)\mathbf{\Phi}^{p+1} + U\mathbf{\Phi}^p = \mathbf{S}, \tag{6.86}$$

in terms of the diagonal matrix $D$, lower and upper triangular matrices $L$ and $U$, respectively, distinct from the notation employed in Eq. (6.24). The SR algorithm makes successive use of the updated solution for iteration $p + 1$ as the solution is updated:

$$\mathbf{\Phi}^{p+1} = D^{-1}[-L\mathbf{\Phi}^{p+1} - U\mathbf{\Phi}^p + \mathbf{S}]. \tag{6.87}$$

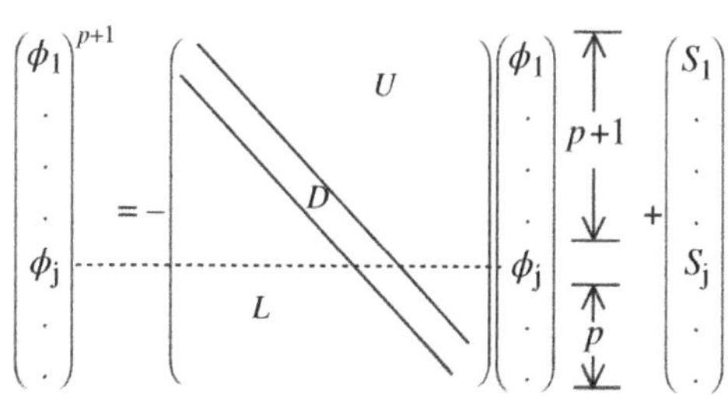


**Figure 6.9** Successive relaxation scheme.

### 6.7.3 Convergence Properties of Matrix Iteration Schemes

Various matrix iteration schemes, e.g. SLOR schemes, for solution of the 3-D diffusion equation may be written in a generic form for solution vector $\mathbf{x}$, iteration matrix $B$, and source vector $\mathbf{z}$ at iteration $p$:

$$\mathbf{x}^{p+1} = B\mathbf{x}^p + \mathbf{z}. \tag{6.88}$$

We will discuss how the rate of convergence of the matrix equation depends on matrix $B$. It should be recognized first that the basic requirement of the converged solution entails

$$\lim_{p\to\infty} \mathbf{x}^p = \mathbf{x}^\infty, \mathbf{x}^\infty = B\mathbf{x}^\infty + \mathbf{z},$$

and further that at the $p$th iteration the residual error in the solution vector $\mathbf{x}$ is written as

$$\epsilon^p = \mathbf{x}^p - \mathbf{x}^\infty = B(\epsilon^{p-1} + \mathbf{x}^\infty) + \mathbf{z} - \mathbf{x}^\infty = B\epsilon^{p-1} = B^p\epsilon^0. \tag{6.89}$$

The convergence of the iterative scheme occurs if and only if

$$\lim_{p\to\infty} \epsilon^p = \lim_{p\to\infty} B^p\epsilon^0 = 0.$$

Assume that the iteration matrix $B$ is simple, i.e. it has a complete set of eigenvector $\mathbf{e}_i$ with eigenvalue $\lambda_i$ such that $B\mathbf{e}_i = \lambda_i\mathbf{e}_i$, making it possible to write the initial error vector in terms of the eigenvectors

$$\epsilon^0 = \sum_i \alpha_i\mathbf{e}_i \text{ and } B^p\epsilon^0 = \sum_i \alpha_i(\lambda_i)^p\mathbf{e}_i$$

for some set of coefficients $\{\alpha_i\}$.

A convergent solution requires that $|\lambda_i| < 1.0$, $\forall i$, and the overall convergence rate is determined by the *spectral radius* $\mu(B) = \max_i|\lambda_i|$. It can be shown [Dah74, Wat02, Saa03] that the convergence rate for the outer or power iteration for the diffusion equation solver may be estimated, for large values of the iteration index $n$

$$|\lambda_i^n - \lambda_i^{n-1}| \propto \rho^n, \quad \forall i,$$

in terms of the *dominance ratio* $\rho = k_1/k_0$, where $k_0$ and $k_1$ are the largest and second largest $k$-eigenvalues, respectively. Pointwise flux convergence is also governed by a similar relationship. Thus, the power iteration convergence is rapid if the dominance ratio is small, i.e. the separation between the first two $k$-eigenvalues is large.

## 6.8 Coarse-Mesh Diffusion Equation Solver

Many formulations exist for performing accurate multigroup 3-D flux calculations using coarse-mesh structures. One such formulation is discussed here in some detail together with a method for reconstructing pin-power distributions within fuel assemblies.

### 6.8.1 Nodal Expansion Method

The *nodal expansion method* (NEM) [Law86] is a well-known coarse-mesh MGD algorithm implemented in the SIMULATE-3 code [DiG95]. The NEM is also known as a polynomial method because it represents flux distributions in each of the 3-D surfaces via polynomials of varying orders. The formulation begins with integrating the 3-D diffusion equation $\nabla \cdot \mathbf{J} + \Sigma\phi = S$ over two surfaces $A_1$ and $A_2$, placed at positions $x_1$ and $x_2$, respectively, to obtain the transverse average current $J_k(x)$, average flux $\phi_k(x)$, and average source $S_k(x)$ at $x_k$, $k = 1, 2$

$$J_k(x) = \frac{1}{A_k}\int\int dydzJ(x,y,z), \tag{6.90a}$$

$$\phi_k(x) = \frac{1}{A_k}\int\int dydz\phi(x,y,z), \tag{6.90b}$$

$$S_k(x) = \frac{1}{A_k}\int\int dydzS(x,y,z), \tag{6.90c}$$

as illustrated in Figure 6.10. The average transverse current is likewise obtained

$$L_k(x) = \frac{1}{A_k}\int\int dydz\left[\frac{\partial}{\partial y} + \frac{\partial}{\partial z}\right]J(x,y,z), \tag{6.91a}$$

$$= \frac{1}{A_k}\left[\int dz(J_4 - J_3) + \int dy(J_6 - J_5)\right], \tag{6.91b}$$

where $J_\ell, \ell = 3, \ldots, 6$, represent the currents across the $y$- and $z$-surfaces.

The transverse leakage term $L_k(x)$ and source term $S_k(x)$ are lumped together as an effective source $Q_k(x)$ to form a 1-D diffusion equation, which is then discretized

**Figure 6.10** NEM geometry for solution in the $x$-direction.

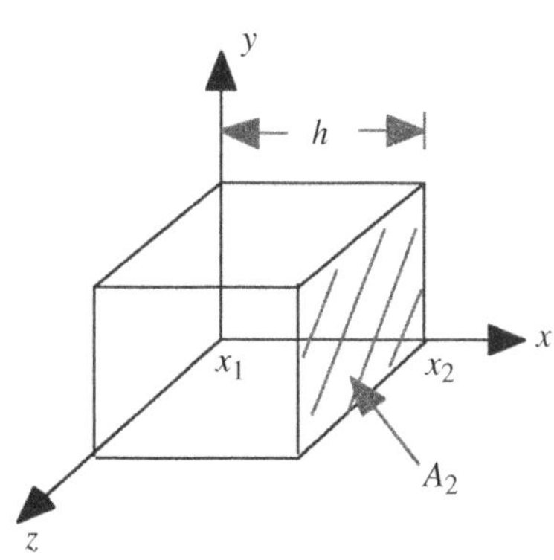

as the volume-centered structure of Eq. (6.69) across cell center $x_0$

$$J_2 - J_1 + \Sigma\phi_0 h = Q_0 h, \tag{6.92}$$

with mesh spacing $h$ and

$$\int_{x_1}^{x_2} \phi_k(x)dx = \phi_0 h, \int_{x_1}^{x_2} Q_k(x)dx = Q_0 h, J_k = J_k^+ - J_k^-, \ k = 1, 2.$$

Introduce a $N$th-order polynomial expansion for $\phi(x)$

$$\phi(x) = \sum_{n=0}^{N} a_n f_n(u), u = \frac{x}{h}, \ u = 0 \text{ at cell center,} \tag{6.93}$$

into Eq. (6.92), and obtain the relationships between the partial currents:

$$a_{11}J_1^- + a_{12}J_2^+ = g_1(\phi_0, J_1^+, J_2^-), \tag{6.94a}$$

$$a_{21}J_1^- + a_{22}J_2^+ = g_2(\phi_0, J_1^+, J_2^-). \tag{6.94b}$$

The set of three equations (6.93) and (6.94), together with the two incoming partial currents $J_1^+$ and $J_2^-$ from neighboring cells, suffice to obtain five unknowns $\{\phi_0, J_1^-, J_2^+, J_1^+, J_2^-\}$. For the 0th order expansion with $N = 2$ in Eq. (6.93), the polynomial is typically chosen as $f_0 = 1, f_1 = u, f_2 = 3u^2 - 0.25$ for 13 unknowns, to be evaluated with seven equations of the type (6.94) and six boundary conditions from adjacent cells. The formulation is invoked in alternate directions to obtain multigroup fluxes in 3-D geometry. In the SIMULATE-3 code, a second-order NEM with $N = 4$ is used, such that the transverse leakage $L_k(x)$ from Eq. (6.91b) forms a quadratic function of $x$ and preserves the continuity of both flux and current

$$\phi(x, y, z) = \sum_{i,j,k=0}^{i+j+k\leq N} c_{ijk} f_i\left(\frac{x}{h_x}\right) f_j\left(\frac{y}{h_y}\right) f_k\left(\frac{z}{h_z}\right), \tag{6.95}$$

with 35 coefficients $c_{ijk}$ obtained through a weighted residual method. The SIMULATE-4 code [Bah05] has incorporated microscopic depletion capability together with a NEM formulation. Another well-known 3-D nodal diffusion equation solver is the ANC code [Liu86] used for PWR core calculations.

### 6.8.2 Pin-Power Reconstruction Algorithm

The coarse-mesh NEM algorithm provides sufficiently accurate flux solutions through polynomial expansions within each 3-D cell but lacks the ability to represent pin-to-pin power distributions for individual fuel assemblies forming heterogeneous fuel-absorber-void arrays. Among several pin-power reconstruction algorithms developed, the approach adopted for the SIMULATE-3 code [DiG95] combines intra-assembly flux calculations $\phi_{form}(x, y)$ from lattice physics

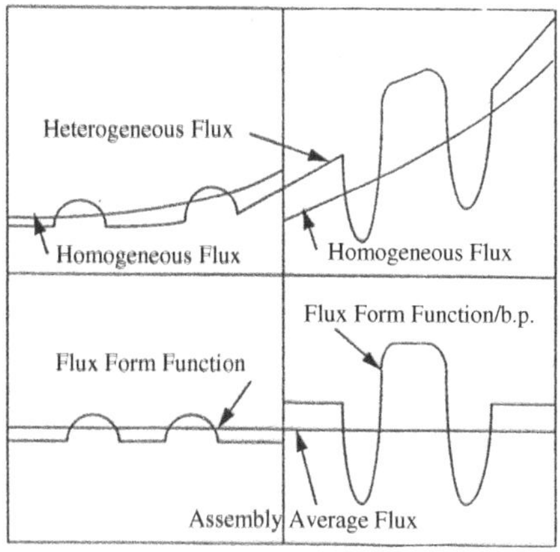


**Figure 6.11** Illustration of the homogenous and heterogeneous fluxes. *Source:* Adapted from DiGiovine et al. [DiG95].

codes, e.g. the CASMO-4 code [Kno95], with global NEM calculations $\phi_{global}(x,y)$ to arrive at an accurate flux distribution for the whole reactor:

$$\phi_{reactor}(x,y) = \phi_{form}(x,y) \cdot \phi_{global}(x,y). \tag{6.96}$$

However, if the global flux $\phi_{global}(x,y)$ is set to be continuous across assembly boundaries based on a homogeneous assembly-average flux, i.e. if $\phi_{global}(x,y) = \phi_{homog}(x,y)$, the actual flux will be discontinuous due to heterogeneities within assemblies, as illustrated in the bottom plot of Figure 6.11. To overcome this problem, determine an *assembly discontinuity factor* (ADF) $\xi$ with the CASMO lattice physics code at the assembly boundary

$$\xi = \frac{\phi_{het}}{\phi_{homog}},$$

so that with the corrected $\phi^*_{homog}$, the continuity of flux at assembly interfaces is satisfied

$$\phi_{reactor}(x,y) = \phi_{form}(x,y) \cdot \phi^*_{homog}(x,y), \tag{6.97}$$

with $\xi^-\phi^{*-}_{homog} = \xi^+\phi^{*+}_{homog}$. For the top plot of Figure 6.11, $\xi^- < 1.0$ and $\xi^+ > 1.0$ would be appropriate. Combination of the coarse-mesh NEM and ADF formulations allow the SIMULATE-3 code [DiG95] to provide accurate global and intra-assembly flux and power distributions for LWR cores without the need to perform pin-resolved fine-mesh calculations for the whole core.

Several diffusion theory codes, including the PARCS code [Dow10] and NESTLE code [Tur95], have been developed in recent years, together with the Denovo 3-D discrete ordinates solver [Eva10] of the SCALE system [Rea18]. The SCALE system, which is under active development, includes the POLARIS lattice physics module and the ORIGEN isotopic depletion module as well as photon transport and shielding routines. Active links between the PARCS and NESTLE codes and the SCALE system have also been developed to facilitate comprehensive steady-state and transient nuclear system simulations for various reactor designs. For 3-D BWR core analysis, the PANACEA code [Gen85] is the latest version used for licensing calculations, although more recent core simulator codes include the AETNA code [Iwa01]. A number of reactor simulation tools have also been developed in the Nuclear Energy Advanced Modeling and Simulation (NEAMS) program [Sta19] through the aegis of the US Department of Energy and may be accessed via the open-source Multiphysics Object-Oriented Simulation Environment (MOOSE) framework [Per20]. Included in the NEAMS system is the Griffin code [Lee21] providing diffusion and transport theory solvers for reactor physics analysis. Other NEAMS software is discussed in Section 13.9.

## 6.9 Krylov Subspace Method as a Diffusion Equation Solver

One of the recent numerical techniques that has been applied to the solution of discretized forms of the neutron diffusion equation is the *Krylov subspace method* [Saa03, Wat02]. We begin with the understanding that a set of vectors $\{\mathbf{v}_1, \mathbf{v}_2, \dots, \mathbf{v}_m\}$ forms a subspace of $n$-dimensional real vector space $\Re^n$ and propose to obtain the solution to the matrix equation $A\mathbf{x} = \mathbf{b}$ with a subspace formed by a polynomial $p(A)$ approximating $A^{-1}$. Thus, for an initial estimate $\mathbf{x}_0$ and residual $\mathbf{r}_0 = \mathbf{b} - A\mathbf{x}_0$, we may form a Krylov subspace $K_m(A, \mathbf{r}_0) = \{\mathbf{r}_0, A\mathbf{r}_0, \dots, A^{m-1}\mathbf{r}_0\}$ or, in general, for a polynomial $q^{m-1}(A)$ of order $m-1$: $A^{-1}\mathbf{b} \simeq \mathbf{x}_m \simeq \mathbf{x}_0 + q^{m-1}(A)\mathbf{r}_0$, which reduces for the case of $\mathbf{x}_0 = 0$ to $A^{-1}\mathbf{b} \simeq q^{m-1}(A)\mathbf{b}$. Through this approach, we can use matrix multiplications in lieu of the computation-intensive inversion of large matrices for the solution of eigenvectors and eigenvalues.

A popular example of the Krylov subspace method is an orthogonal projection method known as *Arnoldi's method*, which is similar to the traditional Gram–Schmidt process for finding an orthogonal basis for a subspace. The technique is particularly useful for the inversion of sparse linear matrices, and other projection techniques have been developed for more complex steady-state or transient solutions. For a matrix $A(n \times n)$, Arnoldi's procedure follows these steps:

(1) Choose an $n$-dimensional normalized vector $\mathbf{v}_1$, i.e. $\|\mathbf{v}_1\|_2 = 1$.

(2) Obtain an orthonormal basis

$$\mathbf{w}_j = A\mathbf{v}_j - \sum_{i=1}^{j} \langle A\mathbf{v}_j, \mathbf{v}_i \rangle \mathbf{v}_i, \tag{6.98}$$

$$\mathbf{v}_{j+1} = \mathbf{w}_j / \|\mathbf{w}_j\|_2, \; j = 1, 2, \dots, m. \tag{6.99}$$

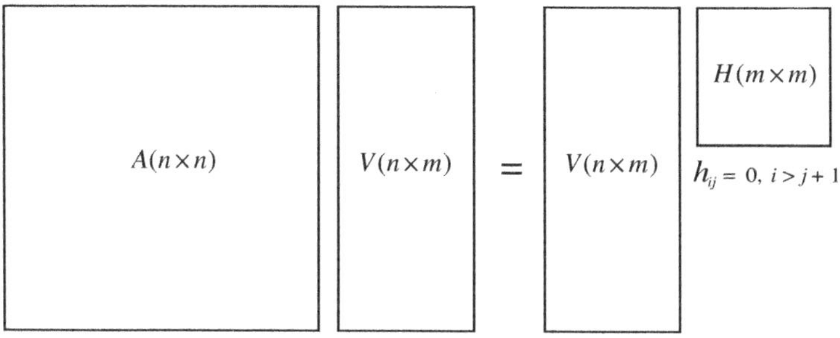


**Figure 6.12** Arnoldi's procedure converting square matrix $A$ to upper Hessenberg matrix $H$, which is generally smaller than $A$ in size.

(3) Defining $h_{ij} = \langle A\mathbf{v}_j, \mathbf{v}_i \rangle$ and $h_{j+1,j} = \|\mathbf{w}_j\|_2$, rewrite Eq. (6.98) as

$$A\mathbf{v}_j = \mathbf{w}_j + \sum_{i=1}^{j} h_{ij}\mathbf{v}_i = \sum_{i=1}^{j+1} \mathbf{v}_i h_{ij}, \tag{6.100}$$

and obtain $V(n \times m) = [\mathbf{v}_1, \mathbf{v}_2, \ldots, \mathbf{v}_m]$ and $H(m \times m)$ such that

$$AV = VH + \mathbf{w}_m \mathbf{e}_m^T, \ \mathbf{e}_m^T(1 \times m) = [0, 0, \ldots, 1]^T. \tag{6.101}$$

Arnoldi's procedure stops when $h_{m+1,m} = \|\mathbf{w}_m\|_2$ vanishes. Figure 6.12 illustrates the matrix structure of Eq. (6.101) involving matrices $V$ and $H$. Remembering that $V(n \times m)$ is a matrix comprising orthogonal basis vectors, we obtain

$$V^T AV = (V^T V)H = H, \text{ with } \langle \mathbf{v}_m, \mathbf{w}_m \rangle = 0. \tag{6.102}$$

The *upper Hessenberg matrix* $H(m \times m)$ is an upper triangular matrix plus non-zero elements in one row below the diagonal such that $h_{ij} = 0, i > j + 1$ for $i = 1, \ldots, m$, and is usually smaller in size than matrix $A$.

Given the eigenvalue equation $A\mathbf{x} = \lambda\mathbf{x}$, construct $\mathbf{x} = V\mathbf{y}$ such that

$$AV\mathbf{y} = \lambda V\mathbf{y} = V(\lambda\mathbf{y}) = V(H\mathbf{y}). \tag{6.103}$$

We then solve the revised eigenvalue equation for a simpler matrix $H$

$$H\mathbf{y} = \lambda\mathbf{y}, \tag{6.104}$$

instead of the original equation $A\mathbf{x} = \lambda\mathbf{x}$. This then finally yields the solution for the eigenvector $\mathbf{x}$ corresponding to the eigenvalue $\lambda$. It should be noted that the eigenvalue solution for matrix $H$ will yield all $m$ modes for $\mathbf{y}$ and hence for $\mathbf{x}$, and associated eigenvalues. In contrast, the usual power iteration algorithm yields only the scalar flux, which is the fundamental mode solution. Finally, note that the solution of Eq. (6.104) involving the simpler upper Hessenberg matrix could be obtained by the traditional $QR$ algorithm [Saa03, Wat02].

**Example 6.2** Illustrate the application of Arnoldi's method to solve linear matrix equations using the eigenvalue equation (6.68) as an example. Compare the results with those of Example 6.1.

For this simple exercise, instead of using the $(3 \times 3)$ matrix equation (6.68) directly to apply Eqs. (6.98) through (6.102), invoke the power-iteration algorithm discussed in Section 6.4 and implemented in the ONED code. In fact, we would need to invert the loss-term matrix $A$ and transform the original matrix equation $A\mathbf{\Phi} = (1/\lambda)\mathbf{Q} = (1/\lambda)F\mathbf{\Phi}$ to $\lambda\mathbf{\Phi} = A^{-1}F\mathbf{\Phi} \equiv A^*\mathbf{\Phi}$. The direct inversion of $A$ is of course to be avoided through Arnoldi's algorithm. Thus, employ orthogonal basis vectors to generate the source vectors $\mathbf{Q}$ sequentially for the code, and use the imbedded Gaussian elimination algorithm to obtain the solution vectors $\mathbf{\Phi}$ without explicitly inverting matrix $A$. This approach demonstrates that we may take similar approaches to obtain solutions to multigroup, multidimensional diffusion theory codes or even neutron transport theory codes that employ a power iteration structure.

We begin with a simple orthogonal basis vector $\mathbf{v}_1 = [1, 0, 0]^T$ to obtain

$$A^*\mathbf{v}_1 = [1.00453, 5.37548 \times 10^{-3}, 2.87646 \times 10^{-5}]^T$$

and

$$\langle A^*\mathbf{v}_1, \mathbf{v}_1 \rangle = h_{11} = 1.004533$$

so that

$$\mathbf{w}_1 = A^*\mathbf{v}_1 - \langle A^*\mathbf{v}_1, \mathbf{v}_1 \rangle \mathbf{v}_1 = [0.0, 5.37548 \times 10^{-3}, 2.87646 \times 10^{-5}]^T.$$

This then generates the second basis vector

$$\mathbf{v}_2 = [0.0, 0.999986, 5.35100 \times 10^{-3}]^T.$$

The entire process of Eqs. (6.98) through (6.102) may be summarized:

$$V = [\mathbf{v}_1, \mathbf{v}_2, \mathbf{v}_3] = \begin{bmatrix} 1.0 & 0.0 & 0.0 \\ 0.0 & 0.999986 & -5.35100 \times 10^{-3} \\ 0.0 & 5.35100 \times 10^{-3} & 0.999986 \end{bmatrix},$$

$$\begin{aligned} A^*V &= [A^*\mathbf{v}_1, A^*\mathbf{v}_2, A^*\mathbf{v}_3] \\ &= \begin{bmatrix} 1.000453 & 5.37556 \times 10^{-3} & -2.64323 \times 10^{-17} \\ 5.37548 \times 10^{-3} & 1.00458 & -4.93963 \times 10^{-15} \\ 2.87646 \times 10^{-5} & 1.07507 \times 10^{-2} & 1.00449 \end{bmatrix} \\ &= VH = [\mathbf{v}_1, \mathbf{v}_2, \mathbf{v}_3] \begin{bmatrix} h_{11} & h_{12} & h_{13} \\ h_{21} & h_{22} & h_{23} \\ 0 & h_{32} & h_{33} \end{bmatrix}, \end{aligned}$$

with

$$H = \begin{bmatrix} 1.000453 & 5.375561 \times 10^{-3} & -2.64323 \times 10^{-17} \\ 5.375561 \times 10^{-3} & 1.004619 & 5.375022 \times 10^{-3} \\ 0.0 & 5.375022 \times 10^{-3} & 1.004475 \end{bmatrix}$$

and $h_{43} = 4.94085 \times 10^{-15} \simeq 0.0$.

Invoking the *eig* function of *MATLAB* for Eq. (6.104) yields the eigenvalues and eigenvectors for the fundamental mode and two higher harmonics of matrix $H$:

$$\lambda = \{\lambda_1, \lambda_2, \lambda_3\} = \{1.0122, 1.0045, 0.9970\}$$

and

$$Y = [\mathbf{y}_1, \mathbf{y}_2, \mathbf{y}_3] = \begin{bmatrix} 0.5000 & 0.7071 & 0.5000 \\ 0.7098 & -0.0038 & -0.7044 \\ 0.4962 & -0.7071 & 0.5038 \end{bmatrix}.$$

Finally, the eigenvector $\mathbf{x}_1$ corresponding to the fundamental eigenfunction $\mathbf{y}_1$ is obtained

$$\mathbf{x}_1 = V\mathbf{y}_1 = [0.5000, 0.7071, 0.5000]^T,$$

together with

$$\mathbf{x}_2 = [0.7071, 0.0, -0.7071]^T \text{ and } \mathbf{x}_3 = [0.5000, -0.7071, 0.5000]^T.$$

The eigenvalue $\lambda_1$ =1.0122 and eigenvector $\mathbf{x}_1$ for the fundamental mode agree with the $k$-eigenvalue = 1.0122 and normalized $\mathbf{\Phi} = [0.0, 0.02513, 0.03550, 0.02513, 0.0]^T$, respectively, obtained via the power-iteration solution of the ONED code.

The main advantage of Arnoldi's method lies in replacing the task to invert a large matrix with a less-involved task to invert a smaller and simpler Hessenberg matrix $H(m \times m)$. Implementing Arnoldi's method may, however, encounter difficulties because the size $m$ of the Hessenberg matrix is unknown and may become large, which will require a large memory to store the orthogonal vectors. In addition, the number of matrix-vector operations also increases linearly as $m$ increases, incurring round-off errors. One approach to avoid this problem with Arnoldi's algorithm involves restarting the process with a new, better estimate after a number of orthogonalization steps through the *implicit restarted Arnoldi method* (IRAM) and related algorithms. As is the case with most iteration methods, the convergence rate of the IRAM and other algorithms depends also on the *condition number* [Dah74, Wat02, Saa03] of the system matrix $A$. Thus, another technique often implemented in subspace algorithms is to *precondition* the matrix with another matrix $M$ such that $MA$ or $AM$ has a better condition number. The supporting solvers may be obtained from the ARPACK [Leh98] or Trilinos package [Lon03].

## References


**[Bah05]** T. Bahadir, S. Lindahl, and S. Palmtag, "SIMULATE-4 Multigroup Nodal Code with Microscopic Depletion Model," *Proc. M&C Conf.* (2005).

**[Bar67]** R.F. Barry, C.C. Emery Jr., and T.D. Knight, "The PANDA Code," WCAP-7048, Westinghouse Electric Corporation (1967).

**[Cad67]** W.R. Cadwell, "PDQ-7 Reference Manual," WAPD-TM-678, Bettis Atomic Power Laboratory (1967).

**[Dah74]** G. Dahlquist and A. Björck, *Numerical Methods*, Prentice-Hall (1974).

**[Der84]** K.L. Derstine, "DIF3D: A Code to Solve One-, Two-, and Three-Dimensional Diffusion Theory Problems," ANL-82-64, Argonne National Laboratory (1984).

**[DiG95]** A.S. DiGiovine, J.D. Rhodes III, and J.A. Umbarger, "SIMULATE-3, Advanced Three-Dimensional Two-Group Reactor Analysis Code, User's Manual," Studsvik/SOA-95/15, Studsvik of America (1995).

**[Dow10]** T. Downar, Y. Xu, and V. Seker, "PARCS V3.0, U. S. NRC Core Neutronics Simulator, User Manual," Department of Nuclear Engineering and Radiological Sciences, University of Michigan (2010).

**[Eva10]** T.M. Evans, A.S. Stafford, R.N. Slaybaugh, and K.T. Clarno, "Denovo: A New Three-Dimensional Parallel Discrete Ordinates Code in SCALE," *Nucl. Technol.* **171**, 171 (2010).

**[Gen85]** General Electric, "Steady-State Nuclear Methods," NEDO-30130-A, General Electric (1985).

**[Iwa01]** T. Iwamoto, M. Tamitani, and B. Moore, "Present Status of GNF New Nodal Simulator," *Trans. Am. Nucl. Soc.* **84**, 55 (2001).

**[Kno95]** D. Knott, B.H. Forssen, and M. Edenius, "CASMO-4, A Fuel Assembly Burnup Program, Methodology," Studsvik/SOA-95/2, Studsvik of America (1995).

**[Law86]** R.D. Lawrence, "Progress in Nodal Methods for the Solution of the Neutron Diffusion and Transport Equations," *Prog. Nucl. Energy* **17**, 271 (1986).

**[Lee74]** J.C. Lee, "Input Instructions for the ONED Code," unpublished lecture notes, University of Michigan (1974, revised 2006).

**[Lee21]** C. Lee, Y.S. Jung, H. Park et al., "Griffin Software Development Plan," ANL/NSE-21/23, Argonne National Laboratory (2021).

**[Leh98]** R.B. Lehoucq, D.C. Sorensen, and C. Yang, "ARPACK Users' Guide: Solution of Large-scale Eigenvalue Problems with Implicitly Restarted

Arnoldi Methods," *Soc. Ind. Appl. Math.* (1998), https://doi.org/10.1137/1.9780898719628.ch1.

**[Lit68]** W.W. Little Jr. and R.W. Hardie, "2DB – A Two-Dimensional Fast Reactor Burnup Code," *Nucl. Sci. Eng.* **32**, 275 (1968).

**[Liu86]** Y.S. Liu, A. Meliksetian, J.A. Rathkopf et al., "ANC: A Westinghouse Advanced Nodal Computer Code," WCAP-10966-A, Westinghouse Electric Company (1986).

**[Lon03]** K.R. Long, R.S. Tuminaro, R.A. Bartlett et al., "An Overview of Trilinos," SAND2003-2927, Sandia National Laboratory (2003).

**[Per20]** C.J. Permann, D.R. Gaston, D. Andrs, R.W. Carlsen, F. Kong, A.D. Lindsay, J.M. Miller, J.W. Peterson, A.E. Slaughter, R.H. Stogner, and R.C. Martineau, "MOOSE: Enabling Massively Parallel Multiphysics Simulation," *SoftwareX* **11**, 100430 (2020).

**[Rea18]** B.T. Rearden and M.A. Jessee, eds., "SCALE Code System," ORNL/TM-2005/39, vers. 6.2.3, Oak Ridge National Laboratory (2018).

**[Saa03]** Y. Saad, *Iterative Methods for Sparse Linear Systems*, 2nd ed., SIAM (2003).

**[Sta01]** W.M. Stacey, *Nuclear Reactor Physics*, Wiley (2001).

**[Sta19]** C. Stanek, "Overview of DOE-NE NEAMS Program," LA-UR-19-22247, Los Alamos National Laboratory (2019).

**[Tur95]** P.J. Turinsky, R.M. Al-Chalabi, P. Engrand, H.N. Sarsour, F.X. Faure, and W. Guo, "NESTLE: A Few-Group Neutron Diffusion Equation Solver Utilizing the Nodal Expansion Method for Eigenvalue, Adjoint, Fixed-Source Steady-State and Transient Problems," *Nucl. Sci. Eng.* **120**, 72 (1995).

**[Wat02]** D.S. Watkins, *Fundamentals of Matrix Computations*, 2nd ed., Wiley (2002).

## Problems

**6.1** Develop a finite-difference solution code in C/C++ for a one-dimensional slab reactor using the face-centered formulation from Section 6.1 for 50 uniform meshes with zero-flux boundary conditions at both reactor boundaries. Compare the discretized flux distribution and eigenvalue with the corresponding analytical solution and four-mesh finite difference solution discussed in Example 6.1.

**6.2** Incorporate the reflecting boundary condition discussed in Section 6.3 with 25 meshes, and compare with the solution obtained in Problem 6.1.

**6.3** Develop a finite-difference solution code for a one-dimensional slab reactor using the body-centered formulation from Section 6.7, and compare with the face-centered solution from Section 6.1 implemented in Problem. 6.1.

**6.4** Develop a finite-difference solution code for a one-dimensional cylindrical reactor using the formulation from Section 6.1, and compare with the Bessel function solution from Table 5.2.

**6.5** Verify the advantages of using the extrapolation factor $\omega$ introduced in Eq. (6.47).

**6.6** Implement Arnoldi's method in the finite-difference code developed in Problem 6.1, and compare with the power-iteration algorithm from Section 6.4, verifying the advantages of the Krylov subspace method.

**6.7** Incorporate Rayleigh's quotient method of eigenvalue determination, discussed in Section 10.5, and compare the enhanced accuracy with that of the source iteration method from Eq. (6.44).

**6.8** Verify that the three-point FD matrix $A$ from Eq. (6.21) can be decomposed into matrices $L$ and $U$ from Eq. (6.24). Show also that the inverse matrices $L^{-1}$ and $U^{-1}$ correctly represent the forward elimination and backward substitution algorithms, respectively.

**6.9** Obtain the expression for the four coefficients $a_{11}, a_{12}, a_{21}, a_{22}$ and functions $g_1, g_2$ for the NEM formulation in Eqs. (6.94) by following these steps: (a) with the specifications

$$\int_{-0.5}^{0.5} du\phi(u) = \phi_0, \phi(-0.5) = \phi_1, \phi(0.5) = \phi_2,$$

obtain the expansion coefficients $a_n, n = 0, 1, 2$, in terms of $\phi_0, \phi_1$, and $\phi_2$; (b) using the relationship for partial current $J^{\pm}(x)$ and defining $J_1 = J(-0.5)$ and $J_2 = J(0.5)$, obtain $a_n, n = 0, 1, 2$, in terms of $\phi_0, J_1^{\pm}$, and $J_2^{\pm}$.

**6.10** Develop a boundary condition formulation for a reflected slab reactor with the reflector represented by albedo $\beta$.

# 7

# Applications of the Two-Group Neutron Diffusion Equation

As the simplest form of the neutron balance statement, the one-group neutron diffusion equation provides many useful results in both steady-state and time-dependent applications. Some analytic solutions to the one-group neutron diffusion equation, together with the concepts of critical buckling and eigenvalue, are discussed in Chapter 5. Chapter 6 presents numerical algorithms for the diffusion equation to obtain the flux and eigenvalue for a multiplying system with arbitrary material distributions. One obvious limitation of the one-group neutron balance statement, however, is that it cannot account for the energy dependence of the neutron flux and hence of the neutron reaction rates that are of interest in many practical reactor physics problems. To remedy this deficiency, we now derive the multigroup neutron diffusion equation as an energy-dependent neutron balance statement and illustrate its applications through the two-group neutron diffusion equation.

Section 7.1 begins with the task of casting the energy-dependent diffusion equation derived in Chapter 4 into a discretized form, i.e. the multigroup diffusion (MGD) equation. In Section 7.2, the MGD equation is simplified somewhat for use in time-independent critical reactor analysis. The equation is further reduced to a two-group representation, which is a particularly useful form of the neutron balance statement for many practical applications, especially for light water reactor (LWR) analysis. Section 7.3 discusses applications of the two-group diffusion equation by obtaining the two-group form of the neutron multiplication factor for a bare reactor. We conclude with qualitative discussions on the numerical solution and use of the two-group neutron diffusion equation for reflected reactors in Section 7.4.

*Nuclear Reactor Physics and Engineering*, Second Edition. John C. Lee.
© 2025 John Wiley & Sons, Inc. Published 2025 by John Wiley & Sons, Inc.
Companion website: www.wiley.com/go/LeeNRPE2

## 7.1 Derivation of Multigroup Neutron Diffusion Equation

Recall the energy-dependent form of the neutron diffusion equation (4.39), where the neutron leakage rate is represented through Fick's law of diffusion and the explicit dependence of various cross sections on space and time is suppressed for notational convenience:

$$\frac{1}{v}\frac{\partial\phi(\mathbf{r},E,t)}{\partial t} = \chi(E)\int_0^\infty dE'\nu\Sigma_f(E')\phi(\mathbf{r},E',t) + Q(\mathbf{r},E,t)$$
$$+\int_0^\infty dE'\Sigma_s(E'\to E)\phi(\mathbf{r},E',t) - \Sigma_t(E)\phi(\mathbf{r},E,t) + \nabla\cdot D(E)\nabla\phi(\mathbf{r},E,t). \tag{7.1}$$

For further convenience in reducing the continuum form of the balance statement into a discrete form, the *lethargy* variable is introduced

$$u = \ln\frac{E_0}{E}, \tag{7.2}$$

where the reference energy $E_0$ is usually chosen at 10.0 MeV. This traditional choice of $E_0$ reflects the upper limit for most of the past cross section libraries, including the Evaluated Nuclear Data File VI, Part B (ENDF/B-VI) [McL96]. Most of the recent cross section libraries, including the ENDF/B-VIII library [Bro18] and JEF-3.3 library [JEF17], have the upper boundary of the energy range extended to at least 20 MeV. As illustrated in Figure 7.1, the lethargy $u$ increases as the neutron energy decreases, which makes the group numbering of discretized energy groups a bit more convenient. That is, we may have the energy groups begin at the upper end energy $E_0$ and continue down the energy or up the lethargy. Since $\phi(E)dE = \phi(u)du$, we readily obtain

$$\phi(u) = \phi(E)\left|\frac{dE}{du}\right| = E\phi(E). \tag{7.3}$$

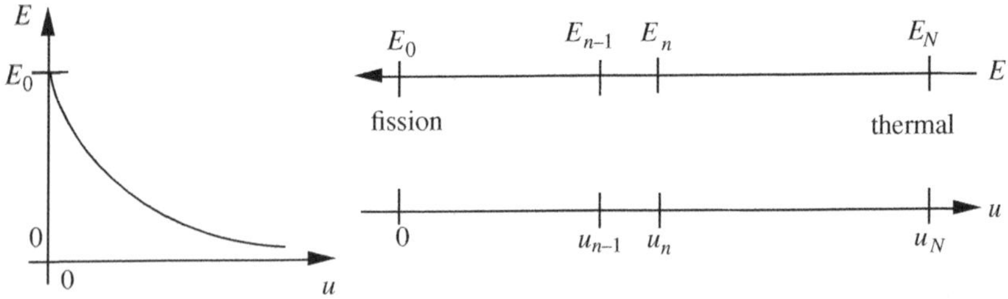


**Figure 7.1** Lethargy variable and energy group structure.

In terms of the lethargy variable, we may now recast the diffusion equation (7.1):

$$\frac{1}{v}\frac{\partial \phi(\mathbf{r},u,t)}{\partial t} = \chi(u)\int_0^\infty du'\, \nu\Sigma_f(u')\phi(\mathbf{r},u',t) + Q(\mathbf{r},u,t) + \int_0^\infty du'\, \Sigma_s(u' \rightarrow u)\phi(\mathbf{r},u',t) - \Sigma_t(u)\phi(\mathbf{r},u,t) + \nabla\cdot D(u)\nabla\phi(\mathbf{r},u,t). \tag{7.4}$$

The discretization of Eq. (7.4) may simply be achieved by integrating it over the $n$th group of lethargy width $\Delta n = u_n - u_{n-1}, n = 1, \ldots, N$, and defining the $n$th group flux:

$$\phi_n(\mathbf{r},t) = \int_{u_{n-1}}^{u_n} du\, \phi(\mathbf{r},u,t) = \int_{\Delta n} du\, \phi(\mathbf{r},u,t). \tag{7.5}$$

Note that $\phi_n(\mathbf{r},t)$ represents an integrated flux, not an average flux, for the $n$th group, and is in units of [neutron $\cdot$ cm$^{-2}$ $\cdot$ s$^{-1}$]. Similarly, we may define the total cross section for the $n$th group:

$$\Sigma_{tn} = \frac{\int_{\Delta n} du\Sigma_t(u)\phi(\mathbf{r},u,t)}{\int_{\Delta n} du\, \phi(\mathbf{r},u,t)} = \frac{\int_{\Delta n} du\Sigma_t(u)\phi(\mathbf{r},u,t)}{\phi_n(\mathbf{r},t)}. \tag{7.6}$$

In this definition of the group-wise cross section, we use the flux as the weighting factor to conserve reaction rates, as we usually have to do whenever an effective cross section is desired. This point was discussed extensively when we considered the two average cross sections, $\overline{\sigma}(v)$ from Eq. (3.26) and $\sigma_{eff}$ from Eq. (3.29). Thus, $\overline{\sigma}(v)$ properly accounts for target nuclei in thermal motion, while $\sigma_{eff}$ accounts for the velocity distributions of both neutrons and nuclei, with the associated reaction rates duly conserved.

Equation (7.6), however, requires the space-, lethargy-, and time-dependent flux $\phi(\mathbf{r},u,t)$, which of course is not known a priori. In fact, the whole idea behind the MGD equation is to obtain a discrete form of the flux described by Eq. (7.4). A group-wise representation of cross sections as in Eq. (7.6) is required before the MGD equation may be solved for $\phi_n(\mathbf{r},t)$. Thus, the flux weighting in Eq. (7.6) is handled through an approximate representation of the flux $\phi(\mathbf{r},u,t)$. Usually, we assume the separability of the flux

$$\phi(\mathbf{r},u,t) = \psi(\mathbf{r},t)\theta(u). \tag{7.7}$$

Assume furthermore a simple form for the lethargy dependence of the flux, e.g. $\theta(u)$ = constant, for the slowing-down region of the flux spectrum. This is one of the important results we obtain in our study of the slowing down of neutrons in Chapter 9. Alternatively, one may assume that $\theta(u)$ is given by the

Maxwell–Boltzmann distribution from Eq. (3.44d) for thermal energy groups, while for high energy regions the fission spectrum may be used for $\theta(u)$. For our immediate purpose, since the MGD equation is often used for the slowing-down region, we assume that the approximation, $\theta(u) = \text{constant}$, is valid whenever necessary. With this representation for the flux, Eq. (7.6) may be simplified to a simple arithmetic average of the lethargy-dependent cross section:

$$\Sigma_{tn} = \frac{\int_{\Delta n} du \Sigma_t(u)}{\Delta n}. \tag{7.8}$$

This simplification is one of the key reasons for employing the lethargy variable in lieu of the more intuitive energy variable, since the averaging process in Eq. (7.6) will then be more involved. This may be understood by noting from Eq. (7.3) that $\theta(u) = \text{constant}$ corresponds to $\theta(E) \propto 1/E$. Similarly, the inverse neutron speed for group $n$ may be defined as

$$\frac{1}{v_n} = \frac{1}{\phi_n} \int_{\Delta n} du \frac{\phi(\mathbf{r}, u, t)}{v}, \tag{7.9}$$

while the discrete fission spectrum is obtained as

$$\chi_n = \int_{\Delta n} du \chi(u). \tag{7.10}$$

The integral of the in-scattering term over the lethargy interval $\Delta n$ may be written

$$\begin{aligned} \int_{\Delta n} du \int_0^{\infty} du' \, \Sigma_s(u' \to u)\phi(\mathbf{r}, u', t) &= \sum_{j=1}^{N} \int_{\Delta n} du \int_{\Delta j} du' \, \Sigma_s(u' \to u)\phi(\mathbf{r}, u', t) \\ &= \sum_{j=1}^{N} \Sigma_{s,j\to n}\phi_j(\mathbf{r}, t), \end{aligned} \tag{7.11}$$

where the integral over the incoming neutron lethargy interval is cast into a summation of the integrals over lethargy groups $j$ and the element $\Sigma_{s,j\to n}$ of the scattering matrix is defined as:

$$\begin{aligned} \Sigma_{s,j\to n} &= \frac{\int_{\Delta n} du \int_{\Delta j} du' \, \Sigma_s(u' \to u)\phi(\mathbf{r}, u', t)}{\int_{\Delta j} du' \, \phi(\mathbf{r}, u', t)} \\ &= \frac{1}{\Delta j} \int_{\Delta n} du \int_{\Delta j} du' \, \Sigma_s(u' \to u). \end{aligned} \tag{7.12}$$

In the last expression in Eq. (7.12), Eq. (7.7) is used again with the assumption $\theta(u) = \text{constant}$.

The leakage term requires the definition of the diffusion coefficient $D_n$ for group $n$

$$\int_{\Delta n} du D(u) \nabla \phi(\mathbf{r}, u, t) = D_n \int_{\Delta n} du \nabla \phi(\mathbf{r}, u, t), \tag{7.13}$$

so that the $n$th group net current may be written as

$$\mathbf{J}_n = -D_n \nabla \phi_n. \tag{7.14}$$

Since no prior knowledge of $\nabla \phi(\mathbf{r}, u, t)$ is available, in general, we have to take somewhat of an ad hoc approach

$$\int_{\Delta n} du D(u) \nabla \phi(\mathbf{r}, u, t) \simeq \nabla \int_{\Delta n} du D(u) \phi(\mathbf{r}, u, t) = \nabla (D_n \phi_n) = D_n \nabla \phi_n, \tag{7.15}$$

with the $n$th group diffusion coefficient defined through conventional flux weighting:

$$D_n = \frac{\int_{\Delta n} du D(u) \phi(\mathbf{r}, u, t)}{\phi_n}. \tag{7.16}$$

Comparing Eqs. (7.13) and (7.16) indicates that the weighted average involving the gradient of flux in Eq. (7.13) is replaced by a simple flux-weighted average in Eq. (7.16), as is done for $\Sigma_{tn}$ in Eq. (7.6), and eventually in Eq. (7.8). This approximate treatment for $D_n$ should be recognized as an indication of the non-uniqueness inherent in determining the diffusion coefficient. Since the neutron transport equation does not require an entity called *transport cross section*, the diffusion coefficient is a concept arising in the diffusion approximation. Hence, even in a continuum representation, the definition for $D$, in general, depends on the problem, and may be operationally defined so that the leakage rate from a region of interest is preserved. Finally, the fission source term may be readily discretized by converting the integral over the lethargy of fissioning neutrons into a summation over the lethargy groups, as was done for the in-scattering term from Eq. (7.11), together with the introduction of a discretized fission spectrum $\{\chi_n, n = 1, \ldots, N\}$.

Collecting terms with $\Sigma_{j \to n} \equiv \Sigma_{s,j \to n}$ and with the discretized external source $Q_n(\mathbf{r}, t)$, the *multigroup diffusion equation* follows:

$$\begin{aligned}\frac{1}{v_n} \frac{\partial \phi_n(\mathbf{r}, t)}{\partial t} &= \chi_n \sum_{j=1}^{N} \nu \Sigma_{fj} \phi_j(\mathbf{r}, t) + Q_n(\mathbf{r}, t) + \sum_{j=1}^{N} \Sigma_{j \to n} \phi_j(\mathbf{r}, t) \\ &\quad - \Sigma_{tn} \phi_n(\mathbf{r}, t) + \nabla \cdot D_n \nabla \phi_n(\mathbf{r}, t).\end{aligned} \tag{7.17}$$

The MGD equation forms the basis of global reactor analysis involving the determination of flux and power distributions in steady-state and transient conditions. The multigroup cross sections appearing in Eq. (7.17) are often called multigroup constants and are generated through the flux-weighted averaging scheme of the type used in Eq. (7.6) or (7.8). In practice, the process of generating the multigroup constants is rather involved and requires the use of lattice physics analysis codes at either the unit cell or unit assembly level. A unit cell typically comprises one fuel rod and the surrounding moderator volume representing an entire (17 × 17) array of fuel rods for a typical PWR fuel assembly. The lattice physics analysis then requires accurate accounting of material heterogeneities present in the unit cell and unit assembly geometries and will be left as a subject of detailed discussion in Chapter 11.

## 7.2 Steady-State Multigroup Diffusion Equation

Using the concept of the system eigenvalue $\lambda$ introduced in our study of the one-group neutron diffusion equation in Chapters 5 and 6, we will simplify the general MGD equation (7.17) for use in the global neutronic analysis of critical reactors. Thus, we drop the time dependence and the external source term, and introduce $\lambda$ into the balance equation. In addition, for notational convenience, we also define the $n$th group *removal cross section*

$$\Sigma_{Rn} = \Sigma_{tn} - \Sigma_{n\to n} = \Sigma_{an} + \sum_{j\neq n}^{N} \Sigma_{n\to j}, \tag{7.18}$$

to represent the probability that neutrons are removed from group $n$ due to either absorption or out-scattering into other groups. With the simplifications and introduction of $\Sigma_{Rn}$, a steady-state form of Eq. (7.17) is obtained:

$$-\nabla\cdot D_n\nabla\phi_n(\mathbf{r}) + \Sigma_{Rn}\phi_n(\mathbf{r}) = \frac{\chi_n}{\lambda}\sum_{j=1}^{N}\nu\Sigma_{fj}\phi_j(\mathbf{r}) + \sum_{j\neq n}^{N}\Sigma_{j\to n}\phi_j(\mathbf{r}), \tag{7.19}$$

The *steady-state MGD equation* (7.19) may readily be written in a matrix from for a $N$-group formulation

$$L\boldsymbol{\Phi} = \frac{1}{\lambda}\chi\mathbf{F}^T\boldsymbol{\Phi} = \frac{1}{\lambda}M\boldsymbol{\Phi}, \tag{7.20}$$

where

$$L = \begin{bmatrix} -\nabla\cdot D_1\nabla + \Sigma_{R1} & -\Sigma_{2\to 1} & -\Sigma_{3\to 1} & \cdot & -\Sigma_{N\to 1} \\ -\Sigma_{1\to 2} & -\nabla\cdot D_2\nabla + \Sigma_{R2} & -\Sigma_{3\to 2} & \cdot & -\Sigma_{N\to 2} \\ -\Sigma_{1\to 3} & -\Sigma_{2\to 3} & -\nabla\cdot D_3\nabla + \Sigma_{R3} & \cdot & -\Sigma_{N\to 3} \\ \cdot & \cdot & \cdot & \cdot & \cdot \\ -\Sigma_{1\to N} & -\Sigma_{2\to N} & -\Sigma_{3\to N} & \cdot & -\nabla\cdot D_N\nabla + \Sigma_{RN}, \end{bmatrix},$$

$$\chi = \begin{bmatrix} \chi_1 \\ \chi_2 \\ \cdot \\ \cdot \\ \chi_N \end{bmatrix}, \mathbf{F} = \begin{bmatrix} \nu\Sigma_{f1} \\ \nu\Sigma_{f2} \\ \cdot \\ \cdot \\ \nu\Sigma_{fN} \end{bmatrix}, \mathbf{\Phi} = \begin{bmatrix} \phi_1 \\ \phi_2 \\ \cdot \\ \cdot \\ \phi_N \end{bmatrix}, \text{ and}$$

$$M = \begin{bmatrix} \chi_1\nu\Sigma_{f1} & \chi_1\nu\Sigma_{f2} & \cdot & \cdot & \chi_1\nu\Sigma_{fN} \\ \chi_2\nu\Sigma_{f1} & \chi_2\nu\Sigma_{f2} & \cdot & \cdot & \chi_2\nu\Sigma_{fN} \\ \cdot & \cdot & \cdot & \cdot & \cdot \\ \cdot & \cdot & \cdot & \cdot & \cdot \\ \chi_N\nu\Sigma_{f1} & \chi_N\nu\Sigma_{f2} & \cdot & \chi_N\nu\Sigma_{f,N-1} & \chi_N\nu\Sigma_{fN} \end{bmatrix}.$$

The steady-state MGD equation (7.20) is widely used to determine global flux and power distributions, as well as the eigenvalue or the *effective multiplication factor* $\lambda = k_{eff} = k$, for steady-state reactor configurations. Equation (7.20) also forms the basis for fuel depletion analysis in multidimensional core geometries. We should also note that Eq. (7.19) for each group is in the form of Eq. (6.1) considered for the numerical solution of the one-group diffusion equation. Thus, as discussed in Section 6.7, the basic finite-difference algorithm developed for the one-group equation may be extended, with coupling via the group-dependent source terms, to the solution of Eq. (7.20). This is discussed further in Section 7.4.

The simplest form of the MGD equation results if we consider a two-group structure, whereby we represent the neutron flux spectrum and slowing down of neutrons in a simple but fully space-dependent manner for an entire reactor core. The division between the groups is usually taken at the *cadmium cutoff energy* $E_{cc} = 0.625$ eV, where the absorption cross section of Cd shows a sharp decrease so that the absorption of neutrons above this cutoff energy may essentially be neglected. With this energy structure adopted, designate the *fast* or *epithermal group* as group 1 and the *thermal group* as group 2. Recognize further that practically all fission neutrons are released into group 1, i.e. $\chi_1 \simeq 1.0$ and $\chi_2 \simeq 0.0$, although both fast and thermal neutrons do induce fission in general. For this simple group structure, *removal cross sections* may be written explicitly as $\Sigma_{R1} = \Sigma_{a1} + \Sigma_{1\to 2}$ and $\Sigma_{R2} = \Sigma_{a2} + \Sigma_{2\to 1}$, and obtain the *two-group neutron diffusion equations*:

$$-\nabla \cdot D_1 \nabla\phi_1(\mathbf{r}) + \Sigma_{R1}\phi_1(\mathbf{r}) - \Sigma_{2\to 1}\phi_2(\mathbf{r}) = \frac{1}{k}[\nu\Sigma_{f1}\phi_1(\mathbf{r}) + \nu\Sigma_{f2}\phi_2(\mathbf{r})], \tag{7.21a}$$

$$-\nabla \cdot D_2 \nabla\phi_2(\mathbf{r}) + \Sigma_{R2}\phi_2(\mathbf{r}) = \Sigma_{1\to 2}\phi_1(\mathbf{r}). \tag{7.21b}$$

The two-group diffusion equations (7.21) may be simplified further by introducing an *effective slowing-down cross section* $\Sigma_r = \Sigma_{1\to 2}\Sigma_{a2}/(\Sigma_{a2} + \Sigma_{2\to 1})$ so that

$$-\nabla \cdot D_1 \nabla\phi_1(\mathbf{r}) + (\Sigma_{a1} + \Sigma_r)\phi_1(\mathbf{r}) = \frac{1}{k}[\nu\Sigma_{f1}\phi_1(\mathbf{r}) + \nu\Sigma_{f2}\phi_2(\mathbf{r})], \tag{7.22a}$$

$$-\nabla \cdot D_2 \nabla\phi_2(\mathbf{r}) + \Sigma_{a2}\phi_2(\mathbf{r}) = \Sigma_r\phi_1(\mathbf{r}). \tag{7.22b}$$

The cross section $\Sigma_r$, introduced in Eqs. (7.22), representing the net slowing down of fast neutrons into the thermal group, is also often referred to as a removal cross section in the PWR literature and should not be confused with $\Sigma_{R1}$. Equations (7.22) may also be obtained without the formal machinery of Eqs. (7.19) through (7.21) by setting up a neutron balance for each of the fast and thermal groups with the loss terms properly representing the leakage, absorption, and net slowing-down rates, together with the fission source. It should also be noted that $\Sigma_r \simeq \Sigma_{1\to2}$ for LWR configurations, since the up-scattering cross section $\Sigma_{2\to1}$ is usually much smaller than the down-scattering cross section $\Sigma_{1\to2}$. With this observation, note also that, for problems where the up-scattering probability is negligible, the terms above the main diagonal in matrix $L$ of Eq. (7.20) may be set to zero.

Simple though they may be, Eqs. (7.22) represent one of the most useful forms of neutron balance statement and are indeed used in a variety of situations where the one-group diffusion equation is found lacking. This is especially the case in many LWR applications, where it becomes necessary to explicitly account for the presence of a significant number of epithermal or fast neutrons, i.e. those with energy above $E_{cc} = 0.625$ eV, in a population of primarily thermal neutrons. This is exemplified by the fact that 20 $\sim$ 25% of fission events in a typical LWR core are caused by neutrons in group 1, although essentially all of the neutrons released from the fission reactions do appear in the fast group. For this reason, the two-group expression for the effective multiplication factor $k_{eff}$ can meaningfully account for all of the key phenomena of the LWR physics, some of which cannot be represented at all by one-group theory. We will illustrate these points clearly in the next section when we consider a two-group diffusion theory model of a bare reactor.

## 7.3 Two-Group Form of Effective Multiplication Factor

As a simple but important application of the two-group diffusion equations (7.22), now we will consider the derivation of the effective multiplication factor $k_{eff} = k$ for a bare reactor. Without specifying the geometrical details of the reactor, we will merely assume that the neutron flux distributions for fast and thermal groups are both described by the Helmholtz or wave equation of the form considered in Eq. (5.62), but with the geometrical bucklings, $B_1^2$ and $B_2^2$, which may in general be unequal:

$$\nabla^2\phi_1(\mathbf{r}) + B_1^2\phi_1(\mathbf{r}) = 0, \tag{7.23a}$$

$$\nabla^2\phi_2(\mathbf{r}) + B_2^2\phi_2(\mathbf{r}) = 0. \tag{7.23b}$$

Equations (7.23) are subjected to the appropriate boundary conditions at the physical boundaries. Since the extrapolation distances vary from group to group, the geometrical buckling would be group-dependent as well. Equations (7.23) would be strictly valid for a bare reactor, although even for a reflected reactor we may obtain an approximate estimate of the geometrical buckling for each group.

With the help of Eqs. (7.23), the neutron balance statement of Eqs. (7.22) may now be converted to a coupled set of algebraic equations:

$$(D_1 B_1^2 + \Sigma_{a1} + \Sigma_r)\phi_1 = \frac{1}{k}(\nu\Sigma_{f1}\phi_1 + \nu\Sigma_{f2}\phi_2), \tag{7.24a}$$

$$(D_2 B_2^2 + \Sigma_{a2})\phi_2 = \Sigma_r \phi_1. \tag{7.24b}$$

Solving for $k$ from the fast-group balance equation (7.24a) yields

$$k = \frac{\nu\Sigma_{f1}\phi_1 + \nu\Sigma_{f2}\phi_2}{(D_1 B_1^2 + \Sigma_{a1} + \Sigma_r)\phi_1}, \tag{7.25}$$

where $k$ is given as a ratio of the neutron production rate to the removal rate for the fast group and hence is a proper representation of the effective multiplication factor. Substituting the flux ratio $\phi_2/\phi_1$ from the thermal-group balance equation (7.24b) into Eq. (7.25) yields the desired expression for the *effective multiplication factor*:

$$k = k_{eff} = \frac{1}{D_1 B_1^2 + \Sigma_{a1} + \Sigma_r}\left(\nu\Sigma_{f1} + \nu\Sigma_{f2}\frac{\Sigma_r}{D_2 B_2^2 + \Sigma_{a2}}\right). \tag{7.26}$$

For an infinitely large reactor with $B_1^2 = B_2^2 = 0$, Eq. (7.26) yields the *infinite multiplication factor*

$$k_\infty = \frac{\nu\Sigma_{f1}}{\Sigma_{a1} + \Sigma_r} + \frac{\nu\Sigma_{f2}}{\Sigma_{a2}}\frac{\Sigma_r}{\Sigma_{a1} + \Sigma_r} \equiv k_1 + k_2, \tag{7.27}$$

where $k_1$ and $k_2$ represent the contributions to $k_\infty$ from fast and thermal fissions, respectively.

By defining the *fast* and *thermal diffusion lengths*, $L_1$ and $L_2$,

$$L_1^2 = \frac{D_1}{\Sigma_{a1} + \Sigma_r}, L_2^2 = \frac{D_2}{\Sigma_{a2}}, \tag{7.28}$$

Eq. (7.26) may be rewritten

$$k = \frac{k_\infty(1 + k_1 L_2^2 B_2^2/k_\infty)}{(1 + L_1^2 B_1^2)(1 + L_2^2 B_2^2)}. \tag{7.29}$$

For a reactor with a small thermal leakage probability, as is the case for LWRs, the term $L_2^2 B_2^2$ is small, typically less than 0.01, and Eq. (7.29) simplifies to

$$k = k_\infty P_{NLT} P_{NLF}, \tag{7.30}$$

with the *fast* and *thermal nonleakage probabilities*

$$P_{NLF} = \frac{1}{1 + L_1^2 B_1^2}, P_{NLT} = \frac{1}{1 + L_2^2 B_2^2}. \tag{7.31}$$

As a final approximation, assume that $B_1^2 = B_2^2 = B^2$, and note that the leakage terms, $L_1^2B_1^2$ and $L_2^2B_2^2$, are both small, with $L_2^2B_2^2 \ll L_1^2B_1^2$, in large reactors, to simplify Eq. (7.30) to

$$k \simeq \frac{k_\infty}{(1 + L_1^2B^2)(1 + L_2^2B^2)} \simeq \frac{k_\infty}{1 + M^2B^2}, \tag{7.32}$$

with the definition for the *neutron migration area*

$$M^2 = L_1^2 + L_2^2 = \frac{D_1}{\Sigma_{a1} + \Sigma_r} + \frac{D_2}{\Sigma_{a2}}. \tag{7.33}$$

Similar to the interpretation of the one-group diffusion length in Eq. (5.14), the migration length may be interpreted as one-sixth of the mean square distance a neutron travels between its birth in the fast group and its capture in the thermal group in an infinite medium. The parameter $L_1^2$ is often called the *Fermi age* $\tau$, which will be discussed further in neutron slowing-down theory in Chapter 9. Finally, Eq. (7.32) may be written as

$$k = k_\infty P_{NL}, \tag{7.34}$$

with the *total nonleakage probability* identified as

$$P_{NL} = P_{NLF}P_{NLT} \simeq \frac{1}{1 + M^2B^2}. \tag{7.35}$$

Remembering that $\Sigma_r$ represents the net probability of neutrons slowing down from the fast to the thermal group, we may, in two-group neutron diffusion theory, define the *resonance escape probability*:

$$p = \frac{\Sigma_r}{\Sigma_{a1} + \Sigma_r}. \tag{7.36}$$

Thus, $p$ may be interpreted as the probability that fast neutrons will escape absorption and slow down into the thermal group. Introducing the *thermal utilization* $f$ as the fraction of thermal neutron absorptions taking place in fuel

$$f = \frac{\Sigma_{a2}^F}{\Sigma_{a2}}, \tag{7.37}$$

and the *number of neutrons released per thermal neutron absorption in fuel*

$$\eta = \frac{\nu\Sigma_{f2}}{\Sigma_{a2}^F}, \tag{7.38}$$

rewrite Eq. (7.27) as

$$k_\infty = k_1 + pf\eta. \tag{7.39}$$

Although Eq. (7.39) can be formally put in the form of the conventional *four-factor formula* from Eq. (5.84) by setting $\varepsilon = 1 + k_1/k_2$, it should be recognized that

Eq. (7.27) is preferable for LWR analysis where $k_1 \simeq k_2/3$. Thus, in this case, $\varepsilon$ cannot be considered a minor correction for fast fissions, as was the intent when it was introduced as the *fast fission factor* in the early days of reactor development. In many practical applications of diffusion theory, such as in the calculation of power coefficients of reactivity, it is often more accurate and meaningful to represent $k_\infty$ through Eq. (7.27) rather than formally using the four-factor formula.

**Example 7.1** Using the two-group constants generated with the Serpent Monte Carlo code [Fri11] for the AP600 core in Example 5.3, obtain collapsed one-group constants, $k_\infty$, and the effective slowing-down cross section $\Sigma_r$.

| Group $n$ | $D$ (cm) | $\Sigma_a$ (cm$^{-1}$) | $\nu\Sigma_f$ (cm$^{-1}$) | $\Sigma_{n\to j}$ (cm$^{-1}$) |
|---|---|---|---|---|
| 1 | 1.3701 | $1.0085 \times 10^{-2}$ | $6.2443 \times 10^{-3}$ | $1.5562 \times 10^{-2}$ |
| 2 | 0.4319 | $9.0798 \times 10^{-2}$ | $1.1695 \times 10^{-1}$ | $1.9456 \times 10^{-3}$ |

Note: $\Sigma_{1\to j} = \Sigma_{1\to 2}$, $\Sigma_{2\to j} = \Sigma_{2\to 1}$.

With the flux ratio $\phi_1/\phi_2 = (\Sigma_{a2} + \Sigma_{2\to 1})/\Sigma_{1\to 2} = 5.960$, obtain the group-average cross sections $\Sigma_a = (\Sigma_{a1}\phi_1 + \Sigma_{a2}\phi_2)/(\phi_1 + \phi_2) = 0.02168\,\text{cm}^{-1}$ and $\nu\Sigma_f = 0.02215\,\text{cm}^{-1}$, together with $\Sigma_r = 0.01524\,\text{cm}^{-1}$. For the one-group diffusion constant, obtain the average $\Sigma_{tr} = 0.3192\,\text{cm}^{-1}$ and $D = 1.0442\,\text{cm}$. This simple example illustrates the steps involved in obtaining flux-weighted cross sections and establishes the one-group constants considered in Examples 5.2 and 5.3. With an estimate of critical buckling $B^2 = 4.056 \times 10^{-4}\,\text{cm}^{-2}$, the flux ratio may be recalculated $\phi_1/\phi_2 = (D_2B^2 + \Sigma_{a2} + \Sigma_{2\to 1})/\Sigma_{1\to 2} = 5.971$, which yields essentially the same one-group constants obtained with $B^2 = 0.0$: $\Sigma_a = 0.02166$ cm$^{-1}$, $\nu\Sigma_f = 0.02212\,\text{cm}^{-1}$, and $D = 1.0446\,\text{cm}$. It is worth noting that $k_\infty = 1.022$ results as a sum of $k_1 = 0.247$ and $k_2 = 0.775$, indicating that thermal fissions generally contribute nearly 80% to $k_\infty$ in LWR cores.

## 7.4 General Two-Group Diffusion Analysis

The application of the two-group neutron diffusion equation in Section 7.3 has been limited to the analysis of a bare reactor, with a simple geometric buckling introduced to represent spatial distributions of the fast and thermal fluxes. The analysis has yielded useful expressions for the effective and infinite multiplication factors, clearly recognizing the key system parameters affecting the multiplication potential of a chain-reacting system.

The application of two-group diffusion theory may be extended to the analysis of reflected reactors of various geometries. A two-group analytical model of a reflected reactor, without an explicit up-scattering representation, entails representing the fast and thermal flux distributions for the core as a linear combination of two distinct eigenfunctions. This is because the fast and thermal fluxes for the core are fully coupled through Eqs. (7.22). For the reflector, Eq. (7.22a) would be in the same form as the one-group diffusion equation for a non-multiplying medium, and the fast flux will be given simply as a function of the fast diffusion length $L_1$ of the reflector material, subject to the proper boundary conditions at the physical boundaries. The thermal group equation (7.22b), even for the reflector, has an inhomogeneous term due to the slowing down of neutrons from the fast group into the thermal group. Thus, the thermal flux distribution in the reflector has to comprise a linear combination of a function of $L_1$ and another function of the thermal diffusion length $L_2$ of the reflector material.

The criticality condition for a reflected reactor evolves from the continuity of the flux and current at the core-reflector interface in each group, requiring the solution of a set of four homogeneous algebraic equations for four unknowns in the flux solutions. Nontrivial solutions to the algebraic equations are possible only if the $(4 \times 4)$ determinant for the coefficients of the algebraic equations vanishes. Thus, the criticality condition for a reflected core requires iteratively searching for the eigenvalue $k$, appearing in the determinant, such that the determinant becomes zero. Such techniques are illustrated in a number of reactor physics textbooks [Lam66,Meg60] for representative geometries.

In practical applications of diffusion theory, however, we turn to numerical solutions of the two-group diffusion equations (7.21) or (7.22) and directly obtain the effective multiplication factor $k$, together with the fast and thermal flux distributions, $\phi_1(\mathbf{r})$ and $\phi_2(\mathbf{r})$. Such a solution typically involves solving for the flux distribution for each group, through an *inner iteration*, and accounting for the coupling between the groups as an integral part of an *outer iteration* for the fast-group source, slowing-down term, and eigenvalue. To gain a better understanding of this approach, we recast Eqs. (7.22) in the general one-group structure of Eq. (6.1)

$$-\nabla \cdot D_n(\mathbf{r})\nabla\phi_n(\mathbf{r}) + \Sigma_n(\mathbf{r})\phi_n(\mathbf{r}) = S_n(\mathbf{r}), \quad n = 1,2, \tag{7.40}$$

by designating

$$\begin{aligned} &\text{Group 1: } \Sigma_1 = \Sigma_{a1} + \Sigma_r = \Sigma_{R1}, \quad S_1 = \left(\nu\Sigma_{f1}\phi_1 + \nu\Sigma_{f2}\phi_2\right)/\lambda, \\ &\text{Group 2: } \Sigma_2 = \Sigma_{a2} = \Sigma_{R2}, \quad S_2 = \Sigma_r\phi_1. \end{aligned} \tag{7.41}$$

With the finite-difference algorithm developed in Sections 6.1 and 6.2, Eq. (7.40) is discretized into matrix form for both groups:

$$A_n\mathbf{\Phi}_n = \mathbf{S}_n, \quad n = 1,2. \tag{7.42}$$

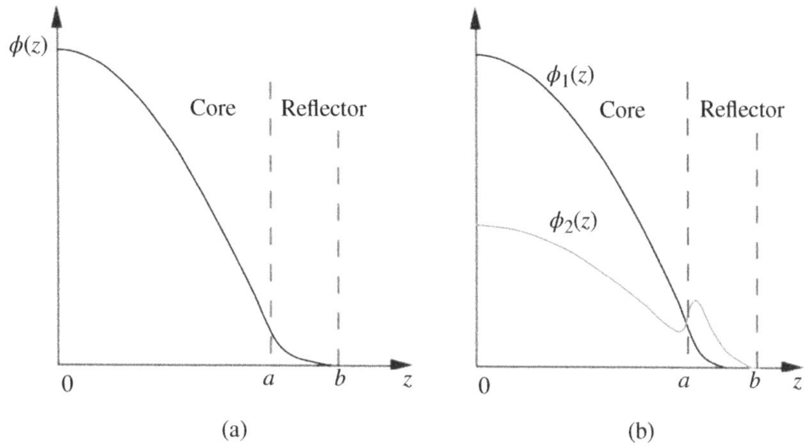


**Figure 7.2** Comparison of one-group and two-group flux distributions for a reflected slab reactor: (a) One-group flux distribution showing a monotonic decrease across the core-reflector interface (b) two-group flux distributions, indicating thermal flux peaking in the reflector.

Starting with an estimate of $\mathbf{S}_1$, we invert the matrix $A_1$ for the fast group first and calculate the flux vector $\mathbf{\Phi}_1$, to determine the source vector $\mathbf{S}_2$ for the thermal group. The inversion of $A_2$ then yields $\mathbf{\Phi}_2$, which is combined with $\mathbf{\Phi}_1$ to yield an improved estimate of $\mathbf{S}_1$ and eigenvalue $\lambda$ through the outer iteration algorithm of Eqs. (6.41) and (6.44). It may also be convenient to represent up-scattering terms explicitly with the matrix $L$ of Eq. (7.20) and proceed with the outer iteration. The rest of the algorithm proceeds in an identical manner as the one-group formulation.

To conclude our discussion on the two-group analysis, we present in Figure 7.2 a comparison of one-group and two-group flux distributions obtained with the ONED code [Lee74] for a reflected slab reactor [Lee16]. In our illustration, we assume that the material composition is homogeneous and uniform in the core, with the extrapolated half thickness $b$ for the reactor and half thickness $a$ for the core. The fast flux distribution $\phi_1(x)$ decreases monotonically from the core midplane and is similar to the one-group flux profile $\phi(x)$. The thermal flux distribution $\phi_2(x)$, however, shows a non-monotonic behavior with a distinct peaking in the reflector region. This is due to fast neutrons leaking out of the core and subsequently reaching the thermal energy in the reflector. In fact, some of the neutrons thermalized in the reflector do actually return to the core, accounting for the positive gradient of $\phi_2(x)$ at the core-reflector interface $x = a$.

Thus, we should now recognize two benefits of surrounding the core with a reflector. The first benefit is readily understood in terms of the *reflector savings*, i.e. the reduction in the critical core size due to the reflection of neutrons leaking out of the core, which is best represented through the albedo $\beta$, defined in Eq. (5.39), for the reflector in one-group diffusion theory. The second benefit of the reflector accrues from the fact that fast neutrons leak out of the core and undergo slowing down in the reflector, with an increased probability of escaping resonance captures in the process, and some of them eventually return to the core as thermal neutrons to induce further fission. It is obvious that this second reflector benefit cannot be represented by one-group theory, demonstrating the need for two-group diffusion theory for realistic analysis of any reactor configurations.

## References

**[Bro18]** D.A. Brown, "ENDF/B-VIII.0: The 8th Major Release of the Nuclear Reaction Data Library with CIELO-project Cross Section, New Standards and Thermal Scattering Data," *Nucl. Data Sheets* **148**, 1 (2018).

**[Fri11]** E. Fridman and J. Leppänen, "On the use of the Serpent Monte Carlo code for few-group cross section generation," *Ann. Nucl. Energy* **38**, 1399 (2011).

**[JEF17]** Nuclear Energy Agency, http://www.oecd-nea.org/dbdata/jeff/jeff33/ (2017).

**[Lam66]** J.R. Lamarsh, *Introduction to Nuclear Reactor Theory*, AddisonWesley (1966).

**[Lee74]** J.C. Lee, "Input Instructions for the ONED Code," unpublished lecture notes, University of Michigan (1974, revised 2006).

**[Lee16]** J.C. Lee, "Nuclear Engineering," *Wiley Encyclopedia of Electrical and Electronics Engineering*, **W5211**, 3rd ed., Wiley (2016).

**[Lit68]** W.W. Little Jr. and R.W. Hardie, "2DB – A Two-Dimensional Fast Reactor Burnup Code," *Nucl. Sci. Eng.* **32**, 275 (1968).

**[McL96]** V. McLane, "ENDF/B-VI Summary Documentation," Cross Section Evaluation Working Group, BNL-NCS-17541, ENDF-201, 4th ed., suppl. I, National Nuclear Data Center, Brookhaven National Laboratory (1996).

**[Meg60]** R.V. Meghreblian and D.K. Holmes, *Reactor Analysis*, McGraw-Hill (1960).

## Problems

**7.1** Derive the expression for the effective slowing-down cross section $\Sigma_r$ introduced in Eqs. (7.22), assuming that the flux ratio $\phi_2/\phi_1$ may be obtained for a core with negligible neutron leakage.

**7.2** In a design analysis for a large PWR core, the two-group diffusion equations are used, where the leakage of thermal neutrons may be neglected and the leakage of fast neutrons is represented in terms of buckling $B^2$. Obtain an expression for the fast-to-thermal flux ratio and the effective multiplication factor $k_{eff}$ for the core.

**7.3** Consider a half space of non-multiplying material surrounded by vacuum. Due to a plane source of fast neutrons located at distance $H$ from the vacuum interface, a steady-state fast neutron flux distribution is established within the medium: $\phi(z) = \phi_0 \exp(-\Sigma_r |z|), -\infty < z < H$. Using the two-group neutron diffusion equation, obtain an expression for the thermal neutron flux distribution $\phi_2(z)$ within the medium.

**7.4** A plane isotropic source of fast neutrons of strength $S$ [neutron $\cdot$ cm$^{-2}$ $\cdot$ s$^{-1}$] is located at the face of the half space of the non-multiplying material. The medium is characterized by two-group constants $D_1, \Sigma_{a1}, \Sigma_r, D_2, \Sigma_{a2}$. Obtain expressions for the fast and thermal neutron flux distributions in the half space.

**7.5** A plane isotropic source of fast neutrons is located at the midplane of a semi-infinite non-multiplying slab of extrapolated thickness $2H$ surrounded by vacuum. The slab material is characterized by two-group constants $D_1, \Sigma_{a1}, \Sigma_r, D_2, \Sigma_{a2}$. (a) Obtain expressions for the fast and thermal neutron flux distributions within the slab. (b) Determine the fraction of the source neutrons that escape the slab as thermal neutrons.

**7.6** Consider a semi-infinite slab of thickness $2H$ consisting of non-multiplying material surrounded by vacuum. Due to a plane source of fast neutrons located at the midplane of the slab, a steady-state fast neutron flux distribution within the slab is established: $\phi_1(z) = S_0 \exp(-\Sigma_r |z|), -H < z < H$, where $\Sigma_r$ is the removal cross section of fast neutrons to the thermal group. Using the two-group diffusion equation, obtain an expression for the thermal flux distribution $\phi_2(z)$ within the slab. The thermal group constants for the medium are given as $D_2$ and $\Sigma_2$, and the neutron extrapolation distance is negligibly small for thermal neutrons.

**7.7** For a large PWR core with a negligible thermal neutron leakage probability, merge the thermal-group equation into the fast group and obtain a one-group diffusion equation (5.58) for the fast-group with $B^2 = (k^* - 1)/M^2$. Identify the parameters $k^*$ and $M^2$.

**7.8** Solve the two-group neutron diffusion equation with the effective removal cross section $\Sigma_r$ for the reflector region for a slab reactor, and determine two-group albedo matrix elements $\alpha_{11}, \alpha_{12}, \alpha_{21}, \alpha_{22}$ so that two-group currents $J_1$ and $J_2$ at the core-reflector interface may be represented in terms of the interface fluxes $\phi_1$ and $\phi_2$:

$$\begin{pmatrix} J_1 \\ J_2 \end{pmatrix} = \begin{pmatrix} \alpha_{11} & \alpha_{12} \\ \alpha_{21} & \alpha_{22} \end{pmatrix} \begin{pmatrix} \phi_1 \\ \phi_2 \end{pmatrix}.$$

Show how the albedo boundary condition may be incorporated into the cell-centered finite-difference formulation of the 2DB code [Lit68].

**7.9** A large PWR core with negligible neutron leakage is just critical with soluble boron concentration $C_b$. If the boron concentration is increased by $\Delta C_b$, show that the resulting reactivity decrease is approximately given by $\Delta k/k \simeq -(\Delta C_b/C_b) f_b k_2$, where $f_b$ is the fraction of thermal neutrons absorbed in boron and $k_2$ is the contribution to $k_\infty$ from the thermal neutron fission. Assume boron is a pure thermal neutron absorber.

**7.10** Exercise a two-group finite-difference diffusion solver, e.g. the ONED code at the University of Michigan, in one-dimensional slab geometry for the axial flux distribution and eigenvalue for a bare cylindrical reactor with a core height of 3.658 m and a 0.3-m reflector each at the top and bottom of the core. Use the two-group constants provided, and represent the radial leakage by radial buckling $B_r^2 = 3.39 \times 10^{-4}\,\text{cm}^{-2}$. Neglect the neutron extrapolation distance.

| Region | Group | $D$ (cm) | $\Sigma_a$ (cm$^{-1}$) | $\nu\Sigma_f$ (cm$^{-1}$) | $\Sigma_r$ (cm$^{-1}$) |
|---|---|---|---|---|---|
| Core | Fast | 1.454 | $9.644 \times 10^{-3}$ | $6.006 \times 10^{-3}$ | $1.631 \times 10^{-2}$ |
| | Thermal | 0.3949 | $9.953 \times 10^{-2}$ | 0.1206 | |
| Reflector | Fast | 1.417 | $4.838 \times 10^{-4}$ | 0.0 | $3.143 \times 10^{-2}$ |
| | Thermal | 0.2780 | $9.818 \times 10^{-3}$ | 0.0 | |

**7.11** A collimated beam of fast neutrons of intensity $I_0$ [neutron $\cdot$ cm$^{-2} \cdot$ s$^{-1}$] is incident normally on a half space of moderating material. The source neutrons become thermalized upon first collisions in the moderator, which has a scattering cross section $\Sigma_r$ for the uncollided fast neutrons and diffusion constants $D$ and $\Sigma_a$ for thermal neutrons. Fast neutron absorption in the moderator may be neglected. Determine the depth into the moderator where the thermal flux is maximized.

**7.12** A sphere of $^{235}U^{19}F_6$ gas of radius $R$ is surrounded by a graphite reflector. Neutrons generated through the fission process are all released as fast neutrons and are neither scattered nor absorbed in the core of $UF_6$ gas. For each fast or thermal neutron leaving the $UF_6$ core, $\beta$ thermal neutrons are returned from the graphite reflector and enter the core. No neutrons entering the reflector can return to the core as fast neutrons. Given thermal group constants $D, \Sigma_a$, and $\nu\Sigma_f$ for the gaseous core, obtain a criticality equation.

# 8

# Nuclear Reactor Kinetics

Analytical and numerical solutions to the one-group neutron diffusion equation covered in Chapters 5 and 6 focus on understanding the spatial distribution of the neutron flux and the concepts of criticality and system eigenvalue. In Chapter 7, with the multigroup neutron diffusion equation, the focus shifts somewhat to the energy dependence of the neutron flux distribution, although the basic interest remains in the spatial flux distribution. We now take up the task of investigating the time-dependent behavior of nuclear reactors in this chapter, before we turn our attention to a more comprehensive study in Chapter 9 of the energy dependence of the neutron population in the reactor core. The approach taken to consider different approximations to the basic neutron balance equation reflects the considerable difficulty encountered in solving the entire energy-dependent diffusion equation (4.39) without separating out the space, energy, and time dependencies of the neutron flux $\phi(\mathbf{r}, E, t)$.

Transient phenomena in nuclear reactor cores can be generally broken up into three different categories according to the time scales associated with them. The first group of transients includes fast transients and accidental conditions, with the time constants typically in the range of milliseconds to seconds. On the opposite end of the time scale are the transients related to fuel depletion, with the time constants in days to months. Between these two groups of transient events lie those associated with fission product poisoning due to $^{135}$Xe and load-follow maneuvers, which have characteristic time constants of hours.

The time constants associated with either fuel burnup or $^{135}$Xe transients are large, in general, compared with those associated with neutron multiplication in the reactor core, and for these transients, the time-independent neutron diffusion equations can be used to represent the reactivity and flux distributions in time. For fast transients, the time-dependent diffusion equation, or some approximation to it, has to be solved explicitly. In this chapter, we begin with the one-group neutron diffusion equation to derive the point kinetics equation in Section 8.1

*Nuclear Reactor Physics and Engineering*, Second Edition. John C. Lee.
© 2025 John Wiley & Sons, Inc. Published 2025 by John Wiley & Sons, Inc.
Companion website: www.wiley.com/go/LeeNRPE2

and indicate how solutions to the point kinetics equation may be obtained for a few simple cases in Section 8.2. The point kinetics equation ignores changes in the spatial flux distribution during a transient and allows us to focus on the time dependence of the transient involved. We consider reactivity-induced transients in Section 8.2 but neglect the effects of temperature feedback. Section 8.3 discusses how recent developments in the state space representation of system dynamics could be applied to the solution of point kinetics equation using the *Simulink* toolbox of *MATLAB*. In Section 8.4, we account for the feedback effects as an example of nonlinear kinetics models useful for reactor power excursion analysis. Section 8.5 presents a brief discussion of various techniques for experimental determination of reactivity in critical and subcritical systems, followed by techniques for stability analysis of reactor systems in Section 8.6. Brief remarks regarding space-dependent kinetics solutions in Section 8.7 close the chapter.

## 8.1 Derivation of Point Kinetics Equation

The simplest form of the reactor kinetics equation is derived here. It can represent the core average or total reactor power evolving as a function of time, without explicit accounting for the time evolution of the spatial distribution of the flux. Even in this simple kinetics formulation, we do need to represent the production of neutrons with certain time delays.

### 8.1.1 Representation of Delayed Neutron Production

Recall that Section 4.3 introduces assumption (7) that all fission neutrons are released instantaneously so that we obtain a simple form of the neutron source term in Eq. (4.39). We now duly recognize that a certain fraction $\beta$ of neutrons in the fission process is typically produced with some time delays through a series of radioactive decays of fission products. One example of such a process involves ${}^{87}_{35}\text{Br}$, which undergoes a $\beta$-decay into a metastable state of ${}^{87}_{36}\text{Kr}^*$, which decays into the ground state of ${}^{86}_{36}\text{Kr}$ with the emission of a neutron. As illustrated in Figure 8.1, the decay of ${}^{87}_{36}\text{Kr}^*$ occurs with half-life $t_{1/2} = 0.1$ ms, which is negligibly short compared with $t_{1/2} = 55.9$ s associated with the $\beta$-decay of ${}^{87}_{35}\text{Br}$ itself. Hence, the effective half-life of the decay chain resulting in the neutron release may be characterized by $t_{1/2} = 55.9$ s. The yield of the fission product ${}^{87}_{35}\text{Br}$ ranges from $2.5 \times 10^{-4}$ or 0.025% for fission of ${}^{235}\text{U}$ induced by thermal neutrons to $8.0 \times 10^{-5}$ for thermal fission of ${}^{239}\text{Pu}$.

The decay chain contributing to the delayed emission of neutrons is summarized as

$$ {}^{87}_{35}\text{Br} \xrightarrow{t_{1/2}=55.9\,\text{s}} {}^{87}_{36}\text{Kr}^* \xrightarrow{t_{1/2}=0.1\,\text{ms}} {}^{86}_{36}\text{Kr} + {}^{1}_{0}n. \tag{8.1} $$

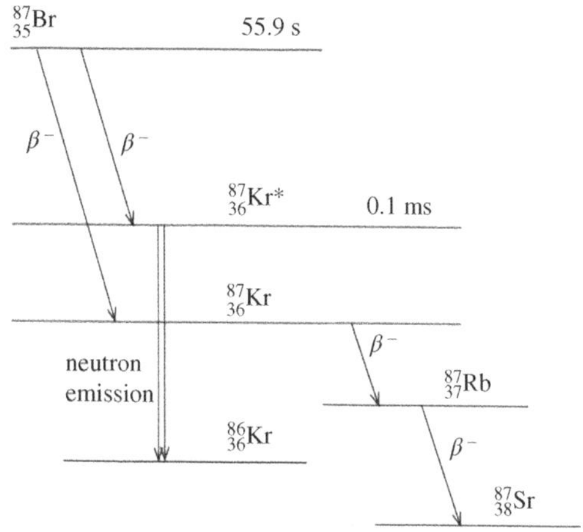


**Figure 8.1** Schematics of decay chains for fission product $^{87}_{35}$Br.

The neutrons produced through this decay chain will contribute to the overall neutron balance essentially with a half-life $t_{1/2} = 55.9$ s associated with the $\beta$-decay of $^{87}_{35}$Br. Thus $^{87}_{35}$Br is known as a precursor for this group of delayed neutrons. Dozens of delayed neutron precursors have been observed [Bra89] to date, and, for convenience, they are usually broken up into six equivalent groups [Kee65]. However, suggestions [Par99] have been made in recent years that different groupings for delayed neutrons should be used. For thermal fission of $^{235}$U, the total delayed neutron yield $\beta = 0.0065$, while for $^{239}$Pu the yield is only 0.0023. Table 2.2 summarizes the total yield and six groups of delayed neutrons for key fissile and fertile nuclides $^{233}$U, $^{235}$U, $^{239}$Pu, $^{241}$Pu, and $^{238}$U from the ENDF/B-VIII file.

### 8.1.2 Point Kinetics Approximation

Considering six groups of delayed neutron precursors with yields $\beta_i$ and decay constants $\lambda_i, i = 1, \dots, 6$, set up a neutron balance in the reactor core in terms of the one-group neutron diffusion equation

$$\frac{1}{v}\frac{\partial \phi(\mathbf{r},t)}{\partial t} = D\nabla^2\phi(\mathbf{r},t) - \Sigma_a\phi(\mathbf{r},t) + (1-\beta)\nu\Sigma_f\phi(\mathbf{r},t) + \sum_{i=1}^{6}\lambda_i C_i(\mathbf{r},t) + S(\mathbf{r},t), \tag{8.2}$$

where the term $S(\mathbf{r},t)$ represents an extraneous source. The delayed neutron precursor concentrations $C_i(\mathbf{r},t)$ in Eq. (8.2) are related to the scalar neutron flux $\phi(\mathbf{r},t)$ through a balance equation

$$\frac{\partial C_i(\mathbf{r},t)}{\partial t} = -\lambda_i C_i(\mathbf{r},t) + \beta_i\nu\Sigma_f\phi(\mathbf{r},t), i = 1, \dots, 6. \tag{8.3}$$

Equation (8.2) indicates that the decay of delayed neutron precursors yields delayed neutrons, not to imply in any way that the precursor concentrations are equated to the neutron number density. Likewise, Eq. (8.3) does not imply that neutrons released from the fission process become precursors but indicates that the precursors in group $i$ are produced at a rate *equivalent* to the fraction $\beta_i$ of the total neutron production rate.

We now introduce the *point kinetics assumption* that the neutron flux, precursor concentrations, and external source exhibit the same spatial dependence

$$\begin{aligned}\phi(\mathbf{r},t) &= vn(t)\psi(\mathbf{r}),\\ C_i(\mathbf{r},t) &= C_i(t)\psi(\mathbf{r}), i = 1,\ldots,6,\\ S(\mathbf{r},t) &= S(t)\psi(\mathbf{r}),\end{aligned} \tag{8.4}$$

with the spatial distribution, or the *shape function*, $\psi(\mathbf{r})$ chosen as the fundamental flux mode. This implies that $\psi(\mathbf{r})$ satisfies

$$\nabla^2\psi(\mathbf{r}) + B^2\psi(\mathbf{r}) = 0, \tag{8.5}$$

where $B^2$ is the critical buckling of the core. Equations (8.4) assume that the spatial distribution of the flux, precursor concentrations, and source will not change during the transient. Hence the point kinetics equation is not strictly applicable to space-time transients involving significant coupling between space and time evolutions and, in general, for short times after the initiation of a transient, where spatial evolutions of the flux would be significant. Substituting the simplifying assumptions from Eqs. (8.4) into the neutron balance equation (8.2) yields

$$\frac{dn(t)}{dt} = -v(\Sigma_a + DB^2)n(t) + (1-\beta)\nu\Sigma_f vn(t) + \sum_{i=1}^{6}\lambda_i C_i(t) + S(t). \tag{8.6}$$

The neutron balance equation (8.6) is now expressed in terms of the neutron number density or the amplitude of the flux. To cast the balance equation in a more tractable form, recall the *effective multiplication factor* $k_{eff}$ defined in Eq. (5.71)

$$k_{eff} = \frac{\nu\Sigma_f}{\Sigma_a + DB^2} \tag{8.7}$$

and introduce the *neutron generation time*

$$\ell = \frac{1}{v(\Sigma_a + DB^2)}. \tag{8.8}$$

Note that the term $1/(\Sigma_a + DB^2)$ represents the *effective mean free path* of neutrons, accounting for both the absorption and leakage, so that the mean free path divided by the neutron speed in Eq. (8.8) yields the average time a neutron spends between its birth and loss due to either absorption or leakage. In terms of $k_{eff} = k$ and $\ell$, we may also define the *reactivity* $K$ and the *normalized generation time* $\Lambda$

$$K = \frac{k-1}{k\beta}, \tag{8.9}$$

$$\Lambda = \frac{\ell}{k\beta}, \tag{8.10}$$

together with the *fractional yield* of delayed neutrons for group $i$

$$a_i = \frac{\beta_i}{\beta}. \tag{8.11}$$

Equations (8.9) and (8.10) reduce Eq. (8.6) to

$$\frac{dn(t)}{dt} = \frac{K(t)-1}{\Lambda}n(t) + \sum_{i=1}^{6}\lambda_i C_i(t) + S(t). \tag{8.12}$$

Substituting Eqs. (8.4) into Eq. (8.3) similarly yields

$$\frac{dC_i(t)}{dt} = -\lambda_i C_i(t) + \frac{a_i}{\Lambda}n(t), \quad i = 1, \ldots, 6. \tag{8.13}$$

Equations (8.12) and (8.13) constitute the *point reactor kinetics equations* (PKEs), expressed here in terms of the reactivity and neutron generation time *normalized* by the delayed neutron fraction $\beta$. We recall from Eq. (5.73) that the reactivity $K$ from Eq. (8.9) expresses the degree of departure from criticality, normalized by $\beta$, and is said to be in units of dollar. The form of the point kinetics equations (8.12) and (8.13) follows that of Ash's book [Ash79] and is one of several forms available in the nuclear engineering literature. One other popular form of the point kinetics equations is written in terms of reactivity $\rho$ from Eq. (5.72) in units of $[\%\Delta k/k]$.

It is worth mentioning here that the point kinetics equations (8.12) and (8.13) could also be derived through an integral formulation [Ash79] using either the Boltzmann transport equation or the multigroup neutron diffusion equation. The point kinetics equations, however, can only represent the magnitude or the core average of an evolving neutron population and cannot explicitly represent the spatial evolution of the neutron flux in time. Note also that, although the point kinetics equations are formally written in terms of the neutron number density $n(t)$ introduced in Eqs. (8.4), we could multiply $n(t)$ by a constant to convert it to core power level $P(t)$ or neutron flux $\phi(t)$. Hence, without loss of generality, we may consider $n(t)$ and $C_i(t)$ in Eqs. (8.12) and (8.13) are expressed in units of either number density, neutron flux, or power level, depending on which interpretation is most convenient or useful.

When $K = 0$ or $k_{eff} = 1.0$, we describe the system as *delayed critical*, to make a distinction from *prompt criticality*, which corresponds to $K = 1.0$ dollar. At prompt criticality, a reactor could conceptually retain a self-sustaining chain reaction even with all the delayed neutron precursors removed from the system. This would correspond to a state where the reactor power or flux can increase exponentially and should be avoided in any normal reactor operation.

Figure 2.6 indicates that the $^{235}$U fission neutron energy spectrum $\chi(E)$ for the delayed neutron emission is substantially shifted away from the average neutron emission of approximately 2.0 MeV for prompt neutrons. Because the fission cross section increases in general as neutrons slow down in a core, the importance of delayed neutrons tends to be higher than indicated by the physical yield fraction $\beta$, requiring the determination of an effective yield fraction $\beta_{eff}$.

### 8.1.3 One-Group Delayed Neutron Model

For many transients of interest, sufficiently accurate results may be obtained via one equivalent group of delayed neutron precursors, with the one-group mean life $\tau$ defined as

$$\tau = \frac{1}{\beta}\sum_{i=1}^{6}\beta_i\tau_i = \sum_{i=1}^{6}a_i\tau_i = \sum_{i=1}^{6}\frac{a_i}{\lambda_i} \equiv \frac{1}{\lambda}. \tag{8.14}$$

Equations (8.12) and (8.13) then simplify to

$$\frac{dn(t)}{dt} = \frac{K(t)-1}{\Lambda}n(t) + \lambda C(t) + S(t), \tag{8.15}$$

$$\frac{dC(t)}{dt} = -\lambda C(t) + \frac{n(t)}{\Lambda}. \tag{8.16}$$

The mean life $\tau = 11.07$ s applies for thermal reactors fueled with $^{235}$U, compared with $\tau = 13.29$ s for $^{239}$Pu in fast reactors. The normalized neutron generation time $\Lambda$ would lie between $10^{-4}$ and $10^{-2}$ s, covering typical fast and thermal reactor configurations, corresponding to the un-normalized generation time $\ell = 3\times10^{-7} \sim 6\times10^{-5}$ s.

The equivalent one-group form of the decay constant defined in Eq. (8.14) is generally applicable for long times after a transient. For short times following a transient, a direct weighting of the decay constants

$$\lambda = \sum_{i=1}^{6}a_i\lambda_i, \tag{8.17}$$

provides more accurate results than Eq. (8.14). Consult Problem 8.1 for further clarifications.

## 8.2 Solution of Point Kinetics Equation without Feedback

To gain some basic understanding of transient reactor behavior, we begin with a few simple solutions to the point kinetics equation following a step reactivity insertion, which can be considered an idealized representation of a rapid reactivity insertion. Assume that the effects of any thermal and hydraulic feedback on the system reactivity may be neglected such that the reactivity will remain constant throughout the transient. Assume also that the transient is initiated from a critical system, i.e. $K(0-) = 0$ and $S(t) = 0$ for all time $t > 0$. The one-group delayed neutron model from Eqs. (8.15) and (8.16) are used whenever possible for notational and algebraic convenience, but the full-blown six-group formulation is also used as necessary.

### 8.2.1 Step Insertion of Reactivity

We are interested in a general solution to Eqs. (8.15) and (8.16) for a rapid insertion of reactivity $K_0$ into an initially critical system, i.e. subject to the initial conditions

$$\frac{dn(0-)}{dt} = \frac{dC(0-)}{dt} = 0,$$

which yield

$$n(0+) = \lambda\Lambda C(0+) \equiv \lambda\Lambda C_0 \equiv n_0. \tag{8.18}$$

Note that there will in general be a discontinuity in the time derivative $dn(t)/dt$ at $t = 0$, although there cannot be any discontinuity in the neutron number density or power level $n(t)$ and hence in $C(t)$ either. The point kinetics equations (8.15) and (8.16) then simplify to a set of ODEs with constant coefficients:

$$\frac{dn(t)}{dt} = \frac{K_0 - 1}{\Lambda} n(t) + \lambda C(t), \tag{8.19a}$$

$$\frac{dC(t)}{dt} = -\lambda C(t) + \frac{n(t)}{\Lambda}. \tag{8.19b}$$

Since the solution of ODEs of the type given in Eqs. (8.19) may be most conveniently obtained through Laplace transforms, basic Laplace transform techniques discussed in Appendix D are employed. Invoking formula (5) of Table D.1, take Laplace transform of Eqs. (8.19) to obtain

$$s\overline{n}(s) - n(0+) = \frac{K_0 - 1}{\Lambda}\overline{n}(s) + \lambda\overline{C}(s), \tag{8.20a}$$

$$s\overline{C}(s) - C(0+) = -\lambda\overline{C}(s) + \frac{\overline{n}(s)}{\Lambda}. \tag{8.20b}$$

Solving for $\overline{C}(s)$ from Eq. (8.20b) and substituting it into Eq. (8.20a), together with the initial conditions of Eq. (8.18), yields the Laplace transform $\overline{n}(s)$ of the power level:

$$\overline{n}(s) = \frac{n_0\left(s + \lambda + \dfrac{1}{\Lambda}\right)}{\left(s - \dfrac{K_0 - 1}{\Lambda}\right)(s + \lambda) - \dfrac{\lambda}{\Lambda}}. \tag{8.21}$$

Remembering that $\lambda \simeq 0.09\,\text{s}^{-1}$ and $\Lambda = 10^{-4} \sim 10^{-2}\,\text{s}$ simplifies the transform of the power in Eq. (8.21) to

$$\overline{n}(s) = \frac{n_0\left(s + \dfrac{1}{\Lambda}\right)}{\left(s - s_1\right)\left(s - s_2\right)}, \tag{8.22}$$

with two roots $s_1$ and $s_2$ of the quadratic equation

$$\left(s - \frac{K_0 - 1}{\Lambda}\right)(s + \lambda) - \frac{\lambda}{\Lambda} = 0. \tag{8.23}$$

For $|K_0 - 1| \gg \lambda\Lambda$ and $(K_0 - 1)^2 \gg 4\lambda\Lambda K_0$, we obtain explicitly

$$s_1 \simeq \frac{K_0 - 1}{\Lambda}, \tag{8.24}$$

$$s_2 \simeq \frac{1}{2}\left[\frac{K_0 - 1}{\Lambda} - \sqrt{\left(\frac{K_0 - 1}{\Lambda}\right)^2 + \frac{4\lambda K_0}{\Lambda}}\right] \simeq \frac{\lambda K_0}{1 - K_0}. \tag{8.25}$$

Furthermore, for many transients of practical interest, $|s_2| \ll |s_1|$ and $|s_2| \ll 1/\Lambda$. Invoking formula (9) from Table D.1 readily inverts Eq. (8.22) into the time domain:

$$n(t) = \frac{n_0}{K_0 - 1}\left[K_0 \exp\left(\frac{K_0 - 1}{\Lambda}t\right) - \exp\left(\frac{\lambda K_0}{1 - K_0}t\right)\right]. \tag{8.26}$$

For sub-prompt critical transients, i.e. for $K_0 < 1.0$ dollar, the first term in the bracket in Eq. (8.26) is significant for short times after the initiation of the reactivity insertion and is known as the *transient solution* of the point kinetics equations. Similarly, the second term plays a dominant role after the initial transient dies out and hence is called the *stable* or *dominant solution*. We refer to the parameter $T = 1/s_2 = (1 - K_0)/\lambda K_0$, representing the *e*-folding time of the dominant solution, as the *stable reactor period* for $K_0 < 1.0$ dollar. On the other hand, for $K_0 > 1.0$ dollar, i.e. super-prompt critical transients, the transient behavior is governed by the first term in the bracket, and the dominant reactor period is $T = \Lambda/(K_0 - 1)$. If we were to use six groups of delayed neutron precursors, we would have altogether seven exponential terms in our time-domain solution for reactor power $n(t)$ instead of the two terms in Eq. (8.26). This point is discussed further in Section 8.2.3.

**Example 8.1** Illustrate the time dependence of the transient and stable terms of Eq. (8.26), with $\lambda = 0.09\,\text{s}^{-1}$ and $\Lambda = 0.01\,\text{s}$, for (a) positive and (b) negative reactivity insertions.

(a) Positive step reactivity insertion with $K_0 = 0.5$ dollar. Using Eqs. (8.24) and (8.25) yield one negative and one positive time constant

$$s_1 = -50\,\text{s}^{-1} \text{ and } s_2 = 0.09\,\text{s}^{-1},$$

and the normalized power level evolves as in Figure 8.2.

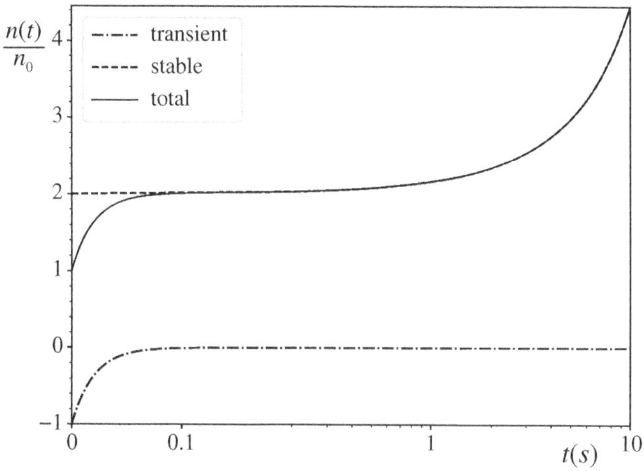


**Figure 8.2** Transient and stable solutions of the point kinetic equations with $K_0 = 0.5$ dollar.

(b) Negative step reactivity insertion with $K_0 = -1.0$ dollar. For this case, the PKE solution yields two negative time constants, and both terms decay exponentially:

$$s_1 = -200\,\text{s}^{-1} \text{ and } s_2 = -0.045\,\text{s}^{-1}.$$

Since the transient term dies out shortly after the reactivity perturbation for both cases, note that, for any reactivity perturbation with $K_0 < 1.0$ dollar, the transient will look, at large times, as though it has started with the amplitude of the stable solution instead of $n(0)$. Thus, the power level will look as if it has suffered a *prompt jump* (PJ) in the amount of

$$\text{PJ} = -\frac{n_0}{K_0 - 1} - n_0 = n_{pj} - n_0 = \frac{n_0 K_0}{1 - K_0}, \tag{8.27}$$

where $n_{pj}$ is the apparent power level attained after the jump. Of course, there will never be any discontinuity in the power level $n(t)$ following a step insertion of reactivity, but in practice, the power level would appear to have gone through a jump equal to that given by Eq. (8.27). For some limiting cases, Eq. (8.26) can further be approximated:

$$\text{(a) } n(t) \simeq n_0 \left[\exp\left(\lambda K_0 t\right) - K_0 \exp\left(\frac{K_0 - 1}{\Lambda} t\right)\right], |K_0| \ll 1.0 \text{ dollar}, \tag{8.28a}$$

$$\text{(b) } n(t) \simeq n_0 \exp\left(\frac{K_0 - 1}{\Lambda} t\right) \simeq n_0 \exp\left(\frac{K_0}{\Lambda} t\right), K_0 > 1.0 \text{ dollar}, \tag{8.28b}$$

$$\text{(c) } n(t) \simeq -\frac{n_0}{K_0} e^{-\lambda t}, K_0 \ll -1.0 \text{ dollar}. \tag{8.28c}$$

Note that Eq. (8.28c) correctly represents a general trend, but the actual time constant, corresponding to a shutdown reactor period, should be properly determined in terms of the decay constant $\lambda_1$ of the longest-lived delayed neutron precursor group, rather than the one-group constant $\lambda$. This is discussed further in Section 8.2.3.

### 8.2.2 Prompt Jump or Zero-Lifetime Approximation

We have observed in Section 8.2.1 that for large times after a reactivity insertion, the neutron flux behaves essentially like the stable solution, i.e. the second term in Eq. (8.26). The stable solution can also be obtained from the point kinetics equations by assuming that the neutron generation time $\Lambda$ is infinitesimally small.

Differentiating Eq. (8.15) with respect to time and dropping the external source term yields

$$\frac{d^2n(t)}{dt^2} = \frac{dK(t)}{dt}\frac{n(t)}{\Lambda} + \frac{K(t)-1}{\Lambda}\frac{dn(t)}{dt} + \lambda\frac{dC(t)}{dt}, \tag{8.29}$$

which, with Eqs. (8.15) and (8.16), is converted to

$$\Lambda\frac{d^2n(t)}{dt^2} = \frac{dK(t)}{dt}n(t) + \frac{dn(t)}{dt}[K(t) - 1 - \lambda\Lambda] + \lambda K(t)n(t). \tag{8.30}$$

With the approximation $\Lambda = 0$, Eq. (8.30) is simplified to

$$\frac{1}{n(t)}\frac{dn(t)}{dt} \simeq \frac{\lambda K(t) + dK(t)/dt}{1 - K(t)}. \tag{8.31}$$

Using the initial conditions for reactivity $K$

$$K(0-) = 0 \text{ and } K(0+) = K_0,$$

integrate Eq. (8.31) to obtain

$$n(t) = n_0\left[\frac{1 - K(0-)}{1 - K(0+)}\right]\exp\left(\frac{\lambda K_0}{1 - K_0}t\right),$$

or

$$n(t) = \frac{n_0}{1 - K_0}\exp\left(\frac{\lambda K_0}{1 - K_0}t\right). \tag{8.32}$$

This is of course the stable solution obtained in Eq. (8.26) and does not provide anything new in itself. Equation (8.31) is, however, valid in general for any arbitrary time variation of the inserted reactivity $K(t)$ and may be integrated to yield a sufficiently accurate solution for large times after the initiation of the transient, for cases where the step insertion solution of Eq. (8.26) is invalid.

Invoking the concept of the prompt jump discussed in Section 8.2.1, assume $dn(t)/dt \simeq 0$ after the "jump" with $K = K_0$, and readily obtain Eqs. (8.32) and

(8.27). To illustrate, approximately set

$$\frac{K_0 - 1}{\Lambda} n(t) = -\lambda C(t) \tag{8.33}$$

and substitute it into Eq. (8.16) to obtain

$$\frac{dC(t)}{dt} = -\lambda C(t) + \frac{\lambda C(t)}{1 - K_0} = \frac{\lambda K_0}{1 - K_0} C(t). \tag{8.34}$$

Integrating Eq. (8.34) and reusing Eq. (8.33), together with the initial condition of Eq. (8.18), we recover Eq. (8.32). Furthermore, to obtain the power after the jump, recast Eq. (8.33), noting that the delayed neutron precursor concentration remains essentially unperturbed during the jump:

$$\frac{K_0 - 1}{\Lambda} n(t) + \lambda C_0 = \frac{K_0 - 1}{\Lambda} n(t) + \frac{n_0}{\Lambda} = 0. \tag{8.35}$$

This then allows us to obtain the power level $n(t)$ after the prompt jump

$$n_{pj} = \frac{n_0}{1 - K_0}. \tag{8.36}$$

This is the amplitude of Eq. (8.32) and of the stable solution of Eq. (8.26) and was introduced in Eq. (8.27).

### 8.2.3 Inhour Equation

Section 8.2.1 discussed solutions to the point kinetics equations with one equivalent group of delayed neutrons through the Laplace transform technique. Now we will obtain a similar solution for six groups of delayed neutron precursors, assuming that the neutron flux level and the precursor concentrations will vary with the same time constant $s = 1/T$. After the initial perturbations have died out, following a step reactivity insertion of $K$, assume an asymptotic time behavior:

$$n(t) = n_0^* e^{st} = n_0^* e^{t/T}, \tag{8.37}$$

$$C_i(t) = C_{i0}^* e^{st} = C_{i0}^* e^{t/T}, \quad i = 1, \ldots, 6. \tag{8.38}$$

Note that the parameters $n_0^*$ and $C_{i0}^*$ do not correspond to the actual power and precursor concentrations, respectively, at $t = 0$, but rather are equal to the amplitude of the stable solution of Eq. (8.26) and the corresponding amplitude for the $i$th-group precursor concentration, respectively. Substituting Eqs. (8.37) and (8.38) into Eqs. (8.12) and (8.13) yields

$$K = s\left[\Lambda + \sum_{i=1}^{6} \frac{a_i}{s + \lambda_i}\right], \tag{8.39}$$

$$K = \frac{\Lambda}{T} + \sum_{i=1}^{6} \frac{a_i}{1 + \lambda_i T}. \tag{8.40}$$

The algebraic steps leading to Eqs. (8.39) and (8.40) are essentially identical to the steps taken for Eqs. (8.20) through (8.23), and in fact, we may readily show that Eq. (8.39), simplified for one group of delayed neutron precursors with $K = K_0$, is identical to Eq. (8.23). Indeed, if we were to start off with the full-blown point kinetics equations (8.12) and (8.13) and follow through the same steps as in Eqs. (8.20) through (8.23), we would obtain Eq. (8.39) as the characteristic equation that may be solved for seven roots, instead of two roots $s_1$ and $s_2$ from Eqs. (8.24) and (8.25). Using the simplified forms of Eqs. (8.37) and (8.38) has allowed us to save a bit in the algebra to obtain the equation characterizing the kinetic behavior of the reactor. Given a measurement of the stable reactor period $T = 1/s$, after the transient solutions have died out, we may conveniently determine the step reactivity $K$ inserted via Eq. (8.39) or (8.40).

1. **Inhour of reactivity**

   For a small reactivity insertion, $|K| \ll 1.0$ dollar, the stable period $T$ will be large and $\lambda_i T \gg 1$, which allows us to approximate Eq. (8.40)

$$K \simeq \frac{\Lambda}{T} + \frac{\tau}{T} \simeq \frac{\tau}{T}, \tag{8.41}$$

   where one-group delayed neutron mean life $\tau$ is given by Eq. (8.14). If we now define one *inhour of reactivity* as the reactivity that yields a stable period of $T = 1$ hr, then the reactivity $K$, corresponding to a measured period $T$ in units of hour, can be represented as

$$K(\text{inhr}) \simeq \frac{\dfrac{\Lambda}{T} + \dfrac{\tau}{T}}{\Lambda + \tau} = \frac{1}{T}. \tag{8.42}$$

   Thus, by measuring the stable reactor period $T$, one can simply determine the reactivity $K$ through Eq. (8.42). Since $K$ from Eq. (8.41) expresses the reactivity in units of dollar, with the period $T$ measured in units of second, we may establish a relationship between the two reactivity units

$$K(\$) = K(\text{inhr}) \times \frac{\text{hr}}{3,600\,\text{s}} \times (\Lambda + \tau), \tag{8.43}$$

   where the neutron generation time $\Lambda$ is usually negligibly small compared with mean life $\tau$ expressed in units of second. Equation (8.39) or (8.40) is known as the *inhour equation* and serves as one of the primary means of reactivity measurement.

2. **Roots of the inhour equation**

   The inhour equation (8.39) provides valuable insights into the dynamic behavior of a reactor, in addition to providing an important means to determine inserted reactivity $K$ from the measurement of stable period $T$. In terms of the seven roots $s = s_n, n = 0, \ldots, 6$, from Eq. (8.40), we may write general solution

$n(t)$ to the point kinetics equation:

$$n(t) = \sum_{n=0}^{6} A_n e^{s_n t}. \tag{8.44}$$

The coefficients $A_n$ can be determined [Ash79] in a manner similar to the techniques used in deriving Eq. (8.26) for the case of one equivalent group of delayed neutrons:

$$A_n = \frac{n(0)[\Lambda + \bar{f}(s_n)]}{\Lambda + \bar{f}(s_n) + s_n \bar{f}'(s_n)}, \quad \bar{f}(s_n) = \sum_{i=1}^{6} \frac{a_i}{s_n + \lambda_i}. \tag{8.45}$$

Thus, Eq. (8.44) may be contrasted to the one-group counterpart, Eq. (8.26)

$$n(t) = \sum_{n=1}^{2} A_n e^{s_n t}, \tag{8.46}$$

with the recognition that

$$s_1 = \frac{K_0 - 1}{\Lambda}, \; s_2 = \frac{\lambda K_0}{1 - K_0}, \; A_1 = \frac{n_0 K_0}{K_0 - 1}, \; A_2 = -\frac{n_0}{K_0 - 1}. \tag{8.47}$$

Indeed, Eq. (8.47) may be obtained from the one-group form of the inhour equation:

$$K = s\left[\Lambda + \frac{1}{s + \lambda}\right]. \tag{8.48}$$

A comparison with the one-group results indicates

(a) The first root $s_0 \geq 0$ for $K \geq 0$ and $s_0 < 0$ for $K < 0$.
(b) All remaining roots are negative, i.e. $s_n < 0, n = 1, \ldots, 6$.

3. **Characteristics of the inhour equation**
   The general behavior of the seven roots of the inhour equation (8.39) is illustrated in Figure 8.3, and we note a few key features:
   (a) For $K < 1$ dollar, six roots, $s_0, \ldots, s_5$, may be obtained from Figure 8.3, and the largest negative root $s_6$ can be calculated from Eq. (8.39) in the limit of a large magnitude of $s$. This yields $K = s_6\Lambda + 1$, or $s_6 = (K - 1)/\Lambda$, which is the transient time constant obtained for the equivalent one group of delayed neutron precursors in Eq. (8.24).
   (b) For $K > 1$ dollar, i.e. for super-prompt critical transients, $s_0 = (K - 1)/\Lambda$ is the dominant period $s_1$ of Eq. (8.24), and the remaining six negative roots, $s_1, \ldots, s_6$, can be obtained from the inhour plot of Figure 8.3.
   (c) The reactor period $T$ obtained earlier in Eq. (8.41) for small reactivity, $|K| \ll 1.0$ dollar, can now be identified as the stable period $T = \tau/K = 1/(\lambda K) \simeq 1/s_2$ of Eq. (8.25) or (8.47).
   (d) For $K \ll -1.0$ dollar, i.e. when the inserted reactivity is large negative, as is the case when the entire control rod banks are inserted to shut down or

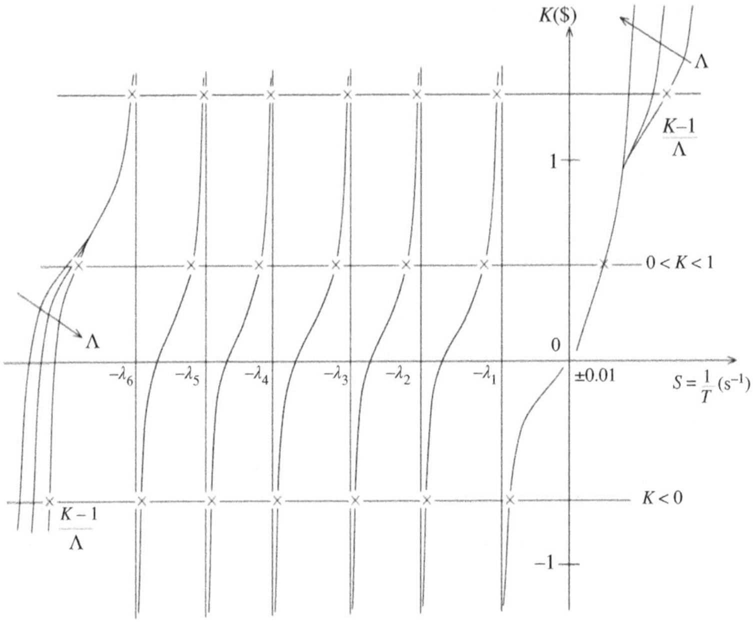


**Figure 8.3** Reactivity versus roots of the inhour equation

*scram* the reactor, we would also have $|s_0| < \lambda_1 < \lambda_2 < \cdots < \lambda_6$ and

$$s_0 \simeq \frac{\lambda_1 K}{a_1 - K} \simeq -\lambda_1. \tag{8.49}$$

Equation (8.49) implies that a reactor cannot be shut down any faster than with a period equal to the mean life of the longest-delayed precursor group or $T \simeq 75.0\,\text{s}$. This should be contrasted with the corresponding result obtained from the approximate solution, Eq. (8.28c), with one equivalent group of delayed neutrons, where the shutdown period is given as $T = \tau = 11.07\,\text{s}$ for thermal fission of $^{235}$U. This indicates a key limitation of the approximation inherent in the one equivalent group formulation studied extensively in Sections 8.2.1 and 8.2.2.

### 8.2.4 Linearized Kinetics Equation and Transfer Function

We have obtained and studied the inhour equations (8.39) and (8.40), which may be used to determine the step reactivity inserted. Consider now another form of the characteristic equation to derive a general relationship for power-level changes due to a reactivity insertion of arbitrary time dependence; the reactivity may not be a step insertion but would be of a sufficiently small magnitude.

Introduce a small reactivity perturbation $\Delta K(t)$ into an equilibrium state, resulting in a perturbed state for $n(t)$ and $C(t)$:

$$n(t) = n_0 + \Delta n(t), \tag{8.50a}$$

$$C(t) = C_0 + \Delta C(t), \tag{8.50b}$$

$$K(t) = K_0 + \Delta K(t), \tag{8.50c}$$

$$S(t) = S_0. \tag{8.50d}$$

For generality, we allow the case where the reactor may initially be subcritical, $K_0 < 0$, but is at steady state with the help of an external neutron source, $S_0 > 0$. Substituting Eqs. (8.50) into the point kinetics equations (8.15) and (8.16) and neglecting a higher-order term involving the product $\Delta n(t)\Delta K(t)$, yields

$$\frac{d\Delta n(t)}{dt} = \frac{K_0 - 1}{\Lambda}\Delta n(t) + \frac{n_0}{\Lambda}\Delta K(t) + \lambda\Delta C(t), \tag{8.51a}$$

$$\frac{d\Delta C(t)}{dt} = -\lambda\Delta C(t) + \frac{\Delta n(t)}{\Lambda}. \tag{8.51b}$$

The linearized form of the point kinetics equations (8.51) apply strictly only for small reactivity perturbations from an equilibrium state but for general temporal variations of the reactivity perturbation $\Delta K(t)$.

Taking Laplace transform of Eqs. (8.51) and recognizing that $\Delta n(0) = \Delta C(0) = 0$ allows us to combine the transformed equations and obtain

$$\overline{\Delta n}(s) = \frac{n_0\overline{\Delta K}(s)}{s\left(\Lambda + \dfrac{1}{s+\lambda}\right) - K_0}. \tag{8.52}$$

Starting from an initially critical system, i.e. $K_0 = 0$, simplifies Eq. (8.52) to

$$\frac{\overline{\Delta n}(s)}{n_0\overline{\Delta K}(s)} = \frac{1}{s\left(\Lambda + \dfrac{1}{s+\lambda}\right)} \equiv \overline{G}(s). \tag{8.53}$$

The function $\overline{G}(s)$ defined by Eq. (8.53) represents the relationship

$$\overline{G}(s) = \frac{\text{normalized change in power}}{\text{unit change in reactivity}} \text{ in Laplace domain,} \tag{8.54}$$

and is known as the *zero-power reactor transfer function* or *open-loop transfer function*. This is in contrast to the closed-loop transfer function that includes thermal-hydraulic feedback effects, which is discussed as part of the system stability analysis later in the chapter. The transfer function $\overline{G}(s)$ can also be shown to be equal to the power transform normalized by the product of neutron source $\overline{S}(s)$ and generation time $\Lambda$, or equal to the power transform due to a neutron

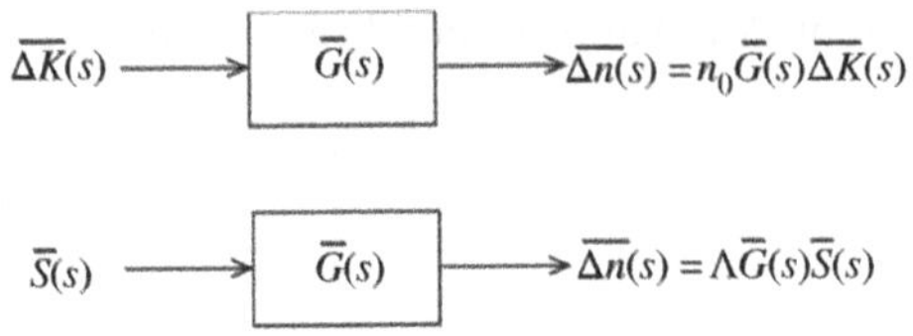


**Figure 8.4** Transfer function representing the output-to-input ratio.

source of strength $\delta(t)/\Lambda$. A general interpretation follows:

$$\overline{G}(s) = \frac{\text{output}}{\text{input}} \text{ in Laplace domain,} \tag{8.55}$$

both for reactivity and source perturbations and hence it is called a *transfer function*. It is called a *zero-power transfer function* because it does not include the effects of temperature feedback on reactivity, which are present invariably in all operating reactors with a sufficient or sensible power output. Figure 8.4 illustrates the output-to-input ratio embodied in the reactor transfer function, representing the dynamic reactor characteristics.

An extension of Eq. (8.53) to the case of six delayed neutron groups shows that the inhour equation (8.39) can be written as

$$K\overline{G}(s) = 1, \tag{8.56}$$

with the *zero-power transfer function* now given as

$$\overline{G}(s) = \frac{1}{s\left(\Lambda + \sum_{i=1}^{6} \frac{a_i}{s + \lambda_i}\right)}. \tag{8.57}$$

For the zero-power case, i.e. when the power level is so low that we may ignore sensible heat generated in the fission process, the roots of Eq. (8.56), with the transfer function of Eq. (8.57), are equal to the roots of the inhour equation (8.39). The reactor transfer function can in general be measured to yield information on the dynamic characteristics of a reactor. We should recognize, however, that the time constant $s$ in Eq. (8.57) is a complex variable, and the transfer function has to be characterized in terms of both its amplitude and phase as a function of the frequency, often plotted in the form of a Bode plot [Ash79].

Since $\overline{G}(s)$ represents the dynamic characteristics of the reactor system, the long-time behavior of the system may be represented equivalently by evaluating $G(t)$ for large values of $t$, with Eqs. (8.53) and (8.57), and invoking the final value theorem of Laplace transform:

$$\lim_{t\to\infty} G(t) = \lim_{s\to 0+} s\overline{G}(s) = \frac{1}{\Lambda + \sum_{i=1}^{6} \frac{a_i}{\lambda_i}} = \frac{1}{\Lambda + \frac{1}{\lambda}}. \tag{8.58}$$

Equation (8.58) now provides a simple justification for the equivalent one-group form of the decay constant and mean life introduced in Eq. (8.14). A similar use of Eqs. (8.53) and (8.57) yields Eq. (8.17), applicable for short times following a transient.

### 8.2.5 Infinite Delayed Approximation

The study of the point kinetics equations without feedback so far has provided considerable insights into the characteristic behavior of chain-reacting systems. We have studied the equations in several different ways and obtained simplified solutions. We conclude this section with one more approximate method for solving the kinetics equations.

For reactor transients that are fast, compared with the mean lives of delayed neutron precursors, we may introduce an approximation that the delayed neutron precursor concentrations do not change in time during the transient. This is equivalent to assuming that the delayed neutrons are produced with infinite mean lives, and the simple approach is known as the *infinite delayed approximation*. Assuming further that the source term $S(t)$ is negligibly small approximates Eq. (8.15) as

$$\frac{dn(t)}{dt} = \frac{K(t)-1}{\Lambda}n(t) + \lambda C_0,$$

or

$$\frac{dn(t)}{dt} = \frac{K(t)-1}{\Lambda}n(t) + \frac{n_0}{\Lambda}, \tag{8.59}$$

for an initially steady-state configuration, satisfying Eq. (8.18). Equation (8.59) is a simple first-order ordinary differential equation for $n(t)$ and can be integrated readily over time for any arbitrary behavior of $K(t)$, resulting in a relatively rapid transient.

## 8.3 State Space Representation of Point Kinetics Equation

We now pause and introduce the *state space representation*, which is a general formulation of system equations developed by the systems and control theory community over the past three decades [Mor01,Doy89]. We begin with a brief introduction to the key ideas behind the state space representation and then illustrate how the linearized point kinetics equations (8.51) may be cast in the state space framework so that we can make efficient use of the *Simulink* toolbox of *MATLAB*. We may employ either the time domain or transfer function representation with *Simulink* to obtain system response as a function of time and eventually perform system stability and control studies. Our discussion in this section is, however, limited to time-domain solutions of the linearized point kinetics equation.

The *state space model* for a dynamical system is represented by a pair of equations connecting *system state vector* $x$ and *system output* $z$ to *control input* $u$:

$$\frac{dx}{dt} = Ax + Bu, \tag{8.60a}$$

$$z = Cx + Du. \tag{8.60b}$$

The system dynamics is expressed in terms of

$$A = \text{internal system dynamics model,} \tag{8.61a}$$

$$B = \text{model representing the effect of control on state vector } x, \tag{8.61b}$$

$$C = \text{model connecting system } x \text{ to output or measurement } z, \text{ and} \tag{8.61c}$$

$$D = \text{model representing the effect of control on measurement } z. \tag{8.61d}$$

All of the variables $x, z,$ and $u$ are time-dependent vectors in general, but in recent systems and control theory literature neither the vector notation nor the time dependence is explicitly carried around. Likewise, all of the system and measurement models, $A, B, C,$ and $D$, are matrices in general, and assumed so without explicit matrix notation. The output variable $z$ also represents a set of *system measurements*. The matrices $A, B, C,$ and $D$ are, however, time-independent, making Eqs. (8.60) linear, and thereby amenable to Laplace transform:

$$s\overline{x}(s) = A\overline{x}(s) + B\overline{u}(s), \tag{8.62a}$$

$$\overline{z}(s) = C\overline{x}(s) + D\overline{u}(s). \tag{8.62b}$$

As in the solution of the linearized point kinetics equations (8.52) and (8.53), we have assumed that the state $x$ represents a perturbation from a steady state or a target state, allowing us to set $x(0) = 0$. In systems and control theory literature, a simplified notational convention is again adopted so that the over-bar indicating the transform and the argument $s$ indicating the transform variable are not explicitly shown:

$$sx = Ax + Bu, \tag{8.63a}$$

$$z = Cx + Du. \tag{8.63b}$$

Equation (8.63a) yields an expression for the transform of the system variable, in terms of identity matrix $I$

$$(sI - A)x = Bu \quad \text{or} \quad x = (sI - A)^{-1}Bu, \tag{8.64}$$

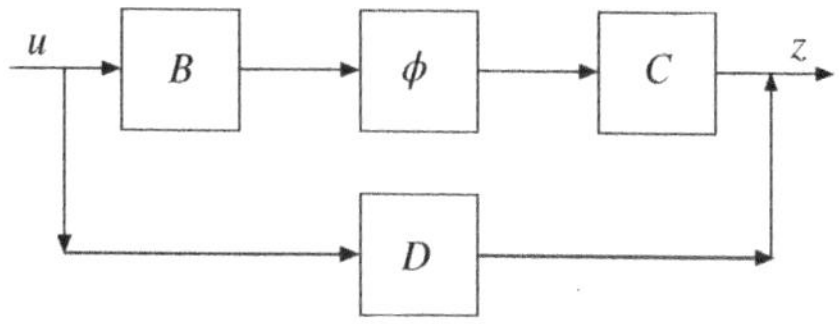


**Figure 8.5** Open-loop transfer function $G = C\phi B + D$ connecting control input $u$ to output $z$ together with controller $D$.

which yields, when combined with Eq. (8.63b), the overall relationship between the output and control input

$$z = [C(sI - A)^{-1}B + D]u = (C\phi B + D)u \equiv Gu, \tag{8.65}$$

with the transform $\phi = (sI - A)^{-1}$. The function $G$ representing the output-to-input ratio is obviously equivalent to the reactor transfer function $\overline{G}(s)$ of Eqs. (8.53) and (8.57) and is simply written as

$$G = \frac{z}{u}, \tag{8.66}$$

with the understanding that all of the functions and the vector-matrix algebra are represented in the Laplace domain. The overall relationship between the control input $u$ and output $z$ is illustrated in Figure 8.5.

Another notation that has taken a root in systems and control theory follows from the matrix form of the system equations (8.60):

$$\begin{bmatrix} \dot{x} \\ z \end{bmatrix} = \begin{bmatrix} A & B \\ C & D \end{bmatrix} \begin{bmatrix} x \\ u \end{bmatrix}. \tag{8.67}$$

With the observation that Laplace transform of Eq. (8.67) yields Eq. (8.65), Eq. (8.67) is simply written as

$$z = \left[ \begin{array}{c|c} A & B \\ \hline C & D \end{array} \right] u. \tag{8.68}$$

The shorthand notation in Eq. (8.68) is known as a *linear fractional transformation* (LFT). The LFT takes a form and structure similar to a block matrix but does not possess normal matrix properties at all. It should be consistently interpreted as a shorthand notation for Eqs. (8.65) and (8.67), and properly treated as such [Mor01,Doy89].

Now cast the linearized point kinetics equations (8.51) into the structure of Eq. (8.67) with

$$x = \begin{bmatrix} x_1 \\ x_2 \end{bmatrix} = \begin{bmatrix} \Delta n \\ \Delta C \end{bmatrix}, A = \begin{bmatrix} \dfrac{K_0 - 1}{\Lambda} & \lambda \\ \dfrac{1}{\Lambda} & -\lambda \end{bmatrix}, B = \begin{bmatrix} \dfrac{1}{\Lambda} \\ 0 \end{bmatrix}, u = n_0 \Delta K, \tag{8.69a}$$

$$C = [1 \quad 0], D = 0. \tag{8.69b}$$

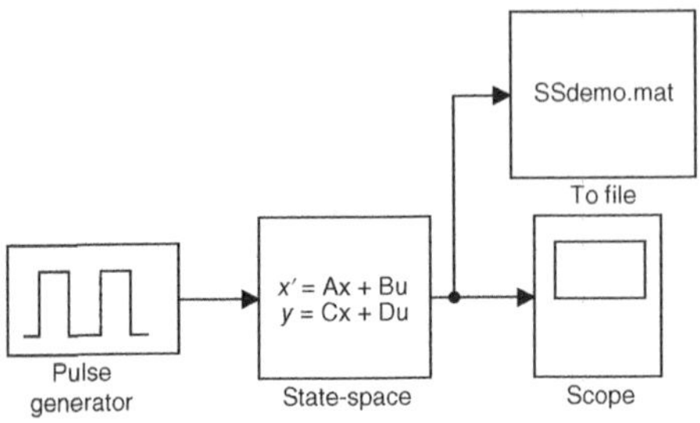


**Figure 8.6** Simulink setup for the pulse source solution of Example 8.2.

Equation (8.69b) renders the output $z$ to equal $\Delta n$ so that Eq. (8.65) yields the transfer function obtained in Eq. (8.53). We may input Eqs. (8.69) into the state space module of *Simulink* to efficiently obtain a time-solution $\Delta n(t)$ due to a step insertion of reactivity $\Delta K(t)$. We may alternatively input the transfer function from Eq. (8.53) into *Simulink* and obtain the time-domain solution $\Delta n(t)$ in much the same way. The one-group form of the state space representation, Eqs. (8.69), may also be readily extended to a six-group delayed neutron structure and likewise represent the transfer function from Eq. (8.57). The state space representation of Eq. (8.67) is thus a general formulation for linear system dynamics and often called the *canonical representation* in systems and control theory literature.

**Example 8.2** Illustrate the application of the *Simulink* toolbox by considering a pulse insertion of neutron source into a critical reactor to obtain an asymptotic power level $n(\infty)/n(0) = 1.1$ with a pulse of neutron source activated for a period of $\tau = 10\,\text{s}$.

With the point kinetics equation model from Eqs. (8.69), a *Simulink mdl* is set up as illustrated in Figure 8.6, comprising a state space model, a pulse generator to represent the source, a scope to display the power trace, and a *mat* file to store the output data. The output file copied from the scope in Figure 8.7 indicates that the power pulse has to overshoot the desired asymptotic power level $n(\infty)$ significantly during the time $\tau = 10\,\text{s}$, before the transient power level converges to $n(\infty) = 1.1 \cdot n(0)$.

## 8.4 Point Kinetics Equation with Feedback

Solutions to the point kinetics equations in Section 8.2 have been limited to low-power operations where thermal and hydraulic feedback in reactivity may be ignored. With the feedback effects included, reactivity $K(t)$ will in general be a function of neutron number density or power $n(t)$. This then means that the point kinetics equations will form a set of nonlinear differential equations for $n(t)$. To illustrate the nonlinear kinetics formulations, consider two such models,

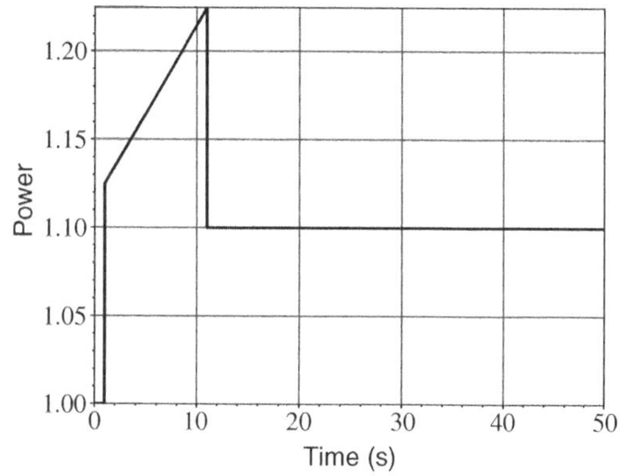


**Figure 8.7** Simulink pulse source output of Example 8.2.

the Ergen-Weinberg model and Nordheim–Fuchs model, both of which may be used to represent the time response of a reactor with a negative temperature coefficient of reactivity subject to a step insertion of positive reactivity. The nonlinear power excursion models were originally developed for analysis of power excursions in bare critical experiments but may also be applied to first-order calculations of reactivity-induced transients or accidents in nuclear power plants. Specific examples could include analysis of the Chernobyl accident of 1986 and a postulated ejection of control rods in pressurized water reactors (PWRs) due to a sudden rupture of the control rod housing.

### 8.4.1 The Ergen–Weinberg Model

In the Ergen–Weinberg model [Wei58], we represent the core as a single homogeneous volume and assume: (a) the reactivity insertion results in a sufficiently prompt super-critical transient so that we may safely neglect contributions from delayed neutrons, (b) the temperature coefficient of reactivity $\alpha$ is independent of the core temperature, and (c) the heat generated in the power transient will be dissipated at a constant rate equal to the initial power level $n_0$ of the system. With the delayed neutron contributions neglected, the point kinetics equation is written simply as

$$\frac{dn(t)}{dt} = \frac{K(t)}{\Lambda} n(t), \tag{8.70}$$

where the temperature feedback is explicitly included in reactivity $K(t)$:

$$K(t) = K_0 - \alpha T(t), \quad \alpha > 0. \tag{8.71}$$

The inserted reactivity is assumed large, i.e. $K_0 \gg 1$ dollar, and the core temperature $T$ is defined as the change from the equilibrium steady-state value. The temperature change $T$ may be calculated from an energy conservation equation, written in terms of the heat capacity $C_p$ of the system:

$$C_p \frac{dT(t)}{dt} = n(t) - n_0. \tag{8.72}$$

In the energy conservation equation, the heat capacity $C_p$ will be in units of $[\text{kJ} \cdot \text{K}^{-1}]$ if the system power is expressed in units of [kW].

Solution of the energy conservation equation (8.72) yields $T(t)$ as an integral of core power $n(t)$ over time. When this expression is used in the reactivity equation (8.71), we conclude quickly that the reactor kinetics equation (8.70) will be a nonlinear differential equation for power $n(t)$:

$$\frac{dn(t)}{dt} = \frac{1}{\Lambda}\left[K_0 - \frac{\alpha}{C_p}\int_0^t \{n(t') - n_0\}\, dt'\right] n(t). \tag{8.73}$$

The nonlinear ordinary differential equation is not amenable to an exact, analytical solution in time, although it can be integrated numerically without too much effort. Since analytic solutions, even approximate ones, often provide valuable insights to the physics of complex problems, we now turn to the task of obtaining an approximate solution to the Ergen–Weinberg model described by Eqs. (8.70) through (8.72). Our approach here consists of an exact solution of the equations in the $(n, T)$ phase-plane, which provides us with an approximate estimate of the key system behavior, including the peak power and temperature expected in the transient. The analytical estimates can then be used to determine approximately the time behavior of system parameters $\{K(t), n(t), T(t)\}$.

The nonlinear phase-plane solution of the Ergen–Weinberg model is achieved by dividing Eq. (8.70) by Eq. (8.72), thereby eliminating the time variable, and substituting Eq. (8.71) for $K(t)$:

$$\frac{dn}{C_p dT} = \frac{Kn}{\Lambda(n - n_0)} = \frac{(K_0 - \alpha T)\, n}{\Lambda(n - n_0)}.$$

Now separate the terms containing $n$ and $T$, and perform a simple integration of the resulting equation

$$\left(\frac{n}{n_0} - 1\right) - \ln\frac{n}{n_0} = \frac{C_p}{\Lambda n_0}\left(K_0 T - \frac{\alpha T^2}{2}\right), \tag{8.74}$$

where we make a judicious use of initial conditions: $n(0) = n_0$ and $T(0) = 0$. Although Eq. (8.74) yields the power level $n$ as a function of temperature change $T$, it is still an exact solution of the Ergen-Weinberg model.

The phase-plane solution of Eq. (8.74) may now be used to obtain an exact solution for the maximum and minimum temperatures expected during the transient. For this purpose, note that the extremum temperatures will occur when $dT/dt = 0$ in Eq. (8.72), or when $n(t) = n_0$. This condition corresponds to setting the left-hand side of Eq. (8.74) to zero, yielding two roots of the quadratic equation on the right-hand side:

$$T_{min} = 0 \quad \text{and} \quad T_{max} = \frac{2K_0}{\alpha}. \tag{8.75}$$

Similarly, the extremum values of power are given by $K(t) = 0$ or $\alpha T = K_0$. For an estimate of the maximum power $n_{max}$, we note that $\ln(n_{max}/n_0)$ will generally be much smaller than the ratio $n_{max}/n_0$ itself in Eq. (8.74), which yields

$$\frac{n_{\max}}{n_0} \simeq 1 + \frac{C_p K_0^2}{2\Lambda n_0 \alpha}. \tag{8.76}$$

For an estimate of the minimum power $n_{min}$, we may safely neglect the term $n_{min}/n_0$ in Eq. (8.74) to obtain

$$\frac{n_{\min}}{n_0} \simeq \exp\left[-\left(1 + \frac{C_p K_0^2}{2\Lambda n_0 \alpha}\right)\right] \simeq \exp\left[-\frac{n_{\max}}{n_0}\right]. \tag{8.77}$$

The solution of the Ergen–Weinberg model illustrated in Figures 8.8 and 8.9 indicates that the system goes through cycles, each consisting of a power burst followed by a cooling period. This cyclic trajectory continues as long as contributions from delayed neutrons can be neglected, which is one of the key assumptions of the model. The period of such oscillation can also be approximately estimated by noting that the total energy generated in a cycle is essentially equal to

$$Q_{tot} \simeq C_p T_{max} = \frac{2K_0 C_p}{\alpha}, \tag{8.78}$$

and that this accumulated energy is dissipated approximately at a rate $n_0$ until the excess energy is drained off. With the duration of the power burst neglected compared with the cooling period, an estimate for the oscillation period $\tau_{osc}$ follows:

$$\tau_{osc} = \frac{2K_0 C_p}{\alpha n_0}. \tag{8.79}$$

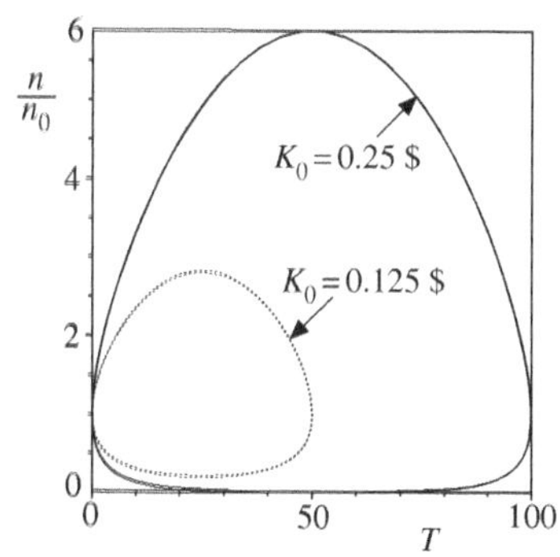


**Figure 8.8** Phase plane solution of the Ergen–Weinberg model.

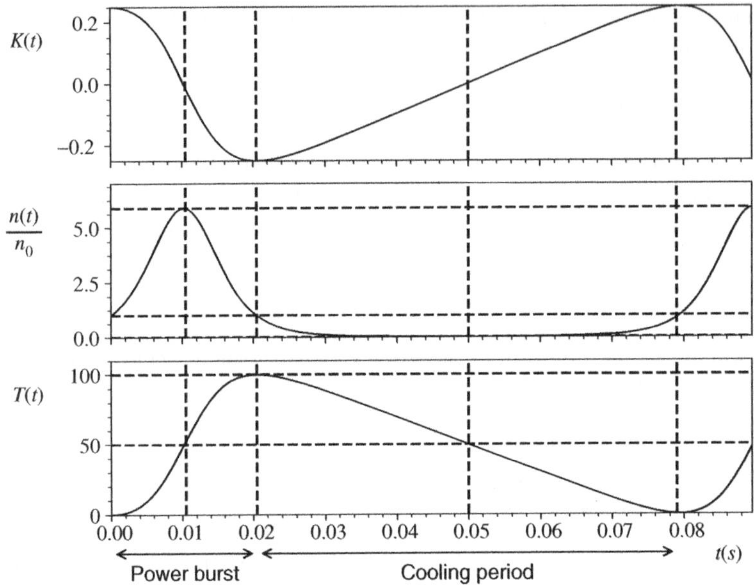


**Figure 8.9** Time-domain behavior of the Ergen–Weinberg model.

**Example 8.3** Illustrate the $(n, T)$ phase-plane solution of the Ergen–Weinberg model with these system parameters: $n_0 = 5.0\,\text{MW}$, $K_0 = 0.25$ dollar, $C_p = 2.5\,\text{kJ} \cdot \text{K}^{-1}$, $\alpha = 5.0 \times 10^{-3}$ dollar $\cdot\,\text{K}^{-1}$, and $\Lambda = 1.0 \times 10^{-3}\,\text{s}$.

Equations (8.70) through (8.72) are integrated numerically in time with a Crank-Nicolson scheme [Dah74] and plotted in Figure 8.8 for the $(n, T)$ phase-plane trajectory and in Figure 8.9 for the time-domain trajectory. The peak relative power $n_{max}/n_0 = 5.88$ and $T_{max} = 100\,\text{K}$ obtained from the numerical integration agree well with the phase-plane solution of Eqs. (8.74) and (8.75). The peak relative power $n_{max}/n_0 = 4.1$ and the oscillation period $\tau_{osc} = 0.05\,\text{s}$ obtained from Eqs. (8.76) and (8.79), respectively, are, however, somewhat approximate compared with the Crank–Nicolson solution for $n_{max}/n_0 = 5.88$ and $\tau_{osc} = 0.080\,\text{s}$. The estimate of the peak relative power $n_{max}/n_0 = 4.1$ is similar to the first-order estimate obtained from a singular perturbation analysis [War87]. The agreement between the simplified phase-plane estimates and direct numerical solutions is expected to improve for larger values of inserted reactivity $K_0$.

Although we have obtained only an approximate estimate of the oscillation period $\tau_{osc}$, the phase-plane solution is quite useful for getting an order-of-magnitude estimate as well as understanding parametric dependence of a postulated rod-ejection accident in a PWR core. The Ergen–Weinberg model may also be used to get a quick assessment of the severity of the Chernobyl accident, including the total energy generated and total radioactivity released, given an estimate of the peak power.

### 8.4.2 The Nordheim–Fuchs Model

Another prompt-power transient formulation is obtained as the Nordheim–Fuchs model [Het71] by adopting an adiabatic assumption for the heat dissipation so that Eq. (8.72) is modified to

$$C_p \frac{dT(t)}{dt} = n(t). \tag{8.80}$$

With the simplification, the set of equations (8.70), (8.71), and (8.80) may be combined to yield explicit solutions in the time domain. Taking the time derivative of Eq. (8.71) yields

$$\frac{dK(t)}{dt} = -\alpha \frac{dT(t)}{dt} = -\frac{\alpha}{C_p} n(t), \tag{8.81}$$

which is combined with Eq. (8.70) to generate phase-plane equations

$$\frac{dn}{dK} = -\frac{C_p}{\Lambda\alpha} K, \frac{dn}{dT} = \frac{C_p}{\Lambda}(K_0 - \alpha T), \tag{8.82}$$

and phase-plane trajectories:

$$n(K) = \frac{C_p}{2\Lambda\alpha}(K_0^2 - K^2), n(T) = \frac{C_p}{\Lambda}\left(K_0 T - \frac{\alpha T^2}{2}\right). \tag{8.83}$$

The phase-plane solutions provide the peak power $n_{max}$ corresponding to $K(t) = 0$ and peak temperature $T_{max}$ corresponding to $n(t) = 0$:

$$n_{max} = \frac{C_p K_0^2}{2\Lambda\alpha}, T_{max} = \frac{2K_0}{\alpha}. \tag{8.84}$$

The phase-plane solutions may also be used to obtain explicit time-domain solutions first by substituting $n(K)$ from Eq. (8.83) into Eq. (8.81)

$$\frac{dK}{dt} = -\frac{1}{2\Lambda}(K_0^2 - K^2), \tag{8.85}$$

which is integrated to yield the time variation of reactivity:

$$K(t) = -K_0 \tanh \frac{K_0}{2\Lambda} t. \tag{8.86}$$

Equation (8.81) finally yields the solution for the time-dependent power

$$n(t) = n_{max}\left(1 - \tanh^2 \frac{K_0}{2\Lambda} t\right) = n_{max} \operatorname{sech}^2 \frac{\omega t}{2}, \omega = \frac{K_0}{\Lambda}. \tag{8.87}$$

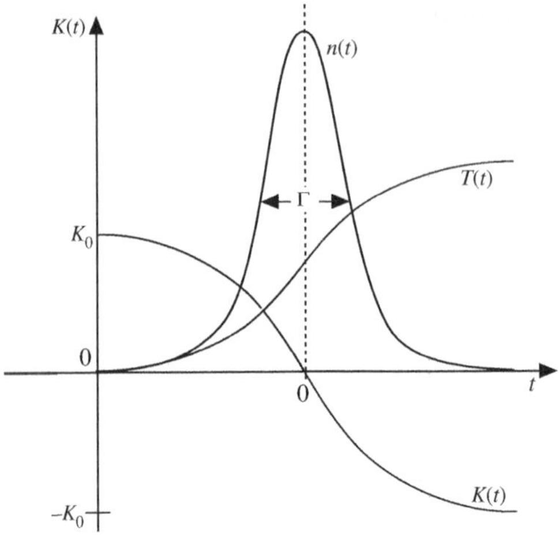


**Figure 8.10** Time-domain behavior of the Nordheim–Fuchs power excursion model.

As a measure of the power pulse width, the full width at half maximum (FWHM) $\Gamma$ of the pulse is readily calculated from Eq. (8.87):

$$\Gamma = \frac{4}{\omega}\cosh^{-1}\sqrt{2} = \frac{3.524}{\omega}. \tag{8.88}$$

The FWHM may be compared with another measure by calculating the total energy $Q_{tot}$ generated over a pulse divided by the maximum power $n_{max}$

$$\frac{Q_{tot}}{n_{max}} = \frac{C_p T_{max}}{n_{max}} = \frac{4}{\omega}, \tag{8.89}$$

where $1/\omega = \Lambda/K_0$ is the initial reactor period. The time-domain behavior of $n(t)$, $T(t)$, and $K(t)$ is illustrated in Figure 8.10, which clearly displays the relationship $K(t) = K_0 - \alpha T(t)$ and $n_{max}$ occurring at $K = 0$. The Nordheim–Fuchs model provides the same total energy released in a pulse $Q_{tot} = 2C_pK_0/\alpha$ as that of Eq. (8.78).

## 8.5 Reactivity Measurements

Since a reactivity perturbation results in a change in the flux or power level of a reactor, we may deduce the reactivity variation by monitoring the flux variation over time. In fact, the measurement of reactivity in any chain-reacting system invariably involves the measurement of time-dependent flux variations coupled with the use of a neutron balance equation in some form. Once a *reactivity scale* is established through the calibration of a control rod, however, we may use this calibrated scale to determine the reactivity of other control devices through an equivalence relationship or substitution technique. We discuss in this section key methods available for experimental reactivity determination.

1. **Inhour equation**
   For small reactivity insertions, we may determine the reactor period $T$ after initial transients have died out. This period may then be substituted into the inhour equation (8.39) or (8.40) to yield reactivity $K$. This is one of the oldest techniques but is still used often as the first choice whenever applicable. One limitation of the technique is that the inserted reactivity has to be small in magnitude so that we can wait until the transient terms in the point kinetics solution have died out. To stay within this limitation, reactivity measurement of a control rod may have to be broken down into multiple steps, each with a sufficiently small reactivity worth.
2. **Inverse multiplication technique**
   This is a technique that may be used during an approach to the critical fuel loading in a reactor when the effective multiplication factor $k = k_{eff} < 1.0$. In such a core, the total neutron population $n$, after an infinite number of generations, due to a neutron source of strength $S_0$ may be determined as
   $$n = S_0 + kS_0 + k^2S_0 + \cdots = \frac{S_0}{1-k} \equiv MS_0. \tag{8.90}$$
   Hence, the reciprocal of the *multiplication* $M$ is plotted as a function of fuel loading, and the critical configuration is estimated through an extrapolation to $1/M = 0$, illustrated schematically in Figure 8.11. This is, of course, the technique that Enrico Fermi and his team adopted in establishing the feasibility of a self-sustaining chain-reacting system in 1942. This old technique is still used to determine the initial criticality in new cores of any kind, including small research reactors and large power reactors. This technique is also routinely used in existing reactor cores whenever sufficient changes are made to the fuel loading pattern.
   The $1/M$ parameter is monitored as a core is reloaded in a power reactor. Criticality is not actually achieved by loading, but $1/M$ is monitored to ensure that the core remains subcritical when fully loaded. Criticality is not, in fact, allowed until the core is fully loaded, the reactor is fully assembled, all Technical Specification requirements are met, the reactor coolant system is at nominal pressure and temperature, and all required surveillance testing is completed. In a typical PWR plant, while the shutdown banks are fully withdrawn, the approach to criticality proceeds with the control rod banks withdrawn sequentially with sufficient overlaps between the control rod groups.
3. **Rod drop technique**
   This technique involves rapidly inserting the neutron absorber of unknown reactivity worth into a critical, steady-state system and is useful to determine the integral or total reactivity worth of a control rod or rod bank. By integrating

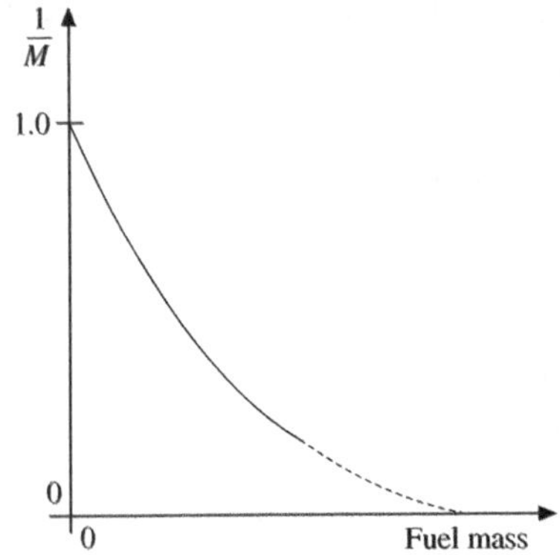


**Figure 8.11** Inverse multiplication as a function of fuel mass.

the time-dependent core power following the negative reactivity insertion, the reactivity worth of the control rod may be established:

$$K[\text{dollar}] = -\frac{\tau n(0)}{\int_0^\infty n(t)dt}. \tag{8.91}$$

Because the technique involves a time integral, it is relatively free from the errors associated with multiple steps of rod-worth measurements subject to a somewhat arbitrary waiting time in the inhour method.

4. **Substitution method**
When there is more than one type of reactivity control device in the system, the reactivity of one control device can be determined indirectly by balancing its reactivity effects against those of another control device of known reactivity worth. One primary example is the determination of soluble boron worth $\partial\rho/\partial C_B$ in units of [pcm/ppm of soluble boron] in a PWR core by finding the movement $\Delta z$ [step] of a calibrated control rod bank with *differential reactivity worth* $\partial\rho_{CR}/\partial z$ [pcm/step] that will cancel the reactivity effects of variation $\Delta C_B$ [ppm] in the soluble boron concentration:

$$\frac{\partial\rho}{\partial C_B}\Delta C_B + \frac{\partial\rho_{CR}}{\partial z}\Delta z = 0. \tag{8.92}$$

5. **Inverse kinetics method**
Given a measurement of the power variation $n(t)$, the point kinetics equations may be solved inversely [Ash79] to determine the reactivity $K(t)$ corresponding to the measured $n(t)$:

$$K(t) = \Lambda\frac{d\ln n(t)}{dt} + \frac{1}{n(t)}\int_0^t \frac{dn(\tau)}{d\tau}\sum_{i=1}^{6} a_i \exp[-\lambda_i(t-\tau)]d\tau. \tag{8.93}$$

This is the primary tool currently used in power reactors. It offers the ability to determine the reactivity as an explicit function of time and does not depend on the assumption of a step reactivity insertion subject to a waiting time, which is the inherent limitation of the inhour method. For power reactors, where reactivity feedback effects are significant, the inhour equation is not very useful, and we are essentially forced to use the inverse kinetics method, implemented as a *reactivity meter*.

6. **Noise analysis techniques**
   From the analysis of noise characteristics in neutron flux variations, we may also obtain reactivity information. This passive method is especially useful in subcritical systems, including experimental subcritical assemblies and storage facilities for fissile materials.
7. **Neutron pulse method**
   With an input of repetitive pulses of neutrons [Lew78], particularly in subcritical systems, the system reactivity may be determined. A variation of this method may also be used as an active interrogation technique to identify any clandestine presence of fissile material in nuclear material safeguards and monitoring.

## 8.6 System Stability Analysis

The open-loop transfer function $G(s)$ with control input $u(s)$ in Figure 8.5 may now be extended to include thermal-hydraulic feedback effects of the type considered in Eq. (8.71) via a transfer function $H(s)$. With the feedback path illustrated in Figure 8.12, a *combined transfer function* or *closed-loop transfer function* $T(s)$ is obtained:

$$\frac{\Delta n}{n_0 \Delta K} = \frac{G}{1 + GH} = T. \tag{8.94}$$

For the stability analysis of a critical or subcritical reactor core, when all roots of the inhour equation are negative or zero (when $K = 0$), it suffices to consider singularities of $1/(1 + GH)$ or zeros of the denominator $F(s) = 1 + G(s)H(s)$. The system with feedback $H$ will be stable if $F(s)$ has no zeros in the right-hand half of the Laplace transform plane so that only negative exponential terms are possible in the time domain. For linear stability analysis of a combined system $T(s)$, as the contour $\Gamma$ is traversed in the right-hand $s$-plane in Figure 8.13a, a *Nyquist*

**Figure 8.12** Schematic diagram of reactor transfer function $G$ with thermal-hydraulic feedback function $H$.

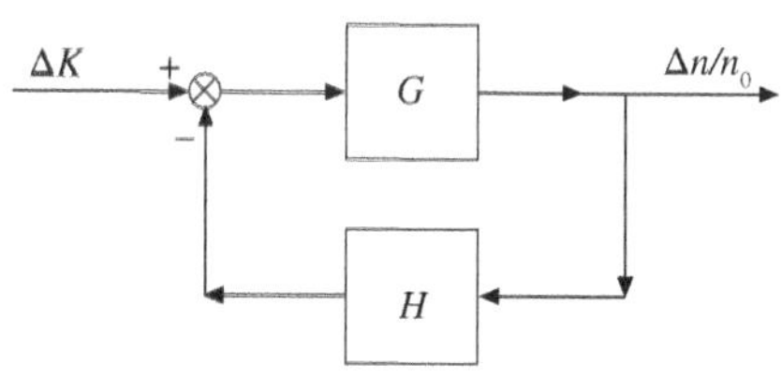

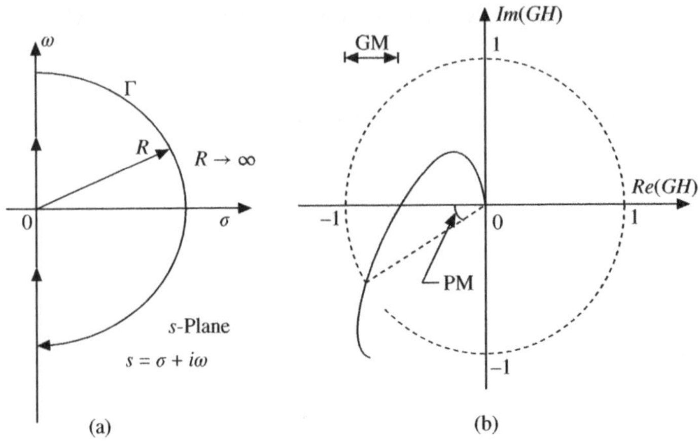


**Figure 8.13** Generation of Nyquist diagram via traversing contour Γ in the right-hand half of the $s$-plane. Contour Γ in diagram (a) is extended with the radius $R \to \infty$, while the phase margin (PM) and gain margin (GM) in diagram (b) indicate the system is stable with sufficient margins.

*diagram* is constructed for the trajectory of either $F(s)$ or, equivalently, $G(s)H(s)$, as in Figure 8.13b.

The stability of the combined system is then determined by the Nyquist contour of $G(s)H(s)$ relative to the instability threshold $(-1, 0)$: if $G(s)H(s)$ has no zeros with real part $Re(GH) < -1$, then the system is stable. For the $GH$ contour in Figure 8.13b, the system is stable, and the margin to the instability threshold is illustrated by the *phase margin* (PM) when the magnitude $|GH| = 1$ and the *gain margin* (GM) when the phase is $-\pi$.

The system stability may also be analyzed via the *Nyquist stability criterion* [Mor01] representing the number of encirclements of origin $(0, 0)$ by the Nyquist contour $F(s)$ in the Laplace plane

$$E = Z - P, \tag{8.95}$$

where $Z$ and $P$ are the number of zeros and poles with positive real parts, respectively, of $F(s)$ along the contour Γ. With $P = 0$ for $G(s)H(s)$, if $E = 0$, then $Z = 0$, i.e. there are no zeros of $F(s) = 1 + G(s)H(s)$ with positive real parts. Thus, if the Nyquist contour for $G(s)H(s)$ does not encircle $(-1, 0)$, then the system $T(s)$ is stable.

**Example 8.4** For $G(s) = 1/s(s + a)$ and $H(s) = C/(s + b)$, with $a = 1, b = 2$, $C = 1$, construct a Nyquist diagram for $G(s)H(s)$ and evaluate the system stability.

Evaluate $GH$ along the imaginary axis with $s = \sigma + i\omega$

$$G(i\omega)H(i\omega) = \frac{C}{i\omega(i\omega + a)(i\omega + b)} = \frac{-C(a+b)\omega + iC(\omega^2 - ab)}{\omega(\omega^2 + a^2)(\omega^2 + b^2)},$$

which indicates that along the real axis, i.e. when $Im(GH) = 0$ or $\omega^2 = ab$,

$$Re(GH) = \frac{-C}{ab(a+b)} = -\frac{1}{6},$$

in agreement with the Nyquist diagram in Figure 8.14. Note that the contour along the imaginary axis in the $s$-plane is distorted around the origin to avoid the pole $s = 0$ for $GH$. The diagram is obtained via the *nyquist* function of MATLAB but special effort is made to zoom around the origin in the $GH$ space. Note also that point $C$ on the distorted contour around the origin in the $s$-plane transforms to point $(\infty, 0)$ on the $GH$-plane. The stability of $GH$ is also confirmed via the Nyquist stability criterion where $Z = P = 0$ yielding $E = 0$, i.e. $GH$ does not encircle $(-1, 0)$.

Another method often utilized for system stability analysis is the *Bode diagram*, where the magnitude and phase of $GH$ are plotted as a function of $\log_{10}\omega$

$$\text{Gain: } N(\omega) = 20\log_{10}|GH| \text{ [decibel]},$$
$$\text{Phase: } \delta(\omega) = \tan^{-1}\frac{Im(GH)}{Re(GH)}, \tag{8.96}$$

and the PM and GM are determined for stability evaluation. The Bode diagram is also used to evaluate various transfer functions introduced for model-based control formulations in Chapter 16.

**Example 8.5** Construct the Bode diagram for $G(s)H(s)$ from Example 8.4, and determine the PM and GM.

Calculate the magnitude and phase angle of $GH$

$$|GK| = \frac{C[\omega^2(a+b)^2 + (\omega^2 - ab)^2]^{1/2}}{\omega(\omega^2 + a^2)(\omega^2 + b^2)} = \frac{C}{\omega(\omega^2 + a^2)^{1/2}(\omega^2 + b^2)^{1/2}},$$

$$\delta(\omega) = -\tan^{-1}\left|\frac{\omega^2 - ab}{\omega(a+b)}\right| = \begin{cases} -\pi/2 \text{ as } \omega \to 0-, \\ -3\pi/2 \text{ as } \omega \to -\infty, \end{cases}$$

and determine the gain and phase margins

$$\text{GM} = -20\log_{10}(1/6) = 15.6 \text{ db at } Im(GH) = 0,$$
$$\text{PM} = \delta(\omega = 0.446) = \tan^{-1} 1.346 = 53.4^\circ \text{ at } |GH| = 1.0,$$

in agreement with the Nyquist diagram of Figure 8.14 and the Bode diagram of Figure 8.15. The Bode diagram, together with the GM and PM edits, is generated via the *bode* function of MATLAB, and the slopes for low, medium $(1 \sim 2\,\text{rad} \cdot \text{s}^{-1})$, and high frequencies are added to the gain plot. Note that $\omega = 1.0$

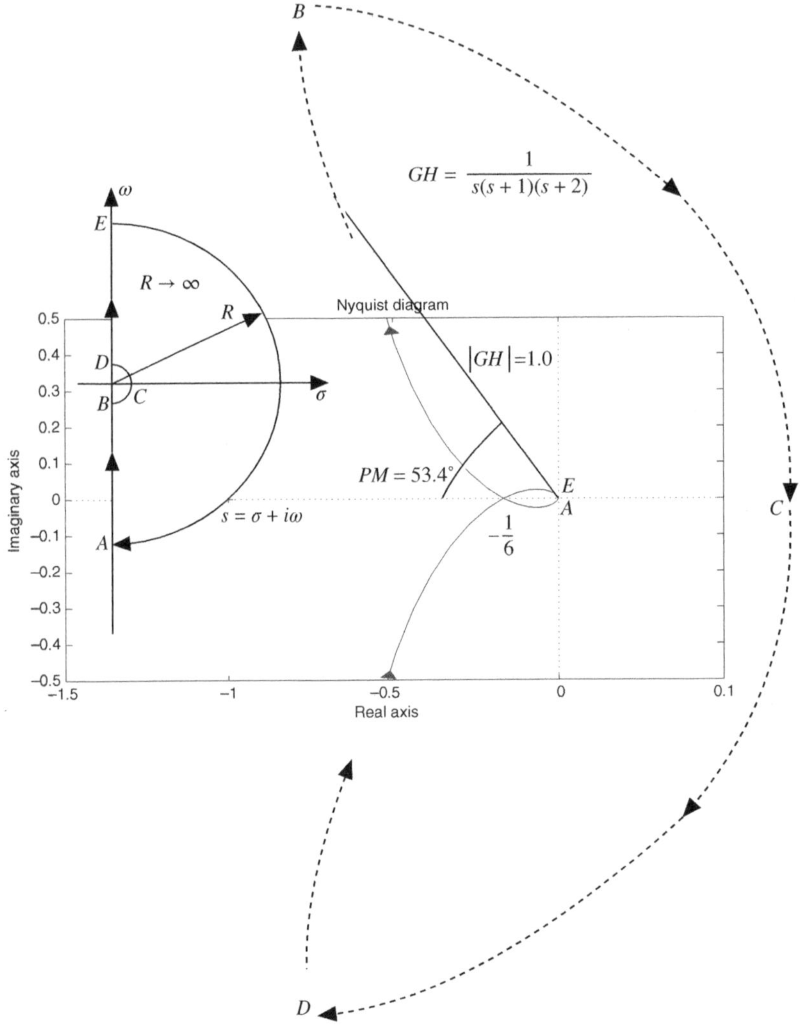


**Figure 8.14** Nyquist diagram for $GH = 1/s(s+1)(s+2)$, Example 8.4.

and $\omega = 2.0\,\text{rad}\cdot\text{s}^{-1}$ are known as the *break* or *corner frequencies* representing changes in the slopes of the plot.

Open-loop reactor transfer function $G(s)$ from Eq. (8.57) is illustrated as a Bode diagram in Figure 8.16 [Ash79] for several different values of neutron lifetime $\Lambda = \ell/\beta$. Note that $\lim_{\omega\to\infty} \delta(\omega) = -\pi/2$, while the magnitude $|G(\omega)| \simeq 1.0$ over

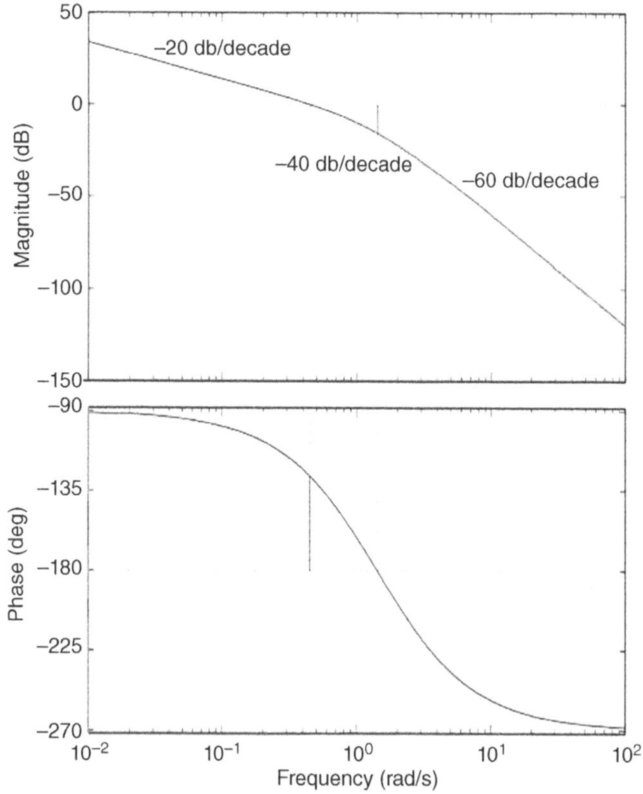


**Figure 8.15** Bode diagram for $GH = 1/s(s + 1)(s + 2)$, Example 8.4, indicating $GM = 15.6$ db and $PM = 53.4°$.

the bulk of the frequency range and $\lim_{\omega\to\infty} |G(\omega)| = 1/(\omega\Lambda)$ with a break frequency $\omega = 1/\Lambda = \beta/\ell$.

## 8.7 Point Reactor and Space-Dependent Reactor Kinetics

Our discussion of fast transients in nuclear reactor cores has been limited to the point kinetics equations with both six groups and one equivalent group of delayed neutron precursors. Solution of the point kinetics equations has provided, however, valuable understanding of the time evolution of flux or power as a function of the reactivity inserted. We have also derived a few simple approximations to the point kinetics equations that could determine power-level

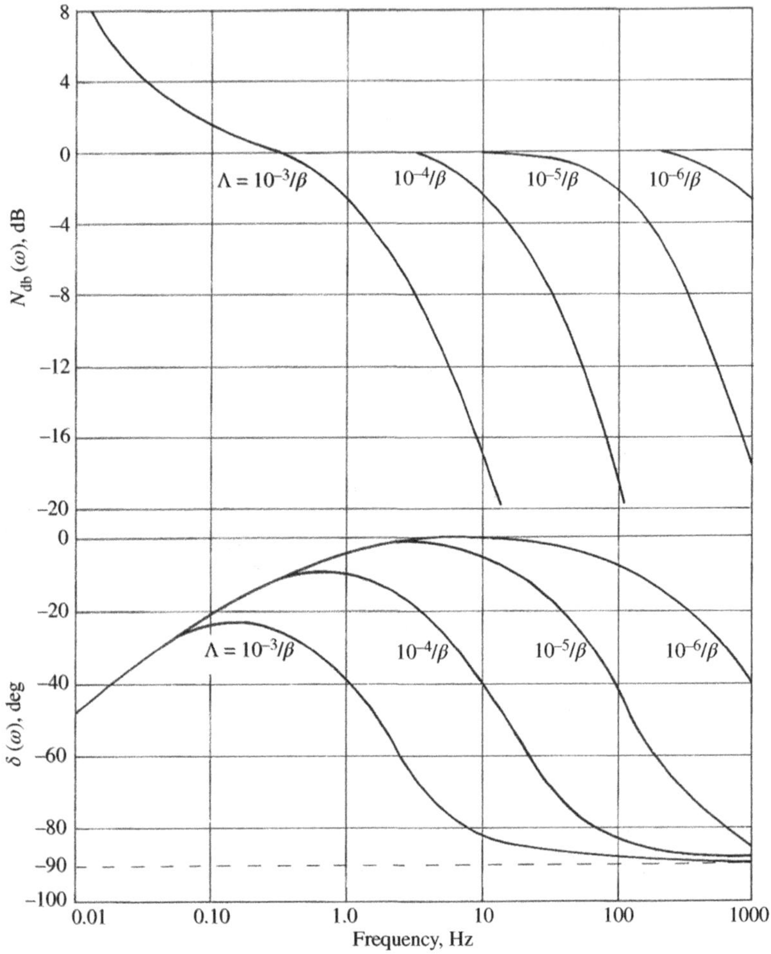


**Figure 8.16** Bode diagram for open-loop reactor transfer function $G(s)$. *Source:* Ash [Ash79]/McGraw-Hill.

variations subject to an arbitrary time-dependent reactivity. Canonical state-space representations have been introduced to solve the point reactor kinetics equation via the *Simulink* toolbox of *MATLAB*. The Ergen–Weinberg and Nordheim–Fuchs models have been studied as simple examples of the nonlinear kinetics equation and reactor excursion model. A number of experimental techniques for reactivity measurement have been discussed, together with techniques for stability analysis of nuclear systems.

Space-dependent kinetics equations have to be used in general for more accurate transient analyses, especially in large power reactor cores. The few-group time-dependent diffusion equations are the primary tool for space-time analyses in large reactor cores. Even today, due to large computational requirements, however, only limited use is made of full-blown three-dimensional calculations for routine design calculations. Techniques and applications of space-time kinetics formulation are discussed in Chapter 16.

## References


**[Ash79]** M. Ash, *Nuclear Reactor Kinetics*, 2nd ed., McGraw-Hill (1979).

**[Bra89]** M.C. Brady and T.R. England, "Delayed Neutron Data and Group Parameters for 43 Fissioning Systems," *Nucl. Sci. Eng.* **103**, 129 (1989).

**[Dah74]** G. Dahlquist and A. Björck, *Numerical Methods*, Prentice-Hall (1974).

**[Doy89]** J.C. Doyle, K. Glover, P.P. Khargonekar, and B.A. Francis, "State-Space Solution to Standard $H_2$ and $H_\infty$ Control Problems," *IEEE Trans. Autom. Control* **34**, 831 (1989).

**[Het71]** D.L. Hetrick, *Dynamics of Nuclear Reactors*, University of Chicago Press (1971).

**[Kee65]** G.R. Keepin, *Physics of Nuclear Kinetics*, Addison-Wesley (1965).

**[Lew78]** J. Lewins, *Nuclear Reactor Kinetics and Control*, Pergamon Press (1978).

**[Mor01]** K. Morris, *Introduction to Feedback Control*, Harcourt/Academic Press (2001).

**[Par99]** T.A. Parish, W.S. Charlton, N. Shinohara, et al. "Status of Six-Group Delayed Neutron Data and Relationships Between Delayed Neutron Parameters from the Macroscopic and Microscopic Approaches," *Nucl. Sci. Eng.* **131**, 208 (1999).

**[War87]** M.E. Ward and J.C. Lee, "Singular Perturbation Analysis of Relaxation Oscillations in Reactor Systems," *Nucl. Sci. Eng.* **95**, 47 (1987).

**[Wei58]** A. Weinberg and E.P. Wigner, *The Physical Theory of Neutron Chain Reactors*, University of Chicago Press (1958).


## Problems

**8.1** Using the expressions for the reactor transfer function expressed in terms of six groups and one group of delayed neutron precursors, show that Eq. (8.17) should apply for short times after a transient.

**8.2** Using the Laplace transform technique, obtain the strength of the source that should be activated to obtain the solution obtained in Example 8.2 with one equivalent group of delayed neutron precursors.

**8.3** Using the results of Problem 8.2, show that the transform $\Delta n(s)$ of the power variation $\Delta n(t)$ may be represented in terms of the zero-power reactor transfer function $G(s)$ from Eq. (8.53): $\Delta n(s) = \Lambda G(s)S(s)$.

**8.4** Repeat the *Simulink* application from Example 8.2 with six groups of delayed neutron precursors with the $^{235}$U data from Table 2.2 to determine a realistic power-level variation with the source activated for the same period $\tau = 10$ s.

**8.5** Obtain the reactivity meter formulation from Eq. (8.93) starting from Eqs. (8.12) and (8.13).

**8.6** Derive the rod-drop expression of Eq. (8.91) starting from Eq. (8.21).

**8.7** A step insertion of reactivity $K_0 = 0.25$ dollar is made at $t = 0$ into an initially critical core with steady-state power level $n_0$. For short times following the reactivity insertion, assume that the concentration of delayed neutron precursors remains unchanged from the steady-state level. (a) Set up a point kinetics equation representing the time-dependent power level $n(t)$, and (b) solve the point kinetics equation of part (a) for $n(t)$ via Laplace transform for short times following the reactivity insertion.

**8.8** A subcritical reactor is operating at steady-state power level $n_0$ with a steady-state source of strength $S_0$ [neutron $\cdot$ s$^{-1}$]. (a) Determine an expression for the reactivity $K_0$ in units of dollar for the system. (b) An accelerator-driven source (ADS) of strength $S_1$ [neutron $\cdot$ s$^{-1}$] is turned on for a period of $\tau$ s. Obtain an expression for the Laplace transform $n(s)$ of the power, and determine the asymptotic power level $n_1$ that will be obtained a long time after the ADS is turned off. Compare the result with that of Problem 8.2 and discuss.

**8.9** The power-level variation $n_1(t)$ is given for a step insertion of reactivity $K_1$ at $t = 0$ into a critical reactor operating at a steady-state power. If a step reactivity insertion is made at $t = 0$ into the reactor operating at steady state, but with a time-independent source of strength $S_0$ so that the resulting reactivity is also $K_1$, the resulting transient power variation is given as $n_2(t)$. Using the point kinetics equation with one equivalent group of delayed neutrons, show that, for either $K_1 < 0$ or $K_1 > 0$, $n_2(t) = C_1 n_1(t) + C_2$, and obtain the constants $C_1$ and $C_2$.

**8.10** Rewrite the point reactor kinetics equation in the form

$$\frac{d\mathbf{\Psi}(t)}{dt} = A(t)\mathbf{\Psi}(t), \mathbf{\Psi}(t) = [n(t), C_1(t), \ldots, C_6(t)]^T,$$

and show that the roots of inhour equations are the eigenvalues of the matrix $A$ for a step insertion of reactivity.

**8.11** In the Dragon subcritical experiment [Het71], a positive reactivity is rapidly inserted into an initially subcritical assembly, with a steady-state power level $n_0$, for a short duration $\tau$. Assume that (i) $K(t) = K_0 < 0$ for $t < 0$ and $t > \tau$, and (ii) $K(t) = K_1 > 1.0$ dollar for $0 \leq t \leq \tau$. Using the point kinetics equation with one equivalent group of delayed neutrons, obtain an expression for the power variation $n(t)$ for the rising $(0 < t < \tau)$ and falling $(t > \tau)$ portions of the pulse, and calculate the energy released in the prompt burst.

**8.12** With the Nordheim–Fuchs model for power excursions, consider an experimental fast reactor consisting of a sphere of pure $^{235}$U metal weighing 20 kg. No cooling is provided, and the reactor is initially at zero power. When the reactor is subcritical, a piece of $^{235}$U metal is shot into a hole in the sphere, suddenly making the effective multiplication factor $k = 1.02$. The temperature coefficient of the reactivity is $d\ln k/dT = -1.0 \times 10^{-5}$ K$^{-1}$, and the neutron generation time $\ell = \Lambda k\beta = 10^{-5}$ s. The effective delayed neutron fraction $\beta = 0.0065$. The heat capacity of the uranium metal is $130\,\text{J}\cdot\text{kg}^{-1}\text{K}^{-1}$. The resulting transient is so rapid that it is completed before any delayed neutrons are formed. Obtain (a) maximum power $n_{max}$, (b) maximum temperature rise $T_{max}$, (c) total energy released $Q_{tot}$ in the pulse, (d) total number of fissions occurring in the power pulse, and (e) FWHM for the power pulse.

**8.13** Given the LFT representation for two transfer functions

$$G_1 = \left[\begin{array}{c|c} A_1 & B_1 \\ \hline C_1 & D_1 \end{array}\right] \quad \text{and} \quad G_2 = \left[\begin{array}{c|c} A_2 & B_2 \\ \hline C_2 & D_2 \end{array}\right],$$

obtain the compound transfer function $G = G_2 G_1$.

**8.14** The dynamics of a system with control input $u$ is described by LFT

$$G = \left[\begin{array}{cc|c} 2 & 0 & 1 \\ 0 & -1 & 1 \\ \hline 1 & 1 & 0 \end{array}\right].$$

Obtain an expression for the output $y = Gu$ in the Laplace domain.

**8.15** A sharp pulse of positive reactivity is introduced into a $^{235}$U-fueled reactor core at time $t = 0$. The reactor core is initially critical with an equilibrium power level $n_0$. The reactivity pulse is so sharp that it may be represented by $K(t) = K_0\delta(t)$. Using the point kinetics equation with one equivalent group of delayed neutrons, obtain the asymptotic power level at long times after the reactivity pulse with $K_0 = 1.28$ \$ and $\lambda = 0.090\,\text{s}^{-1}$ and $\Lambda = 10^{-3}$ s.

**8.16** In the Ergen–Weinberg model representing power excursions, we may approximate the power variation by $n(t) = n(0)e^{Q(t)}$ in terms of the energy $Q(t)$ produced over time $t$. (a) Linearize the E-W model with the assumption that $Q(t)$ remains small throughout the transient for relatively mild excursions, and obtain the solution for $Q(t)$ through Laplace transform. (b) What is the oscillation period $\tau$ of $n(t)$ for the linearized E-W model?

**8.17** Consider, for the point kinetics equation, the task of estimating the effective decay constant $\lambda$ for the one-group delayed neutron precursor representation. Generate simulated experimental data for the reactor power using the full-blown six-group PKE model, linearize the one-group PKE at some frequent intervals, and treat the decay constant $\lambda$ as a system parameter to be estimated through a Kalman filter. With judicious estimates of the covariance matrices for the system and measurement noises, begin with the value of $\lambda$ valid for short times after a reactivity input is introduced into the reactor and show that the filter gradually provides the analytical estimate for $\lambda$ for long times.

**8.18** The transfer function $GH = C[(s+a)(s+b)]^{-1}, a > 0, b > 0$, is given for a closed-loop system. (a) Determine key cross-over points for the Nyquist diagram and compare with the *MATLAB* solution and (b) determine the phase margin for $a = 1, b = 2$, and $C = 2$.

# 9

# Fast Neutron Spectrum Calculation

We discuss in this chapter methods of calculating the distribution of neutrons in a reactor core at energies above the thermal range. The first step in a typical reactor core design analysis consists of estimating the energy distribution of neutrons both in the fast and thermal ranges, usually at the fuel assembly level, through a lattice physics code. In such lattice physics analysis, the slowing down of fast neutrons may be treated separately from the calculation of the thermal neutron spectrum. As discussed in Chapter 7 for the derivation of the multigroup neutron diffusion equation (7.17), some estimate of the energy distribution of neutrons in the slowing-down process is necessary to generate fine-group cross sections from the available experimental data and to collapse them into few-group constants. The few-group constants may then be used in the multigroup neutron diffusion equation to represent the global geometrical configuration and to calculate the flux and power distributions throughout the core. In general, the fast spectrum calculation has to be iteratively coupled to the global analysis if we are to properly represent the effects of neutron leakage in the lattice physics analysis. For analysis of light water reactor (LWR) cores, where the spectral-spatial coupling is rather weak, the iteration between the fine-group spectral calculation and the few-group global spatial calculation is not usually performed.

A key result we derive in this chapter is the approximation introduced in the first-order estimate of the lethargy-dependent flux spectrum $\theta(u)$ in Eq. (7.7). With this in mind, our discussion focuses on the basic methods of calculating the energy distribution of fast neutrons in a homogeneous mixture of fuel and moderator. We will not explicitly consider detailed approaches generally necessary to handle material heterogeneities in typical fuel lattices even in the slowing down range. The representation of heterogeneities both for thermal and fast spectrum calculations will be the subject for Chapter 11. Furthermore, our study of the energy distribution of neutrons in the slowing down range comprises primarily the analysis of the flux spectrum in a large medium where the neutron leakage may be

*Nuclear Reactor Physics and Engineering*, Second Edition. John C. Lee.
© 2025 John Wiley & Sons, Inc. Published 2025 by John Wiley & Sons, Inc.
Companion website: www.wiley.com/go/LeeNRPE2

either neglected or represented approximately. Only a first-order treatment will be included for the analysis of the flux spectrum in a finite medium.

As part of the fast neutron spectrum analysis, an improved description is also necessary for the resonance escape probability $p$ introduced in Eq. (7.36) in terms of two-group diffusion theory and the related concept of effective resonance integral. This will provide a better understanding of parametric dependencies of the infinite multiplication factor represented in terms of the four-factor formula. Furthermore, the effective resonance integral often plays an important role in the calculation of the flux spectrum itself in lattice physics codes.

Section 9.1 presents a derivation of general balance equations that may be used for analysis of the fast neutron spectrum for a homogeneous mixture of fuel and moderator. In particular, we introduce the concept of slowing down density, which plays a key role in slowing down theory. Section 9.2 reviews the concept of scattering kernel obtained in Section 2.6 for elastic scattering of neutrons with nuclei making up the reactor core, highlighting the relationship between the lethargy and energy variables. Section 9.3 presents solutions to the balance equations derived in Section 9.1 for an infinite medium, accounting for both the hydrogenous and non-hydrogenous scattering materials. With expressions obtained for the slowing down density, Section 9.4 moves to the task of calculating the probability that neutrons escape absorption collisions during slowing down, i.e. the resonance escape probability $p$ introduced in Eq. (7.36). In Section 9.5, we discuss the effects of fuel temperature increases on resonance absorbers, usually known as the Doppler broadening of resonances. With Section 9.6 presenting a simple method of accounting for the effects of neutron leakage on the fast neutron flux spectrum, general remarks regarding lattice physics analysis conclude Section 9.7. This will prepare us for Chapter 11 which discusses approaches to incorporate both thermal and fast spectrum calculations in the overall lattice physics analysis.

## 9.1 Neutron Balance Equation and Slowing Down Density

In flux spectrum calculations, as well as in global flux and power distribution calculations, we are often interested in accounting for the first-order anisotropy in the angular flux distribution. This often entails representing the energy-dependent neutron balance equation in terms of the $P_1$ approximation, although in many lattice physics codes a variant of the $P_1$ formulation, known as the $B_1$ method, is often used. The $B_1$ method is discussed in Chapter 11. Representing the energy dependence explicitly in the angular flux $\psi(z, u, \mu)$, the 1-D form of the transport equation (4.44) is rewritten

$$\Sigma_t(u)\psi(z,u,\mu) + \mu\frac{\partial\psi(z,u,\mu)}{\partial z} = S(z,u,\mu) + \rho(z,u,\mu), \tag{9.1}$$

with the scattering integral

$$\rho(z,u,\mu) = \int_0^{\infty} du' \int_{-1}^{1} d\mu' \Sigma_s(u' \to u, u' \to \mu)\psi(z,u',\mu'). \tag{9.2}$$

With the two-term $P_n$ expansion of Eq. (4.49) extended to $\psi(z,u,\mu)$

$$\psi(z,u,\mu) = \sum_{n=0}^{1}\left(\frac{2n+1}{2}\right)\phi_n(z,u)P_n(\mu),$$

the scattering integral of Eq. (9.2) is generalized from Eq. (4.45) to

$$\rho(z,u,\mu) = \sum_{n=0}^{1}\left(\frac{2n+1}{2}\right)\rho_n(z,u)P_n(\mu)$$

in terms of the $P_n$ components of the scattering integral

$$\rho_n(z,u) = \int_{u-\Delta}^{u} du'\Sigma_{sn}(u' \to u)\phi_n(z,u'),\ n = 0, 1. \tag{9.3}$$

Here, $\Sigma_{sn}(u' \to u)$ is the $P_n$ component of the scattering kernel and $\Delta = \ln(1/\alpha)$, with $\alpha = (A-1)^2/(A+1)^2$, for a nucleus of mass number $A$. Recall from Eq. (2.26) that the parameter $\alpha$ represents the minimum fractional energy a neutron may have following an elastic scattering collision with the nucleus. The definition of lethargy $u = \ln(E_0/E)$ in Eq. (7.2) shows that $\Delta$ is the *maximum lethargy gain per collision*.

Evaluate the scattering integral

$$\rho_n(z,u) \simeq \int_{u}^{u+\Delta} du'\Sigma_{sn}(u \to u')\phi_n(z,u), \tag{9.4}$$

assuming that the rate of neutrons scattered from the interval $[u-\Delta, u]$ into unit energy interval around $u$ is equal to the rate of neutrons scattered out of the unit energy interval around $u$ into the lethargy interval $[u, u+\Delta]$. This assumption is certainly only approximate and may be meaningful if the neutron absorption and leakage rates are negligible so that the scattering rate does not change much over the lethargy intervals under consideration. With the approximation introduced into Eq. (9.3), and recognizing that in elastic scattering collisions, there is a one-to-one correspondence between the lethargy transfer $u' \to u$ and directional change $\mathbf{\Omega}' \to \mathbf{\Omega}$ or $\mu_0 = \mathbf{\Omega}' \cdot \mathbf{\Omega}$, evaluate the integrals over $u'$

$$\rho_0(z,u) \simeq \int_{u}^{u+\Delta} du'\Sigma_s(u \to u')\phi_0(z,u) = \Sigma_s(u)\phi_0(z,u), \tag{9.5a}$$

$$\rho_1(z,u) \simeq \int_u^{u+\Delta} du' \mu_0 \Sigma_s(u \to u')\phi_1(z,u),$$
$$= \int_{-1}^{+1} d\mu_0 \mu_0 \Sigma_s(u;\mu_0)\phi_1(z,u) = \overline{\mu}_0 \Sigma_s(u)\phi_1(z,u), \tag{9.5b}$$

with Eq. (4.60b) providing $\Sigma_{s1}(u' \to u) = \overline{\mu}_0 \Sigma_s(u' \to u)$ and $\overline{\mu}_0$ is the average cosine of the scattering angle defined in Eq. (4.66). Generalize Eqs. (4.62) to the lethargy-dependent $P_1$ equations:

$$\Sigma_t(u)\phi_0(z,u) + \frac{\partial \phi_1(z,u)}{\partial z} = S_0(z,u) + \Sigma_s(u)\phi_0(z,u), \tag{9.6a}$$

$$\Sigma_t(u)\phi_1(z,u) + \frac{1}{3}\frac{\partial \phi_0(z,u)}{\partial z} = S_1(z,u) + \overline{\mu}_0 \Sigma_s(u)\phi_1(z,u). \tag{9.6b}$$

Dropping the $P_1$ component $S_1(z,\mu)$ of the source term provides the lethargy-dependent Fick's law of diffusion.

The simple exercise involving Eqs. (9.4) and (9.5) clearly illustrates the need to solve Eqs. (9.6) in a fully coupled manner for an accurate solution of the flux spectrum $\phi_0(z,u)$. Recognizing that this *consistent* $P_1$ *formulation* is often quite involved, we take an expedient approach in this section and assume that Eqs. (9.5) are valid, and simply replace the net current $\phi_1(z,\mu)$ in Eq. (9.6a) with Fick's law. This allows us to adopt the steady-state diffusion equation as the neutron balance statement:

$$\nabla \cdot \mathbf{J}(\mathbf{r},u) + \Sigma_t(u)\phi(\mathbf{r},u) = \int_0^\infty du' \, \Sigma_s(u' \to u)\phi(\mathbf{r},u') + S(\mathbf{r},u), \tag{9.7a}$$

$$\nabla \cdot \mathbf{J}(\mathbf{r},E) + \Sigma_t(E)\phi(\mathbf{r},E) = \int_0^\infty dE' \, \Sigma_s(E' \to E)\phi(\mathbf{r},E') + S(\mathbf{r},E). \tag{9.7b}$$

The effort to solve Eq. (9.7b) for the energy-dependent flux $\phi(\mathbf{r},E)$ is somewhat simplified by introducing the concept of *slowing down density*:

$$q(\mathbf{r},E) = \text{net number of neutrons slowing down past } E \text{ per unit volume at } \mathbf{r} \text{ per unit time.} \tag{9.8}$$

We derive a neutron balance equation in the phase space of $\{\mathbf{r},E\}$, where the slowing down of neutrons is considered together with the production and loss due to leakage. Figure 9.1 illustrates schematically a phase volume $d\mathbf{r}dE$ into which neutrons of energy $E + dE$ arrive at a rate $q(\mathbf{r},E+dE)d\mathbf{r}$ per unit time and out of which neutrons of reduced energy $E$ leave. We also include the production of neutrons in the phase volume at a rate $S(\mathbf{r},E)$ per unit volume and per unit energy interval. The difference between the total influx of neutrons, including the neutron source, and outflux of neutrons has to be equal to the loss of neutrons in the phase volume due to the leakage and absorption of neutrons.

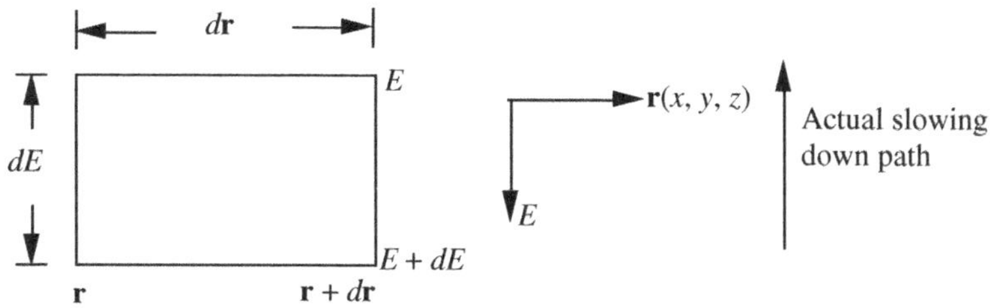


**Figure 9.1** Phase volume representing the neutron balance.

This provides a balance equation for the phase volume $d\mathbf{r}dE$

$$q(\mathbf{r}, E + dE)d\mathbf{r} - q(\mathbf{r}, E)d\mathbf{r} + S(\mathbf{r}, E)dEd\mathbf{r} = \nabla \cdot \mathbf{J}(\mathbf{r}, E)dEd\mathbf{r} + \Sigma_a \phi(\mathbf{r}, E)dEd\mathbf{r}. \tag{9.9}$$

Dividing Eq. (9.9) by phase volume $d\mathbf{r}dE$, and putting it into a differential equation, yields the desired balance equation written in terms of the slowing down density $q(\mathbf{r}, E)$:

$$\frac{\partial q(\mathbf{r}, E)}{\partial E} + S(\mathbf{r}, E) = \nabla \cdot \mathbf{J}(\mathbf{r}, E) + \Sigma_a \phi(\mathbf{r}, E). \tag{9.10}$$

Recognizing that the lethargy $u$ of neutrons increases as neutrons slow down and lose energy, rewrite Eq. (9.10) in terms of $u$:

$$-\frac{\partial q(\mathbf{r}, u)}{\partial u} + S(\mathbf{r}, u) = \nabla \cdot \mathbf{J}(\mathbf{r}, u) + \Sigma_a \phi(\mathbf{r}, u). \tag{9.11}$$

It is instructive to derive Eq. (9.10) alternatively by formally expressing the slowing down density $q(\mathbf{r}, E)$ as the difference between the down-scattering and up-scattering rates across $E$:

$$\begin{aligned} q(\mathbf{r}, E) = &\int_E^\infty dE' \int_0^E dE'' \,\Sigma_s(E' \to E'')\phi(\mathbf{r}, E') \\ &- \int_E^\infty dE' \int_0^E dE'' \,\Sigma_s(E'' \to E')\phi(\mathbf{r}, E''). \end{aligned} \tag{9.12}$$

For now, we have kept the upper limit of the integral over $E'$ at $\infty$ and the lower limit of the integral over $E''$ at 0, although both limits will have to be more restrictive if we wish to account for limited energy degradations possible for neutrons suffering collisions with nuclei with mass number $A > 1$. Applying Leibnitz's general differentiation formula

$$\frac{d}{d\alpha} \int_{a(\alpha)}^{b(\alpha)} f(x, \alpha)dx = f(b, \alpha)\frac{db}{d\alpha} - f(a, \alpha)\frac{da}{d\alpha} + \int_a^b \frac{\partial f}{\partial \alpha}dx \tag{9.13}$$

twice to Eq. (9.12) yields an expression for the rate of change in the slowing down density:

$$\frac{\partial q(\mathbf{r},E)}{\partial E} = \int_0^\infty dE'\Sigma_s(E' \to E)\phi(\mathbf{r},E') - \Sigma_s(E)\phi(\mathbf{r},E). \tag{9.14}$$

Equation (9.14) represents a simple neutron balance statement that the difference between the in-scattering rate and out-scattering rate is equal to the rate of change of the slowing down density with respect to energy. Substituting Eq. (9.14) into the diffusion equation (9.7b) yields Eq. (9.10), obtained by setting up a neutron balance in the phase volume $d\mathbf{r}dE$. Note also that Eq. (9.14) combined with Eq. (9.7a) yields Eq. (9.11), obtained via an intuitive physical approach instead of the mathematical approach employed in Eqs. (9.12) and (9.14). Thus, only two of the balance equations (9.7), (9.10), and (9.14) are independent.

Finally, define the *collision density* as the total collision rate per unit volume at $\mathbf{r}$ per unit energy interval at $E$:

$$F(\mathbf{r},E) = \Sigma_t(E)\phi(\mathbf{r},E). \tag{9.15}$$

For an infinite medium, the balance equations (9.10) and (9.7) may be written in terms of the collision density:

$$\frac{dq(E)}{dE} + S(E) = \Sigma_a(E)\phi(E) = \frac{\Sigma_a(E)}{\Sigma_t(E)}F(E), \tag{9.16a}$$

$$F(E) = \Sigma_t(E)\phi(E) = \int_0^\infty dE'\Sigma_s(E' \to E)\phi(E') + S(E). \tag{9.16b}$$

These are the equations that will form the basis for the bulk of our study of the neutron flux spectrum in the slowing down range in Section 9.3.

## 9.2 Elastic Scattering and Lethargy Variable

Through the study of elastic scattering of neutrons, Section 2.5 provides and clarifies the concept of the scattering kernel in Eq. (2.52), with the observation that neutrons are scattered isotopically in the center-of-mass (CM) system. Recall that the scattering kernel $\sigma_s(E_0 \to E)$ is constant over the interval $[\alpha E_0, E_0]$ and vanishes everywhere else, as displayed in Figure 2.12. Recall also the relationship between the total scattering cross section and the scattering kernel, rewritten in terms of the macroscopic cross section:

$$\Sigma_s(E_0) = \int_0^\infty dE\,\Sigma_s(E_0 \to E) = \int_{\alpha E_0}^{E_0} dE\,\Sigma_s(E_0 \to E). \tag{9.17}$$

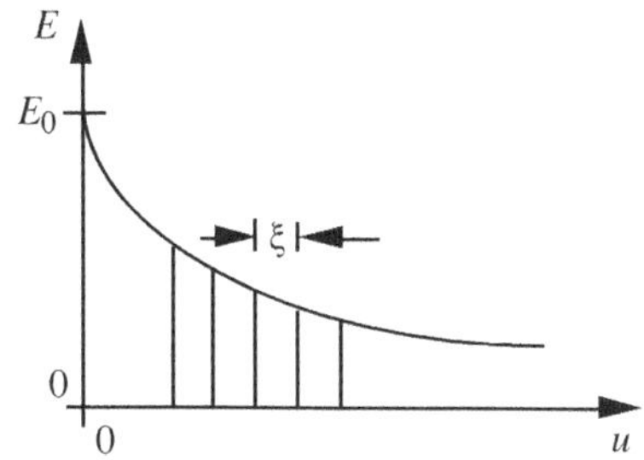


**Figure 9.2** Lethargy variable and average lethargy increase per collision.

For elastic scattering with hydrogen nuclei, $\alpha = 0$, and Eq. (2.52) implies that the scattering kernel $\Sigma_s(E_0 \rightarrow E)$ is constant for all energies below $E_0$.

Since lethargy $u = \ln(E_0/E)$, we may write

$$E = E_0 e^{-u}$$

and, as shown schematically in Figure 9.2, the lethargy of neutrons increases monotonically as neutrons slow down and lose energy. The scattering kernel from Eq. (2.52) may also be written in terms of lethargy

$$\begin{aligned} \Sigma_s(u' \rightarrow u) &= \Sigma_s(E' \rightarrow E)\left(-\frac{dE}{du}\right) \\ &= \Sigma_s(u')p(u' \rightarrow u) = \begin{cases} \dfrac{\Sigma_s(u')e^{u'-u}}{1-\alpha}, & u - \Delta \leq u' \leq u, \\ 0, & \text{otherwise,} \end{cases} \end{aligned} \tag{9.18}$$

where $\Delta = \ln(1/\alpha)$ is again the maximum lethargy increase per collision and $p(u' \rightarrow u)$ represents the conditional probability that a neutron of lethargy $u'$ will emerge from the collision process with lethargy in unit lethargy interval around $u$. The conditional probability is equivalent to the conditional probability of Eq. (2.52) expressed in terms of the energy variable:

$$p(E' \rightarrow E) = \frac{1}{E'(1-\alpha)}. \tag{9.19}$$

Recalling Eq. (7.3), note

$$\phi(u) = E\phi(E) \text{ and } F(u) = EF(E). \tag{9.20}$$

In contrast, since the slowing down density $q(E)$ is not a density function, but represents the net number of neutrons slowing down past $E$ per unit volume and unit time, which is equal to the net number of neutrons scattered past lethargy $u$ corresponding to $E$, yielding

$$q(u) = q(E). \tag{9.21}$$

The average lethargy increase per collision $\xi$ may also be calculated:

$$\begin{aligned}\xi = \langle \Delta u \rangle &= \int_{u-\Delta}^{u} (u-u')p(u' \to u)du' \\ &= \int_{u-\Delta}^{u} (u-u')\frac{e^{u'-u}}{1-\alpha}du' = \frac{1}{1-\alpha}\int_{0}^{\Delta} xe^{-x}dx.\end{aligned} \tag{9.22}$$

The averaging process involves, as a weighting factor, the *conditional probability* $p(u' \to u)$ of Eq. (9.18). By definition of the conditional probability, the integral of $p(u' \to u)$ over the possible collision interval $[u-\Delta, u]$ is unity so that the integral over $u'$ in Eq. (9.22) is properly normalized. Performing the integral in Eq. (9.22) yields

$$\xi = 1 + \frac{\alpha}{1-\alpha}\ln\alpha, \tag{9.23}$$

which is independent of lethargy as illustrated in Figure 9.2. For $A \gg 1, \xi$ can be approximated as

$$\xi \simeq \frac{2}{A+\frac{2}{3}}. \tag{9.24}$$

The integral for $\xi$ in Eq. (9.22) may also be evaluated as an average of the logarithmic energy decrement, $\ln(E_0/E)$, with the conditional probability of Eq. (9.19).

Before returning to the task of solving the slowing down equations, we will note here some of the advantages of using the lethargy variable $u$ in fast spectrum calculations. One immediate advantage follows from the basic property that $u$ increases as neutrons slow down so that, in a discretized formulation of the slowing down equations, one may naturally number the groups along the direction of increasing $u$, as noted in Section 7.1. Another, and perhaps more substantial, advantage is that for asymptotic slowing down regions, $\phi(E) \propto 1/E$ while $\phi(u) \simeq$ constant. The relationship, as discussed in Section 9.3, is a key result obtained in slowing down theory and renders the flux-weighted cross section collapsing process simpler when performed in the lethargy space. The property is used in Eq. (7.8) for the calculation of multigroup cross sections. In addition, the fact that the mean lethargy increase per collision $\xi$ is independent of $u$ provides generally simpler formulation of fast spectrum calculations and slowing down of neutrons.

## 9.3 Neutron Slowing Down in Infinite Medium

To obtain a basic understanding of the behavior of neutrons in the slowing down range, we now turn our attention to the simple neutron balance equations (9.16) applicable to a large medium where neutron leakage and hence the spatial dependence of neutron flux may be neglected. As a reasonable approximation in the slowing down range, neglect up-scattering of neutrons. Furthermore, to simplify the algebra, we consider the idealized case of a mono-energetic source of neutrons at energy $E_0$ of strength $Q$ [neutron · cm$^{-3}$ s$^{-1}$] uniformly distributed throughout the infinite medium. That is, the source term in Eqs. (9.16) is represented as

$$S(E) = Q\delta(E - E_0). \tag{9.25}$$

Since the scattering kernel of Eq. (2.52) is finite only over an energy interval $(1 - \alpha)E_0$, we need to break up the entire energy range below the source energy into an interval down to $\alpha E_0$ and that below $\alpha E_0$ for separate analyses. The energy boundary is chosen because source neutrons suffering first collisions cannot have energy below $\alpha E_0$. For neutrons undergoing scattering with hydrogen nuclei, the first collision interval, i.e. $\alpha E_0 < E < E_0$, covers the whole energy range below $E_0$, and the analysis for the first collision interval will suffice for a scattering medium consisting of a hydrogenous material.

### 9.3.1 Slowing Down in the First Collision Interval

Since the source term of Eq. (9.25) includes a Dirac delta function, it is reasonable to expect that the solution to Eqs. (9.16) for the scalar flux $\phi(E)$ should contain a delta function term as well as a regular, continuous function. Thus, consider $\phi(E)$ in the form

$$\phi(E) = \phi_c(E) + A\delta(E - E_0), \tag{9.26}$$

where the well-behaved function $\phi_c(E)$ and the constant $A$ are yet to be determined.

1. **Collision density**
   Substituting Eq. (9.26) into Eq. (9.16b) yields

$$\Sigma_t(E)\phi_c(E) + A\Sigma_t(E)\delta(E - E_0) = \int_E^{E_0} dE' \frac{\Sigma_s(E')\phi_c(E')}{E'(1-\alpha)} + \frac{A\Sigma_s(E_0)}{E_0(1-\alpha)} + Q\delta(E - E_0). \tag{9.27}$$

   To handle the delta function term, integrate Eq. (9.27) over the energy interval $[E_0 - \epsilon, E_0 + \epsilon]$ and let $\epsilon \to 0$. Since the function $\phi_c(E)$ remains finite for all $E$, the integrals involving the regular functions vanish as $\epsilon \to 0$, and the integrals involving the delta function are equated to obtain

$$A\Sigma_t(E_0) = Q. \tag{9.28}$$

   We may also obtain Eq. (9.28) simply by arguing that the delta function terms cannot be equated to any regular, well-behaved functions and hence that the two delta function terms in Eq. (9.27) have to be equated to each other.

Furthermore, since $\delta(E - E_0)$ vanishes except when $E = E_0$, the total cross section $\Sigma_t(E)$ has to be evaluated at $E = E_0$ in Eq. (9.28).

With the delta function terms in Eq. (9.27) separated out, consider the balance equation for $\phi_c(E)$:

$$\Sigma_t(E)\phi_c(E) = \int_E^{E_0} dE' \frac{\Sigma_s(E')\phi_c(E')}{E'(1-\alpha)} + \frac{Q\Sigma_s(E_0)}{\Sigma_t(E_0)E_0(1-\alpha)}. \tag{9.29}$$

The last term in Eq. (9.29) represents the rate at which source neutrons undergo first scattering collisions and emerge in unit energy interval around $E$, while the integral represents the rate at which neutrons scatter somewhere below the source energy and end up in unit energy interval around $E$. In an infinite medium, the sum of these two in-scattering terms has to equal the total rate of neutrons suffering either scattering or absorption collisions in the unit energy interval around $E$. Thus, $\phi_c(E)$ may now be interpreted as the *collided flux*, i.e. the flux due to neutrons that have undergone at least one collision, while the delta function term in Eq. (9.26) represents the *uncollided flux*, i.e. the flux due to source neutrons yet to suffer collisions. With these interpretations, rewrite Eq. (9.26)

$$\phi(E) = \phi_c(E) + \frac{Q}{\Sigma_t(E_0)}\delta(E - E_0) \tag{9.30}$$

and similarly obtain an expression for the total collision density:

$$F(E) = F_c(E) + Q\delta(E - E_0). \tag{9.31}$$

Both Eqs. (9.30) and (9.31) indicate that, for $E < E_0$, the distinction between the collided and total flux vanishes such that $\phi(E) = \phi_c(E)$ and $F(E) = F_c(E)$, but that the presence of source neutrons undergoing collisions at $E_0$ should be explicitly represented. This complication comes from our idealized representation of the source neutrons as a delta function and is a small penalty to pay to obtain simple analytic solutions to a problem that may otherwise be intractable.

In terms of the *collided component* $F_c(E)$ of the collision density of Eq. (9.31), the balance equation (9.29) is rewritten

$$F_c(E) = \int_E^{E_0} dE' \frac{\Sigma_s(E')F_c(E')}{\Sigma_t(E')E'(1-\alpha)} + F_c(E_0), \tag{9.32}$$

where the rate of the *source neutrons undergoing first scattering collisions* and emerging in unit energy interval around $E$ is formally recognized as

$$F_c(E_0) = \frac{Q\Sigma_s(E_0)}{\Sigma_t(E_0)E_0(1-\alpha)}. \tag{9.33}$$

Note that the ratio $Q\Sigma_s(E_0)/\Sigma_t(E_0)$ in Eq. (9.33) yields the *rate of source neutrons suffering scattering collisions*, rather than getting absorbed, at the source energy $E_0$, i.e. slowing down density $q(E_0)$, which is multiplied by the term $1/E_0(1-\alpha)$, the conditional probability $p(E_0 \rightarrow E)$ from Eq. (9.19), i.e. the probability of scattered neutrons emerging in unit energy interval below $E_0$ but above $\alpha E_0$.

Since Eq. (9.32) is an integral equation, differentiate it with respect to $E$ to cast it into a differential equation:

$$\frac{dF_c(E)}{dE} = -\frac{\Sigma_s(E)}{\Sigma_t(E)E(1-\alpha)} F_c(E). \tag{9.34}$$

Together with the simple boundary condition that $F_c(E)$ at $E_0$ is $F_c(E_0)$ from Eq. (9.33), Eq. (9.34) is integrated back:

$$\begin{aligned} F_c(E) &= F_c(E_0) \exp\left[\int_E^{E_0} \frac{\Sigma_s(E')dE'}{\Sigma_t(E')E'(1-\alpha)}\right], \\ &= F_c(E_0)\left(\frac{E_0}{E}\right)^{1/(1-\alpha)} \exp\left[-\int_E^{E_0} \frac{\Sigma_a(E')dE'}{\Sigma_t(E')E'(1-\alpha)}\right]. \end{aligned} \tag{9.35}$$

The expression for the collided component of the collision density may be combined with the uncollided component to yield the total collision density $F(E)$ of Eq. (9.31). The uncollided component of $F(E)$ represents the collision density associated with source neutrons suffering first collisions.

2. **Slowing down density**
   Neglect up-scattering in Eq. (9.12) to obtain the corresponding solution for the slowing down density

$$q(E) = \int_E^{E_0} dE' \int_{\alpha E'}^{E} dE'' \frac{\Sigma_s(E')\phi(E')}{E'(1-\alpha)}, \tag{9.36}$$

with particular attention given to the lower limit $\alpha E'$ for energy $E''$ of the neutrons landing below the energy $E$ of our interest. Substituting Eq. (9.30) for $\phi(E')$ into Eq. (9.36) yields

$$\begin{aligned} q(E) &= \int_E^{E_0} dE' \frac{F_c(E')\Sigma_s(E')}{\Sigma_t(E')} \frac{E-\alpha E'}{E'(1-\alpha)} + \frac{Q\Sigma_s(E_0)(E-\alpha E_0)}{\Sigma_t(E_0)E_0(1-\alpha)} \\ &= \int_E^{E_0} dE' \frac{F_c(E')\Sigma_s(E')}{\Sigma_t(E')} \frac{E-\alpha E'}{E'(1-\alpha)} + F_c(E_0)(E-\alpha E_0), \end{aligned}$$

which, with the help of Eq. (9.34), is rewritten

$$q(E) = EF_c(E) - \alpha E_0 F_c(E_0) + \alpha \int_E^{E_0} dE' E' \frac{dF_c(E')}{dE'}.$$

Finally, an integration by parts yields a general expression for the slowing down density:

$$q(E) = (1-\alpha)EF_c(E) - \alpha \int_E^{E_0} F_c(E')dE'. \tag{9.37}$$

In particular, note that the slowing down density at the source energy is given by

$$q(E_0) = (1-\alpha)E_0 F_c(E_0) = \frac{Q\Sigma_s(E_0)}{\Sigma_t(E_0)}, \tag{9.38}$$

as recognized in connection with Eq. (9.33) for $F_c(E_0)$. Note also that Eqs. (9.35) and (9.37) are general expressions valid in the first collision interval, $\alpha E_0 < E < E_0$, for both hydrogenous and non-hydrogenous media.

3. **Resonance escape probability**

   In terms of the slowing down density $q(E)$, define the probability of neutrons escaping absorption collisions in slowing down from energy $E_1$ to energy $E_2$:

$$p(E_1 \to E_2) = \frac{q(E_2)}{q(E_1)}, E_2 < E_1. \tag{9.39}$$

   The probability is usually called the *resonance escape probability*, although all absorption reactions, not merely resonance absorptions, should be considered in general.

4. **Specific examples**

   We now illustrate applications of general expressions, Eqs. (9.35) and (9.37), derived for the collision density and slowing down density, respectively. We consider first hydrogenous media with and without capture and then simplify the equations for non-hydrogenous media without capture. Through these specific examples, we will interpret further the expressions obtained so far.

   (a) Hydrogenous medium with capture

   In this case, for all $E < E_0$, merely set $\alpha = 0$ in various governing equations:

$$F_c(E) = \frac{Q\Sigma_s(E_0)}{\Sigma_t(E_0)E} \exp\left[-\int_E^{E_0} \frac{\Sigma_a(E')dE'}{\Sigma_t(E')E'}\right], \tag{9.40a}$$

$$q(E) = EF_c(E) = q(u) = F_c(u) = q(E_0)p(E_0 \to E), \tag{9.40b}$$

$$p(E_1 \to E_2) = \frac{q(E_2)}{q(E_1)} = \exp\left[-\int_{E_2}^{E_1} \frac{\Sigma_a(E)dE}{\Sigma_t(E)E}\right], E_1 > E_2, \tag{9.40c}$$

$$\phi(E) = \frac{1}{\Sigma_t(E)}\left[\frac{q(E)}{E} + Q\delta(E-E_0)\right]. \tag{9.40d}$$

   From the resonance escape probability of Eq. (9.40c), note that the exponential term in Eq. (9.40a) represents the probability $p(E_0 \to E)$ of neutrons escaping absorptions in slowing down from the source energy all the way to energy $E$. Thus, recognize that collision density $F_c(E)$ in an infinite medium should be equal to the rate of neutrons slowing down into unit energy interval around $E$, which is given as the {slowing down density $q(E_0)$ at $E_0$} × {probability $1/E$ of neutrons emerging in unit energy interval around $E$} × {resonance escape probability $p(E_0 \to E)$}. Note also the general $1/E$ dependence in the collided component of the flux $\phi(E)$, which is discussed further in Section 9.3.2.

(b) Hydrogenous medium without capture
In this idealized case, for all $E < E_0$, set $\Sigma_a(E) = 0$ as well as $\alpha = 0$:

$$F_c(E) = \frac{Q}{E}, F_c(u) = Q, \tag{9.41a}$$

$$q(E) = EF_c(E) = Q = q(u) = F_c(u), \tag{9.41b}$$

$$p(E_1 \rightarrow E_2) = 1.0, E_2 < E_1, \tag{9.41c}$$

$$\phi(E) = \frac{Q}{\Sigma_s(E)}\left[\frac{1}{E} + \delta(E - E_0)\right]. \tag{9.41d}$$

With the resonance escape probability trivially set to unity, the simplest expressions are obtained for the collision density, slowing down density, and flux spectrum. In particular, note that the slowing down density at all $E$ has to be merely equal to the source strength $Q$.

(c) Non-hydrogenous medium without capture
Here the resonance escape probability is again unity, and we obtain

$$F_c(E) = \frac{Q}{E_0(1-\alpha)}\left(\frac{E_0}{E}\right)^{1/(1-\alpha)}, \tag{9.42a}$$

$$q(E) = q(u) = Q, \tag{9.42b}$$

$$p(E_1 \rightarrow E_2) = 1.0, E_2 < E_1, \tag{9.42c}$$

$$\phi(E) = \frac{Q}{\Sigma_s(E)}\left[\frac{(E_0/E)^{1/(1-\alpha)}}{E_0(1-\alpha)} + \delta(E - E_0)\right]. \tag{9.42d}$$

### 9.3.2 Slowing Down Below the First Collision Interval

We note that, for neutrons undergoing scattering collisions with hydrogen nuclei, $\alpha = 0$ and the expressions for the flux, slowing down density, collision density, and resonance escape probability derived in Section 9.3.1 for the first collision interval, $\alpha E_0 < E < E_0$, are valid for all energies below the source energy. For non-hydrogenous materials, for which $\alpha \neq 0$, neutron slowing down characteristics below $\alpha E_0$ have to be studied separately.

For $E < \alpha E_0$, with the mono-energetic source at $E_0$, there is no direct contribution from the neutron source, and the distinction between the collided and total flux disappears:

$$\phi(E) = \phi_c(E) \text{ and } F(E) = F_c(E).$$

Thus, the balance equation (9.16b) is simplified to

$$F(E) = \int_E^{E/\alpha} dE' \frac{\Sigma_s(E')F(E')}{\Sigma_t(E')E'(1-\alpha)} \tag{9.43}$$

and may be used to evaluate the collision density for neutron energy just below the first collision boundary

$$F(\alpha E_0 - \varepsilon) = F_c(\alpha E_0 - \varepsilon) = \int_{\alpha E_0}^{E_0} dE' \frac{\Sigma_s(E')F_c(E')}{\Sigma_t(E')E'(1-\alpha)}, \tag{9.44}$$

for a small positive number $\epsilon$. We also use Eq. (9.32) to calculate the collision density for $E = \alpha E_0 + \epsilon$, i.e. neutron energy just above the boundary:

$$F_c(\alpha E_0 + \varepsilon) = \int_{\alpha E_0}^{E_0} dE' \frac{\Sigma_s(E')F_c(E')}{\Sigma_t(E')E'(1-\alpha)} + F_c(E_0). \tag{9.45}$$

Comparing Eqs. (9.44) and (9.45) shows that the collision density $F(E)$ suffers a discontinuity of magnitude $F_c(E_0)$ at the first collision boundary $E = \alpha E_0$. Since $F_c(E_0)$ represents the contribution to $F(E)$ due to source neutrons suffering first collisions, the discontinuity in $F(E)$ at $E = \alpha E_0$ simply reflects the fact that source neutrons may be scattered, in first collisions, only down to the first collision boundary $E = \alpha E_0$ but not below that. This phenomenon results in discontinuities in $F(E)$ or $dF(E)/dE$ at $E = \alpha^n E_0, n = 1, 2, \ldots,$ but the discontinuities decrease in magnitude as $n$ increases or as the energy decreases. For $n > 3$, or $E < \alpha^3 E_0$, the discontinuities become sufficiently small and may appear to have vanished, yielding more or less smooth, asymptotic behavior for $F(E)$. This approach to the asymptotic behavior of the collision density is illustrated in Figure 9.3. Since $\alpha$ is an increasing function of mass number $A$, this asymptotic collision density is attained at a higher energy for heavy scatterers than for light scatterers. These observations are established through analytical solutions for the collision density, which are possible only for pure scatterers.

At the asymptotic energy range, since the mean logarithmic energy decrement or mean lethargy increase per collision $\xi$ is independent of energy, we expect that the number of collisions per unit lethargy will be $1/\xi$ for each neutron. Hence,

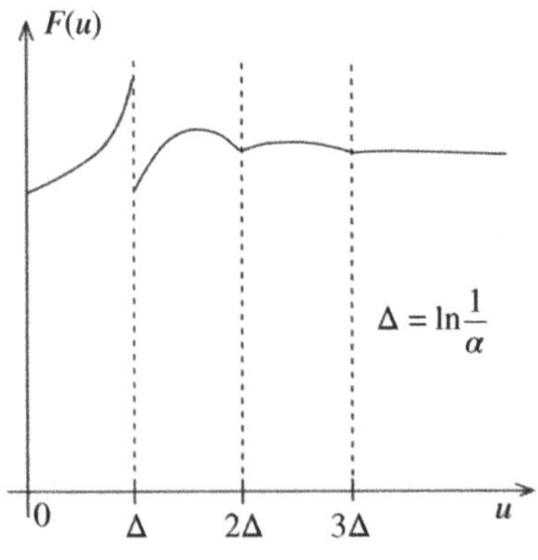


**Figure 9.3** Collision density approaching the asymptotic behavior.

**Figure 9.4** Collision intervals before and after the collision.

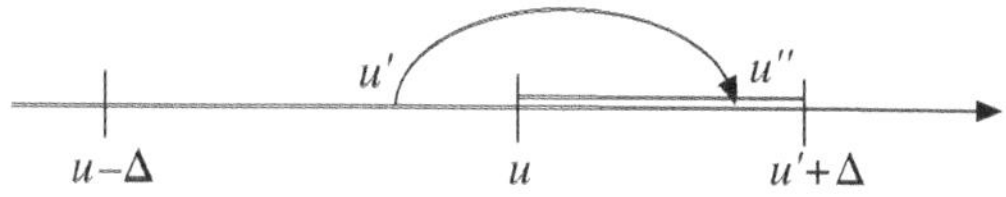


with $q(u)$ [neutrons · cm$^{-3}$ s$^{-1}$] slowing down past $u$, corresponding to energy $E = E_0 e^{-u}$, note that the asymptotic collision density $F(u)$, i.e. the number of collisions per cm$^3$ per s per unit lethargy should be given by

$$\begin{aligned} F(u) &= q(u)\left(\frac{\text{number of neutrons}}{\text{cm}^3\cdot\text{s}}\right)\times\frac{1}{\xi}\left(\frac{\text{number of collisions}}{\text{lethargy}\cdot\text{neutron}}\right) \\ &= \frac{q(u)}{\xi}\left(\frac{\text{number of collisions}}{\text{cm}^3\cdot\text{s}\cdot\text{lethargy}}\right). \end{aligned} \tag{9.46}$$

This intuitive relationship is one of the key results of slowing down theory, and we will now formally derive it, beginning with the definition of the slowing down density, Eq. (9.12). In the process, a number of approximations are introduced and thereby establish the conditions for the validity of Eq. (9.46).

With the up-scattering of neutrons neglected, for $E < \alpha E_0$, Eq. (9.12) may be simplified

$$q(E) = \int_E^{E/\alpha} dE' \int_{\alpha E'}^{E} dE''\,\Sigma_s(E' \to E'')\phi(E'), \tag{9.47}$$

where the minimum fractional energy after collision is duly recognized as $\alpha$. In addition, note that the lower limit in the inner integral covering energy $E''$ after the collision has to be set to $\alpha E'$, not $\alpha E$, as noted in Eq. (9.36). The slowing down density may be written equivalently in terms of lethargy

$$q(u) = q(E) = \int_{u-\Delta}^{u} du' \int_{u}^{u'+\Delta} du''\,\Sigma_s(u' \to u'')\phi(u'), \tag{9.48}$$

with the limits of integrations before and after the collision illustrated in Figure 9.4.

Substituting the scattering kernel from Eq. (9.18) into Eq. (9.48) and integrating over $u''$ yields

$$q(u) = \int_{u-\Delta}^{u} du' \frac{\Sigma_s(u')\phi(u')}{1-\alpha}\left(e^{u'-u} - \alpha\right). \tag{9.49}$$

Analytical solutions to Eq. (9.43) for the collision density and Eq. (9.49) for the slowing down density have been obtained for the idealized case of neutrons slowing down in infinite media without capture. Since our primary interest is in diffusing media with capture at the asymptotic energy range, we obtain an approximate solution to Eq. (9.49), connecting the slowing down density $q(u)$ to the flux $\phi(u)$. For this purpose, the scattering collision density in Eq. (9.49) is expressed in terms of a two-term Taylor series expansion

$$\Sigma_s(u')\phi(u') \simeq \Sigma_s(u)\phi(u) + (u' - u)\frac{d}{du}\left[\Sigma_s(u)\phi(u)\right], \tag{9.50}$$

which may be considered a mathematical statement that the scattering rate is slowly varying in the lethargy space. Upon substituting Eq. (9.50) into Eq. (9.49), we obtain, with a bit of algebra

$$q(u) = \xi\Sigma_s(u)\phi(u) - \gamma\,\xi\,\frac{d}{du}[\Sigma_s(u)\phi(u)], \tag{9.51}$$

where $\xi$ is the mean lethargy gain per collision of Eq. (9.23) and

$$\gamma = 1 - \frac{\alpha \ln^2\alpha}{2(1-\alpha)\xi}. \tag{9.52}$$

Substituting the first approximation $q(u) = \xi\Sigma_s(u)\phi(u)$ from Eq. (9.51) back into the second term yields

$$q(u) \simeq \xi\Sigma_s(u)\phi(u) - \gamma\,\frac{dq(u)}{du}. \tag{9.53}$$

For a source-free problem under consideration, Eq. (9.11) simplifies to

$$\frac{dq(u)}{du} = -\Sigma_a(u)\phi(u), \tag{9.54}$$

which represents a simple statement that the decrease in the slowing down density in an infinite medium, away from the source, is due to neutron absorptions. Substituting Eq. (9.54) into Eq. (9.53) yields a simple expression for the slowing down density:

$$q(u) \simeq \left[\xi\Sigma_s(u) + \gamma\Sigma_a(u)\right]\phi(u). \tag{9.55}$$

Finally, as illustrated in Figure 9.5, approximate the parameter $\gamma$ in Eq. (9.55) by $\xi$, the mean lethargy gain per collision, to obtain

$$q(u) \simeq \xi\Sigma_t(u)\phi(u), \tag{9.56a}$$

$$q(E) \simeq \xi E\Sigma_t(E)\phi(E). \tag{9.56b}$$

Equation (9.56a) is equal to the expression for the collision density heuristically obtained in Eq. (9.46) and is expected to be valid for the asymptotic energy range, when the scattering rate is weakly varying across a scattering interval so that the two-term Taylor's expansion in Eq. (9.50) is a valid approximation. Thus, Eqs. (9.56) may be used for scattering media with (a) weak captures or (b) slowly varying captures. Equations (9.56) may also be used (c) if mass number $A \gg 1$ so that the scattering intervals, indicated by $\Delta = \ln(1/\alpha)$, are small compared with the rate of cross-section variations, and (d) in the energy regions away from absorption resonances. For a weakly absorbing medium, the slowing down density $q(u)$, as well as $\Sigma_t(u)$, does not vary rapidly as a function of $u$, since

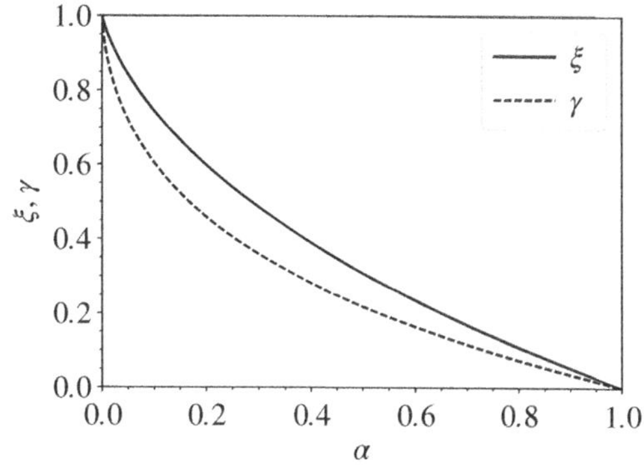


**Figure 9.5** Comparison of parameters $\gamma$ and $\xi$ as a function of $\alpha$.

$\Sigma_s(u)$, primarily comprising the potential scattering cross section, is often nearly constant. Hence,

$$\phi(u) \simeq \frac{Q}{\xi\Sigma_s} \simeq \text{constant, or } \phi(E) \simeq \frac{Q}{\xi\Sigma_s E} \propto \frac{1}{E}. \tag{9.57}$$

This first-order estimate of flux spectrum often serves as the starting point in generating fine-group cross sections for unit-cell or unit-assembly representations through a lattice physics code. The fine-group cross sections are then used in an accurate calculation of the fast spectrum, which is used in turn to collapse fine-group cross sections into few-group constants for multigroup diffusion theory calculations for the whole core. In fact, Eq. (9.57) is used as $\theta(u) = \text{constant}$ in Eqs. (7.7) and (7.8) in the derivation of the multigroup diffusion equation. This rather simple relationship is indeed one of the key results representing the slowing down of neutrons and plays an important role in processing cross-section libraries in general. Equations (9.56) are often known as the *Fermi approximation* or *continuous slowing down model*, although in early days of reactor physics development, typically around the time of Enrico Fermi's involvement in reactor physics, no explicit accounting was made of neutron absorptions in relationships in the form of Eqs. (9.56).

## 9.4 Resonance Escape Probability

Having obtained a general understanding of the neutron flux spectrum in the slowing down range both for hydrogenous and non-hydrogenous media in Section 9.3, we turn our attention now to evaluating the resonance escape probability for the asymptotic slowing down range. Since resonances of our interest usually are located at energies far below that of neutrons released from the fission process, we find it useful to derive an expression for the resonance escape probability based on the Fermi approximation for the slowing down density, Eqs. (9.56), and the effective resonance integral.

### 9.4.1 Effective Resonance Integral

With this purpose in mind, we divide both sides of the neutron balance equation (9.54) by $q$ and integrate over $u$ to obtain

$$q(u) = q(0) \exp\left[-\int_0^u du' \frac{\Sigma_a(u')\phi(u')}{q(u')}\right],$$
$$p(u_1 \to u_2) = \frac{q(u_2)}{q(u_1)} = \exp\left[-\int_{u_1}^{u_2} du \frac{\Sigma_a(u)}{\Sigma_t(u)} \frac{F(u)}{q(u)}\right], u_1 < u_2, \tag{9.58}$$

where Eq. (9.39) is used to determine the resonance escape probability $p(u_1 \to u_2)$ as the fraction of neutrons slowing down past energy $E_1$, lethargy $u_1$, escaping absorption in reaching energy $E_2$, lethargy $u_2$. Since $F(u)$ and $q(u)$ both involve integrals of $\phi(u)$ over $u$, they are, in general, more smoothly varying than the flux $\phi(u)$ itself. Hence, assume, even in the vicinity of absorption resonances, that the ratio $F(u)/q(u)$ may be approximated by the Fermi approximation of Eq. (9.56a). With this assumption, Eq. (9.58) is rewritten as

$$p(u_1 \to u_2) = \exp\left[-\int_{u_1}^{u_2} du \frac{\Sigma_a(u)}{\xi\Sigma_t(u)}\right], u_1 < u_2. \tag{9.59}$$

Since we have to invoke the Fermi approximation of Eqs. (9.56), the resonance escape probability just obtained is expected to be valid, in general, only for

(a) Narrow resonances, widely separated,
(b) Slowly varying captures, and
(c) Weak captures.

Before we proceed further to discuss Eq. (9.59), we need a qualitative understanding of the general behavior of neutron cross sections in the resonance range. For a mixture of fuel and moderating materials in a typical reactor core configuration, the resonance range covers approximately from 1 eV to a few keV. The subscripts $F$ and $M$ are used to denote the fuel and moderator, respectively, and both the resonance absorption and scattering, including potential and inelastic scattering reactions, are considered:

(a) $\sigma_a(E) \simeq \sigma_{aF}(E) \simeq \sigma_{a,res,F}(E)$,
(b) $\sigma_{sF}(E) = \sigma_{pF}(E) + \sigma_{s,res,F}(E) + \sigma_{in,F}(E) \simeq \sigma_{pF}(E)$,
(c) $\sigma_{sM}(E) = \sigma_{pM}(E) + \sigma_{s,res,M}(E) + \sigma_{in,M}(E) \simeq \sigma_{pM}(E)$,
(d) $\sigma_{aF}(E) \gg \sigma_{pF}(E) \gg \sigma_{s,res,F}(E)$.

Thus, we obtain an expression for the total scattering cross section for the fuel-moderator mixture, with the number densities $N_F$ and $N_M$

$$\Sigma_s(E) = N_F\sigma_{sF}(E) + N_M\sigma_{sM}(E) \simeq \Sigma_{pF}(E) + \Sigma_{pM}(E) = \Sigma_p(E), \tag{9.60}$$

which is usually a slowly varying function of $E$. Similarly, the total absorption cross section is given by

$$\Sigma_a(E) \simeq N_F\sigma_{aF}(E) \simeq N_F\sigma_{a,res,F}(E) = N_F\sigma_a(E). \tag{9.61}$$

With Eq. (9.60) and the last expression in Eq. (9.61), take the terms $N_F$ and $\xi\Sigma_s$ outside the integral in Eq. (9.59) to obtain

$$p\left(u_1 \rightarrow u_2\right) = \exp\left[-\frac{N_F}{\xi\Sigma_s}\int_{u_1}^{u_2} du \frac{\sigma_a(u)}{1+\sigma_a(u)N_F/\Sigma_s}\right] = \exp\left[-\frac{N_F}{\xi\Sigma_s}I\right]$$

or

$$p\left(u_1 \rightarrow u_2\right) = \exp\left[-\frac{1}{\xi\rho}\int_{u_1}^{u_2} du \frac{\sigma_a(u)}{1+\sigma_a(u)/\rho}\right], \rho = \frac{\Sigma_s}{N_F}, \tag{9.62}$$

where $I \equiv I_{eff}$ is the *effective resonance integral* and $\rho$ is the scattering cross section per fuel atom, often called the *dilution factor* or *background cross section*. The expressions for the resonance escape probability in Eqs. (9.59) and (9.62) are known as the *narrow resonance (NR) approximation*, since they are primarily applicable when the resonances are narrow and widely separated so that the resonances have negligible effects on the asymptotic slowing down density of Eqs. (9.56). In general, for a mixture of several nuclides, the mean lethargy gain per collision $\xi$ appearing in Eq. (9.62) has to be replaced by an average over the mixture

$$\bar{\xi} = \frac{\sum_i \xi_i \Sigma_{si}}{\sum_i \Sigma_{si}}. \tag{9.63}$$

### 9.4.2 Energy Self-Shielding Factor

To provide an alternative interpretation of the NR approximation of Eq. (9.62), we first introduce a normalization on the asymptotic flux $\phi_{as}(u) = 1.0$ outside the resonance and assume that the slowing down density $q(u)$ is nearly independent of $u$ and is given by the asymptotic expression $q_{as}(u) = \xi\Sigma_s$ throughout the interval $[u_1, u_2]$. With this approximation, valid for a relatively weak absorber, Eq. (9.58) indicates that the effective resonance integral defined in Eq. (9.62) may be recognized as

$$I = \int_{u_1}^{u_2} \sigma_a(u)\phi(u)\,du = \int_{u_1}^{u_2} \sigma_a(u)\frac{\Sigma_s}{\Sigma_t(u)}du. \tag{9.64}$$

Equation (9.64) indicates that $I$ is a flux-weighted effective absorption cross section, with the flux normalization $\phi_{as}(u) = 1.0$, for the absorption resonances for the fuel included in the interval $[u_1, u_2]$. Furthermore, under the NR approximation, the normalized flux in the resonance interval is given by

$$\phi_{NR}(u) = \frac{\Sigma_s}{\Sigma_t(u)} = \frac{\Sigma_{pM} + \Sigma_{pF}}{\Sigma_{aF} + \Sigma_{pM} + \Sigma_{pF}} = \frac{\rho}{\rho + \sigma_{aF}}, \tag{9.65}$$

where $\rho$ is the dilution factor introduced in Eq. (9.62). For an infinitely dilute system, i.e. as $\rho \to \infty$, $\phi_{NR}(u)$ approaches the asymptotic flux $\phi_{as}(u) = 1.0$. Since the resonance integral $I$ represents a *flux-weighted effective absorption cross section*, the NR flux $\phi_{NR}(u)$ indicates a depression in the flux in the neighborhood of a resonance due to a finite presence of resonance absorbers, which results in a decrease in the absorption rate per absorber atom. Thus, the resonance absorption tends to *shield the absorber itself from neutrons of resonance energy*, and the term $\phi_{NR}(u)$ is called the *energy self-shielding factor*, applicable to a homogeneous mixture of fuel and moderator under consideration here.

The effective resonance integral for an infinitely dilute system

$$I_\infty = I\,(\rho = \infty) = \int_{u_1}^{u_2} \sigma_a\,(u)\,du$$

will be larger than that for a finitely dilute system. In general, if $\rho_1 > \rho_2$, then $I(\rho_1) > I(\rho_2)$. The term $N_F/\xi\Sigma_s = 1/\xi\rho$ will, however, decrease more rapidly as $\rho$ increases. Hence, in general

$$p(\rho_1) > p(\rho_2) \text{ for } \rho_1 > \rho_2 \tag{9.66}$$

and a dilute system is preferable as far as the resonance escape probability is concerned.

The normalized NR flux in Eq. (9.65) may be formally derived by simplifying the neutron balance statement from Eq. (9.7a) for an infinite homogeneous medium in the slowing down range with negligible source production

$$\Sigma_t(u)\phi(u) = \int_{u-\Delta}^{u} \Sigma_s(u' \to u)\phi(u')du' = \Sigma_s, \tag{9.67}$$

where the scattering kernel $\Sigma_s(u' \to u)$ is given by Eq. (9.18). The integral is simplified to $\Sigma_s$ with the approximation that the normalized flux is not perturbed much in the interval $[u - \Delta, u]$ due to the presence of a resonance at $u$, and hence that $\phi(u) \simeq 1.0$, together with the observation that the lethargy dependence of $\Sigma_s$, mostly associated with potential scattering, may be suppressed.

### 9.4.3 Wide Resonance Approximation

The NR approximations for the effective resonance integral of Eq. (9.62) and energy self-shielding factor of Eq. (9.65) are expected to be valid if the resonances are narrow. A figure of merit in this regard is the *practical width* $\Gamma_p$ of an absorption resonance, defined in Eq. (2.40). For a fuel-moderator mixture, the practical width should account for the presence of moderator nuclei and should be evaluated as the energy interval over which the resonance cross section is at least as large as the total potential scattering cross section $\Sigma_p$ for the mixture

$$\Gamma_p = \Gamma\sqrt{\frac{\Sigma_{a0}}{\Sigma_p}} = \Gamma\sqrt{\frac{N_F\sigma_{a0}}{\Sigma_p}} \tag{9.68}$$

where $\sigma_{a0}$ is the peak resonance absorption cross section for the fuel resonance. Note that resonances are usually narrow compared with the energy loss due to scattering collisions with moderator nuclei, i.e.

$$\Gamma_p \ll (1 - \alpha_M)E_0. \tag{9.69}$$

A similar condition does not often hold for fuel materials, where $\Gamma_p$ may often be larger than the energy loss due to scattering with fuel nuclei:

$$\Gamma_p > (1 - \alpha_F)E_0. \tag{9.70}$$

This is primarily because $\alpha$ becomes close to one for heavy nuclei.

To account for the presence of broad resonances and thereby remove the inaccuracies introduced by the NR approximation, consider another approach. In this *narrow resonance infinite mass (NRIM)* or *wide resonance (WR) approximation*, allow the resonances to be broad but take the limit as fuel mass number $A_F \to \infty$ or as $\alpha_F \to 1.0$. In this limit, the scattering integral in Eq. (9.67) is broken up into one for the fuel region and another for the moderator region, with the NR approximation for the moderator scattering integral:

$$\Sigma_t(u)\phi(u) = \int_{u-\Delta_F}^{u} \Sigma_{sF}(u' \to u)\phi(u')du' + \Sigma_{pM}. \tag{9.71}$$

With the limit $\alpha_F \to 1.0$ applied to the fuel-scattering integral, the WR flux is represented by

$$\Sigma_t(u)\phi(u) = \Sigma_{pF}\phi(u) + \Sigma_{pM}, \tag{9.72}$$

or

$$\phi_{WR}(u) = \frac{\Sigma_{pM}}{\Sigma_{pM} + \Sigma_{aF}(u)} = \frac{\Sigma_s - \Sigma_{pF}}{\Sigma_t(u) - \Sigma_{pF}} = \frac{s}{s + \sigma_{aF}(u)},\ s = \frac{\Sigma_{pM}}{N_F}, \tag{9.73}$$

where $s$ is the *moderator potential scattering cross section per absorber atom*. This suggests that the WR approximation is equivalent to the NR approximation with fuel-scattering contributions neglected. Comparison of Eqs. (9.65) and (9.73) also suggests [Gol62, Liu15] an intuitive merging of the NR and WR approximations with the introduction of a parameter $\lambda \in [0,1]$:

$$\Sigma_t(u)\phi(u) = \int_{u-\Delta_F}^{u} (1-\lambda)\Sigma_{pF}(u' \to u)\phi(u')du' + \left[\int_{u-\Delta_F}^{u} \lambda\Sigma_{pF}(u' \to u) + \int_{u-\Delta_M}^{u} \Sigma_{pM}(u' \to u)\right]\phi(u')du'. \quad (9.74)$$

Applying the WR approximation to the first integral and the NR approximation to the second set of two integrals, respectively, yields the *intermediate resonance (IR) approximation*

$$\phi_{IR}(u) = \frac{s + \lambda\sigma_{pF}}{s + \sigma_{aF} + \lambda\sigma_{pF}}, \quad (9.75)$$

where the IR parameter $\lambda \in [0,1]$ is obtained semi-empirically [Gol62]. With a proper choice of $\lambda$, Eq. (9.75) provides an accurate representation of resonances in general. We may use Eq. (9.75) to represent the NR approximation of Eq. (9.65) with $\lambda = 1.0$ as well as the WR approximation of Eq. (9.73) with $\lambda = 0.0$. With the designation of $\sigma^*_{pF} = \lambda\sigma_{pF}$ as the effective fuel potential cross section, we may use the NR expression from Eq. (9.65) with $\Sigma_{pF}$ replaced by $\Sigma^*_{pF} = N_F\sigma^*_{pF}$. This is the approach actually implemented effectively in modern lattice physics codes, e.g. CPM-3 and CASMO-4 [Jon87, Ede93].

### 9.4.4 Probability Table or Subgroup Method

The traditional method to represent the energy dependence of effective resonance integrals has been a multigroup approach where the energy range for the fast spectrum calculation is divided into a number of energy or lethargy groups. With the increasing ability to perform fine-group calculations in recent years, as discussed further in Chapter 11, another approach known as the *subgroup* or *probability table method* has been adopted in advanced lattice physics calculations. With rapid cross section variations around resonances, including unresolved resonances, subdividing the fast spectrum range into more-or-less uniform energy or lethargy intervals tends to smear sharp resonance peaks over the energy intervals making up the energy groups. To remedy this tendency, the *probability table method* was introduced [Lev72, Heb07] so that the cross section data are processed according to the cross section values rather than averaged over energy intervals.

Probability tables for the microscopic cross section $\sigma(u)$ are obtained by the probability density function (PDF) $p(\sigma)$ such that $p(\sigma)d\sigma$ represents the probability that the cross section lies in the interval $d\sigma$ around $\sigma$. With the PDF $p(\sigma)$, an integral $I^*$ of function $f[\sigma(u)]$ over lethargy group g, $\Delta u = u_g - u_{g-1}$, considered as a

Riemann integral over the domain of variable $u$, may be equated to a Lebesque integral [Fri56] over the range of variable $\sigma$

$$I^* = \frac{1}{\Delta u}\int_{u_{g-1}}^{u_g} du f[\sigma(u)] = \int_0^{\sigma_m} d\sigma p(\sigma) f(\sigma), \tag{9.76}$$

where $\sigma_m$ is the maximum value of the cross section $\sigma(u)$ in the interval $\Delta u$. The Lebesque integration may be visualized as summing up the function $f(\sigma)$ over areas $p(\sigma)d\sigma$ at discrete values of $\sigma$. The transformation, as illustrated in Figure 9.6, allows a discrete representation of sharp resonances, and $p(\sigma)$ may be cast as a collection of $L$ Dirac delta functions centered at discrete *subgroups* $\sigma_\ell$, with weights $\omega_\ell, \ell = 1, \dots, L$:

$$p(\sigma) = \sum_{\ell=1}^{L} \delta(\sigma - \sigma_\ell)\omega_\ell. \tag{9.77}$$

The discrete representation of the PDF $p(\sigma)$ yields the integral

$$I^* = \frac{1}{\Delta u}\int_{u_{g-1}}^{u_g} du f[\sigma(u)] = \sum_{\ell=1}^{L} \omega_\ell f(\sigma_\ell), \tag{9.78}$$

in terms of the parameters $\{\sigma_\ell, \omega_\ell, \ell = 1, \dots, L\}$ making up a *probability table* for the resonance cross section $\sigma$ for lethargy group $g$. The probability table method facilitates the use of fewer resonance integrals for discrete cross sections $\sigma_\ell$ preselected around sharp resonance peaks than traditional methods that require

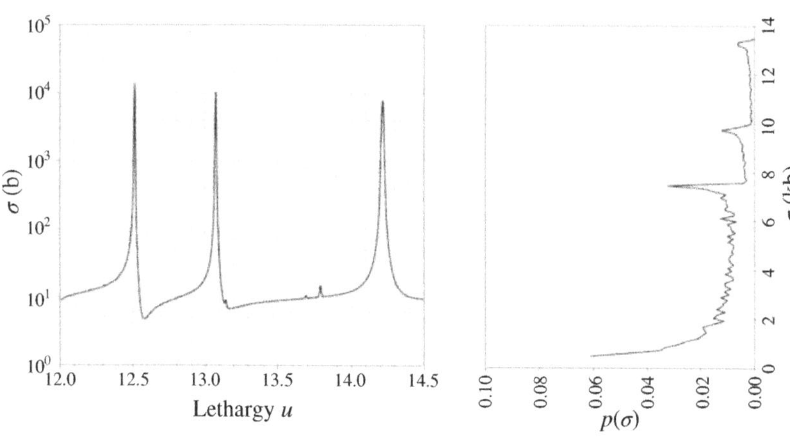


**Figure 9.6** Illustration of the probability table method for absorption cross section $\sigma_a(u)$ for $^{238}$U over the lethargy interval $u = [12.0, 14.5]$ corresponding to the energy interval $E = [37.3, 61.4]$ eV. Probability density function $p(\sigma)$ in the RHS plot represents a distribution of cross section $\sigma_a(u)$ according to its value.

averaging over lethargy intervals. Cullen [Cul74] and Nikolaev [Nik76] discussed the probability table method as the *multiband* and *subgroup method*, respectively.

**Example 9.1** Obtain a subgroup expression for the effective resonance integral for lethargy group g using the NR approximation.

The effective resonance integral of Eq. (9.64) with the normalized NR flux of Eq. (9.65) yields

$$I_{eff} = \int_{\Delta u} \frac{\rho \sigma_a(u)}{\rho + \sigma_a(u)} du = \Delta u \sum_{\ell=1}^{L} \omega_\ell \frac{\sigma_{a\ell}}{1 + \sigma_{a\ell}/\rho}, \quad \sigma_a(u) = \sigma_{aF}(u),$$

which, in the limit of infinite dilution, simplifies to

$$\lim_{\rho \to \infty} I_{eff} = I_\infty = \Delta u \sum_{\ell=1}^{L} \omega_\ell \sigma_{a\ell}.$$

**Example 9.2** Using the four-point probability table for $\sigma_a(u)$ for $^{238}U$ over the lethargy interval $\Delta u = [12.0, 14.5]$ illustrated in Figure 9.6, calculate the effective resonance integral and average absorption cross section $\langle \sigma_a \rangle$ for $\Delta u = 2.5$.

| Group | $\omega_i$ | $\sigma_i$ (b) |
|---|---|---|
| 1 | 0.441 | 4.86 |
| 2 | 0.429 | 11.58 |
| 3 | 0.056 | 33.43 |
| 4 | 0.074 | 1771.68 |

Evaluating the effective resonance integral with an infinite dilution factor $\rho = \infty$ yields $I_\infty = 351.4$ b with a 4-group probability table integral in good agreement with $I_\infty = 357.9$ b obtained via a 200-group conventional integration over the lethargy interval $\Delta u = 2.5$. With a dilution factor $\rho = 106.7$ b, the resonance integral is reduced to $I = 39.2$ b for the subgroup integration, again in good agreement with $I = 38.6$ via conventional integration. The subgroup integration thus yields average absorption cross section $\langle \sigma_a \rangle = 140.6$ b and 15.7 b for $\rho = \infty$ and 106.7 b, respectively.

## 9.5 Doppler Broadening of Resonances

As the temperature of the fuel material increases, thermal motion of the fuel nuclei increases, and the absorption rate of neutrons in fuel resonances increases.

This change in the resonance absorption rate is the result of changes in the relative motion between the neutrons and target nuclei, which broadens the resonance line shape given by the Breit–Wigner formula in Eq. (2.34). This phenomenon is somewhat analogous to the common Doppler effect in acoustics and is known as the *Doppler broadening of resonances*.

### 9.5.1 Qualitative Description of Doppler Broadening

The broadening of resonances can be qualitatively understood if we remember that the cross sections considered so far represent effective cross sections accounting for the thermal motion of target nuclei. For neutrons of velocity $\mathbf{v}$ and nuclei in motion with velocity distribution $N(\mathbf{V})$ at temperature $T$, recall that the reaction rate per neutron is given by Eqs. (3.24) through (3.26)

$$P(v) = N_0\overline{\sigma}(v)v = \int_{\mathbf{V}} d\mathbf{V}\ |\mathbf{v} - \mathbf{V}|\ \sigma(|\mathbf{v} - \mathbf{V}|)N(\mathbf{V}), \tag{9.79}$$

where $\sigma(|\mathbf{v} - \mathbf{V}|)$ is the *true* absorption cross section given as a function of the relative speed between the neutron and nucleus. Note that, for a $1/v$ absorber, the reaction rate is independent of the nuclear velocity distribution and, hence, temperature $T$. For general cases, and especially for sharp resonances, the thermal motion of the target nuclei represented through $N(\mathbf{V})$ in Eq. (9.79) can have a marked effect on the reaction rate. In fact, as $T$ increases, the resonance absorbers become thermally more agitated, and the Breit–Wigner line shape of Eq. (2.34) becomes broader, with the result that the effective resonance absorption cross section $\overline{\sigma}_a(v)$ gets smeared in a manner similar to the Doppler effect in acoustics.

Although the area under the cross section curve remains essentially constant, the cross section smearing results in a decrease in the flux depression represented by the energy self-shielding factor from Eq. (9.65). The broadened resonance also increases the range over which the resonance is felt and increases the energy-integrated *effective absorption rate* represented by the resonance integral $I$. The increase in $I$ in turn decreases the resonance escape probability $p$ of Eq. (9.62) and effective multiplication factor $k_{eff}$ of the system. Thus, the Doppler broadening of resonances provides an inherent negative reactivity feedback mechanism for nuclear reactors in general. Figure 9.7 illustrates how the Breit–Wigner absorption cross section $\overline{\sigma}_a(E)$ at resonance energy $E_0$, simply represented as $\sigma_a(E)$, broadens as a function of temperature $T$, which then reduces the depression in flux $\phi(E)$ around the absorption resonance. The $1/E$ dependence of $\phi(E)$ schematically illustrated represents Eq. (9.57) for the case of an infinitely diluted system, i.e. in the limit of dilution factor $\rho \to \infty$, where the energy self-shielding factor $\phi_{NR}(u)$ of Eq. (9.65) approaches unity; there is no flux depression due to the presence of resonance absorbers infinitely diluted with moderator atoms.

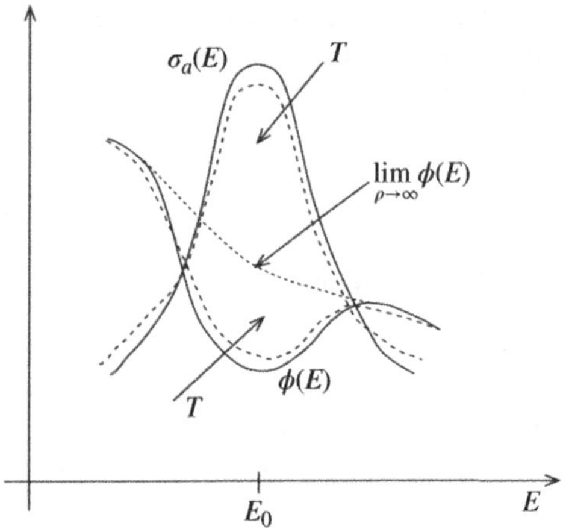


**Figure 9.7** Doppler broadening of absorption resonance at energy $E_0$.

### 9.5.2 Analytical Treatment of Doppler Broadening

We present a simplified analytical description of the temperature dependence of the reaction rate for a single isolated absorption resonance, to gain a better understanding of the statement that the area under the single-level Breit–Wigner line shape remains essentially constant under Doppler broadening as well as obtaining the general temperature dependence of the effective resonance integral. The resonance is assumed sharp so that the capture resonance cross section of Eq. (2.34) is approximated by

$$\sigma_a(E_c) \simeq \frac{\sigma_{a0}}{1+x^2}, \sigma_{a0} = \sigma_0 \frac{\Gamma_\gamma}{\Gamma}, x = \frac{E_c - E_0}{\Gamma/2}, \tag{9.80}$$

where $E_c$ is the CM energy of the neutron–nucleus pair undergoing a resonance reaction at $E_0$. In terms of the reduced mass $\mu = mM/(m+M)$ for the neutron mass $m$ and nuclear mass $M$, write the CM energy in terms of the neutron Lab energy $E$

$$\begin{aligned} E_c &= \frac{1}{2}\mu|\mathbf{v}-\mathbf{V}|^2 = \frac{1}{2}\mu v^2 - \mu \mathbf{v}\cdot\mathbf{V} + \frac{1}{2}\mu V^2 \simeq \frac{1}{2}mv^2 - mvV \\ &= E - \sqrt{2mEV}, \end{aligned} \tag{9.81}$$

with (a) the recognition that $V/v = \sqrt{mkT/ME} \simeq [(0.1\,\text{eV}/4.0\,\text{eV})m/M]^{1/2} \simeq 0.01 \ll 1.0$ for fuel temperature $T = 1200$ K, neutron energy $E = 4.0$ eV, $M/m = A = 238$ for $^{238}$U, and $\mu \simeq m$; and (b) the assumption, without loss of generality, that $\mathbf{v}$ is parallel to $\mathbf{V}$. With the additional approximation that $|\mathbf{v}-\mathbf{V}| \simeq v$, the reaction rate from Eq. (9.79) is simplified to

$$N_0\overline{\sigma}(v) = \int_0^\infty \sigma_a(E_c)N(V)dV, \tag{9.82}$$

written in terms of the Maxwell–Boltzmann distribution of particles moving around with speed $V$:

$$N(V) = N_0\left(\frac{M}{2\pi kT}\right)^{1/2} \exp\left(-\frac{MV^2}{2kT}\right). \tag{9.83}$$

Substituting Eqs. (9.80), (9.81), and (9.83) into Eq. (9.82), with $E \simeq E_0$ for a sharp resonance, yields

$$\overline{\sigma}_a(v) = \sigma_{a0}\psi(\xi, y) = \overline{\sigma}_a(E), \tag{9.84}$$

where

$$\psi(\xi, y) = \frac{\xi}{2\sqrt{\pi}}\int_{-\infty}^{\infty} \exp\left[-\frac{\xi^2(y-x)^2}{4}\right]\frac{dx}{1+x^2}, \tag{9.85}$$

$$y = \frac{E-E_0}{\Gamma/2}, \xi = \frac{\Gamma}{\Delta},$$

and the *Doppler width* is defined as

$$\Delta = \left(\frac{4kTE_0}{A}\right)^{1/2}. \tag{9.86}$$

With the Doppler-broadened effective cross section $\overline{\sigma}_a(v)$ from Eq (9.84) substituted into Eq. (9.64) comprising the dilution factor $\rho = \Sigma_s/N_F$ provides an expression for the effective resonance integral for an isolated *sharp* resonance

$$I = \int_0^{\infty} \overline{\sigma}_a(E)\phi_{NR}(E)dE = \frac{1}{E_0}\int_0^{\infty}\frac{\overline{\sigma}_a(E)dE}{1+\overline{\sigma}(E)/\rho}$$
$$= \frac{1}{E_0}\int_{-2E_0/\Gamma}^{\infty}\frac{\sigma_{a0}\psi(\xi, y)}{1+\sigma_0\psi(\xi, y)/\rho}\frac{\Gamma}{2}dy,$$

or

$$I \simeq \frac{\Gamma}{2E_0}\int_{-\infty}^{\infty}\frac{\sigma_{a0}\psi(\xi, y)}{1+\sigma_0\psi(\xi, y)/\rho}dy = \frac{\Gamma_\gamma\rho}{E_0}J(\xi, \beta), \tag{9.87}$$

with the definition

$$J(\xi, \beta) = \int_0^{\infty}\frac{\psi(\xi, y)}{\psi(\xi, y)+\beta}dy, \tag{9.88}$$

and

$$\beta = \frac{\rho}{\sigma_0} = \frac{\Sigma_s\Gamma_\gamma}{N_F\sigma_{a0}\Gamma} = \frac{\rho}{\sigma_{a0}}\frac{\Gamma_\gamma}{\Gamma}.$$

It should be noted that the peak resonance cross section $\sigma_0$ includes the resonance scattering cross section, while $\Sigma_s$ represents a sum of moderator and fuel potential cross sections introduced in Eq. (9.65). The resonance escape probability is written as

$$p = \exp\left[-\frac{\Gamma_\gamma}{\overline{\xi}E_0}J(\xi, \beta)\right], \ \overline{\xi} = \frac{\sum_i \xi_i \Sigma_{si}}{\sum_i \Sigma_{si}}. \tag{9.89}$$

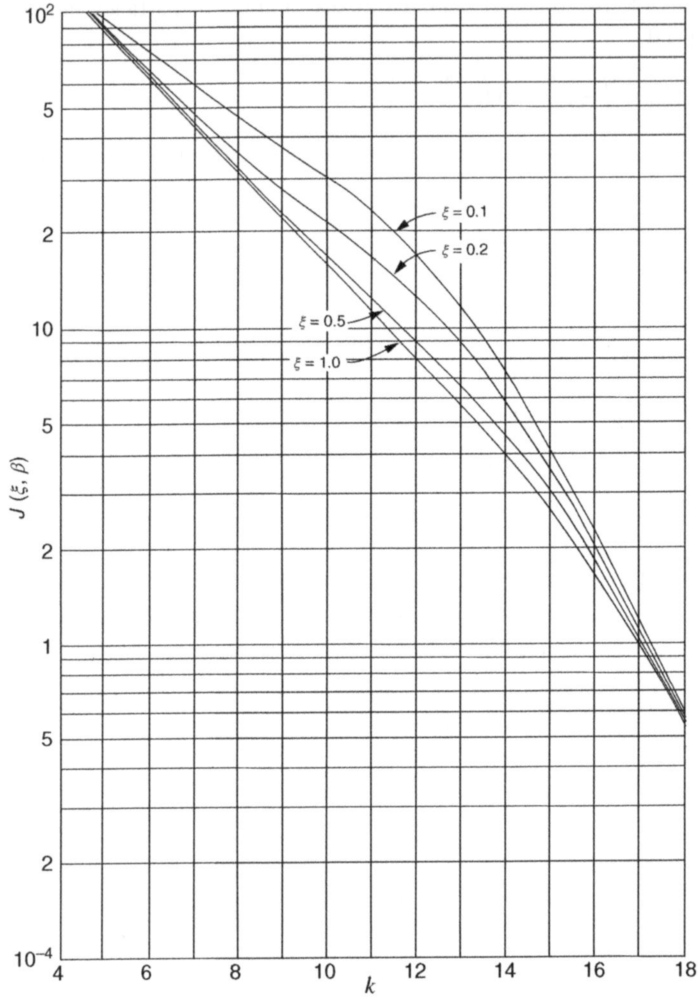


**Figure 9.8** Resonance integral function $J(\xi, \beta)$, with $\beta = 2^k \times 10^{-5}$, representing Doppler broadening. *Source:* Adapted from Dresner [Dre56].

The functions $\psi(\xi, y)$ and $J(\xi, \beta)$ are tabulated [Dre56] and the temperature dependence of $J(\xi, \beta)$ is illustrated in Figure 9.8. In the limit as $T \to 0$, Eq. (9.85) reduces to

$$\lim_{T\to 0} \psi(\xi, y) = \lim_{\xi\to\infty} \int_{-\infty}^{\infty} dx \frac{\delta(x-y)}{1+x^2} = \frac{1}{1+y^2}, \tag{9.90}$$

together with

$$\lim_{T\to 0} J(\xi,\beta) = \int_0^\infty \frac{dx}{1+\beta(1+x)^2} = \frac{\pi}{2\beta}\left(1+\frac{1}{\beta}\right)^{-1/2},$$

and

$$\lim_{T\to 0} I = \frac{\pi}{2}\frac{\Gamma\sigma_{a0}}{E_0}\left(1+\frac{1}{\beta}\right)^{-1/2}. \tag{9.91}$$

Furthermore, in the limit as the dilution factor $\rho \to \infty$, the effective resonance integral of Eq. (9.87) reduces to

$$\begin{aligned} I_\infty &= \frac{\Gamma\sigma_{a0}}{2E_0}\int_{-\infty}^{\infty} dy\psi(\xi,y) \\ &= \frac{\Gamma\sigma_{a0}}{2E_0}\frac{\xi}{2\sqrt{\pi}}\int_{-\infty}^{\infty}\frac{dx}{1+x^2}\int_{-\infty}^{\infty} dy \exp\left[-\frac{\xi^2}{4}(y-x)^2\right] \end{aligned}$$

or

$$I_\infty = \frac{\pi\Gamma\sigma_{a0}}{2E_0}. \tag{9.92}$$

Equation (9.90) shows that the Doppler-broadened effective cross section $\overline{\sigma}_a(E)$ reduces to the original Breit–Wigner line shape of Eq. (9.80) in the absence of Doppler broadening at $T = 0$, as it should. Furthermore, Eq. (9.92) indicates that there is no Doppler broadening effect in an infinitely dilute mixture, i.e. in the absence of energy self-shielding, and hence that the area under a broadened resonance equals that under the un-broadened Breit–Wigner line shape. It has been, however, emphasized [Cul74] that this statement is only meaningful under the approximations introduced in deriving the function $\psi(\xi,y)$ and another associated function $\chi(\xi,y)$ representing the scattering-absorption interference term and is only approximate for realistic representations of Doppler broadening of resonances.

The idealized analysis leading to Eq. (9.92) nonetheless provides valuable insights into the issues associated with Doppler broadening of resonances. For the temperature range of interest for nuclear reactor analysis, $\xi \in [0.1, 1]$, and the function $J(\xi,\beta)$ is a function of $\xi$ or $T$ practically only for the range $\beta \in [1.6\times 10^{-4}, 2.6]$. For the dilution factor $\rho \in [8, 2000]$ b of scattering cross section per $^{238}U$ atom, $\partial I/\partial T \simeq 10^{-4}\ \mathrm{b}\cdot\mathrm{K}^{-1}$ and $\partial p/\partial T \simeq -10^{-5}\ \mathrm{K}^{-1}$. Hence, in LWR cores, where the Doppler effect occurs predominantly in $^{238}U$ absorption resonances, any increase in the fuel temperature results in a prompt, negative reactivity feedback. For this reason, Doppler temperature feedback is an important self-regulating, inherently safe feedback mechanism in LWR cores. In sodium-cooled fast reactor (SFR) cores with a relatively high fissile Pu enrichment and a fast neutron spectrum, the Doppler broadening of resonances in Pu results in a competition between the increases in capture and fission reactions. For current SFR designs featuring either a metallic or (U,Pu) oxide mixture, the Doppler fuel temperature coefficient of reactivity is expected to be relatively small but still negative.

It is instructive to return to Eq. (9.88) and study how the function $J(\xi, \beta)$ increases as a function of $\xi$ or $T$, resulting in a corresponding increase in the effective resonance integral:

$$\frac{\partial I}{\partial \Delta} = -\frac{\Gamma_\gamma \Gamma \rho}{\Delta^2 E_0} \frac{\partial J(\xi, \beta)}{\partial \xi} = -\frac{I_\infty \Gamma \beta^2}{\Delta^2 \pi} \int_{-\infty}^{\infty} \frac{\partial \psi(\xi, y)/\partial \xi}{[\psi(\xi, y) + \beta]^2} dy. \tag{9.93}$$

Substituting

$$\frac{\partial \psi(\xi, y)}{\partial \xi} = -\frac{2}{\xi^3} \frac{\partial^2 \psi(\xi, y)}{\partial y^2}$$

into Eq. (9.93) and integrating by parts yields

$$\frac{\partial I}{\partial \Delta} = \frac{4 I_\infty \beta^2 \Delta}{\pi \Gamma^2} \int_{-\infty}^{\infty} \frac{[\partial \psi(\xi, y)/\partial y]^2}{[\psi(\xi, y) + \beta]^3} dy > 0, \tag{9.94}$$

which clearly indicates that $\partial I/\partial T > 0$ and $\partial p/\partial T < 0$. The steps taken with Eqs. (9.93) and (9.94) also provide a natural path to represent the temperature dependence in the form of the Doppler width $\Delta$ from Eq. (9.86), or $\sqrt{T}$ for the metal-oxide effective resonance correlation in Chapter 11.

## 9.6 Fermi Age Theory

As discussed briefly in Section 9.1, in order to account properly for the leakage effects in the slowing down process, one has to solve the $P_1$ approximation, or preferably the $B_1$ equations, in a fully coupled manner. Such a calculation has to be done invariably as part of a lattice physics code, which accounts for the spectral-spatial coupling in the scalar flux and neutron current for a unit-cell or unit-assembly representation of the reactor core. In this section, we first consider a simple analytical approach that illustrates the basic spectral-spatial coupling effects. More detailed techniques for lattice physics analysis are discussed in Chapter 11.

Assume that the lethargy-dependent current may be represented by Fick's law

$$\mathbf{J}(\mathbf{r}, u) = -D(u) \nabla \phi(\mathbf{r}, u) \tag{9.95}$$

and assume also that the asymptotic slowing down density from Eqs. (9.56) obtained for an infinite medium is valid for a finite medium:

$$q(\mathbf{r}, u) \simeq \xi \Sigma_t(u) \phi(\mathbf{r}, u). \tag{9.96}$$

In addition, write the slowing down density with capture as a product of the slowing down density $q_0(\mathbf{r}, u)$ without capture and the resonance escape probability $p(u) = p(0 \to u)$

$$q(\mathbf{r}, u) = q_0(\mathbf{r}, u)p(u), \tag{9.97}$$

where $p(u)$ is given by Eq. (9.58), obtained again for an infinite medium.

Taking a derivative of Eq. (9.97) with respect to lethargy $u$ provides

$$\frac{\partial q(\mathbf{r}, u)}{\partial u} = p(u)\frac{\partial q_0(\mathbf{r}, u)}{\partial u} - \Sigma_a(u)\phi(\mathbf{r}, u). \tag{9.98}$$

Substituting Eqs. (9.95) and (9.98) into the balance equation (9.11), with $S(\mathbf{r}, u) = 0$, yields

$$\frac{\partial q_0(\mathbf{r}, u)}{\partial u} = \frac{D(u)}{\xi\Sigma_t(u)}\nabla^2 q_0(\mathbf{r}, u), \tag{9.99}$$

which may be recast into

$$\frac{\partial q_0(\mathbf{r}, \tau)}{\partial \tau} = \nabla^2 q_0(\mathbf{r}, \tau), \tag{9.100}$$

with a new variable $\tau$ defined as

$$\tau(u) = \int_0^u du\frac{D(u)}{\xi\Sigma_t(u)}. \tag{9.101}$$

The relationship derived in Eq. (9.100) for the slowing down density without capture, $q_0(\mathbf{r}, u) = q_0(\mathbf{r}, \tau)$, is in the same form as the time-dependent heat conduction equation. In fact, the variable $\tau$ plays the role of the time variable or the chronological age and increases as the neutron lethargy increases or as neutrons slow down from the source energy at $E_0$. Hence, $\tau$ is called the *Fermi age*, although it is in units of area. The governing equation (9.100) is known as the *Fermi age equation* or the *age-diffusion equation*.

The basic idea behind Fermi age theory is to solve Eq. (9.100) for a capture-free slowing down density $q_0(\mathbf{r}, \tau)$ for a finite medium, accounting approximately for the effect of neutron leakage. This is then combined with an estimate of the resonance escape probability obtained for an infinite medium, to arrive at the slowing down density $q(\mathbf{r}, \tau)$ for a finite medium with a finite capture probability for neutrons. The applicability of the Fermi age equation for $q_0(\mathbf{r}, \tau)$ depends on the validity of the basic assumptions introduced in Eqs. (9.95) through (9.97). This suggests that the solution for the slowing down density $q(\mathbf{r}, u)$ for a finite medium through Fermi age theory is expected to be valid primarily for a medium with (a) small leakage and (b) weak or slowly varying captures, typically involving heavy nuclei such that $A \gg 1.0$.

The application of the Fermi age equation is illustrated through the analysis of a critical reactor with the slowing down source explicitly represented by one-group diffusion theory. Assume that the capture-free slowing down density $q_0(\mathbf{r}, \tau)$ is separable in space and energy or age

$$q_0(\mathbf{r}, \tau) = \psi(\mathbf{r})\theta(\tau) \tag{9.102}$$

and substitute it into Eq. (9.100) to obtain

$$\frac{1}{\theta(\tau)}\frac{d\theta(\tau)}{d\tau} = \frac{1}{\psi(\mathbf{r})}\nabla^2\psi(\mathbf{r}) = -B^2. \tag{9.103}$$

The constant that equates the separated terms involving $\theta(\tau)$ and $\psi(\mathbf{r})$ is set here to $-B^2$, based on our recognition that the spatial flux distribution $\psi(\mathbf{r})$ has to satisfy the eigenvalue equation characteristic of a critical reactor. That is, only for certain values of $B_n^2$, we may obtain nontrivial solutions $\psi_n(\mathbf{r})$ to the wave or Helmholtz equation (5.62)

$$\nabla^2\psi_n(\mathbf{r}) + B_n^2\psi_n(\mathbf{r}) = 0 \tag{9.104}$$

subject to the proper boundary conditions at the physical boundary of the reactor. Hence, the general solution to the Fermi age equation (9.100) can be written as

$$q_0(\mathbf{r}, \tau) = \sum_{n=0}^{\infty} A_n\psi_n(\mathbf{r})e^{-B_n^2\tau}. \tag{9.105}$$

Note that Eq. (9.105) is similar to the space- and time-dependent flux of Eq. (5.65) and the Fermi age $\tau$ plays the role of physical time variable of Eq. (5.65).

The slowing down density from Eq. (9.105) serves as the neutron source for a one-group neutron diffusion equation for the thermal group comprising neutrons of energy $E = E_{th}$ or age $\tau = \tau_{th}$:

$$-D\nabla^2\phi(\mathbf{r}) + \Sigma_a\phi(\mathbf{r}) = q(\mathbf{r}, \tau_{th}). \tag{9.106}$$

For a core, with the effective multiplication factor $k$, the slowing down density at the fission energy is equal to the fission source:

$$q(\mathbf{r}, 0) = q_0(\mathbf{r}, 0) = \frac{\nu\Sigma_f}{k}\phi(\mathbf{r}). \tag{9.107}$$

An expression for the flux follows from combining Eqs. (9.105) and (9.107):

$$\phi(\mathbf{r}) = \frac{k}{\nu\Sigma_f}\sum_{n=0}^{\infty} A_n\psi_n(\mathbf{r}).$$

From the study of the time-dependent one-group neutron diffusion equation in Section 5.3, recall that the nontrivial solution for a critical system should be described by a fundamental mode, i.e. $\phi(\mathbf{r}) \propto \psi_0(\mathbf{r})$, corresponding to the lowest eigenvalue $B_0^2 = B_g^2$ from Eq. (9.104) or, more specifically,

$$\phi(\mathbf{r}) = \frac{A_0 k}{\nu\Sigma_f}\psi_0(\mathbf{r}). \tag{9.108}$$

Hence, Eq. (9.105) is simplified to

$$q_0(\mathbf{r}, \tau) = A_0 \psi_0(\mathbf{r}) e^{-B_g^2 \tau}. \tag{9.109}$$

Substituting Eqs. (9.108) and (9.109) into Eq. (9.97), together with Eq. (9.106), yields a criticality condition

$$k = \frac{\nu \Sigma_f p(\tau_{th}) \exp\left(-B_g^2 \tau_{th}\right)}{\Sigma_a + D B_g^2} = \frac{k_\infty \exp\left(-B_g^2 \tau_{th}\right)}{1 + L_2^2 B_g^2}, \tag{9.110}$$

with the infinite multiplication factor $k_\infty$ and the thermal diffusion length $L_2$ defined as

$$k_\infty = \frac{\nu \Sigma_f p(\tau_{th})}{\Sigma_a} \text{ and } L_2^2 = \frac{D}{\Sigma_a}.$$

With the recognition that the thermal non-leakage probability $P_{NLT}$ is represented by Eq. (7.31)

$$P_{NLT} = \frac{1}{1 + L_2^2 B_g^2},$$

identify the term $\exp(-B_g^2 \tau_{th})$ as the fast non-leakage probability:

$$P_{NLF} = \exp(-B_g^2 \tau_{th}). \tag{9.111}$$

For a large reactor with small leakage, define the *neutron migration area*

$$M^2 = L_2^2 + \tau_{th} \tag{9.112}$$

and approximate Eq. (9.110) in the form of Eq. (7.32):

$$k = \frac{k_\infty}{1 + M^2 B_g^2}. \tag{9.113}$$

Comparison with two-group diffusion theory immediately suggests that the Fermi age $\tau_{th}$ for thermal neutrons is equal to the square of the diffusion length for the fast group defined in Eq. (7.28):

$$\tau_{th} = L_1^2 = \frac{D_1}{\Sigma_{a1} + \Sigma_r}. \tag{9.114}$$

The age $\tau_{th}$ may be physically interpreted as one-sixth of the mean square distance that a fast neutron travels from its birth until its removal from the fast group due to either absorption or slowing down into the thermal group. Thus, the migration area $M^2$ provides a useful measure of the distance a neutron travels in its lifetime in a reactor core and, for historical reasons, is often written in the form of Eq. (9.112), rather than Eq. (7.33) in terms of $L_1^2$.

The simple exercise of Fermi age theory has resulted in an expression for the effective multiplication factor that is fairly similar to the corresponding two-group expression. The main difference is in the determination of $k_\infty$, and in some sense we may regard Fermi age theory as a 1-1/2 group model. Although somewhat inferior to two-group diffusion theory, the Fermi age model still provides a simple way of accounting for neutron leakage in the slowing down density in a finite reactor core. The Fermi model for the asymptotic slowing down density, Eq. (9.96), provides useful insight into how the neutron flux spectrum $\phi(\mathbf{r}, u)$ is affected by the neutron leakage.

## 9.7 Comments on Lattice Physics Analysis

In this chapter, we have investigated basic approaches for determining the neutron flux in the slowing down range, primarily focusing on simple solutions for an infinite medium. We have obtained basic relations for the slowing down density, collision density, neutron flux, and resonance escape probability, which serve as the starting point for more detailed fast spectrum calculations in lattice physics analysis. Our investigation has been primarily limited to a mono-energetic source of neutrons, with the explicit purpose of obtaining simple analytic solutions for both hydrogenous and non-hydrogenous media. Indeed, this approach has yielded some important results: in particular, the Fermi model for the slowing down density for the asymptotic energy range. It is obvious, however, that for a more realistic representation of neutron slowing down, we need to explicitly account for the energy distribution of neutrons released in the fission process.

Our study has been limited to a homogeneous mixture of fuel and moderator materials. This often entails a substantial approximation for actual reactor configurations, especially in determining the resonance escape probability. In a lumped fuel configuration, where fuel atoms are separated from moderator atoms, the flux in the resonance range is depressed in the fuel region due to the large absorption cross section of fuel, essentially shielding the fuel atoms from neutrons of resonance energy. This depression of flux in the fuel lump, in a heterogeneous layout of fuel and moderator materials, is known as the *spatial self-shielding factor*, and usually represents a larger effect than the energy self-shielding effect. This point was effectively utilized by E. Fermi and his colleagues in the design and demonstration of the first self-sustaining fission system at the University of Chicago in 1942. They were able to achieve a critical configuration with a heterogeneous assembly of natural uranium fuel and graphite moderator, which otherwise would have remained substantially subcritical even for an infinitely large assembly. This concept was classified during the Manhattan Project.

As briefly discussed in Section 9.1, more realistic calculations of the fast spectrum often entail numerical solutions to the coupled $P_1$ or $B_1$ equations. More recently, collision probability or method-of-characteristics (MOC) algorithms are employed directly in various lattice physics codes, e.g. the CASMO5 code [Fer17] and the POLARIS module of the SCALE 6.2.3 package [Rea18], to account for material heterogeneities at the unit-cell or unit-assembly level, coupled with the $B_1$ treatment to represent leakage effects on the overall spectrum calculation. Collision probability and MOC formulations solve essentially an integral form of the neutron transport equation (4.44) without the spherical harmonics expansion of Eq. (4.42) or the Legendre polynomial expansion of Eq. (4.49). In the $B_1$ formulation, the spatial dependence of the 1-D slab-geometry form of the angular flux is represented via the fundamental mode or buckling mode from Eq. (5.62), thereby transforming the $P_1$ equations (9.6) into algebraic equations. They are then solved in a coupled manner for both the scalar flux $\phi_0(E)$ and net current $\phi_1(E)$ to accurately represent the neutron leakage out of a unit assembly. Even in these production lattice physics codes, often the concepts of the resonance escape probability and effective resonance integral from Section 9.4 are used as an integral part of the detailed fast spectrum calculations. One particular feature we discussed is the IR treatment from Eq. (9.75), which forms the basis for accurate resonance treatments, especially with an extension made effective for the spatial self-shielding effect. Thus, the approximate analytical results obtained in this chapter serve as a useful starting point for the study of detailed methods for fast spectrum calculations in Chapter 11.

## References


**[Cul74]** D.E. Cullen, "Application of the Probability Table Method to Multigroup Calculations of Neutron Transport," *Nucl. Sci. Eng.* **55**, 387 (1974).

**[Dre56]** L. Dresner, "The Effective Resonance Integrals of U-238 and Th-232," *Nucl. Sci. Eng.* **1**, 68 (1956).

**[Ede93]** M. Edenius, K. Ekberg, B.H. Forssén, and D. Knott, *CASMO-4, A Fuel Assembly Burnup Program*, Studsvik/SOA-93/1, Studsvik of America (1993).

**[Fer17]** R. Ferrer, J. Hykes, and J. Rhodes, "CASMO5, A Fuel Assembly Burnup Program: Methodology Manual," SSP-08/405, rev. 7, Studsvik Scandpower (2017).

**[Fri56]** B. Friedman, *Principles and Techniques of Applied Mathematics*, Wiley (1956).

**[Gol62]** R. Goldstein and E.R. Cohen, "Theory of Resonance Absorption of Neutrons," *Nucl. Sci. Eng.* **13**, 132 (1962).

**[Heb07]** A. Hébert "A Review of Legacy and Advanced Self-Shielding Models for Lattice Calculations," *Nucl. Sci. Eng.* **155**, 310 (2007).

**[Jon87]** D.B. Jones, *CPM-3 Computer Code Manual*, EPRI EP-P1295/C507, Vol. 2, Rev. A, Electric Power Research Institute (1987).

**[Lev72]** L.B. Levitt, "The Probability Tables Method for Treating Unresolved Neutron Resonances in Monte–Carlo Calculations," *Nucl. Sci. Eng.* **49**, 450 (1972).

**[Liu15]** Y. Liu, W.R. Martin, M. Williams, and K.-S. Kim, "A Full-Core Resonance Self-Shielding Method Using a Continuous Energy Quasi-One-Dimensional Slowing-Down Solution that Accounts for Temperature-Dependent Fuel Subregions and Resonance Interference," *Nucl. Sci. Eng.* **180**, 247 (2015).

**[Nik76]** M.N. Nikolaev, "Comments on the Probability Table Method," *Nucl. Sci. Eng.* **61**, 286 (1976).

**[Rea18]** B.T. Rearden and M.A. Jessee, eds., "SCALE Code System," vers. 6.2.3, ORNL/TM-2005/39 (2018).

## Problems

**9.1** A source of neutrons of energy $E_0$ and strength $Q$ [neutron $\cdot$ cm$^{-3}$ s$^{-1}$] is distributed uniformly throughout an infinite homogeneous medium. The medium consists of hydrogen and a purely absorbing material. Assume that the total absorption cross section $\Sigma_a$ of the medium and the scattering cross section $\Sigma_{sH}$ of hydrogen alone are both independent of energy. Obtain the collision density $F(E)$ and slowing down density $q(E)$ for all energies below the source energy $E_0$.

**9.2** A source of neutrons of energy $E_0$ and strength $Q$ [neutron $\cdot$ cm$^{-3}$ s$^{-1}$] is distributed uniformly throughout an infinite homogeneous medium. The medium has a zero absorption cross section for $E_1 < E < E_0$, where $E_1 < \alpha^3 E_0$, and an infinite absorption cross section for $E_2 < E < E_1$. (a) Calculate the fraction of source neutrons that slow down past $E_2$ when $\alpha E_1 < E_2 < E_1$. (b) Compare the result of part (a) with the corresponding result with the NR approximation. Discuss for narrow resonances with small values of $\alpha$ and for wide resonances.

**9.3** Absorption cross section $\overline{\sigma}(E)$ of a nuclide is inversely proportional to neutron energy $E$ and is equal to $\sigma_1$ at neutron energy $E_1$. (a) If the neutron flux $\phi(E)$ [neutron $\cdot$ cm$^{-2}$ s$^{-1}$ eV$^{-1}$] is constant over the interval $[E_1, E_2]$, with $E_2 > E_1$, determine the average cross section for the nuclide for the interval $[E_1, E_2]$. (b) Over the same interval $[E_1, E_2]$, if the flux $\phi(u)$ [neutron $\cdot$ cm$^{-2}$ s$^{-1}$ (lethargy)$^{-1}$] is constant, determine the average cross section for the nuclide for the interval. (c) Without numerically comparing the results of part (a) and (b), determine which average cross section is larger. Justify your answer.

**9.4** Starting from the neutron balance equation (9.71), complete the derivation for wide resonance energy shielding factor from Eq. (9.73). You may use L'Hospital's theorem in the limiting process.

**9.5** A source of neutrons of energy $E_0$ and strength $Q$ [neutron $\cdot$ cm$^{-3}$ s$^{-1}$] is distributed uniformly throughout an infinite homogeneous medium. The medium has $\Sigma_s$ that is independent of energy and $\Sigma_a = 0$ for $E_1 < E < E_0$, where $E_1 > \alpha E_0$, and $\Sigma_a = \Sigma^*$ for $E < E_1$. Determine the collision density $F(E)$ for $E > \alpha E_0$.

# 10

# Perturbation Theory and Adjoint Flux

With the primary purpose of deriving expressions for the differential and integral worths of control rods in nuclear reactors, in this chapter, we present the basic concepts behind perturbation theory and its applications to the determination of reactivity perturbations due to cross section changes. Perturbation theory may be used to evaluate, to the first order of accuracy, variations in the eigenvalue without the need to evaluate the associated changes in the neutron flux distribution. Perturbation theory may also be formulated to obtain first-order estimates of flux perturbations without numerically performing the full-blown eigenvalue calculations comprising inner and outer iterations.

We begin by offering a brief introduction to the operator notation with two simple examples in Section 10.1, followed by the definition of the adjoint operator and the associated adjoint flux in Section 10.2. Then, in Section 10.3, we present a formal derivation of the general perturbation expression for eigenvalue changes and indicate how the first-order perturbation formulation is obtained. The specific expression for the eigenvalue changes in one-group neutron diffusion theory is also discussed. Section 10.4 presents a few first-order perturbation examples, including the expressions for the differential and integral rod worths, together with interpretations of the adjoint flux. Other applications of adjoint formulations are discussed in Sections 10.5–10.7, including the method to obtain flux perturbations due to operator perturbations without explicitly solving the perturbed neutron balance equation. Some concluding remarks are presented in Section 10.8.

## 10.1 Operator Notation for Neutron Diffusion Equation

For the purpose of developing general perturbation expressions, we introduce the concept of neutron production and destruction operators for a general neutron

*Nuclear Reactor Physics and Engineering*, Second Edition. John C. Lee.
© 2025 John Wiley & Sons, Inc. Published 2025 by John Wiley & Sons, Inc.
Companion website: www.wiley.com/go/LeeNRPE2

balance equation

$$L\phi = \left(L_1 - \frac{L_2}{\lambda}\right)\phi = 0, \tag{10.1}$$

where $L_1$ and $L_2$ are the operators representing the neutron destruction and production terms, respectively, together with the eigenvalue $\lambda$. Two simple examples will illustrate the use of operators.

**Example 10.1** Obtain an operator structure for the steady-state one-group diffusion equation:

$$-\nabla \cdot D\nabla\phi + \Sigma_a\phi - \frac{\nu\Sigma_f}{k}\phi = 0. \tag{10.2}$$

The one-group diffusion equation is simply written in the form of Eq. (10.1)

$$\left(L_1 - \frac{L_2}{k}\right)\phi = 0, \tag{10.3}$$

provided we define

$$\begin{aligned} L_1 &= -\nabla \cdot D\nabla + \Sigma_a, \\ L_2 &= \nu\Sigma_f, \\ \lambda &= k. \end{aligned} \tag{10.4}$$

Thus, the operators may be regarded as a shorthand notation for handling both differential and multiplicative operations.

**Example 10.2** Obtain an operator form of the steady-state two-group diffusion equation (7.24):

$$\begin{aligned} -\nabla \cdot D\nabla\phi_1 + (\Sigma_{a1} + \Sigma_r)\phi_1 &= \frac{1}{k}(\nu\Sigma_{f1}\phi_1 + \nu\Sigma_{f2}\phi_2), \\ -\nabla \cdot D_2\nabla\phi_2 + \Sigma_{a2}\phi_2 &= \Sigma_r\phi_1. \end{aligned} \tag{10.5}$$

The two-group equation takes the form of a matrix equation

$$\begin{aligned} &\begin{pmatrix} -\nabla \cdot D_1\nabla + \Sigma_{a1} + \Sigma_r & 0 \\ -\Sigma_r & -\nabla \cdot D_2\nabla + \Sigma_{a2} \end{pmatrix}\begin{pmatrix} \phi_1 \\ \phi_2 \end{pmatrix} \\ &= \frac{1}{k}\begin{pmatrix} \nu\Sigma_{f1} & \nu\Sigma_{f2} \\ 0 & 0 \end{pmatrix}\begin{pmatrix} \phi_1 \\ \phi_2 \end{pmatrix}, \end{aligned} \tag{10.6}$$

again with $\lambda = k$.

## 10.2 Adjoint Operator and Adjoint Flux

We define the adjoint operator $L^\dagger$ corresponding to the operator $L$ from Eq. (10.1) through an inner product involving a function $f$

$$\langle f, L\phi\rangle = \langle L^\dagger f, \phi\rangle, \tag{10.7}$$

where the inner product involves an integral over the entire volume for which the operator equation and the neutron flux are defined. Thus, the evaluation of $L^\dagger$ entails performing the left-hand inner product via integration by parts and imposing the proper boundary condition (BC) on the general function $f$ so that the resulting boundary terms, usually known as the *conjunct*, vanish, to yield the right-hand inner product involving the adjoint operator $L^\dagger$. Once the adjoint operator is identified, the adjoint flux $\phi^\dagger$ may be obtained to satisfy the adjoint operator equation

$$L^\dagger\phi^\dagger = L_1^\dagger\phi^\dagger - \frac{1}{\lambda^\dagger}L_2^\dagger\phi^\dagger = 0, \tag{10.8}$$

subject to the proper boundary conditions. If $L^\dagger = L$, the operator $L$ is called *self-adjoint*.

**Example 10.3** Determine the adjoint operator for the one-group neutron diffusion equation (10.2) for a slab reactor of thickness $2H$, where the destruction operator from Eq. (10.4) may be written as

$$L_1 = -\nabla\cdot D\nabla + \Sigma_a = -\frac{d}{dx}D\frac{d}{dx} + \Sigma_a. \tag{10.9}$$

For the solution of the forward operator equation (10.3), the simple zero-flux boundary condition is imposed

$$\phi(H) = \phi(-H) = 0, \tag{10.10}$$

and the inner product of Eq. (10.7) yields

$$\begin{aligned}
\langle f, L_1\phi\rangle &= -\int_{-H}^{H} f\frac{d}{dx}D\frac{d\phi}{dx}dx + \int_{-H}^{H} f\,\Sigma_a\phi dx\\
&= -fD\left.\frac{d\phi}{dx}\right|_{-H}^{H} + \int_{-H}^{H} D\frac{df}{dx}\frac{d\phi}{dx}dx + \langle\Sigma_a f, \phi\rangle\\
&= -fD\left.\frac{d\phi}{dx}\right|_{-H}^{H} + D\frac{df}{dx}\phi\Big|_{-H}^{H} - \int_{-H}^{H}\frac{d}{dx}D\frac{df}{dx}\phi dx + \langle\Sigma_a f, \phi\rangle\\
&= \langle L_1^+ f, \phi\rangle.
\end{aligned} \tag{10.11}$$

Invoking the BC from Eq. (10.10) shows that the final inner product of Eq. (10.11) is equal to the inner product involving the forward operator $L_1$

$$\langle L_1^\dagger f, \phi \rangle = \langle L_1 f, \phi \rangle, \tag{10.12}$$

provided the same zero-flux boundary condition is also applied to the general function $f$

$$f(H) = f(-H) = 0. \tag{10.13}$$

Since the product operator $L_2$ is obviously self-adjoint, we recognize that the one-group diffusion operator $L = [L_1 - (1/\lambda)L_2]$ is self-adjoint, provided the adjoint neutron flux satisfies the same zero-flux boundary condition from Eq. (10.10) or (10.13). It can be proven that the one-group diffusion theory operator is in general self-adjoint for other boundary conditions for the flux, including the more-general boundary condition that the return current at a vacuum boundary vanishes.

We now show that the adjoint eigenvalue $\lambda^\dagger$ for the adjoint operator equation

$$L_1^\dagger \phi^\dagger = \frac{1}{\lambda^\dagger} L_2^\dagger \phi^\dagger \tag{10.14}$$

is equal to the forward eigenvalue $\lambda$. For this purpose, evaluate the inner product

$$\langle L_1^\dagger \phi^\dagger, \phi \rangle = \langle \phi^\dagger, L_1 \phi \rangle = \frac{1}{\lambda}\langle \phi^\dagger, L_2 \phi \rangle = \frac{1}{\lambda}\langle L_2^\dagger \phi^\dagger, \phi \rangle = \frac{1}{\lambda^\dagger}\langle L_2^\dagger \phi^\dagger, \phi \rangle. \tag{10.15}$$

One may show that the adjoint eigenvalue is in general equal to the forward eigenvalue for multigroup diffusion equations.

**Example 10.4** Obtain the adjoint form of the two-group neutron diffusion equation.

Apply the definition of the adjoint operator in Eq. (10.7) to the two-group equation (10.6)

$$\begin{aligned}(\phi_1^\dagger \;\; \phi_2^\dagger)\begin{pmatrix} -\nabla \cdot D_1 \nabla + \Sigma_{a1} + \Sigma_r & 0 \\ -\Sigma_r & -\nabla \cdot D_2 \nabla + \Sigma_{a2} \end{pmatrix}\begin{pmatrix} \phi_1 \\ \phi_2 \end{pmatrix} \\ = \tfrac{1}{k}(\phi_1^\dagger \;\; \phi_2^\dagger)\begin{pmatrix} \nu\Sigma_{f1} & \nu\Sigma_{f2} \\ 0 & 0 \end{pmatrix}\begin{pmatrix} \phi_1 \\ \phi_2 \end{pmatrix},\end{aligned} \tag{10.16}$$

and perform the integrations by part, which is equivalent to the steps taken in Eq. (10.11). This naturally yields the transpose of the $(2 \times 2)$ operator matrices for the adjoint two-group equation

$$\begin{pmatrix} -\nabla \cdot D_1 \nabla + \Sigma_{a1} + \Sigma_r & -\Sigma_r \\ 0 & -\nabla \cdot D_2 \nabla + \Sigma_{a2} \end{pmatrix}\begin{pmatrix} \phi_1^\dagger \\ \phi_2^\dagger \end{pmatrix}$$

$$= \frac{1}{k}\begin{pmatrix} \nu\Sigma_{f1} & 0 \\ \nu\Sigma_{f2} & 0 \end{pmatrix}\begin{pmatrix} \phi_1^\dagger \\ \phi_2^\dagger \end{pmatrix}, \tag{10.17}$$

which indicates that the adjoint neutrons would gain energy from $\phi_2^\dagger$ to $\phi_1^\dagger$ as regular neutrons slow down from the fast group to the thermal group.

## 10.3 First-Order Perturbation Theory

Consider now a perturbed operator $L'$ so that the neutron balance equation representing a perturbed reactor configuration may be written as

$$L'\phi' = \left(L_1' - \frac{1}{\lambda'}L_2'\right)\phi', \tag{10.18}$$

where each of the perturbed quantities is written as the sum of the original unperturbed quantity and a perturbation term:

$$\begin{aligned} L_1' &= L_1 + \delta L_1, \\ L_2' &= L_2 + \delta L_2, \\ \lambda' &= \lambda + \delta\lambda, \\ \phi' &= \phi + \delta\phi. \end{aligned} \tag{10.19}$$

The key objective for perturbation theory is to obtain an expression for the eigenvalue perturbation due to operator perturbations, without explicitly solving for the flux perturbations.

For this purpose, multiply the perturbed equation (10.18) by the adjoint flux $\phi^\dagger$, and obtain the required inner product

$$\langle \phi^\dagger, L'\phi' \rangle = 0, \tag{10.20}$$

which leads to

$$\langle \phi^\dagger, L_2'\phi' \rangle = \lambda'\langle \phi^\dagger, L_1'\phi' \rangle = \lambda\langle \phi^\dagger, L_1'\phi' \rangle + \delta\lambda\langle \phi^\dagger, L_1'\phi' \rangle. \tag{10.21}$$

Equation (10.21) is then solved for $\delta\lambda$ and divided by the eigenvalue $\lambda$ to obtain the fractional perturbation in the eigenvalue:

$$\begin{aligned} \frac{\delta\lambda}{\lambda} &= \frac{\langle \phi^\dagger, L_2'\phi' \rangle - \lambda\langle \phi^\dagger, L_1'\phi' \rangle}{\lambda < \langle \phi^\dagger, L_1'\phi' \rangle} \\ &= \frac{\langle \phi^\dagger, L_2\phi' \rangle + \langle \phi^\dagger, \delta L_2\phi' \rangle - \lambda\langle \phi^\dagger, L_1\phi' \rangle - \lambda\langle \phi^\dagger, \delta L_1\phi' \rangle}{\lambda\langle \phi^\dagger, L_1'\phi' \rangle}. \end{aligned} \tag{10.22}$$

Now apply the adjoint equation (10.14) with $\lambda^\dagger = \lambda$ to the numerator of Eq. (10.22) to cancel out terms involving unperturbed operators $L_1$ and $L_2$ and obtain a simplified expression:

$$\frac{\delta\lambda}{\lambda} = \frac{\langle\phi^\dagger, \delta L_2\phi'\rangle - \lambda\langle\phi^\dagger, \delta L_1\phi'\rangle}{\lambda\langle\phi^\dagger, L_1\phi'\rangle + \lambda\langle\phi^\dagger, \delta L_1\phi'\rangle}. \tag{10.23}$$

This is an exact expression known as the *general* or *strong perturbation equation* for the fractional change in the eigenvalue due to operator perturbations $\delta L_1$ and $\delta L_2$. Note that the simplification is possible through the use of the adjoint equation (10.14). Indeed, this is the main reason for introducing the adjoint operator and adjoint flux in the first place.

Evaluation of Eq. (10.23), however, requires knowledge of the perturbed flux $\phi'$. To avoid the calculation of $\phi'$, now approximate it with the unperturbed flux $\phi$ and retain only the first-order terms to reduce Eq. (10.23) to

$$\frac{\delta\lambda}{\lambda} = \frac{\langle\phi^\dagger, \delta L_2\phi\rangle - \lambda\langle\phi^\dagger, \delta L_1\phi\rangle}{\lambda\langle\phi^\dagger, L_1\phi\rangle + \lambda\langle\phi^\dagger, L_1\delta\phi\rangle + \lambda\langle\phi^\dagger, \delta L_1\phi\rangle}. \tag{10.24}$$

Finally, we note that both of the first-order terms involving perturbed quantities in the denominator of Eq. (10.24) result in second-order terms when applied to the two first-order terms in the numerator. The *first-order perturbation* expression for the fractional change in the $k$-eigenvalue or the reactivity is finally obtained:

$$\frac{\delta k}{k} = \frac{\delta\lambda}{\lambda} = \frac{\left\langle\phi^\dagger, \left(\frac{1}{k}\delta L_2 - \delta L_1\right)\phi\right\rangle}{\frac{1}{k}\langle\phi^\dagger, L_2\phi\rangle} \tag{10.25}$$

For the one-group neutron diffusion equation (10.2), the operator perturbations may be explicitly written as

$$\delta L_1 = -\nabla\cdot\delta D\nabla + \delta\Sigma_a, \tag{10.26a}$$

$$\delta L_2 = \delta(\nu\Sigma_f), \tag{10.26b}$$

so that Eq. (10.25) may be recast:

$$\frac{\delta k}{k} = \frac{\int_V \phi^\dagger\left[\frac{1}{k}\delta(\nu\Sigma_f)\phi + \nabla\cdot\delta D\nabla\phi - \delta\Sigma_a\phi\right]d\mathbf{r}}{\frac{1}{k}\int_V \nu\Sigma_f\phi^\dagger\phi d\mathbf{r}}. \tag{10.27}$$

With the recognition that the one-group neutron diffusion operator is self-adjoint, i.e. $\phi^\dagger = \phi$, Eq. (10.27) is finally simplified to

$$\frac{\delta k}{k} = \frac{\int_V \left[\frac{1}{k}\delta(\nu\Sigma_f)\phi^2 - \delta D(\nabla\phi)^2 - \delta\Sigma_a\phi^2\right]d\mathbf{r}}{\frac{1}{k}\int_V \nu\Sigma_f\phi^2 d\mathbf{r}}, \tag{10.28}$$

where Green's theorem is also used for the leakage term in the numerator together with the BC that the flux vanishes on the outside boundary of the core. It should

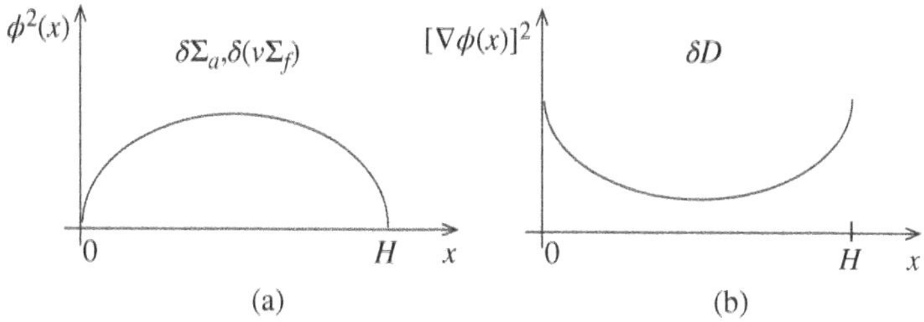


**Figure 10.1** Weighting factors for perturbations (a) $\delta\Sigma_a$ and $\delta(\nu\Sigma_f)$ and (b) $\delta D$.

be emphasized that Eqs. (10.27) and (10.28) provide first-order expressions for the reactivity change, due to changes in the cross sections for the core, based on the unperturbed flux distribution. It should also be recognized that the perturbations in $\nu\Sigma_f$ and $\Sigma_a$ are represented with weighting factors proportional to $\phi^2$ but that the weighting factor for the perturbations in $D$ is $(\nabla\phi)^2$, as illustrated in Figure 10.1. This means perturbations in $\nu\Sigma_f$ and $\Sigma_a$ would have the largest impact where the flux is highest, while perturbations in $D$ would have the largest impact near the periphery of the core where $\nabla\phi$ has the largest magnitude.

## 10.4 Adjoint Flux for Control Rod Worth Calculation

A few examples of the first-order perturbation formulation from Eqs. (10.27) and (10.28) are presented in this section to illustrate the application of the basic formulation and to provide some physical insights to the adjoint flux introduced primarily for mathematical convenience in deriving Eq. (10.23). One of the examples derives the differential control rod worth formulation, which is a key objective for studying perturbation theory in nuclear reactor physics.

**Example 10.5** For a reactor core with a uniform $\nu\Sigma_f$, determine an expression for the reactivity perturbation due to a uniform perturbation $\delta\Sigma_a$.

Setting $\delta(\nu\Sigma_f) = \delta D = 0$ simplifies Eq. (10.28) to

$$\frac{\delta k}{k} = \frac{-k\delta\Sigma_a \int_V \phi^2 d\mathbf{r}}{\nu\Sigma_f \int_V \phi^2 d\mathbf{r}} = -\frac{\delta\Sigma_a}{\Sigma_a + DB^2}, \tag{10.29}$$

with the one-group expression for the effective multiplication factor

$$k = \frac{\nu\Sigma_f}{\Sigma_a + DB^2}. \tag{10.30}$$

By taking a direct variation of Eq. (10.30), we also obtain

$$\delta \ln k = -\delta \ln \left(\Sigma_a + DB^2\right), \tag{10.31}$$

which verifies Eq. (10.29).

**Example 10.6** For the same reactor configuration as in Example 10.5, determine an expression for the reactivity perturbation due to a localized perturbation $\delta\Sigma_a(\mathbf{r}) = A\delta(\mathbf{r} - \mathbf{r}_0)$ of magnitude $A$ at position $\mathbf{r}_0$.

Equation (10.28) yields

$$\frac{\delta k}{k} = \frac{-k\phi^+(\mathbf{r}_0)A\phi(\mathbf{r}_0)}{\int_V \nu\Sigma_f \phi^2 d\mathbf{r}} \equiv \Delta\rho, \tag{10.32}$$

from which we obtain an expression for the adjoint flux:

$$\phi^+(\mathbf{r}_0) = \frac{-\int_V \nu\Sigma_f \phi^2 d\mathbf{r}}{k} \frac{\Delta\rho}{A\phi(\mathbf{r}_0)} \equiv C\frac{\Delta\rho}{A\phi(\mathbf{r}_0)}. \tag{10.33}$$

With a lumped parameter $C$ introduced in Eq. (10.33), note that the adjoint flux $\phi^\dagger(\mathbf{r}_0)$ is proportional to the reactivity change per absorption rate $A\phi(\mathbf{r}_0)$ at position $\mathbf{r}_0$. The adjoint flux $\phi^\dagger(\mathbf{r}_0)$ may be interpreted as the importance of neutrons absorbed at position $\mathbf{r}_0$.

**Example 10.7** Determine the reactivity worth of a control rod bank inserted to position $x$ from the bottom of a slab reactor core of height $H$. The reactor is initially critical before the rod insertion, with uniform $\nu\Sigma_f$ throughout the core. Assume $\delta(\nu\Sigma_f) = \delta D = 0$ for the control rod insertion, and consider the control rod as a pure absorber with absorption cross section $\Delta\Sigma_a$:

$$\delta\Sigma_a(x') = \begin{cases} \Delta\Sigma_a, & 0 < x' < x, \\ 0, & x < x' < H. \end{cases} \tag{10.34}$$

Substituting the unperturbed flux

$$\phi(x) = \phi_0 \sin \frac{\pi x}{H}, \tag{10.35}$$

into Eq. (10.28) yields

$$\frac{\delta k}{k} = \rho(x) = \frac{-\int_0^H \delta\Sigma_a(x')\phi^2(x')dx'}{\nu\Sigma_f \int_0^H \phi^2(x)dx} = -\frac{\Delta\Sigma_a}{\nu\Sigma_f}\left[\frac{x}{H} - \frac{1}{2\pi}\sin\frac{2\pi x}{H}\right]. \tag{10.36}$$

Equation (10.36) represents the *integral rod worth* for a control road inserted to position $x$. Differentiating Eq. (10.36) yields the *differential rod worth* for a control

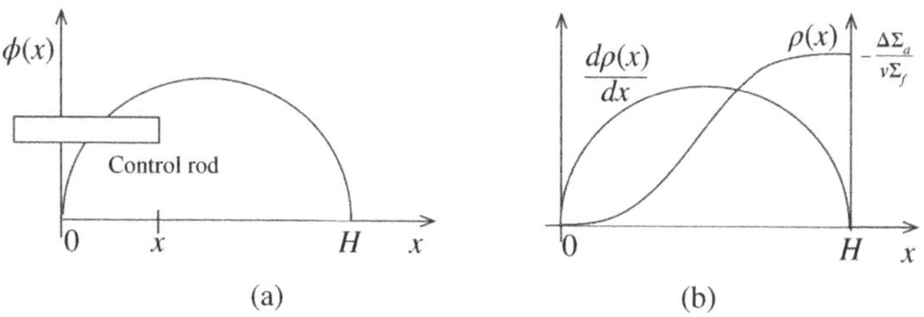


**Figure 10.2** Control rod worth curves: (a) Control rod insertion to position $x$ and (b) corresponding rod worth curves.

rod at position $x$:

$$\frac{d\rho(x)}{dx} = -\frac{\Delta\Sigma_a}{\nu\Sigma_f H}\left[1 - \cos\frac{2\pi x}{H}\right]. \tag{10.37}$$

The rod worth curves are illustrated in Figure 10.2 together with the position of the control rod inserted from the LHS or bottom of the core.

Since the differential rod worth $d\rho(x)/dx = 0$ at the bottom of the core $x = 0$, the rate of reactivity control will be zero as the rod begins to be inserted into the core. Thus, the control rods are typically inserted a few steps into the core, known as the *bite position*, during normal operation for PWRs so that a negative reactivity insertion is immediately available whenever required. It should of course be noted that control rods are typically inserted from the top of the core, but the rod is illustrated here as inserted from the bottom of the core for notational convenience.

## 10.5 Adjoint Flux for Variational Formulation

Somewhat related to the use of adjoint flux in the eigenvalue perturbation calculation considered in Section 10.3, a variational formulation for the eigenvalue $\lambda$ in Eq. (10.1) is developed here with the adjoint flux used as a weighting function. It will be demonstrated that, through the variational formulation, the eigenvalue calculated is free from first-order errors in the eigenfunction, i.e. the flux distribution $\phi$.

A simple method to determine the eigenvalue $\lambda$ is the volume integral approach used in the outer iteration in Eq. (6.43), which is, however, subject to errors or inaccuracies in flux $\phi$ to the first order. The solution $\phi^\dagger$ to the adjoint equation (10.8) may instead be used as a weighting function to determine $\lambda$ from the weighted volume integrals

$$\left\langle \phi^\dagger, \left(L_1 - \frac{L_2}{\lambda}\right)\phi \right\rangle = 0,$$

or

$$\lambda = \frac{\langle \phi^\dagger, L_2 \phi \rangle}{\langle \phi^\dagger, L_1 \phi \rangle}. \tag{10.38}$$

Taking the first-order variation of Eq. (10.38) with respect to the variation $\delta\phi$ in the eigenfunction yields

$$\delta\lambda(\delta\phi) = \frac{\partial \lambda}{\partial \phi}\delta\phi = \frac{\langle \phi^\dagger, L_2 \delta\phi \rangle - \lambda\langle \phi^\dagger, L_1 \delta\phi \rangle}{\langle \phi^\dagger, L_1 \phi \rangle} = \frac{\langle (L_2^\dagger - \lambda L_1^\dagger)\phi^\dagger, \delta\phi \rangle}{\langle \phi^\dagger, L_1 \phi \rangle} = 0,$$

which shows that Eq. (10.38) is free from first-order errors in $\phi$. One could likewise show that Eq. (10.38) is free from first-order errors in $\phi^\dagger$. Thus, the eigenvalue $\lambda$ obtained via Eq. (10.38) provides second-order accuracy in the presence of first-order uncertainties in the forward and adjoint flux.

Taking another step to evaluate the *total* first-order variation in $\lambda$, including variations in the operators $L_1$ and $L_2$, yields the first-order perturbation equation (10.24), with the second-order terms in the denominator removed. The ratio in Eq. (10.38) is an example of the Rayleigh quotient [Wat02], which is used in numerical analysis for a variety of applications. Several other variational formulations have been developed over the years for various reaction rates or ratios of reaction rates, all of which employ adjoint flux structures [Sta01].

## 10.6 Adjoint Flux for Detector Response Calculation

Another useful application of the adjoint formulation is demonstrated in this section by considering a neutron balance equation for neutron flux $\phi$ associated with an external source $f$

$$L\phi = f, \tag{10.39}$$

together with an adjoint balance equation:

$$L^\dagger \phi^\dagger = g^\dagger. \tag{10.40}$$

Assigning a reaction cross section to the adjoint source $g^\dagger$ allows the volume integral $R = \langle g^\dagger, \phi \rangle$ to represent a detector response or signal. Invoking the properties of the adjoint operator, evaluate the detector signal effectively with the adjoint flux $\phi^\dagger$:

$$R = \langle g^\dagger, \phi \rangle = \langle L^\dagger \phi^\dagger, \phi \rangle = \langle \phi^\dagger, L\phi \rangle = \langle \phi^\dagger, f \rangle. \tag{10.41}$$

Consider a specific case where the function $f$ represents a point source of unit magnitude at position $\mathbf{r}_1$

$$f = \delta(\mathbf{r} - \mathbf{r}_1) \tag{10.42}$$

and the adjoint source $g^{\dagger}$ represents a point detector at $\mathbf{r}_0$ with reaction cross section $\Sigma_d$:

$$g^{\dagger} = \Sigma_d \delta(\mathbf{r} - \mathbf{r}_0). \tag{10.43}$$

The response of the detector at $\mathbf{r}_0$ due to the source at $\mathbf{r}_1$ is obtained with Eq. (10.41)

$$R(\mathbf{r}_1 \rightarrow \mathbf{r}_0) = \Sigma_d \phi(\mathbf{r}_0) = \phi^{+}(\mathbf{r}_1), \tag{10.44}$$

which indicates that the adjoint flux in this case represents the detector response. In fact, when the response of a detector at $\mathbf{r}_0$ due to multiple point sources $\mathbf{r}_n, n = 1, \ldots, N$, is desired, solve the fixed-source adjoint equation (10.40) once and then merely look up the values of the adjoint flux $\phi^{\dagger}(\mathbf{r}_n)$ for various points $\mathbf{r}_n, n = 1, \ldots, N$. This is obviously much more efficient than solving the forward equation (10.39) $N$ times to determine $N$ values of $\phi(\mathbf{r}_0)$ and determine $N$ detector responses one by one.

This approach was taken to determine the detector weighting function for an excore neutron detector system for an operating PWR core [Cru78]. As is typically the case in PWR plants, the excore detectors are located outside the reactor pressure vessel in the concrete shield of the containment building. Excore detectors usually comprise uncompensated ion chambers intended to provide information about the core power in each quadrant and in the top and bottom halves of the core. Through periodic calibrations through incore neutron detectors, the excore detector system serves as the primary instrumentation system for continuous monitoring of total power output and spatial power distribution. In particular, the excore instrumentation system provides the axial offset (AO) and quadrant tilt (QT) of power as key measures of the 3-D power distribution in PWR cores. The AO and QT of power in turn serve as measures of core stability against spatial power fluctuations associated with space-time evolutions of $^{135}$I-$^{135}$Xe concentrations in a reactor core. This issue is discussed further in Chapter 16, together with control measures for $^{135}$Xe-induced transient phenomena.

With the core and detector geometry presented in Figure 10.3, it is not possible to determine through direct measurements the contribution of neutrons produced in various fuel assemblies to the excore detectors. Representation of detector readings also requires sufficiently accurate neutron transport calculations for the 193-assembly four-loop PWR core at the Indian Point Unit 2 plant with an equivalent diameter of 3.37 m, surrounded by the water reflector, core barrel, thermal shield, and reactor vessel with thickness of 0.43, 0.11, 0.13, and 0.40 m, respectively. The actual deep-penetration calculation was performed with a semi-empirical point transport kernel validated by 1-D ANISN discrete-ordinate [Eng67] calculations. An adjoint transport calculation representing Eq. (10.43) was performed with the detector $g^{\dagger}(\mathbf{r}_0)$ at the excore detector location to yield $\phi^{\dagger}(\mathbf{r})$. The adjoint flux is used together with the measured relative assembly

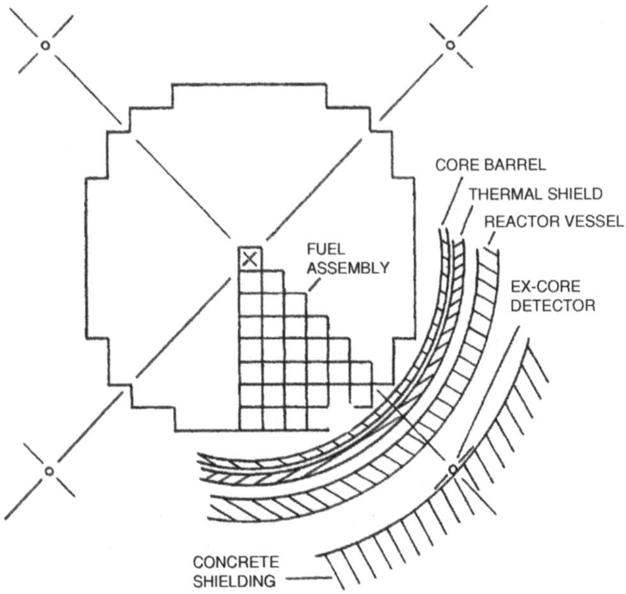


**Figure 10.3** Core and detector geometry for Indian Point Unit 2 plant. *Source*: Crump and Lee [Cru78]/U.S. Department of Energy/Public domain.

power map in Figure 10.4a serving as the source distribution $f(\mathbf{r})$ to generate the detector response distribution $\langle \phi^\dagger, f \rangle = R(\mathbf{r} \rightarrow \mathbf{r}_0)$ summarized in Figure 10.4b. The detector response distribution in Figure 10.4b indicates that 91% of the detector signal comes from the five adjacent fuel assemblies, indicating the limited applicability of using the excore detectors for global core power monitoring in PWR cores.

## 10.7 Adjoint Formulation for Flux Perturbation Calculation

As the final application of the adjoint operator and flux distribution, consider the task of obtaining the perturbation $\delta\phi(\mathbf{r})$ in flux due to perturbation $\delta L$ in the operator representing the neutron balance equation (10.1), without explicitly recalculating the perturbed flux $\phi'(\mathbf{r})$ with $L'\phi' = 0$. The flux perturbation $\delta\phi(\mathbf{r})$ is obtained as a sum of high-order harmonics considered for the solution of the time-dependent neutron diffusion equation in Section 5.3.

| | | | | | |
|---|---|---|---|---|---|
| 1.10 | | | | | |
| 1.03 | 1.11 | | | | |
| 1.10 | 0.971 | 1.15 | | | |
| 1.06 | 1.14 | 1.05 | 1.09 | | |
| 1.19 | 1.09 | 1.16 | 1.03 | 1.26 | |
| 1.07 | 1.16 | 1.21 | 1.15 | 0.902 | 1.07 |
| 1.04 | 1.08 | 1.05 | 1.02 | 0.873 | 0.502 |
| 0.755 | 0.884 | 0.702 | 0.597 | | |

(a) Relative assembly power map

| | | | | | |
|---|---|---|---|---|---|
| 7.77[a]<br>−9 | | | | | |
| 3.05<br>−8 | 1.81<br>−7 | | | | |
| 1.25<br>−7 | 6.62<br>−7 | 4.48<br>−6 | | | |
| 4.66<br>−7 | 2.98<br>−6 | 1.76<br>−5 | 1.10<br>−4 | | |
| 2.04<br>−6 | 1.18<br>−5 | 7.73<br>−5 | 4.73<br>−4 | 3.76<br>−3 | |
| 6.21<br>−6 | 4.90<br>−5 | 3.48<br>−4 | 2.33<br>−3 | 1.39<br>−2 | 1.19<br>−1 |
| 1.25<br>−5 | 1.19<br>−4 | 1.08<br>−3 | 8.92<br>−3 | 6.54<br>−2 | 3.29<br>−1 |
| 2.15<br>−5 | 2.06<br>−4 | 1.60<br>−3 | 1.54<br>−2 | | |

[a]Denotes $7.77 \times 10^{-9}$

(b) Detector reponse distribution

**Figure 10.4** Relative assembly power map and excore detector response distribution for the octant of Indian Point Unit 2 core. (a) Relative assembly power map and (b) detector response distribution. *Source*: Crump and Lee [Cru78]/Taylor & Francis Group.

Returning to Eqs. (10.18) and (10.19), with $L'(\mathbf{r}) = L(\mathbf{r}) + \delta L(\mathbf{r})$ and $\phi'(\mathbf{r}) = \phi(\mathbf{r}) + \delta\phi(\mathbf{r})$, and dropping the high-order term $\delta L(\mathbf{r})\delta\phi(\mathbf{r})$ yield

$$L(\mathbf{r})\delta\phi(\mathbf{r}) = -\delta L(\mathbf{r})\phi(\mathbf{r}) = -\left[\delta L_1(\mathbf{r}) - \frac{1}{\lambda}\delta L_2(\mathbf{r}) + \frac{\delta\lambda}{\lambda^2}L_2(\mathbf{r})\right]\phi(\mathbf{r}), \tag{10.45}$$

into which is substituted an expansion for the perturbed flux $\delta\phi(\mathbf{r})$:

$$\delta\phi(\mathbf{r}) = \sum_{l=1}^{N} a_l\phi_l(\mathbf{r}). \tag{10.46}$$

Given $\delta L(\mathbf{r})$, the expansion coefficients, $\{a_l, i = 1, \ldots, N\}$, depend on the choice of high-order harmonics. The simplest choice would be the geometric modes considered in Figure 5.8, which usually require a large number of terms. One popular choice is the *lambda modes*, also known as the *natural modes*, which are defined by

$$L\phi_i = \left(L_1 - \frac{L_2}{\lambda_i}\right)\phi_i = 0, \; i = 0, \ldots, N, \tag{10.47}$$

with a corresponding set of adjoint modes with the set of eigenvalues $\{\lambda_i^\dagger = \lambda_i, i = 0, \ldots, N\}$. The adjoint lambda modes may be normalized to provide the orthonormality property:

$$\langle \phi_i^\dagger, L_1\phi_j \rangle = \left\langle \phi_i^\dagger, \frac{L_2}{\lambda_j}\phi_j \right\rangle = \delta_{ij} = \langle L_1^\dagger \phi_i^\dagger, \phi_j \rangle = \left\langle \frac{L_2^\dagger}{\lambda_i^\dagger}\phi_i^\dagger, \phi_j \right\rangle, \ \forall\, i,j. \tag{10.48}$$

It should be recognized that the eigenvalue $\lambda_0 = \lambda = k_{eff}$ for the fundamental mode and that $\lambda_i^\dagger = \lambda_i, i \geq 1$, as demonstrated for the fundamental mode in Eq. (10.15).

Multiplying the perturbed equation (10.45) by $\phi_i^\dagger$ on the LHS and taking an inner product yields

$$\langle \phi_i^\dagger, L\delta\phi \rangle = -\left[ \langle \phi_i^\dagger, \delta L_1 \phi \rangle - \frac{1}{\lambda}\langle \phi_i^\dagger, \delta L_2 \phi \rangle + \frac{\delta\lambda}{\lambda^2}\langle \phi_i^\dagger, L_2 \phi \rangle \right] \tag{10.49a}$$

$$= \sum_{j=1}^{N} a_j \left[ \langle \phi_i^\dagger, L_1 \phi_j \rangle - \frac{1}{\lambda}\langle \phi_i^\dagger, L_2 \phi_j \rangle \right] \tag{10.49b}$$

$$= \sum_{j=1}^{N} a_j \left(1 - \frac{\lambda_j}{\lambda}\right)\delta_{ij} = a_i \left(1 - \frac{\lambda_i}{\lambda}\right), \tag{10.49c}$$

where the lambda mode expansion from Eq. (10.46) is substituted for $\delta\phi(\mathbf{r})$ in Eq. (10.49b), and the orthonormality of Eq. (10.48) is invoked to single out the expansion coefficient $a_i$ corresponding to the $i$th lambda mode $\phi_i$ in Eq. (10.49c). The expansion coefficients $a_i, i = 1, \ldots, N$, are obtained with the operator perturbations $\delta L_1$ and $\delta L_2$ applied to the unperturbed fundamental mode flux $\phi$, together with the eigenvalue perturbation $\delta\lambda$ represented in Eq. (10.49a). Thus, the determination of the lambda modes and the corresponding eigenvalues via Eq. (10.47) is essentially all that is required to calculate the flux perturbation $\delta\phi(\mathbf{r})$ to the order $N$ of desired accuracy.

In the TerraPower Advanced Reactor Modeling Interface (ARMI) framework [Tou15], high-order modal expansion algorithms were developed through the Krylov routines in the Trinilos package for Traveling Wave Reactor core design and fuel cycle studies. For the representation of localized perturbations in fast reactor cores loaded with 178 fuel assemblies, up to 400 harmonics were calculated in 3-D 33-group diffusion theory models with one-third core symmetry. Figure 10.5 illustrates fractional errors in point-wise flux distributions in offset fuel assemblies obtained with modal expansion algorithms relative to direct 33-group diffusion theory calculations, as a function of varying levels of $^{235}$U enrichment perturbations. Sufficiently accurate flux calculations with relative errors of O(1.0%) can be obtained with 20 ~ 30 harmonics even for localized enrichment perturbations of up to 20%. With an initial investment made to obtain

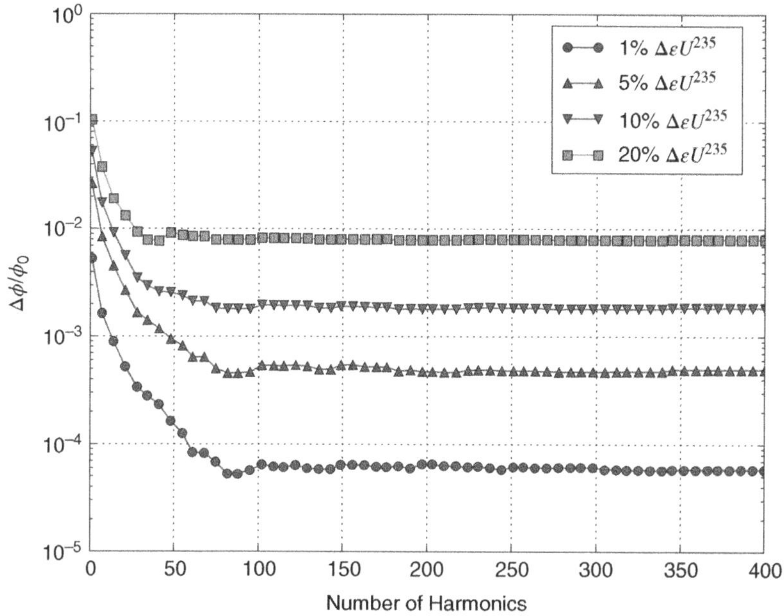


**Figure 10.5** Fractional pointwise flux errors in modal expansion calculations for offset fuel assemblies relative to direct 3-D diffusion theory calculations, subject to varying levels of $^{235}$U enrichment perturbations in a fast reactor. *Source*: Touran and Lee [Tou15]/Taylor & Francis.

high-order modal expansions, subsequent core and fuel cycle optimization studies were performed with a minimal computational effort in the ARMI structure.

In another application of the modal expansion technique for the space-time kinetics study, a somewhat different approach [Mor84] was taken for the KAHTER experimental critical assembly for a pebble-bed gas-cooled reactor. In this study, performed when the computational assets were still rather limited, a combination of a modal expansion technique and a direct perturbation calculation was used to represent the effects of control rod movements on flux distributions. The time-dependent flux distribution is represented through perturbations in the shape function $\psi(\mathbf{r}, E, t)$, discussed briefly in Section 8.1

$$\delta\psi(\mathbf{r}, E, t) = \sum_{i=1}^{N} a_i \phi_i(\mathbf{r}, E) + f(\mathbf{r}, E, t), \tag{10.50}$$

where the localized perturbations due to control rod movements are represented by a time-dependent balance equation for the local function $f(\mathbf{r}, E, t)$, primarily related to the neutron destruction operator $L_1$. The local function $f(\mathbf{r}, E, t)$ is obtained through a fixed-source diffusion solver and is reflected in the overall neutron balance equation for the global perturbation function comprising the modal expansion terms. This is discussed in more detail as an example of the quasi-static space-dependent kinetics formulation in Chapter 16.

Another important application of the modal expansion technique is in the analysis of the reactor core stability associated with spatial power oscillations due to $^{135}Xe$ buildup and poisoning, which is also discussed in detail in Chapter 16. The modal technique is invaluable in this case for gaining theoretical understanding of the phenomena involved and deriving stability criteria. Optimal methods to control the space-time xenon oscillation could also be derived by combining modal analysis and optimal control techniques.

## 10.8 Concluding Remarks on Adjoint Flux

A few examples considered in Sections 10.4 and 10.6 have provided valuable insights into the usefulness of adjoint formulations in reactor physics, in particular Eqs. (10.36) and (10.44). The fixed-source formulation from Eq. (10.44) may be also extended to efficiently obtain uncertainties in integral quantities like $R$ without having to make direct parametric perturbations. In fact, this is one of the major approaches taken currently to quantify uncertainties in various nuclear energy system parameters obtained with complex computer models. An effort has been made to develop adjoint formulations for dynamic system models similar to the RELAP5 code [NRC01].

Calculations of the adjoint neutron flux also serve various useful purposes, including the formulation for the $k$-eigenvalue and flux perturbations discussed in Sections 10.5 and 10.6. The adjoint flux is, however, a mathematical construct in a dual space and may not offer direct physical interpretations. For example, adjoint neutrons may gain energy as the real, forward neutrons slow down in the energy domain, as indicated by Eq. (10.17). Similarly, adjoint neutrons may travel in directions opposite to the regular neutrons and may travel backward in time so that adjoint neutrons may not leak out into vacuum but return freely from vacuum.

## References

**[Cru78]** M.W. Crump and J.C. Lee, "Calculation of Spatial Weighting Functions for Ex-Core Neutron Detectors," *Nucl. Technol.* **41**, 87 (1978).

**[Eng67]** W.W. Engle, Jr., "*A User's Manual for ANISN, A One-Dimensional Discrete Ordinates Transport Code with Anisotropic Scattering*," K-1697, Oak Ridge Gaseous Diffusion Plant (1967).

[Mor84] J. Moreira and J.C. Lee, "Space-Time Analysis of Reactor-Control-Rod-Worth Measurements," *Nucl. Sci. Eng.* **86**, 91 (1984).

[Nic64] R.B. Nicholson, "Methods for Determining the Energy Release in Hypothetical Fast Reactor Meltdown Accidents," *Nucl. Sci. Eng.* **18**, 207 (1964).

[NRC01] U. S. Nuclear Regulatory Commission, "RELAP5/MOD3.3 Code Manual, Volume 1: Code Structure, Systems Models, and Solution Methods," NUREG/CR-5535, rev. 1, U. S. Nuclear Regulatory Commission (2001).

[Sta01] W.M. Stacey, *Nuclear Reactor Physics*, Wiley (2001).

[Tou15] N.W. Touran and J.C. Lee, "A Modal Expansion Equilibrium Cycle Perturbation Method for Optimizing Fast Reactors," *Nucl. Sci. Eng.* **179**, 1 (2015).

[Wat02] D.S. Watkins, *Fundamentals of Matrix Computations*, 2nd ed., Wiley (2002).

## Problems

**10.1** Show that the one-group neutron diffusion equation is self-adjoint, subject to the boundary condition that the incoming partial current at either end of a slab reactor of thickness $a$ is zero, i.e. $J_+(0) = J_-(a) = 0$.

**10.2** Consider a bare critical slab reactor fueled with a uniform mixture of fuel and moderator. In the analysis of a postulated accident in the reactor, it is assumed that the central 5% of the core suffers a 5% decrease in the density of the fuel-moderator mixture. Determine the change in the effective multiplication factor due to the localized density change. Use one-group neutron diffusion theory, and neglect neutron extrapolation distance. For the unperturbed core material, $k_\infty = 1.03$.

**10.3** In a bare spherical reactor that is critical with homogenized density $\rho$ and volume $V$, a perturbation $\delta\rho(\mathbf{r})$ is introduced. The resulting reactivity change may be represented by $\delta K = \int_V \delta\rho(\mathbf{r}) w(\mathbf{r}) d\mathbf{r}$, in terms of the reactivity worth function (density worth function) $w(\mathbf{r})$. For a spherical reactor with radius $R$, the neutron flux profile $\phi(r) = \sin(Br)/r$, with buckling $B^2$, may be approximated by $\phi(r) = 1 - q(r/R)^2$. (a) Obtain an expression for the parameter $0 < q \leq 1$. (b) Derive the worth function $w(r) = w(0)[1 - \gamma(r/R)^2]$ via first-order perturbation theory. (c) If a uniform change in density is made to the spherical core such that $\delta\rho/\rho = 0.05$, with $q = 0.6, \Sigma_a = 2.652 \times 10^{-3}\,\text{cm}^{-1}, \nu\Sigma_f = 3.788 \times 10^{-3}\,\text{cm}^{-1}, D = 1.463\,\text{cm}$, $\beta = 3 \times 10^{-3}$, and $V = 6.0\,\text{m}^3$, determine the resulting reactivity change

$\delta K$. The result obtained here constitutes a basis for the Bethe-Tait model for a hypothetical core disruptive accident (HCDA) analysis [Nic64].

**10.4** Perturbation in the absorption cross section $\delta\Sigma_a(x) = \Sigma^* \exp(-\gamma x)$ is introduced over $0 < x < H$ into a critical homogeneous slab reactor of thickness $H$. Obtain the resulting change in the effective multiplication factor.

**10.5** A thickness $\Delta x$ of the fuel-moderator mixture is replaced by pure moderator at position $x$ in a bare slab reactor of thickness $H$. Obtain an expression for the change in reactivity, assuming that the moderator has the same diffusion coefficient $D$ but has $\Sigma_a = \nu\Sigma_f = 0$.

**10.6** A step-wise perturbation in $\Sigma_a$

$$\delta\Sigma_a(x) = \begin{cases} \Delta\Sigma_a, & -H/2 \le x \le 0, \\ -\Delta\Sigma_a, & 0 < x \le H/2, \end{cases}$$

is introduced into a critical bare slab reactor of height $H$. With $\Delta\Sigma_a/\Sigma_a = 0.01$ and a neutron leakage probability of 0.03, determine the amplitude of the first harmonic mode relative to the fundamental mode.

# 11

# Lattice Physics Analysis of Heterogeneous Cores

In most operating reactor cores, various material heterogeneities are introduced for thermal and mechanical design considerations as well as for neutronic reasons. A typical example is the fuel material lumped separately in the form of thin plates or pins, encapsulated with some cladding material, and separated from other fuel elements by coolant or moderator materials. Material heterogeneities in a reactor core have to be explicitly considered if the mean free path of neutrons in the core is comparable to the characteristic dimensions of such heterogeneities. Thus, in LWR cores where typically about 75% of the neutron production is due to thermal neutron fission, such material heterogeneities are of major importance in calculating the reactivity and spatial power distributions. In contrast, the average energy of neutrons in fast reactor cores, typically cooled by liquid sodium, is on the order of $100 \sim 200$ keV, resulting in a much longer mean free path of neutrons. This suggests that, in fast reactor cores, a detailed treatment of spatial-spectral coupling will play a more important role than material heterogeneities. In fact, for fast reactor analysis, flux spectrum calculations are typically coupled with global spatial flux calculations and iteratively performed, whereas for LWR cores, neutron spectrum calculations are normally decoupled from global spatial calculations and performed on a unit cell or unit assembly basis.

An LWR core may be approximately represented as an infinite array of unit cells, each consisting of a fuel element, cladding, coolant/moderator, and often structural or parasitic absorber materials. The fuel elements are typically arranged in fuel assembly structures, illustrated for a PWR core in Figure 1.6. In a unit-cell arrangement, no neutron is assumed lost across the radial boundary of the cell, and axial leakage of neutrons is represented simply in terms of an axial buckling. Thus, a unit-cell spectrum calculation is typically performed in a cylindrical geometry, where the material heterogeneities are explicitly represented to generate the cell-average flux spectrum $\phi(E)$. The energy-dependent scalar flux is then used to collapse fine-group neutron cross sections into a set of multi- or

*Nuclear Reactor Physics and Engineering*, Second Edition. John C. Lee.
© 2025 John Wiley & Sons, Inc. Published 2025 by John Wiley & Sons, Inc.
Companion website: www.wiley.com/go/LeeNRPE2

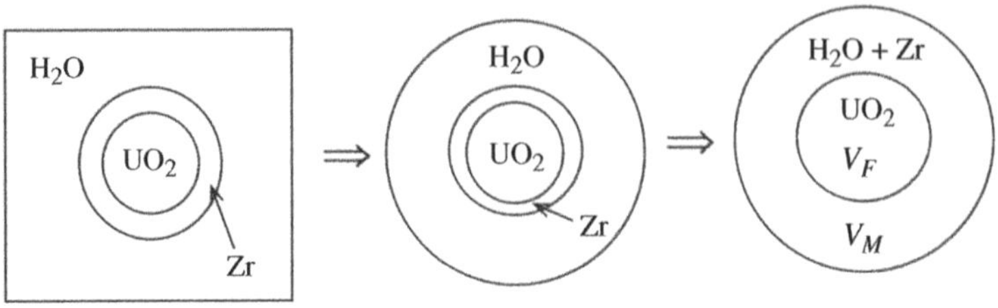


**Figure 11.1** Unit-cell representations.

few-group cross sections. This process is sketched in Figure 11.1. For other types of fuel arrangements, e.g. hexagonal fuel assemblies for sodium-cooled fast reactors or He-cooled high-temperature reactors, modifications have to be made to the basic unit-cell geometry, but the concept of a unit cell still applies.

In recent years, collision probability (CP) and method of characteristics (MOC) formulations have been used to solve an integral form of the neutron transport equation for the entire energy spectrum, explicitly accounting for material heterogeneities at a unit-assembly level. In an assembly CP or MOC calculation, the presence of lattice positions not containing the fuel is accurately represented in a full-blown 2-D geometry, but with boundary conditions similar to those for unit-cell calculations.

In addition to thermal and mechanical reasons, there is a strong neutronic incentive for lumping fuel materials separately rather than homogeneously mixing them with coolant or moderator materials. This is because fuel lumping increases the resonance escape probability considerably, even at the expense of some reduction in the thermal utilization. Thus, a homogeneous mixture of carbon with natural uranium has $k_\infty < 0.85$, and Enrico Fermi's CP-1 assembly could not have achieved criticality without the fuel-lumping technique.

We begin with a simple description of the basic heterogeneous unit-cell representation in Section 11.1, with the spatial flux distribution represented in terms of a fuel-moderator two-region model. This will be followed in Section 11.2 by a discussion of the neutronic advantages of fuel lumping through a two-group diffusion theory model. With a simple unit-cell arrangement, physical effects of heterogeneous fuel arrangements may be studied effectively, and this is the main reason for focusing our study on simple unit-cell structures rather than delving quickly into assembly arrangements.

We present in Section 11.3 a basic diffusion theory formulation of flux distribution in the unit cell, leading to an expression for the thermal utilization. Section 11.4 introduces a number of improvements that may be made to the diffusion theory results of Section 11.3, in particular the well-known Amouyal–Benoist–Horowitz model. Section 11.5 presents approaches that may be used to evaluate the resonance escape probability in a heterogeneous unit-cell configuration. We discuss in Section 11.6 the Wigner–Wilkins model that has been used for PWR thermal flux calculations for a number of years. We then turn our attention in Section 11.7 to the integral transport techniques that form the basis for lattice physics analysis of LWR cores in recent years. Section 11.8 presents the $B_1$ formulation that is used for neutron leakage calculations in unit-cell and unit-assembly lattice physics models. We conclude this chapter with a discussion of lattice physics models more tailored for fast-spectrum nuclear reactors in Section 11.9, followed by a brief introduction to the Monte Carlo lattice physics analysis in Section 11.10. An overview of overall reactor physics analysis is presented in Section 11.11.

## 11.1 Material Heterogeneity and Flux Distribution in Unit Cell

The primary product we desire to get out of a neutron spectrum calculation for a unit cell is spectrum-averaged neutron cross sections in a suitable group structure. In a multigroup structure, the spectrum-averaged cross section for the cell in the gth group may be defined in terms of space- and energy-dependent scalar flux $\phi(\mathbf{r}, E)$ as

$$\Sigma_g = \frac{\int_V d\mathbf{r} \int_{E_g}^{E_{g-1}} dE \Sigma(\mathbf{r}, E)\phi(\mathbf{r}, E)}{\int_V d\mathbf{r} \int_{E_g}^{E_{g-1}} dE \phi(\mathbf{r}, E)}, \tag{11.1}$$

where the volume integral is carried over the entire volume $V$ of the unit cell. With the multigroup flux $\phi_g(\mathbf{r})$ and flux-weighted cross section $\Sigma_g(\mathbf{r})$ defined as

$$\phi_g(\mathbf{r}) = \int_{E_g}^{E_{g-1}} dE \phi(\mathbf{r}, E), \tag{11.2}$$

$$\Sigma_g(\mathbf{r}) = \frac{\int_{E_g}^{E_{g-1}} dE \Sigma(\mathbf{r}, E)\phi(\mathbf{r}, E)}{\phi_g(\mathbf{r})}, \tag{11.3}$$

the cell-average cross section of Eq. (11.1) follows:

$$\Sigma_g = \frac{\int_V d\mathbf{r} \Sigma_g(\mathbf{r})\phi_g(\mathbf{r})}{\int_V d\mathbf{r} \phi_g(\mathbf{r})}. \tag{11.4}$$

In performing the spectral weighting indicated in Eq. (11.3), the scalar flux is often assumed separable in the space and energy variables:

$$\phi(\mathbf{r}, E) = \psi(\mathbf{r})\phi(E). \tag{11.5}$$

This then allows us to obtain multigroup microscopic constants

$$\sigma_g = \frac{\int_{E_g}^{E_{g-1}} dE\sigma(E)\phi(\mathbf{r}, E)}{\psi(\mathbf{r})\int_{E_g}^{E_{g-1}} dE\phi(E)} = \frac{\int_{E_g}^{E_{g-1}} dE\sigma(E)\phi(E)}{\int_{E_g}^{E_{g-1}} dE\phi(E)}, \tag{11.6}$$

which are independent of position **r**. The macroscopic group constants for a nuclide with number density $N(\mathbf{r})$ of Eq. (11.3) may then be simply obtained as $\Sigma_g(\mathbf{r}) = N(\mathbf{r})\sigma_g$. For a mixture comprising several nuclides, e.g. for $UO_2$ fuel, a summation has to be made over the nuclides of the mixture to arrive at the macroscopic cross sections for each region of the unit cell.

Once the macroscopic group constants are generated in this way for each region, spatial flux distribution $\phi_g(\mathbf{r})$ may be calculated for each group and cell-average group constants obtained through Eq. (11.4). Thus, our main task in treating material heterogeneities is to calculate $\phi_g(\mathbf{r})$, with due account given for large differences in cross sections between the fuel and moderator regions. The effects of flux depression in the fuel region are manifest dominantly for thermal neutrons and resonance energy neutrons. We will study the basic concepts required for treating the effects of material heterogeneities in terms of a two-region unit cell consisting of a fuel region of volume $V_F$ and a moderator region of volume $V_M$, illustrated in Figure 11.1. For our two-region unit cell, the cell-average cross section defined in Eq. (11.4) may be written

$$\Sigma_g = \frac{\Sigma_{gF}\overline{\phi}_{gF}V_F + \Sigma_{gM}\overline{\phi}_{gM}V_M}{\overline{\phi}_{gF}V_F + \overline{\phi}_{gM}V_M}, \tag{11.7}$$

where

$$\overline{\phi}_{gF} = \frac{1}{V_F}\int_{V_F}\phi_g(\mathbf{r})d\mathbf{r} \quad \text{and} \quad \overline{\phi}_{gM} = \frac{1}{V_M}\int_{V_M}\phi_g(\mathbf{r})d\mathbf{r}$$

are the average flux for the fuel and moderator regions, respectively, and $\Sigma_{gF}$ and $\Sigma_{gM}$ are the fuel and moderator cross sections, respectively, for group g. Extension of our simple two-region unit-cell model may be made, without undue effort, to represent the fuel cladding or other *non-lattice materials*, including instrumentation guide tubes and control materials present in a fuel assembly.

## 11.2 Neutronic Advantages of Fuel Lumping

To gain physical understanding of the neutronic effects of fuel lumping, we set up a neutron balance in a reactor core via two-group neutron diffusion theory discussed in Chapter 7. In particular, recall Eq. (7.27) for the infinite multiplication factor

$$k_\infty = \frac{\nu\Sigma_{f1}}{\Sigma_{a1} + \Sigma r} + \frac{\nu\Sigma_{f2}}{\Sigma_{a2}}\frac{\Sigma_r}{\Sigma_{a1} + \Sigma_r} = k_1 + k_2 = k_1 + pf\eta, \tag{11.8}$$

where $k_1$ and $k_2$ represent the contributions to $k_\infty$ from fast and thermal fissions, respectively, and $k_2$ is further broken up into resonance escape probability $p$, thermal utilization $f$, and the parameter $\eta$ representing the number of neutrons released per thermal neutron absorption in the fuel.

The main effect of fuel lumping in a heterogeneous geometry is the depression of spatial flux distribution in the fuel region. Such flux depression phenomena are qualitatively shown in Figures 11.2 and 11.3. Due to the variation of the fuel absorption cross section sketched in Figure 11.2 over the energy range covering thermal to fission neutrons, the flux distribution in a unit cell may show the trends illustrated in Figure 11.3. At fission energy $E_{fiss}$, since neutrons are produced in the fuel, the flux distribution tends to be slightly peaked in the fuel region. As neutrons slow down and the fuel absorption cross section $\Sigma_{aF}(E)$ increases, the neutrons coming back into the fuel from the moderator are absorbed increasingly at or near the surface of the fuel element. Hence, the flux distribution, for neutrons in the slowing-down range $[E_1, E_2]$, is suppressed in the fuel region compared with that in the moderator region, as illustrated in Figure 11.3. At a resonance

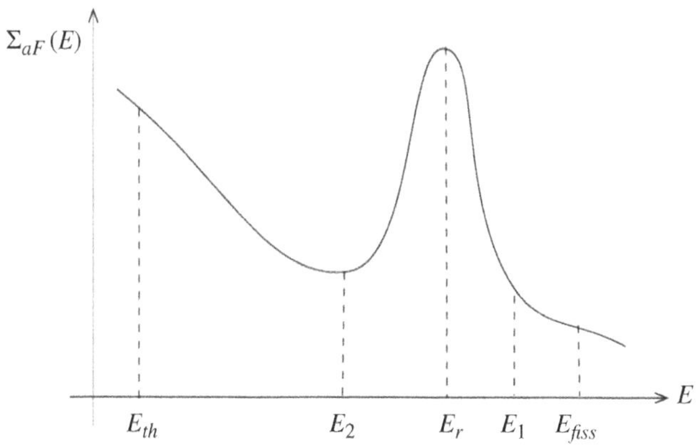


**Figure 11.2** Energy dependence of a fuel absorption cross section.

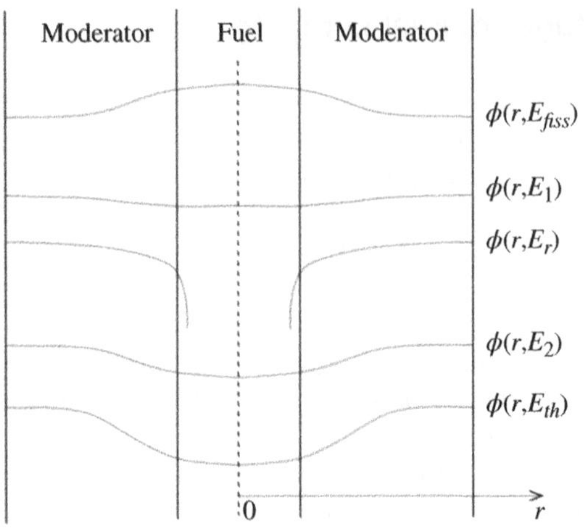


**Figure 11.3** Spatial flux distribution as a function of neutron energy.

energy $E_r$, the flux distribution suffers a sharp decline near the surface of the fuel element and essentially vanishes in the interior region of the fuel element. Finally, the neutrons, escaping absorption resonances and slowing down into the thermal group, may show some flux depression in the fuel region, reflecting the general trend that $\Sigma_{aF} > \Sigma_{aM}$.

The depression of spatial flux distribution in the fuel region over the bulk of the energy spectrum is similar to the *energy self-shielding* phenomenon discussed in Section 9.4.2, which is of importance in the calculation of resonance escape probability for a homogeneous mixture of fuel and moderator. As illustrated in Figure 11.3, because of the presence of highly absorbing material in the fuel region, the neutron flux is depressed there, and the absorption rate per fuel atom is reduced compared with the case without flux depression. Thus, a strong resonance absorption in fuel tends to shield the absorber from neutrons of resonance energy and is known as the *spatial self-shielding* of resonances. This is the primary reason why the resonance escape probability is increased in a heterogeneous unit cell, compared with the corresponding homogeneous mixture of fuel and moderator, and is the main advantage of fuel lumping. This advantage of fuel lumping is partly offset, however, by a decrease in the thermal utilization, due to the depression of thermal neutron flux in the fuel region, also illustrated in Figure 11.3. We will study the effects of fuel lumping further in terms of various parameters summarized in Eq. (11.8) for the two-group form of the infinite multiplication factor $k_\infty$.

1. **Resonance escape probability**

   Recall that, for a homogeneous cell, the *narrow resonance* (NR) approximation of Eq. (9.62) may be used to obtain probability $p(E)$ of neutrons escaping resonance captures in slowing down from source energy $E_0$ to $E < E_0$

$$p(E_0 \rightarrow E) = \exp\left[-\frac{N_F}{\xi\Sigma_s}\int_0^u du\,\frac{\sigma_a(u)}{1+\sigma_a(u)N_F/\Sigma_s}\right] = \exp\left[-\frac{N_F}{\xi\Sigma_s}I\right],$$

or

$$p(0 \rightarrow u) = \exp\left[-\frac{N_F}{\xi\Sigma_s}\int_0^u du\sigma_a(u)\phi_{NR}(u)\right] = \exp\left[-\frac{N_F}{\xi\Sigma_s}\int_0^u du\frac{\sigma_a(u)\Sigma_s}{\Sigma_t(u)}\right], \tag{11.9}$$

where $I = I_{eff}$ is the *effective resonance integral*, physically representing a flux-weighted effective absorption cross section for the resonance region, with the normalized NR flux $\phi_{NR}(u) = \Sigma_s/\Sigma_t(u)$ defined as the *energy self-shielding factor* in Eq. (9.65). Extending this result to a heterogeneous configuration with a large resonance absorption cross section in the lumped fuel region, we anticipate that the neutron flux is suppressed in the fuel region, resulting in a smaller effective resonance integral for fuel and hence a larger resonance escape probability. The flux depression in the lumped fuel region due to the presence of large resonance absorbers is the spatial self-shielding of absorbers, discussed further in Section 11.5. In a heterogeneous arrangement, there is an increased probability that neutrons produced in the fuel region are able to migrate into the moderator region and slow down there to sufficiently low energies without undergoing resonance captures. This is, however, a much smaller, secondary effect, compared with the primary advantage of fuel lumping due to the spatial flux self-shielding of resonance absorbers.

2. **Thermal utilization**
   In terms of the cell-average macroscopic cross section defined in Eq. (11.7), for the thermal group, i.e. $g = 2$ in a two-group model, rewrite Eq. (7.37) for the *thermal utilization*, i.e. the fraction of thermal neutron absorptions that take place in fuel

$$f = \frac{\Sigma_{a2}^F\overline{\phi}_{2F}V_F}{\Sigma_{a2}^F\overline{\phi}_{2F}V_F + \Sigma_{a2}^M\overline{\phi}_{2M}V_M} = \frac{\Sigma_{a2}^F}{\Sigma_{a2}^F + \Sigma_{a2}^M\zeta_2 V_M/V_F}, \tag{11.10}$$

with the *thermal disadvantage factor* introduced

$$\zeta_2 = \frac{\overline{\phi}_{2M}}{\overline{\phi}_{2F}}. \tag{11.11}$$

Due to the flux depression in fuel, as illustrated in Figure 11.3, it is natural to expect $\zeta_2 > 1$ in a heterogeneous configuration. The larger $\zeta_2$ is, the smaller $f$ is, and hence the name *thermal disadvantage factor*. For a homogeneous mixture of fuel and moderator, recover Eq. (7.37) for the thermal utilization by setting $\zeta_2 = 1$ in Eq. (11.10), remembering that $\Sigma_{a2}^F$ and $\Sigma_{a2}^M$ are absorption cross sections for the separated fuel and moderator materials, respectively, and that the volume ratio simply accounts for the homogeneous mixing of fuel and moderator. Hence, the thermal utilization $f$ is smaller in a heterogeneous system compared with an equivalent homogeneous system. This is obviously the penalty we have to pay in a heterogeneous arrangement to achieve a much larger gain in the form of an increased resonance escape probability.

3. **The number $\eta$ of neutrons produced per absorption in fuel**
   The parameter $\eta$ is essentially given by a ratio of fuel cross sections and is nearly unaffected by fuel lumping. This is particularly true under the assumption that the scalar flux is separable in energy and space.
4. **Fast fission term $k_1$**
   Using the cell-average macroscopic cross section from Eq. (11.7), we may write the fast-group contribution $k_1$ to the infinite multiplication factor $k_\infty$

$$k_1 = \frac{\nu\Sigma_{f1}}{(\Sigma_{a1}^F + \Sigma_r^F) + (\Sigma_{a1}^M + \Sigma_r^M)\zeta_1 V_M/V_F}, \tag{11.12}$$

   where the parameter $\zeta_1 = \overline{\phi}_1^M / \overline{\phi}_1^F$ may be called the fast advantage factor, Since fast neutrons have, on the average, a higher chance of interacting with fuel atoms than with moderator atoms in a heterogeneous system, we may get $\zeta_1 \leq 1$, and $k_1$ could increase just slightly as compared with an equivalent homogeneous system. This increase in $k_1$ is usually a minor effect.
5. **Neutron leakage**
   To assess the effects of fuel lumping on thermal neutron leakage, write the thermal absorption cross section $\Sigma_{a2}$ averaged over a heterogeneous cell

$$\Sigma_{a2} = \frac{\Sigma_{a2}^M}{1-f} \cdot \frac{\overline{\phi}_{2M} V_M}{\overline{\phi}_{2F} V_F + \overline{\phi}_{2M} V_M}. \tag{11.13}$$

   For a small fuel lump such that $\overline{\phi}_{2F} V_F \ll \overline{\phi}_{2M} V_M$, Eq. (11.13) may be approximated as

$$\Sigma_{a2} \simeq \frac{\Sigma_{a2}^M}{1-f}. \tag{11.14}$$

   The thermal diffusion coefficient $D_2$, which is on the order of 1.0 cm, is essentially equal to $D_{2M}$ of the moderator for small fuel lumps. Hence, we may represent the thermal diffusion length $L_2$ for the cell as

$$L_2 \simeq L_{2M}\sqrt{1-f}. \tag{11.15}$$

   Equation (11.15) indicates that, with $L_{2M}$ of the moderator nearly equal to $L_2$ of a homogeneous mixture, the thermal neutron leakage will increase slightly in a heterogeneous arrangement. Thermal neutron leakage is, however,

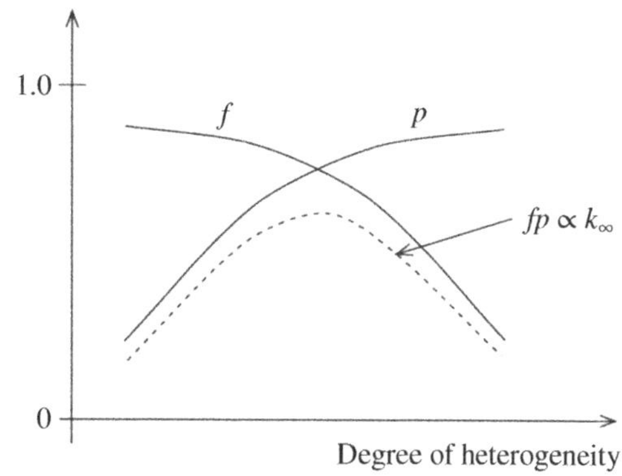


**Figure 11.4** Parameters $f$, $p$, and $k_\infty$ vs. degree of heterogeneity.

usually small in large LWR cores and the effect of heterogeneities on $k_{eff}$ through the increase in thermal neutron leakage is insignificant. On the other hand, the spatial flux distribution for fast neutrons is nearly flat, as illustrated in Figure 11.3, and hence fast neutron leakage is hardly affected by fuel lumping.

In summary, the primary advantage of lumping fuel in a heterogeneous geometry is a substantial increase in the resonance escape probability due to the spatial self-shielding of the resonance absorbers. As an example, the effective resonance integral for a mixture of natural uranium and graphite could decrease from 280 b for a homogeneous mixture to a mere 9 b for a heterogeneous system. The resulting increase in $p$ in a heterogeneous geometry is substantial and can usually override the accompanying decrease in $f$, yielding a net increase in $k_\infty$ and also in $k_{eff}$ itself. As a function of degree of heterogeneity, we sketch in Figure 11.4 the overall trend of the variations in two key parameters, thermal utilization $f$, and resonance escape probability $p$, affecting $k_\infty \propto fp$. Thus, we may anticipate that some effort would be required to obtain optimal fuel configurations in various LWR design studies.

## 11.3 Diffusion Theory Model for Thermal Utilization

As discussed in Section 11.2, the thermal disadvantage factor $\zeta_2$ has to be evaluated in order to calculate the thermal utilization $f$ and the cell-average group constants for the thermal group. As an example of various techniques that may be used for heterogeneous core calculations, we begin with a one-group diffusion theory model [Gla52] for the calculation of the thermal flux distribution in a unit cell. We discuss in Section 11.4 improvements to this simple diffusion theory formulation.

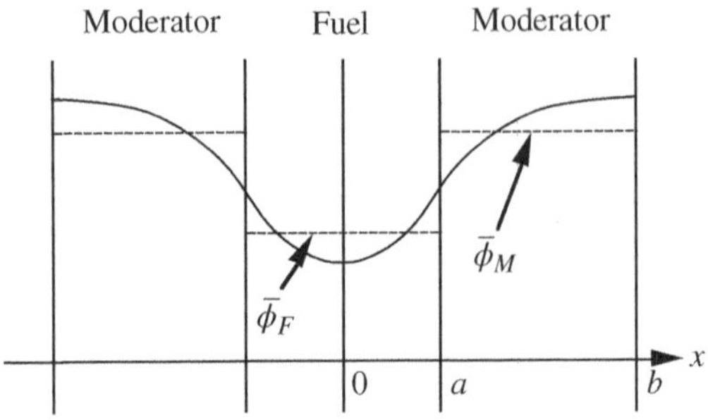


**Figure 11.5** Two-region unit cell in slab geometry.

To minimize the algebraic complexities required, we transform the two-region cylindrical unit cell from Figure 11.1 into a unit cell in slab geometry, represented schematically in Figure 11.5. Extension to a more realistic cylindrical unit cell can be made with some additional algebra and will be discussed in due course.

Together with the slab geometry approximation, we introduce the following simplifying assumptions:

(1) The overall thickness of the reactor core is much larger than the half thickness $b$ of the slab unit cell so that we may consider the reactor core consisting of an infinite lattice of unit cells of half thickness $b$ and fuel plates of half thickness $a$. Thus we also assume that there is *no net neutron leakage across the unit cell boundaries* at $x = \pm b$.

(2) The one-group neutron diffusion equation is valid in both the fuel and the moderator, although the fuel is highly absorbing and diffusion theory is a poor approximation in the fuel region. In Section 11.4, we present approaches to remedy this limitation of diffusion theory for heterogeneous core analysis. At the moment, we are interested in obtaining a simple expression for the thermal disadvantage factor $\zeta_2$, although somewhat approximate, that may give us a good physical understanding of the effects of material heterogeneities.

(3) A volumetric source of thermal neutrons of strength $Q$ [neutron $\cdot$ cm$^{-3}$ s$^{-1}$] is distributed uniformly in the moderator region, and no thermal neutrons are produced in the fuel region. This is a reasonable approximation because fast neutrons born in the fuel region have to migrate to the moderator region, where they suffer a number of scattering collisions before they are thermalized. Hence, the slowing-down density in the moderator region tends to be flat spatially. Likewise, no significant amount of slowing down takes place in fuel, and it is reasonable to assume that the slowing-down density in the fuel region is negligibly small.

With the necessary simplifying assumptions introduced, we write the one-group diffusion equations for the two-region unit cell, suppressing the subscript 2 that was used to denote the thermal group in Chapter 7:

$$-D_F \frac{d^2\phi_F(x)}{dx^2} + \Sigma_{aF}\phi_F(x) = 0, \quad \text{fuel,} \tag{11.16a}$$

$$-D_M \frac{d^2\phi_M(x)}{dx^2} + \Sigma_{aM}\phi_M(x) = Q, \quad \text{moderator.} \tag{11.16b}$$

Equations (11.16a) and (11.16b) may be solved, subject to the usual boundary conditions of the continuity of flux and current at the interface between the fuel and moderator, and symmetry properties at $x = 0$ and $x = b$:

$$\text{(i) } \phi_F(a) = \phi_M(a), \ \text{(ii) } D_F\phi_F'(a) = D_M\phi_M'(a), \ \text{(iii) } \phi_F'(0) = 0, \ \text{(iv) } \phi_M'(b) = 0.$$

The actual solution of Eqs. (11.16) follows standard procedure to yield

$$\phi_F(x) = \frac{Q\cosh\kappa_F x}{\Sigma_{aM} C D_F \kappa_F \sinh\kappa_F a}, \tag{11.17}$$

$$\phi_M(x) = \frac{Q}{\Sigma_{aM}}\left[1 - \frac{\cosh\kappa_M(b-x)}{CD_M\kappa_M\sinh\kappa_M(b-a)}\right], \tag{11.18}$$

where

$$\kappa_F = \sqrt{\Sigma_{aF}/D_F}, \quad \kappa_M = \sqrt{\Sigma_{aM}/D_M},$$

and

$$C = \frac{\coth\kappa_F a}{D_F\kappa_F} + \frac{\coth\kappa_M(b-a)}{D_M\kappa_M}. \tag{11.19}$$

The flux solutions provide the thermal disadvantage factor $\zeta_2$ of Eq. (11.11)

$$\zeta_2 = \frac{\overline{\phi}_M}{\overline{\phi}_F} = \frac{a\displaystyle\int_a^b dx\phi_M(x)}{(b-a)\displaystyle\int_0^a dx\phi_F(x)} = aC\Sigma_{aF}\left[1 - \frac{1}{C\Sigma_{aM}(b-a)}\right]. \tag{11.20}$$

In the limit of homogeneous mixture, i.e. as both $a$ and $b$ approach zero, while the ratio $a/b$ is kept constant, $\zeta_2$ approaches unity. Figure 11.6 qualitatively indicates how the thermal disadvantage factor $\zeta_2$ varies as a function of the degree of heterogeneity, i.e. the half thickness $a$ of fuel. Substituting Eq. (11.19) for $C$ into Eq. (11.20), write $\zeta_2$ as

$$\zeta_2 = \frac{a\Sigma_{aF}}{(b-a)\Sigma_{aM}}\left[\frac{(b-a)\Sigma_{aM}}{a\Sigma_{aF}}\kappa_F a\coth\kappa_F a + (b-a)\kappa_M\coth\kappa_M(b-a) - 1\right]. \tag{11.21}$$

Noting that, in our geometry, $V_F/V_M = a/(b-a)$ yields

$$\zeta_2 = \frac{\Sigma_{aF}V_F}{\Sigma_{aM}V_M}\left[\frac{\Sigma_{aM}V_M}{\Sigma_{aF}V_F}F + E - 1\right], \tag{11.22}$$

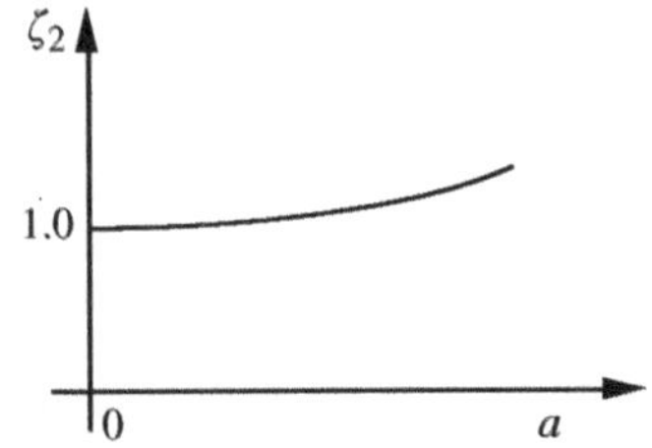


**Figure 11.6** Thermal disadvantage factor as a function of fuel half thickness $a$ with the ratio $a/b$ kept constant.

with the definition for two *lattice functions*

$$F = a\kappa_F \coth \kappa_F a, \tag{11.23}$$

$$E = (b - a)\kappa_M \coth \kappa_M (b - a). \tag{11.24}$$

It can be readily recognized that the lattice function $F$ is equal to the *surface-to-average flux ratio in the fuel*

$$F = \frac{\phi_F(a)}{\overline{\phi}_F}, \tag{11.25}$$

while $(E - 1)$ is a measure of the *excess absorption due to the non-flat flux distribution in the moderator* as compared with a flat flux distribution

$$E - 1 = \frac{\Sigma_{aM} V_M [\overline{\phi}_M - \phi_M(a)]}{\Sigma_{aF} V_F \overline{\phi}_F} = \frac{\text{excess absorption in moderator}}{\text{absorption in fuel}}. \tag{11.26}$$

The lattice function $E$ represents the fact that the flux in the moderator is not completely uniform due to the finiteness of $D_M$, and hence that the neutron absorption in the moderator is slightly higher than it would have been if $\phi_M(x) = \phi_M(a)$ everywhere in the moderator region. The interpretations of the two lattice functions may be clarified by rewriting Eq. (11.22)

$$\zeta_2 = \frac{\Sigma_{aF} V_F}{\Sigma_{aM} V_M} \left[ \frac{\Sigma_{aM} V_M}{\Sigma_{aF} V_F} \frac{\phi_F(a)}{\overline{\phi}_F} + \frac{\Sigma_{aM} V_M}{\Sigma_{aF} V_F} \left\{ \frac{\overline{\phi}_M}{\overline{\phi}_F} - \frac{\phi_M(a)}{\overline{\phi}_F} \right\} \right], \tag{11.22a}$$

which simply reproduces the definition of $\zeta_2$ in Eq. (11.20).

The thermal disadvantage factor $\zeta_2$ in Eq. (11.22), expressed in terms of the lattice functions $F$ and $E$, also holds for other geometries. For example, in a cylindrical unit cell with fuel radius $a$ and cell radius $b$, Eq. (11.22) holds, provided [Gla52] Eqs. (11.23) and (11.24) for slab geometry are replaced by

$$F = \frac{\kappa_F a I_0(\kappa_F a)}{2 I_1(\kappa_F a)}, \tag{11.27}$$

$$E = \frac{\kappa_M (b^2 - a^2)}{2a} \left[ \frac{I_0(\kappa_M a) K_1(\kappa_M b) + K_0(\kappa_M a) I_1(\kappa_M b)}{I_1(\kappa_M b) K_1(\kappa_M a) - K_1(\kappa_M b) I_1(\kappa_M a)} \right], \tag{11.28}$$

where $I_n$ and $K_n$ are the $n$th-order modified Bessel functions of the first and second kinds, respectively [Appendix C]. Finally, Eqs. (11.10) and (11.22) provide an expression for the thermal utilization

$$\frac{1}{f} - 1 = \frac{\Sigma_{aM} V_M}{\Sigma_{aF} V_F} \zeta_2,$$

or equivalently,

$$\frac{1}{f} = \frac{\Sigma_{aM} V_M \overline{\phi}_M}{\Sigma_{aF} V_F \overline{\phi}_F} + 1 = \frac{\Sigma_{aM} V_M}{\Sigma_{aF} V_F} F + E. \tag{11.29}$$

For numerical evaluation, the lattice function $E$ may be approximated [Arg63] for $\kappa_M b \ll 1$

$$E - 1 = \begin{cases} \dfrac{\kappa_M^2 (b-a)^2}{3} : & \text{slab} \\ \dfrac{(\kappa_M b)^2}{2} \left[ \dfrac{\ln(b/a)}{1-(a/b)^2} - \dfrac{3}{4} + \dfrac{1}{4}\left(\dfrac{a}{b}\right)^2 \right] : & \text{cylinder.} \end{cases} \tag{11.30}$$

Although derived for slab geometry, Eq. (11.29) is a general expression that is also valid for other unit-cell configurations, provided the lattice functions $F$ and $E$ properly represent the particular geometry. Furthermore, the slab geometry lattice functions, Eqs. (11.23) and (11.24), may be used as a good approximation for the corresponding cylindrical expressions, Eqs. (11.27) and (11.28), provided the mean chord lengths of both the fuel and moderator regions are preserved. The *mean chord length*, defined as $\overline{s} = 4V/A$ for a lump with volume $V$ and surface area $A$, is an effective mean free path of neutrons for collision probabilities, as discussed further in Chapter 17. The use of the mean chord length to find a slab geometry unit-cell equivalent to a cylindrical unit cell is an empiricism, which is illustrated in Example 11.1.

**Example 11.1** Calculate the thermal disadvantage $\zeta_2$ and thermal utilization $f$ for a two-region unit cell with the zircalloy clad homogenized into the moderator region. For a HFP PWR configuration with a fuel pellet radius $a = 0.4692$ cm and lattice pitch $p = 1.4194$ cm, obtain the cell-average cross section with the thermal-group constants:

$$D_F = 0.5201 \text{ cm}, \quad \Sigma_{aF} = 0.2641 \text{ cm}^{-1}, \quad D_M = 0.3062 \text{ cm}, \quad \Sigma_{aM} = 0.0454 \text{ cm}^{-1}.$$

Determine the radius $b = 0.8008$ cm for an equivalent cylindrical cell by setting $p^2 = \pi b^2$ and obtain $\kappa_F = \sqrt{\Sigma_{aF}/D_F} = 0.7126\,\text{cm}^{-1}$ and $\kappa_M = \sqrt{\Sigma_{aM}/D_M} = 0.3851\,\text{cm}^{-1}$. Obtain the lattice functions $F = 0.3343 \times 1.0281/(2 \times 0.1695) = 1.0138$ and $E = 1.00864$, respectively, with Eqs. (11.27) and (11.30). With

$V_F/V_M = a^2/(b^2 - a^2) = 0.5228$, Eq. (11.22) yields $\zeta_2 = 1.0402$, together with $f = 0.7452$ from Eq. (11.29). Equation (11.7) finally yields

$$\langle\Sigma_a\rangle_{cell} = \frac{\Sigma_{aF} + \Sigma_{aM}V_M\zeta_2/V_F}{1 + V_M\zeta_2/V_F} = 0.1186\ \text{cm}^{-1}.$$

**Example 11.2** Repeat the calculations of $\zeta_2$, $f$, and $\langle\Sigma_a\rangle_{cell}$ using the slab geometry expressions by preserving the mean chord lengths for both the fuel and moderator regions.

The equivalent half thickness $a_s$ for the fuel region and the half thickness $b_s$ for the unit cell itself is obtained by setting $\bar{s}_F = 4a_s = 4\pi a^2/(2\pi a) = 2a$, and $\bar{s}_M = 4(b_s - a_s) = 4\pi(b^2 - a^2)/(2\pi a)$ so that

$$a_s = \frac{a}{2},\quad b_s = a_s + \frac{b^2 - a^2}{2a}.$$

Note that the mean chord lengths $\bar{s}_F$ and $\bar{s}_M$ are evaluated with surfaces with vanishing net current excluded, as is the case for the hydraulic diameter in Chapter 13; the wetted perimeter has to be limited to the surfaces where the fluid is in actual contact with the channel wall. With $a_s = 0.2346$ cm and $b_s = 0.6835$ cm, the lattice functions are readily evaluated, with Eqs. (11.23) and (11.24): $F = 1.0093$ and $E = 1.0099$, yielding $\zeta_2 = 1.0410$, $f = 0.7450$, and $\langle\Sigma_a\rangle_{cell} = 0.1185\ \text{cm}^{-1}$, in excellent agreement with that of Example 11.1. This demonstrates the power of mean chord length in reaction-rate calculations.

## 11.4 Improved Method for Thermal Disadvantage Factor

The derivation of the thermal disadvantage factor and thermal utilization is presented in terms of one-group diffusion theory in Section 11.3 to illustrate the basic concepts of the unit cell model. For highly absorbing fuel elements in the form of thin plates or rods, however, the probability of scattering is small in fuel. Hence, the use of diffusion theory is a rather poor approximation, usually underestimating the thermal disadvantage factor and overestimating the thermal utilization. Improved methods that better account for the large absorption rate in fuel are introduced in this section.

### 11.4.1 Blackness or Simplified Collision Probability Method

We introduce the concept of CP, discussed in detail in Chapter 17, to build an improved method for the surface-to-average flux ratio $F$ in Eq. (11.25). With the *blackness* $\beta^*$ representing the net absorption probability of neutrons incident on a diffusing medium after any number of collisions, an improved neutron balance statement is set up for the neutron absorption rate for the fuel region:

$$\Sigma_{aF}\overline{\phi}_F V_F = -J(a)A_F = J^-(a)A_F\beta_F^*. \tag{11.31}$$

Equation (11.31) represents the correct neutron balance statement that the net current of neutrons into the fuel should be equal to the absorption rate in the fuel and also, by definition of blackness, equal to the fuel blackness $\beta_F^*$ times the incoming current. Equation (11.31) yields

$$\beta_F^* = -\frac{J(a)}{J^-(a)} = \frac{J^-(a) - J^+(a)}{J^-(a)} = 1 - \frac{J^+(a)}{J^-(a)}$$

or

$$\frac{J^+(a)}{J^-(a)} = 1 - \beta_F^*. \tag{11.32}$$

For neutrons isotropically incident on the fuel region, $\phi_F(a)$ is equal to twice the *total* neutron current at the surface. Equation (11.31) is rewritten for the lattice function

$$F = \frac{\phi_F(a)}{\overline{\phi}_F} = \frac{2[J^+(a) + J^-(a)]}{J^-(a)A_F\beta_F^*/(\Sigma_{aF}V_F)},$$

which, with the help of Eq. (11.32), simplifies to

$$F = \left(\frac{1}{\beta_F^*} - \frac{1}{2}\right)\overline{s}_F\Sigma_{aF}. \tag{11.33}$$

Finally, the blackness relationship of Eq. (17.43), $\beta_F^* = \overline{s}_F\Sigma_{aF}P_{0F}^*$, allows introducing the net escape probability $P_{0F}^*$ for the fuel region to obtain the desired improvement for the lattice function:

$$F = \frac{\phi_F(a)}{\overline{\phi}_F} = \frac{1}{P_{0F}^*} - \frac{\overline{s}_F\Sigma_{aF}}{2}. \tag{11.34}$$

### 11.4.2 Amouyal-Benoist-Horowitz Method

In a collision-probability approach developed by Amouyal, Benoist, and Horowitz (ABH) [Amo57,Str65], in addition to the improved expression for the lattice function in Eq. (11.34), an effort is made to improve the accuracy of calculating the excess absorption term $(E - 1)$ defined in Eq. (11.26). Through the combination, the ABH method offers a sufficiently accurate formulation to calculate the thermal disadvantage factor in LWR unit cell analysis and forms a basis for the SOFOCATE module [Shu62,Ams57] of the LEOPARD code [Bar63].

Write the absorption rate in the moderator as a sum of the rate at which neutrons born in the moderator are absorbed there without escaping and the rate at which neutrons returning from the fuel are absorbed in the moderator

$$\Sigma_{aM}\overline{\Phi}_M V_M = QV_M(1 - P^*_{0M}) + J^+(a)S_F\beta^*_M, \tag{11.35}$$

where

$P^*_{0M}$ = net escape probability for the moderator,
$J^+(a)$ = partial current of neutrons entering the moderator, and
$\beta^*_M$ = blackness for the moderator.

This supplements the expression for the fuel absorption rate from Eq. (11.31). Since the thermal neutrons born in the moderator will eventually be absorbed somewhere in the unit cell, we have

$$QV_M = \Sigma_{aM}\overline{\Phi}_M V_M + \Sigma_{aF}\overline{\Phi}_F V_F. \tag{11.36}$$

Solve for $Q$ by combining Eqs. (11.35) and (11.31) with Eq. (11.36), and substitute it back into Eqs. (11.35) and (11.31) to obtain the thermal utilization of Eq. (11.10)

$$\frac{1}{f} - 1 = \frac{1 - P^*_{0M}}{P^*_{0M}} + \frac{\beta^*_M}{P^*_{0M}}\left(\frac{1}{\beta^*_F} - 1\right), \tag{11.37}$$

where Eq. (11.32) is also used. Together with the improved expression for the lattice function $F$ from Eq. (11.34), an alternate expression for thermal utilization $f$ follows:

$$\frac{1}{f} - 1 = \frac{\Sigma_{aM}V_M}{\Sigma_{aF}V_F}F + \frac{1 - P^*_{0M}}{P^*_{0M}} - \frac{1}{2}\bar{s}_M\Sigma_{aM}. \tag{11.38}$$

If we compare Eq. (11.38) with the corresponding expression obtained from one-group diffusion theory, i.e. Eq. (11.29), we note that the last two terms in Eq. (11.38) correspond to the excess absorption term $(E - 1)$ from Eq. (11.26). Since usually $E - 1 \ll 1.0$, we now propose to obtain an estimate of the moderator escape probability $P^*_{0M}$ based on a simple model and thereby improve our determination of $f$. The ABH method treats the fuel as a *black absorber*, i.e. a large and strong absorber, and solves the one-group neutron diffusion equation only in the moderator. Assume that the flux vanishes at an extrapolated distance $d$ into the fuel region

$$\frac{1}{\phi_M(x)}\left.\frac{d\phi_M(x)}{dx}\right|_{x=a} = \frac{1}{d}, \tag{11.39}$$

together with the ordinary unit-cell boundary condition that the net current vanishes at the cell boundary:

$$\left.\frac{d\phi_M(x)}{dx}\right|_{x=b} = 0. \tag{11.40}$$

From the solution of the flux for the moderator region, evaluate the average moderator flux. Noting that some of the neutrons leaking out of the moderator

**Table 11.1** Extrapolation distance $d$ in units of transport mean free path $\lambda_{tr,M}$ for a black cylinder.

| $a/\lambda_{tr,F}$ | 0.0 | 0.25 | 0.5 | 1.0 | 2.0 | 3.0 | 4.0 |
|---|---|---|---|---|---|---|---|
| $d/\lambda_{tr,M}$ | 1.28 | 1.13 | 1.03 | 0.94 | 0.84 | 0.80 | 0.77 |

The moderator surrounds a black cylinder of radius $a$ with transport mean free path $\lambda_{tr,F}$.
*Source*: Adapted from Lamarsh [Lam66].

are eventually captured in the moderator, obtain an expression for the moderator escape probability as the fraction of the source neutrons escaping absorption in the moderator region

$$1 - P^*_{0M} = \frac{\Sigma_{aM}\overline{\phi}_M V_M}{QV_M}, \tag{11.41}$$

or equivalently

$$\frac{1}{P^*_{0M}} = \frac{d}{4D_M}\bar{s}_M\Sigma_{aM} + E, \tag{11.42}$$

where $E$ is the lattice function obtained from diffusion theory in Eqs. (11.24) and (11.28) for slab and cylindrical geometries, respectively.

Substituting Eq. (11.42) into Eq. (11.38) finally yields

$$\frac{1}{f} - 1 = \frac{\Sigma_{aM}V_M}{\Sigma_{aF}V_F}F + \frac{1}{2}\left(\frac{d}{2D_M} - 1\right)\bar{s}_M\Sigma_{aM} + E - 1. \tag{11.43}$$

Note that Eq. (11.43) derived for slab geometry is also valid for cylindrical geometry, provided the lattice functions $F$ and $E$ from Eqs. (11.27) and (11.28) are used.

The thermal disadvantage factor $\zeta_2$ can then be determined from Eqs. (11.43) and (11.10):

$$\zeta_2 = \frac{\overline{\phi}_M}{\overline{\phi}_F} = \frac{1}{P^*_{0F}} + \bar{s}_F\Sigma_{aF}\left(\frac{d}{4D_M} - 1 + \frac{E-1}{\bar{s}_M\Sigma_{aM}}\right). \tag{11.44}$$

The fuel escape probability $P^*_{0F}$ can be obtained with the first-flight CP $P_{cF}$ [Cas53], and extrapolation length $d$ can be obtained from numerical estimates [Str65,Lam66] in Table 11.1. For a large rod diameter, $d$ approaches the familiar slab geometry expression $d = 0.7104\ \lambda_{tr,M}$. In the ABH method, a refinement is also made to account for a non-flat distribution of neutrons in each generation of collisions in the fuel lump [Amo57,Str65].

**Example 11.3** Repeat the calculation of $\zeta_2, f$, and $\langle\Sigma_a\rangle_{cell}$ using the blackness model of Eq. (11.34) and finally ABH model of Eq. (11.44), with $\Sigma_{sF} = 0.3932\,\text{cm}^{-1}$ and $\Sigma_{sM} = 3.1302\,\text{cm}^{-1}$.

**Table 11.2** Comparison of thermal-group lattice parameters calculated in Examples 11.1 through 11.3.

| Parameter | Diffusion theory (cylinder) | Diffusion theory (slab) | Blackness method | ABH method | Homogeneous model |
|---|---|---|---|---|---|
| $F$ | 1.0138 | 1.0093 | 1.0397 | 1.0397 | – |
| $E$ | 1.00864 | 1.00990 | 1.00864 | 1.03394 | – |
| $\zeta_2$ | 1.0402 | 1.0410 | 1.0657 | 1.1429 | 1.0 |
| $f$ | 0.7452 | 0.7450 | 0.7404 | 0.7268 | 0.7525 |
| $\langle\Sigma_a\rangle_{cell}$ (cm$^{-1}$) | 0.1186 | 0.1185 | 0.1174 | 0.1140 | 0.1205 |

Obtain the first-flight CP for the fuel region $P_{cF} = 0.2893$ with $a\Sigma_{tF} = 0.3084$, which yields, together with $\gamma = \Sigma_{sF}/\Sigma_{tF} = 0.5982$, the net escape probability $P^*_{0F} = (1 - P_{cF})/(1 - \gamma P_{cF}) = 0.8594$. The lattice parameter $F = 1/0.8594 - 0.2478/2 = 1.0397$ is obtained from Eq. (11.34). Combining the updated value of $F$ with $E = 1.00864$ from Example 11.1 yields improved estimates of $f = 0.7404$ and $\zeta_2 = 1.0657$, together with $\langle\Sigma_a\rangle_{cell} = 0.1174\,\text{cm}^{-1}$.

Finally, for the ABH formulation, we obtain the extrapolation distance $d = 1.08/1.089 = 0.992\,\text{cm}$ from Table 11.1 with $a/\lambda_{tr,F} = 0.301$ for the fuel rod, leading to an estimate of the ABH correction term $0.5(d/2D_M - 1)\bar{s}_M\Sigma_{aM} = 0.5 \times [0.992/(2 \times 0.3062) - 1] \times 0.08153 = 0.0253$. Combining this correction term with that considered for the blackness model for $F$ and $E$ yields $f = 0.7268$, $\zeta_2 = 1.1429$, and $\langle\Sigma_a\rangle_{cell} = 0.1140\,\text{cm}^{-1}$, which compares quite well with the actual LEOPARD result of $0.1134\,\text{cm}^{-1}$. The good agreement between the one-group ABH calculation and 172-group LEOPARD result for the cell-average cross section should be considered rather fortuitous, reflecting the effects of several approximations. The thermal-group lattice parameters summarized in Table 11.2 indicate, however, the improvements associated with higher-order methods. We note in summary that thermal utilization $f$ decreases due to fuel lumping in a fuel assembly.

## 11.5 Resonance Escape Probability for Heterogeneous Cell

Having discussed in Sections 11.3 and 11.4 various approaches to account for material heterogeneities in the thermal spectrum, including the concept of collision probabilities, we now turn to the task of determining the resonance escape probability for a heterogeneous unit cell. We concentrate on a two-region fuel-moderator geometry and consider the problem of determining the average flux $\overline{\phi}_F(E)$ or $\overline{\phi}_F(u)$ in the fuel so that we may extend the resonance escape probability of Eq. (11.9) to heterogeneous cell calculations. We discuss in Section 11.5.1 the concept of spatial self-shielding referred to in Section 11.2. This is followed by a discussion of the engineering approaches to handle the resonance escape probability calculation for unit cell geometry in Section 11.5.2.

### 11.5.1 Spatial Self-Shielding for Heterogeneous Unit Cell

Since the resonance escape probability in Eq. (11.9) for a homogeneous cell is expressed in terms of a flux-weighted effective absorption cross section $I$, all that is necessary now is to replace the cell-average flux $\phi_{NR}(u) = \Sigma_s/\Sigma_t(u)$ with the fuel-average flux $\overline{\phi}_F(u)$

$$I = \int_{u_1}^{u_2} du\sigma_{aF}(u)\overline{\phi}_F(u), \tag{11.45}$$

which clearly indicates that the absorption cross section is evaluated for the fuel region only. With this purpose in mind, we return to the neutron balance statements for a homogeneous medium discussed in Chapter 9 and set up a similar balance statement for a fuel-moderator two-region heterogeneous cell.

**1. Neutron Balance and Spatial Self-Shielding Factor**

When the concept of the energy self-shielding factor was discussed in Section 9.4, we introduced three equivalent approaches. The first approach involves assuming that the ratio $F(u)/q(u)$ in Eq. (9.58) may be approximated by the Fermi approximation of Eq. (9.56a) in and around a narrow resonance. In Eq. (9.64), another approach is taken to introduce the asymptotic expression $q_{as}(u) = \xi\Sigma_s$ around the resonance with normalization $\phi_{as}(u) = 1.0$. These two approaches allowed us to interpret the NR flux $\phi_{NR}(u) = \Sigma_s/\Sigma_t(u)$ as a normalized flux representing the depression due to the presence of the resonance, i.e. the energy self-shielding factor.

We now modify the third construct from Eq. (9.67) that is used to derive the $\phi_{NR}(u)$ using the lethargy-dependent neutron diffusion equation (9.7a) but for the asymptotic slowing down range, with the spatial dependence suppressed for notational convenience. This provides a simple neutron balance statement, with due consideration given to the lethargy interval $[u - \Delta, u]$ from which neutrons would scatter into a unit lethargy interval around $u$

$$\Sigma_t(u)\phi(u) = \int_{u-\Delta}^{u} \Sigma_s(u' \to u)\phi(u')du', \tag{11.46}$$

where the scattering kernel $\Sigma_s(u' \to u)$ is given by Eq. (9.18). The scattering integral is simplified to $\Sigma_s$, and the expression for the NR flux $\phi_{NR}(u) = \Sigma_s/\Sigma_t(u)$ is retrieved again.

In light of Eq. (11.46) representing the neutron balance for a homogeneous cell, we now set up for the heterogeneous cell a neutron balance in terms of the collision density for the fuel and break up the fuel collision density into contributions due to neutrons that slow down in the fuel and suffer collisions in the fuel and those that enter the fuel from the moderator

$$\Sigma_{tF}(u)\overline{\phi}_F(u) = P_{cF}\int_{u-\Delta_F}^{u} du'\Sigma_{sF}(u' \rightarrow u)\overline{\phi}_F(u') + \Sigma_{tF}(u)\overline{\phi}_{FM}(u), \tag{11.47}$$

where $P_{cF}$ is the first-flight CP for the fuel and $\overline{\phi}_{FM}(u)$ is the average flux in the fuel due to transport of neutrons from the moderator. With the first-flight escape probability from Eq. (17.40) to represent the flux depression in the fuel region, we obtain

$$\overline{\phi}_{FM}(u) = P_{0F}\overline{\phi}_M(u), \tag{11.48}$$

in terms of the average moderator flux $\overline{\phi}_M(u)$.

To obtain the heterogeneous counterpart of the NR approximation, Eq. (9.65), apply the *NR approximation*, i.e. $q(u) = \xi\Sigma_s$, or $\overline{\phi}_F(u) = \overline{\phi}_M(u) = 1.0$, in the RHS of Eq. (11.47), which is equivalent to assuming that the resonance in the slowing down range $[u - \Delta_F, u]$ does not materially perturb the asymptotic flux attained in the interval. Recalling that the remaining integral yields the total scattering cross section of fuel as in Eq. (11.46), we obtain a simple neutron balance statement for the fuel region

$$\Sigma_{tF}(u)\overline{\phi}_F(u) = P_{cF}\Sigma_{sF} + P_{0F}\Sigma_{tF}(u), \tag{11.49}$$

which yields the desired expression for the average flux in fuel:

$$\begin{aligned}\overline{\phi}_F(u) &= P_{cF}\frac{\Sigma_{sF}}{\Sigma_{tF}(u)} + P_{0F} = \frac{(1-P_{0F})\Sigma_{sF} + P_{0F}\Sigma_{tF}(u)}{\Sigma_{tF}(u)}\\ &= \frac{\Sigma_{aF}(u)P_{0F} + \Sigma_{pF}}{\Sigma_{tF}(u)} = \frac{\sigma_{aF}(u)P_{0F} + \sigma_{pF}}{\sigma_{tF}(u)} \equiv \phi_{NR}^{het}(u).\end{aligned} \tag{11.50}$$

The moderator-to-fuel transfer term in Eq. (11.47), represented through Eq. (11.48), can also be derived a bit more formally by explicitly considering the collision rate for the moderator region with volume $V_M$ and first-flight escape probability $P_{M\rightarrow F}$ and converting it into the collision rate for the fuel region with volume $V_F$ and $P_{F\rightarrow M}$ via the reciprocity relationship of Eq. (17.20):

$$\begin{aligned}&(\Sigma_{pM}\overline{\phi}_M V_M P_{M\rightarrow F})/V_F = \Sigma_{tF}\overline{\phi}_M P_{F\rightarrow M},\\ &\qquad \text{with}\,\Sigma_{tM} \simeq \Sigma_{pM}, P_{F\rightarrow M} = P_{0F}.\end{aligned} \tag{11.51}$$

Note also that the expression in Eq. (11.50) written in terms of the fuel microscopic cross sections is valid strictly for the case of a single fuel nuclide, but is introduced for notational clarity.

The normalized average flux $\overline{\phi}_F(u)$ for the fuel in Eq. (11.50) represents the flux depression in the fuel due to the resonances, illustrated in Figure 11.3, and is known as the *spatial self-shielding factor*. The spatial self-shielding factor clearly depends on the geometry through the term $P_{0F}$ and approaches unity as $P_{0F}$ approaches unity in the limit of an infinitely dilute mixture. For practical heterogeneous cells of interest, the spatial self-shielding effect is much larger than the energy self-shielding effect, and the latter is usually ignored, i.e. $\phi_{NR}^{het}(u) < \phi_{NR}^{homog}(u)$. Indeed, the importance of the spatial self-shielding effect may best be understood by recalling that the resonance integral for a heterogeneous cell consisting of natural uranium and graphite reduces to ~9 b from a homogeneous value of ~280 b, which is a key factor that contributed to the success of Fermi's Chicago Pile.

**2. Escape Cross Section and Equivalence Relationship**

The wide resonance (WR) and intermediate resonance (IR) expressions for heterogeneous cells can be obtained in a form similar to Eq. (11.50), and the Doppler broadening of fuel resonances can likewise be included in Eq. (11.50) or the equivalent WR or IR results. This approach, however, involves considerable complications, and hence, for many engineering applications, an equivalence relationship between heterogeneous and homogeneous cells is established so that the homogeneous results may be used with equivalent heterogeneous corrections introduced. The derivation of the equivalence relationship begins with the concept of a fictitious *escape cross section* $\sigma_e$ so that the first-flight escape probability for the fuel may be written as

$$P_{0F} = \frac{\sigma_e}{\sigma_e + \sigma_{tF}}. \tag{11.52}$$

Substituting Eq. (11.52) into Eq. (11.50) yields a NR spatial self-shielding factor for a heterogeneous cell:

$$\phi_{NR}^{het}(u) = \frac{\sigma_{aF}\sigma_e + \sigma_{pF}(\sigma_e + \sigma_{tF})}{\sigma_{tF}(\sigma_e + \sigma_{tF})} = \frac{\sigma_e + \sigma_{pF}}{\sigma_e + \sigma_{tF}}. \tag{11.53}$$

This expression may be compared with the homogeneous NR expression for the energy self-shielding factor:

$$\phi_{NR}^{homog}(u) = \frac{\Sigma_s}{\Sigma_t(u)} = \frac{\rho}{\rho + \sigma_{aF}} = \frac{\Sigma_{pM} + \Sigma_{pF}}{\Sigma_t(u)} = \frac{s + \sigma_{pF}}{s + \sigma_{tF}}, \rho = \frac{\Sigma_s}{N_F}, s = \frac{\Sigma_{pM}}{N_F}. \tag{11.54}$$

Recall that the dilution factor $\rho$ is defined in Eq. (9.62) as the total scattering cross section per fuel nucleus, while $s$ is introduced as the *moderator* scattering cross section per fuel nucleus for the homogeneous cell in Eq. (9.73). Comparing Eqs. (11.53) and (11.54), note that $\sigma_e$ for a heterogeneous cell plays the role of $s$ for a

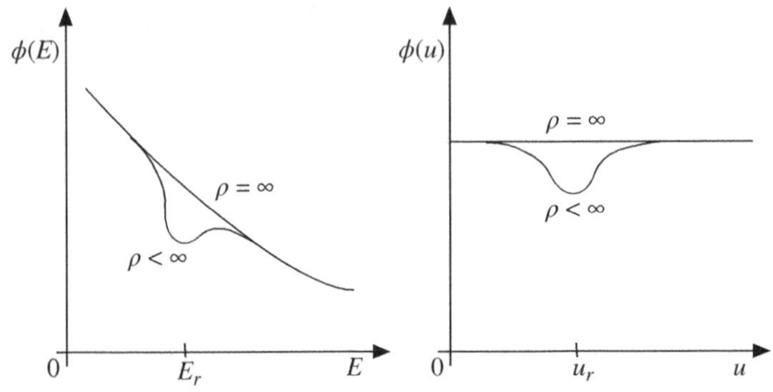


**Figure 11.7** Flux self-shielding around resonance energy $E_r$ corresponding to $u_r = \ln(E_r/E_0)$ for finite dilution factor $\rho < \infty$.

homogeneous cell and that a heterogeneous cell with $\sigma_e = s$ would yield the same resonance integral. This also suggests that a heterogeneous cell can be represented as a homogeneous cell, provided we replace the homogeneous dilution factor

$$\rho_{homog} = s + \sigma_{pF}, \tag{11.55}$$

with the heterogeneous dilution factor

$$\rho_{het} = \sigma_e + \sigma_{pF}. \tag{11.56}$$

The *equivalence relationship* is often used effectively to convert the resonance integral for a heterogeneous cell into a simplified, equivalent homogeneous configuration. Two heterogeneous cells with different configurations are naturally equivalent to one another if $\rho_{het}$ is the same. Self-shielding of the flux is compared in Figure 11.7 for either a homogeneous or heterogeneous configuration for a finite dilution factor $\rho < \infty$ with that without any flux self-shielding for an infinite dilution factor $\rho = \infty$. The illustration also provides a reminder that in the slowing down range, $\phi(E) \propto 1/E$ while $\phi(u) \simeq$ constant away from absorption resonances.

**3. Spatial and energy self-shielding factors**

Together with the equivalence relationship, the first-flight escape probability is often expressed by *Wigner's rational approximation*

$$P_0 = \frac{1}{1 + \bar{s}\Sigma_t}, \tag{11.57}$$

which yields

$$\sigma_e = \frac{1}{\bar{s}_F N_F}. \tag{11.58}$$

Note that for a typical LWR unit cell, $\sigma_e$ is larger than $\sigma_{pF}$ in Eq. (11.56) but smaller than $s$, and hence smaller than the volume-average dilution factor for the cell $\langle \rho \rangle_{cell}$

$$\sigma_{pF} \ll [\rho_{het} = \sigma_e + \sigma_{pF}]$$
$$< \rho_{homog} = \langle \rho \rangle_{cell} = \frac{\Sigma_{pM} V_M + \Sigma_{pF} V_F}{N_F V_F} \simeq \frac{\Sigma_{pM} V_M}{N_F V_F}, \tag{11.59}$$

where we recognize that the scattering cross section $\Sigma_{pM}$ for the moderator region is much larger than the scattering cross section $\Sigma_{pF}$ for the fuel region. Thus, a heterogeneous cell appears to be less dilute than a homogeneous cell of the identical physical mixture, and the resonance integral for the heterogeneous layout is less than the corresponding homogeneous integral. This is another interpretation of the importance of the spatial self-shielding factor obtained in Eq. (11.50). Note, however, that this equivalent dilution factor interpretation should apply only to the resonance integral and not to the factor $N_F/\xi\Sigma_s$ in the exponential of Eq. (11.9) for the resonance escape probability itself. In fact, for a two-region heterogeneous cell, the factor $N_F/\xi\Sigma_s$ has to be replaced by a volume-weighted quantity:

$$\frac{N_F}{\xi\Sigma_s} \Rightarrow \frac{N_F V_F}{\overline{\xi}_F \Sigma_{pF} V_F + \overline{\xi}_M \Sigma_{pM} V_M}. \tag{11.60}$$

In general, $\overline{\xi}_F$ and $\overline{\xi}_M$ represent the mean lethargy increase per collision averaged, with scattering cross sections as weighting factors, over the fuel and moderator regions, respectively. The resonance escape probability and effective multiplication factor for a typical LWR fuel cell are larger than the corresponding values for a homogeneous cell of the same composition:

$$\rho_{het} < \rho_{homog} \Rightarrow \phi_{NR}^{het}(u) < \phi_{NR}^{homog}(u) \Rightarrow I_{het} < I_{homog}$$
$$\Rightarrow p_{het} > p_{homog} \Rightarrow k_{het} > k_{homog}. \tag{11.61}$$

### 11.5.2 Engineering Approaches for Resonance Integral

In addition to the equivalence relationship approach discussed in Section 11.5.1, other approximate treatments are often used to evaluate the resonance integral for heterogeneous cells. To understand the physical reasons behind some of these approximate models, substitute Eq. (11.50) for the NR flux into the definition for the resonance integral of Eq. (11.45):

$$I = \int_{u_1}^{u_2} du \frac{\sigma_{aF}(u)}{1 + \sigma_{aF}(u)/\sigma_{pF}} + \int_{u_1}^{u_2} du\, \sigma_{aF}(u) P_{0F} \frac{\sigma_{aF}(u)}{\sigma_{tF}(u)}. \tag{11.62}$$

Note that the first term is equal to the effective resonance integral for the fuel only and hence is independent of the fuel geometry. In contrast, the second term represents the net neutron transport across the fuel-moderator interface $A_F$ and is a function of the fuel geometry through the escape probability $P_{0F}$. Hence, it was suggested, and experimentally verified, that the resonance integral $I$ can be written [Gla52] as the sum of a term depending on the fuel volume only and another term depending on the fuel surface area-to-mass ratio

$$I = a + b\frac{A_F}{M_F}, \tag{11.63}$$

where $M_F$ is the fuel mass. Equation (11.63) is known as the *standard formula*, with the first term treated as the volume absorption term and the second as the surface absorption term.

Up to now, we have implicitly assumed that a unit cell comprising the fuel and moderator is entirely isolated. This assumption is actually not very accurate in a close-packed lattice, because the probability of resonance neutron captures in a fuel rod decreases due to neutron captures in other fuel rods in the lattice. This effect may be expressed by the *Dancoff-Ginsberg factor* $C$ and an effective surface area $A_{eff}$ of the fuel rod:

$$A_{eff} = A_F(1 - C). \tag{11.64}$$

The fuel surface appears to be shadowed, as far as the resonance neutron interactions are concerned, by neighboring fuel rods. Hence the Dancoff effect is also known as the *rod-shadowing effect*, discussed further in Eq. (17.47). The rod-shadowing effect also implies that the *effective mean chord length* is increased:

$$\bar{s}_{eff} = \frac{\bar{s}_F}{1 - C}. \tag{11.65}$$

It can be shown that the Dancoff factor is equal to $1 - \beta_M$, where $\beta_M$ is the first-flight blackness of the moderator region. Several variations from the standard formula in Eq. (11.63) have been also proposed. One such empirical resonance integral formula is Strawbridge's *metal-oxide correlation* [Str65] for $^{238}$U

$$I^{28} = 2.16x + 2.56 + (0.0279x - 0.0537)\sqrt{T_{eff}}, \tag{11.66}$$

where the *heterogeneous correction factor* $x$ is given by

$$x = \left[\frac{\Sigma_{sF}}{N_{28}}P_{0F} + \frac{1 - C}{\bar{s}_F N_{28}}\right]^{1/2}, \tag{11.67}$$

with the number density $N_F$ of the resonance absorber $^{238}$U written as $N_{28}$. Rewriting Eq. (11.67) as

$$x = \left[\sigma_{pF} P_{0F} + (1 - C)\sigma_e\right]^{1/2}$$

indicates that $x$ is approximately equal to $\sqrt{\rho_{het}}$, with the heterogeneous dilution factor defined in Eq. (11.56). The effective temperature $T_{eff}$ in Eq. (11.66) may be chosen to represent the Doppler effect due to a distributed temperature profile within a fuel rod. The metal-oxide correlation applies to fuel rods consisting of either uranium metal or dioxide, and the temperature dependence is discussed in connection with Eq. (9.94).

**Example 11.4** Calculate the resonance escape probability $p^{28}$ for $^{238}U$ with the metal-oxide correlation $I^{28}$ for a PWR unit cell studied in Examples 11.1 through 11.3 and evaluate the effects of a heterogeneous fuel layout on the resonance escape probability for a PWR core. The unit cell represents $UO_2$ fuel with $U^{235}$ enrichment of 2.78 wt% and 98.8% of theoretical density surrounded by water with density $\rho_M = 0.727\,\text{g}\cdot\text{cm}^{-3}$ containing 2210 ppm by weight of natural boron. A typical unit cell configuration is simplified into a two-region setup with a pellet of radius $a = 0.4692$ cm and a square pitch $p = 1.4194$ cm. The effective fuel temperature $T_{eff} = 1033$ K, and the scattering cross sections $\sigma_s = \sigma_p$ for the resonance energy neutrons from the Serpent code [Fri11] are tabulated here:

| Nuclide | $^{238}U$ | $^{235}U$ | $^{16}O$ | $^{1}H$ |
|---|---|---|---|---|
| $\sigma_s$ (b) | 12.9 | 11.9 | 3.9 | 16.8 |

With the number density $N_F = 0.02415\,(\text{b}\cdot\text{cm})^{-1}$ for $UO_2$ and U, converting wt% to at% yields the number density $N_{28}$ for $^{238}U$ and $N_{25}$ for $^{235}U$, respectively

$$N_{28} = \frac{N_F \times 0.9722/238.0}{0.9722/238.0 + 0.0278/235.0} = 0.02347(\text{b}\cdot\text{cm})^{-1},$$

$$N_{25} = 6.787 \times 10^{-4}(\text{b}\cdot\text{cm})^{-1},$$

and total fuel scattering cross section $\Sigma_{sF} = 0.4992\,\text{cm}^{-1}$ and mean chord length for the fuel pellet $\bar{s}_F = 2a = 0.9382$ cm. The first-flight escape probability for the fuel region $P_{0F} = 0.7766$ and Dancoff factor $C = 0.316$ yield the heterogeneous correction factor $x = 6.90$ for Eq. (11.67) and the effective resonance integral $I^{28} = 21.9$ b. With the moderator number density $N_M = 0.0243\,(\text{b}\cdot\text{cm})^{-1}$, obtain $(\xi\Sigma_s)_M = 0.828\,\text{cm}^{-1}$, which is substituted into Eq. (11.60), together with $(\xi\Sigma_s)_F = 0.0252\,\text{cm}^{-1}$ and $V_F/(V_M + V_F) = 0.3433$, finally providing $\overline{\xi\Sigma_s} = 0.5524\,\text{cm}^{-1}$ and $p^{28} = \exp[-0.02347 \times 0.3433 \times 21.9/0.5524] = 0.727$.

For an equivalent homogeneous configuration, together with $\Sigma_{sF} = 0.4992\,\text{cm}^{-1}$, obtain $\Sigma_{sM} = 0.911\,\text{cm}^{-1}$, which yields $\Sigma_{s,cell} = 0.7696\,\text{cm}^{-1}$ and $x = \sqrt{\rho} = (\Sigma_{s,cell}/N_{28})^{1/2} = 9.77$, with $\rho$ representing the usual dilution factor, i.e. scattering cross section per $^{238}U$ nucleus for a homogeneous mixture. Substituting $x = 9.77$ into the metal-oxide correction yields $I^{28} = 30.69$ b and

$p^{28} = 0.640$, which is significantly less than 0.727 obtained for a heterogeneous configuration, clearly indicating the main neutronic advantage of heterogeneous fuel cells. The effects of heterogeneity on the resonance escape probability may also be evaluated through comparing the self-shielding factor and dilution factor, as suggested in Problems 11.5 and 11.7, respectively.

With absorptions in other nuclides, including $^{235}$U, accounted for, the net resonance escape probability $p$ is reduced from $p^{28}$. We may now assess, however, the overall impact of heterogeneous fuel arrangements by comparing $k_2$, the contribution to $k_\infty$ from thermal neutron fissions, with the values for $f$ obtained from Table 11.2:

$$\frac{k_{2,het}}{k_{2,homog}} \simeq \frac{(fp)_{het}}{(fp)_{homog}} \simeq \frac{(fp^{28})_{het}}{(fp^{28})_{homog}} = \frac{0.7268 \times 0.727}{0.7525 \times 0.640} = 1.10.$$

The approximate estimate of a 10% increase in $k_2$ associated with heterogeneity could be translated into a 7% increase in $k_\infty$, in good agreement with Serpent calculations for a PWR unit cell.

The MUFT module [Boh57,Str65] combines the metal-oxide correlation together with the NR approximation to calculate the effective resonance integral for typical unit-cell geometry. Since it is difficult to account accurately for all individual resonances, let alone the Doppler broadening and heterogeneous effects, an effort is made to preserve the total resonance integral, rather than individual ones, through an iterative scheme. With a self-shielding factor $L$ to be applied to the resonance integral $I$ iteratively, use the homogeneous NR integral for $T = 0$ K. In step 1 of the two-step $\omega$-search, the resonance integral $I$ is evaluated only for $^{238}$U captures, with the NR approximation and buckling $B^2 = 0$. The $L$ factor is iteratively updated, as a multiplicative correction factor to $I$, until the resonance escape probability $p^{28}$ for uranium fuel rods agrees with that predicted by the metal-oxide correlation $I^{28}$ of Eq. (11.66). In the actual MUFT algorithm, a parameter $\omega$

$$\omega = \frac{1 - p^{28}}{p^{28}}$$

is matched rather than $p^{28}$. Once a converged value of the $L$ factor is obtained, the MUFT fast-spectrum calculation is repeated in step 2, with resonance captures in other nuclides included, together with either an input value of $B^2 \neq 0$ or a critical buckling iteratively searched for. The two-step $\omega$-search assumes that an empirical correction factor $L$ obtained for $^{238}$U applies approximately for other resonance absorbers as well.

### 11.5.3 Implementation in the CPM-3 Code

In the CPM-3 code [Jon00], a homogeneous IR model is used together with the equivalence relationship to account for the Doppler and spatial self-shielding effects. For the first-flight escape probability $P_{0F}$ from Eq. (17.40), in place of the traditional Wigner's approximation, Carlvik's two-term rational approximation is used

$$P_{0F} = \frac{\beta\alpha_1}{x+\alpha_1} + \frac{(1-\beta)\alpha_2}{x+\alpha_2} \equiv \sum_{i=1}^{2} \frac{\beta_i\alpha_i}{x+\alpha_i}, \tag{11.68}$$

with $x = \bar{s}_F \Sigma_{tF}^* = (\sigma_{aF} + \sigma_{pF}^*)/\sigma_e$,

where $\sigma_{pF}^* = \lambda\sigma_{pF}$ is the *effective potential scattering cross section*, including the IR parameter $\lambda$, introduced in Eq. (9.75). When Eq. (11.68) is applied to the heterogeneous NR flux from Eq. (11.50), the resonance integral can be obtained as

$$I = \sum_{i=1}^{2} \beta_i I(\rho_i), \tag{11.69}$$

where

$$I(\rho_i) = \int du \frac{\sigma_{aF}(u)}{1+\sigma_{aF}(u)/\rho_i}, \quad \rho_i = \sigma_{pF}^* + \alpha_i\sigma_e. \tag{11.70}$$

The construct $\sigma_{pF}^* = \lambda\sigma_{pF}$ allows the use of the NR formulation to represent the general IR formulation. With flux $\phi(\rho_i)$ written in terms of the effective dilution factor $\rho_i$

$$\phi(\rho_i) = \frac{\rho_i}{\rho_i + \sigma_{aF}(u)}, \tag{11.71}$$

obtain a resonance-shielded cross section

$$\langle\sigma_{aF}(\rho_i)\rangle = \frac{I(\rho_i)}{\Delta u - I(\rho_i)/\rho_i}, \tag{11.72}$$

and a combined shielded cross section for the lethargy interval $\Delta u$

$$\langle\sigma_{aF}\rangle = \frac{\sum_{i=1}^{2} \beta_i I(\rho_i)}{\sum_{i=1}^{2} \beta_i[\Delta u - I(\rho_i)/\rho_i]}. \tag{11.73}$$

The actual CPM-3 implementation of $\langle\sigma_{aF}\rangle$ in terms of the *resonance shielding factor* $f(\rho_i) = \langle\sigma(\rho_i)\rangle/\langle\sigma(\infty)\rangle$ is left as an exercise [Problems 11.9 and 11.10]. The Doppler broadening is also represented through a tabulation of the resonance integrals $I(\rho_i)$ as a function of the effective resonance temperature $T_{eff}$ and the dilution factor $\rho_i$. The code also has special provisions to handle overlaps between resonances and the Dancoff effect. A qualitative discussion of the effect of temperature rise on the resonance absorption is presented in Section 9.5, together with a simplified analytical treatment for the single-level Breit–Wigner line shape. Figure 9.7 illustrates the broadening of the resonance due to Doppler effect, representing the relative motion of the neutron and nucleus involved.

## 11.6 Thermal Spectrum Calculation

In Section 3.3, we had a brief discussion of the behavior of thermal neutrons in LWR cores, where we assumed that the thermal flux may be approximately represented by the Maxwell–Boltzmann distribution from Eq. (3.30) or (3.44d). In actual reactor cores, the thermal flux deviates substantially from the idealized Maxwell–Boltzmann distribution due to a number of reasons. We sketch in Section 11.6.1 one method that accounts for neutron absorptions in the thermal range of the neutron flux and qualitatively discuss in Section 11.6.2 the departure from thermal equilibrium characterizing the Maxwell–Boltzmann distribution.

### 11.6.1 Wigner–Wilkins Model

A thermal spectrum model developed by Wigner and Wilkins [Wig44] is based on the free gas kernel and serves as the basis for the TEMPEST code [Shu62] and the SOFOCATE code [Ams57], which have been subsequently incorporated into the LEOPARD code [Bar63]. The Wigner–Wilkins model assumes that the moderator can be approximated by an ideal gas of free protons that are in thermal equilibrium at temperature $T$ in an infinite medium. The model also assumes that

(a) The microscopic scattering cross section $\sigma_s$ of a proton is independent of the relative speed $v_r = |\mathbf{v} - \mathbf{V}|$ between neutrons and protons,
(b) The absorption cross section $\sigma_a$ of proton is inversely proportional to the relative speed $v_r$, and
(c) There is no direct contribution from fission to the thermal energy range.

Under these assumptions, the neutron diffusion equation (4.39) may be simplified to a simple steady-state neutron balance statement for a medium without leakage and source

$$v\left[\Sigma_a(E) + \Sigma_s(E)\right] n(E) = \int_0^{\infty} v'\Sigma_s(E' \to E)n(E')dE', \tag{11.74}$$

together with the free-gas kernel

$$\Sigma_s(E' \to E) = \begin{cases} \dfrac{\Sigma_{fr}}{E'} \exp\left(-\dfrac{E-E'}{kT}\right) \operatorname{erf}\sqrt{\dfrac{E'}{kT}}, & E' < E, \\ \dfrac{\Sigma_{fr}}{E'} \operatorname{erf}\sqrt{\dfrac{E}{kT}}, & E' \geq E, \end{cases} \tag{11.75}$$

and the free-atom scattering cross section $\Sigma_{fr}$ at 0 K, i.e. for $\mathbf{V} = 0$, expressed in terms of the proton number density $N_H$ and scattering cross section $\sigma_{sH}$

$$\Sigma_{fr} = N_H\,\sigma_{sH}. \tag{11.76}$$

Wigner and Wilkins obtained a solution to Eq. (11.74) by converting the integral equation into a nonlinear differential equation of the Ricatti type. The differential equation is solved in terms of a normalized speed variable $x$

$$x = \frac{v}{\sqrt{2kT/m}} = \frac{v}{v_0} = v\beta, \tag{11.77}$$

by a combination of 29th-order asymptotic expansion and Milne's five-point predictor-corrector numerical integration [Car69] for 172 energy groups covering the interval [0.1 meV, 0.625 eV]. In Eq. (11.77), $v_0$ is the most-probable speed for the Maxwell–Boltzmann distribution given in Eq. (3.38). The numerical technique turned out to be quite accurate and served as a key module for the LEOPARD code for many years during the development of the PWR technology.

Figure 11.8 presents the Wigner–Wilkins plots of $n(x)$ vs. $x$ as a function of $\beta\gamma = \Gamma = \Sigma_{a0}/\Sigma_{fr}$, indicating that the spectrum shifts toward the higher speed, or the spectrum becomes *harder* as $\Gamma$ increases, where $\Sigma_{a0}$ is the absorption cross section

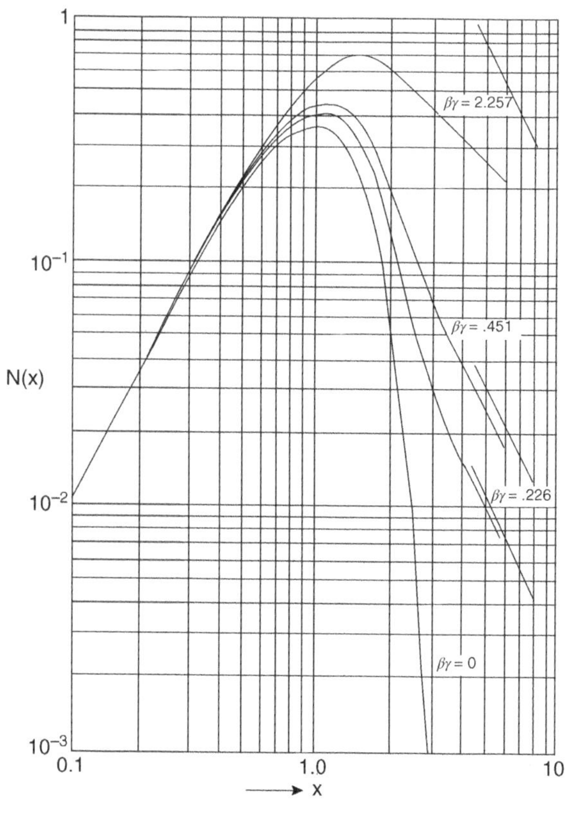


**Figure 11.8** Neutron flux spectrum $n(x)$ vs. $x$, as a function of $\beta\gamma = \Gamma = \Sigma_{a0}/\Sigma_{fr}$. *Source*: Wigner and Wilkins [Wig44]/United States Atomic Energy Commission.

of proton at speed $v_0$. The asymptotic behavior of $n(x)$ for large values of $x$ is also shown in Figure 11.8.

Although the Wigner-Wilkins model was originally developed for free proton gas in thermal equilibrium at temperature $T$, it was later extended in the SOFOCATE code [Ams57] to account for the presence of other nuclides by modifying the parameter $\Gamma$

$$\Gamma = \frac{x[\Sigma_a(x) + D(x)B^2]}{\bar{\xi}\Sigma_s} \simeq \frac{\Sigma_{a0} + xD(x)B^2}{\bar{\xi}\Sigma_s}, \tag{11.78}$$

where all nuclides in the system are included in the cross sections. This extension of the parameter $\Gamma$ has a rather weak theoretical justification, but its usefulness has been borne out by the reasonable accuracy of the LEOPARD code. Zero-dimensional spectrum calculations for LWR cells may be performed via the Wigner-Wilkins model with the extension of $\Gamma$ in Eq. (11.78) and the 172-group TEMPEST or SOFOCATE library. For this calculation, for cell-averaged cross sections for each of the 172 groups are obtained via the flux- and volume-weighted scheme from Eq. (11.7) and the ABH thermal disadvantage factor from Eq. (11.44).

### 11.6.2 Qualitative Behavior of Thermal Neutron Spectrum

The Wigner–Wilkins thermal spectrum plotted in Figure 11.8 indicates that the peak energy of the spectrum increases as the absorption cross section $\Sigma_{a0}$ of the proton increases. We may gain physical understanding of this phenomenon by plotting in the LHS plot of Figure 11.9 the general trend of the absorption cross section decreasing as a function of neutron energy $E$ in the thermal range. This implies that at lower energy, there is an increasing depletion of the neutron population, shifting the peak of the flux spectrum to a higher energy. This trend

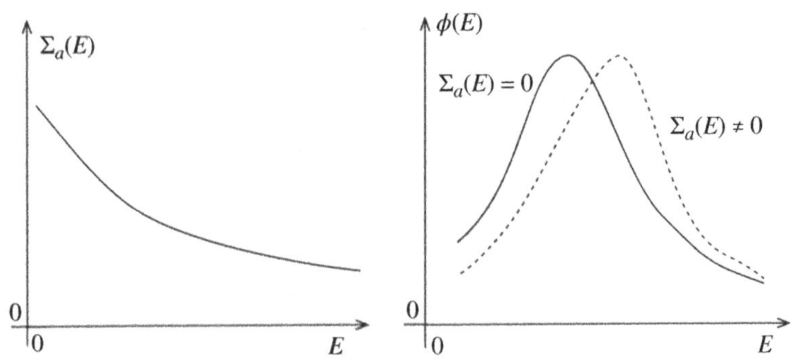


**Figure 11.9** Absorption cross section and normalized flux plots in the thermal range.

is known as *hardening of the spectrum* and is sketched via *normalized* flux profiles in the RHS plot of Figure 11.9. The shifting of the spectrum to a higher energy is equivalent effectively to a Maxwell–Boltzmann distribution at a higher temperature, and hence this dependence of the flux spectrum on neutron absorption is referred to as *spectral hardening due to absorption heating.*

Another cause for the departure of the thermal neutron spectrum from the idealized Maxwell–Boltzmann distribution is the leakage of neutrons. In this case, which is quite opposite the effect of neutron absorption, neutrons of higher energy tend to leak out more freely, thereby depleting the high-energy end of the spectrum and shifting the flux peak to a lower energy. This trend is known as *diffusion cooling of the spectrum.*

The Wigner–Wilkin's thermal spectrum displayed in Figure 11.8 may then be approximately represented by a Maxwell–Boltzmann distribution $M(E, T)$ obtained from $n(\mathbf{v})$ of Eq. (3.30), provided the physical temperature $T$ is replaced by an *effective neutron temperature*

$$\phi(E) \simeq vM(E, T_n), \tag{11.79}$$

where $T_n$ may be expressed as

$$T_n = T(1 + A\Gamma). \tag{11.80}$$

in terms of $\Gamma$ of Eq. (11.78) and an empirical parameter $A$. Corresponding to this representation of the thermal flux, the effective temperature $T_n$ may be used together with Westcott's $g$-factor, introduced in Eq. (3.50) for the effective thermal cross section.

## 11.7 Integral Transport Methods

One of the major approximations of the LEOPARD code [Bar63] is that the spectrum calculation, in either the MUFT module [Boh57] or the SOFOCATE module [Ams57,Shu62] of the code, is performed for a zero-dimensional unit cell. Through the ABH method, the spatial flux distribution is accounted for in the thermal spectrum, to a reasonable degree of accuracy, for a normal LWR fuel cell. However, it is rather a poor approximation for calculating the flux spectrum near sizable water gaps and in or near strong, localized absorbers. For general spectrum calculations, space-dependent formulations have to be directly used. One such lattice physics model widely used in recent years for core design calculations is the integral transport or CP method of the CPM and CASMO assembly-level lattice physics codes [Jon87,Ede92,Jon00].

The CP method was originally developed for the THERMOS code [Hon61] to solve the integral neutron transport equation in 1-D geometry. The code uses

Nelkin's scattering kernel [Nel60] to explicitly represent molecular effects in thermal neutron scattering with water molecules, and accurately accounts for material heterogeneities associated with water gaps and strong lumped absorbers.

The basic CP formulation begins with an integral form of the steady-state neutron transport equation, discussed more formally in Section 17.1,

$$\phi(\mathbf{r}, E) = \int_V S(\mathbf{r}', E) T(\mathbf{r}' \rightarrow \mathbf{r}, E) d\mathbf{r}', \tag{11.81}$$

where the *transport kernel* $T(\mathbf{r}' \rightarrow \mathbf{r}, E)$ is given as

$$T(\mathbf{r}' \rightarrow \mathbf{r}, E) = \frac{\exp\left[-\int_0^{|\mathbf{r}-\mathbf{r}'|} ds\, \Sigma_t\left(\mathbf{r} - s\frac{\mathbf{r}-\mathbf{r}'}{|\mathbf{r}-\mathbf{r}'|}\right)\right]}{4\pi\,|\mathbf{r}-\mathbf{r}'|^2}, \tag{11.82}$$

and the source is assumed isotropic, i.e. $S(\mathbf{r}', E, \boldsymbol{\Omega}) = S(\mathbf{r}', E)/4\pi$. For a spatially uniform medium, the transport kernel $T(\mathbf{r}' \rightarrow \mathbf{r}, E)$ may be simplified to

$$T(\mathbf{r}' \rightarrow \mathbf{r}, E) = \frac{\exp\left[-\Sigma_t(E)|\mathbf{r}-\mathbf{r}'|\right]}{4\pi\,|\mathbf{r}-\mathbf{r}'|^2}. \tag{11.83}$$

Equation (11.83) is a simple mathematical statement that the contribution to the scalar flux of neutrons of energy $E$ at position $\mathbf{r}$, from a point isotropic source of the same energy located at $\mathbf{r}'$, is given by {the flux on the surface of a sphere of radius $|\mathbf{r}-\mathbf{r}'|$ centered at source $\mathbf{r}'$ in vacuum} × {the probability of neutrons escaping collisions in traveling the distance $|\mathbf{r}-\mathbf{r}'|$ in a medium with total cross section $\Sigma_t$}. Thus, without going through a formal derivation of Eq. (11.81) from the neutron transport equation, we may invoke the concept of Green's function from Eq. (5.53) and the physical interpretation of Eq. (11.83) just presented to justify or explain Eq. (11.81). Note that Eq. (11.82) represents a transport kernel, in contrast to a diffusion kernel in Eq. (5.52).

The source distribution can be written as a sum of the in-scattering term and fission source $Q(\mathbf{r}, E)$

$$S(\mathbf{r}, E) = \int_0^\infty dE' \phi(\mathbf{r}, E') \Sigma(\mathbf{r}, E' \rightarrow E) + Q(\mathbf{r}, E), \tag{11.84}$$

where $\Sigma(\mathbf{r}, E' \rightarrow E)$ represents the scattering kernel of Eq. (2.52). Substituting Eq. (11.84) into Eq. (11.81) yields

$$\phi(\mathbf{r}, E) = \int_V d\mathbf{r}'\, T(\mathbf{r}' \rightarrow \mathbf{r}, E) \left[\int_0^\infty dE' \phi(\mathbf{r}', E') \Sigma(\mathbf{r}', E' \rightarrow E) + Q(\mathbf{r}', E)\right]. \tag{11.85}$$

Integrating Eq. (11.85) over subvolume $V_n$ and energy group $g$, and defining discretized forms of the transport and scattering kernels, flux, and source with the transformation of variables

$$\mathbf{r} \to n,\ \mathbf{r}' \to n';\quad E \to g,\ E' \to g', \tag{11.86}$$

we obtain

$$\phi_{ng} = \sum_{n'} T_{n'\to n,g} V_{n'} \left[ \sum_{g'} \left( \Sigma_{n',g'\to g} \phi_{n'g'} \right) + Q_{n'g} \right]. \tag{11.87}$$

The CPM-3 code [Jon87,Jon00] solves Eq. (11.87) in matrix notation

$$\mathbf{\Phi} = T(\Sigma\mathbf{\Phi} + \mathbf{Q}), \tag{11.88}$$

for space- and energy-dependent flux vector $\mathbf{\Phi}$ through an iterative technique.

Actual evaluation of the transport kernel in a discretized form for 2-D geometry requires considerable computational effort for realistic geometries. Hence, a sequential combination of 1-D and 2-D CP formulations is used for the CPM-3 and CASMO-3 codes for both fast and thermal spectrum calculations at the assembly level. The first step involves a 1-D fine-mesh, micro-group calculation, along the lines of the original THERMOS approach, for each of the distinct fuel and absorber rod types. Fine-group fluxes from the micro-group calculations are then used to generate macro-group unit-cell average cross sections for each rod in the assembly. The code then performs a 2-D CP calculation using these coarse-mesh, coarse-group constants, representing the actual locations of fuel rods and non-lattice regions of the assembly in $(x - y)$ geometry. Between the unit-cell and 2-D assembly-level CP calculations, a 1-D cylindrical calculation is optionally performed to represent the entire assembly, and location-dependent corrections are made to the unit-cell results for the subsequent 2-D calculation. In addition, an empiricism is introduced in many CP formulations to replace the total cross section $\Sigma_t$ with the corresponding transport cross section so that anisotropic scattering is represented to the first order in the transport kernel from Eq. (11.82).

Through a synthesis of fine-mesh, fine-group unit-cell calculations and a coarse-mesh, coarse-group assembly calculation, the CASMO-3, CPM-3, and PHOENIX-P codes represent material heterogeneities, explicitly and with sufficient accuracy, at both the unit-cell and 2-D assembly levels. The CASMO-3 code introduces somewhat of an approximate representation of neutron transport at the 2-D assembly level to expedite the overall CP calculation. In passing, we also note that the methodology is quite similar to that of the proprietary GEBLA lattice physics code [Mar76], updated to DGEBLA, in use at General Electric Company for the design analysis of boiling water reactor (BWR) cores.

Some of the CP lattice physics codes, e.g. CPM-3 [Jon00], as well as the CASMO5, LANCR02, and POLARIS codes [Fer17,Glo09,Rea18], provide alternate integral transport representations via the MOC algorithm. In the MOC formulation, essentially identical spatial discretizations as for the CP formulation are used to solve the neutron transport equation along the trajectories of neutron travel. For notational convenience, consider the 1-D transport equation (4.44) with the in-scattering term included into the source term and $\Sigma \equiv \Sigma_t$:

$$\mu \frac{\partial \psi(z,\mu)}{\partial z} + \Sigma \psi(z,\mu) = S(z,\mu). \tag{11.89}$$

Rewriting the angular flux $\psi(z,\mu) = \psi(s)$ along a trajectory $s = z/\mu$ in direction $\theta = \cos^{-1}\mu$ transforms Eq. (11.89) into

$$\frac{d\psi(s)}{ds} + \Sigma \psi(s) = S(s). \tag{11.90}$$

The ODE may be conveniently integrated to yield

$$\psi(s) = e^{-\Sigma s}\left[\int_0^s S(s')e^{\Sigma s'}ds' + \psi(0)\right], \tag{11.91}$$

together with the boundary value $\psi(0)$ of the flux at the start of the trajectory. The characteristic equation (11.90) can represent a 3-D path along $\mathbf{\Omega}$ by properly including the azimuthal angle $\varphi$ together with the polar angle $\theta$ into the path length $s$. Equation (11.91) may then be discretized via schemes similar to that in Eq. (11.87) and integrated to provide the scalar flux distribution.

The CASMO5 code [Fer17], the HELIOS code [Stu00], the PARAGON module [Wes10] of the Westinghouse APA package, the LANCR02 code [Glo09] for BWR analysis, and the POLARIS module of the SCALE-6.2.3 package [Rea18] represent more recent enhancements to the lattice physics methodology. The CASMO-4 [Rho04] and CASMO5 [Fer17] codes incorporate multi-assembly 2-D MOC calculations with direct microscopic depletion of burnable absorbers, e.g. Gd-admixed fuel rods, instead of a separate Microburn routine used in earlier versions of the codes. The POLARIS code has also adopted a MOC solver, which replaces the TRITON solver used in earlier versions of the SCALE package, while the LANCR02 code combines 1-D pin-cell calculations with 2-D MOC assembly representations. Primary examples of the global MOC code are the DeCART code [Joo04] and the MPACT code [Koc13], which was developed for the Consortium for Advanced Simulations of Light Water Reactors (CASL). The MPACT code may be used efficiently for 2-D lattice physics analysis as well as for pin-resolved whole-core analysis.

## 11.8 $B_1$ Formulation for Spectrum Calculation

With an explicit intent to minimize mathematical complexities in representing the transport of neutrons in space, energy, and time, we have limited ourselves to the neutron diffusion equation as the balance statement as much as possible. We now return to a brief discussion we had on the neutron transport equation in Section 4.5 so that we may develop the $B_1$ method that is used to calculate the flux spectrum in a finite medium. As part of the unit-cell and unit-assembly formulations, including the assembly-level CP formulation in Section 11.7, the neutron balance statement excludes the leakage of neutrons across the unit-cell or assembly boundaries. Recalling that Fick's law of Eq. (4.36) for the neutron leakage and current is derived through a number of approximations, we now need to develop a formulation that can represent the migration of neutrons more accurately.

### 11.8.1 Basic Structure of $B_1$ Formulation

We return to the transport equation (4.44) in steady-state, 1-D form with the lethargy variable $u$ suppressed, and review the steps taken with the $P_1$ expansion of the equation:

$$\Sigma_t \psi(z,\mu) + \mu \frac{\partial \psi(z,\mu)}{\partial z} = S(z,\mu) + \rho(z,\mu). \tag{11.92}$$

The key approximation in the $P_1$ formulation of the neutron balance equations (4.62) is truncating the Legendre polynomial expansion of the angular flux after the first two terms, i.e. Eq. (4.49). For a fuel lattice with significant leakage in or out of the lattice, the angular flux $\psi(z,\mu)$ may exhibit a higher-order dependence on the angular variable. Thus, in the $B_1$ formulation, the truncation in the angular flux is not made, but instead, an alternate approximation is introduced to represent the spatial dependence of the angular flux.

The departure from the $P_1$ formulation is to assume the spatial and angular dependence in $\psi(z,\mu)$ may be separated

$$\psi(z,\mu) = Z(z)\psi(\mu) = e^{iBz}\psi(\mu), \tag{11.93}$$

so that

$$\frac{\partial \psi(z,\mu)}{\partial z} = iB\psi(z,\mu). \tag{11.94}$$

By taking the second derivative of the spatial component $Z(z)$, we obtain

$$\frac{d^2 Z(z)}{dz^2} + B^2 Z(z) = 0, \tag{11.95}$$

which represents the fundamental-mode eigenvalue equation (5.62) with buckling $B^2$. Equation (11.93) may also be considered a Fourier transform of the angular flux and, in a formal derivation of the $B_1$ formulation, the angular component $\psi(\mu)$ in Eq. (11.93) is often written as $\psi(B,\mu)$, with the parameter $B$ treated as the Fourier transform variable. Thus, assuming the fundamental mode shape $Z(z)$ for the $B_1$ formulation, we sacrifice the spatial details that the $P_1$ approximation provides and instead focus on a more accurate treatment for the angular component $\psi(\mu)$. Introducing the *fundamental mode* representation from Eq. (11.93) into the source and scattering terms, cancel out the spatial component $Z(z) = e^{iBz}$ from both sides of Eq. (11.92) and obtain an algebraic equation for the transport equation

$$iB\mu\psi(\mu) + \Sigma_t\psi(\mu) = S(\mu) + \rho(\mu), \tag{11.96}$$

with the isotropic source and scattering terms written via the same $P_\ell$ expansions as Eqs. (4.61) and (4.56)

$$S(\mu) = \frac{S_0}{2}, \tag{11.97a}$$

$$\rho(\mu) = \frac{\rho_0}{2} + \frac{3\mu}{2}\rho_1. \tag{11.97b}$$

*Without invoking the $P_1$ approximation* for the angular component $\psi(\mu)$, now integrate Eq. (11.96) over $\mu \in [-1, 1]$

$$iB\phi_1 + \Sigma_t\phi_0 = S_0 + \rho_0, \tag{11.98}$$

recalling the definitions of scalar flux $\phi_0$ and net current $\phi_1$. Multiplying back both sides of Eq. (11.98) by the fundamental mode $Z(z)$, and recalling Eq. (11.94), we immediately recognize that Eq. (11.98) is simply equal to Eq. (4.62a), the $P_0$-component of the transport equation. Next, multiply both sides of Eq. (11.96) by $\Sigma_t\mu/(iB\mu + \Sigma_t)$ and carry on the integrals again over $\mu \in [-1, 1]$

$$\Sigma_t\phi_1 = A_{01}(S_0 + \rho_0) + 3A_{11}\rho_1, \tag{11.99}$$

with the definition

$$A_{ij} = \frac{1}{2}\int_{-1}^{1} \frac{P_i(\mu)P_j(\mu)d\mu}{1 + iB\mu/\Sigma_t} = \frac{1}{2}\int_{-1}^{1} \frac{P_i(\mu)P_j(\mu)d\mu}{1 + iy\mu}, \quad y = \frac{B}{\Sigma_t}, i, j = 0, 1, \ldots. \tag{11.100}$$

Note for low-order values of indices $i, j$

$$A_{00} = \frac{1}{2}\int_{-1}^{1} \frac{d\mu}{1 + iy\mu} = \frac{\tan^{-1}y}{y}, \tag{11.101a}$$

$$A_{01} = \frac{1}{2}\int_{-1}^{1} \frac{\mu\, d\mu}{1 + iy\mu} = \frac{1}{2iy}\int_{-1}^{1} \frac{iy\mu + 1 - 1}{1 + iy\mu}d\mu = \frac{i}{y}(A_{00} - 1) = A_{10}, \tag{11.101b}$$

$$A_{11} = \frac{1}{2y^2}\int_{-1}^{1} \frac{y^2\mu^2\, d\mu}{1 + iy\mu} = \frac{1}{2y^2}\int_{-1}^{1} \frac{1 + y^2\mu^2 - 1}{1 + iy\mu}d\mu = \frac{1 - A_{00}}{y^2}. \tag{11.101c}$$

Now substitute Eq. (11.98) into Eq. (11.99) to obtain

$$\Sigma_t\phi_1 = A_{01}(iB\phi_1 + \Sigma_t\phi_0) + 3A_{11}\rho_1, \tag{11.102}$$

which simplifies to

$$\gamma \Sigma_t \phi_1 + \frac{1}{3} iB\phi_0 = \rho_1, \tag{11.103}$$

with the help of Eqs. (11.101) and the definition

$$\gamma = \frac{A_{00}}{3A_{11}} = \frac{y^2 \tan^{-1} y}{3(y - \tan^{-1} y)} \simeq 1 + \frac{4}{15}\left(\frac{B}{\Sigma_t}\right)^2. \tag{11.104}$$

Substituting Eq. (4.60b) for the $P_1$ component of the scattering integral $\rho_1 = \overline{\mu}_0 \Sigma_s \phi_1$ into Eq. (11.103) again yields Fick's law but with the transport cross section now defined as

$$\Sigma_{tr} = \gamma \Sigma_t - \overline{\mu}_0 \Sigma_s. \tag{11.105}$$

With the expansion given for the parameter $\gamma$ in Eq. (11.104), note that the correction to the conventional $\Sigma_{tr}$ would become significant as the leakage increases.

The enhanced angular treatment of the $B_1$-formulation may be understood if we note that the scattering term of Eq. (11.92) involves an integral over the angular flux $\psi(z, \mu)$ and hence is in general more isotropic than the flux itself. Thus, less error is introduced by truncating, up to the same order, the scattering integral rather than the angular flux itself. This is also the case for $\phi_j$ for all orders of $j$, including the $P_0$- and $P_1$-components of the angular flux, if the formulation is obtained for higher orders of $j$. This also means that if the scattering is linearly anisotropic in the laboratory, $B_1$ formulation preserves the scattering source and hence determines exactly $\phi_j$ for all orders of $j$. In the actual implementation of the $B_1$ formulation in lattice physics codes, e.g. CPM, CASMO, and PHOENIX, the lethargy dependence is restored to Eqs. (11.98) and (11.103). The resulting equations are then discretized and solved in a multigroup structure, typically between 70 and 97 groups, to account accurately for the neutron leakage out of the unit assembly.

### 11.8.2 Numerical Solution of $B_1$ Equations

The energy-independent $B_1$ formulation from Eqs. (11.102) and (11.103) is now rewritten to include the energy or lethargy dependence and the slowing down density $q(u)$ of Eq. (9.48) extended to the $P_1$-component $\phi_1(B, u)$ of neutron flux:

$$q_\ell(B, u) = \int_{u-\Delta}^{u} du' \int_{u}^{u'+\Delta} du'' \Sigma_{s\ell}(u' \to u'') \phi_\ell(B, u'), \quad \ell = 1, 2. \tag{11.106}$$

Extending likewise the steps taken to derive Eq. (9.53) yields the *Greuling–Goertzel approximation*

$$\lambda_\ell \frac{\partial q_\ell(B, u)}{\partial u} + q_\ell(B, u) = \xi_\ell \Sigma_s(u) \phi_\ell(B, u), \quad \ell = 1, 2, \tag{11.107}$$

where $\xi_0 = \xi$ is the mean lethargy gain per collision $\langle \Delta u \rangle$ of Eq. (9.23), $\lambda_0 = \gamma$ of Eq. (9.52), and

$$\xi_1 = \langle \mu_0 \xi \rangle, \quad \lambda_1 = \frac{\langle \mu_0 \xi^2 \rangle}{2\xi_1}.$$

Taking the derivative of Eq. (11.106), using Leibnitz's differentiation formula from Eq. (9.13) also generalizes Eq. (9.14) to arrive at

$$\frac{\partial q_\ell(B,u)}{\partial u} = \phi_\ell(B,u) \int_u^{u+\Delta} du'' \Sigma_{s\ell}(u \to u'') - \int_{u-\Delta}^{u} du' \Sigma_{s\ell}(u' \to u) \phi_\ell(B,u'),$$

or

$$\frac{\partial q_\ell(B,u)}{\partial u} = \Sigma_{s\ell}(u)\phi_\ell(B,u) - \rho_\ell(B,u). \tag{11.108}$$

Note that Greuling–Goertzel equation (11.107) connects $q_\ell(B,u)$ to $\phi_\ell(B,u)$, while Eq. (11.108) relates $q_\ell(B,u)$ to $\phi_\ell(B,u)$ and slowing down integrals $\rho_\ell(B,u)$. Thus, we now need a relationship that connects $\phi_\ell(B,u)$ and $\rho_\ell(B,u)$ to the neutron source $S(B,\mu)$, which is provided by the $B_1$ equations (11.98) and (11.102).

In actual implementation of the Greuling–Goertzel formulation in the MUFT code, the anisotropic component of the slowing down density for heavy nuclides is assumed negligibly small to represent the two moments of the slowing down density

$$\begin{aligned} q_0(B,u) &= q_{0H}(B,u) + q_{0A}(B,u), \\ q_1(B,u) &= q_{1H}(B,u), \end{aligned} \tag{11.109}$$

where $q_{0A}(B,u)$ is the $P_0$ slowing down density for heavy scatterers. Grouping the $B_1$ equations (11.98) and (11.102) with Eq. (11.108) for $\ell = 0$ for heavy scatterers and $\ell = 0, 1$ for protons finally yields a set of differential equations that MUFT solves for $\phi_0(B,u)$ and $\phi_1(B,u)$, with the fission source $\chi$ and inelastic scattering cross section $\Sigma_{in}$ and scattering integral $\rho_{in}$:

$$(\Sigma_a + \Sigma_{in})\phi_0 + iB\phi_1 = \chi + \rho_{in} - \frac{\partial q_{0H}}{\partial u} - \frac{\partial q_{0A}}{\partial u}, \tag{11.110a}$$

$$\Sigma_{tr}\phi_1 + \frac{iB}{3}\phi_0 = -\frac{\partial q_{1H}}{\partial u}, \tag{11.110b}$$

$$\frac{\partial q_{0H}}{\partial u} + q_{0H} = \Sigma_{sH}\phi_0, \tag{11.110c}$$

$$\frac{2}{3}\frac{\partial q_{1H}}{\partial u} + q_{1H} = \frac{4}{9}\Sigma_{sH}\phi_1, \tag{11.110d}$$

$$\lambda\frac{\partial q_{0A}}{\partial u} + q_{0A} = \xi_A \Sigma_{sA}\phi_0, \tag{11.110e}$$

where

$$\lambda = \frac{\sum_{i \neq H} \xi_i \Sigma_{si} \lambda_i}{\sum_{i \neq H} \xi_i \Sigma_{si}}, \quad \xi_A = \frac{\sum_{i \neq H} \xi_i \Sigma_{si}}{\sum_{i \neq H} \Sigma_{si}} = \frac{\sum_{i \neq H} \xi_i \Sigma_{si}}{\Sigma_{sA}}.$$

It should be noted that Eqs. (11.110c) and (11.110d) follow from Eq. (11.108) with $\xi_0 = \lambda_0 = 1, \xi_1 = 4/9$, and $\lambda_1 = 2/3$. The proton scattering equations may, however, be derived from Eq. (11.108) without the approximations introduced in Eq. (9.53). Thus, Eqs. (11.110c) and (11.110d) form an exact balance statement for protons. The MUFT code obtains the two flux moments $\phi_0(B, u)$ and $\phi_1(B, u)$ by integrating the set of five differential equations over each of 54 groups covering the fast neutron spectrum, with a typical cutoff energy $E = 0.625$ eV. With some further approximations related to the in-group inelastic scattering, the set of five discretized equations for each of the 54 groups is simplified into a $(2 \times 2)$ matrix equation for $\phi_0(B, u)$ and $\phi_1(B, u)$, which is solved group by group, without iterations. The adoption of the Greuling–Goertzel equations avoided the use of integral equations for the slowing down density covering the entire fast spectrum for a hydrogenous medium typical of LWR cores, and the MUFT formulation for the sequential solution of discretized equations served as the fast-spectrum module of the popular LEOPARD code during the early days of PWR development.

In contrast to the MUFT formulation, which requires a group-by-group inversion of a simple matrix equation for the calculation of the discretized neutron flux and current, a more direct approach is taken in modern lattice physics codes, including the CPM-3 code. The actual discretization of the $B_1$ equations (11.98) and (11.102) closely follows the group-transfer scattering matrix formulation in Eqs. (7.12) and (7.20) for the multigroup diffusion (MGD) equation. Thus, the inelastic scattering integral $\rho_{in}(B, u)$ and elastic scattering integrals $\rho_0(B, u)$ are represented by a combined $P_0$ scattering kernel $\Sigma_{s0}(u' \to u)$ and discretized for transfer from lethargy group $\Delta_j$ to $\Delta_n$:

$$\Sigma_{s0,j\to n} = \frac{\int_{\Delta n} du \int_{\Delta j} du' \, \Sigma_{s0}(u' \to u)\phi_0(B, u')}{\int_{\Delta j} du' \, \phi_0(B, u')} \simeq \frac{1}{\Delta j} \int_{\Delta n} du \int_{\Delta j} du' \, \Sigma_{s0}(u' \to u).$$

A similar discretization of $\rho_1(B, u)$ for $\Sigma_{s1,j\to n}$, together with discretized cross sections $\Sigma_{tj}$ and $\gamma\Sigma_{tj}$, generates matrices $G$ and $H$ with elements

$$G_{jn} = \Sigma_{tj}\delta_{jn} - \Sigma_{s0,j\to n} \quad \text{and} \quad H_{jn} = \gamma\Sigma_{tj}\delta_{jn} - \Sigma_{s1,j\to n},$$

which transforms Eqs. (11.98) and (11.102) for multigroup flux vector $\mathbf{\Phi}$ and current vector $\mathbf{J}$ to

$$G\mathbf{\Phi} + B\mathbf{J} = M\mathbf{\Phi} \quad \text{and} \quad H\mathbf{J} = B\mathbf{\Phi}, \tag{11.111}$$

with the matrix $M$ representing the fission cross section and fission yield defined in Eq. (7.20). Combining the two discretized $B_1$ equations yields

$$(G + B^2 H^{-1})\mathbf{\Phi} \equiv L\mathbf{\Phi} = M\mathbf{\Phi} = \chi, \tag{11.112}$$

with the proper source normalization as in Eq. (7.20). Inversion of matrix $L$ in Eq. (11.112) finally provides the desired flux solution $\mathbf{\Phi}$ in terms of the fission yield vector $\chi$ for a given buckling $B^2$. Typically, a buckling search is performed with a single- or few-assembly $B_1$ setup to arrive at $k_{eff} = 1.0$ for each assembly, thereby effectively representing the flow of neutrons across the assembly boundaries.

## 11.9 Lattice Physics Methodology for Fast Reactor

The bulk of the discussion so far in this chapter has been devoted to the lattice physics methodology applicable to LWR core design calculations. For the lattice physics analysis for fast-spectrum nuclear reactors, e.g. the sodium-cooled fast reactor (SFR) under development with the Generation-IV initiative, the leakage fraction of neutrons from the core is an order of magnitude larger than that for a LWR core. At the same time, since the average energy of neutrons in an SFR core is around $0.1 \sim 0.2$ MeV, as illustrated in the neutron flux spectra, $\phi(u) = E\phi(E)$ for typical SFR and LWR cores in Figure 11.10. Because SFR fuel rods are also thinner than the LWR counterpart, it is much less important to represent the detailed flux distribution within a fuel cell. The mean free path of neutrons is also significantly larger than that in the LWR core. Thus, it becomes much more important for fast reactor to account for spectral-spatial coupling at the global level than is required for LWR core physics analysis. Although resonance neutron absorptions play a nearly equally important role in fast reactor lattice physics analysis, the spatial self-shielding effect is not as important as in LWR analysis. With these perspectives, we discuss a few representative lattice physics methods for fast reactor analysis.

### 11.9.1 Bondarenko Formulation for Self-Shielding Factor

Bondarenko [Bon64] suggested a simple method to account for energy self-shielding of resonances, based on the NR approximation from Eq. (9.65), $\phi(u) = \Sigma_s/\Sigma_t(u)$. Noting that the scattering cross section $\Sigma_s$ is nearly constant over the fast spectrum, we determine the resonance shielded cross section through standard flux weighting

$$\overline{\sigma}_k = \langle \sigma_k \rangle = \frac{\displaystyle\int_{\Delta u} \sigma_k(u)\phi(u)du}{\displaystyle\int_{\Delta u} \phi(u)du} = \frac{\displaystyle\int_{\Delta u} \frac{\sigma_k(u)}{\Sigma_t(u)}du}{\displaystyle\int_{\Delta u} \frac{du}{\Sigma_t(u)}}, \tag{11.113}$$

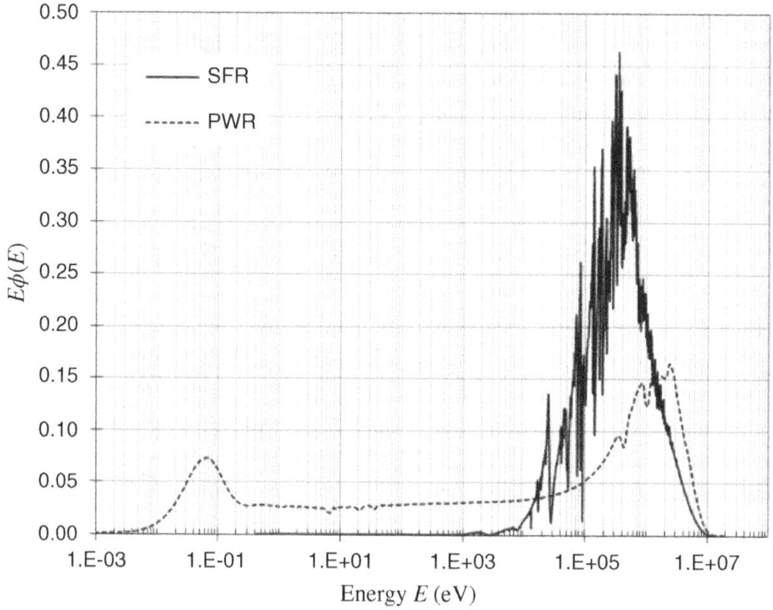


**Figure 11.10** Comparison of flux spectra for typical SFR and PWR cores. *Source*: Yang [Yan17]/with permission of W.S. Yang.

and recognize that the shielded cross section is a function of some *background cross section* making up $\Sigma_t$. The subscript $k$ in Eq. (11.113) is used to designate {*nuclide, energy group, reaction type*}, but for simplicity, we will refer to it as the *nuclide index*. Construct the total cross section

$$\Sigma_t = N_k\overline{\sigma}_{tk} + \sum_{j\neq k} N_j\overline{\sigma}_{tj} \equiv N_k(\overline{\sigma}_{tk} + \overline{\sigma}_{0k}), \tag{11.114}$$

where the *background cross section* is defined as

$$\begin{aligned}\overline{\sigma}_{0k} &= \frac{1}{N_k}\sum_{j\neq k} N_j\sigma_{tj}\\ &= \text{total cross section seen by each nucleus of the } k\text{th nuclide}\\ &\quad\ \text{excluding itself.}\end{aligned} \tag{11.115}$$

This then allows us to represent the resonance-shielded cross section of Eq. (11.113) as a product of the *resonance shielding factor* $f$ and *unshielded cross section* $\sigma_k$:

$$\overline{\sigma}_k = f(\sigma_{0k}, T)\sigma_k. \tag{11.116}$$

The resonance shielding factor is equivalent to the factor $f(\rho_i)$ introduced for the resonance integral $I(\rho_i)$ in Eq. (11.73). To account for Doppler broadening of resonances, the self-shielding factor is calculated as a function of an effective resonance temperature $T$ and the background cross section. Since the total cross section $\overline{\sigma}_{tj}$ has to be calculated, however, with the proper self-shielding, the actual implementation of Eqs. (11.115) and (11.116) require an iteration for the background cross section:

$$\overline{\sigma}_{0k} = \frac{1}{N_k}\sum_{j\neq k} N_j \overline{\sigma}_{tj}. \tag{11.117}$$

Together with the energy self-shielding represented through the shielding factor $f$, Bondarenko's algorithm is implemented in a multigroup structure in global neutron diffusion theory calculations in the 1DX code [Har69] and the TRANSX code [Mac92] to account for the spatial-spectral coupling of importance for fast reactor analysis. In addition, some effort is made to represent the spatial self-shielding through an escape cross section concept. Although Bondarenko's original tabulation of the self-shielding factor was presented in a 26-group structure, recent implementations of the method have been made with $50 \sim 100$ energy groups.

### 11.9.2 MC$^2$-3 Code

A sophisticated lattice physics code specialized for fast reactor analysis is the MC$^2$-3 code [Lee17] developed at the Argonne National Laboratory. In MC$^2$-3, a ultra-fine energy group structure is used to represent each resolved resonance separately for key nuclides, with due accounting for Doppler broadening and accurate treatment for unresolved resonances. The self-shielding of resolved resonances is represented by direct numerical integration of point-wise cross sections, with allowance made for anisotropic scattering sources in hyper-fine group spectrum calculations.

Recent improvements to the code include a direct representation of spectral-spatial coupling at the global level, thereby obviating the need for iterations with the 1-D SDX cell calculations [Sta72] in the earlier MC$^2$-2 code [Hen76]. Region-dependent cross sections are generated with the angular flux obtained from whole-core ultra-fine group transport calculations with the TWODANT code [Alc84].

### 11.9.3 ERANOS System

Another well-known fast reactor physics code is the ERANOS system [Pal91], which combines a lattice physics code with a global analysis code in an integrated package. The ECCO module [Rim95] of the ERANOS system performs the CP calculations for a number of fast reactor configurations in 1-D or 2-D geometries, providing the capability for accurate self-shielding factor calculations. In addition, the code incorporates the subgroup or probability table method from Section 9.4.4, which has the capability to accurately represent both resolved and unresolved resonances. In the subgroup method, resonance cross sections in a given group are processed into a structure listing the distribution of multiple or unresolved resonance cross sections as a function of the cross section value itself in a probability structure, rather than generating point cross section data in traditional multigroup formulations.

The ERANOS system offers the ability to perform fine-group fuel depletion and space-time kinetics analysis, involving as many as 1968 groups for lattice physics analysis. The global calculations are, however, generally limited to 33-group structures. It requires substantial computational resources to fully utilize the integrated system package.

## 11.10 Monte Carlo Lattice Physics Analysis

One computational algorithm for lattice physics analysis that has achieved considerable importance in recent years is the Monte Carlo (MC) method. By selecting a host of random numbers, MC algorithms simulate the life cycle or history of individual particles that follow physical laws of particle interactions and transport, as represented by Eq. (4.40), without the need to discretize any of the spatial, energy, or direction variables. Monte Carlo algorithms offer the potential to provide accurate solutions for transport problems with complex geometries and material heterogeneities, with the solution accuracy limited only by the computing resources available. With rapid advances made in computer hardware, there has been a significant increase in the popularity of MC algorithms in both neutron and photon transport applications.

This increase in popularity of MC algorithms is owed in no small measure to the versatility that the MC codes, including MCNP6 [Goo12] and Serpent [Fri11], offer: (i) simple description of complex geometries using well-defined surfaces, (ii) separate or coupled neutron and gamma transport calculations, and (iii) cross section libraries in a continuous energy structure, rather than in discrete group formulations. Despite these features, the applicability of MC codes for lattice physics analysis had been limited, partly because of the difficulty of representing [Dav05] the critical configuration for assembly-level calculations and accurately determining diffusion coefficients for global MGD theory codes. The recent development of the Serpent code [Fri11] that incorporates the $B_1$ formulation in lattice physics analysis has opened up the possibility to use MC algorithms for a variety of reactor configurations with different levels of material heterogeneities.

## 11.11 Overall Reactor Physics Analysis

In this chapter, we have studied various lattice physics methods for both LWR and SFR cores. Although unit-cell codes exemplified by the well-known LEOPARD code have rarely been used in recent years in actual design calculations, some effort has been made to discuss the unit-cell formulations for material heterogeneities in both the thermal spectrum and resonance integral calculations to gain a physical understanding of the key physical phenomena. We should also remark that the LEOPARD code provides acceptable accuracy for routine PWR spectrum calculations, despite its simplicity, and has made significant contributions to the commercial development of the PWR technology. The validity of the LEOPARD code is, however, usually limited to normal fuel cells. More accurate models, e.g. the CPM-3, PHOENIX-P, PARAGON, and CASMO5 codes, involving space-dependent spectrum calculations have to be used for handling large water gaps or strong localized absorbers. A brief discussion of SFR lattice physics methods highlighted the importance of spectral-spatial coupling for fast reactor analyses. Recent introduction of the Serpent Monte Carlo code indicates the possible future focus for lattice physics algorithms.

We conclude our discussion of lattice physics analysis with a brief overview of the overall reactor physics calculations required for both LWR and fast reactor calculations. The overall structure for reactor physics analysis is indicated in Figure 11.11, with computer code systems illustrated in rectangles and databases or computer output in oval or rounded boxes. Included is the cross section processing code that generates a cross section library in a 30 ~ 2000 group structure for the

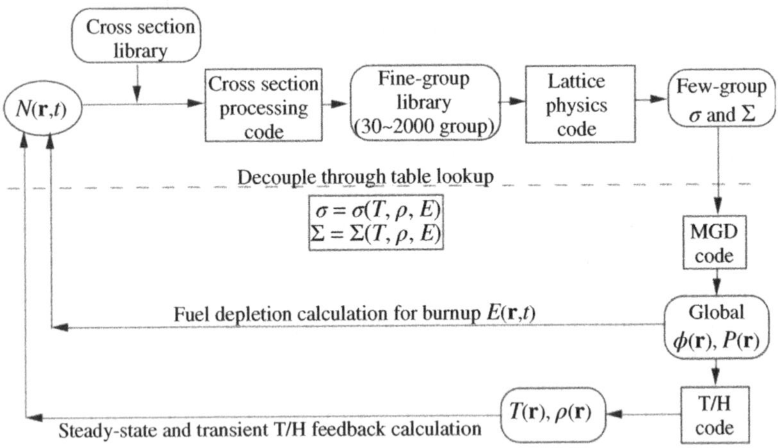


**Figure 11.11** Overall reactor physics calculational procedure.

appropriate lattice physics code. We also indicate the importance of accounting for changes in the cross sections and number densities associated with variations in temperature $T$ and density $\rho$ of various materials at position $\mathbf{r}$ in a reactor core, as well as the effects due to fuel depletion represented by exposure $E(\mathbf{r}, t)$ at time $t$ in a fuel cycle. For many LWR applications, the complexities associated with fully coupled processes involving lattice physics and global MGD calculations are simplified often by generating cross section tables. The cross section tables may be interpolated as a function of $T$, $\rho$, and $E$ in the MGD codes, thereby decoupling the lattice physics calculations from the MGD and thermal-hydraulic (T/H) calculations.

## References

**[Alc84]** R.E. Alcouffe, F.W. Brinkley, D.R. Marr, and R.D. O'Dell, "User's Guide for TWODANT: A Code Package for Two-Dimensional Diffusion-Accelerated, Neutral-Particle Transport," LA-10049-M-Rev.1 (1984).

**[Amo57]** A. Amouyal, P. Benoist, and J. Horowitz, "Nouvelle Methode de Determination du Facteur d'Utilisation Thermique d'une Cellule," *J. Nucl. Energy* **6**, 79 (1957).

**[Ams57]** H. Amster and R. Suarez, "The Calculation of Thermal Constants Averaged Over a Wigner-Wilkins Flux Spectrum: Description of the SOFOCATE Code," WAPD-TM- 39, Westinghouse Electric Corporation (1957).

**[Arg63]** Argonne National Laboratory, *Reactor Physics Constants*, ANL-5800, 2nd ed., Argonne National Laboratory (1963).

**[Bar63]** R.F. Barry, "LEOPARD - A Spectrum Dependent Non-Spatial Depletion Code for the IBM-7094," WCAP-3269-26, Westinghouse Electric Corporation (1963).

**[Boh57]** H. Bohl, E. Gelbard, and G. Ryan, "MUFT-4 - Fast Neutron Spectrum Code for the IBM-704," WAPD-TM-72, Westinghouse Electric Corporation (1957).

**[Bon64]** I.I. Bondarenko et al., *Group Constants for Nuclear Reactor Calculations*, Consultants Bureau, New York (1964).

**[Car69]** B. Carnahan, H.A. Luther, and J.O. Wilkes, *Applied Numerical Methods*, Wiley (1969).

**[Cas53]** K.M. Case, F. de Hoffmann, and G. Placzek, *Introduction to the Theory of Neutron Diffusion*, Los Alamos Scientific Laboratory (1953).

**[Dav05]** J.C. Davis and J.C. Lee, "Comparison of Monte Carlo and Deterministic Depletion Codes for LWR Fuel Cycle Analysis," *Trans. Am. Nucl. Soc.* **92**, 651 (2005).

**[Der84]** K.L. Derstine, "DIF3D: A Code to Solve One-, Two-, and Three-Dimensional Finite Difference Diffusion Theory Problems," ANL-82-64, Argonne National Laboratory (1984).

**[Ede92]** M. Edenius and F. Forssen, "CASMO-3, A Fuel Assembly Burnup Program," Studsvik/NFA-89/3, Rev. 2, Studsvik AB (1992).

**[Fer17]** R. Ferrer, J. Hykes, and J. Rhodes, "CASMO5, A Fuel Assembly Burn-up Program: Methodology Manual," SSP-08/405, rev. 7, Studsvik Scandpower (2017).

**[Fri11]** E. Fridman and J. Leppänen, "On the Use of the Serpent Monte Carlo Code for Few-Group Cross Section Generation," *Ann. Nucl. Energy* **38**, 1399 (2011).

**[Gla52]** S. Glasstone and M.C. Edlund, *The Elements of Nuclear Reactor Theory*, Van Nostrand (1952).

**[Glo09]** Global Nuclear Fuel, "LANCR02 Lattice Physics Model Description," NED-33376-A, rev. 3, Global Nuclear Fuel (2009).

**[Goo12]** T. Goorley, et al., "Initial MCNP6 Release Overview," *Nucl. Technol.* **180**, 298 (2012).

**[Har69]** R.W. Hardie and W. W. Little, Jr., "1DX, A One Dimensional Diffusion Code for Generating Effective Nuclear Cross Sections," BNWL-954, Battelle Northwest Laboratory (1969).

**[Hen76]** H. Henryson II, B.J. Toppel, and C.G. Stenberg, "$MC^2$-2: A Code to Calculate Fast Neutron Spectra and Multigroup Cross Sections," ANL-8144, Argonne National Laboratory (1976).

**[Hon61]** H.C. Honeck, "THERMOS, A Thermalization Transport Theory Code for Reactor Lattice Calculations," BNL-5826, Brookhaven National Laboratory (1961).

**[Jon87]** D.B. Jones, "ARMP-02 Documentation, Part II, Chapter 6 – CPM-2 Computer Code Manual," EPRI NP-4574-CCM, Part II, Ch. 6, Volumes 1, 2, and 3, Electric Power Research Institute (1987).

**[Jon00]** D.B. Jones, K.E. Watkins, and M.L. Williams, "CPM-3 Computer Code Manual, Volume 1: Theory and Numerics Manual," EPRI-CPM-001-M-001, Rev. A, Electric Power Research Institute (2000).

**[Joo04]** H.G. Joo, J.Y. Cho, K.S. Kim, C.C. Lee, and S.Q. Zee, "Methods and Performance of a Three-Dimensional Whole-Core Transport Code DeCART," *PHYSOR 2004* (2004).

**[Koc13]** B. Kochunas, et al., "Overview of Development and Design of MPACT: Michigan Parallel Characteristics Transport Code," *Proc. M&C Sun Valley* **1**, 42 (2013).

**[Lam66]** J.R. Lamarsh, *Introduction to Nuclear Reactor Theory*, AddisonWesley (1966).

**[Lee17]** C. Lee and W.S. Yang,"$MC^2$-3: Multigroup Cross Section Generation Code for Fast Reactor Analysis," *Nucl. Sci. Eng.* **187**, 268 (2017).

**[Mac92]** R.E. MacFarlane, "TRANSX 2: A Code for Interfacing MATXS Cross Section Libraries to Nuclear Transport Codes," LA-12312-MS, Los Alamos National Laboratory (1992).

**[Mar76]** C.L. Martin, "Lattice Physics Methods," NEDE-20913-P, General Electric Company (1976).

**[Nel60]** M. Nelkin, "Scattering of Slow Neutrons by Water," *Phys. Rev.* **119**, 741 (1960).

**[Pal91]** G. Palmiotti, R.F. Burstall, E. Kiefhaber, et al., "New Methods Developments and Rationalization of Tools for LMFBR Design in the Frame of the European Collaboration," *Proc. Int. Conf. Fast Reactors and Related Fuel Cycles*, Kyoto, Japan (1991).

**[Rea18]** B.T. Rearden and M.A. Jessee, eds., "SCALE Code System," vers. 6.2.3, ORNL/TM-2005/39, Oak Ridge National Laboratory (2018).

**[Rho04]** J. Rhodes and M. Edenius, "CASMO-4, A Fuel Assembly Burnup Program User's Manual," SSP-01/400, rev. 4, Studsvik Scandpower (2004).

**[Rim95]** G. Rimpault, "Algorithmic Features of the ECCO Cell Code for Treating Heterogeneous Fast Reactor Subassemblies," *Proc. Intl. Conf. Math. Comp. Reactor Phys. Env. Analyses*, Portland, Oregon (1995).

**[San09]** A. Santamarina, D. Bernard, P. Blaise, P. Leconte, et al., "APOLLO2.8: A validated code package for PWR neutronics calculations," *Proc. ANFM-IV* (2009).

**[Shu62]** R.H. Shudde and J. Dyer, "TEMPEST II - A Neutron Thermalization Code," TID-18284, Atomics International (1962).

**[Sta72]** W.M. Stacey Jr., B.J. Toppel, H. Henryson II, B.A. Zolotar, et al., "A New Space-Dependent Fast Neutron Multigroup Cross-Section Preparation Capability," *Trans. Am. Nucl. Soc.* **15**, 292 (1972).

**[Str65]** L.E. Strawbridge and R.F. Barry, "Criticality Calculations for Uniform Water-Moderated Lattices," *Nucl. Sci. Eng.* **23**, 58 (1965); also L.E. Strawbridge, "Calculation of Lattice Parameters and Criticality for Uniform Water-Moderated Lattices," WCAP-3269-25, Westinghouse Electric Corporation (1963).

**[Stu00]** Studsvik Scandpower, "HELIOS 1.7 Methods," Studsvik Scandpower (2000).

**[Top90]** B.J. Toppel, "A User's Guide for the REBUS-3 Fuel Cycle Analysis Capability," ANL-83-2, Argonne National Laboratory (1983; updated 1990).

**[Wes02]** Westinghouse Electric Company, "PHOENIX-P User Manual," Westinghouse Electric Company (2002).

**[Wes10]** Westinghouse Electric Company, "PARAGON User Manual, Rev. 4," Westinghouse Electric Company (2010).

**[Wig44]** E.P. Wigner and J.E. Wilkins Jr., "Effects of the Temperature of the Moderator on the Velocity Distribution of Neutrons with Numerical Calculations for H as Moderator," AECD-2275, Oak Ridge National Laboratory (1944).

**[Yan17]** W. S. Yang, University of Michigan, private communication, 2017.

## Problems

**11.1** Obtain resonance parameters for the $^{239}$Pu at 0.296 eV from the ENDF/B-VIII.0 library in the NNDC database [www.nndc.bnl.gov], and calculate the total, fission, and capture cross sections at the peak of the resonance at temperature $T = 0$ K. Compare the results with plots available in the NNDC database.

**11.2** Prove that the first-flight escape probability of Eq. (11.48) for a purely absorbing lump represents the flux depression factor using the 1-D transport equation for a purely absorbing slab of thickness $a$.

**11.3** Verify that the thermal disadvantage factor $\zeta_2$ of Eq. (11.21) for the slab-geometry unit cell approaches 1.0 for a homogeneous mixture, as illustrated in Figure 11.6.

**11.4** Consider a unit cell in slab geometry, where the half thickness of the fuel region is $a$ and that of the moderator region is $(b - a)$. Assume that thermal neutrons are uniformly produced at a rate of $Q$ [neutron $\cdot$ cm$^{-3}$ s$^{-1}$] in the moderator region and essentially zero in the fuel region, respectively. Find an expression for the ratio of the average thermal flux in the moderator to that in the fuel, if the moderator absorption cross section is negligibly small. Use the one-group diffusion equation with group constants $D_F$ and $\Sigma_{aF}$ for the fuel and $D_M$ for the moderator.

**11.5** For the AP1000 unit cell, we propose an idealized two-region structure, where the clad region is homogenized with coolant water into one moderator region. Based on spectrum-averaged cross sections for resonance-energy neutrons, the following cross sections are obtained for the *separate* fuel and moderator regions, i.e. not smeared over the entire cell volume: $\Sigma_{sF} = 0.418\,\text{cm}^{-1}$, $\Sigma_{aF} = 5.64\,\text{cm}^{-1}$, $\Sigma_{sM} = 0.983\,\text{cm}^{-1}$, $\Sigma_{aM} = 0.0$, pin radius $a = 0.4096$ cm, and fuel volume fraction $f_F = 0.332$. Using the NR approximation and the escape cross section evaluated through Wigner's rational approximation for the heterogeneous lattice, calculate the spatial self-shielding factor $\phi_{NR}^{het}(u)$. Compare $\phi_{NR}^{het}(u)$ with the energy self-shielding factor $\phi_{NR}^{homog}(u)$ for the corresponding homogeneous lattice, and discuss the results in terms of the effective resonance integral for fuel and the resonance escape probability.

**11.6** For the two-region unit cell considered in Problem 11.4, the moderator absorption cross section is finite now, but the fuel material is a strong absorber so that every thermal neutron incident on the fuel region is absorbed. Use one-group diffusion theory to obtain the thermal flux *only* in the moderator region and determine the thermal utilization factor $f$ for the unit cell. Neutron cross sections are not known for the fuel region.

**11.7** For the PWR unit cell in Example 11.4, calculate the effective dilution factor $\rho_{het}^{28}$ for $^{238}U$ in the two-region heterogeneous cell through the escape cross section concept $\rho_{het}^{28} = (\Sigma_{sF} + \Sigma_e)/N_{28}$, with the escape cross section evaluated through Wigner's rational approximation. Compare the result with the homogeneous dilution factor $\rho_{homog}^{28} = \langle \Sigma_s \rangle_{cell} / \langle N^{28} \rangle_{cell}$, and discuss how this difference between $\rho_{het}^{28}$ and $\rho_{homog}^{28}$ impacts the effective resonance integral and resonance escape probability for the unit cell.

**11.8** Perform the discretization of the five governing equations (11.110) for the MUFT module, and derive the $2 \times 2$ matrix structure for each of the 54 MUFT groups.

**11.9** In a homogeneous mixture of resonance absorber and moderator nuclei with a dilution factor $\rho$, neutron absorption in the moderator and neutron scattering in the absorber may be neglected. Obtain an expression for the resonance-integral self-shielding factor $f_I = I(\rho)/I(\infty)$ in terms of the cross-section self-shielding factor $f_x = \langle \sigma(\rho) \rangle / \langle \sigma(\infty) \rangle$, unshielded cross section $\langle \sigma(\infty) \rangle$, and $\rho$.

**11.10** Perform the step indicated in the CPM-3 implementation of the resonance integral to obtain the expression for the resonance-shielded cross section $\langle \sigma \rangle$.

**11.11** A slab of thickness $H$ consists of a homogeneous mixture of a fissile material and hydrogen, which is described by one-group cross sections $\Sigma_a$ and $\nu\Sigma_f$, and a negligibly small inelastic scattering cross section. The elastic scattering cross section $\Sigma_s^A$ for the fuel material is isotropic in the laboratory, and the elastic cross section $\Sigma_s^H$ for hydrogen is linearly anisotropic in the laboratory, so that $\Sigma_s^H(\mu_0) = 0.5\Sigma_{s0}(1 + 0.5\mu_0)$, where $\mu_0 = \mathbf{\Omega} \cdot \mathbf{\Omega}'$ is the cosine of the scattering angle in the laboratory. Assuming that the neutron extrapolation distance is negligibly small compared with $H$, use the $B_1$ formulation to obtain an expression for the effective multiplication factor $k_{eff}$ for the slab.

**11.12** Consider a monoenergetic, isotropic source of neutrons distributed uniformly in a semi-infinite slab consisting of a purely scattering hydrogenous material. Assume that the scattering cross section is independent of neutron energy and that neutrons suffering more than one collision are distributed isotropically in the laboratory. Solve the $B_1$ equation for the scalar flux $\phi_0(u) = \phi(B, u)$ when the source neutrons are born at 10.0 MeV. You may want to break up the $B_n$ components of the flux and scattering integrals into collided and uncollided components, and follow these steps: (a) Use the $B_1$ equations for $\phi_0(u)$ and $\phi_1(u)$ to obtain the elastic scattering integrals $\rho_n(u) = \rho_n(B, u)$, for $n = 0$ and 1, for this idealized slowing down problem. (b) From the expression for the collided component $\rho_0^c(u)$ derived in part (a), obtain a differential equation connecting $\rho_0^c(u)$ and the collided scalar flux $\phi_0^c(u)$. Compare the result with the slowing down density $q_0(B, u)$ from Eq. (11.106). (c) From the results of parts (a) and (b), derive an ordinary differential equation for $\rho_0^c(u)$, which can be integrated in terms of the functions $A_{jk}$ of Eqs. (11.101). (d) Substituting the integrated solution for $\rho_0^c(u)$ from part (c) into the differential equation of part (b), and utilizing low-order expansions for $A_{jk}$, obtain finally the desired solution for $\phi_0(u)$. Plot the result, compare with the Fermi age approximation for large and small leakage cases as a function of $u$, and discuss the differences between the two models. The Fermi age equation may be obtained from Eqs. (9.101) and (9.109).

**11.13** Derive the matrix $H$ for a two-group representation of the $B_1$ formulation of Eq. (11.111) with the transport cross sections $\Sigma_{tr,1} = \gamma\Sigma_{t1} - \Sigma_{s1}$ and $\Sigma_{tr,2} = \gamma\Sigma_{t2} - \Sigma_{s2}$ with the $P_1$ components of the group transfer cross sections $\Sigma_{1\to2}$ and $\Sigma_{2\to1}$, and obtain the two-group diffusion coefficients $D_1$ and $D_2$ in terms of the two-group flux $\phi_1$ and $\phi_2$. Discuss the limiting form of the diffusion coefficient when the off-diagonal transfer cross sections $\Sigma_{1\to2}$ and $\Sigma_{2\to1}$ are ignored.

# 12

# Nuclear Fuel Cycle Analysis and Management

One of the important tasks for nuclear engineers working at either a reactor vendor or a utility company operating nuclear power plants is to perform fuel cycle analysis for safe and economical operation of the plants involved. Such analysis entails various operations involved in the preparation and utilization of nuclear fuel, and reprocessing and disposal of used nuclear fuel. For an engineer working at a reactor manufacturing company, e.g. Westinghouse, Areva, or General Electric, the task begins with a design analysis for the initial core and eventually continues with a reload analysis for subsequent fuel cycles. For a utility engineer, the task typically begins with a reload core analysis, often in consultation with reactor vendors or fuel suppliers, and may cover a broader scope extending to updating safe analysis reports as necessary.

With the emphasis on developing alternate fuel cycles under the Generation IV Roadmap [DOE02, GIF14] and a number of advanced and modular reactor designs actively pursued in recent years, there is an increasing need to study the basic methods of fuel cycle analysis as well as key aspects of fuel cycle management. In-depth understanding of issues involved with the optimization of fuel cycles is necessary for the continuing development of advanced fuel cycles that can satisfy the goals enunciated in the Generation IV Roadmap. With these perspectives in mind, this chapter covers a broad spectrum of topics related to fuel cycle analysis and management, extending eventually to the management and disposal of used nuclear fuel.

We begin with a brief overview of various steps in a typical nuclear fuel cycle in Section 12.1 and discuss key nuclide chains applicable to power production in Section 12.2. Section 12.3 introduces methods required for fuel transmutation calculations and simple approaches for estimating the depletion and burnup of fuel in a reactor core, together with the concept of equilibrium cycle and the associated mass balance. This is followed by a simplified fuel cycling technique in Section 12.4, which makes direct use of the reactivity variation as a function

*Nuclear Reactor Physics and Engineering*, Second Edition. John C. Lee.
© 2025 John Wiley & Sons, Inc. Published 2025 by John Wiley & Sons, Inc.
Companion website: www.wiley.com/go/LeeNRPE2

of fuel burnup. We then discuss in Section 12.5 the effect on reactor operation of the buildup of the fission product $^{135}$Xe, with a thermal absorption cross section of ~3 Mb. Section 12.6 presents various topics of relevance to incore fuel management, followed by characteristics of fission products and radioactive wastes in Section 12.7. We delve into a discussion on the disposition of used nuclear fuel in Section 12.8 and conclude in Section 12.9 with topics related to geological repository (GR) performance.

## 12.1 Nuclear Fuel Cycle Analysis

Fuel cycle tasks supporting the operation of a nuclear power plant encompass a broad spectrum of processes, starting from the mining of nuclear fuel and ending with the disposal of nuclear waste. The tasks are generally grouped into two major operations: incore and excore fuel managements. *Incore fuel management* refers to tasks directly related to the operation of the reactor, while *excore fuel management* covers the rest of the fuel cycle and provides support to incore fuel management. Incore fuel management includes the selection of fuel enrichment and burnable absorber configurations, the assembly loading pattern, and the cycle length. Various stages of a typical LWR cycle are summarized in Figure 12.1, where we assume a *once-through uranium cycle* resulting in the storage of the irradiated or used nuclear fuel (UNF) in a repository without any reprocessing. The excore fuel management begins with the mining of the uranium ore and milling of the ore into yellow cake $U_3O_8$, which is then converted into uranium hexafluoride $UF_6$. Since the natural fluorine comprises only one isotope, the gaseous compound $UF_6$ renders itself conveniently for the enrichment of $^{235}$U from natural uranium. The actual process of converting yellow cake to $UF_6$ involves three steps, (a) dissolution and purification into $UO_3$, (b) hydrofluorination into $UF_4$, and (c) fluorination into $UF_6$, which then goes through the enrichment process [Vex21].

For the current generation of LWRs, the enriched U contains 2.5 ~ 4.5 wt% of $^{235}$U, which is fabricated into $UO_2$ fuel rods and loaded into fuel assemblies. The series of operations beginning with the uranium ore mining and leading to the fabrication of fuel assemblies is known as the *front-end fuel cycle*. The *back-end fuel cycle* includes the storage of used fuel for the decay of short-lived radioactive material and associated cooling of used fuel in interim storage facilities, transportation, and eventual placement of the used fuel in a permanent repository, such as the Yucca Mountain geological repository that has been under development in the United States.

Together with the development of Generation IV nuclear energy systems, it has become apparent that we should seriously consider closing the fuel cycle by

**Figure 12.1** Flow chart for the once-through uranium cycle.

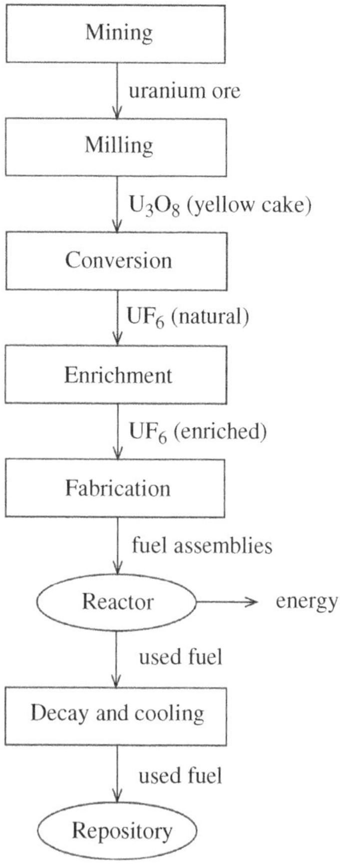


reprocessing and recycling the used fuel. A simple closed fuel cycle, illustrated in Figure 12.2, is what the French have implemented in many of their 58 PWR plants. This involves reprocessing used fuel and recycling Pu only in the mixed oxide (MOX) form of $(U\text{-}Pu)O_2$. The remainder of the *transuranic* (TRU) elements in the UNF, comprising Am, Np, and Cm, are known as the *minor actinides* (MAs). Note that we should refer to all the elements in the periodic table above uranium, including Pu, Am, Np, and Cm, as TRUs rather than actinides, since to be proper actinides should cover all the elements including actinium and above. All of the actinides, usually starting with Th and U, used as fuel in nuclear reactors and those produced during the fuel cycle are referred to as *heavy metal* (HM) elements. As illustrated in Figure 12.2, MAs separated in the reprocessing stream are vitrified together with other structural components, including end fittings and hulls of the fuel assemblies, and stored for eventual disposal in a permanent repository yet to be developed. In fact, some Pu that may not be

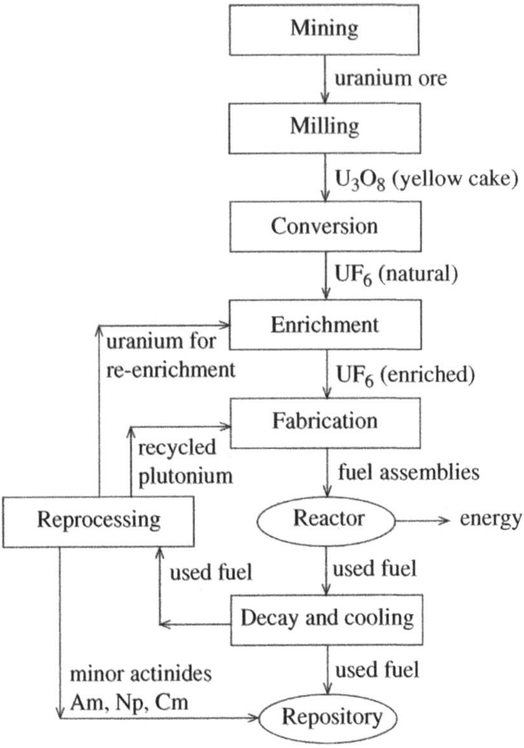


**Figure 12.2** Closed fuel cycle with plutonium recycling.

fully extracted for recycling is also sent to the repository. Currently, even with the recycling of plutonium only, the MOX fuel is used for one recycling only, and the once-used MOX fuel assemblies are stored for future reprocessing and disposal. The limitation placed on the MOX recycling is due to the degradation of the quality of Pu associated with an increased fraction of *fertile* Pu, i.e. $^{240}$Pu and $^{242}$Pu, in the once-recycled MOX fuel. The degraded Pu fuel in turn results in degraded performance of a PWR core, in particular reduction in the reactivity worth of control rods and decreased magnitude of the negative coefficients of reactivity, both of which have deleterious effects on safe operation of the reactor. Further discussion is provided on the actual reprocessing technology in Section 12.8.

One TRU recycling scheme that holds some promise as part of the Gen-IV Roadmap is a two-tier scheme [Van01] that includes synergistic use of LWRs and SFRs. Such a strategy is illustrated in a simplified structure in Figure 12.3. In this two-tier strategy, LWRs may be utilized for initial recycling and utilization of either the entire TRU or Pu only, to be followed by recycling the entire used fuel, including all the remaining MAs, in SFRs. This would allow efficient depletion of

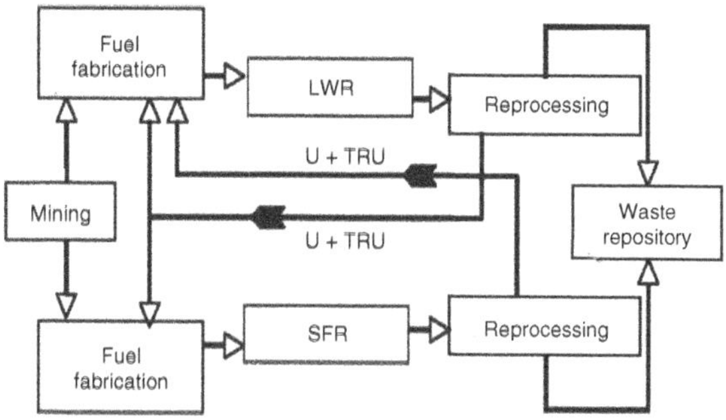


**Figure 12.3** Symbiotic two-tier fuel cycle involving both LWR and SFR transmuters.

the entire TRU inventory in the stockpiled used fuel. A more detailed illustration of a two-tier nuclear fuel cycle is presented in Section 12.8 as part of the fuel cycle discussion for advanced nuclear systems.

## 12.2 Nuclear Transmutation Formulation

### 12.2.1 Key Nuclide Chains for Nuclear Fuel Cycle

For all nuclear reactors in operation and those under development, the nuclear fuel cycle is based on one of three *fissile* materials, $^{235}$U, $^{233}$U, and $^{239}$Pu, and the associated isotopic chains. Figure 12.4 illustrates the U-Pu cycles, comprising the first chain built around $^{235}$U and the second around $^{239}$Pu. The Th-U cycle built around the fissile nuclide $^{233}$U is illustrated in Figure 12.5. Note that we explicitly represent in Figures 12.4 and 12.5 fission events only in fissile nuclides, although all fertile nuclides, including $^{234}$U, $^{236}$U, $^{238}$U, $^{240}$Pu, and $^{242}$Pu, undergo fission, especially upon capture of high-energy neutrons. Branching ratios for $\beta$-decay paths, half-lives, and $\sigma(n,f)$ and $\sigma(n,\gamma)$ for thermal neutrons are indicated in the transmutation diagrams.

The Th cycle ending with $^{235}$U in Figure 12.5 could in principle be connected to the uranium cycle beginning with $^{235}$U in Figure 12.4, but the amount of $^{235}$U produced in a thorium cycle is small and insignificant for the whole cycle. On the other hand, the two nuclide chains in Figure 12.4 are closely linked in LWR cycles, because slightly enriched uranium fuel with a $^{235}$U enrichment of $2.5 \sim 4.5$ wt% is typically used in both PWR and BWR plants, where substantial energy is produced from Pu isotopes. A more detailed nuclide chain for LWR fuel cycles is discussed in Section 12.2, together with the actual depletion schemes adopted for either lattice physics or global depletion calculations.

One important comparison for the three major fissile nuclides, $^{235}$U, $^{233}$U, and $^{239}$Pu, is their breeding potential. A *breeder* is by definition a reactor that produces more fissile material than it consumes and needs a minimum of two neutrons

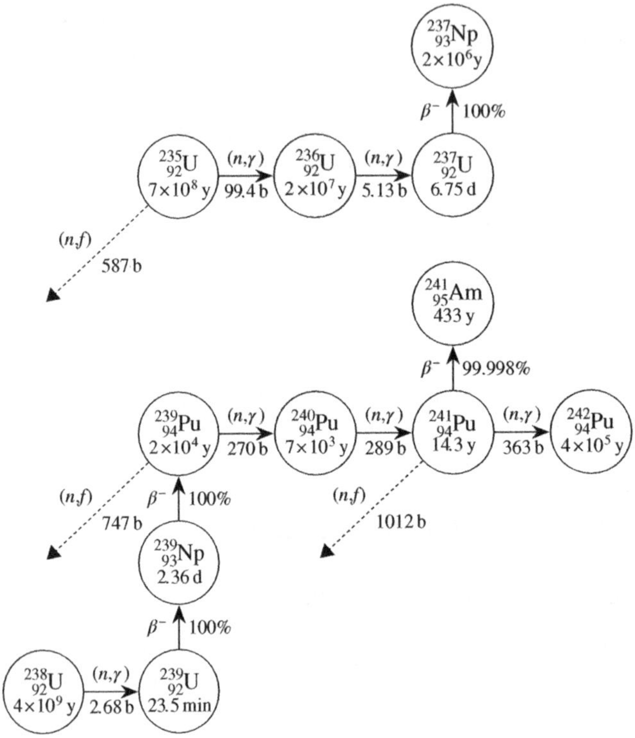


**Figure 12.4** Uranium-plutonium fuel cycle.

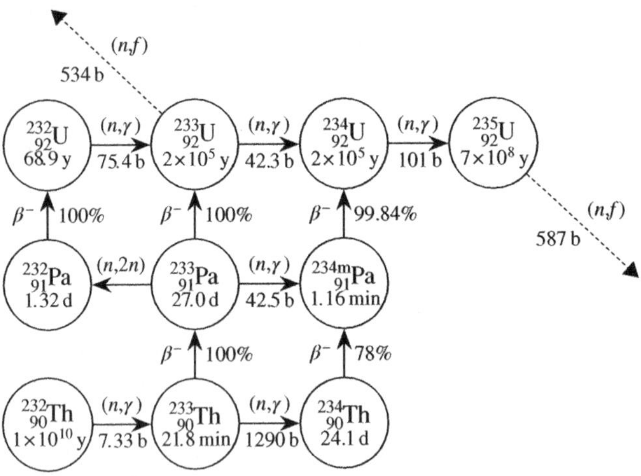


**Figure 12.5** Thorium-uranium fuel cycle.

produced for each neutron absorbed in a fissile nucleus. In terms of the simple four-factor formula discussed in Section 5.4, consider the parameter $\eta_{fissile}$ representing the number of fission neutrons released per absorption in fissile nuclide:

$$\eta_{fissile} = \frac{\nu\sigma_f}{\sigma_a} = \frac{\nu\sigma_f}{\sigma_f + \sigma_\gamma} = \frac{\nu}{1+\alpha}, \quad \alpha = \frac{\sigma_\gamma}{\sigma_f}. \tag{12.1}$$

The breeding potential of the three fissile nuclides requires that the parameter $\eta_{fissile}$ be at least equal to or greater than 2.0 so that one neutron released from fission can sustain the chain reaction, while the other neutron can be captured in a fertile nucleus, e.g. a $^{238}U$ nucleus, to produce one fissile nucleus of $^{239}Pu$. A comparison of $\eta_{fissile}$ for the three key fissile nuclides, $^{235}U$, $^{233}U$, and $^{239}Pu$, for use in both thermal and fast spectrum reactors is given in Table 12.1.

In an actual reactor, we need to account for the fraction $L$ of neutrons lost due to leakage and parasitic captures and for additional fissions induced by fast neutrons released from thermal fissions. Thus, in general, the *conversion ratio* (CR) is defined as the *ratio of fissile material produced to that destroyed* and may be written as

$$CR = \eta_{fissile} \cdot \varepsilon - 1 - L, \tag{12.2}$$

where $\varepsilon$ is the fast fission factor defined in Eq. (5.84). When the CR is greater than 1.0, it is called the *breeding ratio* (BR). Note from Table 12.1 that $^{233}U$ offers the potential for BR > 1.0 in both thermal and fast reactors, while $^{239}Pu$ offers a superior breeding potential for fast reactors. Indeed, significant developmental effort has been made over the past few decades in India for fast breeders utilizing the $^{232}Th$-$^{233}U$ cycle.

The $^{232}Th$-$^{233}U$ cycle also offers significant potential for enhanced transmutation of plutonium from the UNF because the cycle can preferentially consume [Sor06] Pu and produce $^{232}U$, which is highly radioactive, thereby providing proliferation resistance. The nuclide $^{232}U$ has a half-life of 68.9 yr with two strong $\gamma$-rays with 57.6 and 129 keV, respectively, so that 1.0 kg of $^{232}U$ produces approximately 20 kCi of radioactivity. Note, however, that $^{233}U$ is itself a fissile nuclide with a proliferation limit of 12 wt% compared with 20 wt% for $^{235}U$. To maximize the proliferation resistance, denaturing the cycle with a suitable amount of $^{238}U$ could be considered.

**Table 12.1** Comparison of parameter $\eta_{fissile} = \nu/(1+\alpha)$ for key fissile nuclides.

| Reactor spectrum | $^{235}U$ | $^{233}U$ | $^{239}Pu$ |
|---|---|---|---|
| Thermal | 2.0 | 2.2 | 1.9 |
| Fast | 2.1 | 2.3 | 2.6 |

We derive in Section 12.2.2 general isotopic balance equations involving the nuclide chains and discuss numerical schemes for solving the isotopic transmutation equations. This will naturally define the concept of fuel burnup and leads us to a simple method for estimating fuel burnup.

### 12.2.2 Fuel Transmutation Equation

We derive a general isotopic transmutation equation representing the rate of change of number density $N_i(\mathbf{r}, t)$ of nuclide $i$ at position $\mathbf{r}$ and time $t$ in a reactor core as a balance between the destruction and production rates of the nuclide

$$\frac{\partial N_i(\mathbf{r}, t)}{\partial t} = -\left[\sigma_{ai}(\mathbf{r}, t)\phi(\mathbf{r}, t) + \lambda_i\right] N_i(\mathbf{r}, t) + \sum_{j\neq i}\gamma_{ji}N_j(\mathbf{r}, t), \quad i = 1, \ldots, I, \tag{12.3}$$

for a transmutation chain comprising a total of $I$ nuclides. We account for the destruction rate of nuclide $i$, through neutron absorption with an effective one-group cross section $\sigma_{ai}(\mathbf{r}, t)$ and neutron flux $\phi(\mathbf{r}, t)$, combined with the radioactive decay rate with decay constant $\lambda_i$. Note that the absorption cross section $\sigma_{ai}(\mathbf{r}, t)$ is a function of position and time, because it is an effective cross section averaged with the energy-dependent neutron flux as a weighting function, and the flux itself is a function of position and time.

The production rate of nuclide $i$ is written via the parameter $\gamma_{ji}$ to account for three different channels that could produce nuclide $i$ from nuclide $j$ or effectively *transform* nuclide $j$ into nuclide $i$

$$\gamma_{ji} = \begin{cases} \lambda_j: & \text{decay of nuclide } j, \\ \sigma_{\gamma j}(\mathbf{r}, t)\phi(\mathbf{r}, t), \sigma_{sj}(\mathbf{r}, t)\phi(\mathbf{r}, t): & (n, \gamma), (n, 2n) \text{ reactions}, \\ y_i\sigma_{fj}\phi(\mathbf{r}, t): & \text{fission in nuclide } j \text{ yielding nuclide } i, \end{cases} \tag{12.4}$$

where the first channel represents simply the radioactive decay of nuclide $j$, resulting in nuclide $i$. For the second channel, we explicitly represent only the radiative capture reaction and the $(n, 2n)$ reaction in nuclide $j$, among a variety of possible reactions that could yield nuclide $i$. The third channel represents the case where a neutron absorption in fissionable nuclide $j$ induces fission with fission yield $y_i$ for nuclide $i$. Note here that nuclide $j$ could be either a fissile or fertile nuclide in general.

The energy-dependent flux, used in obtaining the effective cross section $\sigma_{ai}(\mathbf{r}, t)$ in Eq. (12.3), should reflect the thermal-hydraulic feedback effects associated with the space- and time-dependent variations in fuel temperature $T_F$ and moderator density $\rho_M$, as well as the fuel-number densities evolving as a function of fuel *burnup* or *exposure* $E$. Thus, we write the effective cross section explicitly as

$$\sigma_{ai}(\mathbf{r},t) = f[E, T_F(P), \rho_M(T_M)], E = E(\mathbf{r},t), P = P(\mathbf{r},t), T_M = T_M(\mathbf{r},t), \quad (12.5)$$

where $T_F$ is written as a function of power density $P(\mathbf{r},t)$ and $\rho_M$ is represented as a function of moderator temperature $T_M(\mathbf{r},t)$. The relationship between burnup $E(\mathbf{r},t)$ and power density $P(\mathbf{r},t)$ is discussed in Section 12.2.4. In general, for BWR cores, we need to determine the moderator density $\rho_M$ as a function of the steam content, known as the *void fraction*, as well as the temperature for the two-phase mixture of steam and liquid water. Furthermore, the simple one-group cross section of Eq. (12.5) should be extended in general to few-group cross sections for both PWR and BWR analysis. The parametric dependence of the absorption cross section represented in Eq. (12.5) forms the basis for the cross section table lookup illustrated in Figure 11.11 for the overall reactor-physics calculational procedure.

Since all LWR cores are loaded with solid fuel elements, the fuel is stationary, and hence we may represent the partial differential equation (12.3) as an ordinary differential equation (ODE) for each position $\mathbf{r}$ in terms of an isotopic number density vector $\mathbf{N}(t)$ of dimension $I$

$$\frac{d\mathbf{N}(t)}{dt} = A(t)\mathbf{N}(t), A = A\left[\phi(t), \sigma_{ai}(t)\right], \quad (12.6)$$

where the $(I \times I)$ matrix $A$ combines all of the reaction rate terms on the RHS from Eq. (12.3). Because of the time dependence of both cross section $\sigma_{ai}(\mathbf{r},t)$ and flux $\phi(\mathbf{r},t)$, $A$ will in general be a sensitive function of time $t$. Nonetheless, the ODE from Eq. (12.6) may be solved numerically for each position $\mathbf{r}$ and is called the *pointwise fuel transmutation equation*. For typical lattice physics calculations, Eq. (12.3) should in general represent $15 \sim 30$ heavy nuclides, i.e. fissionable nuclides, and $25 \sim 200$ fission product (FP) chains, requiring the solution of up to 230 coupled ODEs in the form of Eq. (12.6). In routine fuel cycle calculations, substantial simplifications and the corresponding improvement in the calculational efficiency may be achieved by lumping FP nuclides into one or two groups. Such a lumped FP representation for LWR fuel cycles is usually coupled with separate, specialized treatments for nuclide chains for fission products $^{135}$Xe and $^{149}$Sm, both of which have large thermal absorption cross sections with half-lives ranging from hours to days. The effects of this FP buildup and decay in LWR fuel management and reactor operation are discussed in Section 12.5.

### 12.2.3 Solution of Pointwise Transmutation Equation

Time integration of the pointwise fuel transmutation equation (12.6) may be obtained through a number of different approaches depending on the level of details represented for the time dependence of the transmutation matrix $A$. We discuss two representative methods in this section.

1. **Laplace transform approach**
   In a simple scheme where the time dependence of the elements of $A$ is suppressed, Eq. (12.6) may be integrated directly with a Laplace transform or a Bateman's formula applicable to a set of coupled ODEs. The inversion of the resulting matrix equation may, however, present some difficulty, because the magnitude of the elements of the matrix may differ by orders of magnitude. Thus, this simple method is used typically for simplified depletion equations involving relatively few nuclide chains, as is the case with the LEOPARD and CITATION codes [Bar63, Fow70].
2. **Matrix exponential method**
   The pointwise transmutation equation (12.6) may be integrated using a discrete time step $\Delta t$ directly to advance from time $t$ to $t + \Delta t$ in a fuel cycle

$$\mathbf{N}(t + \Delta t) = e^{A\Delta t}\mathbf{N}(t), \tag{12.7}$$

where the *matrix exponential* is evaluated through a Taylor series expansion

$$e^{A\Delta t} = I + A\Delta t + \frac{A^2\Delta t^2}{2!} + \frac{A^3\Delta t^3}{3!} + \dots . \tag{12.8}$$

The order of the expansion in Eq. (12.8) is determined empirically for the accuracy desired. This is the basis for the integration algorithms used in well-known fuel depletion codes, ORIGEN2.2, CINDER90, and REBUS [Rea18, Wil95, Top83].

In the lattice physics codes, CASMO, CPM, and PHOENIX-P [Ede92, Fer17, Jon87, Jon00, Wes02], a *predictor-corrector algorithm* is implemented to account for the flux dependence of matrix $A$ with relatively long time steps. In the predictor step, determine $A$ based on the flux and cross sections at the beginning of the time step at $t$:

$$\mathbf{N}_p(t + \Delta t) = \exp[A\{\phi(t), \sigma(t)\}\Delta t]\mathbf{N}(t). \tag{12.9}$$

Once a *predictor estimate* $\mathbf{N}_p(t + \Delta t)$ of the number density vector at the end of the time step $t + \Delta t$ is obtained, utilize the new number densities to estimate the flux and cross sections at $t + \Delta t$:

$$\mathbf{N}_p(t + \Delta t) \Rightarrow \{\phi^*(t + \Delta t), \sigma^*(t + \Delta t)\}. \tag{12.10}$$

In the *corrector step*, we use the *intermediate* flux $\phi^*(t + \Delta t)$ and cross sections $\sigma^*(t + \Delta t)$ to get an improved estimate of the number density vector:

$$\mathbf{N}_c(t + \Delta t) = \exp[A\{\phi^*(t + \Delta t), \sigma^*(t + \Delta t)\}\Delta t]\mathbf{N}(t). \tag{12.11}$$

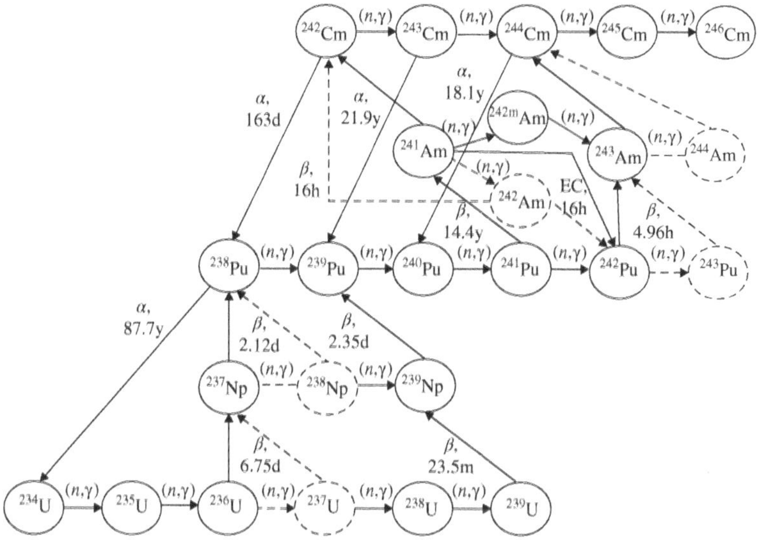


**Figure 12.6** Extended heavy-nuclide transmutation chain of the CASMO-3 code. *Source:* Lee and Du [Lee94]/Taylor & Francis Group.

The final number-density vector at $t + \Delta t$ is then obtained by an arithmetic average of the predictor and corrector estimates:

$$\mathbf{N}_{final}(t + \Delta t) = \frac{1}{2}\left[\mathbf{N}_p(t + \Delta t) + \mathbf{N}_c(t + \Delta t)\right]. \tag{12.12}$$

The cycle then proceeds to the next time step with $\mathbf{N}_{final}(t + \Delta t)$ replacing $\mathbf{N}(t)$ in the predictor step.

The actual HM nuclide chain used in the CASMO-3 code [Ede92], featuring 18 TRU nuclides and six U isotopes, is summarized in Figure 12.6, where dashed circles indicate some of the short-lived nuclides, including $^{237}$U, $^{238}$Np, $^{243}$Pu, and $^{244}$Am, that are not explicitly represented in actual depletion calculations. The dashed arrows indicate the true decay paths for the short-lived nuclides, while the dotted arrows indicate the approximate paths represented in CASMO-3 calculations. The transmutation chains are considerably more detailed and complex than those presented in Figure 12.4.

### 12.2.4 Fuel Transmutation Equation in Global MGD Calculation

For fuel cycle calculations with multigroup diffusion theory (MGD) codes, we need to account for the cross section dependence on fuel exposure $E$, as indicated in Eq. (12.5). For this purpose, the fuel burnup or exposure is defined by equating the energy production rate to power

$$\frac{\partial E(\mathbf{r},t)}{\partial t} = P(\mathbf{r},t), \tag{12.13}$$

with a suitable unit for the *power density* $P(\mathbf{r},t)$. For reactor physics and fuel cycle analysis, $P(\mathbf{r},t)$ is expressed in terms of the mass of heavy nuclides or actinides, usually referred to as HM. Thus, one convenient unit for $P(\mathbf{r},t)$ is MWt/kgHM, which then leads to the unit of MWd/kgHM for fuel burnup $E(\mathbf{r},t)$. Another popular burnup unit is MWd per metric tonne of HM, often written as MWD/MTHM. For the target discharge burnup of AP1000 fuel assemblies, for example, note that it is a bit more convenient to write as 49 MWd/kgHM rather than 49,000 MWD/MTHM.

To utilize the cross section table lookup illustrated in Figure 11.11, we can perform interpolations of either the microscopic or macroscopic cross sections. This leads to two different implementations of the isotopic depletion equations:

1. **Microscopic transmutation scheme**
In this approach, we basically follow the pointwise transmutation scheme discussed in Section 12.2.2 but account explicitly for each point in the reactor core. The *microscopic transmutation scheme* for an interval $t$ to $t + \Delta t$ proceeds

$$\begin{aligned} \{E(\mathbf{r},t),\ \mathbf{N}(\mathbf{r},t), \sigma(E, T_F, \rho_M)\} &\Rightarrow \{\phi(\mathbf{r},t), P(\mathbf{r},t)\} \\ &\Rightarrow \{E(\mathbf{r},t+\Delta t), \mathbf{N}(\mathbf{r},t+\Delta t)\}, \end{aligned} \tag{12.14}$$

where the nuclear number densities $\mathbf{N}(\mathbf{r},t+\Delta t)$ at the end of the time step $t + \Delta t$ are evaluated through the matrix exponential method from Eq. (12.7) or equivalent.

Consider a simple case involving the depletion of a single nuclide with number density $N_F(\mathbf{r},0)$ at the beginning-of-cycle (BOC) and time-independent absorption cross section $\sigma_a(\mathbf{r})$ to obtain the nuclear density at time $t$ during the cycle:

$$\frac{\partial N_F(\mathbf{r},t)}{\partial t} = -\sigma_a(\mathbf{r})\phi(\mathbf{r},t)N_F(\mathbf{r},t). \tag{12.15}$$

The simple balance equation is integrated

$$N_F(\mathbf{r},t) = N_F(\mathbf{r},0)\exp[-\sigma_a(\mathbf{r})\theta(\mathbf{r},t)], \tag{12.16}$$

with the *flux-time* or *fluence*

$$\theta(\mathbf{r},t) = \int_0^t \phi(\mathbf{r},t')dt'. \tag{12.17}$$

If the flux is assumed constant, i.e. $\phi(\mathbf{r},t) = \phi(\mathbf{r})$ over the interval $[0,t]$, then we obtain a simple expression $\theta(\mathbf{r},t) = \phi(\mathbf{r})t$ in units of [neutron $\cdot$ cm$^{-2}$]. Note also that the term *fluence* is often used to denote the time integral of the flux but with the neutron energy above a certain cutoff, e.g. 0.1 MeV, above which the degradation of the properties of structural materials due to neutron irradiation primarily occurs. Simple as the model is, Eqs. (12.15)–(12.17) may be used to obtain an approximate estimate of the consumption of the $^{235}U$ fuel inventory in an LWR cycle, as illustrated in Section 12.2.5.

The microscopic transmutation scheme from Eq. (12.14) has been adopted in the well-known PDQ code [Cad67] as well as the University of Michigan version of the 2DB code [Lit69], UM2DB. The SFR fuel cycle code, REBUS-3 [Top83], also performs fuel transmutation calculations via a multigroup, multi-nuclide microscopic transmutation algorithm with a set of burnup-independent microscopic cross sections. Alternate fuel cycle codes, including MONTEBURNS [Pos99] and MCNPX-CINDER90 [Hen04, Wil95], also employ the microscopic transmutation algorithm. In these codes, the flux calculations are performed by the Monte Carlo codes, MCNP6 [Goo12] and MCNPX, while the isotopic transmutation equations are solved via the ORIGEN2.2 code [Rea18] and CINDER90 code, respectively.

2. **Macroscopic transmutation scheme**
Many popular MGD codes, including SIMULATE [DiG95], ANC [Liu86], and UM2DB, use a simpler algorithm based on the macroscopic cross sections tabulated as a function of $E$, $T_F$, and $\rho_M$, as illustrated in Figure 11.11. In this approach, we perform time-wise explicit integration from Eq. (12.13) over time step $\Delta t$:

$$E(\mathbf{r}, t + \Delta t) = E(\mathbf{r}, t) + P(\mathbf{r}, t)\Delta t. \tag{12.18}$$

Averaging Eq. (12.18) over the core volume $V$ yields

$$\langle E(\mathbf{r}, t + \Delta t)\rangle_V = \langle E(\mathbf{r}, t)\rangle_V + \langle P(\mathbf{r}, t)\rangle_V \Delta t,$$

or equivalently a relationship between the core-average incremental burnup and core-average power density:

$$\langle \Delta E(\mathbf{r}, t)\rangle_V = \langle P(\mathbf{r}, t)\rangle_V \Delta t. \tag{12.19}$$

Substituting Eq. (12.19) into Eq. (12.18) provides

$$\begin{aligned} E(\mathbf{r}, t + \Delta t) &= E(\mathbf{r}, t) + \frac{P(\mathbf{r}, t)}{\langle P(\mathbf{r}, t)\rangle_V}\langle \Delta E(\mathbf{r}, t)\rangle_V \\ &= E(\mathbf{r}, t) + P_{rel}(\mathbf{r}, t)\langle \Delta E(\mathbf{r}, t)\rangle_V, \end{aligned} \tag{12.20}$$

where we recall the normalized or *relative power distribution* $P_{rel}(\mathbf{r}, t)$ introduced in Eq. (6.50). Equation (12.20) indicates that the average of $P_{rel}(\mathbf{r}, t)$ over core volume $V$ is, by definition, equal to unity, i.e. $\langle P_{rel}(\mathbf{r}, t)\rangle_V = 1.0$.

Once $P_{rel}(\mathbf{r}, t)$ is obtained from the converged flux distribution at time $t$, we may update the burnup distribution at every position $\mathbf{r}$ by simply incrementing $E(\mathbf{r}, t)$ by a product of $P_{rel}(\mathbf{r}, t)$ and the core-average burnup step $\langle \Delta E(\mathbf{r}, t)\rangle_V$

corresponding to the time step $\Delta t$. This is another important role of the relative power distribution $P_{rel}(\mathbf{r}, t)$ in MGD codes, in addition to its role in determining power-peaking factors as discussed in Section 6.5. In fact, in actual global MGD theory and lattice physics analysis, fuel depletion steps often are specified in terms of core-average burnup steps $\langle \Delta E(\mathbf{r}, t) \rangle_V$ in units of MWd/kgHM, rather than in terms of time steps in units of hours or days. In contrast to the microscopic transmutation scheme from Eq. (12.14), the *macroscopic transmutation algorithm* proceeds in a simple fashion

$$\{E(\mathbf{r}, t), \Sigma(E, T_F, \rho_M)\} \Rightarrow P(\mathbf{r}, t) \Rightarrow P_{rel}(\mathbf{r}, t) \Rightarrow E(\mathbf{r}, t + \Delta t), \tag{12.21}$$

without the need to determine the isotopic number densities $\mathbf{N}(\mathbf{r}, t)$. If the isotopic data are needed, they may be extracted from lattice physics calculations that generated the macroscopic cross-section table. Finally, for LWR fuel-cycle calculations, the first burnup step should be specified such that an equilibrium $^{135}$Xe concentration is properly built up over the first step, followed by longer burnup steps of either 1.0 or 2.0 MWd/kgHM.

### 12.2.5 Simple Model for Fuel Burnup Estimation

Having discussed the solution of the pointwise isotopic transmutation equation and the implementation of the equation in the microscopic and macroscopic depletion algorithms, we now present a simple model [Lee16] that provides a quick estimate of fuel burnup that may be attained in a given fuel cycle. We limit our discussion here to LWR fuel cycles, although similar formulations can be derived for SFR cycles.

We construct an energy balance expressed in terms of the discharge fuel burnup $E$, either for a batch of fuel assemblies or for the entire core, corresponding to the depletion of the HM inventory releasing fission energy of 200 MeV · fission$^{-1}$:

$$\begin{aligned} E\left(\frac{\text{MWd}}{\text{kgU}}\right) &= fima\left(\frac{\text{number of fissions}}{\text{initial U atom}}\right) \times \left(\frac{6.022 \times 10^{26}\ \text{atoms}}{\text{kg atom of U}}\right) \\ &\quad \times \left(\frac{1\ \text{kg atom}}{238\ \text{kg U}}\right) \times \left(\frac{200\ \text{MeV}}{\text{fission}}\right) \times \left(\frac{1\ \text{MWd}}{5.39 \times 10^{23}\ \text{MeV}}\right) \\ &= 939\, fima. \end{aligned} \tag{12.22}$$

Here, *fima* representing *fissions per initial metal atom*, e.g. per initial uranium atom, may be obtained in terms of *fissions per initial fissile atom (fifa)* and the $^{235}$U enrichment $e$

$$fima = fifa\left(\frac{\text{number of fissions}}{\text{initial fissile atom}}\right) \times e\left(\frac{\text{number of}\ ^{235}\text{U atoms}}{\text{U atom}}\right). \tag{12.23}$$

In turn, recast *fifa* as

$$fifa = \beta \left( \frac{\text{number of}\ ^{235}\text{U atoms fissioned}}{\text{initial}\ ^{235}\text{U atom}} \right) \times F \left( \frac{\text{total number of fissions}}{^{235}\text{U atom fissioned}} \right). \tag{12.24}$$

The *burnup fraction* $\beta$ of the fissile nuclide, in this case $^{235}U$, may finally be obtained using the simple depletion integral from Eq. (12.16) corresponding to irradiation time $T$

$$\beta = \frac{N^{25}(0) - N^{25}(T)}{N^{25}(0)} \frac{\sigma_f^{25}}{\sigma_a^{25}} = \frac{1 - e^{-\sigma_a^{25}\theta}}{1 + \alpha^{25}}, \theta = \int_0^T \phi(t)dt, \tag{12.25}$$

with the superscript 25 representing the last digits of atomic number 92 and mass number 235 for $^{235}U$, as adopted by Glenn Seaborg during the wartime discovery of Pu isotopes. Note that the fission-to-absorption ratio $\sigma_f^{25}/\sigma_a^{25}$ represents the fact that the actual depletion or burnup of a $^{235}U$ nucleus occurs only when it fissions, not merely when it captures a neutron; a radiative capture in $^{235}U$ results in $^{236}U$, as illustrated in Figure 12.4.

As a simple illustration, consider the proposed fuel cycle parameters for the AP1000 design. With a $^{235}U$ enrichment of 4.5 wt%, we estimate $\beta = 0.75$, $F = 1.6$, yielding $fifa = 1.20$ and $fima = 0.054$. This yields a discharge burnup $E = 51$ MWd/kgHM, close to the target discharge burnup of 49 MWd/kgHM.

## 12.3 Equilibrium Cycle and Mass Balance

An *equilibrium cycle* is defined as a fuel cycle that has reached an asymptotic equilibrium configuration such that each cycle is the same in all its characteristics as one before and one after. In actual power plant operation, such an idealized configuration is never reached primarily because of the demand of a particular power grid that does not allow precise outage and startup times for any plant. In addition, during a power plant shutdown and refueling outage, unanticipated events, e.g. leaky fuel assemblies that need replacement, may dictate departure from a planned fuel loading pattern and fuel management scheme. Nonetheless, for the purpose of planning and evaluating various fuel cycle options, it is quite useful to construct and study equilibrium cycles for any operating plants or new plant designs under development.

In addition, equilibrium cycle configurations offer a meaningful comparison between reactor designs of significantly different characteristics, e.g. between an LWR and a SFR. This is because the fuel-cycle characteristics during an approach to the equilibrium cycle, via *transition cycles*, may in general be significantly different for different reactor types. We begin in Section 12.3.1 with a brief discussion of nuclide balance statements that allow for the determination of an equilibrium cycle and present a corresponding mass balance relationship in Section 12.3.2.

### 12.3.1 Nuclide Balance Statement

With the *cycle length* formally defined as the fuel irradiation time $T$ between two consecutive refuelings expressed in a suitable unit, e.g. MWd/kgHM, we can use the pointwise depletion equation (12.6) to represent the core-average fuel burnup over a cycle. This is accomplished by rewriting Eq. (12.7) to cover the cycle length $T$

$$\mathbf{N}(T) = \mathbf{N}_d = \exp\left(\int_0^T A(t)dt\right)\mathbf{N}(0) \equiv B\mathbf{N}(0) = B\mathbf{N}_c, \tag{12.26}$$

with the introduction of *transmutation matrix* $B$ to connect the BOC nuclide vector $\mathbf{N}(0)$ to the end of cycle (EOC) nuclide vector $\mathbf{N}(T)$. Note also that $\mathbf{N}(0)$ represents the *charge nuclide vector* $\mathbf{N}_c$, while $\mathbf{N}(T)$ is equal to the *discharge vector* $\mathbf{N}_d$. For SFR fuel-cycle analysis, where we may assume constant power operation with a set of burnup-independent multigroup microscopic cross sections, we can evaluate $B$ with only a few burnup steps with nearly burnup-independent matrix $A$. For LWR cycle calculations, due to significant changes in the flux spectrum and two-group microscopic and macroscopic cross sections typically used, matrix $A$ is a sensitive function of irradiation time $t$ or fuel burnup.

For the determination of equilibrium cycle configuration, especially with TRU recycling where the equilibrium cycle feed composition has to be determined, an iterative solution of transmutation matrix $B$ is necessary [Sor08]. Such an iterative technique for the determination of an equilibrium cycle configuration is illustrated in Figure 12.7. An initial estimate of the charge vector $\mathbf{N}_c$ yields an initial estimate of the discharge vector $\mathbf{N}_d$ via Eq. (12.26), which is then coupled to the excore cycle entailing the reprocessing operation $R$ to provide a new estimate of $\mathbf{N}_c$. To reduce the computational burden associated with the iteration, an algorithm was developed to generate a simplified matrix representation for $B$, which may be assumed constant to arrive at an approximate estimate of equilibrium $\mathbf{N}_c$ and $\mathbf{N}_d$, together with the feed nuclide vector $\mathbf{N}_b$, through the combined incore and excore loop illustrated in Figure 12.7. Another inner iteration loop may be established, where the flux spectrum and microscopic reaction rates are updated for $\sigma\phi$ and hence the transmutation matrix $B$, iteratively, until the overall iteration converges.

The evolving nuclide vectors for the combined incore and excore fuel-cycle processes are illustrated in Figure 12.8 for a closed fuel cycle, where we consider a Pu feed vector $\mathbf{N}_f$ combined with natural U feed $\mathbf{N}_b$ to be mixed with the recycled

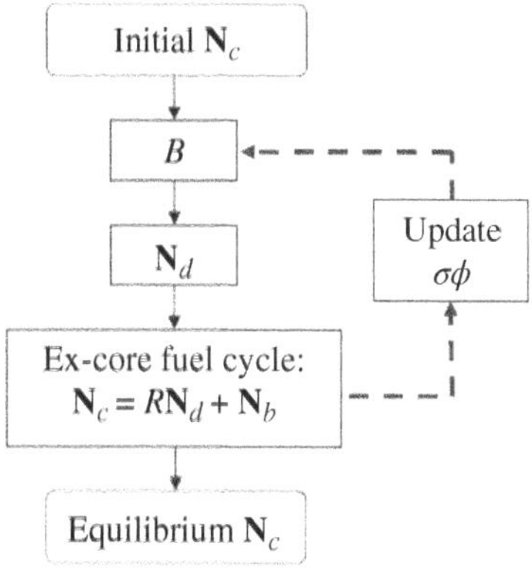


**Figure 12.7** Iterative search for an equilibrium cycle.

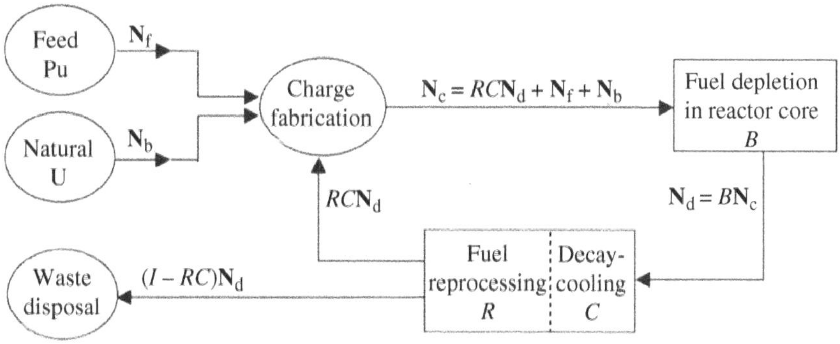


**Figure 12.8** Nuclide vectors for incore and excore fuel cycle processes.

fuel vector $RC\mathbf{N}_d$ making up the total charge $\mathbf{N}_c$ for fuel fabrication. Note also that the discharge fuel goes through the decay and cooling operation $C$, followed by reprocessing $R$, yielding the nuclide vector $RC\mathbf{N}_d$ for fuel fabrication. The final process in the combined incore-excore fuel cycle is the waste disposal process receiving the remnant $(I - RC)\mathbf{N}_d$ of the discharge nuclide vector at the conclusion of the decay-cooling and reprocessing operations. The feed stream represented in Figure 12.8 by the Pu and natural U feeds may obviously be altered for different fuel feeds, e.g. TRU from the UNF.

### 12.3.2 Material Flow Sheet

For the purpose of determining the overall material or mass balance in a closed fuel cycle, it is often convenient to recast the nuclide balance relationships considered in Section 12.3.1 into a *material flow sheet* for fuel material or heavy metal illustrated in Figure 12.9. The burnup mass flow rate $B$ (Mg/cycle) represents the

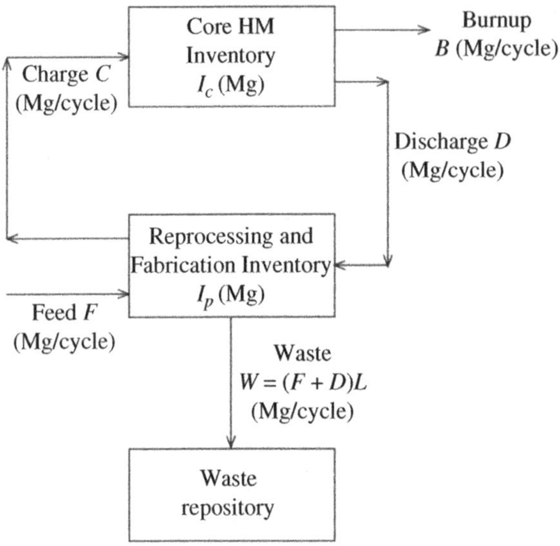


**Figure 12.9** Heavy-metal material flow sheet for a closed fuel cycle.

HM inventory consumed per cycle and hence is equal to the FPs generated in a power production cycle.

Assume that the partitioning and reprocessing of the discharge fuel entails unrecoverable mass flow represented by a loss fraction $L$. This is equivalent to the loss fraction $(I - RC)$ associated with the decay-cooling and reprocessing processes in Figure 12.8. Note also the total inventory $I$ (Mg) of the combined incore-excore system:

$$I = I_c + I_p. \tag{12.27}$$

A simple mass balance for the feed and discharge flow streams yields

$$C = D + B = (F + D)(1 - L), \tag{12.28}$$

where the recoverable fraction $(1 - L)$ of the combined feed-discharge stream $(F + D)$ into the reprocessing and fabrication operations makes up the charge flow $C$, sometimes called the *makeup fuel flow*.

Note that, unlike the vector-matrix notations used for the nuclide vector **N** and associated operations in Section 12.3.1, simple scalar notations are used in this section. This is possible because we are dealing with the total HM mass inventories and flow rates, without any regard for the composition of the HM inventories involved. Note also that we have singled out the FP material flow rate $B$ in Figure 12.9, while the nuclide vector representations in Figures 12.7 and 12.8 do not explicit separate out this component of the total mass flow. At the same time, it is necessary to eventually account for the disposal of the bulk of the FPs, represented by $B$; some of the volatile FPs will be lost during either the reactor operation or partitioning and reprocessing process. Thus, the accumulation rate $W$ at the waste repository in Figure 12.9 represents only the HM waste stream, not the total waste stream that could include volatile FPs.

### 12.3.3 REBUS Equilibrium Inventory Calculation

The REBUS fuel-cycle code [Hos78, Top83] is built to perform equilibrium fuel cycle calculations for SFRs, with the DIF3D code [Der84] serving as its MGD equation solver. Because of the need to represent spectral-spatial coupling effects accurately in fast-spectrum reactors, the REBUS-DIF3D calculations are typically performed with finer group structures than the fast-thermal two-group structure of the Westinghouse APA system or the three-group structure sometimes used for coupled nuclear-thermal-hydraulic calculations for BWRs. Thus, in the $MC^2$-REBUS [Lee17b, Top83] setup at Michigan, a set of 33-group burnup-independent microscopic cross sections is used for SFR design and fuel cycle analysis.

The overall nuclide balance involving both the incore and excore processes summarized in Figure 12.7 closely follows the REBUS equilibrium cycling methodology. As part of the methodology, the code performs an iterative search for the fraction of the primary HM fuel, e.g. Pu in (Pu-U)Zr metallic fuel, or for the fissile enrichment in Pu itself. In a typical SFR cycle, the fuel isotopics do not undergo large changes, and the REBUS cycle calculations may be performed with only three or four burnup steps, unlike a dozen or more steps required for MGD depletion calculations for LWR systems. Using this feature, the enrichment search in REBUS is performed through a simple gradient algorithm to yield a desired $k = k_{eff}$ at EOC. For example, given two calculations $k_0$ and $k_1$ corresponding to enrichments $e_0$ and $e_1$, we determine the gradient $\Delta e/\Delta k = (e_1 - e_0)/(k_1 - k_0)$ and obtain the next estimate for enrichment $e_2$ for the desired $k = k_2$:

$$e_2 = e_1 + \frac{\Delta e}{\Delta k}(k_2 - k_1). \tag{12.29}$$

A similar search procedure would also provide a means to obtain a desired cycle length.

Another important feature of the REBUS equilibrium methodology is to effectively represent the fuel shuffling and reloading in an SFR core based on a cluster of hexagonal fuel assemblies. In an $N$-batch fuel management scheme, we introduce the notion of a macro-cycle consisting of $N$ micro-cycles, where at the end of each micro-cycle, we discharge one out of every $N$ assemblies, each having resided in the core for $N$ cycles, and load fresh fuel assemblies in their locations so that each micro-cycle has a unique loading pattern without allowing for shuffling of fuel assemblies. To clarify the macro-cycle/micro-cycle relationship, we introduce the concept of an assembly cluster, with each cluster containing $N$ assemblies

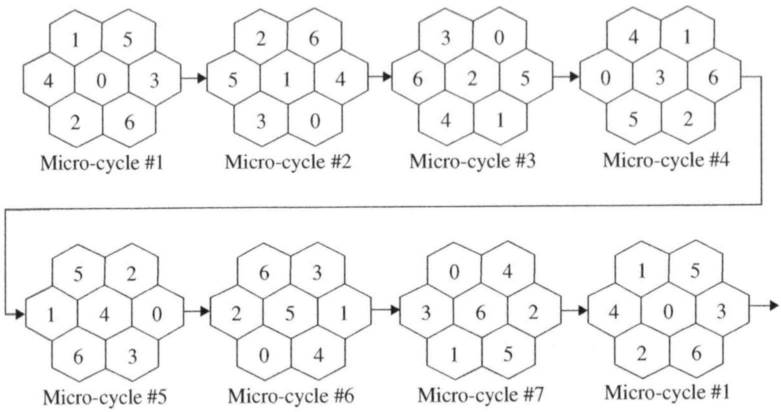


**Figure 12.10** One macro-cycle consisting of seven micro-cycles for a cluster of seven driver assemblies in a honeycomb configuration. The number within each assembly represents the number of previous cycles that the assembly has resided in that position. For example, "0" represents a fresh assembly, whereas "6" represents an assembly that has been irradiated for six micro-cycles.

with a unique incore residence time or color. In Figure 12.10, we show the color evolution of a typical seven-batch cluster during a macro-cycle consisting of seven micro-cycles, i.e. for $N = 7$. For a seven-batch core, in the REBUS equilibrium cycle mode, the macro-cycle becomes fixed and the same micro-cycle repeats seven times, with no explicit representation of distinct loading patterns for the evolving micro-cycles.

For equilibrium cycling, the code calculates the time-averaged macro-cycle equilibrium condition of the reactor, and this extends to the equilibrium composition of each assembly in the reactor. Thus, in our seven-batch fuel management scheme, each assembly at BOC will contain one-seventh fresh fuel, one-seventh fuel that has burned through one micro-cycle, one-seventh that has burned through two micro-cycles, and so on. The equilibrium cycle configuration is thus equivalent to homogenizing the fuel in the seven-batch cluster illustrated in Figure 12.10 such that each assembly in the cluster contains the same fuel composition at BOC. The equilibrium cycle provides fairly accurate estimates of the mass depletion and reactivity information, but may suffer somewhat in the calculation of the power distribution.

In an $N$-batch equilibrium cycling algorithm, we determine the BOC HM inventory of the $n$th batch, i.e. $1/N$ of an assembly that has gone through $n$ irradiation cycles

$$M_n = M_{feed} - (n - 1)\Delta M, \; n = 1, \dots, N, \tag{12.30}$$

where $M_{feed}$ is the inventory of the fresh batch and $\Delta M$ is the inventory decrease per cycle. With Eq. (12.30) representing the inventory for the $1/N$ of the assembly via the approximate *equilibrium cycling method*, we then determine the total BOC inventory $M_{total}$ of an assembly comprising $N$ fuel volumes of inventories $M_n, n = 1, \ldots, N$:

$$M_{total} = \sum_{n=1}^{N} M_n = N\left[M_{feed} - \frac{N-1}{2}\Delta M\right] \equiv N\langle M\rangle. \tag{12.31}$$

For $N = 7$, the *average batch inventory* is given by $\langle M\rangle = M_{feed} - 3\Delta M$, reflecting a decrease from the feed inventory $M_{feed}$ due to an average residence time of three cycles for the six irradiated batches.

## 12.4 Simplified Cycling Model

For the evaluation of alternate design concepts and associated fuel economics, a simple cycling model is often used at utility companies and reactor vendors including General Electric Company and Westinghouse Electric Corporation. It was called the *instant cycling method* at Westinghouse Electric Corporation in the early 1970s, discussed as part of fuel cycle models [Gra79, Ben81], and later published as the *reactivity-based cycling* (RBC) model [Dri90]. The basis of the model is to approximate global MGD fuel depletion calculations with assembly-level lattice physics calculations, with the assumption that the infinite multiplication factor $k_\infty$ of a fuel assembly is a linear function of fuel burnup. This assumption is applicable to the first order in common PWR designs, after an equilibrium $^{135}Xe$ concentration is built up, and provided we do not include the reactivity effects of burnable absorbers, which is discussed in Section 12.6.

### 12.4.1 Reactivity-Based Instant Cycling Method

Consider $N$ enrichment zones in an equilibrium PWR core with a cycle length of $\theta$ [MWd/kgHM], illustrated for the case of $N = 3$ in Figure 12.11. Assume that the effective multiplication factor $k = k_{eff}$ can be approximated as a linear function of the region-average fuel burnup for an $N$-region core and that each fuel region produces $1/N$ of the core power. For an equilibrium cycle with cycle length $\theta$, express the criticality condition at EOC in terms of $k$ averaged over $N$ regions or equivalently in terms of $k$ for the core-average EOC burnup $E^*$:

$$\frac{1}{N}\sum_{n=1}^{N} k(n\theta) = k(E^*) = 1.0, \tag{12.32}$$

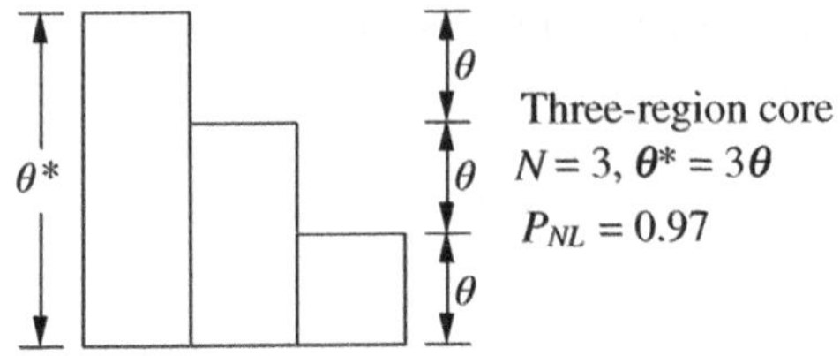


**Figure 12.11** Simplified three-region equilibrium cycle modeling.

with

$$E^* = \frac{1}{N}\sum_{n=1}^{N} n\theta = \frac{N+1}{2}\theta. \tag{12.33}$$

Since the discharge fuel burnup $\theta^* = N\theta$ in an equilibrium cycle, Eq. (12.32) is rewritten in terms of the infinite multiplication factor $k_\infty$

$$k_\infty\left(\frac{N+1}{2N}\theta^*\right) = \frac{1}{P_{NL}}, \tag{12.34}$$

for a core with non leakage probability $P_{NL}$. For the three-zone core, i.e. $N = 3$, illustrated in Figure 12.11, with a leakage probability of 3% or $P_{NL} = 0.97$, Eq. (12.34) suggests that we determine the EOC core-average EOC burnup $E^* = 2\theta$, when $k_\infty = 1.03$, based on assembly-level depletion calculations with a lattice physics code, e.g. the PHOENIX-P code [Wes02] for PWR fuel cycle analysis. This then yields the cycle length $\theta$ and the discharge burnup of the batch that has gone through three cycles of incore irradiation. The fuel isotopics of discharged fuel batches are determined from the lattice physics calculation just performed. Alternately, an equilibrium feed enrichment of $^{235}U$ is determined through Eq. (12.34) for a desired discharge burnup $\theta^*$. Accuracy of the reactivity-based cycling method is discussed [Sor08] via actual 2-D fuel cycle calculations for a realistic PWR configuration.

### 12.4.2 Application of Instant Cycling Method

In a typical three-zone PWR fuel-management scheme, for a fresh core, i.e. the first fuel cycle loaded with batches of fresh fuel assemblies but with different $^{235}U$ enrichments, the highest enrichment is usually chosen as the desired equilibrium feed enrichment, yielding an equilibrium cycle length $\theta$, and the cycle length of the first cycle is determined as $\theta_1 = 3\theta/2$. The basic objective of the instant cycling (IC) method is to achieve for each cycle the core average EOC burnup $E^*$ from Eq. (12.33) with either a uniform or nonuniform first core and to approach the equilibrium cycle discharge burnup of $\theta^* = N\theta$. This approach to an equilibrium cycle configuration is illustrated through two applications for a three-region core with an *out-in* fuel shuffling scheme.

The cycle length $\theta_n$ for cycle $n$, with the BOC burnup distribution $B_{ni}$ for batch number $i = 1, \dots, N$, is obtained by setting the average of the EOC burnup distribution $E_{ni}$ to the target cycle burnup $E^*$ from Eq. (12.33)

$$\begin{aligned}\langle E_{ni}\rangle &= \frac{1}{N}\sum_{i=1}^{N} E_{ni} = \frac{1}{N}\sum_{i=1}^{N}(B_{ni} + \theta_n) \\ &= \frac{1}{N}\left(\sum_{i=1}^{N-1} E_{n-1,i+1} + \theta_n N\right) = \frac{N+1}{2}\theta,\end{aligned} \tag{12.35}$$

which simplifies to

$$\langle E_{2i}\rangle_i = \frac{1}{N}[\theta_1(N-1) + \theta_2 N], \quad \theta_2 = \frac{1}{2}\frac{N+1}{N}\theta,$$

with $B_{11} = 0$ and $\theta_1 = E^*$ for a uniform fresh first core. The discharge burnup for cycle $n$, with batch $i = 1$ to be discharged, is obtained as

$$\theta_n^* = E_{ni} = B_{n1} + \theta_n = E_{n-1,2} + \theta_n, n = 2, \dots, \tag{12.36}$$

again with $B_{11} = 0$ for a fresh first core and $\theta_2^* = \frac{N}{2}\left(\frac{N+1}{N}\right)^2\theta$. Continuing the process provides

$$\theta_n = \frac{1}{2}\left(\frac{N+1}{N}\right)^{n-1}\theta \quad \text{and} \quad \theta_n^* = \frac{N}{2}\left(\frac{N+1}{N}\right)^n\theta, \quad n = 2, \dots, N. \tag{12.37}$$

The peak discharge burnup is achieved at cycle $n = N$

$$\theta_N^* = \frac{N}{2}\left(\frac{N+1}{N}\right)^N\theta,$$

which is the limiting design parameter typically imposed by the allowable fuel cladding fluence. For subsequent cycles $n > N$, when the first batches of fresh fuel elements have all been discharged, the cycle length $\theta_n$ and discharge burnup $\theta_n^*$ are determined by Eqs. (12.35) and (12.36).

**Example 12.1** Set up a systemic approach for multiple cycles approaching an equilibrium configuration starting with a uniform first cycle and equilibrium discharge burnup $\theta$.

A six-cycle evolution is summarized in Table 12.2 beginning with a three-batch core with BOC burnup $B_{1i} = 0.0, i = 1, 2, 3$, indicating fresh fuel for all three batches. At the EOC $n = 1$, with cycle length $\theta_1 = 2\theta$, all three batches accumulate the same burnup of $E_{1i} = 2\theta$, and batch 1 fuel is discharged with discharge burnup $\theta_1^* = 2\theta$, as indicated by Eq. (12.36). For the first reload cycle, $n = 2$, an *out-in* scheme is adopted, moving the batch 2 fuel to region 1 and batch 3 fuel to region 2, with a fresh batch loaded into region 3. The cycle length $\theta_2 = 2\theta/3$ will be much shorter than $\theta_1 = 2\theta$ due to the reactivity requirement that the EOC core average burnup $\langle E_{ni} \rangle_i = (1/3) \sum_1^3 E_{ni} = 2\theta$ for $n = 2$. The region 1 discharge batch for cycle 2 has $\theta_2^* = 2.67\theta$. Cycle $n = 3$, with the same out-in scheme, allows a slightly longer cycle length $\theta_2 = 0.89\theta$ and significantly larger discharge burnup $\theta_3^* = 3.56\theta$. At the end of the third cycle, all fuel batches loaded in the first cycle are discharged, and the evolution toward an equilibrium cycle continues with $\theta_4 = 1.18\theta$ and $\theta_4^* = 2.74\theta$. By the EOC $n = 6$, both the cycle length $\theta_6 = 0.99\theta$ and the discharge burnup $\theta_6^* = 3.09\theta$ indicate that the cycle is fairly close to the desired equilibrium cycle with $\theta_n = \theta$ and $\theta_n^* = 3.0\theta$. The feed enrichment may be specified according to the desired equilibrium cycle length $\theta$, typically for a 12-month or 18-month fuel reloading and outage schedule. The evolution path toward an equilibrium cycle configuration with an out-in scheme indicates, however, significant fluctuations in the cycle length $\theta_n$, suggesting a non-uniform fuel loading to start the fuel cycle in a new reactor core.

**Example 12.2** Reconstruct Table 12.2 with a non-uniform enrichment distribution for the three batches starting with the first cycle.

Batches 1 and 2, loaded in the inner regions of the core, are now assigned lower enrichments than batch 3 so that the cycle-to-cycle fluctuations in the cycle length $\theta_n$ may be reduced together with the average enrichment and associated fuel cycle cost for the first core. The enrichment deficit in batches 1 and 2 relative to that of batch 3 is often referred to as a *pseudo burnup*, which is added to the actual burnup for the first two cycles in Table 12.3. For this example, it is assumed that the cycle length $\theta_1 = 1.5\theta$ for the first cycle for $N = 3$ and the core average EOC burnup $\langle E_{ni} \rangle = E^* = (N+1)\theta/2$ including the pseudo burnup. The progression of the fuel cycles with the out-in scheme is similar to that of the uniform first core in Example 12.1, with the observation that the *effective discharge burnups* $\theta_1^* = 2.5\theta$ and $\theta_2^* = 2.83\theta$, including the pseudo burnup contributions, are larger than those for the uniform core. It should of course be noted that *actual discharge burnup* values for the first two cycles are shorter than those for the uniform first core, but with substantial savings in the enrichment cost. It may be noted that the phrase pseudo burnup is also used often to classify few-group cross section libraries in global diffusion-depletion calculations.

In this non-uniform refueling scheme, the discharge burnup $\theta_1^*$ of the first, lowest-enrichment batch will be half that of the discharge burnup $3\theta$ of the equilibrium cycle, but the cycle length $\theta_1$ itself is 50% larger than the equilibrium cycle length $\theta$. With this general fuel-management practice in mind, one may iteratively determine first the enrichment for the mid-enrichment zone, corresponding to batch 2, for a three-enrichment core. For this purpose, cycle length $\theta_1 = 3\theta/2$ of the first cycle may be determined via Eq. (12.34) to calculate the

**Table 12.2** Evolution of a three-batch core beginning with a uniform first cycle with equilibrium cycle length $\theta$.

| Cycle number $n$ | Batch number $i$ | BOC burnup $B_{ni}$ | EOC burnup $E_{ni}$ | Cycle length $\theta_n$ | Discharge burnup $\theta_n^*$ |
|---|---|---|---|---|---|
| 1 | 1 | 0 | $2\theta$ | $2\theta$ | $2\theta$ |
| 1 | 2 | 0 | $2\theta$ | | |
| 1 | 3 | 0 | $2\theta$ | | |
| 2 | 1 | $2\theta$ | $2.67\theta$ | $0.67\theta$ | $2.67\theta$ |
| 2 | 2 | $2\theta$ | $2.67\theta$ | | |
| 2 | 3 | 0 | $0.67\theta$ | | |
| 3 | 1 | $2.67\theta$ | $3.56\theta$ | $0.89\theta$ | $3.56\theta$ |
| 3 | 2 | $0.67\theta$ | $1.56\theta$ | | |
| 3 | 3 | 0 | $0.89\theta$ | | |
| 4 | 1 | $1.56\theta$ | $2.74\theta$ | $1.18\theta$ | $2.74\theta$ |
| 4 | 2 | $0.89\theta$ | $2.07\theta$ | | |
| 4 | 3 | 0 | $1.18\theta$ | | |
| 5 | 1 | $2.07\theta$ | $2.99\theta$ | $0.92\theta$ | $2.99\theta$ |
| 5 | 2 | $1.18\theta$ | $2.10\theta$ | | |
| 5 | 3 | 0 | $0.92\theta$ | | |
| 6 | 1 | $2.10\theta$ | $3.09\theta$ | $0.99\theta$ | $3.09\theta$ |
| 6 | 2 | $0.92\theta$ | $1.91\theta$ | | |
| 6 | 3 | 0 | $0.99\theta$ | | |

core-average $k_\infty$

$$\langle k_\infty(\theta_1)\rangle = \frac{1}{3}[k_\infty^1(\theta_1) + k_\infty^2(\theta_1) + k_\infty^3(\theta_1)] \simeq k_\infty^2(\theta_1) = \frac{1}{P_{NL}}, \tag{12.38}$$

where $k_\infty^i$ corresponds to batch $i = 1, 2, 3$. This relationship is, of course, strictly valid only if $k_\infty^i$ is a linear function of fuel burnup, but provides a first-order estimate of the desired fuel composition that could yield an equilibrium discharge burnup $\theta^* = 3\theta$. One may then iteratively determine via Eq. (12.34) the enrichment for the highest-enrichment zone so that the cycle length is equal to $3\theta$, the discharge burnup of the equilibrium cycle. The enrichment for the lowest-enrichment zone may likewise be determined to yield cycle length $\theta_1 = 3\theta/2$ for the first cycle. In a

**Table 12.3** Evolution of a three-batch core beginning with a non-uniform first cycle approaching equilibrium cycle length $\theta$.

| Cycle number $n$ | Batch number $i$ | BOC burnup (actual + pseudo) $B_{ni}$ | EOC burnup (actual + pseudo) $E_{ni}$ | Cycle length $\theta_n$ | Discharge burnup (actual + pseudo) $\theta_n^*$ |
|---|---|---|---|---|---|
| 1 | 1 | $0.0\theta + 1.0\theta$ | $1.5\theta + 1.0\theta$ | $1.5\theta$ | $1.5\theta + 1.0\theta$ |
| 1 | 2 | $0.0\theta + 0.5\theta$ | $1.5\theta + 0.5\theta$ | | |
| 1 | 3 | 0.0 | $1.5\theta + 0.0\theta$ | | |
| 2 | 1 | $1.5\theta + 0.5\theta$ | $2.33\theta + 0.5\theta$ | $0.83\theta$ | $2.33\theta + 0.5\theta$ |
| 2 | 2 | $1.5\theta + 0.0\theta$ | $2.33\theta + 0.0\theta$ | | |
| 2 | 3 | 0.0 | $0.83\theta + 0.0\theta$ | | |
| 3 | 1 | $2.33\theta$ | $3.28\theta$ | $0.95\theta$ | $3.28\theta$ |
| 3 | 2 | $0.83\theta$ | $1.78\theta$ | | |
| 3 | 3 | 0 | $0.95\theta$ | | |
| 4 | 1 | $1.78\theta$ | $2.87\theta$ | $1.09\theta$ | $2.87\theta$ |
| 4 | 2 | $0.95\theta$ | $2.04\theta$ | | |
| 4 | 3 | 0 | $1.09\theta$ | | |
| 5 | 1 | $2.04\theta$ | $3.00\theta$ | $0.96\theta$ | $3.00\theta$ |
| 5 | 2 | $1.09\theta$ | $2.05\theta$ | | |
| 5 | 3 | 0 | $0.96\theta$ | | |
| 6 | 1 | $2.05\theta$ | $3.05\theta$ | $1.00\theta$ | $3.05\theta$ |
| 6 | 2 | $0.96\theta$ | $1.96\theta$ | | |
| 6 | 3 | 0 | $1.00\theta$ | | |

All burnup values are actual for cycle 3 and beyond.

typical approach to an equilibrium cycle, it takes three to six transition cycles of varying cycle lengths before a near-equilibrium cycle is attained. Thus, the first task of a core physics design engineer entails finding a satisfactory fuel design and fuel assembly loading pattern that will yield the desired cycle length $\theta_1 = 3\theta/2$ for the first cycle.

For the AP1000 design [Wes03], three $^{235}$U enrichments for the first core are selected as $\{2.35, 3.40, 4.45\}$wt% to yield a cycle length $\theta_1 = 20$ MWd/kgHM for the first core. In a rather complicated fuel-management scheme [Dru07], an average equilibrium feed enrichment of 4.67 wt%, slightly larger than the highest enrichment 4.45 wt% for the first core, is chosen to yield an equilibrium cycle length $\theta = 20$ MWd/kgHM and equilibrium discharge burnup $\theta^* = 49$ MWd/kgHM, corresponding to 18 months of full-power operation with a capacity factor of 0.95. For a tall, slender core design of AP1000, the cycle-average non leakage probability $P_{NL} = 0.95$, smaller than 0.97 for the current generation of PWRs. Furthermore, $\theta \simeq \theta_1 \simeq 20$ MWd/kgHM, and the equilibrium discharge burnup $\theta^*$ is only $2.45\theta$, rather than $3\theta$, as suggested by the simplified cycling model in Figure 12.11 and Eq. (12.38).

With Pu or TRU recycling, Eq. (12.38) may similarly be used to find the first-core composition, but the equilibrium TRU loading depends on the composition of the combined charge considered in Figure 12.8 and generally requires [Lee94, Sor08] another loop of iterations for the proper mix between the feed and recycled fuel. Nonetheless, the RBC method represented by Eq. (12.38) would provide meaningful first estimates before full-blown MGD calculations are performed. Thus, the simplified cycling method discussed in this section is useful for fuel cycle economics as well as fuel cycle optimization tasks.

The out-in refueling scheme illustrated in Tables 12.2 and 12.3 has been used traditionally for PWR and BWR cores so that the power distribution in the $X - Y$ plane is flattened to minimize the power-peaking factors $F_{\Delta h}$ and $F_q$ and to reduce the overall neutron leakage probability. In recent years, however, alternate loading patterns and refueling schemes have been utilized, primarily to reduce the neutron and gamma fluence on the reactor pressure vessel. The rather complicated AP1000 loading pattern scheme is another illustration of departure from traditional three-zone loading pattern schemes. Nonetheless, it is useful to summarize *approximate* reactivity parameters applicable for the fuel cycle analysis of PWR configurations:

(a) Reactivity change as a function of fuel burnup: $0.7 \sim 1.0\%\Delta k/k$ per MWd/kgHM,
(b) Reactivity worth of enrichment variation: $1.0\%\Delta k/k$ per 0.1 wt% $^{235}U$,
(c) Reactivity worth of burnable absorbers: $1.0\%\Delta k/k$ per absorber rod/assembly,
(d) Fuel cycle length as a function of reactivity variation: 30 equivalent full power day (EFPD) per $1.0\%\Delta k/k$.

## 12.5 Fission Product Xenon Buildup

Among the many FPs generated in thermal reactors, $^{135}Xe$ is of particular concern because it has an absorption cross section of approximately 3 Mb for thermal neutrons, the highest cross section of any material known to humanity. The generation of this FP was discovered during World War II at one of the Hanford plutonium production reactors. The reactor was slowly shutting itself down for some unknown reason, and E. Fermi was quickly able to resolve the mystery behind this phenomenon. With reserve fuel elements fortunately available, the reactor was restored to full-power operation within days. Because of its large absorption cross section, the buildup and transient behavior of the $^{135}$Xe concentration present operational and economic issues in LWR fuel management.

We begin in Section 12.5.1 with a discussion of the FP decay chain related to $^{135}$Xe production and buildup and associated balance equations. Section 12.5.2 presents the time-domain solution of the balance equations, with particular attention given to the buildup of $^{135}$Xe following a reactor shutdown. A brief discussion of related FP $^{149}$Sm concludes Section 12.5.3.

### 12.5.1 Mechanism for $^{135}$Xe Production and Balance Equation

The FP $^{135}$Xe is produced directly from the fission process and is also formed through a decay of $^{135}$I, which in turn is produced through a decay of another FP $^{135}$Te:

$$^{135}_{52}\text{Te} \xrightarrow[t_{1/2}=19s]{\beta-} {}^{135}_{53}\text{I} \xrightarrow[t_{1/2}=6.57h]{\beta-} {}^{135}_{54}\text{Xe} \xrightarrow[t_{1/2}=9.1h]{\beta-} {}^{135}_{55}\text{Cs} \xrightarrow[t_{1/2}=2.3\times10^{6}y]{\beta-} {}^{135}_{56}\text{Ba}. \quad (12.39)$$

Since the half-life of $^{135}$Te is short compared with those of $^{135}$Xe and $^{135}$I, we may for all practical purposes assume $^{135}$I is produced directly as a FP. Similarly, since the half-life of $^{135}$Cs is much longer than those of $^{135}$Xe and $^{135}$I, we may treat $^{135}$Cs essentially as the final nuclide in the chain. The fission yield of $^{135}$I from thermal fission of $^{235}$U is $\gamma_I = 0.064$, while the direct fission yield of $^{135}$Xe is $\gamma_X = 0.002$. The thermal absorption cross sections for $^{135}$I and $^{135}$Xe are $\sigma_I = 7$ b and $\sigma_X = 2.65$ Mb, respectively, with the corresponding decay constants, $\lambda_I = 2.89 \times 10^{-5}$ s$^{-1}$ and $\lambda_X = 2.08 \times 10^{-5}$ s$^{-1}$. For the AP1000 core with an average neutron energy estimated as 0.2 eV, the effective one-group cross section $\sigma_X \simeq 1.5$ Mb is calculated.

In terms of the scalar flux $\phi(\mathrm{r}, t)$, set up balance equations representing the production and destruction of the $^{135}$I and $^{135}$Xe concentrations, $I(\mathbf{r}, t)$ and $X(\mathbf{r}, t)$, respectively:

$$\frac{\partial I(\mathbf{r},t)}{\partial t} = \gamma_I \Sigma_f \phi(\mathbf{r},t) - \lambda_I I(\mathbf{r},t) - \sigma_I \phi(\mathbf{r},t) I(\mathbf{r},t), \quad (12.40)$$

$$\frac{\partial X(\mathbf{r},t)}{\partial t} = \gamma_X \Sigma_f \phi(\mathbf{r},t) + \lambda_I I(\mathbf{r},t) - \lambda_X X(\mathbf{r},t) - \sigma_X \phi(\mathbf{r},t) X(\mathbf{r},t). \quad (12.41)$$

Note that Eq. (12.41) includes the production of $^{135}$X through the radioactive decay of $^{135}$I, which is represented as a loss term in Eq. (12.40). Since the absorption cross section of $^{135}$I is small, we may neglect the last term, representing the destruction of $^{135}$I nuclei through neutron absorption, in Eq. (12.40). Because the half-lives of $^{135}$Xe and $^{135}$I are on the order of hours, much longer than the time constants associated with time-dependent neutron flux variations, we may obtain $\phi(\mathbf{r}, t)$ from the steady-state one-group neutron diffusion equation (5.70).

In operating reactors, there are two types of issues we need to address regarding the production and decay of $^{135}$Xe and $^{135}$I nuclei. The first issue deals with the fundamental mode or time-domain evolution of the $^{135}$Xe and $^{135}$I concentrations during startup, during routine operation, and after shutdown of the reactor. One particular concern in this regard is the buildup of $^{135}$Xe after shutdown due to the decay of the $^{135}$I nuclei, decreasing reactivity. Since commercial nuclear power plants have sufficient excess reactivity except perhaps near EOC, this usually is not an operational concern in commercial power plants. In contrast, in small research reactors and naval reactors, where excess reactivity may not be plentiful, the post-shutdown poison-out of the reactor could be a concern. Optimal strategies [Rob65] for shutting down the reactor to minimize the buildup of $^{135}$Xe have been studied, and programmed shutdown maneuvers are apparently used in actual naval reactor operations.

The second concern related to the $^{135}$Xe production and decay is of a different nature and is present in large, high-power commercial reactors. In this case, due to a competition between the production of $^{135}$Xe via $^{135}$I decay (the second RHS term) and the destruction of $^{135}$Xe via neutron capture (the last RHS term) in Eq. (12.41), significant oscillations [Sch80] in the *spatial* flux distribution may occur. Such space-time oscillations can be induced by local perturbations in $\phi(\mathbf{r}, t)$ and may grow in amplitude [Lee71] if not properly controlled. The stability of large LWR cores against xenon-induced oscillations decreases as the core size and power density increase and also as the fuel burnup increases. This phenomenon is not strictly a safety concern, due to the large time constants involved, but may present operational or economic penalties. Allowing for xenon-induced space-time oscillations in flux and power distributions may result in an increase in the power peaking factor, which may reduce the thermal margin and limit the rated power level. In addition, monitoring and controlling space-time power oscillations may require additional instrumentation systems and provisions for control rods or distributed control poisons. With the issues related to xenon-induced space-time flux oscillations deferred to Chapter 16, only the time-domain issues will be discussed here.

### 12.5.2 Time-Domain Solution of Xe-I Balance Equation

For the time-domain study of the $^{135}$Xe-$^{135}$I dynamics, we rewrite Eqs. (12.40) and (12.41) in terms of core-average concentrations $I(t)$ and $X(t)$:

$$\frac{dI(t)}{dt} = \gamma_I \Sigma_f \phi(t) - \lambda_I I(t), \tag{12.42}$$

$$\frac{dX(t)}{dt} = \gamma_X \Sigma_f \phi(t) + \lambda_I I(t) - \lambda^*(t)X(t), \ \ \lambda^*(t) = \lambda_X + \sigma_X \phi(t). \tag{12.43}$$

The equilibrium concentrations corresponding to flux $\phi$ may be obtained first

$$I_{\infty}(\phi) = \frac{\gamma_I \Sigma_f \phi}{\lambda_I}, \tag{12.44}$$

$$X_{\infty}(\phi) = \frac{\gamma \Sigma_f \phi}{\lambda^*}, \gamma = \gamma_I + \gamma_X, \tag{12.45}$$

where $\gamma$ is the total effective fission yield of $^{135}$Xe.

Derive general solutions to Eqs. (12.42) and (12.43) subject to a step change in flux at time $t = 0$, with the initial conditions

$$\phi(0-) = \phi_0, \ \phi(0+) = \phi_1, \tag{12.46}$$

$$X(0-) = X_0, \ I(0-) = I_0. \tag{12.47}$$

With the $^{135}$Xe and $^{135}$I concentrations initially in general non-equilibrium conditions from Eq. (12.47), Eq. (12.42) is simplified to an ODE with constant coefficients

$$\frac{dI(t)}{dt} = \gamma_I \Sigma_f \phi_1 - \lambda_I I(t), \tag{12.48}$$

which is integrated to yield

$$I(t) = e^{-\lambda_I t}\left[\gamma_I \Sigma_f \phi_1 \int_0^t e^{\lambda_I \tau} d\tau + I_0\right] = \frac{\gamma_I \Sigma_f \phi_1}{\lambda_I}\left(1 - e^{-\lambda_I t}\right) + I_0 e^{-\lambda_I t}. \tag{12.49}$$

In terms of Eq. (12.44), Eq. (12.49) simplifies to

$$I(t) = (I_0 - I_1^{\infty})e^{-\lambda_I t} + I_1^{\infty}, \quad I_1^{\infty} = I_{\infty}(\phi_1). \tag{12.50}$$

Similarly, integrating the $^{135}$Xe balance equation

$$\frac{dX(t)}{dt} = \gamma_X \Sigma_f \phi_1 + \lambda_I I(t) - \lambda_1^* X(t), \quad \lambda_1^* = \lambda_X + \sigma_X \phi_1, \tag{12.51}$$

yields

$$X(t) = e^{-\lambda_1^* t}\left[\int_0^t \left\{\gamma_X \Sigma_f \phi_1 + \lambda_I I(\tau)\right\} e^{\lambda_1^* \tau} d\tau + X_0\right]. \tag{12.52}$$

Substituting Eq. (12.50) into Eq. (12.52) and recalling Eq. (12.45), perform the integration

$$\begin{aligned} X(t) &= e^{-\lambda_1^* t}\left[\int_0^t \left\{\lambda_1^* X_1^{\infty} + \lambda_I (I_0 - I_1^{\infty}) e^{-\lambda_I \tau}\right\} e^{\lambda_1^* \tau} d\tau + X_0\right], X_1^{\infty} = X_{\infty}(\phi_1), \\ &= X_1^{\infty}(1 - e^{-\lambda_1^* t}) + \frac{\lambda_I (I_0 - I_1^{\infty})}{\lambda_1^* - \lambda_I}\left(e^{-\lambda_I t} - e^{-\lambda_1^* t}\right) + X_0 e^{-\lambda_1^* t}, \end{aligned}$$

from which we finally obtain

$$X(t) = (X_0 - X_1^{\infty})e^{-\lambda_1^* t} + \frac{\lambda_I (I_0 - I_1^{\infty})}{\lambda_1^* - \lambda_I}\left(e^{-\lambda_I t} - e^{-\lambda_1^* t}\right) + X_1^{\infty}. \tag{12.53}$$

Two transient scenarios are illustrated with Eqs. (12.50) and (12.53):

(1) Reactor startup from a clean, xenon-free core, yielding simple initial conditions

$$\phi_0 = X_0 = I_0 = 0. \tag{12.54}$$

In this case, Eqs. (12.50) and (12.53) yield

$$I(t) = I_1^\infty(1 - e^{-\lambda_I t}) = \frac{\gamma_I \Sigma_f \phi_1}{\lambda_I}(1 - e^{-\lambda_I t}), \tag{12.55}$$

$$X(t) = X_1^\infty \left(1 - e^{-\lambda_1^* t}\right) - \frac{\lambda_I I_1^\infty}{\lambda_1^* - \lambda_I} \left(e^{-\lambda_I t} - e^{-\lambda_1^* t}\right). \tag{12.56}$$

(2) Reactor shutdown from equilibrium $^{135}Xe$ and $^{135}I$ concentrations, corresponding to flux $\phi = \phi_0$

$$I_0 = I_0^\infty, X_0 = X_0^\infty, \phi_1 = 0, \lambda_1^* = \lambda_X, I_1^\infty = X_1^\infty = 0. \tag{12.57}$$

We obtain post-shutdown solutions:

$$I(t) = I_0^\infty e^{-\lambda_I t}, \tag{12.58}$$

$$\begin{aligned} X(t) &= X_0^\infty e^{-\lambda_X t} + \frac{\lambda_I I_0^\infty}{\lambda_I - \lambda_X}\left(e^{-\lambda_X t} - e^{-\lambda_I t}\right) \\ &= \left(X_0^\infty + \frac{\lambda_I I_0^\infty}{\lambda_I - \lambda_X}\right) e^{-\lambda_X t} - \frac{\lambda_I I_0^\infty}{\lambda_I - \lambda_X} e^{-\lambda_I t}, \quad \lambda_I > \lambda_X. \end{aligned} \tag{12.59}$$

Time-dependent $^{135}I$ and $^{135}Xe$ concentrations for the combined startup and shutdown scenarios are illustrated in Figure 12.12. The saturation of the $^{135}I$ and $^{135}Xe$ concentrations following a startup from a clean core to the equilibrium concentrations from Eqs. (12.44) and (12.45) is easily understood from Eqs. (12.55) and (12.56). The decay of the post-shutdown $^{135}I$ concentration is also evident from Eq. (12.58), while the post-shutdown increase of the $^{135}Xe$ concentration may be understood as the result of a competition between the post-shutdown $^{135}Xe$ buildup from the decay of the $^{135}I$ concentrations and the decay of the existing $^{135}Xe$ concentration. In fact, since $\lambda_I > \lambda_X$, the last, negative term in Eq. (12.59) controlled by $\lambda_I$ decays away faster than the positive terms controlled by $\lambda_X$, so that for some time after the shutdown the $^{135}Xe$ concentration actually increases, and then decays away eventually.

The time $t_m$ at which the post-shutdown $^{135}Xe$ concentration peaks is illustrated in Figure 12.12 and is obtained by setting the time-derivative from Eq. (12.59) to zero:

$$t_m = \frac{1}{\lambda_I - \lambda_X} \ln \left[\frac{\lambda_I/\lambda_X}{1 + \left(1 - \lambda_X/\lambda_I\right) X_0^\infty / I_0^\infty}\right]. \tag{12.60}$$

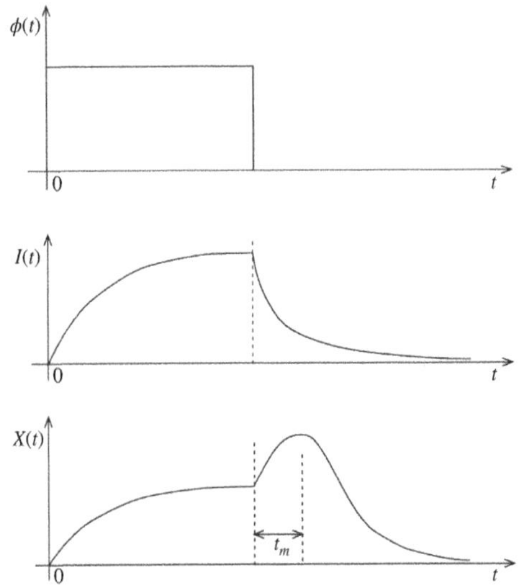


**Figure 12.12** Evolution of $^{135}$I and $^{135}$Xe concentrations during startup and shutdown.

For high power reactors, in the limit as $I_0^\infty \to \infty$ and $X_0^\infty \to \gamma\Sigma_f/\sigma_X$, the $^{135}$Xe peak time approaches

$$t_m = \frac{1}{\lambda_I - \lambda_X} \ln \frac{\lambda_I}{\lambda_X} \simeq 11.3 \text{ hours.} \tag{12.61}$$

Hence, there is a need to implement a programmed shutdown strategy so that the core will not be poisoned out during the 12 ~ 24 hrs following a reactor shutdown, as discussed in Section 12.5.1.

In one-group diffusion theory, the reactivity effect of the equilibrium $^{135}$Xe concentration may be determined by the effect of macroscopic $^{135}$Xe absorption cross section $\Sigma_{aX}^\infty$ on the total loss term $\Sigma_a + DB^2$

$$\rho_X = \frac{\delta k}{k} = \delta \ln k = -\delta \ln\left(\Sigma_a + DB^2\right) = -\frac{\Sigma_{aX}^\infty}{\Sigma_a + DB^2} = -\frac{\sigma_X X_0^\infty}{\nu\Sigma_f}, \tag{12.62}$$

where the equilibrium $^{135}$Xe concentration $X_0^\infty$ is given by Eq. (12.45) with $\phi = \phi_0$. For high-flux reactors, typical of LWRs, Eq. (12.62) simplifies to

$$\rho_X = -\frac{\gamma\Sigma_f}{\nu\Sigma_f} \simeq 2.7\% \frac{\Delta k}{k}. \tag{12.63}$$

### 12.5.3 Effect of Samarium Buildup

Another FP of importance in LWR incore fuel management is $^{149}$Sm with a thermal absorption cross section of 40 kb. The decay chain for $^{149}$Sm

$$^{149}_{60}\text{Nd} \xrightarrow[t_{1/2}=1.73h]{\beta-} {}^{149}_{61}\text{Pm} \xrightarrow[t_{1/2}=2.2d]{\beta-} {}^{149}_{62}\text{Sm (stable)}, \tag{12.64}$$

indicates that we may neglect $^{149}$Nd and begin the chain with $^{149}$Pm with a fission yield of 1.1% for $^{235}$U fission.

The dynamics of $^{149}$Sm buildup via the decay of $^{149}$Pm is similar to the $^{135}$I-$^{135}$Xe dynamics, except that the $^{149}$Sm concentration will continually approach an asymptotic level following a reactor shutdown. The post-shutdown $^{149}$Sm buildup will eventually be reduced by neutron capture in a subsequent restart of the reactor. The total equilibrium reactivity effect of $^{149}$Sm in thermal reactors is on the order of $0.6\%\Delta k/k$ and has to be accounted for, together with that of $^{135}$Xe, during the first few days of reactor operation.

## 12.6 General Incore Management Considerations

We have discussed in this chapter methods for fuel depletion calculations, equilibrium cycle analysis, and techniques for estimating fuel burnup. This section presents a number of general issues that are important for incore fuel management.

### 12.6.1 Reactivity Variation over Fuel Cycle

Except for a breeder, where more fissile material is produced than consumed, the effective multiplication factor of any reactor decreases as fuel is depleted over a fuel cycle. Even in many SFR designs that focus on the transmutation of TRUs, including the legacy Pu inventory, there is a substantial decrease in reactivity over a fuel cycle with a cycle length on the order of a year. The reactivity decrease over a cycle is yet larger for LWRs fueled with slightly enriched $UO_2$ fuel, since the fissile inventory is depleted significantly over a $12 \sim 18$-month cycle. In fact, in the simple example discussed in Section 12.2.5, for AP1000 $UO_2$ assemblies with a $^{235}$U enrichment of 4.5 wt%, approximately 75% of the initial $^{235}$U may be consumed in an 18-month cycle.

A qualitative sketch of $k = k_{eff}$ as a function of fuel burnup in Figure 12.13 illustrates the issues associated with the reactivity decrease due to fuel burnup in LWRs. The plots indicate the reactivity trends with and without burnable absorbers (BAs). The sharp decrease in reactivity near BOC for both plots is due to the buildup of $^{135}$Xe and $^{149}$Sm discussed in Section 12.5. Following the initial decrease in $k$, the top plot without BAs indicates a continuous decrease in $k$ as fuel depletes, while the bottom plot shows a small increase followed by a decrease toward EOC with $k = 1.0$. Burnable absorbers comprise materials with large thermal absorption cross sections, e.g. $B_4C$, hafnium, $Gd_2O_3$, and erbium, in LWRs so that the depletion rate of BAs is larger than that of fuel itself. This then allows a rapid decrease in neutron captures in BAs, resulting in a small increase

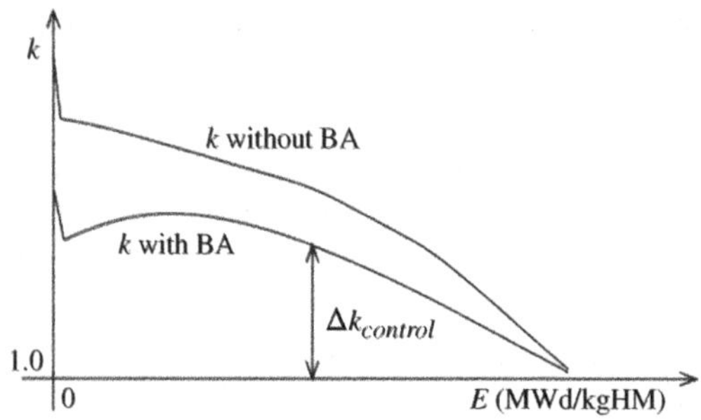


**Figure 12.13** Reactivity as a function of fuel burnup.

in $k$ around the burnup of 5 ~ 8 MWd/kgHM. Eventually, BAs are consumed, and the reactivity follows the general trend associated with fuel burnup. The excess reactivity that has to be controlled to bring the core to criticality is marked as $\Delta k_{control}$ in Figure 12.13.

The excess reactivity $\Delta k_{control}$ is primarily controlled by boric acid dissolved in water in PWRs and by the movement of control blades in BWRs. Thus, the first benefit of implementing BAs in LWRs is to reduce the requirements on reactivity control devices. Since BAs are typically introduced as lumped absorbers or admixed in fuel rods and distributed within each fuel assembly, BAs also contribute toward flattening the power distribution and reducing the power-peaking factor discussed further in Chapter 13. Furthermore, in BWR cores, the introduction of $Gd_2O_3$ as BAs reduces the movement of control blades over the cycle and thereby reduces local perturbations in the three-dimensional power distribution. One final and important contribution of BAs in PWRs is to make the moderator temperature coefficient larger negative. This is discussed in Chapter 14, dealing with reactivity coefficients. One potential problem with BAs, however, especially with $Gd_2O_3$ admixed with fuel, is that they may induce sharp variations in the power distribution as they deplete, and may cause slow oscillations in the axial power distribution over a fuel cycle. This trend amplifies as the core height increases and hence is the reason for rather complex heterogeneous axial fuel configurations in the 14-foot AP1000 core design [Wes03].

### 12.6.2 Thermal-Hydraulic Feedback and Power Distribution

The power distribution in a reactor core with a sufficient power density, producing sensible heat, shows the effects of temperature and density distributions in the core. Thus, the power distribution calculation in any power reactor should reflect the effects of *pointwise thermal-hydraulic (T/H) feedback*, in addition to the core-wide feedback effects on reactivity.

The fuel temperature feedback effect, primarily associated with Doppler broadening of absorption resonances, increases as the power density and the associated fuel temperature increase. Hence, the fuel temperature feedback in any reactor core tends to flatten the power distribution.

In LWR cores, the flux and power density increase locally as the density of water, serving as the neutron moderator, increases. Thus, in a PWR core, as the coolant heats up along the length of the core, the average coolant density in the top half of the core is lower than that in the bottom half of the core. This results in the axial power distribution shifting toward the bottom of the core at BOC. This is usually represented as a negative *axial offset* (AO) of power defined as the normalized difference in power between the top and bottom halves of the core

$$\mathrm{AO} = \frac{P_T - P_B}{P_T + P_B}, \tag{12.65}$$

where $P_T$ and $P_B$ represent the power in the top half and bottom half, respectively. As the fuel depletes preferentially in the bottom half of the core, due to a higher power density, the axial power distribution gradually shifts toward the top of the core, resulting in a positive AO toward EOC in a typical PWR core. During the middle of the cycle, the fuel burnup distribution peaks near the central region of the core and can induce a *double-hump* power distribution with two peaks located away from the mid-point of the core height.

### 12.6.3 Control Requirements for Light Water Reactor

We now reconsider the reactivity control requirements discussed in Section 12.6.1 to account for the T/H feedback effects on reactivity explicitly. Figure 12.14 displays general dependence of the core reactivity on the moderating ratio in LWRs expressed in terms of the hydrogen-to-$^{235}$U atom ratio $N_H/N_{25}$. As discussed further in connection with reactivity coefficients in Chapter 14, thermal utilization $f$ decreases while resonance escape probability $p$ increases as the ratio $N_H/N_{25}$ increases, resulting in a peak in the plot of $k$ vs. $N_H/N_{25}$ for each fissile enrichment or fuel burnup in the family of three representative curves. All LWRs operate in the under-moderated region indicated by solid dots for the cold zero power (CZP) and hot full power (HFP) operation points both for the top and bottom curves. The top curve corresponds to the fresh BOC configuration and the bottom curve to the EOC state. Thus, the difference in reactivity between the CZP BOC condition and the HFP EOC condition is the excess reactivity $\Delta k_{cycle}$ for an operating cycle indicated in Figure 12.14. Note here that the HFP EOC is typically determined when $k = 1.005$, not 1.0 as casually assumed so far in this chapter,

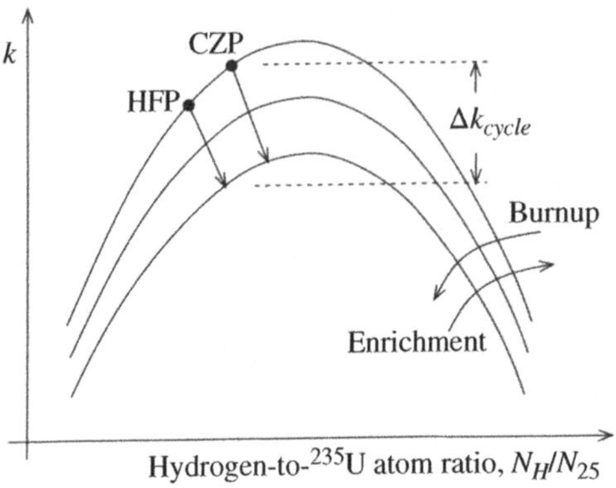


**Figure 12.14** Core reactivity vs. hydrogen-to-$^{235}$U atom ratio.

so that we retain some transient maneuver capability via an excess reactivity of $0.5\%\Delta k/k$.

The excess reactivity for LWR cycles with a target cycle length of 18 months and discharge burnup of 49 MWd/kgHM is estimated to be $17 \sim 19\%\Delta k/k$ and may be broken down into several components, as summarized in Table 12.4. For both PWR and BWR plants, each of the moderator temperature and fuel temperature feedback accounts approximately for a reactivity decrease of $1.0 \sim 1.5\%\Delta k/k$ in moving from a CZP to HFP condition. In addition, in BWR cores, another $2\%\Delta k/k$ decrease in reactivity is estimated due to the generation of approximately 40% core-average void fraction at HFP. The buildup of fission products, including $^{135}$Xe and $^{149}$Sm, accounts for about a $5\%\Delta k/k$ decrease. Finally, a decrease of $10\%\Delta k/k$ is estimated due to fuel burnup of 20 MWd/kgHM over an 18-month cycle, resulting in a total excess reactivity requirement of $\Delta k_{cycle} = 17 \sim 19\ \%\Delta k/k$. The control of the excess reactivity $\Delta k_{cycle}$ for PWR plants is split between $6 \sim 7\ \%\Delta k/k$ for BAs and $11 \sim 12\%\Delta k/k$ for soluble boron. The *soluble boron letdown curve* takes the shape of the bottom plot of $k$ vs. fuel burnup, presented in Figure 12.13. For BWR plants, the excess reactivity $\Delta k_{cycle} = 19\%\Delta k/k$ is split more or less evenly between BAs and control blades actively managed as a function of fuel burnup. This is discussed further in Section 12.6.4.

For SFR configurations, the reactivity swing $\Delta k_{cycle}$ over the cycle is much smaller in absolute magnitude compared with those for LWRs just discussed. Since the effective delayed neutron fraction $\beta$ is, however, only 0.3%, less than half that for LWRs, it also becomes necessary to provide provisions for properly controlling excess reactivity throughout the cycle. Furthermore, many of the traditional BA materials, including boron, would have a much-reduced reactivity worth in a fast spectrum. At the same time, to reduce the possibility of transient overpower (TOP) conditions and avoid concerns about hypothetical core

**Table 12.4** Components of excess reactivity ($\%\Delta k/k$) over cycle for LWRs.

| Reactivity components | PWR | BWR |
|---|---|---|
| Moderator temperature feedback | 1.0 ~ 1.5 | 1.0 |
| Fuel temperature feedback | 1.0 | 1.0 |
| Void feedback | 0 | 2.0 |
| Fission product poisoning | 5.0 | 5.0 |
| Fuel burnup effects | 10.0 | 10.0 |
| Total excess reactivity required | 17 ~ 18 | 19 |

disruptive accidents (HCDAs), modern SFR designs [Boa00] call for a secondary control and shutdown system, coupled with a restriction on control rod insertions in the core.

### 12.6.4 Power Distribution Control

Another key task of incore fuel management involves the safe and efficient control of the power distribution in the core. For PWR plants, active power distribution controls are needed during occasional load variations, where any space-time xenon-induced oscillations in power should be controlled judiciously so that no compromise is made in the core thermal margin. This typically requires some form of constant AO control [Sch80], whereby the AO of power is retained within a narrow band, e.g. ±10%, through control rod maneuvers. For this purpose, part-length control rods were installed in some plants but have not seen much use. In the latest design for the 14-foot AP1000 plants, however, gray control rods with reduced reactivity worths have been introduced with this purpose in mind. As we envision the deployment of new nuclear power plants, eventually sharing a larger fraction of the nation's electric supply, we need to develop efficient and reliable methods for load-follow maneuvers in both PWR and BWR plants. Indeed, nearly all PWR plants routinely undergo load-follow maneuvers in France, where nuclear power plants generate ~75% of the electricity. Nuclear plant control issues are discussed further in Chapter 16.

In BWR plants, control blades in the form of a cross, with the wings loaded with rods of neutron absorbers (typically $B_4C$), are actively used to control both excess reactivity, as discussed in Section 12.6.3, and power distributions during transient maneuvers. In a BWR core, the average void fraction within the assembly channel box is about 40% and provides a major feedback mechanism in both reactivity and flux and power distributions. The *cruciform* control blades are inserted from the bottom of the core up to variable depths or steps in the core. Some control blades are inserted deep into the core, and these *deep control rods* primarily control the reactivity, while the *shallow control rods* are inserted only partially into the core, primarily to shape the power distribution. A typical arrangement for the cruciform blade inserted into the *wide-wide (W-W) gap* of a cluster of (2 × 2) BWR bundles, as they are often called, is presented in Figure 1.10. The diagonal opposite of the W-W corner is the *narrow-narrow (N-N) gap*, where an incore instrumentation tube marked *R* is located.

We may replicate the (2 × 2) bundle cluster shown in Figure 1.10 into a cluster of (4 × 4) bundles to form a unit control cluster, which contains four W-W gaps. The power distribution control in a typical BWR core begins with an approximately 25% full-length equivalent of control blades inserted into the core with a varying pattern of deep and shallow control rods at BOC. Thus, at BOC, some blades are inserted all the way into the top of the core, while others are inserted only a few steps out of a total of 48 steps for a 12-foot BWR core. As the fuel burnup progresses, to obtain as uniform a fuel burnup distribution as possible throughout the core, the control blades are swapped in and out in a set of four sequences A, B, C, and D. Each of the four sequences has different combinations of deep and shallow control rods inserted into two alternating W-W gaps out of a total of four such gaps: deep-shallow and shallow-deep sequence in one diagonal pair of W-W gaps, followed by deep-shallow and shallow-deep sequence in the other pair of W-W gaps, in a unit control cluster comprising 4 × 4 bundles. A basis for BWR power distribution control strategies is to maintain a burnup-independent power shape, known as the *Haling power distribution* [Hal63].

The basis for the Haling power distribution may be established via a proof by contradiction. Suppose that we have an alternate power distribution $P_A(z)$ that is flatter than the Haling distribution $P_H(z)$, i.e. the axial power peaking factor $F_z$ is smaller for $P_A(z)$ than for $P_H(z)$ throughout the cycle, as illustrated for BOC in Figure 12.15a. The burnup distributions $E(z)$ at EOC reflect mostly the BOC power distributions indicated in Figure 12.15b. Since $k_\infty$ is a decreasing function of in $E(z)$ at EOC translate into $k_\infty$ distributions in Figure 12.15c, which finally result in a higher $F_z$ for $P_A(z)$ than for $P_H(z)$, as summarized in Figure 12.15d. This contradicts the assumption that $P_A(z)$ will provide flatter distributions throughout the cycle, and hence the constant Haling power shape $P_H(z)$ is the optimal distribution. Such constant power distribution may be considered an optimal distribution provided the resulting power-peaking factor is equal to the licensed limit, which is only approximately the case in actual operating BWRs [Jac82]. Nonetheless, the constant Haling power shape usually serves as a basis for the control-blade sequencing with 3D BWR core simulators.

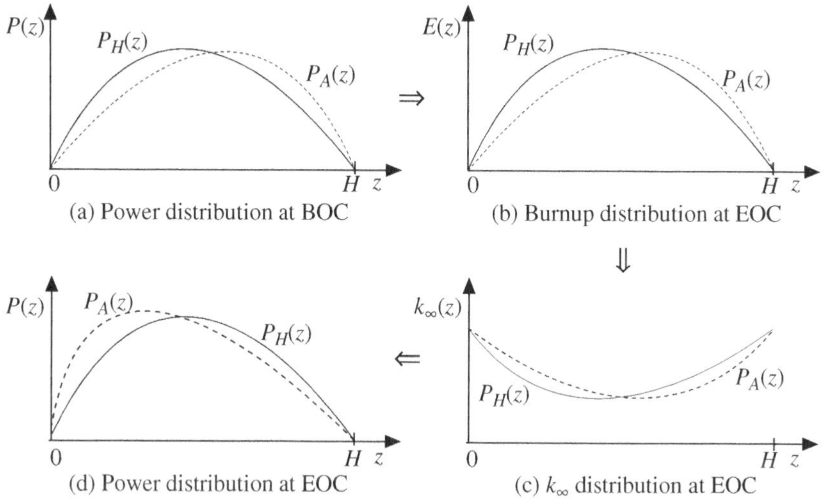


**Figure 12.15** Evolution of power and burnup distributions over a cycle for the Haling power distribution. (a) Power distribution at BOC, (b) burnup Power distribution at EOC, (c) $k_\infty$ distribution at EOC, and (d) Power distribution at EOC.

## 12.7 Fission Products and Radioactive Waste

One of the long-standing issues associated with the continued development of nuclear energy has been the disposition of the used nuclear fuel (UNF) stockpiled and associated radioactive wastes as the critical back-end fuel cycle task. We briefly discuss first in this section the classification of radioactive wastes and key characteristics of the fission products and wastes. This will be followed by a review of current status of UNF management issues in Section 12.8 and considerations for geological disposal of UNF in 12.9.

### 12.7.1 Classification of Radioactive Waste

Since the bulk of the radioactive waste accrues from UNF, the classification of wastes in the United States follows the Nuclear Waste Policy Act (NWPA) of 1982 and the NWPA amended (NWPAA) of 1987, 1992, and 1997. The NWPA system represents in three categories various wastes resulting from the disposition of UNF and other wastes generated in processes involving the use of ionizing radiation:

(1) **High-Level Waste (HLW):**
Highly radioactive liquid and solid material resulting from the reprocessing of UNF that contains FPs in sufficient concentrations are included in this category. This means unprocessed UNF is also included as HLW.

(2) **TRU Waste:**
Any $\alpha$-emitting nuclides with atomic number $Z > 92$, with a half-life greater than 5 yr and activity greater than $0.1\,\text{mCi} \cdot \text{kg}^{-1}$, are treated as TRU waste. Both civilian and weapons-related wastes are included in this category.

(3) **Low-Level Waste (LLW):**
Essentially, all radwastes not included in the previous two categories are grouped into this category and include activation products generated in the operation of nuclear reactors and other wastes generated in the industry, hospitals, and research laboratories. Thus, LLW comprises (i) collection of liquids from equipment leak, vents, and drains, (ii) solid wastes including contaminated rags, tools, clothing, filter cartridges, and demineralizer resin, and (iii) gaseous or evaporated wastes collected from vents and other systems.

LLW packages are further subdivided into four classes. Class A wastes represent LLWs that will become non-hazardous during the institutional control of the repository, Class B wastes are more radioactive than Class A wastes and will require monitoring up to 300 yr, Class C wastes will require monitoring beyond 300 yr, and Greater-than Class C wastes may require geological disposal.

Other countries, including France and Korea, have established another category, Intermediate-level waste (ILW), which is equivalent to Greater-than-class C in the US system. In addition, uranium mill tailings containing radium and other decay products are also considered radwastes, and some of the radwastes presenting both radioactive and chemical hazards are classified as mixed wastes.

### 12.7.2 Characteristics of Fission Products and Radioactive Waste

Hazards due to ionizing radiation and heat generated through the decay of fission products are two important characteristics of radwastes and UNF and obviously have to be considered in their disposal and management. The radiation hazards of radionuclides are represented by their radiological toxicity as well as their activity.

The *radiological toxicity* of radioactive substances may be defined either by (i) their biological effects represented in units of *rem* or *sievert* when the material is ingested or inhaled or (ii) the volume of air (or water) with which the radioactive substance must be diluted such that the resulting dose rate is less than the maximum allowed occupational limit of $5\,\text{rem} \cdot \text{yr}^{-1}$.

1. **Radiological dose coefficient**
The biological effects of radionuclides inhaled or ingested are summarized in terms of radiological dose coefficients and tabulated in the International Committee on Radiological Protection report, ICRP Publication 119 (2011). The dose coefficients are presented for both occupational and public exposures and in terms of the rate of exposure to the individual radionuclides

involved. We consider both ingestion and inhalation dose coefficients for three representative radionuclides.

(1) $^{3}H$: for ingested and inhaled particulate in the form of organic bound tritium, the dose coefficient $e = 4.2 \times 10^{-11}$ $(Sv \cdot Bq^{-1})$ is provided, while for tritiated water $e = 1.8 \times 10^{-11}$ $(Sv \cdot Bq^{-1})$ for occupational doses is suggested. For public doses, the ICRP 119 recommends $e = 6.2 \times 10^{-12}$ $(Sv \cdot Bq^{-1})$ for fast and $e = 2.6 \times 10^{-10}$ $(Sv \cdot Bq^{-1})$ for slow exposures, respectively.

(2) $^{137}Cs$: for occupation exposures, $e = 4.8 \times 10^{-9}$ $(Sv \cdot Bq^{-1})$ for 1.0 $\mu$m particles and $e = 6.7 \times 10^{-9}$ $(Sv \cdot Bq^{-1})$ for 5.0 $\mu$m particles, respectively, are suggested for inhalation doses, together with $e = 1.38 \times 10^{-8}$ $(Sv \cdot Bq^{-1})$ for ingestion doses. For public doses, $e = 4.6 \times 10^{-9}$ $(Sv \cdot Bq^{-1})$ for fast exposures, $9.7 \times 10^{-9}$ $(Sv \cdot Bq^{-1})$ for medium exposures, and $3.9 \times 10^{-8}$ $(Sv \cdot Bq^{-1})$ for slow exposures, respectively, are suggested.

(3) $^{239}Pu$: for occupational ingestion exposures, $e = 2.5 \times 10^{-7}$ $(Sv \cdot Bq^{-1})$ for medium and $e = 5.3 \times 10^{-8}$ $(Sv \cdot Bq^{-1})$ for slow exposures, respectively, while for inhalation exposures, $e = 3.2 \times 10^{-5}$ $(Sv \cdot Bq^{-1})$ and $e = 8.3 \times 10^{-6}$ $(Sv \cdot Bq^{-1})$ for medium and slow exposures, respectively, are obtained. For public ingestion, the dose coefficient $e = 2.5 \times 10^{-7}$ $(Sv \cdot Bq^{-1})$, equal to that for medium occupational exposures, is suggested.

The dose coefficients have been developed to account for distinct time-dependent effects of radioactive substances affecting various organs via ingestion or inhalation. As an example, we note that the $^{239}Pu$ ingestion coefficient $e = 2.5 \times 10^{-7}$ $(Sv \cdot Bq^{-1})$ for medium exposures is considerably less than $e = 3.2 \times 10^{-5}$ $(Sv \cdot Bq^{-1})$ for inhalation exposures that could significantly affect respiratory systems. The dose coefficients from ICRP Publication 119 are used to calculate UNF radiological toxicity for five representative fuel cycles in Section 12.8.

2. **Radiological toxicity**

The *radiological toxicity* is also defined in terms of the amount of air required to dilute the contaminated air to a permissible level. Given the activity $A$ [Ci] of a radionuclide, the toxicity $Tx$ is defined in terms of the *derived air concentration* (DAC)

$$Tx = \frac{A}{DAC}, \tag{12.66}$$

where DAC in units of $[\mu Ci \cdot m^{-3}]$ may in turn be obtained from the *annual limit on intake* (ALI)

$$DAC = \frac{ALI\,[\mu Ci]}{\text{annual inhalation volume of air } [m^3]}. \tag{12.67}$$

The annual inhalation volume of air is set to $2.4 \times 10^3$ m$^3$ based on the inhalation rate of 0.02 m$^3 \cdot$ min$^{-1}$ and 2,000 working hours per year. Both the *DAC* and *ALI* values are listed in Table B.1 of Appendix B to 10 CFR 20 [NRC12].

**Example 12.3** Illustrate the relationship between *ALI*, *DAC*, and *Tx* for $^3$H and $^{137}$Cs with a reference activity $A = 1.0$ Ci using the tabulations from 10 CFR 20, Appendix B. Compare the relative toxicity with that obtained with the ICRP 119 dose coefficients.

For $^3$H, $ALI = 8 \times 10^4 \mu$Ci, and we obtain DAC = 33 $\mu$Ci $\cdot$ m$^{-3}$, which compares with 20 $\mu$Ci $\cdot$ m$^{-3}$ listed in Table B.1 of Appendix B. The difference is apparently due to the single-digit significant figures used in the table. Equation (12.66) yields $Tx(^3\text{H}) = 3.0 \times 10^4$ m$^3$. For $^{137}$Cs, $ALI = 200$ $\mu$Ci, yielding $DAC =$ 0.083 $\mu$Ci $\cdot$ m$^{-3}$, which again compares with $DAC = 0.06$ $\mu$Ci $\cdot$ m$^{-3}$ in Table B.1 of Appendix B. We obtain $Tx(^{137}\text{Cs}) = 1.2 \times 10^7$ m$^3$. The ratio of toxicity

$$\frac{Tx(^{137}\text{Cs})}{Tx(^3\text{H})} = \frac{1.2 \times 10^7}{3.0 \times 10^4} = 400, \qquad (12.68)$$

indicates that $^{137}$Cs is 400 times more hazardous than $^3$H for the same activity. This should be compared with a somewhat lower ratio based on ICRP 119 dose coefficients for public exposures:

$$\frac{e(^{137}\text{Cs})}{e(^3\text{H})} = \frac{6.7 \times 10^{-9}}{4.2 \times 10^{-11}} = 160. \qquad (12.69)$$

This simple comparison suggests different assumptions possibly used in various databases.

For the general public, the *DAC* for effluent is reduced by a factor of 300 to account for the reduction by (a) a factor of 50 in permissible dose rate from 5 rem $\cdot$ yr$^{-1}$ to 0.1 rem $\cdot$ yr$^{-1}$, (b) a factor of 3 to represent the differences in exposure duration and inhalation rate, and (c) a factor of 2 to account for the general age group. Thus, for the example of $^3$H and $^{137}$Cs, we obtain $Tx(^3\text{H}) =$ $9.0 \times 10^6$ m$^3$ and $Tx(^{137}\text{Cs}) = 3.6 \times 10^9$ m$^3$, respectively.

3. **Fission product decay heat**
Heat generated through the decay of fission products is typically calculated with the ANS 5.1 standard [ANS14], which represents the rate $f(t)$ [MeV $\cdot$ fission$^{-1}$s$^{-1}$] of energy emitted in the form of $\beta$- and $\gamma$-rays at $t$ s after the fission of one nucleus

$$f(t) = \sum_{i=1}^{23} \alpha_i \exp\left(-\lambda_i t\right), \qquad (12.70)$$

where 23 pairs of empirical coefficients $\{\alpha_i, \lambda_i\}$ are tabulated for each of the fissionable nuclides $^{235}$U, $^{238}$U, $^{239}$Pu, and $^{241}$Pu. We obtain an expression

[Lee17a] for the FP decay power $P_d(t, T)$ generated at $t$ s after the reactor is shut down following the operation at thermal power $P$ for a period of time $T$ s. With recoverable energy $Q$ [MeV] generated per fission, Eq. (12.70) yields

$$P_d(t,T) = \int_{-T}^{0} dt' \frac{P}{Q} f(t - t') = \frac{P}{Q} F(t,T), \tag{12.71}$$

where $P/Q$ represents the number of fissions per second that take place during the reactor operation and $t - t'$ is the elapsed time after the fission events in interval $dt'$ around $t'$. The integration in Eq. (12.71) is performed over the interval $[-T, 0]$ of reactor operation, and the function $F(t, T)$ represents the decay heat generated in units of MeV · fission$^{-1}$ for cooling time $t$ s. Breaking up the integral over the operating time $[-T, 0]$ into two integrals over the intervals $[-\infty, 0]$ and $[-T, -\infty]$ yields

$$F(t,T) = F(t,\infty) - F(t+T,\infty). \tag{12.72}$$

The ANS 5.1 standard provides, for each of the four fissionable nuclides, a one-dimensional tabulation $F(t, \infty)$, which may be conveniently used twice, once for the cooling time $t$ (s) and a second time for the sum $t + T$ (s) of the cooling and operating times to yield $F(t, T)$. This approach avoids the laborious task of managing two-dimensional tabulations for $F(t, T)$ in terms of $t$ and $T$ separately. Additional guidelines are given in the standard on accounting for the effects of neutron capture in FPs and stepwise variations in the operating power level, together with estimates for uncertainties in the tabulated data. According to the standard, following a long period of reactor operation, 6 ~ 7% of rated thermal power $P$ is generated through the FP decay immediately after the reactor is shut down, with the decay heat decreasing to ~0.5% of $P$ in a day.

An estimate of the radionuclide inventory in a batch of UNF assemblies may be obtained based on a simple physical analysis [Lee17a], with the approximation that every fission event is a binary fission yielding two fission products and that every FP undergoes one radioactive decay in an equilibrium operating condition. With this simple but reasonable approximation, together with a recoverable energy $Q = 200$ MeV released per fission, 1.0 W of thermal energy generated requires approximately (1.0 J per s)/($3.2 \times 10^{-11}$ J per fission) = $3.1 \times 10^{10}$ fission · s$^{-1}$, which then produces approximately $6.2 \times 10^{10}$ Bq = 1.7 Ci of radioactivity. Thus, a 1.0-GWe nuclear plant with a thermal efficiency of 33% produces 3.0 GWt, which then yields an equilibrium radioactivity of 5 BCi. This simple estimate compares favorably with a total radioactivity inventory of 5.6 BCi, including 3.8 BCi of FP radioactivity in a 3.56-GWt reactor [Lee17a]. Given another approximate relationship, verifiable through the ANS 5.1 Standard, that 5 BCi of radioactivity corresponds to 6 ~ 7% of the total thermal power output, or 200 MWt FP power, in a 1.0 GWe reactor, the

ANS 5.1 Standard may be used to estimate the decay heat generation rate and radioactivity inventory in UNF assemblies stored in either a UNF pool or a dry storage cask.

## 12.8 Management of Used Nuclear Fuel

### 12.8.1 Overview of UNF Issues for the United States

As the most viable approach for the disposition of the used nuclear fuel stockpiled in the United States from the operation of LWRs over the past 55 yr, an underground repository had been under development at Yucca Mountain (YM) in Nevada. The YM project was discontinued in 2009, and the Blue Ribbon Commission was formed to study future options for the management of the UNF from nuclear plants. The Commission report [BRC12] suggests that additional effort should be made to develop more than one underground repository and interim consolidated storage facilities through a consent-based licensing approach, and that the administration of the waste management fund of $49 billion collected as of 2012 should be assigned to a non-governmental entity. The fund has been and is being collected as a user fee from nuclear utilities at a rate of $1.0/MWh of electricity sold. In 2013, following a specific request from the US Congress, the US Nuclear Regulatory Commission reopened the review of the risk assessment of the YM project that the US Department of Energy had prepared before the project was discontinued. The YM project review has not, however, progressed in recent years for mostly political issues.

The YM repository was developed originally for the disposition of 63,000 MgHM of UNF together with 8.0 Mg of weapons-grade Pu. For the UNF stockpiled with an average fuel burnup of 39.6 and 33.0 MWd/kgHM [Car11] for PWR and BWR fuel assemblies, respectively, we may consider a rough estimate of the UNF composition: 3 ~ 5% fission products, 1% TRU, 0.1% MA, and 94 ~ 95% U. A 1.0-GWe LWR plant operates with a fuel inventory of approximately 100 Mg, with one third of fuel discharged and reloaded every 18 months. With continued generation of 2,000 ~ 2,400 MgHM/yr of UNF from the approximately 95 LWR plants operating in the United States, approximately 86,000 MgHM of UNF with a radioactivity inventory of 30 BCi is stored as of 2022 [NAS23]. In anticipation of the eventual placement of unprocessed UNF assemblies in the YM repository, significant effort was made to develop multipurpose transportation, aging, and disposal (TAD) casks [Fer12] as well as transportation and storage casks currently in use. The total UNF inventory would occupy a volume roughly equal to a football field with a depth of 4 m, although actual disposal would, of course, require a much larger space to allow for the proper heat dissipation and engineered barriers.

The difficulty associated with the UNF disposition in the YM repository would suggest the importance of the technology for successful partitioning and transmutation (P&T) of TRUs from the legacy LWR used fuel. A study of P&T issues by the Separations Technology and Transmutation Systems (STATS) panel [Ras96] of the US National Academy of Sciences (NAS) indicates that it would cost at least $100 billion to reprocess and transmute the 63,000 MgHM of UNF targeted for the YM repository in the early 1990s. The high price tag for the P&T options that the STATS panel suggested may be understood by recognizing that the reprocessing of the UNF would require at least $1,000/kgHM, even in the 1990s, and the transmutation of TRUs would require more than one pass through the transmuters. The P&T cost estimate of $100 billion, however, did not account for revenues from the sale of electricity generated in the transmuters. Indeed, the French experience with one-pass recycling of Pu from PWR used fuel suggests that the reprocessing and recycling increases the electricity generation cost by approximately 5%, as discussed further in Chapter 15. This additional cost may be justified in light of the potential reduction in the long-term UNF stewardship burden. There are, however, various technical issues that should be developed further before UNF reprocessing could seriously be considered and some of the issues associated with UNF reprocessing as well as geological disposal are discussed in this section.

A geological repository for radwastes, Waste Isolation Pilot Plant (WIPP), has been in operation since 1999 in Carlsbad, NM. This repository was developed for the purpose of storing TRU wastes generated as byproducts from the weapons program over the past five decades or so. It is to date the only radwaste repository in the United States and is constructed in a salt formation more than 667 m below the earth's surface. The WIPP project involved two other candidate sites that were studied before the decision was made to develop the repository in the current site. To placate concerns raised by surrounding communities, the US Department of Energy originally promised that in 2030 it would begin to close the repository and transfer the wastes to national laboratories as required. It appears, however, that the WIPP will continue to serve as a repository for TRU wastes for the foreseeable future. The repository was temporarily closed in 2014, due to a fire involving an underground vehicle and a radiological incident involving a TRU waste drum, but it reopened in 2017.

For LLW management, a considerable premium is placed on decreasing its volume, to reduce both storage space and disposal charges. Volume reduction of LLW up to a factor of 10 may be achieved through a number of techniques including compaction, evaporation, and incineration. The processed LLW is stored in above-ground facilities, in the form of either covered trenches or tumulus. Licensed LLW disposal facilities in the United States are currently located at (a) Barnwell, SC, (b) Richland, WA, (c) Clive, UT, and (d) Andrews, TX.

Considerable effort will be, however, required in the future to clean up and manage mixed chemical-nuclear waste, including TRU waste, from the nuclear weapons program, as illustrated by a recent assessment. Vitrification of fuel reprocessing waste at the Hanford Waste Management Plant was considered in 2007, with a projected cost of $4.3 B, which is now projected to cost $19–30 B, in addition to $11 B already spent. The MOX fuel fabrication at the Savannah River Site for handling 34 Mg of Pu from dismantled nuclear weapons could cost $30 B [NAS23].

A NAS committee [NAS23] was convened in 2020 to evaluate various advanced and modular reactors under development and to provide guidance on UNF management for the emerging reactor designs. Chapter 1 summarizes the new reactor designs that were presented to the NAS committee and studied, while we discuss in the balance of this chapter the UNF management issues that are summarized in the 2023 NAS report, beginning with material flow diagrams in Section 12.8.2. The NAS committee recognized the importance of managing the UNF inventory continually increasing and recommended the establishment of a dedicated entity to develop geological repositories as soon as possible. The US Department of Energy initiated in 2021 a consensus-based siting approach to involve various local communities to study possible geological repository facilities and active study is underway with funding made available.

### 12.8.2 Fuel Cycle Characteristics of Advanced Reactors

Neutron reaction cross sections representing neutron-nucleus reaction probabilities, in particular, fission cross section $\sigma_f$ and radiative capture cross section $\sigma_\gamma$, depend heavily on the energy of neutrons and hence the neutron flux spectrum as well as on the type of nuclides that the neutrons interact with. Table 12.5 compares key spectrum-average neutron cross sections for thermal and fast-spectrum reactors, with the flux spectrum compared in Figure 11.10. We note that the fission cross section $\sigma_f$ is generally much larger in the thermal spectrum, while the ratio of $^{239}$Pu fission cross section to $^{238}$U capture cross section, $\sigma_f^{239}/\sigma_\gamma^{238}$, is 8.14 in the fast spectrum compared with 34 ~ 98 for the thermal spectrum. This indicates that a fast-spectrum reactor requires a larger fissile enrichment of either $^{239}$Pu or $^{235}$U to attain a critical system than for a thermal-spectrum reactor. This feature, at the same time, also allows enhanced neutron captures in fertile nuclides, e.g. converting fertile $^{238}$U to fissile $^{239}$Pu, serving as the basis for a fast-spectrum breeder reactor.

Representative reactor concepts are analyzed with a focus on material flow sheets in an effort to present and compare fuel cycle characteristics of various advanced reactor designs under development. The material flow sheets are constructed primarily with the data presented in the 2014 Nuclear Fuel Cycle Evaluation and Screening (NFCE&S) report [Wig14]. The flow sheets reflect

**Table 12.5** Spectrum-average cross sections for thermal and fast-spectrum reactors.

| Reaction | Thermal spectrum PWR | Thermal spectrum VHTR | Fast spectrum SFR |
|---|---|---|---|
| $^{238}$U capture $\sigma_\gamma^{238}$ (b) | 0.9 | 4.8 | 0.20 |
| $^{239}$Pu fission $\sigma_f^{239}$ (b) | 89.2 | 165.0 | 1.65 |
| $\sigma_f^{239}/\sigma_\gamma^{238}$ | 97.7 | 34.3 | 8.14 |

*Source*: Adapted from Yang [Yan12].

three once-through cycles contrasted with a design requiring continuous fuel processing and a two-tier system combining thermal and fast-spectrum reactors. The analysis covers primarily three major concepts that have been supported by the US Department of Energy following the release of the Generation IV Roadmap.

Fuel cycle characteristics are represented primarily in terms of the fuel composition and coolant that play a dominant role in determining the neutron flux spectrum in the reactor core. The flux spectrum, together with the fuel composition, in turn characterizes the depletion and isotopic evolution of nuclear fuel materials. This observation suggests classifying advanced reactor designs as (a) fast spectrum reactor, (b) gas cooled reactor, and (c) molten salt reactor, together with (d) modular light water reactor, as summarized in Table 1.2.

1. **Pressurized Water Reactor**

   The PWR design represents the current generation of LWRs utilizing low enrichment uranium (LEU) fuel with $^{235}$U enrichment less than 5.0 wt%. The AP1000 plant, representing Generation III+ designs, features fuel assemblies with an effective length of 4.27 m and rated power generation of 1.1 GWe, with a heat-to-electricity conversion efficiency of 33%. The NuScale Small Modular Reactor design under development features 12 smaller PWR units with a fuel length reduced to 2.0 m and a rated power of 77 MWe each. The average energy of neutrons in a PWR core is approximately 0.25 eV.

   A typical PWR core is loaded with ~200 $UO_2$ fuel assemblies with a total inventory of 90 MgU or 90 MgHM at the beginning of cycle as indicated in Figure 12.16. One third of the fuel elements with an inventory of 30 MgHM are discharged and new fuel elements are loaded and shuffled with partially used fuel elements with a cycle length of 18 months. This explains the charge and discharge fuel inventory of 21.9 MgHM per yr and the nominal fuel residence time of 4.5 yr, reduced to an effective fuel residence time of 4.1 yr accounting for routine maintenance. This then yields a discharge fuel burnup of 50 MWd/kgHM or 50 GWd/MTU for fuel elements operating at a

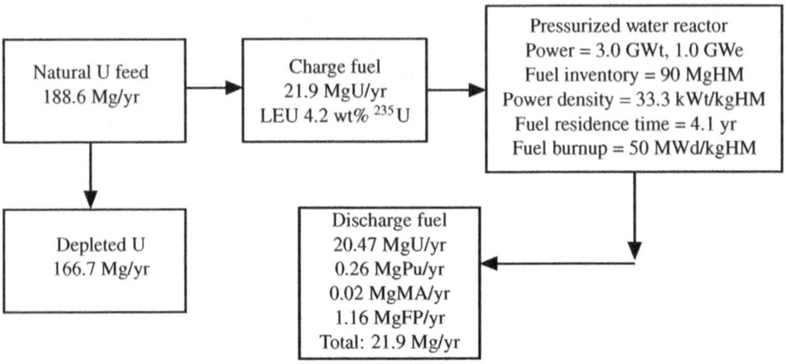


**Figure 12.16** Material flow for a pressurized water reactor design with once-through LEU fuel. Adapted from NFCE&S Evaluation Group 1. *Source:* Wigeland et al. [Wig14].

power density of [3.0 GWt/90 MgHM = 33.3 kWt/kgHM]. The discharged fuel elements contain 4 ~ 5% of FPs, ~1.2% TRU, including 0.1% MA, and 94 ~ 95% unused uranium. The material flow diagram of Figure 12.16 represents a once-through simplification of the closed fuel cycle illustrated in Figure 12.9.

An estimate of the FP generation rate may be obtained from a simplified fuel cycle relationship for *fima* of Eq. (12.22), which indicates that the FP inventory in a reactor system should primarily be proportional to the fuel burnup $E$ [MWd/kgHM]. With a discharge burnup of 50 MWd/kgHM for the PWR configuration, the energy balance statement yields *fima* = 0.053 or a FP yield of 1.2 Mg/yr, in agreement with the detailed NFCE&S calculation included in Figure 12.16. This estimate of FP inventory reflects essentially a simple rule of thumb that a 1.0-GWe nuclear power plant consumes approximately 1.0 Mg/yr of fissionable material.

The UNF inventory or discharge rate of HM for the PWR configuration of Figure 12.16 may also be obtained from a material balance relationship using the discharge fuel burnup of 50 MWd/kgHM as well as from the fuel inventory of 90 MgHM and fuel residence time of 4.1 yr:

$$\begin{aligned}\text{UNF inventory} &= \left(\frac{50\,\text{MWd}}{\text{kgHM}}\right)^{-1} \times \frac{3.0\,\text{GWt}}{1.0\,\text{GWe}} \times \frac{365\,\text{MWd}}{\text{MWt}\cdot\text{yr}} \\ &= \frac{90\,\text{MgHM}}{4.1\,\text{GWe}\cdot\text{yr}} = \frac{21.9\,\text{MgHM}}{\text{GWe}\cdot\text{yr}}.\end{aligned} \tag{12.73}$$

For the AP1000 design with a rated power of 1.09 GWe and 3.4 GWt, yielding a discharge fuel burnup of 49 MWd/kgHM, we obtain a slightly larger fuel discharge rate of 23.4 MgHM/GWe · yr.

Approximately two-thirds of 94 LWRs operating in the United States in 2024 are PWRs and the rest of the fleet are BWRs. The fuel elements or bundles are a bit smaller in a typical BWR core, which contains a slightly larger number of fuel bundles. Thus, it is reasonable to assume that the fuel cycle characteristics of Figure 12.16 are applicable also to BWR plants. Several modular LWR plants under development, including the NuScale modular reactor, feature LEU oxide fuel similar to that used in LWR plants and fuel cycle characteristics will not change much from those of Figure 12.16, subject to possible improvements with increased thermal efficiency.

In fact, as discussed in connection with Table 1.2 and contrary to the suggestion in [Kra22], each of 12 modular NuScale cores, designed to produce 77 MWe with 13 PWR fuel elements, is expected to have a discharge fuel burnup of 41 ~ 60 MWd/kgHM [Rey22]. With the NuScale discharge fuel burnup comparable to 49 MWd/kgHM for the 1.09-GWe AP1000 core, the NuScale UNF discharge rate is also expected to be comparable to that of AP1000 and 22 MgHM/yr illustrated for the reference PWR core in Figure 12.16. It is possible, at the same time, that a smaller reactor core could increase the neutron leakage into the excore regions, including the neutron shields and reflectors, resulting in increased production of ILW or LLW products [NAS23].

2. **Sodium-Cooled Fast Reactor**

The SFR design features one of the key Generation IV reactors fueled with uranium and TRU elements in the form of metallic U-Zr or U-Pu-Zr fuel rods. The SFR core typically features tight fuel assembly designs with the assembly length much shorter than that for LWR designs. Liquid-sodium coolant produces a hard neutron spectrum with an average neutron energy around 0.1 ~ 0.2 MeV, providing flexibility for using a variety of fuel materials including TRUs from used LWR fuel assemblies. In the fast neutron spectrum, the fission-to-capture cross section ratio $\sigma_f/\sigma_\gamma$ for $^{235}U$ or $^{239}Pu$ is smaller than that in the thermal spectrum of the LWRs by an order of magnitude, which requires higher fissile enrichments for fast reactors, represented by the HALEU fuel composition planned for the demonstration core of the TerraPower Natrium reactor [Hej21]. This feature is discussed in connection with Table 12.5. The higher average energy of the fast spectrum reactors would allow efficient production of fissile Pu from $^{238}U$ and could work efficiently with minor actinides Am, Np, and Cm, allowing for the breeding of fissile materials for subsequent fast reactor cycles or as feed material for LWRs. With the HALEU fuel with 16.7 wt% fissile $^{235}U$ enrichment, the charge fuel inventory per year and HM inventory in the Natrium Advanced Type 1B Metal core, summarized in Figure 12.17, are both reduced significantly from the PWR counterparts in Figure 12.16. TerraPower plans to complete the development of fuel element designs without sodium bonding for the Type 1B

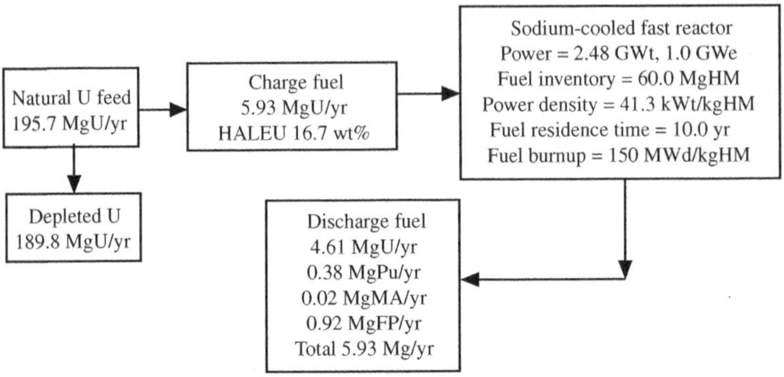


**Figure 12.17** Material flow for a sodium-cooled fast reactor design with once-through HALEU fuel. Adapted from the Natrium Advanced Type 1B Metal core design. *Source:* Hejzlar [Hej21].

Metal Core. With an increased thermal efficiency, the Natrium core design as a fast spectrum reactor, indicates a three-fold increase in the discharge fuel burnup and a four-fold reduction in the UNF inventory discharged, compared with an equivalent 1.0-GWe LWR plant. It should, however, be noted that the depleted U inventory increases by ~15% from the LWR counterpart.

3. **High-Temperature Gas-Cooled Reactor**

The design utilizes sub-mm diameter tristructural-isotropic (TRISO) particles with multiple pyrolytic graphite coatings that form the basic building block. The prismatic design loads TRISO particles, often featuring uranium oxycarbide (UCO) fuel, into graphite pin cells or compacts, while in the pebble bed design the TRISO particles are packed into graphite spheres with a diameter of 40 ~ 60 mm. The high-temperature gas-cooled reactor (HTGR) or VHTR designs provide mostly thermal spectrum but may be configured to provide an epi-thermal spectrum with an average neutron energy in the eV range.

The fuel characteristics summarized in Figure 12.18 illustrate a HALEU design with a combined fissile U enrichment of 15.5 wt% in a thermal neutron spectrum of a graphite-moderated HTGR, similar to the Xe-100 pebble bed reactor design. Utilizing a HALEU design allows the charge and discharge rates of fuel similar to those for the HALEU-fueled SFR of Figure 12.17, again indicating about a third of the PWR fuel data. The core fuel inventory is also reduced to about a third of the PWR inventory. With a high He coolant temperature possible with the HTGR, a higher thermal efficiency of 50% could be attained, compared with 41% for the Natrium design. With a higher power density of 67.1 kWt/kgHM, compared with 41.3 kWt/kgHM for the Type 1B Natrium core, a discharge burnup of 120 MWd/kgHM could be obtained for the HTGR in 4.9 yr compared with a discharge burnup of 150 MWd/kgHM in

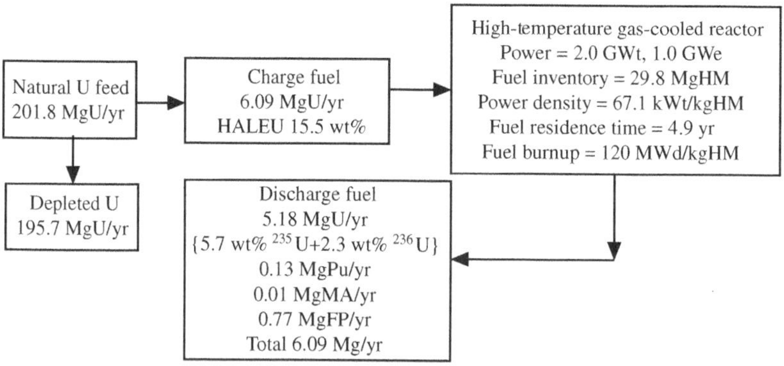


**Figure 12.18** Material flow for a high-temperature gas-cooled reactor design with HALEU TRISO fuel. Adapted from NFCE&S Evaluation Group 2. *Source:* Wigeland et al. [Wig14].

10 yr for the Natrium design. The FP production rate is also reduced for the HTGR design with a higher thermal efficiency compared with the SFR and PWR designs.

4. **Molten Salt Reactor**

The MSR design illustrates the use of various salts, including LiF-$BeF_2$ molten fluoride salt, as a coolant for the core as well as the salt admixed with fuel. The material flow sheet in Figure 12.19 represents a molten fuel design that allows continuous circulation of the fuel salt with an equivalent period of three days or 121.67 circulations a year, indicating a large flow rate of fuel salt summed over a year as high as 8,862 Mg/yr. The fuel residence time of 76.5 yr is estimated from the limited recycle MSR case, represented as Evaluation Group 10, and essentially indicates that the fuel salt circulates through the system indefinitely resulting in a large fuel burnup.

The MSR system provides a continuous on-line fuel salt treatment with low processing loss so that the material processed and discharged consists primarily of FPs produced during the reactor operation. The circulating salt comprises Th as the feed material producing $^{233}U$ as the fissile material, as exemplified in the ThorCon design, listed in Table 1.3, demonstrating the significant potential of a thorium converter operating with merely 0.93 Mg/yr of Th as feed material; this may be compared with 21.9 Mg/yr of LEU fuel for the once-through PWR cycle in Figure 12.16. In the thermal-spectrum MSR design featuring a Th-U fuel cycle in Figure 12.19, the production of TRU elements is very small so that the sum of the $^{233}U$, Th, and MA inventories discharged through the fuel salt treatment is merely 0.02 Mg/yr. The fissile U content, comprising $^{233}U$ and $^{235}U$, is 69.0 wt% in the recirculating fuel stream, and may require denaturing the fuel cycle with the addition of $^{238}U$.

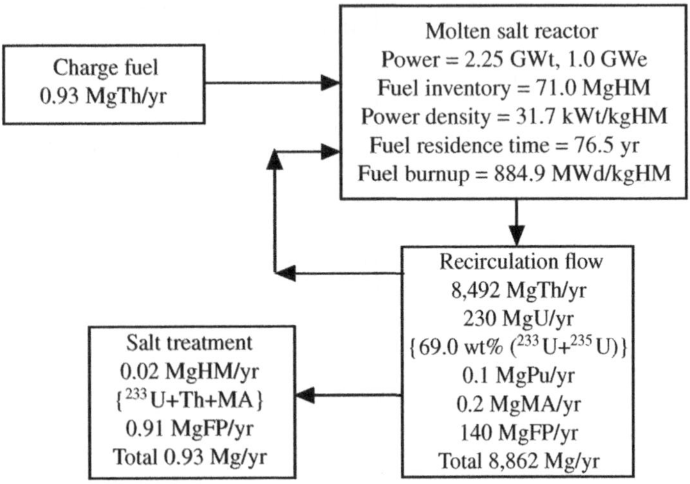


**Figure 12.19** Material flow for a molten salt reactor design with circulating HALEU and thorium fuel. Adapted from NFCE&S Evaluation Group 26. *Source:* Wigeland et al. [Wig14].

Depending on the salt utilized, a MSR could be configured either as a thermal or fast spectrum reactor. With KCl salt as the coolant and $UPuCl_3$ as the fuel, the Moltex design indeed suggests a fast spectrum reactor configuration with an online refueling strategy and spent CANDU fuel used as the feed material. Molten salt fuel syntheses will involve electrochemical processes under inert atmosphere, as the salts form corrosive acids in the presence of water. Radiolysis could present problems for aqueous processes in high radiation fields involving fluoride or chloride salts [McF21].

5. **Two-Tier Coupled SFR-PWR System**
   Coupling a SFR with a full breeding capability to a PWR in a symbiotic manner, the system could recycle the Pu generated in both the PWR and SFR cores. The flow sheet of Figure 12.20 represents a synergistic arrangement where only Pu, together with the uranium fuel, is recycled and MAs and FPs are sent to a repository. The only feed material required then is a small amount of natural uranium for the combined system replenishing the FPs and MAs to be sent to a repository.

   The coupled system is structured so that the combined power output is 1.0 GWe, with a 60%–40% split between the SFR and the PWR. For the SFR itself, the power production is split 90%–10% between the driver and blanket regions. The PWR configuration is nearly equivalent to that of Figure 12.16, if the power output is normalized to 1.0 GWe, with the fuel residence time of 3.9 yr compared with 4.1 yr in Figure 12.16 and the identical discharge fuel

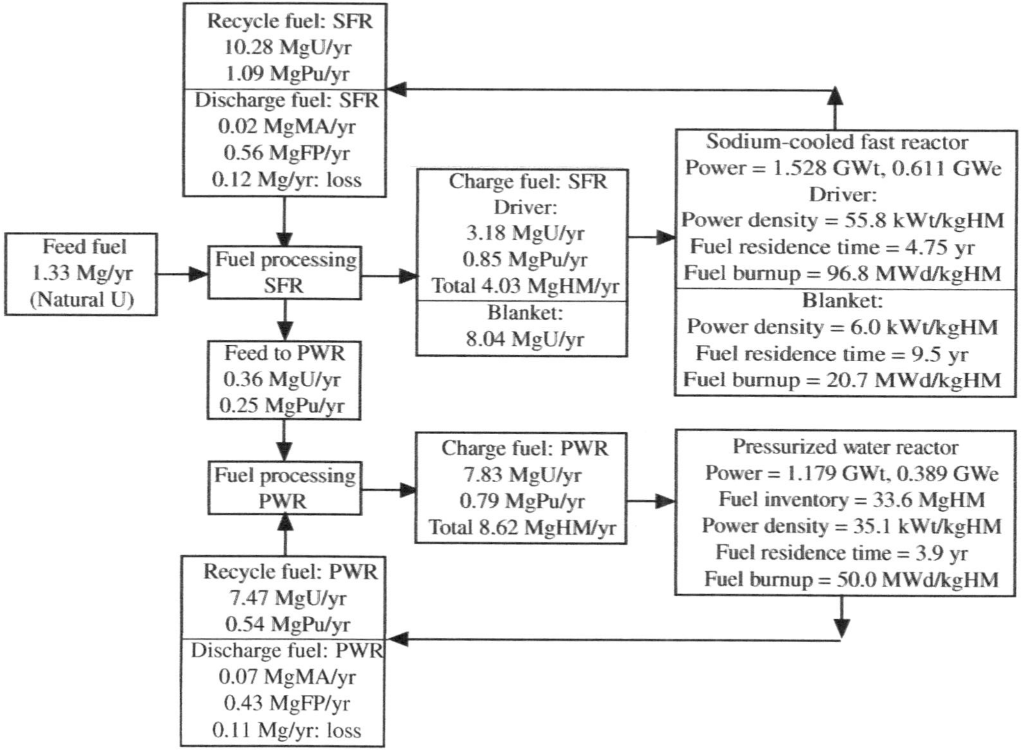


**Figure 12.20** Material flow for a coupled SFR-PWR system, utilizing Pu-U-Zr fuel for the SFR and U-Pu mixed oxide fuel for the PWR. Adapted from NFCE&S Evaluation Group 29. *Source:* Wigeland et al. [Wig14].

burnup of 50 MWd/kgHM. The main difference, of course, is the PWR use of Pu remaining from the previous PWR cycle combined with Pu produced in the SFR. With full recycling, the inventory of U and Pu from the previous SFR cycle is split between the charge to the SFR and additional mix to the PWR to bring up the required fissile contents for the PWR charge fuel. It should be also noted that the discharged fuel from both the SFR and PWR comprise MA and FP, and a total of 0.23 Mg of inventory, out of a total feed inventory of 1.33 Mg of natural U, i.e. equivalent to 17% of the total feed stream, is assumed lost during the recycling. Furthermore, it is useful to note that the 60%–40% power split between the SFR and PWR is approximately represented in the FP production split of 0.56 Mg vs. 0.43 Mg. The total discharge fuel inventory, including the reprocessing loss, is naturally equal to the total feed rate of 1.33 Mg/yr, which compares with the discharge rates of 21.9 Mg/yr for the open-cycle PWR in Figure 12.16 and 5.9 Mg/yr for the open-cycle SFR in Figure 12.17.

Two important final observations relate to: (a) net Pu *production* of $(1.09-0.85=0.24)$ Mg in the SFR vs. net Pu *consumption* of $(0.79-0.54=0.25)$ Mg in the PWR and (b) MA production of 0.02 Mg in the SFR vs. 0.07 Mg in the PWR, despite the 60%–40% power split between the two reactors. These observations translate to a key potential of the SFR serving as a breeder of fissile Pu despite a minor negative impact due to tracer quantities of MA present in the recycled fuel. These features are not explicitly illustrated in the Natrium design summarized in Figure 12.17, because the design features U-Zr HALEU fuel.

### 12.8.3 Comparison of Material Flow and Repository Metrics

The charge and discharge fuel material flows illustrated in Figures 12.6–12.20 are now reflected in a comparison of UNF radioactivity and radiotoxicity over a geological repository life in Tables 12.6–12.9. The repository metrics including radioactivity at 100 and $10^5$ yr and ingestion radiotoxicity at $10^5$ yr are compared for the three advanced reactor designs, together with the reference PWR and two-tier SFR-PWR recycle system. In the NFCE&S report, an adjustment is made to the discharge fuel inventories and the corresponding repository metrics to normalize the thermal efficiency to 33% selected for the reference PWR design. For the analysis of repository metrics, the NAS Committee decided not to adopt the normalization process, because the increased thermal efficiencies of the advanced reactor designs represent a unique feature of the advanced concepts and hence should be retained.

As discussed earlier in connection with Figure 12.16 for a reference PWR configuration, a 1.0-GWe plant produces approximately 1.0 Mg of FPs in 1 yr, as quantified via *fima* of Eq. (12.22). The yields for short- and long-lived FPs of

**Table 12.6** Used nuclear fuel radioactivities at 100 yr (kCi/GWe-yr).

| PWR (EG01) | | SFR (Natrium) | | HTGR (EG02) | | MSR (EG26) | | SFR-PWR (EG29) | |
|---|---|---|---|---|---|---|---|---|---|
| **Total:** | **1,340** | **Total:** | **530** | **Total:** | **940** | **Total:** | **959** | **Total:** | **1,010** |
| $^{137}$Cs | 341 | $^{137}$Cs | 262 | $^{137}$Cs | 231 | $^{137}$Cs | 245 | $^{137}$Cs | 292 |
| $^{137}$Ba* | 322 | $^{90}$Sr | 168 | $^{137}$Ba* | 219 | $^{90}$Sr | 235 | $^{137}$Ba* | 277 |
| $^{90}$Y | 215 | $^{238}$Pu | 30 | $^{90}$Sr | 161 | $^{90}$Y | 235 | $^{241}$Am | 167 |
| $^{90}$Sr | 215 | $^{239}$Pu | 21 | $^{90}$Y | 161 | $^{137}$Ba* | 231 | $^{90}$Y | 99 |
| $^{241}$Am | 122 | $^{240}$Pu | 8 | $^{241}$Am | 87 | | | $^{90}$Sr | 99 |

*Source:* Wigeland et al. [Wig14] and Hejzlar [Hej21].

**Table 12.7** Used nuclear fuel radioactivities at $10^5$ yr (Ci/GWe-yr).

| PWR (EG01) | | SFR (Natrium) | | HTGR (EG02) | | MSR (EG26) | | SFR-PWR (EG29) | |
|---|---|---|---|---|---|---|---|---|---|
| **Total:** | **1,650** | **Total:** | **1,580**[a)] | **Total:** | **1,350** | **Total:** | **1,020** | **Total:** | **818** |
| $^{239}$Pu | 494 | $^{239}$Pu | 1,180 | $^{239}$Pu | 228 | $^{99}$Tc | 245 | $^{99}$Tc | 289 |
| $^{99}$Tc | 304 | $^{99}$Tc | 254 | $^{99}$Tc | 210 | $^{93}$Zr | 62 | $^{239}$Pu | 137 |
| $^{243}$Pu | 61 | $^{93}$Zr | 48 | $^{234}$U | 77 | $^{93}$Nb* | 59 | $^{233}$Pa | 43 |
| $^{93}$Zr | 56 | $^{135}$Cs | 40 | $^{222}$Rn | 52 | $^{209}$Pb | 53 | $^{237}$Np | 43 |
| $^{93}$Nb* | 54 | $^{234}$U | 20 | $^{218}$Po | 52 | $^{213}$Bi | 53 | $^{135}$Cs | 39 |
| $^{234}$U | 53 | $^{237}$Np | 12 | $^{226}$Ra | 52 | $^{217}$At | 53 | $^{93}$Zr | 36 |
| $^{233}$Pa | 37 | $^{236}$U | 10 | $^{214}$Pb | 52 | $^{221}$Fr | 53 | $^{93}$Nb* | 35 |
| $^{237}$Np | 37 | | | $^{214}$Bi | 52 | $^{229}$Th | 53 | $^{126}$Sb* | 16 |
| $^{222}$Rn | 35 | | | $^{210}$Po | 52 | $^{225}$Ra | 53 | $^{126}$Sn | 16 |
| $^{214}$Pb | 35 | | | $^{210}$Pb | 52 | $^{225}$Ac | 53 | $^{235}$U | 15 |
| $^{218}$Po | 35 | | | $^{214}$Po | 52 | $^{213}$Po | 51 | $^{213}$Bi | 14 |
| $^{214}$Bi | 35 | | | $^{210}$Bi | 52 | $^{233}$U | 50 | $^{217}$At | 14 |
| $^{226}$Ra | 35 | | | $^{230}$Th | 52 | $^{234}$U | 16 | $^{221}$Fr | 14 |
| $^{210}$Pb | 35 | | | $^{242}$Pu | 43 | $^{126}$Sb* | 16 | $^{225}$Ra | 14 |
| $^{210}$Po | 35 | | | $^{93}$Zr | 38 | $^{126}$Sn | 16 | $^{225}$Ac | 14 |
| $^{210}$Bi | 35 | | | $^{93}$Nb* | 37 | $^{210}$Po | 11 | | |
| $^{214}$Po | 35 | | | $^{233}$Pa | 27 | $^{214}$Po | 11 | | |
| $^{230}$Th | 35 | | | $^{237}$Np | 27 | $^{222}$Rn | 11 | | |
| $^{135}$Cs | 15 | | | $^{135}$Cs | 17 | | | | |
| $^{233}$U | 13 | | | | | | | | |

a) Excludes contributions from isometric $^{235}_{92}U^*$.
*Source:* Wigeland et al. [Wig14] and Hejzlar [Hej21].

**Table 12.8** Ingestion radiotoxicity of UNF at $10^5$ yr (MSv/GWe-yr).

| PWR (EG01) | | SFR (Natrium) | | HTGR (EG02) | | MSR (EG26) | | SFR-PWR (EG29) | |
|---|---|---|---|---|---|---|---|---|---|
| **Total:** | **8.84** | **Total:** | **11.0** | **Total:** | **7.63** | **Total:** | **2.29** | **Total:** | **2.01** |
| $^{239}$Pu | 4.57 | $^{239}$Pu | 10.9 | $^{210}$Po | 2.32 | $^{229}$Th | 0.95 | $^{239}$Pu | 1.27 |
| $^{218}$Po | 1.55 | $^{237}$Np | 0.05 | $^{239}$Pu | 2.11 | $^{210}$Po | 0.48 | $^{229}$Th | 0.25 |
| $^{210}$Pb | 0.89 | $^{234}$U | 0.04 | $^{210}$Pb | 1.33 | $^{210}$Pb | 0.28 | $^{237}$Np | 0.17 |
| $^{242}$Pu | 0.54 | $^{236}$U | 0.02 | $^{226}$Ra | 0.54 | $^{225}$Ra | 0.19 | $^{210}$Po | 0.08 |
| $^{226}$Ra | 0.36 | $^{99}$Tc | 0.01 | $^{230}$Th | 0.40 | $^{226}$Ra | 0.11 | $^{225}$Ra | 0.05 |
| $^{230}$Th | 0.27 | | | $^{242}$Pu | 0.38 | $^{233}$U | 0.10 | $^{210}$Pb | 0.04 |
| $^{229}$Th | 0.22 | | | $^{229}$Th | 0.16 | $^{230}$Th | 0.08 | $^{242}$Pu | 0.03 |
| | | | | $^{234}$U | 0.14 | | | $^{233}$U | 0.03 |

*Source:* [Wig14, Hej21].

**Table 12.9** UNF inventory and repository metrics of advanced nuclear reactors.

| Fuel cycle and reactor type (Reference) | UNF Inventory (Mg/GWe-yr) | Radioactivity at 100 yr (MCi/GWe-yr) | Radioactivity at $10^5$ yr (kCi/GWe-yr) | Radiotoxicity at $10^5$ yr (MSv/GWe-yr) |
|---|---|---|---|---|
| Light water reactor LEU PWR (NFCE&S EG01) | 21.92 | 1.34 | 1.65 | 8.8 |
| Fast spectrum reactor HALEU (TerraPower Natrium) | 5.93 | 0.53 | 1.58[a)] | 11.0 |
| Gas cooled reactor HALEU TRISO (NFCE&S EG02) | 6.09 | 0.94 | 1.35 | 7.6 |
| Molten salt reactor HALEU Th-$^{233}$U (NFCE&S EG26) | 0.93 | 0.96 | 1.02 | 2.3 |
| Two-tier SFR-PWR Recycle (NFCE&S EG29) | 1.33 | 1.01 | 0.82 | 2.0 |

a) Excludes contributions from isometric $^{235}_{92}$U*.
*Source:* [Wig14, Hej21].

interest in the used fuel management and geological repository analysis depend somewhat on the neutron flux spectrum and fissile nuclides, but the sum of FP yields of interest appear nearly independent of the fuel cycle. The comparison of UNF radioactivity for five fuel cycles was presented in the NAS 2023 report, but the radiotoxicity comparison, in particular, for the Natrium design, has been introduced in this chapter. We offer a few perspectives:

(1) Effective use of HALEU with a thermal efficiency of 40 ~ 50% for the SFR and HTGR, compared with 33% for the PWR, reduces the fuel discharge rates to ~30% of the PWR counterpart. The HTGR fuel inventory is also reduced to ~30% of the PWR inventory. The continuous circulating MSR and coupled SFR–PWR system requiring fuel reprocessing reduce the UNF inventory to 5% of the PWR discharge fuel.
(2) Despite the decrease in the UNF and FP inventories for the four advanced systems analyzed, the differences in the FP and HM compositions evolve to yield the radioactivity at 100 yr fairly similar to that for the LWR. In the coupled SFR-PWR system, U and Pu are recycled and MAs and FPs are to be sent to a geological repository. The UNF radioactivity at 100 yr depends heavily on FP radioactivity, which evolves gradually so that, at $10^5$ yr following the discharge, the radioactivity depends significantly on the Pu and Tc inventories.
(3) Table 12.8 indicates that the ingestion radiotoxicity at $10^5$ yr comes mostly from the Pu and Ra toxicity which for the MSR and coupled SFR-PWR system is reduced to 25% of the LWR radiotoxicity. The evolution of key metrics up to $10^5$ yr, summarized in Table 12.9, is illustrated in Figure 12.21, indicating graphically the complexity involved in the evolution of various HM and FP inventories for the five representative fuel cycles analyzed. Figure 12.22 succinctly illustrates the key contributors to the ingestion radiotoxicity, evolving from short-lived FPs early on to Pu much later in the SKB-3 repository concept [Hed97].
(4) For the MSR project or other designs utilizing U-Th cycle, $^{233}$U ($t_{1/2} = 1.59 \times 10^5$ yr) produces {$^{229}$Th, $^{225}$Ac, $^{225}$Ra, $^{221}$Fr, $^{217}$At, $^{213}$Po, $^{213}$Pt, $^{209}$Pb} and decay products are more radioactive than $^{239}$Pu products for $10^5$ yr.

### 12.8.4 Fuel Reprocessing Technology

We begin with a review of general nuclear fuel reprocessing technology, followed by the particular issues that should be considered for advanced nuclear systems, featuring non-LWR coolant systems.

1. **Representative Reprocessing Technology**
   The UNF reprocessing technology that has been developed may be classified either as aqueous or non-aqueous pyroprocessing technology.

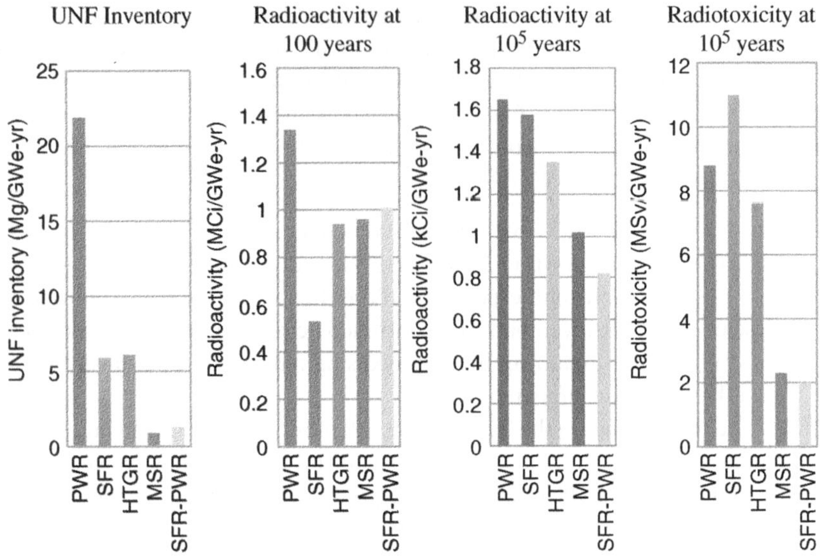


**Figure 12.21** Evolution of UNF ingestion radiotoxicity over time for five representative fuel cycles.

(1) Aqueous Reprocessing Technology

The plutonium uranium reduction extraction (PUREX) process is the best developed process for UNF processing, which produces two separate product streams: (a) combined (Pu + U) stream and (b) nitric acid solution comprising MA and FP, together with solid wastes comprising metal hull. The nitric acid solution is mixed with glass matrix and vitrified into a stainless steel container. Volatile $^{129}$I is captured in trap, with 1% retained in vitrified logs for verification purposes. Long-lived FPs could be also extracted by modifying the PUREX process, as developed in the French CEA technology. In this regard, transmutation of $^{129}$I and $^{99}$Tc has been considered, but requires further study [NAS23].

Conditioning/processing of waste for disposal typically involves vitrification in borosilicate glass. The short-lived FPs, $^{137}$Cs ($t_{1/2} = 30.2$ yr) and $^{90}$Sr ($t_{1/2} = 28.8$ yr), account for 90% of FP decay heat and, if the UNF is allowed to decay in storage for 60 yr, the heat load could decrease by a factor of two.

A coupled two-tier SFR-LWR system could recycle and destroy MA, but further study is needed for reprocessing and recycling used SFR-MOX fuel via aqueous processes. Recycling Np is manageable, but Am presents more difficulty during fabrication due to the $^{239}$Np $\gamma$-activity from the decay of $^{243}$Am.

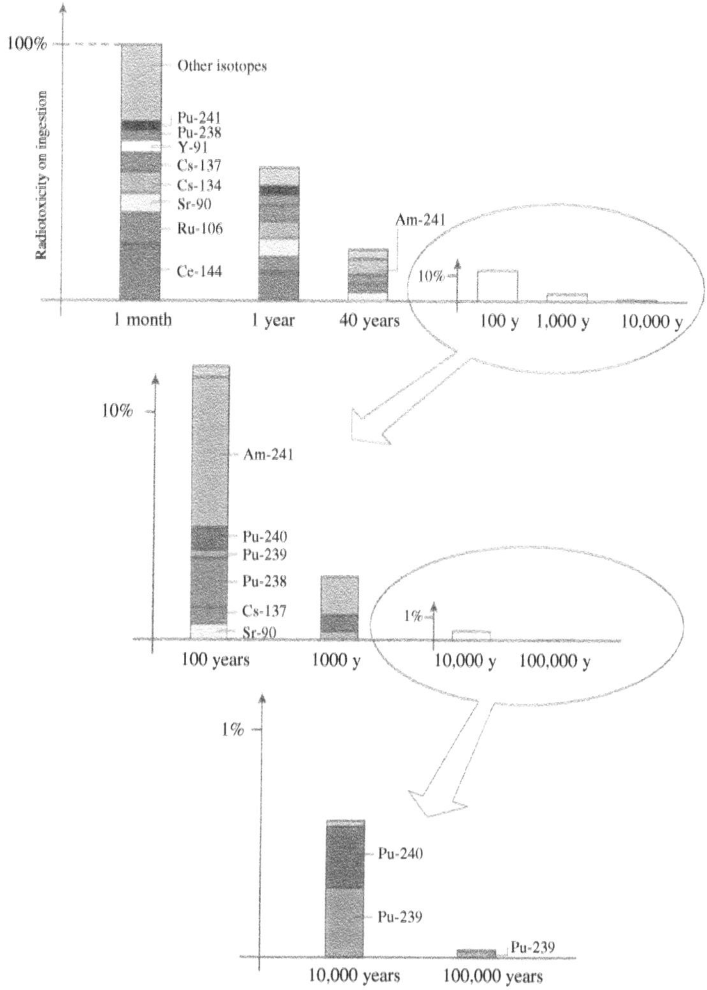


**Figure 12.22** Evolution of UNF ingestion radiotoxicity over time for the SKB-3 repository concept. *Source:* Hedin [Hed97]/SKB.

Advanced PUREX aqueous reprocessing features two different strategies illustrated in Figure 12.23:

(a) GANEX homogeneous strategy [Poi17a] incorporates a solvent extraction process to separate all TRUs, (Pu/Np/Am/Cm), from FPs in a solution of dissolved UNF for oxide fuel, but it could be also applied for metallic fuel.

(b) Heterogeneous reprocessing features the COEX process [Poi17a] developed in France or the UREX process [Van04] developed in

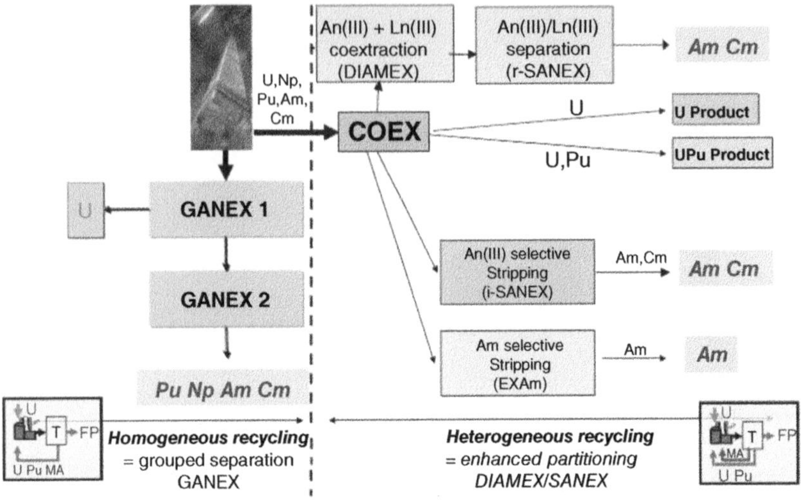


**Figure 12.23** Advanced aqueous reprocessing schemes: homogenous recycling of entire TRU via GANEX and heterogeneous recycling via SANEX that separates Am/Cm from Pu/U and ExAm that separates Am from Cm. *Source:* Poinssot et al. [Poi17a]/HAL/Public domain; AREVA S.A / https://radioactivity.eu.com/articles/nuclearenergy/mox_fuels / Last accessed on July 31, 2024.

the United States and proceeds through the DIAMEX-SANEX or TRUEX-TALSPEAK process, where (i) all trivalent actinides Am and Cm, written as An(III) in Figure 12.23, together with lanthanides are separated from U/Pu or U/Pu/Np in DIAMEX and (ii) subsequently an Am/Cm stream is generated, separate from lanthanides, in SANEX. In the COEX/UREX process, all MOX (U/Pu or U/Pu/Np) remain together in a separate stream, thereby addressing potential proliferation concerns. In the EXAm process [Poi17a] developed in France, Am is further separated from the Am/Cm solution, with Cm allowed to decay in storage for 100 yr, resulting mostly in Pu [NAS23].

The evolution of UNF radiotoxicity in Figure 12.24 indicates the toxicity of Pu, MA, and FP over $1.0 \times 10^6$ yr, clearly suggesting the desirability of P&T of TRUs [Poi17b]. If the TRU radiotoxicity is removed, the UNF radiotoxicity is expected to reach that of natural U in a few hundred years. In France, where Pu recycling has been practiced for a number of years, a strategy was adopted, as a national effort via the 2006 Waste Management Act, known as Bataille's Law, to study separate partitioning of minor actinides, Am, Cm, and Np, from Pu and U. This has resulted in significant progresses made with the DIAMEX-SANEX-EXAm processes, together with continuing Pu recycling in the current generation of PWRs in France.

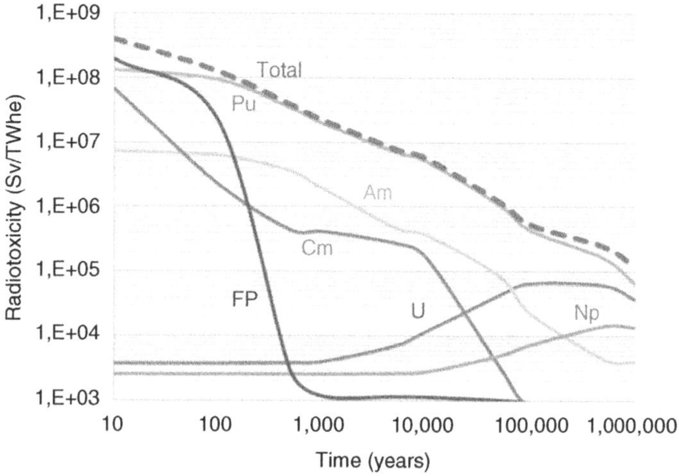


**Figure 12.24** UNF radiotoxicity as a function of time over $10^6$ yr. *Source:* Poinssot et al. [Poi17b]/HAL/Public domain.

All commercial reprocessing facilities have used the PUREX, UREX, or COEX process. Orano has reprocessed to date 37,000 MgHM LEU and 2,900 MgHM MOX fuel [Gay21]. This has allowed 20 French PWRs operating with 1/3 of the fuel assemblies loaded with MOX fuel. It is estimated that the fuel recycling has reduced the UNF volume by a factor of 5 and the radiotoxicity by an order of magnitude. La Hague can reprocess up to 1,300 MgHM/yr of UNF [Wor21]. B. Cipiti suggests [Cip21] that an investment of $60B may be required to establish a meaningful PUREX reprocessing facility in the United States.

(2) Pyroprocessing Technology

Pyroprocessing is the primary non-aqueous reprocessing technology developed to date [Lai97] and involves electrorefining techniques outlined in Figure 12.25 and implemented with LiCl-KCl molten salt at 773 ~ 1,173 K, forming anodic dissolution, followed by metal deposition at the cathode. Three waste streams are produced: (a) glass-bonded sodalite comprising TRU, some U, and FP, (b) mostly U, and (c) metal hull. In this process, the salt plus zeolite forms sodalite and glass is added. The iron–phospate glass may yield better performance than the ceramic waste and alternate liquid-liquid reductive extraction process is possible for oxide fuel [NAS23].

The EBR-II metallic fuel was processed via the electrometallurgical treatment (EMT) process since 1996 using LiCl-KCl eutectic mixture. In addition, oxide fuel has been reprocessed via the pyroprocssing technology at the RIAR facility in Russia [Vav04].

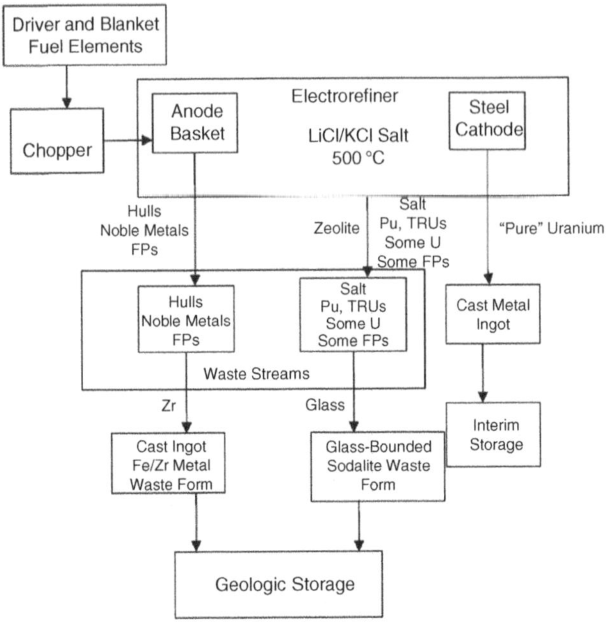


**Figure 12.25** Schematic flowsheet for pyroprocessing of EBR-II metallic fuel. *Source:* Lee et al. [NAS23]/National Academy of Sciences.

Since salts tend to be hygroscopic and metals pyrophoric, the EMT processes should be performed in inert atmosphere. In addition, CsO may be left in the crucible as slag, resulting in a highly radioactive FP remnant. Pyroprocessing is a compact batch process but may be difficult to scale up. A 2018 Argonne National Laboratory study [Cha19] estimates an expenditure of \$900M capital and \$90M/yr operating cost for a 400 MgHM/yr pyroprocessing facility.

It should be noted that the UNF reprocessing reduces the HLW volumes but usually increases ILW/LLW inventories. For example, off-gas $^{3}$H, $^{14}$C, $^{95}$Kr, $^{129}$I, $^{135}$Cs, $^{79}$Se, and $^{99}$Tc require separation and additional treatment; it is naturally much easier to handle ILW than HLW. Reprocessing Pu can be performed with a glove box, while reprocessing MA requires a shielded facility. Na-bonded fuel requires chemical treatment for Na and is not suitable for direct disposal in geological repository [NAS23].

2. **Reprocessing of UNF for Advanced Reactors**
With higher enrichments, HALEU allows a smaller UNF mass and a higher fuel burnup. Used fuel recycling reduces the total UNF mass but may not reduce heat load of the waste stream significantly. As summarized in Table 12.6, FPs dominate the UNF radioactivity at 100 yr after the fuel discharge and advanced

reactors are expected to produce only slightly reduced levels of radioactivity and decay heat per unit energy generated. Similarly, radioactivity and radiotoxicity at $10^5$ yr are reduced, as summarized in Tables 12.7 and 12.8, but the effects may not be significant for geological disposal, if geochemical and geological conditions are properly selected to limit the mobility of radionuclides.

Advanced reactors and SMRs will have little impact on the waste management issues and nuclear industry will still need geological repositories. The main challenge is the containment of long-lived FPs contributing to the long-term dose and hence TRU recycling does not significantly impact geological repository (GR) performance. Integrated or modular reactors produce UNF inventories per energy generated similar to those of current LWRs, while the SMRs could produce increased ILW due to increased neutron leakage from the core. This is discussed in connection with the modular NuScale features in Section 1.5.2 and with Figure 12.16.

The high-temperature gas-cooled reactors may produce activation products from impurities $^{14}N$, $^{13}C$, and $^{17}O$, while He gas may be contaminated by radionuclides from failed fuel particles, together with radioactive C and Si remaining in the waste stream. A TRISO-fueled PBMR, e.g. the Xe-100 design, produces $160\,m^3 \cdot yr^{-1}$ of pebbles, compared with $6.8\,m^3 \cdot yr^{-1}$ of UNF from a PWR plant [Mul21]. Graphite blocks exposed to fast neutron irradiation require special care and may suffer distortion and cracking. The TRISO designs require (a) development of non-destructive techniques to separate the fuel kernels from the graphite matrix and (b) disposal of large volumes of radioactive graphite as ILW/LLW [NAS23].

The Natrium SFR design plans to use metal fuel, sodium-bonded to HT9 cladding, for enhanced heat transfer, producing 80 Mg ($800\,m^3$) of irradiated sodium reflectors and internals, 70 Mg of control rods for processing at the time of decommissioning [Hej21]. The EBR-II decommissioning produced 327 Mg of Na and Fast Flux Test Facility (FFTF) 900 Mg, requiring special attention to $^{23}Na$, $^{3}H$, $^{60}Co$, and $^{137}Cs$, especially since Na reacts with moisture and air, posing fire hazard. A total inventory of 1.7 Mg of EBR-II salt waste from the EMT will generate 51 Mg of ceramic waste, a factor of 30 increase in waste mass [Fra14]. Na-bonded fuel requires additional chemical treatment before geological disposal, while annular metal fuel designs, which could allow for direct disposal, are under development,

The MSR designs require multiple, complex waste streams, some including continuous treatment of circulating fuel and off-gas waste. While particulates, aerosols, and reactive gases ($I_2$, $Cl_2$, $F_2$) can be immobilized in ceramic waste form, the $^{3}H$ waste is more complex [McF21]. It may be possible to solidify fuel salt and dispose it directly in a geological repository, but this process requires further research. A significant increase in waste mass is possible, as experienced for the EBR-II salt treatment process. It has been observed that the MSRE fuel salt in storage at the Oak Ridge National Laboratory generates F and $UF_6$ gas via radiolysis, posing potential criticality concerns and ORNL is not actively considering online reprocessing at MSR plants at this point. The MSRE used fluoride salts, but some molten fuel designs propose to use chloride salts for harder spectrum for breeding, which requires enrichment in $^{37}$Cl. Such designs need to limit the production of $^{36}$Cl ($t_{1/2} = 3.0 \times 10^5$ yr), and desire $^7$Li to reduce $^3$H production. Managing volatile FPs in the molten salt system offgas is a significant design issue [McF21]. Finally, one possible path forward for the MSR used fuel treatment could feature processing into stable ceramic Synroc [NAS23].

## 12.9 Key Considerations for Geological Disposal

A defense-in-depth approach, similar to that adopted for the safety analysis of nuclear power plants involving a combination of {prevention, mitigation, protection}, should be adopted also for the UNF repository. For the UNF disposal, the defense-in-depth approach entails {waste matrix, UNF canister, buffer seals, host rock, geological retention/immobility}.

A geological repository (GR) should thus account for (a) radionuclide (RN) source term including (i) thermal pulses from short-lived FP (SLFP), (ii) $\alpha$-decay of TRU for radiolysis of water, (iii) long-lived mobile FP, (iv) release of long-lived TRU [Gra08], (b) behavior of engineered barriers including the durability and heat load, (c) geophysical, hydrological properties of the repository, (d) RN transport through geological media, and (e) exposure pathways. GR performance should also consider chemical and radiological complexity: (a) SLFP $^{131}$I, $^{137}$Cs, $^{90}$Sr, (b) long-lived FP (LLFP) $^{129}$I, $^{99}$Tc, $^{79}$Se, (c) long-lived HM $^{235}$U, $^{238}$U, $^{239}$Pu, Am, Cm, and (d) activation products $^{60}$Co, $^{63}$Ni [Ewi21, Gra08, NAS23].

### 12.9.1 Essential Issues for Geological Disposal

Eight chemical processes for the evolution of a multitude of chemical species and corrosion of UNF in a GR are illustrated in Figure 12.26: (a) dissolution of the fuel, (b) precipitation of secondary phases onto the fuel surface, (c) colloid formation, (d) sorption of RNs onto corrosion products of the metal canister, (e) formation of gases, e.g. $H_2$, during corrosion of metal, (f) formation of chemical species in solution, (g) cation exchange/sorption between RNs and host rocks, and (h) formation of $H_2O_2$, $O_2$, and OH due to radiolysis [Ewi21, NAS23]. Processes controlling the RN release could also be viewed in terms of (a) kinetics of waste form corrosion, (b) formation of secondary, alternate phases, (c) reduction and

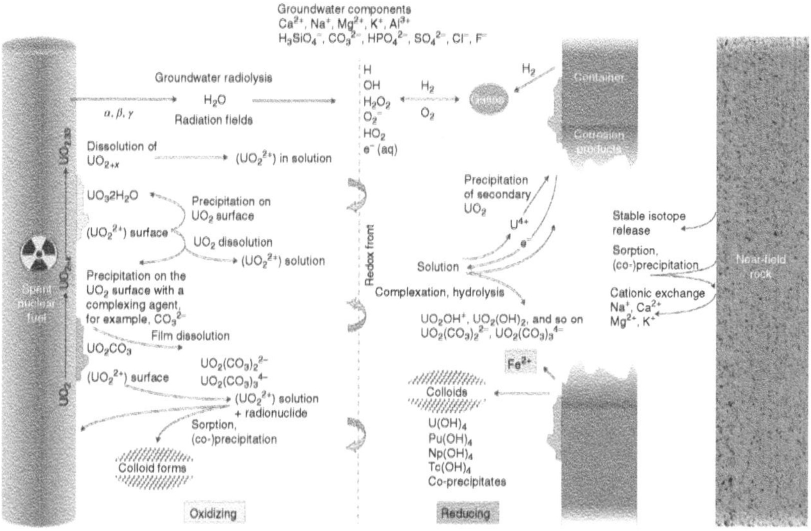


**Figure 12.26** Chemical processes for evolution of species and corrosion of UNF in geological repository. *Source:* Lee et al. [NAS23]/National Academy of Sciences, Originally published by R.C. Ewing (2015)/With permission of Springer Nature.

sorption into near-field materials. Rates for the processes depend on thermal gradient between the UNF and host rock; long-lived mobile FPs travel in ground water as negative-charged and are not sorbed onto mineral surfaces, which are also charged negative [NAS23].

Transuranics are not highly mobile compared with FPs in reducing GR environment; the redox or reduction-oxidation state of the GR can significantly affect the mobility of multivalent actinides. Used nuclear fuel reprocessing reduces long-lived TRUs, but retains SLFP, LLFP, and activation products. The LLFP $^{129}I$, $^{79}Se$, $^{36}Cl$ are mobile in all GR environments [Swi20]. Reducing environment reduces mobility of long-lived TRUs, e.g. organic clay can limit the TRU mobility to tens of meters up to a million year. The UNF dissolution rate in the GR also depends heavily on the geological characteristics: $10^{-8} \sim 10^{-6}$/yr in reducing media vs. $> 10^{-5}$/yr in oxidizing media [Swi10]. The release of $^{129}I$ is faster in MOX fuel than in $UO_2$ and, for vitrified waste from reprocessing, borosilicate glass is not effective for retention of $^{129}I$. Iodine, however, can be fabricated into durable apatite (phosphate mineral) [Ewi21].

Radionuclides contributing significantly to the radiation doses in the near-term and late-term GR performance are illustrated in Figure 12.27, where long-term doses from $^{98}Tc$ and $^{129}I$ are noted for the YM repository [Swi08]. Figure 12.28 illustrates radiotoxicity release rates into the buffer, Opalinus clay bed, and eventually into the biosphere over a period of $10^7$ yr from a Swiss GR performance

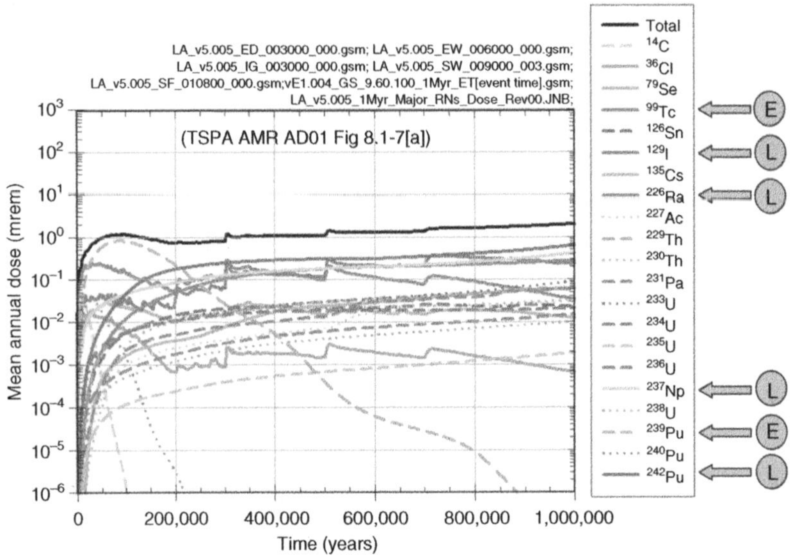


**Figure 12.27** Radionuclides important to Yucca Mount repository mean dose, early (E) and late (L). *Source:* Swift and Knowles [Swi08]/U.S. Department of Energy.

analysis [Zui21]. As discussed in connection with Figure 12.24, if the TRU radiotoxicity is removed, the UNF radiotoxicity is expected to reach that of natural U in a few hundred years. In addition, TRU recycling reduces long-term heat load in reducing, saturated clay GR proposed in France; direct disposal of 80,000 MgHM UNF requires 180,000 $m^3$ of GR space, while vitrified HLW disposal may require 13,000 $m^3$ [Wor21].

The disposal of UNF at the YM repository was studied by solubility limits and sorption of RNs, indicating that the reduction of HM inventory has little effect on disposal risk. The performance assessment suggests (a) some fraction of waste package could fail by $O(10^3$ yr) and (b) most of the waste forms are exposed to geochemical conditions by $O(10^5$ yr), with risk significance indicated for both TRU and FP: $^{36}$Cl, $^{79}$Se, $^{99}$Tc, $^{129}$I, $^{233}$Pa, $^{233}$U, $^{238}$U, $^{237}$Np, $^{239}$Pu, $^{240}$Pu, $^{241}$Pu, $^{241}$Am [Pet07]. In particular, the dominant risk is associated with $^{99}$Tc, $^{14}$C, $^{239}$Pu, $^{129}$I up to $10^4$ yr and with $^{99}$Tc, $^{129}$I up to $10^6$ yr [Han10].

### 12.9.2 Key Features of Geological Repository

Engineered barriers developed and proposed to date include: (a) processed borosilicate glass with low corrosion rate, (b) metal canister waste package comprising steel and copper, (c) bentonite backfill for heat-producing canisters, and (d) MgO

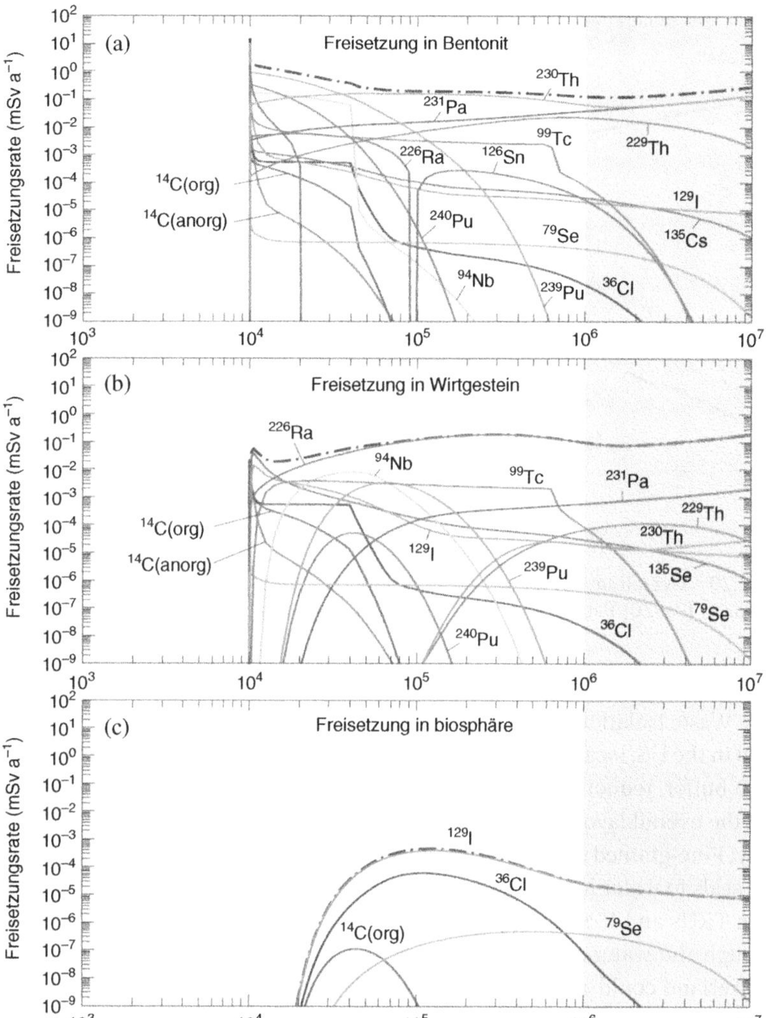


**Figure 12.28** Major contributors to dose in (a) buffer (top), (b) Opalinus clay (middle), (c) biosphere (bottom) associated with UNF disposal in Switzerland. *Source:* Adapted from Zuidema [Zui21].

emplaced with TRU waste. Geological barriers including (a) salt dome, with low water content, (b) clay and shale with little water flow, (c) crystalline rock, with water contained in fracture, and (d) volcanic tuff, with substantial water content have been considered.

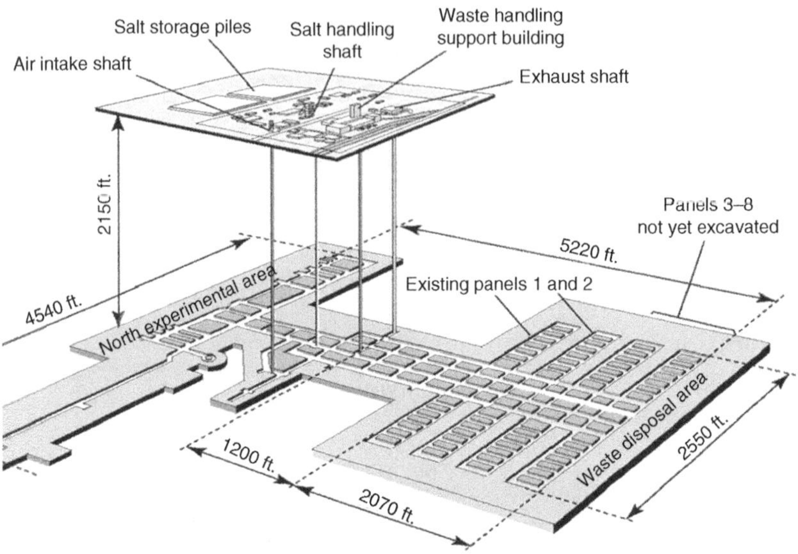


**Figure 12.29** Overall layout for the Waste Isolation Pilot Plant at Carlsbad, New Mexico. *Source:* Paviet [Pav20]/Pacific Northwest National Laboratory (PNNL).

Actual GRs developed or proposed include various geological formations:

(1) **Salt**: Waste Isolation Pilot Plant (WIPP) at Carlsbad, NM, is the only GR developed in the US, located in a 910-m thick bedded salt, 600 m below surface with MgO buffer, reducing TRU solubility. The repository stores TRU, mostly Pu, and the overall layout for the facility is illustrated in Figure 12.29.

(2) **Clay**: Fine-grained silicate with a sheet structure, e.g. bentonite provides effective seals to water ingress. Figure 12.28 illustrates that the dose contributions from TRUs and FPs are significant in the near-field buffer, but the transport through the water-saturated reducing environment of the Opalinus clay in Switzerland could mostly eliminate the TRU contributions to the biosphere, indicating the important role of geosphere in containing RNs.
A similar concept has been developed for the Cigéo Andra facility scheduled to open by 2040. The GR proposes to store 240 $m^3$ HLW and 3,700 $m^3$ ILW, and features reducing clay, 130 ~ 170 m thick, at a depth of 420 ~ 580 m, arranged with 2.0-km drifts and $10^3$ boreholes. Anionic RNs, $^{129}I$, $^{79}Se$, $^{14}C$, and $^{36}Cl$, play a significant role for the repository, with some RN migration to accessible environment possible over $10^5$ yr [Lan21].

(3) **Crystalline rock** represented by granite and basalt formations: The KBS-3 design implemented in the GR set to receive up to 6,500 Mg of UNF at

Onkalo, Finland, in 2024 or 2025 features UNF assemblies placed in Fe inserts and Cu canisters, surrounded by bentonite, deposited at 430 m in depth in groundwater-saturated reducing granite (gneise) rock; bentonite is porous clay formed from volcanic acid [Els22]. Figure 12.30 illustrates the key barrier

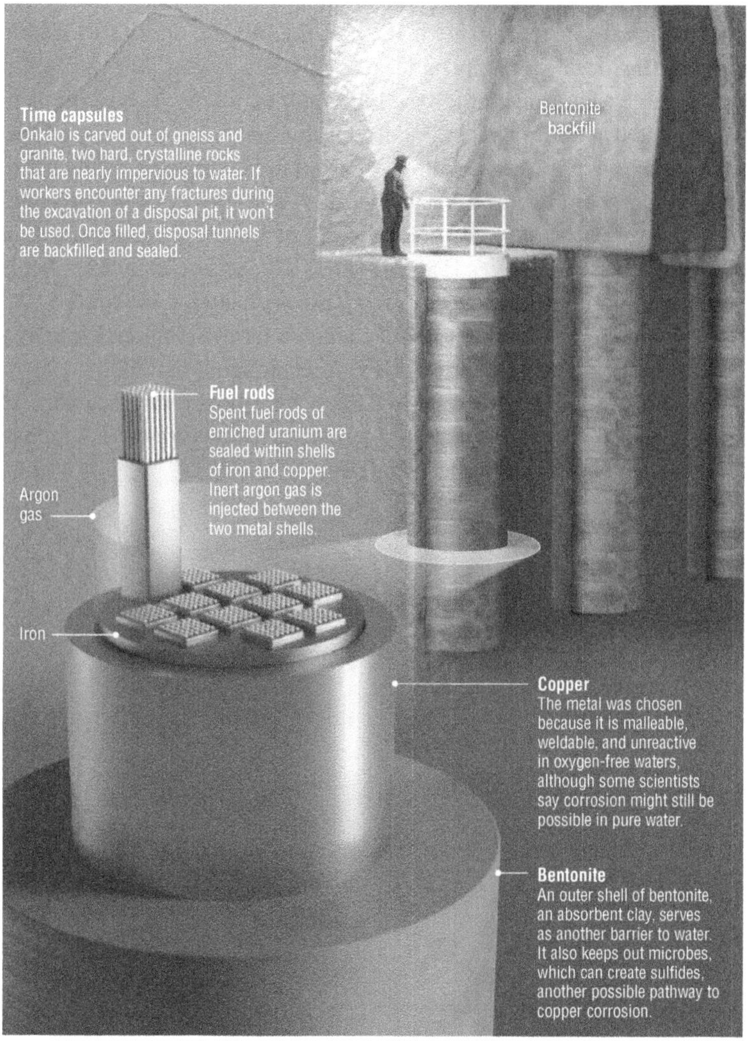


**Figure 12.30** Internal structures of the Onkalo geological repository. *Source:* Reproduced from Graphic: V. Altounian/Science. From Science, Vol. 375, Issue 6583 (2022)/with permission of American Association for the Advancement of Science – AAAS.

steps of the KBS-3 concept, with an actual copper canister visualized, together with the bentonite backfill in the GR layout. The GR will feature 100 tunnels.

The Forsmark repository in Sweden likewise follows the KBS-3 approach, with a plan to store 12,000 Mg UNF, and features a saturated granite arrangement with reducing environment such that the dissolution rate of $10^{-8} \sim 10^{-6}$/yr would control the radiation dose. Radionuclides are not expected to move more than 10 m for $10^6$ yr in the Onkalo repository [Swi10].

4. **Volcanic tuff**: Yucca Mountain is located in a dry region, but the tuff, a form of volcanic ash, proposed for the repository is not dry. Repository is to be located at 200 m above water table in unsaturated zone, with engineered barriers comprising Ni-Cd based alloy, 15-mm thick Ti drip shield, and stainless steel containment over the waste packages. The surface of the packages is designed to remain above the boiling point of water, to delay contact of water with the waste packages [Far03]. The UNF package including the Ti drip shield to minimize the seepage of moisture is illustrated in Figure 12.31. Fractured oxidizing environment raised concerns for the long-term integrity of the GR concept, with suggestions for modified designs [Car05].
5. **Deep borehole** disposal in crystalline rock for horizontal placement of waste packages <0.5 m diameter at 3 ~ 5 km depth is also proposed. Borehole disposal may not depend on waste forms [Nuc18].

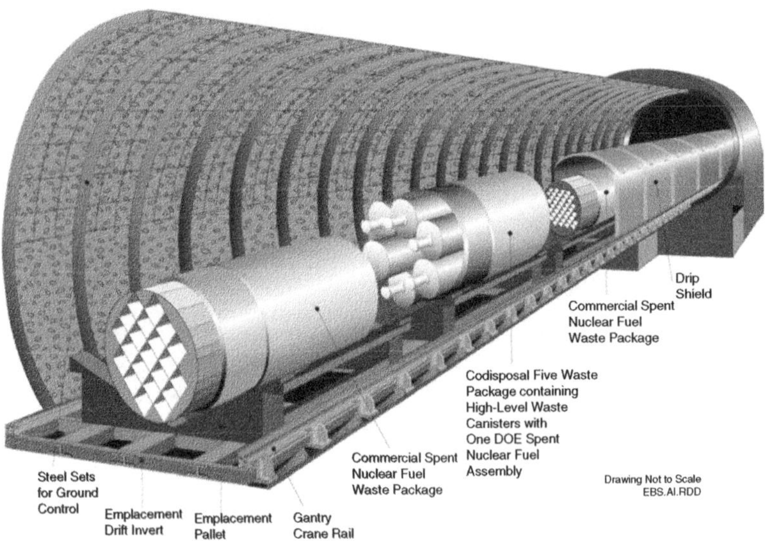


**Figure 12.31** UNF Engineered barrier system planned for the Yucca Mountain GR, Nevada. *Source:* Farmer et al. [Far03]/U.S. Department of Energy/Public domain.

### 12.9.3 General Comments on Reprocessing and Repository

Having discussed UNF characteristics for advanced reactors and fuel reprocessing technology in Section 12.8 and geological repository characteristics and features in Sections 12.9.1 and 12.9.2, respectively, we now offer brief considerations for the steps forward for fuel reprocessing and repository technology with continuing deployment of nuclear power plants in the United States and around the world. We note first that advanced reactors, with HALEU and high thermal efficiencies, could reduce the UNF inventory significantly, but the overall impact on radiotoxicity over time is more difficult to generalize. As discussed with the radiotoxicity for five representative fuel cycles in Table 12.9 and Figure 12.21, and for the SKB-3 concept in Figure 12.22, it is difficult to correlate the UNF radiotoxicity at $10^5$ yr after the discharge simply with the UNF inventory or radioactivity at discharge. It is also noted that the Th-$^{233}$U cycle generates radiotoxicity similar to those of the traditional U-Pu cycle.

Furthermore, radiotoxicity may not be the most significant issue for repository designs, if suitable arrangements are secured with reducing geological environments. Vitrified HLW volume could, however, be reduced by an order of magnitude via TRU recycling. Significant heat load reductions appear also possible for (a) short-lived FP for $50 \sim 70$ yr, (b) Pu and MA with $\alpha$-emissions $>100$ yr, and (c) thermal pulses due to FPs over $300 \sim 400$ yr, with suitable P&T processes adopted. Potential radiological dose reductions achievable via UNF recycling could be also significant for an oxidizing geological repository. In this regard, reducing environments are certainly preferred for a repository apart from the UNF reprocessing and recycling considerations [NAS23].

Small modular plants may certainly be designed so that the UNF inventory per energy generated could be comparable to those of large plants. Fission product inventories in the UNF will in general be proportional to the fission energy generated, with the actual isotopic composition depending somewhat on the fuel type and neutron flux spectrum. With a higher fuel burnup, the UNF from advanced reactors would generate a higher short-term heat load with FP ($^{90}$Sr, $^{137}$Cs) and a larger long-term mobile inventory of FP ($^{129}$I, $^{99}$Tc). Long-lived mobile FPs are essentially independent of reactor type and power rating. Advanced fuel cycles require significant investments and once-through cycle should continue, together with further P&T study conducted as new nuclear power plants are deployed.

TRU recycling with MA removal helps waste disposal, but it should be recognized that MA could be efficiently recycled in fast spectrum reactors and coupled two-tier fuel cycles should be considered, together with TRU recycling. Used MOX fuel, however, has a higher heat content than used $UO_2$ fuel and fuel recycling should be considered together with the overall objective of reducing the disposal volume and radiotoxicity and shortening the storage time in the repository as the motivation for the P&T of UNF. Together with a priority given for the development of a geological repository for the legacy UNF in the United States, additional effort effectively utilizing fuel reprocessing could help reduce the future accumulation of UNF from advanced reactors under development.

## References


**[ANS14]** American Nuclear Society, "American National Standard Decay Heat Power in Light Water Reactors," ANSI/ANS-5.1-2014, American Nuclear Society (2014).

**[Bar63]** R.F. Barry, "LEOPARD - A Spectrum Dependent Non-Spatial Depletion Code for the IBM-7094," WCAP-3269-26, Westinghouse Electric Corporation (1963).

**[Ben81]** M. Benedict, T.H. Pigford, and H.W. Levi, *Nuclear Chemical Engineering*, McGraw-Hill (1981).

**[Boa00]** C.E. Boardman, "A Description of the S-PRISM Plant," *ICONE 8*, paper 8168 (2000).

**[BRC12]** Blue Ribbon Commission on America's Nuclear Future, "Report to the Secretary of Energy," Blue Ribbon Commission on America's Nuclear Future (2012).

**[Cad67]** W.R. Cadwell, "PDQ-7 Reference Manual," WAPD-TM-678, Bettis Atomic Power Laboratory (1967).

**[Car05]** L.J. Carter and T.H. Pigford, "Proof of Safety at Yucca Mountain," *Science* **310**, 447 (2005).

**[Car11]** J.T. Carter, R.H. Jones Jr., and A.J. Luptak, "US Radioactive Waste Inventory and Characteristics Related to Potential Future Nuclear Energy Systems," FCRD-USED-2011-000068, rev. 1, US Department of Energy (2011).

**[Cha19]** Y.I. Chang, R.W. Benedict, M.D. Bucknor, et al. "Conceptual Design of a Pilot-Scale Pyroprocessing Facility," *Nucl. Technol.* **205**, 708 (2019).

**[Cip21]** B. Cipiti, "Advanced Reactor Safeguards and Security," presentation to the US National Academy of Sciences Committee, 17 May 2021, https://doi.org/10.17226/26500 (2023).

**[Der84]** K.L. Derstine, "DIF3D: A Code to Solve One-, Two-, and Three-Dimensional Diffusion Theory Problems," ANL-82-64, Argonne National Laboratory (1984).

**[DiG95]** A.S. DiGiovine, J.D. Rhodes III, and J.A. Umbarger, "SIMULATE-3, Advanced Three Dimensional Two-Group Reactor Analysis Code, User's Manual," Studsvik/SOA-95/15, Studsvik of America (1995).

**[DOE02]** *A Technology Roadmap for the Generation IV Nuclear Energy Systems*, https://www.gen-4.org/gif/jcms/c40473/a-technology-roadmap-for-generation-iv-nuclear-energy-systems (2002).

**[Dri90]** M.J. Driscoll, T.J. Downar, and E.E. Pilat, *The Linear Reactivity Model for Nuclear Fuel Management*, American Nuclear Society (1990).

**[Dru07]** K. Drudy, Westinghouse Electric Company, primate communication, March 2007.

**[Ede92]** M. Edenius and B. Forssen, "CASMO-3, A Fuel Assembly Burnup Program," Studsvik/NFA-89/3, Rev. 2, Studsvik AB (1992).

**[Els22]** S. El-Showk, "Final Resting Place," *Science* **375**, 806 (2022).

**[Ewi21]** R.C. Ewing and S. Park, "The Concept of Geological Disposal of Highly Radioactive Nuclear Waste," *Encyclopedia of Nuclear Energy*, Elsevier (2021).

**[Far03]** J. Farmer, V. Pasupathi, P. Nair, et al., "Technical Basis Document No. 6: Waste Package and Drip Shield Corrosion," UCRL-LR-155288, Lawrence Livermore National Laboratory (2003).

**[Fer12]** M.S. Fertel, "Industry's Safety Record and the Blue Ribbon Recommendations: The Way Ahead for the Management of Used Nuclear Fuel," *The Bridge* 41, National Academy of Engineering (Summer 2012).

**[Fer17]** R. Ferrer, J. Hykes, and J. Rhodes, "CASMO5, A Fuel Assembly Burn-up Program: Methodology Manual," SSP-08/405, rev. 7, Studsvik Scandpower (2017).

**[Fow70]** T.B. Fowler and D.R. Vondy, "Nuclear Reactor Analysis Code: CITATION," ORNL-TM-2496, rev. 1, Oak Ridge National Laboratory (1970).

**[Fra14]** S.M. Frank and M.N. Patterson, "Electrorefiner Salt Disposition Technical Evaluation," TEV-2201, Idaho National Laboratory (2014).

**[Gay21]** A. Gay, "Back-End of the Nuclear Fuel Cycle," presentation to the US National Academy of Sciences committee, 15 September 2021, https://doi.org/10.17226/26500 (2023).

**[GIF14]** *Technology Roadmap Update for Generation IV Nuclear Energy Systems*, https://www.gen-4.org/gif/upload/docs/application/pdf/2014-03/gif-tru2014.pdf (2014).

**[Goo12]** T. Goorley, M. James, T. Booth, et al."Initial MCNP6 Release Overview," *Nucl. Technol.* **180**, 298 (2012).

**[Gra08]** B. Granbow, "Mobile Fission and Activation Products in Nuclear Waste Disposal," *Radiogeochemical Aspects of Nuclear Waste Disposal* **102**, 180 (2008).

**[Gra79]** H.W. Graves Jr., *Nuclear Fuel Management*, Wiley (1979).

**[Hal63]** R.K. Haling, "Operating Strategy for Maintaining an Optimal Power Distribution throughout Life," TID-7672, General Electric Company (1963).

**[Han10]** C.W. Hansen, M.K. Knowles, R.J. MacKinnon, P.N. Swift, and S.D. Sevougian, "Summary of the Total System Performance Assessment for the Yucca Mountain License Application," SAND2010-3754C, Sandia National Laboratories (2010).

**[Hed97]** A. Hedin, "Spent nuclear fuel – how dangerous is it?" TR-97-13, Swedish Nuclear Fuel and Waste Management Co. (1997).

**[Hej21]** P. Hejzlar, "TerraPower: Fuel Cycle and Waste Management," presentation to the US National Academy of Sciences committee, 22 February 2021, https://doi.org/10.17226/26500 (2023).

**[Hen04]** J.S. Hendricks, ed., "MCNPX User's Manual Version 2.5.e," LACP-04-0569, Los Alamos National Laboratory (2004).

**[Hos78]** R.P. Hosteny, "Fuel Cycle Analysis Capability, REBUS-2," ANL-7721, Argonne National Laboratory (1978).

**[Jac82]** J. Jachic and J.C. Lee, "Optimal Fuel Depletion Strategy," *Trans. Am. Nucl. Soc.* **40**, 230 (1982).

**[Jon87]** D.B. Jones, "ARMP-02 Documentation, Part II, Chapter 6 – CPM-2 Computer Code Manual," EPRI NP-4574-CCM, Part II, Ch. 6, Volumes 1, 2, and 3, Electric Power Research Institute (1987).

**[Jon00]** D.B. Jones and K.E. Watkins, "CPM-3 Computer Code Manual, Volume 2: User's Manual," EPRI-CPM-001-M-002, Revision A, Electric Power Research Institute (2000).

**[Kra22]** L.M. Krall, A.M. Macfarlane, and R.C. Ewing, "Nuclear Waste From Small Modular Reactors," *Proc. Natl. Acad. Sci. U.S.A.* **119**, 2111833119 (2022).

**[Lai97]** J.J. Laidler, J.E. Batles, W.E. Millwer, J.P. Ackerman, and E.L. Carls, "Development of Pyroprocessing Technology," *Prog. Nucl. Energy* **31**, 131 (1997).

**[Lan21]** P. Landais, "Franes Strategy on the Back-End of the Fuel Cycle and the Management and Disposal of Radioactive Waste," presentation to the US National Academy of Sciences committee, 28 September 2021, https://doi.org/10.17226/26500 (2023).

**[Lee71]** J.C. Lee, K.A. Jones, and W.I. McCoy, "Axial Stability Measurements at the Rochester Gas and Electric Reactor," *Trans. Am. Nucl. Soc.* **14**, *Suppl.* 2, 15 (1971); J.C. Lee, K.A. Jones, D.P. Felentzer, and D. Rawle, "Axial Xenon Transient Tests at the Rochester Gas and Electric Reactor," WCAP-7964, Westinghouse Electric Corporation (1971).

**[Lee94]** J.C. Lee and J. Du, "Equilibrium Cycle Calculations for Transuranics Recycling in Pressurized Water Reactor," *Nucl. Technol.* **108**, 61 (1994).

**[Lee16]** J.C. Lee, "Nuclear Engineering," *Wiley Encyclopedia of Electrical and Electronics Engineering*, W5211, 3rd ed., Wiley (2016).

**[Lee17a]** J.C. Lee and N.J. McCormick, *Risk and Safety Analysis of Nuclear Systems*, Wiley (2011/2017).

**[Lee17b]** C. Lee and W.S. Yang,"MC$^2$-3: Multigroup Cross Section Generation Code for Fast Reactor Analysis," *Nucl. Sci. Eng.* **187**, 268 (2017).

**[Lit69]** W.W. Little Jr. and R.W. Hardie, "2DB - User's Manual," BNWL-831, rev. 1, Battelle Pacific Northwest Laboratory (1969).

**[Liu86]** Y.S. Liu, A. Meliksetian, J.A. Rathkopf, et al., "ANC: A Westinghouse Advanced Nodal Computer Code," WCAP-10966-A, Westinghouse Electric Company (1986).

**[McF21]** J. McFarlane, "Hazards Associated with Molten Salt Reactor Systems," presentation to the US National Academy of Sciences committee, 15 October 2021, https://doi.org/10.17226/26500 (2023)..

**[Mul21]** E.J. Mulder, "Overview of X-Energy's 200 MWth Xe-100 Reactor," presentation to the US National Academy of Sciences committee, 13 January 2021, https://doi.org/10.17226/26500 (2023).

**[NAS23]** J.D. Lee, P.A. Baisden, R.C. Ewing, et al., "Merits and Viability of Different Nuclear Fuel Cycles and Technology Options and the Waste Aspects of Advanced Nuclear Reactors," US National Academy of Sciences, https://doi.org/10.17226/26500 (2023).

**[NRC12]** US Nuclear Regulatory Commission, "Standards for Protection Against Radiation," Title 10, *Code of Federal Regulation*, Part 20, US Nuclear Regulatory Commission (2012).

**[Nuc18]** NuclearSAFE, "Technology," http://nstusa.net/technology (2018).

**[Pav20]** P. Paviet, "Sustainability of the Nuclear Fuel Cycle," PNNL-SA 156416, Pacific Northwest National Laboratory (2020).

**[Pet07]** M.T. Peters and R.C. Ewing, "A Science-Based Approach to Understanding Waste From Durability in Open and Closed Nuclear Fuel Cycles," *J. Nucl. Mater.* **362**, 395 (2007).

**[Poi17a]** C. Poinssot, S. Bourg, S. Brandjean et al., "French perspective on the respective interests of hydro- and pyro-chemical processes for recycling future spent nuclear fuels," *Proc. GLOBAL 2017*, Seoul (2017); "Discussion Points for Used Fuel Reprocessing," presentation to the US National Academy of Sciences committee, 20 July 2021, https://doi.org/10.17226/26500 (2023).

**[Poi17b]** C. Poinssot, B. Boullis, J.M. Adnet et al., "From the Bataille's Law to the Am-recycling integral experiments, overview of 25 years of R&D in CEA on Minor Actinides Partitioning Processes," *Proc. GLOBAL 2017*, Seoul (2017); "Discussion Points for Used Fuel Reprocessing," presentation to the US National Academy of Sciences committee, 20 July 2021, https://doi.org/10.17226/26500 (2023).

**[Pos99]** D.I. Poston and H.R. Trellue, "User's Manual, Version 2.0, for Monteburns, Version 1.0," LA-UR-99-4999, Los Alamos National Laboratory (1999).

**[Ras96]** N.C. Rasmussen, J. Buckham, T.J. Burke, et al., *Nuclear Wastes: Technologies for Separations and Transmutation*, US National Academy of Sciences (1996).

**[Rea18]** B.T. Rearden and M.A. Jessee, "Scale Code Systen, 6.2.3.," ORNL/TM-2005/39, Oak Ridge National Laboratory (2018).

**[Rey22]** J. Reyes, "Overview of Design Optimization Process for a Commercial Nuclear Core," transmittal to the US National Academy of Sciences committee, 26 August 2022, https://doi.org/10.17226/26500 (2023).

**[Rob65]** J.J. Roberts and H.P. Smith Jr., "Time Optimal Solution to the Reactivity-Xenon Shutdown Problem," *Nucl. Sci. Eng.* **22**, 470 (1965).

**[Sch80]** E.J. Schulz and J.C. Lee, "Time-Optimal Control of Spatial Xenon Oscillations to a Generalized Target," *Nucl. Sci. Eng.* **73**, 140 (1980).

**[Sor06]** R.T. Sorensen, J.C. Davis, and J.C. Lee, "Thorium-based Fuels for Enhancing Plutonium Transmutation in Light Water Reactors," *Trans. Am. Nucl. Soc.* **94**, 87 (2006).

**[Sor08]** R.T. Sorensen and J.C. Lee, "LWR Equilibrium Cycle Search Methodology for Assembly-Level Fuel Cycle Analysis," *Nucl. Sci. Eng.* **158**, 213 (2008).

**[Swi08]** P.N. Swift and M.K. Knowles, "Long-Term Performance of the Proposed Yucca Mountain Repository USA," Sandia National Laboratories, SAND2008-5879C (2008).

**[Swi10]** P.N. Swift, C.W. Hansen, E. Hardin et al., "Potential Impacts of Alternative Waste Forms on Long-Term Performance of Geological Repositories for Radioactive Waste," *PSAM-10* (2010).

**[Swi20]** P.N. Swift and D.C. Sassani, "Impacts of Nuclear Fuel Cycle Choices on Permanent Disposal of High Activity Radioactive Waste," IAEA-CN-272-185, International Atomic Energy Agency (2020).

**[Top83]** B.J. Toppel, "A User's Guide for the REBUS-3 Fuel Cycle Analysis Capability," ANL-83-2, Argonne National Laboratory (1983).

**[Van01]** G.J. Van Tuyle, "Candidate Approaches for an Integrated Nuclear Waste Management Strategy - Scoping Evaluations," LA-UR-01-5572, Los Alamos National Laboratory (2001).

**[Van04]** G.F. Vandegrift, M.C. Regalbuto, S.B. Aase et al., "Large-Scale Demonstration of the UREX+ Process," *Waste Management 2004 International Symposium, Tucson, AZ* (2004).

**[Vav04]** S. Vavilov, T. Kobayashi, and M. Myochin, "Principles and Test Experience of the RIAR's Oxide Pyro-Process," *J. Nucl. Sci. Technol.* **41**, 1018 (2004).

[Vex21] A. Vexler, "HALEU," presentation to the US National Academy of Sciences committee, 15 September 2021, https://doi.org/10.17226/26500 (2023).

[Wes02] Westinghouse Electric Company, "PHOENIX-P User Manual," Westinghouse Electric Company (2002).

[Wes03] Westinghouse Electric Company, "AP1000 Design Control Document," APP-GW-Gl-700, rev. 3, Westinghouse Electric Company (2003).

[Wig14] R. Wigeland, T. Taiwo, H. Ludewig et al., "Nuclear Fuel Cycle Evaluation and Screening – Final Report," INL/EXT-14-31465, Idaho National Laboratory (2014).

[Wil95] W.B. Wilson, T.R. England, D.C. George, et al., "Recent Development of the CINDER'90 Transmutation Code and Data Library for Actinide Transmutation Studies," *Proc. Global '95 Int. Conf. on Evaluation of Emerging Nuclear Fuel Cycle*, Versailles, France (1995).

[Wor21] A. Worral, "The Rationale for Reprocessing and Recycling," presentation to the US National Academy of Sciences committee, 2 June 2021, https://doi.org/10.17226/26500 (2023)

[Yan12] W.S. Yang, "Fast Reactor Physics and Computational Methods," *Nucl. Eng. Technol.* **44**, 183 (2012).

[Zui21] P. Zuidema, "Radioactive Waste Management Issues to be Considered when Evaluating Different Nuclear Fuel Cycle & Reactor Variants," presentation to the US National Academy of Sciences committee, 7 June 2021, https://doi.org/10.17226/26500 (2023).

## Problems

**12.1** A simplified fuel cycle analysis is to be performed with a core-average unit cell representing the average power density for a PWR core loaded with a fresh batch of $UO_2$ fuel containing 3.5 wt% of $^{235}$U. One-group constants, $\sigma_a = 450$ b and $\sigma_f = 310$ b, are given for $^{235}$U and assume that the neutron flux $\phi = 5.0 \times 10^{13}$ neutron $\cdot$ cm$^{-2}$s$^{-1}$ remains constant during the cycle. (a) Solve the balance equation for the $^{235}$U number density $N(t)$ as a function of time $t$ during the cycle and, obtain the fraction of initial $^{235}$U atoms that are destroyed over cycle length $T = 3$ yr. (b) Using the result of part (a), determine the fraction $\beta$ of initial $^{235}$U fissioned during the cycle.

**12.2** The AP1000 plant produces 3,400 MWt in a core with an active fuel length of 4.27 m and an equivalent active diameter of 3.04 m. (a) Obtain the core-average fission rate $\Sigma_f \phi$ and core-average thermal flux for the AP1000 core, using an estimate of thermal fission cross section

$\Sigma_f = 0.065\,\text{cm}^{-1}$ and neglecting fast-group fissions. With $^{135}$Xe thermal absorption cross section $\sigma_X = 1.5\,$Mb, determine the core-average $^{135}$I and $^{135}$Xe concentrations $I_0^\infty$ and $X_0^\infty$, respectively, at the HFP equilibrium condition. (b) Using the results of part (a), calculate the $^{135}$Xe concentration at its post-shutdown peak and the corresponding reactivity poisoning.

**12.3** For a batch of irradiated $UO_2$ fuel rods from a BWR core, it is determined that 70% of the initial $^{235}$U inventory is fissioned during the irradiation. If the fresh fuel had a $^{235}$U enrichment $e = 4.0\,$wt%, and one half of $^{239}$Pu produced during the fuel cycle fissioned, determine the total number of fissions per initial fissile atom (fifa) for the irradiated fuel, using the one-group constants: $\sigma_a^{25} = 400\,$b, $\sigma_f^{25} = 330\,$b, $\sigma_a^{28} = 2.0\,$b, $\sigma_f^{28} = 0.6\,$b.

**12.4** Derive the relationship of Eq. (12.72) for the FP decay heat function $F(t, T)$ from Eq. (12.71).

**12.5** The average energy $f(t)$ emitted per s in the form of $\beta$- and $\gamma$-rays, at $t$ s after the fission of one $^{235}$U nucleus, can be estimated as $f(t) = 2.66\,t^{-1.2}$ [MeV $\cdot$ fission$^{-1}$s$^{-1}$]. This equation is known as the Way-Wigner formula and is approximately valid for the range $t \in [1, 10^6]$ s for thermal neutron fission of $^{235}$U. (a) What is the time after the fission of a $^{235}$U nucleus over which we can expect the release of one half of the total amount of energy that can be ultimately released by the decay of fission products? (b) Obtain an expression for the thermal power released due to the decay of FPs in a PWR at $T$ s of operation at a constant power level of 1,000 MWe. (c) Determine the equilibrium FP decay power for the reactor in units of MWt. What fraction of the rated reactor power does this amount to?

**12.6** Suppose a nuclear fission reactor has been operating with $^{235}$U as the fuel at a steady-state thermal power of $P$ W for a period of $T$ s. It is then shut down essentially instantaneously. Use the Way-Wigner formula to determine the heat generation rate $P_d(t, T)$ due to FP decay at $t$ s after shutdown.

**12.7** The FP decay heat data may be approximated as the heat generation rate $f(t)$ at $t$ s following the fission of one $^{235}$U nucleus: $f(t) = \alpha\lambda\exp(-\lambda t)$ [MeV $\cdot$ fission$^{-1}$s$^{-1}$], with $\alpha = 13.0\,$MeV $\cdot$ fission$^{-1}$ and $\lambda = 3 \times 10^{-7}$ s$^{-1}$. Calculate (a) the equilibrium FP decay power for a 1.0-GWe PWR and (b) the fraction $\xi$ of the equilibrium FP decay power reached after 1.0 day of full-power operation.

**12.8** In a TRU transmutation study [Lee94] for the AP600 plant with a typical three-batch cycle operation, unit-assembly equilibrium cycle calculations for a 3.0 wt% fuel assembly indicate a total TRU inventory $I = 36.65$ Mg out of a total HM inventory of 80.8 Mg and TRU production rate $S = 0.257\ \text{Mg} \cdot \text{yr}^{-1}$ for a 1.0 GWe BOC configuration with a thermal efficiency of 33% and capacity factor of 0.80. With a total TRU makeup rate of $0.644\ \text{Mg} \cdot \text{yr}^{-1}$, TRU feed rate $F = 12.22\ \text{Mg} \cdot \text{yr}^{-1}$, and reprocessing loss fraction $L = 0.001$, a net TRU depletion rate $B = 0.64$ $\text{Mg} \cdot \text{yr}^{-1}$ is obtained. Obtain the TRU inventory reduction (TIR) factor $\Psi$ [Ras96] representing the ratio of the TRU accumulation rate, without the transmutor, to the TRU accumulation rate with the transmutor.

**12.9** Determine the radiological toxicity for $^{14}$C in the form of carbon monoxide and $^{135}$Cs. Compare the results with those of $^{3}$H and $^{137}$Cs.

# 13

# Thermal-Hydraulic Analysis of Reactor Systems

Due to large heat generation rates in a nuclear reactor core, careful analysis has to be performed to determine temperature and density distributions of various materials in the core both for steady-state and transient conditions. The maximum power that can be generated in a reactor core is normally determined by limitations on core materials due to distributed temperature, radiation effects, thermal stress, and other mechanical considerations. Hence, accurate determination of the temperature distribution in a reactor core is important both for economic and safety reasons. For safe operation of the reactor, we would like to be assured that there is an ample margin between the actual and limiting conditions of fuel and structural materials. Given an inventory of fuel, coupled nuclear-thermal-hydraulic calculations are required to determine an optimum fuel loading pattern so that the maximum possible power generation is achieved. Such coupled calculations are necessary because the neutron flux spectrum and nuclear reaction rates are influenced by thermal-hydraulic (T/H) feedback including thermal expansion of the coolant or moderator and changes in the resonance escape probability in the fuel due to the Doppler broadening of absorption resonances.

In current light water reactor (LWR) designs, core performance limits are usually imposed in two ways, one on fuel temperature and the other one on heat flux through the fuel cladding. The limitation on fuel temperature requires that the fuel temperature anywhere in the core at operating conditions will not exceed its melting point, e.g. 3,123 K for $UO_2$. It is desirable to avoid fuel melting in the core for several reasons, including

(a) Fuel melting increases the rate of fission gas release out of the $UO_2$ ceramic structure,
(b) The prompt negative reactivity feedback due to the Doppler effect is reduced in a partially molten fuel, and

*Nuclear Reactor Physics and Engineering*, Second Edition. John C. Lee.
© 2025 John Wiley & Sons, Inc. Published 2025 by John Wiley & Sons, Inc.
Companion website: www.wiley.com/go/LeeNRPE2

(c) Fuel melting increases the fuel volume and could put additional stress on the cladding material.

The T/H limitation on heat flux is required so that the fuel cladding will not fail due to temperature- and radiation-induced stress and strain. Integrity of the fuel cladding is important for containing fission gases within the fuel element and avoiding fuel-coolant interactions. The heat flux limitation is expressed in terms of either a limiting clad temperature or a limiting heat flux into the coolant, below which the fuel cladding is expected to maintain its integrity. The design limits on the fuel cladding performance are represented in the form of the Final Acceptance Criteria [Lee17] for the emergency core cooling system of LWR plants.

In this chapter, we derive general T/H equations representing the conservation of mass, momentum, and energy, and discuss how they can be used to determine distributions of fuel temperature and coolant temperature and density in typical reactor cores. Our study will not include the analysis dealing with thermal stress and mechanical vibrations, which has to be performed in actual design of fuel elements. Although the T/H conservation equations are derived with time dependence represented, our effort will be focused on steady-state T/H analysis applicable to the fuel rod and coolant channel in a reactor core.

The T/H equations we develop in this chapter represent various processes involved with the transport of momentum and energy in fluid flow. We begin in Section 13.1 with a brief discussion of empirical relations governing fluid flow and energy transfer, followed by the derivation of general conservation equations for fluid flow in Section 13.2. Sample solutions of the fluid conservation equations for idealized conditions are then presented in Section 13.3, primarily to clarify some fundamental concepts and to provide perspectives on the T/H analysis of nuclear systems. Calculation of the radial temperature distribution within a nuclear fuel rod is included as a sample problem. The general conservation equations are simplified in Section 13.4 for the case of channel flow, characteristic of fluid flow in reactor coolant channels or feedwater flow in steam generators. The channel flow model is then used in Section 13.5 to determine axial fuel and coolant temperature distributions in reactor coolant channels. We also discuss in Section 13.5 hot channel factors that characterize the limiting heat flux and power distributions in a reactor core. Extension of the fluid conservation equations to handle two-phase flow and boiling heat transfer is presented in Section 13.6, together with approaches to represent the limiting heat flux in a channel flow via the critical heat flux (CHF), i.e. the heat flux corresponding to the departure from nucleate boiling (DNB). We indicate in Section 13.7 how the concept of DNB heat flux is used as a figure of merit, together with due considerations for the limiting fuel temperature, to determine the power generation capability of a reactor core.

Thermal-hydraulic computer codes developed for nuclear plant system analysis are discussed in Section 13.8, followed by concluding remarks in Section 13.9.

## 13.1 Empirical Laws for Energy and Momentum Transport

Among many textbooks on fluid mechanics and heat transfer, we rely heavily on a combined treatment of mass, energy, and momentum transport by Bird, Stewart, and Lightfoot [Bir07] for the derivation of empirical laws governing viscous flow and heat transport as well as basic fluid conservation equations. These empirical laws define key T/H parameters, viscosity, thermal conductivity, and heat transfer coefficient, and hence are essential for meaningful derivation and use of the conservation equations for fluid flow.

### 13.1.1 Fourier's Law of Heat Conduction

One of many empirical observations with which we are familiar is related to the transfer of heat through a solid object subject to a temperature gradient. Consider a semi-infinite slab of solid material of thickness $Y$, and assume that the upper surface is at temperature $T_0$ and lower surface at $T_1$, where $T_1 > T_0$. When the heat flux through the slab has reached an equilibrium distribution, we obtain a linear temperature profile shown in Figure 13.1 and obtain the heat flux $q_y$ in the $y$-direction

$$q_y = k\frac{T_1 - T_0}{Y},$$

or, in general for 3-D geometry,

$$\mathbf{q} = -k\nabla T, \tag{13.1}$$

where $k$ is a constant of proportionality known as *thermal conductivity*. A negative sign is introduced because heat flows against a temperature gradient. Equation (13.1) is known as *Fourier's law of heat conduction*.

**Figure 13.1** Temperature gradient in a slab.

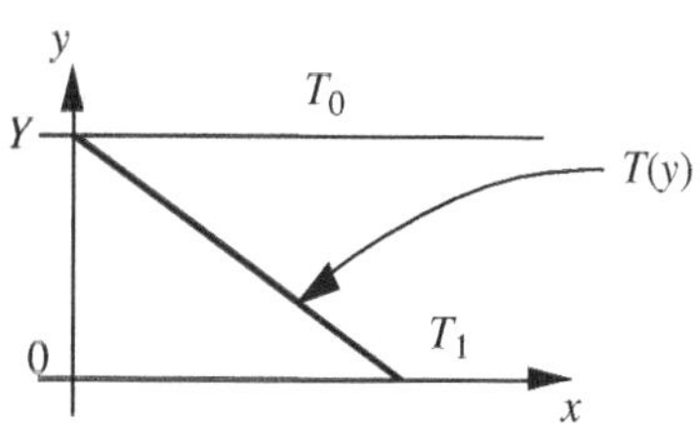

The thermal conductivity $k$ is given in units of $[\mathrm{W} \cdot \mathrm{m}^{-1}\,\mathrm{K}^{-1}]$, while the heat flux $\mathbf{q}$ is given in units of $[\mathrm{W} \cdot \mathrm{m}^{-2}]$. Together with the conductivity $k$, we also define the *thermal diffusivity* $\alpha = k/\rho C_p$, where $\rho$ is the density and $C_p$ is the heat capacity per unit mass at constant pressure. Fourier's law of heat conduction may be considered, in practice, as a definition for the thermal conductivity $k$. Indeed, Eq. (13.1) is used to measure $k$ for various materials as a function of temperature. Temperature dependence of thermal conductivity $k$ is especially important for fuel material because we need to account accurately for large temperature gradients across thin fuel rods, as we discuss further in Sections 13.3.1 and 13.3.2.

### 13.1.2 Newton's Law of Viscosity

The next empirical law we derive governs the momentum transport through a velocity gradient. Somewhat similar to the heat conduction case of Section 13.1.1, consider two flat plates of infinite extent separated by distance $Y$ and filled with a viscous flow. We assume the lower plate is moving at a constant speed $V$ and the upper plate is held stationary. When the fluid has reached an equilibrium velocity profile, the fluid velocity $v_x(y)$ in the $x$-direction becomes a linear function of $y$, as shown in Figure 13.2. Define $\tau_{yx}$ as the *shearing* or *tangential stress in the fluid plane normal to y due to fluid motion in the x-direction* and obtain

$$\tau_{yx} = \mu \frac{V}{Y}.$$

Generalizing this equation to 3-D geometry, we obtain *Newton's law of viscosity*

$$\tau_{yx} = -\mu \frac{\partial v_x}{\partial y}, \tag{13.2}$$

where $\mu$ is a constant of proportionality called *viscosity*. A negative sign is included in Eq. (13.2) because the shear force in the fluid is opposite to the direction of motion of the plate. The *shear stress* $\tau_{yx}$ can also be interpreted as the *viscous flow of x-momentum in the y-direction due to the velocity gradient*. In general, the $x$-component $v_x$ of the fluid velocity vector can be a function of $x$, $y$, and $z$ and Eq. (13.2) can be generalized for the $ij$th component of the $(3 \times 3)$ *shear stress*

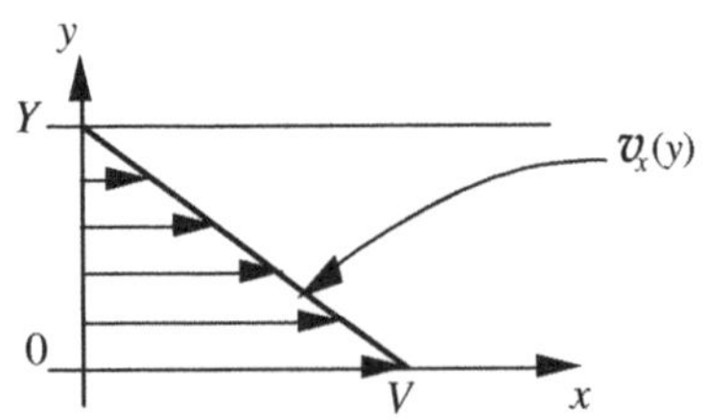


**Figure 13.2** Velocity gradient in fluid between flat plates.

tensor $\tau$

$$\tau_{ij} = -\mu \left[ \frac{\partial v_i}{\partial x_j} + \frac{\partial v_j}{\partial x_i} - \left( \frac{2}{3} - \frac{\kappa}{\mu} \right) (\nabla \cdot \mathbf{v}) \delta_{ij} \right], \tag{13.3}$$

where $\kappa$ is the *bulk viscosity* or the *modulus of elasticity*. Equation (13.3) is known as the *generalized Newton's law of viscosity*, and any fluid obeying Eq. (13.3) is known as a *Newtonian* or *Stokes fluid*. At this point in our discussion of basic concepts, assume that a tensor is a quantify defined in terms of vector operations and may be represented as a matrix.

Viscous force is obviously due to interactions or friction among fluid molecules, and many attempts have been made to calculate viscosity from microscopic theory of gases and liquids. Unfortunately, due to the complex nature of molecular interactions, such attempts have met with only limited success, and invariably we have to rely on empirical correlations for viscosity. For this purpose, Eq. (13.2) can be considered a definition for viscosity. The common unit for viscosity $\mu$ is *Poise* = $[0.1\ \mathrm{kg} \cdot \mathrm{m}^{-1}\ \mathrm{s}^{-1}]$ or *centipoise* = $[\mathrm{mPa} \cdot \mathrm{s}]$, with the stress tensor $\tau$ expressed in units of [Pa]. Note also the definition of the *kinematic viscosity* $\nu = \mu/\rho$.

### 13.1.3 Newton's Law of Cooling

Another common observation is related to the energy transfer at a solid–fluid interface, schematically shown in Figure 13.3, where a heated surface is cooled by a fluid flow. Given the temperature $T_s$ at the interface, the heat flux $q$ may be represented by *Newton's law of cooling*

$$q = h(T_s - T_b), \tag{13.4}$$

where $h$ is the *heat transfer coefficient* and $T_b$ an effective temperature of the fluid called the *bulk fluid temperature*. Analogous to the interpretation that Eqs. (13.1) and (13.2) define thermal conductivity $k$ and viscosity $\mu$, respectively, we may consider Eq. (13.4) as a defining equation for the heat transfer coefficient $h$ given a suitable definition for $T_b$. In many engineering applications, the bulk fluid temperature $T_b$ is defined as a flow rate-average fluid temperature. For a fluid flow in a heated pipe or duct, this temperature may be visualized as the average temperature

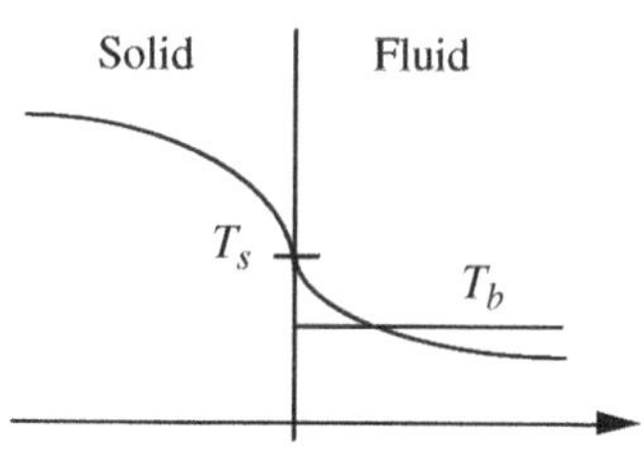


**Figure 13.3** Temperature distribution at a solid–fluid interface.

of the fluid poured into a cup and thoroughly mixed. Hence, $T_b$ is also known as the *cup-mixing temperature*, as discussed further in Section 13.3.3.

The scalar heat flux $q$ in Eq. (13.4) may be considered equivalent to the normal component of the vector heat flux $\mathbf{q}$ in Eq. (13.1). Hence, the heat transfer coefficient $h$ is given in units of the thermal conductivity $k$ divided by that of length, e.g. $[\mathrm{W \cdot m^{-2}\,K^{-1}}]$. In spite of this dimensional relationship between $k$ and $h$, the heat transfer coefficient cannot in general be obtained directly from measurements of $k$ and has to be empirically determined for a particular flow pattern, flow channel geometry, and other system parameters. It is particularly difficult to measure heat transfer coefficients for time-dependent fluid flows applicable to nuclear power plant analysis. Given empirical correlations for $h$, Eq. (13.4) serves as a key boundary condition in solving fluid conservation equations.

## 13.2 Derivation of Fluid Conservation Equations

### 13.2.1 Equation of Continuity

To derive the equation of continuity governing the conservation of mass in a fluid flow, we consider a differential volume element at $\mathbf{r}$ with volume $d\mathbf{r}$, surface area $dS$, and unit outward normal vector $\mathbf{n}$, shown in Figure 13.4. Similar to the approach we have taken in deriving the neutron leakage term in Eqs. (4.15) and (4.16), begin with a small volume element $\Delta V = \Delta x \Delta y \Delta z$ and small surface area $\Delta A$, make use of the Gauss divergence theorem, and take the limit as $\Delta V$ and $\Delta A$ approach differential elements.

Construct a mass balance over a unit volume element

$$\begin{aligned}&(\text{Time rate of density change})\\&\quad = (\text{Time rate of change of mass in unit volume})\\&\quad = -(\text{Rate of leakage of mass out of unit volume}),\end{aligned} \tag{13.5}$$

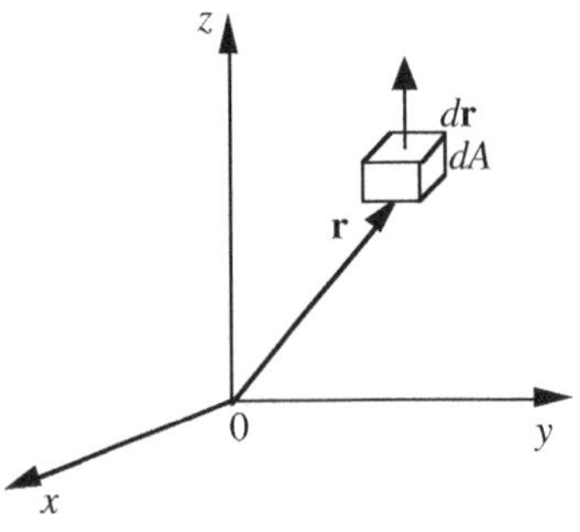


**Figure 13.4** Differential fluid volume.

and obtain the rate of mass leakage out of the differential volume

$$\mathbf{n}\cdot[\rho(\mathbf{r},t)\mathbf{v}(\mathbf{r},t)]\Delta A = \int_{\Delta A} dA\mathbf{n}\cdot[\rho(\mathbf{r},t)\mathbf{v}(\mathbf{r},t)] = \int_{\Delta V} d\mathbf{r}\nabla\cdot[\rho(\mathbf{r},t)\mathbf{v}(\mathbf{r},t)]$$
$$= \Delta V\nabla\cdot[\rho(\mathbf{r},t)\mathbf{v}(\mathbf{r},t)] = \nabla\cdot[\rho(\mathbf{r},t)\mathbf{v}(\mathbf{r},t)]d\mathbf{r}, \quad (13.6)$$

where the Gauss divergence theorem is used to convert the surface leakage rate to volume leakage rate. Since Eq. (13.6) represents the leakage rate out of a volume $d\mathbf{r}$, the rate of mass leakage per unit volume is $\nabla\cdot(\rho\mathbf{v})$ and the mass balance statement of Eq. (13.5) yields the *equation of continuity*

$$\frac{\partial\rho(\mathbf{r},t)}{\partial t} = -\nabla\cdot[\rho(\mathbf{r},t)\,\mathbf{v}(\mathbf{r},t)]. \quad (13.7)$$

Define the *substantial derivative*

$$\frac{D}{Dt} = \frac{\partial}{\partial t} + v_x\frac{\partial}{\partial x} + v_y\frac{\partial}{\partial y} + v_z\frac{\partial}{\partial z} = \frac{\partial}{\partial t} + \mathbf{v}\cdot\nabla, \quad (13.8)$$

and rewrite the mass conservation equation

$$\frac{D\rho}{Dt} = -\rho\nabla\cdot\mathbf{v}. \quad (13.9)$$

The substantial derivative defined in Eq. (13.8) is essentially a total derivative accounting for the parametric dependence on space and represents a *time-derivative for a path following the fluid motion.* For a *steady-state flow*, Eq. (13.7) yields

$$\nabla\cdot(\rho\,\mathbf{v}) = 0, \quad (13.10)$$

while, for an *incompressible flow*, Eq. (13.9) implies

$$\nabla\cdot\mathbf{v} = 0. \quad (13.11)$$

### 13.2.2 Equation of Motion and Navier–Stokes Equation

In analogy to the derivation of the continuity equation, consider the rate of change of linear momentum $\rho\mathbf{v}$ in the differential volume of Figure 13.4 to derive the equation of motion generally applicable to fluid flow. By invoking Newton's law of motion, the momentum balance requires

(Time rate of change of linear momentum in unit volume)
= – (Rate of momentum leakage per unit volume)
+ (Sum of body forces acting on the fluid volume). (13.12)

The momentum leakage should include both leakage through bulk fluid motion, i.e. convective fluid motion, and molecular or viscous transport due to velocity gradient. Likewise, the body forces should include gravitational force and effect of any hydrostatic pressure gradient acting on the fluid volume. In terms of hydrostatic pressure $p$ and acceleration of gravity $\mathbf{g}$, and the viscous momentum transport represented through shear stress tensor $\boldsymbol{\tau}$, we may represent the individual terms for the differential volume of Figure 13.4:

(1) Rate of convective momentum loss = $\int_{dA} dA\, \mathbf{n} \cdot (\rho \mathbf{v}\mathbf{v}) = d\mathbf{r}\, \nabla \cdot (\rho \mathbf{v}\mathbf{v})$
(2) Rate of viscous transport loss = $\int_{dA} dA\, \mathbf{n} \cdot \boldsymbol{\tau} = d\mathbf{r}\, \nabla \cdot \boldsymbol{\tau}$
(3) Force due to hydrostatic pressure = $-\int_{dA} dA\, \mathbf{n}p = -d\mathbf{r}\, \nabla p$, and
(3) Gravitational force = $d\mathbf{r}\, \rho \mathbf{g}$.

We have used a generalized form of the divergence theorem for terms (1) through (3). A negative sign is included in term (3) because $\mathbf{n}$ is the outward unit normal vector, and the pressure acting on the volume element should be represented as a positive contribution to the momentum of the system.

Summing up terms (1) through (4) and noting that $\rho \mathbf{v}$ represents the linear momentum of the fluid per unit volume generates the *equation of motion*, usually known as the *Navier–Stokes equation*

$$\begin{aligned} \frac{\partial}{\partial t} \rho \mathbf{v} &= -\nabla \cdot (\rho \mathbf{v}\mathbf{v}) - \nabla \cdot \boldsymbol{\tau} - \nabla p + \rho \mathbf{g}, \\ &= -\nabla \cdot (\rho \mathbf{v}\mathbf{v}) - \nabla \cdot (\boldsymbol{\tau} + p\boldsymbol{I}) + \rho \mathbf{g}, \end{aligned} \tag{13.13}$$

where the term $\mathbf{v}\mathbf{v}$ is a tensor or dyad similar to $\boldsymbol{\tau}$, and can be considered a quantity defined through its vector operational properties [Bir07]. The $j$th component of $\nabla \cdot \boldsymbol{\tau}$ is defined as

$$(\nabla \cdot \boldsymbol{\tau})_j = \sum_i \frac{\partial}{\partial x_i} \tau_{ij} \tag{13.14}$$

and the $j$th component of $\nabla \cdot (\rho \mathbf{v}\mathbf{v})$ as

$$[\nabla \cdot (\rho \mathbf{v}\mathbf{v})]_j = \sum_i \frac{\partial}{\partial x_i} \left( \rho v_i v_j \right). \tag{13.15}$$

Likewise, the parameter $\boldsymbol{I}$ is a unit tensor.

Using the continuity equation (13.7), rewrite the LHS of Eq. (13.13) as

$$\frac{\partial}{\partial t} \rho \mathbf{v} = \rho \frac{\partial \mathbf{v}}{\partial t} - \mathbf{v} \nabla \cdot (\rho \mathbf{v}),$$

while Eq. (13.15) indicates the convective momentum flow term $\nabla \cdot (\rho \mathbf{v}\mathbf{v})$ in Eq. (13.13) can be written as

$$\nabla \cdot (\rho \mathbf{v}\mathbf{v}) = \rho \mathbf{v} \cdot \nabla \mathbf{v} + \mathbf{v} \nabla \cdot (\rho \mathbf{v}),$$

where $\nabla \mathbf{v}$ is a dyad such that

$$(\mathbf{v} \cdot \nabla \mathbf{v})_j = \sum_i v_i \frac{\partial v_j}{\partial x_i}.$$

With the substantial derivative notation

$$\rho\frac{D\mathbf{v}}{Dt} = \rho\frac{\partial\mathbf{v}}{\partial t} + \rho\mathbf{v}\cdot\nabla\mathbf{v}.$$

Equation (13.13) is rewritten as

$$\rho\frac{D\mathbf{v}}{Dt} = \frac{\partial}{\partial t}\rho\mathbf{v} + \nabla\cdot(\rho\mathbf{v}\mathbf{v}) = -\nabla\cdot(\boldsymbol{\tau} + p\boldsymbol{I}) + \rho\mathbf{g}. \tag{13.16}$$

For constant $\rho$ and $\mu$, Eq. (13.16) can be simplified to

$$\rho\frac{D\mathbf{v}}{Dt} = \mu\nabla^2\mathbf{v} - \nabla p + \rho\mathbf{g}, \tag{13.17}$$

which is another form of the *Navier–Stokes equation*. For $\nabla\cdot\tau = 0$ or for an *inviscid fluid*, Eq. (13.17) further reduces to *Euler's equation*:

$$\rho\frac{D\mathbf{v}}{Dt} = -\nabla p + \rho\mathbf{g}. \tag{13.18}$$

The momentum conservation equation (13.13) can, in principle, be solved for velocity $\mathbf{v}$, provided we use Eq. (13.3) for shear stress $\boldsymbol{\tau}$ and an equation of state for fluid density $\rho$ together with fluid pressure $p$. For real fluids, however, the shear stress, as well as the density, is a complicated nonlinear function of fluid parameters, and it is often difficult even to obtain approximate solutions to the equation of motion for either steady-state or transient conditions.

### 13.2.3 Equations of Energy Conservation

The equation of conservation of energy for fluid flow can be similarly derived by setting up an energy balance for unit volume. We need to derive, however, a balance equation for each type of energy we wish to consider, e.g. mechanical energy, internal energy, enthalpy, and total energy. Since the potential energy is not of much concern for T/H analysis of nuclear systems, concentrate on deriving the conservation equation for the total energy $E$ per unit volume representing a sum of internal energy density $U$ and kinetic energy density:

$$E = \rho\left(U + \frac{1}{2}v^2\right). \tag{13.19}$$

With $U$ given in units of energy per mass, set up an energy balance

$$\begin{aligned}
&(\text{Time rate of change of total energy in unit volume})\\
&\quad = -(\text{Rate of energy leakage by convection from the volume})\\
&\qquad + (\text{Rate of heat addition by conduction to the volume})\\
&\qquad - (\text{Rate of work done by the fluid volume on the surroundings})\\
&\qquad + (\text{Rate of energy produced in the volume}).
\end{aligned} \tag{13.20}$$

With the understanding that the work done by the fluid volume on the surroundings should include the work associated with the gravitational force, hydrostatic pressure, and viscous force, obtain various terms contributing to the total energy balance:

(1) Rate of convective loss = $dA\,\mathbf{n}\cdot(\mathbf{v}E) = d\mathbf{r}\nabla\cdot(\mathbf{v}E)$,
(2) Rate of heat conduction = $-dA\,\mathbf{n}\cdot\mathbf{q} = -d\mathbf{r}\nabla\cdot\mathbf{q}$,
(3) Rate of work done against the gravitational force = $-d\mathbf{r}\mathbf{v}\cdot(\rho\mathbf{g}) = -d\mathbf{r}\rho(\mathbf{v}\cdot\mathbf{g})$,
(4) Rate of work done against the pressure force = $dA\,\mathbf{n}\cdot\mathbf{v}p = d\mathbf{r}\nabla\cdot(p\mathbf{v})$,
(5) Rate of work done against the viscous force = $dA\,\mathbf{n}\cdot(\boldsymbol{\tau}\cdot\mathbf{v}) = d\mathbf{r}\nabla\cdot(\boldsymbol{\tau}\cdot\mathbf{v})$,
(6) Volumetric heat source = $d\mathbf{r}S(\mathbf{r})$.

A negative sign is introduced in term (2) because heat flow out of the volume makes a negative contribution to the energy inventory. Similarly, a negative sign is introduced in term (3) because fluid motion in the same direction as the gravitational acceleration **g** should result in a net increase in the energy inventory. A short-hand application of the divergence theorem has been made in terms (1), (2), (4), and (5), as in the derivation of the continuity and momentum conservation equations.

Collecting terms yields the *equation of conservation of total energy*:

$$\frac{\partial E}{\partial t} = \frac{\partial}{\partial t}\left[\rho\left(U + \frac{v^2}{2}\right)\right] = -\nabla\cdot(E\mathbf{v}) - \nabla\cdot\mathbf{q} + \rho\mathbf{v}\cdot\mathbf{g} - \nabla\cdot(p\mathbf{v}) - \nabla\cdot(\boldsymbol{\tau}\cdot\mathbf{v}) + S. \qquad (13.21)$$

Here the term $\boldsymbol{\tau}\cdot\mathbf{v}$ is a vector, whose $i$th component is given as

$$(\boldsymbol{\tau}\cdot\mathbf{v})_i = \sum_j \tau_{ij}v_j.$$

In terms of the substantial derivative, rewrite Eq. (13.21)

$$\frac{DE}{Dt} = \frac{\partial E}{\partial t} + \mathbf{v}\cdot\nabla E = -E\nabla\cdot\mathbf{v} - \nabla\cdot\mathbf{q} + \rho\mathbf{v}\cdot\mathbf{g} - \nabla\cdot(p\mathbf{v}) - \nabla\cdot(\boldsymbol{\tau}\cdot\mathbf{v}) + S$$

and, with the continuity equation (13.9), obtain the conservation equation for total energy

$$\rho\frac{D}{Dt}\left(U + \frac{v^2}{2}\right) = \rho\frac{\partial}{\partial t}\left(U + \frac{v^2}{2}\right) + \rho\mathbf{v}\cdot\nabla\left(U + \frac{v^2}{2}\right) = -\nabla\cdot\mathbf{q} + \rho\mathbf{v}\cdot\mathbf{g} - \nabla\cdot(p\mathbf{v}) - \nabla\cdot(\boldsymbol{\tau}\cdot\mathbf{v}) + S. \qquad (13.22)$$

An equation representing the *conservation of mechanical energy* can be simply derived by taking a scalar product of fluid velocity **v** with the momentum conservation equation (13.16):

$$\rho \frac{D}{Dt}\left(\frac{1}{2}v^2\right) = -\mathbf{v}\cdot\nabla p - \mathbf{v}\cdot(\nabla\cdot\boldsymbol{\tau}) + \rho\mathbf{v}\cdot\mathbf{g}. \tag{13.23}$$

Subtracting Eq. (13.23) from the total energy conservation equation (13.22) provides an *equation of internal energy conservation*

$$\rho\frac{DU}{Dt} = -\nabla\cdot\mathbf{q} - p\nabla\cdot\mathbf{v} + \mathbf{v}\cdot(\nabla\cdot\boldsymbol{\tau}) - \nabla\cdot(\boldsymbol{\tau}\cdot\mathbf{v}) + S,$$

or equivalently, with the help of the continuity equation (13.7)

$$\rho\frac{DU}{Dt} = \frac{\partial}{\partial t}\rho U + \nabla\cdot(\rho U\mathbf{v}) = -\nabla\cdot\mathbf{q} - p\nabla\cdot\mathbf{v} - \boldsymbol{\tau} : \nabla\mathbf{v} + S \tag{13.24}$$

and the tensorial notation

$$\boldsymbol{\tau} : \nabla\mathbf{v} = \nabla\cdot(\boldsymbol{\tau}\cdot\mathbf{v}) - \mathbf{v}\cdot(\nabla\cdot\boldsymbol{\tau}) = \sum_{i,j}\frac{\partial}{\partial x_i}(\tau_{ij}v_j) - \sum_{i,j} v_j\frac{\partial\tau_{ij}}{\partial x_i} = \sum_{i,j}\tau_{ij}\frac{\partial v_i}{\partial x_j}.$$

We may express Eq. (13.24) also in terms of fluid temperature $T$ recalling the thermodynamic relationship

$$dU = \left(\frac{\partial U}{\partial V}\right)_T dV + \left(\frac{\partial U}{\partial T}\right)_V dT = \left[T\left(\frac{\partial p}{\partial T}\right)_V - p\right] dV + C_v dT,$$

where $C_v$ is the heat capacity at constant specific volume $V = 1/\rho$. The equation can be simply rewritten in terms of the substantial derivative

$$\rho\frac{DU}{Dt} = \left[T\left(\frac{\partial p}{\partial T}\right)_V - p\right]\rho\frac{DV}{Dt} + \rho C_v\frac{DT}{Dt}, \tag{13.25}$$

where the derivative $DV/Dt$ can be replaced by the continuity equation (13.9)

$$\rho\frac{DV}{Dt} = \rho\frac{D}{Dt}\left(\frac{1}{\rho}\right) = -\frac{1}{\rho}\frac{D\rho}{Dt} = \nabla\cdot\mathbf{v}. \tag{13.26}$$

Substituting Eqs. (13.25) and (13.26) into Eq. (13.24) yields an alternate form of the internal energy conservation equation expressed in terms of fluid temperature $T$:

$$\rho C_v\frac{DT}{Dt} = -\nabla\cdot\mathbf{q} - T\left(\frac{\partial p}{\partial T}\right)_V \nabla\cdot\mathbf{v} - \boldsymbol{\tau} : \nabla\mathbf{v} + S. \tag{13.27}$$

The terms on the RHS of Eq. (13.27) represent the changes in internal energy due to heat conduction, expansion, viscous effect, and volumetric source, respectively.

For an inviscid, ideal gas, $(\partial p/\partial T)_V = p/T$, and the equation of energy reduces to

$$\rho C_v\frac{DT}{Dt} = -\nabla\cdot\mathbf{q} - p(\nabla\cdot\mathbf{v}) + S. \tag{13.28}$$

Similarly, for an inviscid, incompressible fluid, $C_p = C_v$, and we obtain

$$\rho C_v \frac{DT}{Dt} = \rho C_p \frac{DT}{Dt} = -\nabla \cdot \mathbf{q} + S. \tag{13.29}$$

This further reduces for solids to the standard *time-dependent heat conduction equation*

$$\rho C_p \frac{\partial T}{\partial t} = -\nabla \cdot \mathbf{q} + S = \nabla \cdot k\nabla T + S, \tag{13.30}$$

where we have used Fourier's law of heat conduction, Eq. (13.1).

Another form of the energy conservation equation follows from Eq. (13.24), if fluid enthalpy $h = U + pV$ is introduced:

$$\frac{\partial}{\partial t}\rho h = -\nabla \cdot (\rho h \mathbf{v}) - \nabla \cdot \mathbf{q} - \boldsymbol{\tau} : \nabla \mathbf{v} + \frac{Dp}{Dt} + S. \tag{13.31}$$

Assuming further that the kinetic energy, viscous heating, and gravitational potential energy are negligibly small yields $\mathbf{v} \cdot \nabla p = 0$, and we obtain

$$\frac{\partial}{\partial t}\rho h = -\nabla \cdot (\rho h \mathbf{v}) - \nabla \cdot \mathbf{q} + \frac{\partial p}{\partial t} + S. \tag{13.32}$$

Equations (13.30) and (13.32) perhaps are the two most useful forms of the energy conservation equation for a large class of T/H problems in nuclear reactor analysis. One case where Eq. (13.32) may not be sufficiently accurate is rapid T/H transients associated with postulated loss-of-coolant accidents. In such transients, the kinetic energy of flashing liquid could make non-negligible contributions to the overall energy balance. Finally, for *steady-state heat conduction in solids*, obtain from Eq. (13.30)

$$\nabla \cdot \mathbf{q} = -\nabla \cdot k\nabla T = S. \tag{13.33}$$

Equation (13.33) is a simple statement of energy balance that the heat generated in unit volume of any solid material will be dissipated through heat conduction to maintain a steady-state temperature distribution. Similarly, for *steady-state fluid flow*, Eq. (13.32) reduces to

$$\nabla \cdot (\rho h \mathbf{v}) = -\nabla \cdot \mathbf{q} + S, \tag{13.34}$$

which represents the corresponding energy balance accounting for convective as well as conductive flow of energy.

### 13.2.4 Comments on Fluid Conservation Equations

Table 13.1 summarizes various forms of conservation equations derived in this section in terms of partial and substantial derivatives, together with a generalized structure of the fluid transport equations in Table 13.2. It should

**Table 13.1** Equations of mass, momentum, and energy conservation.

| Conservation type | Eq. No. |
|---|---|
| Mass conservation | |
| $\frac{\partial \rho}{\partial t} = -\nabla \cdot (\rho \mathbf{v})$ | (13.7) |
| $\frac{D\rho}{Dt} = -\rho \nabla \cdot \mathbf{v}$ | (13.9) |
| Momentum conservation | |
| $\frac{\partial}{\partial t}\rho \mathbf{v} = -\nabla \cdot (\rho \mathbf{v}\mathbf{v}) - \nabla \cdot \boldsymbol{\tau} - \nabla p + \rho \mathbf{g}$ | (13.13) |
| $\rho \frac{D\mathbf{v}}{Dt} = -\nabla \cdot \boldsymbol{\tau} - \nabla p + \rho \mathbf{g}$ | (13.16) |
| Conservation of internal and kinetic energy | |
| $\frac{\partial E}{\partial t} = -\nabla \cdot (E\mathbf{v}) - \nabla \cdot \mathbf{q} + \rho \mathbf{v} \cdot \mathbf{g} - \nabla \cdot (p\mathbf{v}) - \nabla \cdot (\boldsymbol{\tau} \cdot \mathbf{v}) + S$ | (13.21) |
| $\rho \frac{D(E/\rho)}{Dt} = -\nabla \cdot \mathbf{q} + \rho \mathbf{v} \cdot \mathbf{g} - \nabla \cdot (p\mathbf{v}) - \nabla \cdot (\boldsymbol{\tau} \cdot \mathbf{v}) + S$ | (13.22) |
| Conservation of mechanical energy | |
| $\rho \frac{D}{Dt}\left(\frac{1}{2}v^2\right) = -\mathbf{v} \cdot \nabla p - \mathbf{v} \cdot (\nabla \cdot \boldsymbol{\tau}) + \rho \mathbf{v} \cdot \mathbf{g}$ | (13.23) |
| Conservation of internal energy | |
| $\rho \frac{DU}{Dt} = -\nabla \cdot \mathbf{q} - p\nabla \cdot \mathbf{v} - \boldsymbol{\tau} : \nabla \mathbf{v} + S$ | (13.24) |
| $\rho C_v \frac{DT}{Dt} = -\nabla \cdot \mathbf{q} - T\left(\frac{\partial p}{\partial T}\right)_V \nabla \cdot \mathbf{v} - \boldsymbol{\tau} : \nabla \mathbf{v} + S$ | (13.27) |
| Heat conduction equation | |
| $\rho C_p \frac{\partial T}{\partial t} = -\nabla \cdot \mathbf{q} + S = \nabla \cdot k\nabla T + S$ | (13.30) |
| Conservation of fluid enthalpy | |
| $\frac{\partial}{\partial t}\rho h = -\nabla \cdot (\rho h \mathbf{v}) - \nabla \cdot \mathbf{q} - \boldsymbol{\tau} : \nabla \mathbf{v} + \frac{Dp}{Dt} + S$ | (13.31) |

**Table 13.2** Generalized form of fluid transport equations.

$$\frac{\partial(\rho\psi)}{\partial t} + \nabla \cdot (\rho\psi\mathbf{v}) = -\nabla \cdot \mathbf{J} + \phi$$

| Parameter | $\psi$ | $\mathbf{J}$ | $\phi$ | Eq. No. |
|---|---|---|---|---|
| Mass | 1 | 0 | 0 | (13.7) |
| Momentum | $\mathbf{v}$ | $\boldsymbol{\tau} + p\boldsymbol{I}$ | $\rho\mathbf{g}$ | (13.13) |
| Energy | | | | |
| Internal + kinetic | $U + \frac{v^2}{2}$ | $\mathbf{q} + (\boldsymbol{\tau} + p\boldsymbol{I}) \cdot \mathbf{v}$ | $\rho\mathbf{v} \cdot \mathbf{g} + S$ | (13.22) |
| Internal | $U$ | $\mathbf{q}$ | $-p\nabla \cdot \mathbf{v} - \boldsymbol{\tau} : \nabla\mathbf{v} + S$ | (13.24) |
| Enthalpy | $h$ | $\boldsymbol{q}$ | $\frac{Dp}{Dt} - \boldsymbol{\tau} : \nabla\mathbf{v} + S$ | (13.31) |

be noted that the conservation equations are written in terms of parameters $\psi = \left[1, \mathbf{v}, U + v^2/2, U, h\right]$ and a substantial derivative

$$\rho\frac{D\psi}{Dt} = \frac{\partial}{\partial t}(\rho\psi) + \nabla \cdot (\rho\psi\mathbf{v}).$$

Solution of the fluid conservation equations, in general, requires a number of boundary conditions representing the continuity of momentum or energy flux. Some of the common boundary conditions are

(1) At a solid–fluid interface, the fluid velocity is equal to the velocity with which the solid surface itself is moving. This condition is essentially based on the assumption that viscous flow will cling to any solid surface with which it is in contact.
(2) At a liquid–gas interface, the momentum flux, and hence the velocity gradient, can be assumed to be zero.
(3) At a liquid–liquid interface of two immiscible fluids, the momentum flux perpendicular to the interface and the velocity are continuous across the interface.
(4) At a solid–fluid interface, the heat flux is usually specified in terms of Newton's law of cooling, Eq. (13.4), together with the interface temperature.
(5) At an interface between two different media, the temperature and heat flux are continuous.

## 13.3 Simple Solutions of Fluid Conservation Equations

As illustrative examples of analytical methods that can be used to solve the fluid conservation equations derived in Section 13.2, we present solutions for 1-D fluid flow and heat transfer problems in this section, mostly borrowed from [Bir07]. The sample problems will introduce a number of key concepts in fluid mechanics and heat transfer, including dimensionless numbers and friction factor, which are useful for general T/H analysis of reactor systems.

**Example 13.1** Solve the fluid conservation equation for the *Couette flow* representing the 1-D flow of a viscous fluid between two flat plates, illustrated in Figure 13.2, for the derivation of Newton's law of viscosity.

For a Newtonian fluid at steady state, the Navier–Stokes equation (13.17) reduces to

$$\rho\left(v_x\frac{\partial v_x}{\partial x}+v_y\frac{\partial v_x}{\partial y}\right)=\mu\left(\frac{\partial^2 v_x}{\partial x^2}+\frac{\partial^2 v_x}{\partial y^2}\right).$$

Since the flow is uniform and finite only in the $x$-direction, $v_x = v_x(y)$ and $v_y = 0$, and the momentum conservation equation further simplifies to

$$\mu\frac{d^2 v_x}{dy^2}=0.$$

The general solution is

$$v_x(y) = C_1 y + C_2,$$

with two constants $C_1$ and $C_2$. Invoking the boundary condition at solid–fluid interfaces discussed in Section 13.2.4 requires that the fluid moves with the bottom plate at $y = 0$ with speed $V$ and remains stationary at the top plate at $y = Y$, resulting in the boundary conditions

(i) $v_x(0) = V$,
(ii) $v_x(Y) = 0$.

Obtain $C_2 = V$ and $C_1 = -V/Y$, and the velocity distribution

$$v_x(y) = V\left(1-\frac{y}{Y}\right)$$

and the shear stress

$$\tau_{yx} = -\mu\frac{dv_x}{dy} = \frac{\mu V}{Y} = \text{constant}.$$

We have essentially re-derived Newton's law of viscosity, and for the simple Couette flow, the shear stress $\tau_{yx}$ is constant across the flow cross section.

**Example 13.2** Obtain the axial fluid velocity distribution $v_z$ for the *Hagen–Poiseuille flow* describing a steady-state fully developed laminar flow of a Newtonian fluid with constant $\rho$ and $\mu$ in a long cylindrical tube of radius $R$ and length $L$.

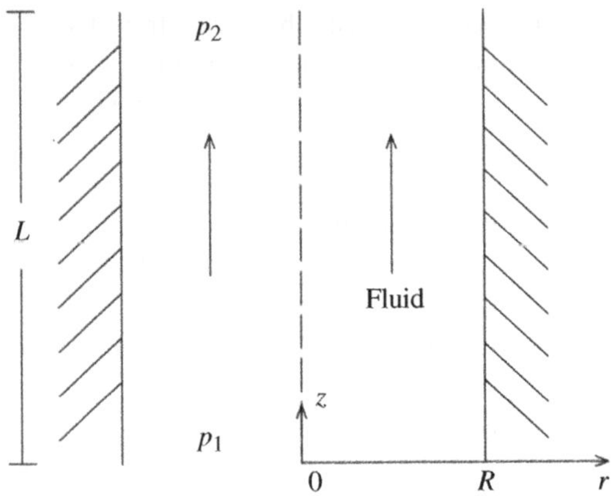


**Figure 13.5** Hagen–Poiseuille flow.

For a fully developed flow, consider $v_z$ at distances far from the inlet and outlet so that $v_z = v_z(r)$. The idealized 1-D flow is illustrated in Figure 13.5.

The equation of motion for this steady-state flow in a pipe

$$\rho v_z \frac{\partial v_z}{\partial z} = -\frac{\partial p}{\partial z} - \frac{1}{r}\frac{\partial}{\partial r}(r\tau_{rz}) - \rho g$$

reduces to

$$\frac{1}{r}\frac{d}{dr}(r\tau_{rz}) = -\frac{dp}{dz} - \rho g = -\frac{d}{dz}(p + \rho g z).$$

Define the total pressure $P$ as the sum of static pressure and gravitational pressure

$$P = p + \rho g z, \tag{13.35}$$

with $p_1 = p(0)$ and $p_2 = p(L)$, and recast the equation of motion

$$\frac{1}{r}\frac{d}{dr}(r\tau_{rz}) = -\frac{dP}{dz} = \frac{P_1 - P_2}{L} = \frac{(p_1 - p_2) - \rho g L}{dz}.$$

Applying the boundary conditions

(i) Symmetry or finiteness of $\tau_{rz}$ at $r = 0$,
(ii) $v_z(R) = 0$ at the fluid–solid interface,

provides the solution for the shear stress and velocity profiles

$$\tau_{rz}(r) = \left(\frac{P_1 - P_2}{2L}\right) r = -\mu \frac{dv_z(r)}{dr} \tag{13.36}$$

and

$$v_z(r) = \left(\frac{P_1 - P_2}{4\mu L}\right) R^2 \left(1 - \frac{r^2}{R^2}\right), \tag{13.37}$$

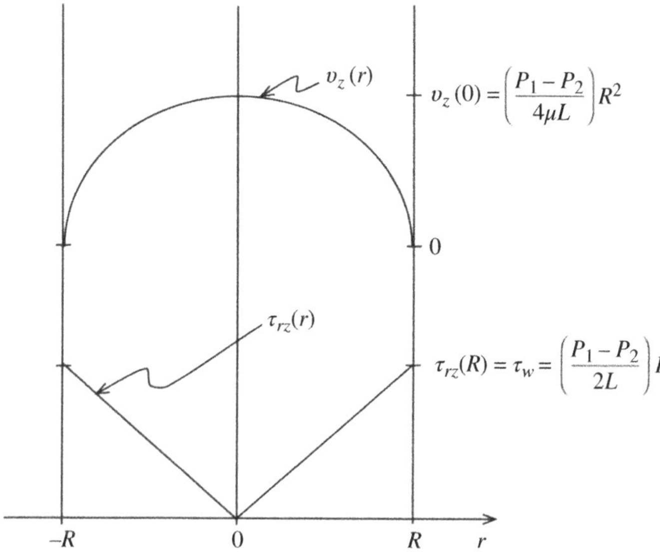


**Figure 13.6** Shear stress and fluid velocity for the Hagen–Poiseuille flow.

schematically illustrated in Figure 13.6. From Eq. (13.37), note that

$$v_{z,max} = v_z(0) = \left(\frac{P_1 - P_2}{4\mu L}\right) R^2$$

and

$$\langle v_z \rangle = \frac{2\pi \int_0^R v_z(r) r dr}{\pi R^2} = \left(\frac{P_1 - P_2}{8\mu L}\right) R^2 = \frac{1}{2} v_{z,max}.$$

The *mass flow rate* $W$ is obtained for this channel flow

$$W = \rho \langle v_z \rangle \pi R^2 = \frac{\pi \rho (P_1 - P_2) R^4}{8\mu L} \tag{13.38}$$

and the force $F_z$ of the fluid on the wetted surface at the tube wall is given by

$$F_z = 2\pi R L \tau_{rz}(R) = 2\pi R L \tau_w = \pi R^2 (P_1 - P_2),$$

where $\tau_w$ is the *wall shear stress*. The equation expresses the force balance on the fluid in the tube, i.e. the shear force $F_z$ should equal the total pressure force on the fluid $\pi R^2 (P_1 - P_2)$. Similarly, the net static pressure drop $\Delta p$ is given as

$$\Delta p = p_1 - p_2 = (P_1 - P_2) + \rho g L = \frac{8\mu L}{\pi R^4} \frac{W}{\rho} + \rho g L. \tag{13.39}$$

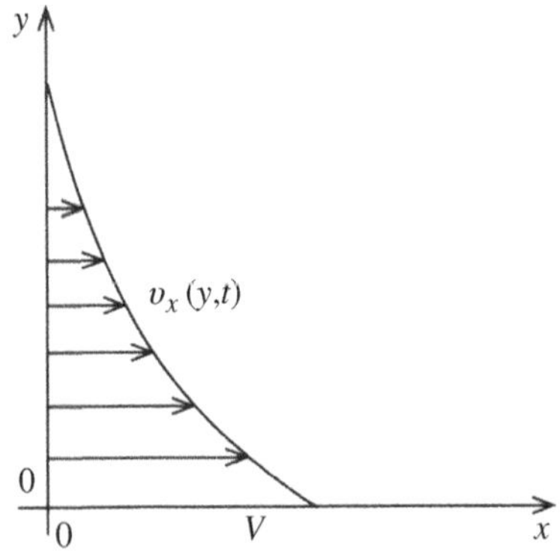


**Figure 13.7** Time-dependent one-dimensional velocity profile.

Equation (13.39) shows that the net pressure drop $\Delta p$ to be provided by external force, e.g. a mechanical pump, is the sum of the frictional pressure drop and gravitational or elevation loss. For the Hagen–Poiseuille flow, define a dimensionless number *Re*, called the *Reynolds number*

$$Re = \frac{\rho \left\langle v_z \right\rangle D}{\mu} \tag{13.40}$$

as a measure of the ratio of the inertial force to viscous force acting on the fluid. Usually a viscous flow with $Re < 2{,}100$ is classified as a laminar flow; beyond $Re = 2{,}100$, some turbulence would usually begin to appear in the flow. The Reynolds number, marking the boundary between laminar and turbulent flows, is an example of dimensionless numbers frequently used in fluid mechanics and heat transfer.

**Example 13.3** Obtain the solution to the time-dependent fluid conservation equations for a semi-infinite volume of fluid with constant $\rho$ and $\mu$ in contact with a flat plate at $y = 0$. Assume that the effects of hydrostatic pressure and gravity are negligible. When the plate is suddenly set in motion in the $x$-direction at speed $V$, a 1-D flow with speed $v_x(y, t)$ is established in the $x$-direction, as illustrated in Figure 13.7.

For this 1-D flow, we write the Navier–Stokes equation (13.17) for the $x$-component $v_x$ of fluid velocity, neglecting the terms $\nabla p$ and $\rho \mathbf{g}$:

$$\rho \frac{\partial v_x}{\partial t} = -\rho \left( v_x \frac{\partial v_x}{\partial x} + v_y \frac{\partial v_x}{\partial y} \right) + \mu \frac{\partial^2 v_x}{\partial y^2}.$$

Since $v_x \neq f(x)$ and $v_y = 0$, the equation of motion reduces to

$$\frac{\partial v_x}{\partial t} = \nu \frac{\partial^2 v_x}{\partial y^2}, \tag{13.41}$$

with the *kinematic viscosity* $\nu = \mu/\rho$, and subject to the initial and boundary conditions:

I.C. $v_x(y, t) = 0$, for $t \leq 0$,
B.C. (i) $v_x(0, t) = V$, for $t > 0$,
B.C. (ii) $v_x(\infty, t) = 0$, for $t > 0$.

Although Eq. (13.41) can be readily solved through separation of variables and other standard techniques, introduce a method that combines independent variables into a single variable called the *similarity variable*. This technique is applicable only when two boundary conditions can be combined into one condition in terms of the similarity variable, and hence its applicability is limited. When it is applicable, however, the method provides elegant solutions to partial differential equations, yielding physical insights to the problem under consideration with minimal effort.

Introduce a dimensionless variable $\eta = y/\sqrt{4\nu t}$ as the similarity variable, and consider the velocity profile as $v_x/V = \phi(\eta)$. The initial and boundary conditions are rewritten in terms of $\eta$:

B.C. (i) $\phi(0) = 1$,
I.C. + B.C. (ii) $\phi(\infty) = 0$.

In terms of the new variables, obtain

$$\frac{\partial(v_x/V)}{\partial t} = \frac{d\phi}{d\eta}\frac{\partial\eta}{\partial t} = -\frac{\eta}{2t}\phi'(\eta),$$

$$\frac{\partial^2(v_x/V)}{\partial y^2} = \frac{\eta^2}{y^2}\phi''(\eta),$$

and rewrite Eq. (13.41) as

$$\phi'' + 2\eta\phi' = 0,$$

which becomes a simple first-order differential equation for $\psi = \phi'$

$$\psi' + 2\eta\psi = 0.$$

Solving for $\psi = \phi'$

$$\psi = \phi' = C_1 \exp(-\eta^2)$$

and integrating once more for $\phi$ yields

$$\phi(\eta) = C_1 \int_0^\eta \exp(-u^2)du + C_2.$$

Applying the boundary conditions for $\phi(\eta)$ to evaluate the integration constants $C_1$ and $C_2$, finally obtain a dimensionless velocity profile

$$\phi(\eta) = 1 - \frac{2}{\sqrt{\pi}} \int_0^\eta \exp(-u^2)du = 1 - \text{erf}(\eta), \tag{13.42}$$

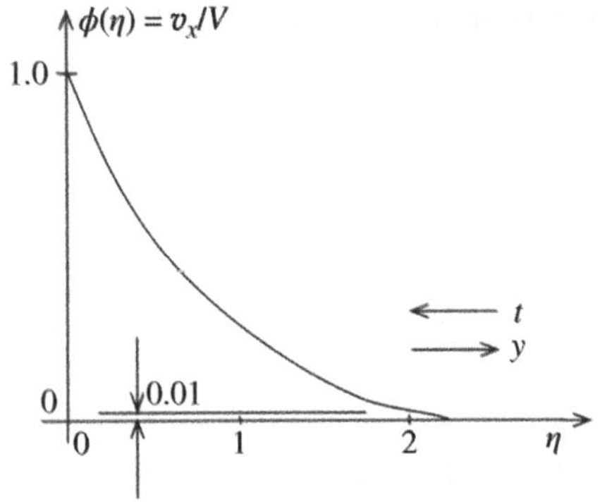


**Figure 13.8** Dimensionless velocity profile.

or

$$\frac{v_x(y,t)}{V} = 1 - \text{erf}\left(\frac{y}{\sqrt{4\nu t}}\right).$$

As shown in Figure 13.8, the velocity $v_x$ decreases as $\eta$ increases, due to either an increase in distance $y$ or a decrease in time $t$, which correctly represents the evolution of the fluid flow. The usefulness of the dimensionless velocity profile of Eq. (13.42) can be simply illustrated by considering a *boundary layer thickness* $\delta$ of the fluid at which the fluid speed reaches 1% of the asymptotic speed $V$. With $\phi(2) \simeq 0.01$, we obtain $\delta = 4\sqrt{\nu t}$ as a measure of the extent to which the momentum of the plate motion penetrates the semi-infinite body of fluid. The concept of the boundary layer thickness $\delta$ is further illustrated in Figure 13.9, where $\delta$ increases as a function of time $t$. This is a simple indication that the plate motion is felt deeper into the fluid volume as the motion continues.

The similarity variable approach can be applied also to other partial differential equations, e.g. the time-dependent heat conduction equation [Bir07]. Since such solutions are expressed in terms of dimensionless similarity variables, they are often valuable for comparing the performance of systems of similar geometrical

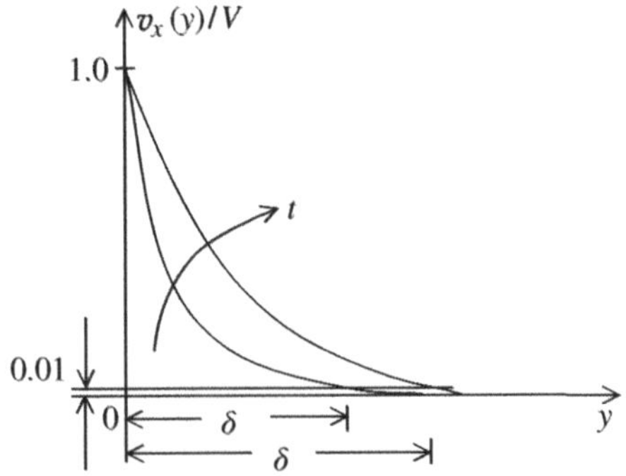


**Figure 13.9** Boundary layer thickness $\delta$ as a function of time.

shape but of different scale. This is another good example of the usefulness of dimensionless analysis in fluid mechanics, heat transfer, and in general systems analysis.

### 13.3.1 Heat Conduction in Cylindrical Fuel Rod

As a sample solution of the energy conservation equations derived in Section 13.2.3, consider the problem of calculating the steady-state distribution of temperature in a long, thin cylindrical fuel rod of radius $a$ and length $L$, subject to a uniform volumetric heat source distribution $S$ and a uniform surface temperature $T_s$. For distances away from the ends in a long rod, axial heat conduction can be neglected compared with the radial heat conduction, and the heat conduction equation (13.30) can be written in 1-D cylindrical coordinates:

$$\frac{1}{r}\frac{d}{dr}(rq_r) = S,\ q_r(r) = -k\frac{dT(r)}{dr}.$$

With the boundary conditions

(i) Symmetry of heat flux or $q_r(0) = 0$,
(ii) $T(a) = T_s$,

integrate the 1-D heat conduction equation to get

$$q_r(r) = \frac{Sr}{2} = -k\frac{dT(r)}{dr}$$

and

$$\int_{T_s}^{T(r)} k(T')dT' = \frac{S}{2}\int_r^a rdr = \frac{Sa^2}{4}\left(1 - \frac{r^2}{a^2}\right). \tag{13.43}$$

Defining $P$ as the total heat generated in a fuel rod of length $L$ suggests $\pi a^2 S = P/L$, and Eq. (13.43) can be rewritten as

$$\int_{T_s}^{T(r)} k(T')dT' = \frac{P/L}{4\pi}\left(1 - \frac{r^2}{a^2}\right). \tag{13.44}$$

In terms of the fuel centerline temperature $T_c = T(0)$, Eq. (13.44) indicates

$$\int_{T_s}^{T_c} k(T)dT = \frac{P/L}{4\pi}. \tag{13.45}$$

If the temperature dependence of thermal conductivity $k(T)$ can be neglected over the temperature range between $T_s$ and $T_c$, Eq. (13.45) simplifies to

$$k(T_c - T_s) = \frac{P/L}{4\pi}. \tag{13.46}$$

Equation (13.45) indicates that the *linear heat generation rate* $P/L$ of a fuel rod is determined entirely by limitations on fuel temperatures, $T_c$ and $T_s$, and is independent of the fuel radius $a$, provided the volumetric heat generation rate $S$ is spatially uniform. In general, the volumetric heat source $S$ is not uniform across the fuel cross section, and Eq. (13.45) has to be corrected, often through a multiplying factor on $P/L$, which then depends to some extent on radius $a$.

For calculation of the temperature distribution in nuclear fuel rods, the temperature dependence of $k(T)$ is significant, as shown for $UO_2$ at 95% of theoretical density in Figure 13.10, and is often expressed in terms of empirical correlations, e.g.

$$\int_0^T k(T')dT' = \sum_{n=0}^{N} a_n T^n. \tag{13.47}$$

The data for $k(T)$ of $UO_2$ plotted in Figure 13.10 are represented for the T/H design calculations of the AP1000 core through the correlation [Hon12]

$$k\left[\frac{\mathrm{W}}{\mathrm{m}\cdot\mathrm{K}}\right] = \frac{18.86}{1+4.49\times10^{-3}T} + 8.775\times10^{-11}(T-273)^3, \quad T\,[\mathrm{K}]. \tag{13.48}$$

Actual calculation of $T(r)$ through Eqs. (13.44) and (13.47) involves conceptually inverting Eq. (13.44) to obtain radius $r$ for a given temperature $T$. Thus, Eq. (13.47) is used to determine, for an assumed temperature $T$, an integral of $k$ over temperature via Eq. (13.44), which then yields radius $r$ corresponding to the assumed

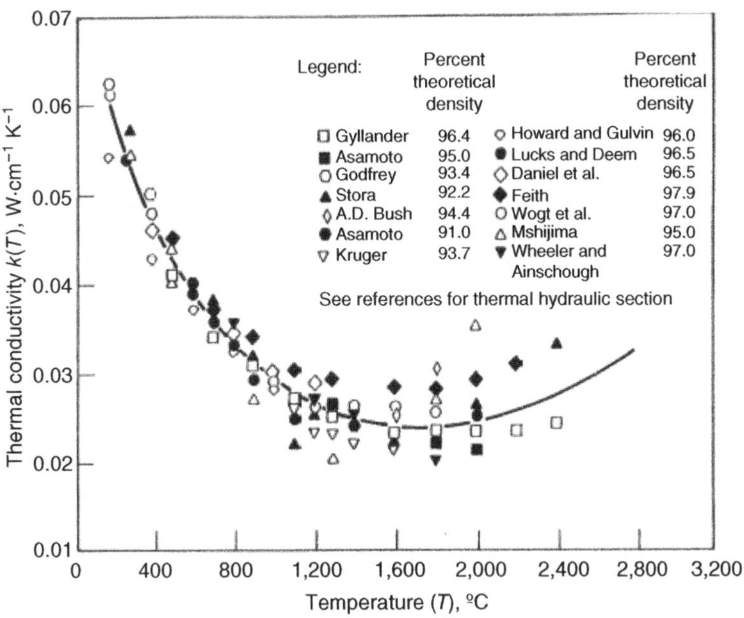


**Figure 13.10** Thermal conductivity of $UO_2$ at 95% of theoretical density vs. temperature. *Source:* Hone et al. [Hon12].

temperature $T$. We may repeat this process as many times as necessary to obtain pairs of $[T(r), r]$. Because the fuel temperature distribution plays a key role in determining the overall power output of a reactor core, this approach to accurately calculate the fuel temperature distribution is very much necessary.

### 13.3.2 Heat Conduction Through Composite Wall

Extend the steady-state 1-D heat conduction solution for a bare cylindrical fuel rod of Section 13.3.1 to a more realistic fuel rod geometry including a clad and a fuel-clad gap. To save on the algebra involved, however, we consider in detail steady-state heat conduction through a composite wall in 1-D slab geometry and derive the concept of overall heat transfer coefficient. For the actual cylindrical geometry of fuel rods in nuclear reactor cores, we simply present the result in terms of an overall heat transfer coefficient. For the slab geometry shown in Figure 13.11, consider three semi-infinite slabs of solid materials of different thicknesses and thermal conductivity values. Assume that inside the slabs there is no heat source but that the temperature is maintained at $T_0$ at the left surface $x = x_0 = 0$, while the right surface at $x = x_3$ is cooled by a fluid flow. We specify the bulk fluid temperature $T_b$ and the heat transfer coefficient $h$ for the fluid. For this simple slab geometry problem, the heat conduction equation (13.30) reduces to

$$\frac{dq}{dx} = 0,$$

subject to the boundary conditions

(i) $T(x_0) = T_0$,
(ii) $q(x_3) = h[T(x_3) - T_b] = q_0$.

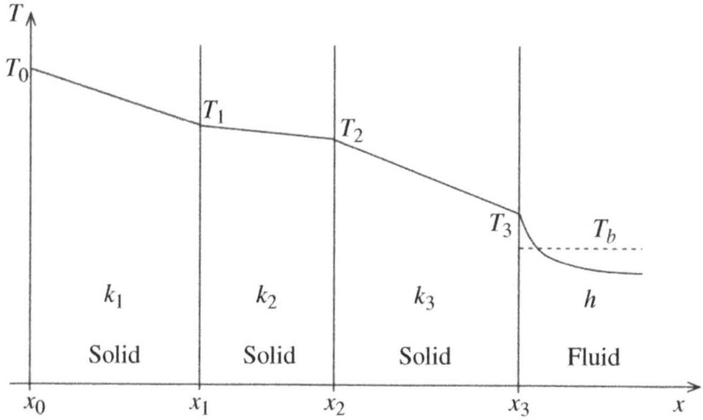


**Figure 13.11** Heat conduction through three slabs.

With the continuity of heat flux at material interfaces, write

$$q_0 = -k_1\frac{T_1 - T_0}{x_1 - x_0} = -k_2\frac{T_2 - T_1}{x_2 - x_1} = -k_3\frac{T_3 - T_2}{x_3 - x_2} = h(T_3 - T_b),$$

from which the temperature difference follows:

$$T_0 - T_b = \sum_{i=1}^{3}(T_{i-1} - T_i) + (T_3 - T_b) = q_0\left[\sum_{i=1}^{3}\frac{x_i - x_{i-1}}{k_i} + \frac{1}{h}\right]. \tag{13.49}$$

An *overall heat transfer coefficient* $U$ can be defined in terms of the overall temperature difference $(T_0 - T_b)$ over the composite slab

$$q_0 = U(T_0 - T_b),$$

or explicitly in terms of Eq. (13.49)

$$U = \left[\sum_{i=1}^{3}\frac{\Delta x_i}{k_i} + \frac{1}{h}\right]^{-1}, \quad \Delta x_i = x_i - x_{i-1}. \tag{13.50}$$

In analogy to electric circuitry, Eq. (13.50) for $U$ can be readily interpreted as the inverse of the sum of heat transfer resistances over various heat transfer paths under consideration.

We may simply extend the results obtained here to multi-region problems in slab geometry and to different geometries. For example, consider a cylindrical fuel rod of radius $a$ surrounded by a clad of thickness $t_c$, with a gap of thickness $t_g$ between the fuel rod and clad, with the temperature distribution $T(r)$ shown in Figure 13.12. Assume that a uniform volumetric source $S$ of heat is produced in the

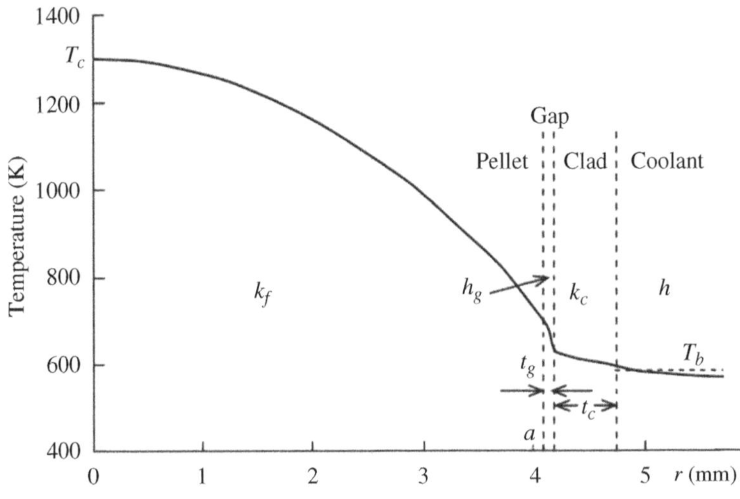


**Figure 13.12** Radial temperature distribution $T(r)$ in a fuel rod cooled by fluid flow.

fuel rod and that the clad outer surface is cooled by a coolant flow, whose bulk fluid temperature is $T_b$. In addition, assume temperature-independent thermal conductivity $k_f$ and $k_c$ for the fuel and clad, respectively, and heat transfer coefficient $h_g$ and $h$ for the fuel-clad gap and coolant, respectively. Then, the heat flux $q$ into the coolant can be represented in terms of the overall temperature difference as

$$q = U_c(T_c - T_b), \tag{13.51}$$

where $T_c$ is the fuel centerline temperature and the *overall heat transfer coefficient* $U_c$ is obtained as

$$U_c = \left[\frac{a}{2k_f} + \frac{1}{h_g} + \frac{a}{k_c}\ln\frac{a + t_g + t_c}{a + t_g} + \frac{a}{a + t_g + t_c}\frac{1}{h}\right]^{-1}. \tag{13.52}$$

The heat transfer coefficient $h_g$ of the fuel-clad gap is often referred to as the *gap conductance* and is a sensitive function of the size of the gap and hence of the overall fuel rod configuration [Ton96]. In addition, actual calculation of the temperature distribution in fuel rods in a reactor core requires a number of detailed engineering considerations including those for spatial distribution of heat flux $S$ and temperature dependence of fuel thermal conductivity, discussed in Section 13.3.1. Other engineering considerations required for accurate fuel temperature calculations include thermal expansion and restructuring of the fuel rods. As a result of repetitive thermal cycling, radial and circumferential cracks may occur in the fuel rods. At the same time, micro-pores retained during the fuel pellet manufacturing process and volatile fission products produced during the irradiation may migrate along temperature gradients, and may result in central voids later in fuel life. This is known as the *coring* of fuel pellets.

In contrast, early in fuel life, due to in-pile sintering of the pellets, the micro-pores may be expelled, resulting in an increased density for the pellets. The *fuel densification* results in reduced fuel volume and may produce axial gaps within fuel rods and flattening of the clad around the gaps. Thermal flux and power spikes are then produced due to extra moderation of neutrons around the flattened clad regions. The fuel densification phenomenon occurred rather unexpectedly in LWR fuel rods in the early 1970s and resulted in temporary derating of a number of nuclear power plants until the fuel pellet design and manufacturing process were altered to minimize the potential for fuel densification. The power spike penalty due to fuel densification was minimized through a Monte Carlo analysis [Lee72] of the effects due to gaps between pellets and clad flattening.

### 13.3.3 Forced Convection in Laminar Flow

Heat transfer by convection in a fluid flow can, in general, be divided into two types: forced convection and free convection. In forced convection, the flow pattern is determined by external forces, and the temperature profile within the fluid may be determined based on the velocity profile. In contrast, in free convection or natural circulation flow, the fluid velocity profile itself is determined by the buoyancy effect of the heated fluid. Hence, in free convection, velocity and temperature profiles are closely interrelated and not separable.

As an example of forced convection heat transfer problems, consider the flow of a viscous fluid with constant properties, $\rho$, $\mu$, $k$, and $C_p$, in a long circular pipe of radius $R$ and length $L$, illustrated in Figure 13.5 for the Hagen–Poiseuille flow. Assume that the fluid flow attains a steady-state, fully developed velocity profile due to an external pressure drop $\Delta p = p_1 - p_2$, and is subject to a spatially uniform heat flux $q_0$ at the tube wall. There is no volumetric heat source in the fluid. The energy conservation equation (13.32) is written for this steady-state convection problem

$$\rho C_p v_z \frac{\partial T(r,z)}{\partial z} = k\left[\frac{1}{r}\frac{\partial}{\partial r}\left(r\frac{\partial T}{\partial r}\right) + \frac{\partial^2 T}{\partial z^2}\right].$$

In a typical forced convection problem, axial fluid speed $v_z$ is large enough or thermal conductivity $k$ is small enough so that the axial heat conduction in the fluid can be neglected compared with the axial convection, which simplifies the energy conservation equation to

$$\rho C_p v_z \frac{\partial T(r,z)}{\partial z} = k\frac{1}{r}\frac{\partial}{\partial r}\left(r\frac{\partial T}{\partial r}\right), \tag{13.53}$$

where $v_z(r)$ is given by the Hagen–Poiseuille velocity profile of Eq. (13.37). With the proper boundary conditions

(i) Symmetry of $T(r,z)$ around $r = 0 \ \forall z$,
(ii) Inlet temperature $T(r,0) = T_0 \ \forall r$,

we may obtain a general solution [Bir07] to Eq. (13.53). However, the limiting form of the solution for large $z$, i.e. at distances away from the inlet and outlet, will be of more interest, because a fully developed temperature profile is established in that limit.

For a fully developed temperature profile that would exist in the heated section of the pipe, the temperature $T(r,z)$ increases as a linear function of $z$, due to the constant surface heat flux $q_0$, whereas the radial temperature profile $\psi(r)$ remains constant as a function of $z$. Thus, write $T(r,z)$ as

$$T(r,z) = C_0 z + \psi(r), \tag{13.54}$$

where $C_0$ is a constant. This rather intuitive observation can further be understood by defining the *bulk fluid temperature* $T_b(z)$ as the flow rate-weighted fluid temperature, averaged over the cross sectional area $A = \pi R^2$ of the pipe

$$T_b(z) = \frac{\langle \rho v_z(r) T(r,z) \rangle_r}{\langle \rho v_z(r) \rangle_r} = \frac{2\pi \int_0^R \rho v_z(r) T(r,z) r dr}{\rho \pi R^2 \langle v_z \rangle_r}. \tag{13.55}$$

Thus, $T_b$ is the average fluid temperature, which would result in a fluid sample collected from the channel and thoroughly mixed in a cup, and is also known as the *cup-mixing temperature* or the *flow-average temperature*. From Eq. (13.55), note that a fully developed temperature profile can be represented by the relationship

$$\frac{\partial T(r,z)}{\partial z} = \frac{dT_b(z)}{dz}.$$

The relationship also implies that the temperature difference $[T(r,z) - T_b(z)]$ is a function only of $r$ and not of $z$. This also follows from Eq. (13.54) and provides an additional interpretation of its functional form.

Define a dimensionless temperature profile $\theta(r)$ as a function of $r$ only

$$\theta(r) = \frac{T(0,z) - T(r,z)}{T(0,z) - T_b(z)} = 1 + \frac{T_b(z) - T(r,z)}{T(0,z) - T_b(z)}. \tag{13.56}$$

Instead of B.C. (ii) on inlet temperature $T_0$, use the definition of the bulk fluid temperature $T_b$ in Eq. (13.55) and write a heat balance over a differential length $dz$ of the tube

$$WC_p dT_b = 2\pi R dz \cdot q_0 = M q_0 dz$$

or

$$\frac{dT_b(z)}{dz} = \frac{Mq_0}{WC_p} = \frac{\partial T(r,z)}{\partial z} = \text{constant}, \tag{13.57}$$

in terms of mass flow rate $W$ and the *wetted perimeter* $M = 2\pi R$. Substituting Eqs. (13.37) for $v_z(r)$ together with Eq. (13.57) into Eq. (13.53) yields

$$k\frac{1}{r}\frac{\partial}{\partial r}\left(r\frac{\partial T}{\partial r}\right) = 2\rho C_p \langle v_z \rangle \left(1 - \frac{r^2}{R^2}\right)\frac{Mq_0}{WC_p} = \frac{4q_0}{R}\left(1 - \frac{r^2}{R^2}\right).$$

With the symmetry of $T(r,z)$ around $r = 0$, the energy conservation equation can be readily integrated for the temperature distribution

$$T(r,z) = C_1(z) + \frac{q_0 r^2}{kR}\left(1 - \frac{r^2}{4R^2}\right),$$

where $C_1(z)$ is a constant of integration given as a function of $z$. Thus, we get

$$T(r,z) - T(0,z) = \frac{q_0 r^2}{kR}\left(1 - \frac{r^2}{4R^2}\right)$$

and

$$T_b(z) - T(0,z) = \frac{q_0}{kR} \frac{4\pi \left\langle v_z \right\rangle \int_0^R r^3 \left(1 - \frac{r^2}{4R^2}\right)\left(1 - \frac{r^2}{R^2}\right) dr}{\pi R^2 \left\langle v_z \right\rangle} = \frac{7}{24}\frac{q_0 R}{k}.$$

Hence, the dimensionless temperature profile $\theta(r)$ can be written as

$$\theta(r) = \frac{T(0,z) - T(r,z)}{T(0,z) - T_b(z)} = \frac{24}{7}\frac{r^2}{R^2}\left(1 - \frac{r^2}{4R^2}\right). \tag{13.58}$$

For the purpose of dimensional analysis and experimental correlations for forced convection problems, a number of dimensionless numbers are often introduced in addition to the Reynolds number introduced in Eq. (13.40). Perhaps the best-known dimensionless numbers for forced convection are the *Nusselt number Nu* and the *Prandtl number Pr*, which are defined as

$$Nu = \frac{hD}{k} = \frac{\text{convective heat flux}}{\text{conductive heat flux}}, \tag{13.59}$$

$$Pr = \frac{C_p \mu}{k} = \frac{\nu}{\alpha} = \frac{\text{diffusion of momentum}}{\text{diffusion of heat}}, \tag{13.60}$$

where $D$ is the pipe diameter, $\nu = \mu/\rho$ *kinematic viscosity*, and $\alpha = k/\rho C_p$ *thermal diffusivity*. As an example, the Nusselt number $Nu$ is evaluated for the forced convection problem just analyzed. Determine the positive heat flux $q_o$ into the fluid, corresponding to a positive temperature gradient $(\partial T/\partial r)_R$ at the tube wall

$$q_0 = h[T(R,z) - T_b(z)] = k\left.\frac{\partial T}{\partial r}\right|_{r=R}$$

and obtain

$$Nu = 2R\frac{(\partial T/\partial r)_R}{T(R,z) - T_b(z)} = 2R\frac{\partial}{\partial r}\left[\frac{T(r,z) - T_b(z)}{T(R,z) - T_b(z)}\right]_{r=R}$$

$$= 2R\frac{\partial}{\partial r}\left[\frac{T_b(z) - T(r,z)}{T(0,z) - T_b(z)} \cdot \frac{T(0,z) - T_b(z)}{T_b(z) - T(R,z)}\right]_{r=R}.$$

Using Eq. (13.56), rewrite

$$Nu = 2R\left.\frac{d\left[\theta(r) - 1\right]}{dr}\right|_{r=R}\frac{1}{\theta(R) - 1}$$

and, with Eq. (13.58), obtain

$$\theta(R) = \frac{18}{7} \text{ and } \left.\frac{d\theta}{dr}\right|_{r=R} = \frac{24}{7R}$$

to arrive finally at $Nu = 48/11$. One empirical correlation for the Nusselt number often used for fuel channel analysis, when the coolant is heated, is the Dittus–Boelter correlation

$$Nu = 0.023\,Re^{0.8}Pr^{0.4}, \quad 0.7 \leq Pr \leq 100, Re \geq 10^4. \tag{13.61}$$

### 13.3.4 Velocity Distribution in Turbulent Flow

As a simple example of solutions to the Navier–Stokes equation, a steady-state laminar flow in a pipe is considered in Example 13.2. As the fluid velocity increases substantially so that $Re > 2{,}100$, the laminar flow transforms into turbulent flow with its fluctuating eddies. Turbulent flow is often characterized by its high flow rate so that the inertial forces associated with the flow are much larger than the corresponding viscous forces. The equations of continuity and motion derived in Section 13.2 are in general valid for turbulent flow because the average size of the turbulent eddies is usually larger than the mean free path of the molecules of the fluid. The Navier–Stokes equation may, however, not apply if the fluid is non-Newtonian. In addition, solutions of the equation of motion become too complex because of the wildly fluctuating distributions of velocity, pressure, and temperature.

In some analysis of turbulent flow, the instantaneous values of the velocity and temperature are expressed in terms of average components plus turbulent components

$$\begin{aligned} v &= \overline{v} + v', \\ T &= \overline{T} + T', \end{aligned} \tag{13.62}$$

where $\overline{v}$ and $\overline{T}$ are the averages over time corresponding to periods of turbulent eddies, and $v'$ and $T'$ are the turbulent components, respectively. For $\overline{v}$ and $\overline{T}$, the results obtained for laminar flows may be used. Many attempts have been made to find analytic expressions for the turbulent components $v'$ and $T'$, but only with partial success; one has to rely usually on empirical correlations to represent characteristics of turbulent flow.

For example, it has been observed [Bir07] experimentally that for turbulent flow in a circular tube, and for $10^4 \leq Re \leq 10^5$, the time-smoothed velocity distribution $\overline{v}$ can be represented approximately as a power-law profile:

$$\overline{v} = \overline{v}_{max}\left(1 - \frac{r}{R}\right)^{1/7} \text{ and } \frac{\langle \overline{v} \rangle}{\overline{v}_{max}} \simeq \frac{4}{5}. \tag{13.63}$$

The velocity profile is qualitatively compared with the laminar profile of the Hagen–Poiseuille flow in Figure 13.13. Note that the turbulent velocity profile is much flatter, approaching that of a *plug flow*, i.e. a uniform velocity profile.

Simulation of turbulent flows remains very much a challenging field of computational fluid dynamics (CFD) and is often performed in terms of the turbulent components $v'$ and $T'$ of Eqs. (13.62). Included in CFD models are the concepts of the *Reynolds stress*, representing the mean square term $\langle v'v' \rangle$, and the *large eddy simulation* focusing on small scale phenomena. A popular turbulence representation known as the *k–ε model* introduces the kinetic energy $k$ of turbulence

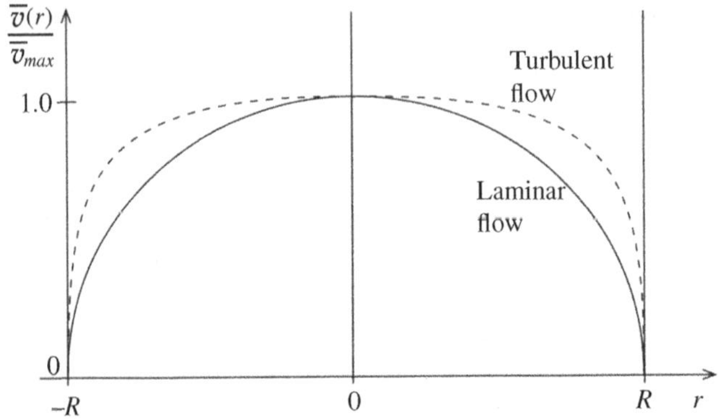


**Figure 13.13** Turbulent velocity profile.

and the turbulent dissipation rate $\varepsilon$ [Dur10]. One of the best known CFD codes is START-CCM+ [CDa16], which offers multi-physics capability built around CFD formulations and is discussed further in Section 13.8.5.

### 13.3.5 Friction Factor and Hydraulic Diameter

With the purpose of extending the pressure drop calculation performed for the Hagen–Poiseuille flow in Example 13.2, define the *friction factor* $f$ as a ratio of the viscous force on the channel wall to the inertial force of the fluid

$$f = \frac{2\pi RL \cdot \tau_w}{2\pi RL \cdot \rho\langle v_z\rangle^2/2} = \frac{\tau_w}{\rho\langle v_z\rangle^2/2} = \frac{1}{2}\left(\frac{R}{L}\right)\frac{P_1 - P_2}{\rho\langle v_z\rangle^2/2}, \tag{13.64}$$

where the *wall shear stress* $\tau_w$ is obtained from Eq. (13.36) in terms of the total pressure drop $(P_1 - P_2)$. For a laminar flow, Eqs. (13.37) and (13.40) yield

$$f = \frac{R}{2L}\frac{P_1 - P_2}{\rho\langle v_z\rangle/2}\frac{8\mu L}{(P_1 - P_2)R^2} = \frac{16\mu}{\rho\langle v_z\rangle D} = \frac{16}{Re}. \tag{13.65}$$

The friction factor $f$ defined in Eq. (13.64) is known as the *Fanning friction factor* and is perhaps the most widely used definition, although alternate definitions are available. For turbulent flows as well as for complex geometries, simple analytical expressions of the type given in Eq. (13.65) are not usually available and empirical correlations have to be employed for $f$. A simple correlation applicable to turbulent flow is the Blasius formula:

$$f = \frac{0.0791}{Re^{1/4}}, 2,100 \le Re \le 10^5. \tag{13.66}$$

Another well-known example is the Moody diagram [Whi86], where $f$ is plotted as a function of $Re$ and roughness of the tube wall. The applicability of empirical correlations for $f$ as a function of $Re$ indicates that complex physical phenomena may be represented succinctly in terms of dimensionless quantities including $f$, $Re$, $Nu$, and $Pr$.

It has also been observed empirically that turbulent flows in channels of different geometries can be represented approximately by an equivalent circular geometry, provided the ratio of the flow area $A$ to the wetted perimeter $M$ is preserved. Thus, define an *equivalent hydraulic diameter* $D_h$ as

$$D_h = \frac{4A}{M}. \tag{13.67}$$

For a circular pipe of diameter $D$, the hydraulic diameter reduces to $D_h = D$, as it should. The concept of the equivalent hydraulic diameter introduced in Eq. (13.67) is an empiricism applicable to turbulent flows, and it is invalid for laminar flows. For turbulent flows, regardless of the geometry, we may determine the *Reynolds number*

$$Re = \frac{\rho \left\langle v_z \right\rangle D_h}{\mu}, \tag{13.68}$$

and the pressure drop $\Delta P$

$$\Delta P = P_1 - P_2 = \frac{2f\rho \left\langle v_z \right\rangle^2 L}{D_h}. \tag{13.69}$$

## 13.4 Conservation Equations for Channel Flow

For analyzing flow in a channel, we are often interested in the average flow behavior across the channel, e.g. $\left\langle v_z \right\rangle$ and $T_b(z)$, rather than $v_z(r)$ and $T(r, z)$. This is the case for coolant channels in a reactor core, where we would be mainly interested in determining the coolant temperature and density along the channel length, averaged over the cross-sectional area of each channel, and the overall pressure drop across each channel. Thus we now consider a channel of cross-sectional area $A$ and wetted perimeter $M$ and reformulate the conservation equations derived in Section 13.2 in a form averaged over the channel cross-sectional area.

### 13.4.1 Equation of Continuity

By taking an average of Eq. (13.7) over the channel cross section of area $A$, we obtain a 1-D form of the equation of continuity

$$\frac{\partial \langle \rho \rangle_r}{\partial t} = -\frac{\partial}{\partial z}\left\langle \rho v_z \right\rangle_r \simeq -\frac{\partial}{\partial z}\left(\langle \rho \rangle_r \left\langle v_z \right\rangle_r\right),$$

where the last expression will be valid for narrow channels, with the density or velocity variation across the channel cross section small compared with that along the channel length. For notational convenience, set $\rho(z) = \langle\rho\rangle_r$ and $v(z) = \langle v_z\rangle_r$ and determine the *mass flow rate* $W$

$$W(z) = \rho(z)v(z)A = G(z)A$$

with $W$ expressed in units of $[\mathrm{kg\cdot s^{-1}}]$ and the *mass velocity* or *momentum flux* $G$ in units of $[\mathrm{kg\cdot m^{-2}\,s^{-1}}]$. In terms of $W$ and $G$, the continuity equation can be written as

$$\frac{\partial\rho}{\partial t} = -\frac{\partial G}{\partial z} = -\frac{1}{A}\frac{\partial W}{\partial z}. \tag{13.70}$$

For steady-state flow, Eq. (13.70) yields $W$ = constant, a simple statement that the total mass of fluid in the channel is conserved along the channel length and that the flow rate at any point along the channel is simply equal to the inlet and outlet flow rates.

### 13.4.2 Equation of Motion and Pressure Drop

To obtain the equation of motion for a channel flow, average the momentum conservation equation (13.13) over the channel cross-sectional area. Similar to the treatment of the leakage term in the continuity equation, the convective term $\nabla\cdot(\rho\mathbf{v}\mathbf{v})$ simplifies to

$$\frac{\partial}{\partial z}\langle\rho v^2\rangle_r \simeq \frac{\partial}{\partial z}\left(\rho\langle v\rangle^2\right) = \frac{\partial}{\partial z}\left(\frac{G^2}{\rho}\right).$$

The 1-D form of the viscous term can be expressed in terms of the wall shear stress $\tau_w$ by performing an integration of $\nabla\cdot\boldsymbol{\tau}$ over the channel volume of differential length $dz$ and dividing by the volume $Adz$:

$$\frac{\int_{Adz}\nabla\cdot\boldsymbol{\tau}d\mathbf{r}}{Adz} = \frac{\int_{Mdz}\boldsymbol{\tau}\cdot\mathbf{n}dA}{Adz} = \frac{\tau_w M}{A}. \tag{13.71}$$

With a similar cross-section averaging of the linear momentum density $\rho\mathbf{v}$ on the LHS, the hydrostatic term $\nabla p$, and the gravitational term $\rho\mathbf{g}$, with the $z$-coordinate taken against the gravity, obtain the equation of motion for channel flow

$$\frac{\partial G}{\partial t} = -\frac{\partial}{\partial z}\left(\frac{G^2}{\rho}\right) - \frac{\tau_w M}{A} - \frac{\partial p}{\partial z} - \rho g = -\frac{\partial p}{\partial z} + R, \tag{13.72}$$

where the term $R$ represents the sum of the pressure gradients due to convective momentum loss, frictional loss, and elevation or gravitational loss.

For a steady-state flow, Eq. (13.72) simply reduces to

$$\frac{dp}{dz} = R = \left(\frac{dp}{dz}\right)_{mom} + \left(\frac{dp}{dz}\right)_{fric} + \left(\frac{dp}{dz}\right)_{elev}. \tag{13.73}$$

To calculate the total pressure drop $\Delta p$, integrate Eq. (13.73) over the channel length $L$ and obtain

$$\Delta p = \Delta p_{mom} + \Delta p_{fric} + \Delta p_{elev}. \tag{13.74}$$

Further break up the momentum loss term $\Delta p_{mom}$ into the reversible pressure drop $\Delta p_{acc}$ due to density changes and hence velocity changes, and the irreversible pressure drop $\Delta p_{local}$ or $\Delta p_{form}$ associated with the flow across contractions, orifices, expansions, and other local effects

$$\Delta p_{mom} = \Delta p_{acc} + \Delta p_{local} = G^2 \Delta \left(\frac{1}{\rho}\right) + \sum_i \frac{G_i^2}{2\rho_i} \left(\sigma_{i+1}^2 - \sigma_i^2\right) + \sum_i K_i \frac{G_i^2}{2\rho_i}, \tag{13.75}$$

where the summations are taken over the area changes, with $\sigma_i$ representing the ratio of the flow area after to that before the $i$th area change, and $K_i$ the empirical irreversible *loss factor* for the $i$th area change. Substituting Eq. (13.75) into Eq. (13.74), and using the Fanning friction factor of Eq. (13.64), obtain the net pressure drop:

$$\Delta p = G^2 \Delta \left(\frac{1}{\rho}\right) + \sum_i \frac{G_i^2}{2\rho_i}\left(\sigma_{i+1}^2 - \sigma_i^2\right) + \sum_i K_i \frac{G_i^2}{2\rho_i} + \frac{2G^2}{D_h} \int_0^L \frac{f}{\rho} dz + \int_0^L \rho g dz. \tag{13.76}$$

Knowledge of the overall pressure drop over coolant channels is necessary to determine the pumping power requirement of the heat removal system of a nuclear power plant. Given the total flow rate of the system, the pressure drop across the core will also determine the flow rate distribution from channel to channel, which in turn provides the coolant density distribution and ultimately the power distribution in a reactor core. The pumping power of the system is the net rate of work per unit time required to pump the flow through the channels. The *pumping power* $F_p$ for a channel is given as

$$F_p = \Delta p A \langle v \rangle_z = \frac{\Delta p W}{\langle \rho \rangle_z}, \tag{13.77}$$

where $\langle v \rangle_z$ and $\langle \rho \rangle_z$ are the fluid velocity and density, respectively, averaged over the channel length, and the pressure drop $\Delta p$ is obtained from Eq. (13.76).

### 13.4.3 Equation of Energy Conservation

For the channel flow form of the energy conservation equation, we limit our derivation to the enthalpy conservation form of Eq. (13.32). Averaging over the channel cross section along the lines of Sections 13.4.1 and 13.4.2 yields

$$\frac{\partial}{\partial t}(\rho h) = -\frac{\partial}{\partial z}(Gh) + \frac{Mq_{\mathrm{w}}}{A} + \frac{\partial p}{\partial t} + S, \tag{13.78}$$

where the wall heat flux $q_w$ is defined to be positive for heat flux into the channel. For a steady-state flow, with the mass continuity $W =$ constant, Eq. (13.78) reduces to

$$W\frac{dh(z)}{dz} = Mq_w(z) + AS(z). \tag{13.79}$$

In terms of the bulk fluid temperature $T_b$ of Eq. (13.55), rewrite Eq. (13.79) as

$$WC_p\frac{dT_b(z)}{dz} = Mq_w(z) + AS(z). \tag{13.80}$$

Equations (13.79) and (13.80) are generalizations of Eq. (13.57) and represent a simple energy balance statement that the rate of increase of the fluid enthalpy across a flow channel is equal to the sum of energy added through the wall and energy produced in the fluid volume within the length of the channel under consideration.

## 13.5 Axial Temperature Distribution in Reactor Core

In order to facilitate a simple calculation of axial temperature distributions in reactor cores, assume that cross flows between coolant channels and between fuel assemblies can be neglected. This simplifying assumption is applicable to most steady-state and mild transient conditions because the power distribution varies slowly from one coolant channel to another. With the cross flows neglected, we may associate with each fuel element the volume of coolant that would typically be contained in a unit cell description of the fuel element. The *single-channel flow model* assumes that no heat is deposited directly in the coolant and neglects axial heat conduction in fuel rods and in coolant channels.

With the understanding that the unit cell boundary should not be included in calculating the hydraulic diameter, we may represent the coolant flow in a typical fuel assembly as a flow in a channel or duct. This is reminiscent of the need to exclude the unit cell boundary in calculating the mean chord length for the moderator in collision probability theory in Section 11.3. Given a heat source distribution for the fuel rod, we use the channel flow model developed in Section 13.4 to calculate the fuel and coolant temperature distributions.

Since the energy conservation equation (13.79) for the channel flow requires the knowledge of wall heat flux $q_w(z)$, we discuss in Section 13.5.1 how the heat flux is obtained from the power distribution in a reactor core. Section 13.5.2 derives the temperature distributions for single-phase flow in a typical PWR channel, and Section 13.5.3 extends the analysis to a typical BWR channel with two-phase flow and boiling heat transfer. Hot channel factors characterizing the temperature and enthalpy distributions in reactor cores are then presented in Section 13.5.4.

### 13.5.1 Power Distribution and Heat Flux in Reactor Core

For notational convenience, use the one-group model with fission cross section $\Sigma_f(\mathbf{r})$ and scalar neutron flux $\phi(\mathbf{r})$ to represent the heat generation rate or power distribution $S(\mathbf{r})$ in a reactor core

$$S(\mathbf{r}) = E_f\Sigma_f(\mathbf{r})\phi(\mathbf{r}), \tag{13.81}$$

where $E_f \simeq 200\,\text{MeV}\cdot\text{fission}^{-1}$ is the recoverable fission energy. Although about 3% of the fission energy is deposited outside the fuel and clad in a typical LWR design, assume that all of the energy generated is deposited in the fuel. This is one of the simplifying assumptions introduced for the single-channel analysis. With heterogeneous fuel arrangements in a reactor core, however, we need to account for the flux depression within the fuel elements. For notational convenience, consider a cylindrical reactor core consisting of an array of fuel rods. At axial position $z$ along a fuel rod located at radius $r$, introduce a normalized flux distribution $F(\rho)$, as a function of radius $\rho$ within the rod, to generalize Eq. (13.81):

$$S(r, z, \rho) = E_f\Sigma_f(r, z)\phi(r, z)F(\rho). \tag{13.82}$$

For a cylindrical core with an extrapolated radius $R'$ and extrapolated height $H'$, and uniform enrichment throughout the core, assume the neutron flux distribution $\phi(r, z)$ is separable in $r$ and $z$ to obtain

$$\phi(r, z) = \phi(0, 0)\psi(r)X(z), \tag{13.83}$$

where

$$\psi(r) = J_0\left(\frac{2.405r}{R'}\right)$$

and

$$X(z) = \cos\frac{\pi z}{H'}.$$

introduced in Eq. (5.81). The surface heat flux $q(r, z)$ at height $z$ into a coolant channel at radius $r$ may now be obtained by integrating the volumetric heat source

$S(r,z,\rho)$ over the cross-sectional area $A = \pi a^2$ of a fuel rod of radius $a$ and dividing by the wetted perimeter $M = 2\pi a$

$$q(r,z) = \frac{2\pi \int_0^a S(r,z,\rho)\rho d\rho}{M} = E_f \Sigma_f(r,z)\phi(r,z) \frac{2\pi \int_0^a F(\rho)\rho d\rho}{M}.$$

Remembering that the flux distribution $F(\rho)$ within the rod is normalized so that the cross section average $\langle F \rangle_\rho = 1$, we get

$$q(r,z) = E_f \Sigma_f(r,z)\phi(r,z)\frac{A}{M} = P(r,z)\frac{A}{M}, \tag{13.84}$$

which indicates that the *wall heat flux into the channel* is simply proportional to the global heat generation rate or *power distribution* $P(r,z)$.

In actual operating reactor cores, due to non-uniform loading of fuel enrichments and various material heterogeneities, and due to spatial distributions of fuel temperature and moderator density, the fission cross section $\Sigma_f$ will not be spatially uniform even in a fresh core. In addition, fuel depletion will complicate the spatial dependence of $\Sigma_f$. Hence, in general, coupled nuclear-T/H calculations have to be performed iteratively in any global diffusion-depletion analysis to obtain the power and heat flux distributions. Consistent with other simplifying assumptions introduced so far, assume, however, that the heat flux distribution is simply given by Eqs. (13.83) and (13.84).

For a reactor core with uniform fuel loading, resulting in a spatially uniform $\Sigma_f$, rewrite Eq. (13.84) for a channel at radial position $r$

$$q(r,z) = q(0,0)X(z)\psi(r) = q(z)\psi(r), \tag{13.85}$$

where the axial heat flux distribution is defined in terms of the axial neutron flux distribution

$$q(z) = q(0,0)X(z). \tag{13.86}$$

Now turn to the task of obtaining fuel and coolant temperature distributions in a coolant channel at radius $r$, subject to an axial heat flux distribution given by Eq. (13.86). Keep in mind that temperature distributions at other radial positions need to account for the radial dependence $\psi(r)$.

### 13.5.2 Axial Temperature Profile in PWR Core

Since the coolant in a PWR core remains a subcooled liquid for normal operating conditions, represent the energy conservation in PWR channels in terms of the bulk fluid temperature $T_b$. The energy conservation equation (13.80), with volumetric heat source $S = 0$, yields

$$WC_p \frac{dT_b(z)}{dz} = Mq(z),$$

where the wall heat flux $q(z)$ is given by Eq. (13.86). For a coolant channel of length $H$, with the inlet and outlet coolant temperatures specified as $T_1$ and $T_2$, respectively, integrate the equation

$$T_b(z) - T_1 = \frac{Mq(0,0)}{WC_p} \int_{-H/2}^{z} X(z')dz',$$

or

$$T_b(z) - T_1 = \frac{MH'q(0,0)}{\pi WC_p} \left( \sin \frac{\pi z}{H'} + \sin \frac{\pi z_0}{H'} \right), \tag{13.87}$$

with the elevation at exit $z_0 = H/2$. Since the bulk fluid temperature $T_b(z_0) = T_2$, at the channel exit, we get

$$T_2 - T_1 = \frac{2MH'q(0,0)}{\pi WC_p} \sin \frac{\pi z_0}{H'}, \tag{13.88}$$

and obtain a dimensionless fluid temperature profile [Pig65] $\theta_b(z)$

$$\theta_b(z) = \frac{T_b(z) - T_1}{T_2 - T_1} = \frac{1}{2} \left( 1 + \frac{\sin \pi z/H'}{\sin \pi z_0/H'} \right). \tag{13.89}$$

If the neutron extrapolation distance or reflector savings $\delta$ is negligible so that $H' = H$, then a simplified equation results:

$$\theta_b(z) = \frac{1}{2} \left( 1 + \sin \frac{\pi z}{H} \right). \tag{13.90}$$

Given the bulk fluid temperature $T_b(z)$, determine the fuel surface temperature $T_s(z)$ by writing the axial heat flux $q(z)$ in terms of Newton's law of cooling, Eq. (13.4)

$$q(z) = h(z)[T_s(z) - T_b(z)], \tag{13.91}$$

where $h(z)$ is the heat transfer coefficient for the convective heat transfer between the fuel rod and coolant. Solve for the fuel surface temperature distribution

$$T_s(z) - T_b(z) = \frac{q(0,0)}{h(z)} \cos \frac{\pi z}{H}, \ \delta = 0, \tag{13.92}$$

and obtain a dimensionless surface temperature profile:

$$\theta_s(z) = \frac{T_s(z) - T_1}{T_2 - T_1} = \frac{T_b(z) - T_1}{T_2 - T_1} + \frac{T_s(z) - T_b(z)}{T_2 - T_1},$$
$$= \frac{1}{2} \left( 1 + \sin \frac{\pi z}{H} \right) + \frac{\pi WC_p}{2MHh(z)} \cos \frac{\pi z}{H}, \ \delta = 0. \tag{13.93}$$

Equation (13.93) indicates that the surface temperature profile consists of a term proportional to the integral of the heat flux $q(z)$ along the channel and a term proportional to the heat flux itself. Similarly, the fuel centerline temperature $T_c$ can be obtained in terms of the overall heat transfer coefficient $U_c$ defined in Eq. (13.51). With

$$q(z) = U_c(z)[T_c(z) - T_b(z)], \tag{13.94}$$

a dimensionless profile for the centerline temperature $T_c(z)$ is identical to that for the surface temperature $T_s(z)$, provided we replace $h(z)$ by $U_c(z)$. Hence, Eq. (13.93) can be generalized to any characteristic fuel temperature $T_f(z)$, with the corresponding overall heat transfer coefficient $U_f(z)$ defined as

$$U_f(z) = \frac{q(z)}{T_f(z) - T_b(z)}, \tag{13.95}$$

where the convective heat flux $q(z)$ is still defined at the fuel element surface. Equation (13.93) can be rewritten for a dimensionless fuel temperature profile

$$\theta_f(z) = \frac{T_f(z) - T_1}{T_2 - T_1} = \frac{1}{2}\left(1 + \frac{\sin \pi z}{H}\right) + A_f(z)\frac{\cos \pi z}{H}, \tag{13.96}$$

in terms of a lumped parameter

$$A_f(z) = \frac{\pi W C_p}{2MHU_f(z)}. \tag{13.97}$$

For uniform geometry and constant fluid properties, the location $z_m$ of the maximum fuel temperature can be obtained by taking a derivative of Eq. (13.96) with respect to axial position $z$ and setting the derivative equal to zero

$$\frac{d\theta_f(z)}{dz} = \frac{1}{2}\cos\frac{\pi z}{H} - A_f(z)\sin\frac{\pi z}{H} = 0$$

or

$$\tan\frac{\pi z_m}{H} = \frac{1}{2A_f} = \frac{MHU_f}{\pi W C_p}. \tag{13.98}$$

Substituting Eq. (13.98) for the location $z_m$ into Eq. (13.96) yields an expression for the value of the dimensionless temperature profile at the maximum fuel temperature:

$$\theta_{fm} = \frac{1}{2}\left(1 + \frac{1}{\sin \pi z_m/H}\right) = \frac{1}{2}\left(1 + \sqrt{1 + 4A_f^2}\right), \delta = 0. \tag{13.99}$$

The total power $P(r)$ produced in a fuel rod located at radius $r$, corresponding to the inlet and outlet coolant temperatures, $T_1$ and $T_2$, respectively, can now be obtained in terms of the *maximum fuel temperature* $T_{fm}$

$$P(r) = WC_p[T_2(r) - T_1]$$

or

$$P(r) = \frac{2WC_p[T_{fm}(r) - T_1]}{1 + \sqrt{1 + 4A_f^2}} = \frac{WC_p[T_{fm}(r) - T_1]}{\theta_{fm}}. \tag{13.100}$$

Thus, any limitation on fuel temperature $T_{fm}$ can be translated into a corresponding limitation on the maximum power production in a fuel rod, and this in turn can determine the maximum total power that can be produced in a reactor core. The general problem of determining the power capability of a given reactor configuration subject to fuel temperature limitations will be discussed in Section 13.7.

Figure 13.14 illustrates the axial temperature profiles represented by Eqs. (13.89) and (13.96); a comparison between $\theta_b(z)$ and $\theta_f(z)$ is given in the LHS plot, and a parametric variation of the location $z_m$ of the maximum fuel temperature is shown in the RHS plot. Since $\theta_b(z)$ is proportional to an integral of the heat flux $q(z)$ along the channel, it increases monotonically along the channel length. In contrast, the characteristic fuel temperature profile $\theta_f(z)$ is a sum of $\theta_b(z)$ and a term proportional to $q(z)$ itself, as noted earlier in connection with Eq. (13.93). Hence, the peak fuel temperature or the *hot spot* in general occurs away from both the outlet and the core midplane, but, as the parameter $A_f$ increases, the hot spot moves toward the core midplane. This reflects the intuitive result that, as the overall heat transfer coefficient $U_f$ decreases, the temperature difference in Eq. (13.95) becomes large compared with the bulk temperature rise across the channel and the fuel temperature profile approaches that of the heat flux.

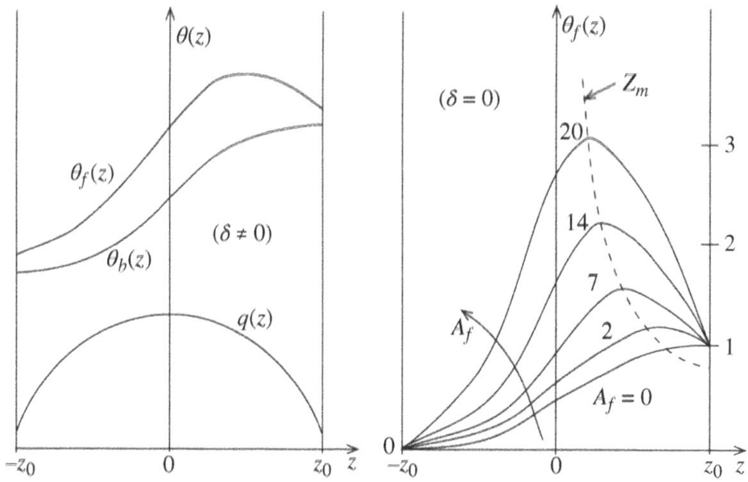


**Figure 13.14** Axial temperature profiles in a PWR core.

### 13.5.3 Axial Temperature Profile in BWR Core

For BWR cores, due to the presence of a significant amount of boiling in coolant channels, the axial temperature profiles obtained for PWR cores in Section 13.5.2 have to be modified. A simple adaptation of the PWR results can be made, however, by considering the energy conservation equation (13.79) expressed in terms of coolant enthalpy $h(z)$, again with volumetric source $S = 0$. Integrating the equation, with the inlet and outlet enthalpies, $h_1$ and $h_2$, respectively, obtain, in analogy to Eqs. (13.87) and (13.88)

$$h(z) - h_1 = \frac{MHq(0,0)}{\pi W}\left(1 + \sin\frac{\pi z}{H}\right), \delta = 0,$$

$$h_2 - h_1 = \frac{2MHq(0,0)}{\pi W},$$

and a dimensionless enthalpy distribution

$$\theta(z) = \frac{h(z) - h_1}{h_2 - h_1} = \frac{1}{2}\left(1 + \sin\frac{\pi z}{H}\right), \tag{13.101}$$

which is functionally equal to Eq. (13.90) for the bulk fluid temperature distribution in a PWR core. The bulk fluid temperature $T_b$ in a BWR core can be determined by Eq. (13.87) until bulk boiling starts at a distance $z_b$ from the inlet. For the two-phase region, the fluid temperature remains constant at its saturation temperature $T_{sat}$, providing a dimensionless fluid temperature distribution

$$\theta_b(z) = \frac{T_b(z) - T_1}{T_{sat} - T_1} = \begin{cases} \dfrac{1 + \sin \pi z/H}{1 + \sin \pi z_b/H}, & z \le z_b, \\ 1, & z > z_b. \end{cases} \tag{13.102}$$

Similarly, following Eq. (13.96), obtain a dimensionless fuel temperature distribution

$$\theta_f(z) = \frac{T_f(z) - T_1}{T_{sat} - T_1} = \frac{T_b(z) - T_1}{T_{sat} - T_1} + \frac{T_f(z) - T_b(z)}{T_{sat} - T_1}$$

or

$$\theta_f(z) = \begin{cases} \dfrac{1 + \sin \pi z/H}{1 + \sin \pi z_b/H} + \dfrac{2A_f(z)\cos \pi z/H}{1 + \sin \pi z_b/H}, & z \le z_b, \\ 1 + \dfrac{2A_f(z)\cos \pi z/H}{1 + \sin \pi z_b/H}, & z > z_b. \end{cases} \tag{13.103}$$

The dimensionless distributions for fluid enthalpy $\theta(z)$, bulk fluid temperature $\theta_b(z)$, and fuel temperature $\theta_f(z)$ of Eqs. (13.101)–(13.103), respectively are illustrated in Figure 13.15. Note that in BWR cores, the maximum fuel temperature will be located somewhat nearer to the core midplane than in PWR cores.

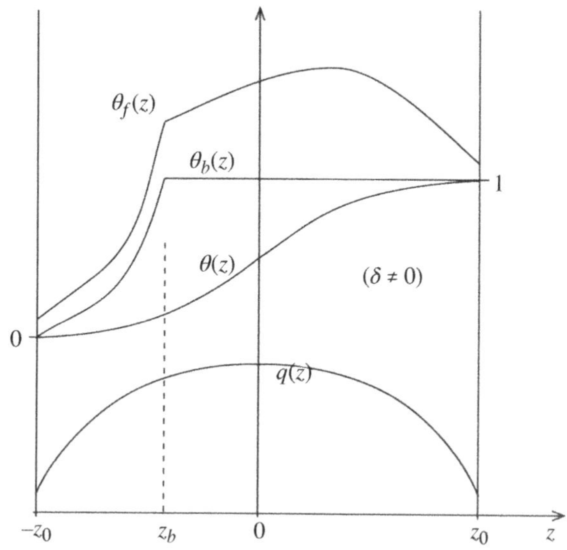


**Figure 13.15** Axial temperature profiles in a BWR core.

### 13.5.4 Hot Channel Factors

In terms of power and heat flux distributions discussed in Section 13.5.1 and axial temperature and enthalpy distributions obtained in Sections 13.5.2 and 13.5.3, we now derive hot channel factors or power peaking factors often used in T/H analysis of reactor cores. We will offer general definitions for the hot channel factors and clarify how they characterize the limiting heat flux distributions, although the analytical results are by necessity limited to those corresponding to the simple cylindrical power distribution of Section 13.5.1.

To characterize the overall heat flux distribution in a reactor core, define the *overall hot channel factor* or *overall power peaking factor* as the ratio of the maximum heat flux to the core-average heat flux:

$$F_q = \frac{\text{maximum heat flux}}{\text{core average heat flux}} = \frac{q_{max}}{\langle q \rangle_{r,z}} = \frac{(\Sigma_f \phi)_{max}}{\left\langle \Sigma_f \phi \right\rangle_{r,z}}. \tag{13.104}$$

Similarly, to characterize the radial heat flux distribution, the *enthalpy rise hot channel factor* is defined as

$$F_{\Delta h} = \frac{\text{maximum enthalpy rise}}{\text{core average enthalpy rise}} = \frac{\Delta h_{max}}{\langle \Delta h \rangle_r}. \tag{13.105}$$

From the energy conservation equation (13.79) employed in Section 13.4.3, obtain the enthalpy rise for a coolant channel

$$\Delta h(r) = \frac{M}{W}\int_{-H/2}^{H/2} q(r,z)dz = \frac{MH\langle q(r,z)\rangle_z}{W} = \frac{P(r)}{W} = \frac{P(0)\psi(r)}{W},$$

where $P(r)$ is the total power generated in a fuel rod at $r$. Hence, Eq. (13.105) is rewritten as

$$F_{\Delta h} = \frac{\langle q \rangle_{z,max}}{\langle q \rangle_{r,z}} = \frac{\left\langle \Sigma_f \phi \right\rangle_{z,max}}{\left\langle \Sigma_f \phi \right\rangle_{r,z}} = \frac{P(0)}{\langle P \rangle_r} = \frac{1}{\langle \psi \rangle}.$$

The enthalpy rise hot channel factor $F_{\Delta h}$ is equal to the ratio of the peak power to the radial-average power on a rod-to-rod basis and is often used synonymously as the *radial power peaking factor* $F_{xy}$. The *axial power peaking factor* $F_z$ is finally defined as

$$F_z = \frac{\text{maximum heat flux in hot channel}}{\text{average heat flux in hot channel}} = \frac{q_{max}}{\langle q \rangle_{z,max}}, \tag{13.106}$$

where the hot channel refers to the channel with the largest value of heat flux $\langle q \rangle_z$ averaged over the length of the channel.

From Eqs. (13.104) through (13.106), a relationship is obtained covering the three hot channel factors:

$$F_q = F_{\Delta h} \cdot F_z. \tag{13.107}$$

The relationship basically represents the separability of the heat flux distribution in the $r$ and $z$ directions. In actual operating reactor cores, heat flux distributions are not in general separable, and full-blown 3-D power and heat flux calculations are necessary to determine the hot channel factor $F_q$. In terms of $F_q$, the total power $P_t$ produced in a cylindrical reactor core, containing $N$ fuel rods with height $H$ and wetted perimeter $M$, is obtained as

$$P_t = NMH\langle q \rangle_{r,z} = NMH\frac{q_{max}}{F_q}. \tag{13.108}$$

If the fission cross section $\Sigma_f$ is spatially uniform, then $q(r, z)$ may be given by Eq. (13.85) and $F_{\Delta h}$ is obtained as the peak-to-average ratio of radial neutron flux $\psi(r)$

$$\frac{1}{F_{\Delta h}} = \frac{\langle P \rangle_r}{P(0)} = \langle \psi \rangle = \frac{2}{R^2}\int_0^R J_0\left(\frac{2.405r}{R}\right) rdr = \frac{2}{2.405}J_1(2.405), \tag{13.109}$$

or $\langle \psi \rangle = 0.4313$ with $\delta = 0$. Similarly, determine $F_z$ as the peak-to-average ratio of axial neutron flux $X(z)$:

$$\frac{1}{F_z} = \langle X \rangle = \frac{1}{H}\int_{-H/2}^{H/2} X(z)dz = \frac{2}{\pi}, \quad \delta = 0. \tag{13.110}$$

Finally, the overall hot channel factor $F_q$ is obtained as the peak-to-average ratio of the $r$–$z$ neutron flux $X(z)\psi(r)$:

$$F_q = \frac{1}{\langle X \rangle \langle \psi \rangle} = \frac{\pi}{2} \cdot 2.32 = 3.64. \tag{13.111}$$

In actual operating reactor cores, the heat flux $q(r,z)$ is not usually separable in $r$ and $z$, partly because the fuel enrichment is not spatially uniform throughout the core, as discussed in Section 13.5.1. The analytical expression obtained for the hot channel factors is not strictly valid for reactor cores with zoned fuel loading. In fact, zoned fuel loading, coupled with distributed burnable absorbers, greatly improves the overall power distribution in LWR cores so that current LWR designs typically have $F_q < 2.5$, compared with our simple estimate of $F_q = 3.64$ for uniform fuel loading. Equation (13.108) indicates that the flatter the power distribution is, the more power we may generate out of a given fuel loading, and hence the incentive for achieving as flat a power distribution as possible in any reactor core design.

The hot channel factors discussed so far are known as *nuclear hot channel factors*, which characterize spatial distributions of nuclear heat generation rate. In addition, two other factors known as the *engineering hot channel factor* $F_q^E$ and the *uncertainty hot channel factor* $F_q^U$ are introduced to account for engineering tolerances in manufacturing fuel and core structures, and for uncertainties in power distribution measurements, respectively. With the nuclear hot channel factor of Eq. (13.107) now designated as $F_q^N$, we may write the overall hot channel factor as

$$F_q = F_q^N \cdot F_q^E \cdot F_q^U. \tag{13.112}$$

The uncertainty factor $F_q^U$ accounts for statistical fluctuations inherent in instrumentation signals and approximations in synthesizing 3-D power distributions from a limited number of detector signals. A typical value for $F_q^U$ is 1.05. Similarly, a value of 1.05 is typically assumed also for $F_q^E$ to account for engineering factors, including variations in coolant flow rate, pellet dimensions, enrichment and density from rated values, and the effects of cross flow between channels not explicitly represented in the analysis.

Since the methods of calculating $F_q^U$ and $F_q^E$ are rather similar, we illustrate the statistical process [Ton67, Ton96] in determining $F_q^E$, assuming that the engineering hot channel factor accounts for variations in fuel enrichment and pellet density only. Assume further that the effects of fractional enrichment and density variations on the power peaking factor are equal and may be linearly superimposed:

$$F_q = F_q^N \left(1 + \frac{\Delta e}{\langle e \rangle} + \frac{\Delta \rho}{\langle \rho \rangle}\right). \tag{13.113}$$

Recall that, according to the law of error propagation, given the standard deviations $\sigma_x$ and $\sigma_y$ for variables $x$ and $y$, respectively, the standard deviation $\sigma_z$ for a variable $z = f(x,y)$, is obtained as

$$\sigma_z^2 = \sigma_x^2 \left(\frac{\partial z}{\partial x}\right)^2 + \sigma_y^2 \left(\frac{\partial z}{\partial y}\right)^2. \tag{13.114}$$

Applying Eq. (13.114) to Eq. (13.113) yields the variance for $F_q$

$$\sigma_{F_q}^2 = \left(F_q^N\right)^2 \left[\frac{\sigma_e^2}{\langle e\rangle^2} + \frac{\sigma_\rho^2}{\langle \rho\rangle^2}\right] \equiv \left(F_q^N \sigma_{F_q^E}\right)^2.$$

Given the standard deviation $\sigma_{F_q}$, we desire to be assured that the actual hot channel factor $F_q$ will be less than that calculated. Hence, choose a one-sided confidence level $P_k$ to determine the confidence parameter $k$ such that

$$\text{Probability}\,\{F_q < F_q^N + k\sigma_{F_q}\} = P_k$$

and obtain the hot channel factor

$$F_q = F_q^N + k\sigma_{F_q} = F_q^N(1 + k\sigma_{F_q^E}) \equiv F_q^N \cdot F_q^E. \tag{13.115}$$

Typically $k = 3$ is chosen corresponding to an upper confidence level of 99.9%, i.e. $P_k = 0.999$, to determine $F_q^E$. The statistical treatment of the uncertainties in the hot channel factors has been replaced by more direct computational approaches in recent years, as discussed further in connection with the CHF issues in Section 13.7.3.

## 13.6 Boiling Heat Transfer and Two-Phase Flow

Heat transfer accompanied by a change in phase from liquid to vapor can take place either in an initially quiescent liquid or in a fluid flow. Two-phase flow and the associated boiling heat transfer play an important role in determining the limiting T/H condition in PWR cores as well as in the coupled nuclear-T/H analysis of BWR cores. We will first study some of the characteristics of boiling heat transfer in a pool of stagnant fluid, i.e. the pool boiling phenomena, in Section 13.6.1, and then extend the result to boiling heat transfer in a forced convective flow in Section 13.6.2. Representative two-phase flow models are then presented in Sections 13.6.3 through 13.6.5.

### 13.6.1 Pool Boiling Regimes

If heat is added to a pool of quiescent liquid through a heated surface, the resulting heat transfer process can be divided into several regimes starting from the natural convection process [Ton97, Ghi17]. The distinct regimes may be illustrated schematically, as in Figure 13.16, by plotting the wall heat flux $q_w$ as a function of *wall superheat* $\Delta T_w = T_w - T_{sat}$, i.e. the difference between the surface temperature $T_w$ of the heated wall and the saturation temperature $T_{sat}$ of the

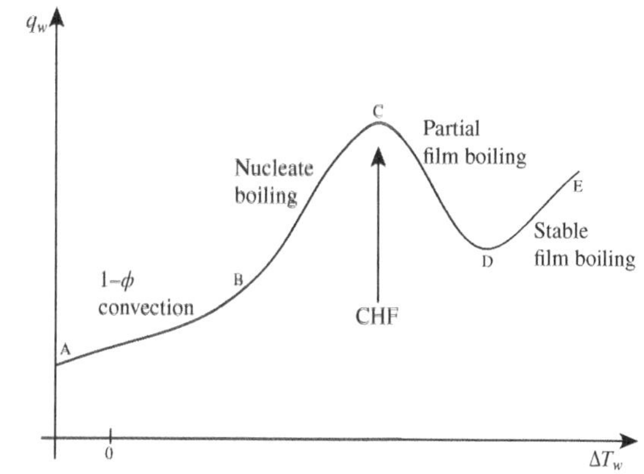


**Figure 13.16** Pool boiling regimes with heat flux represented as a function of wall superheat $\Delta T_w$.

liquid. In regime *AB*, heat is transmitted through natural convection in the single phase $(1-\phi)$ liquid. As the heat flux is increased beyond point *B*, small bubbles are formed near nucleation sites, e.g. impurities or rough spots on the wall. Thus regime *BC* is known as the *nucleate boiling* regime and is characterized by a high heat transfer rate. This is because bubbles carry the latent heat of vaporization with them, and hence the bubble motion tends to agitate the liquid and increase the convective transfer of heat. Nucleate boiling in a subcooled liquid is known as *local boiling* and is characterized by bubbles that tend to collapse locally away from the heated wall. Nucleate boiling in a saturated liquid is known as *bulk boiling*.

Regime *CD* is known as the *partial film boiling* or *transition boiling* regime, where, due to an increased vapor fraction, the heated wall is alternately covered with vapor and liquid, resulting in an unstable, oscillating surface temperature distribution. Since the vapor layers tend to block flow of liquid toward the heated surface, the heat transfer rate also decreases substantially beyond point *C*, resulting in a sharp increase in the wall surface temperature and possible burnout of the wall. Hence point *C* is known as the point of *departure from nucleate boiling* (DNB) or the *boiling crisis*, and the heat flux at *C* is called $q_{DNB}$, the *burnout heat flux* or *CHF*. If the heated surface does not burn out beyond point *D*, a stable film of vapor is formed around the heated surface, and heat flux can increase further. Region *DE* represents a single-phase vapor and is referred to as the *stable film boiling* regime. Beyond point *E*, as the surface temperature is increased further, the heat transfer rate can increase somewhat also due to radiative heat transfer.

In most cases of interest, however, if DNB takes place, the wall temperature may increase beyond the point of material integrity of the heated wall, and it is desirable to limit the heat flux to a value sufficiently below the DNB flux, $q_{DNB}$.

This limitation on surface heat flux of the fuel cladding material forms one of the major design limitations for nuclear fuel elements in current LWR plants.

### 13.6.2 Flow Boiling Regimes and Two-Phase Flow Patterns

Heat transfer in a forced convective fluid flow initially takes place through forced convection in the liquid phase. As the heat flux is increased, bubbles are formed near nucleation sites and thus subcooled nucleate boiling or local boiling initiates. Flow patterns beyond the initiation of nucleate boiling can generally be broken up into four sequential phases, as illustrated in Figure 13.17:

(1) Bubbly flow, where bubbles are formed near heated walls resulting in a discontinuous vapor phase in the midst of a continuous liquid phase,
(2) Slug flow, which is characterized by large separated slugs of liquid and vapor,
(3) Annular flow, where vapor flows in the core of the channel and the walls are covered with liquid films,
(3) Mist or drop flow, where liquid is completely vaporized and a single-phase vapor flow is established in the channel.

Compared to the pool boiling regimes, bulk boiling, i.e. saturated nucleate boiling, could usually initiate somewhere in the bubbly flow regime, and DNB could take place somewhere between the annular flow and the mist flow. Depending on flow conditions, however, not all of these four flow patterns are followed in a flow, and likewise other flow patterns could also develop. For example, under some flow conditions, an annular flow with a liquid core surrounded by a vapor film could exist before DNB takes place. Furthermore, the transitions among the various flow patterns are not usually very distinct. Turbulent eddies also appear frequently in a two-phase flow. In addition to the flow regimes, Figure 13.17 indicates the four heat transfer regimes including the subcooled nucleate boiling regime, where bubbles are nucleated before the fluid temperature reaches the saturation temperature. Likewise, the vapor temperature increases above the saturation temperature in the drop flow regime, despite the presence of liquid droplets in the channel. These are two non-equilibrium phenomena in a boiling channel that require the use of non-equilibrium two-phase flow models discussed later in the chapter.

In a boiling channel, the onset of *local boiling* or *subcooled nucleate boiling* is characterized by a constant wall temperature $T_w$, while bulk boiling or saturated nucleate boiling is characterized by a constant bulk fluid temperature $T_b = T_{sat}$. For a constant wall heat flux $q_w$, Figure 13.18 illustrates the location $z_{LB}$ corresponding to the onset of local boiling and location $z_{BB}$ corresponding to that of bulk boiling. As long as a two-phase flow is maintained, *wall superheat* $\Delta T_w$ remains essentially constant in a boiling channel.

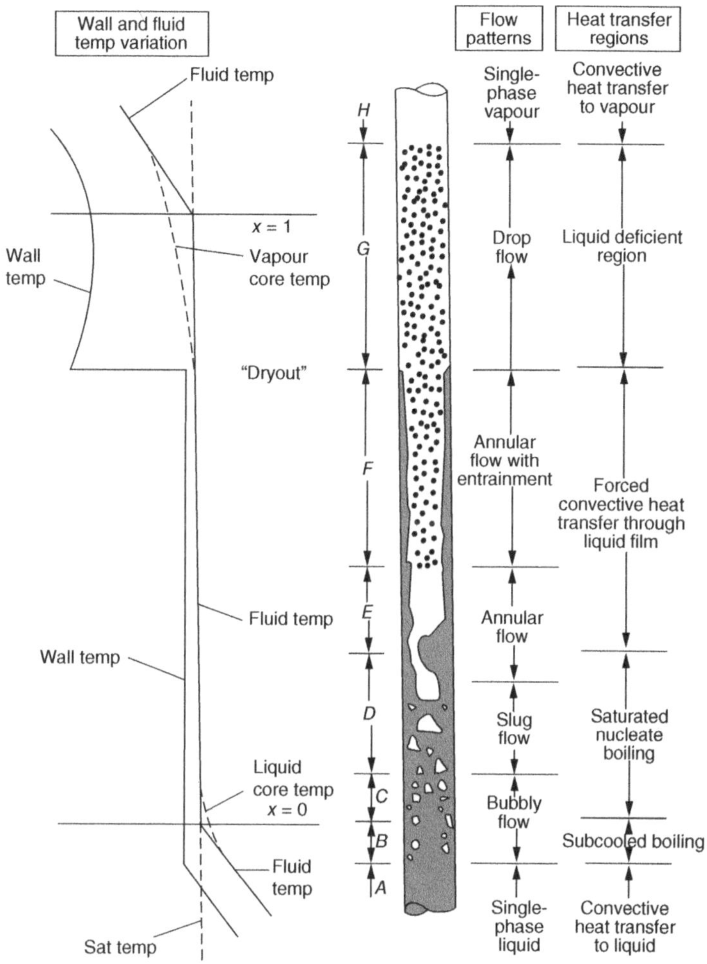


**Figure 13.17** Flow regimes in forced convective flow. *Source:* Adapted from Collier [Col72].

In a typical BWR core, the weight fraction or quality of steam at the coolant channel exit could be on the order of 0.1–0.15. Corresponding to the exit *steam quality*, the volume fraction of vapor, commonly known as *void fraction*, could be on the order of 0.3–0.5, as the density of steam is only about 1/20 of the liquid density. Thus, in order to find the coolant density in a BWR core, an accurate determination of void fraction has to be made. Due considerations have to be given to two-phase flow also in steady-state and transient analyses of limiting conditions in PWR cores, because bulk boiling as well as local boiling is possible in these

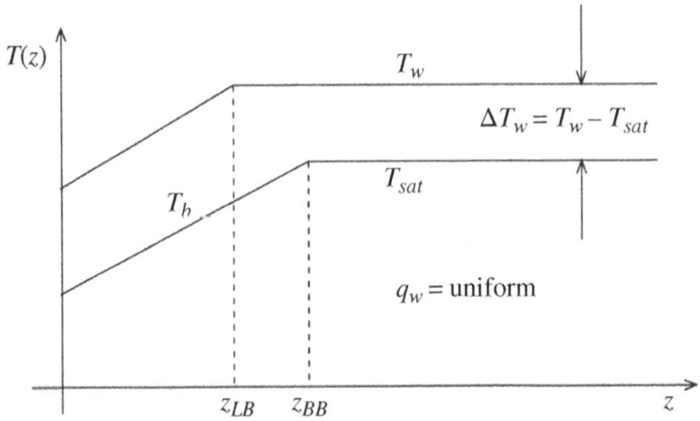


**Figure 13.18** Temperature distribution illustrating the onset of local boiling and bulk boiling.

conditions. The actual flow patterns in two-phase flow are, however, complex and not very distinct, and analysis of convective heat transfer in two-phase flows has to rely significantly on various empirical correlations. We discuss a number of models that can be used to describe two-phase flow and boiling heat transfer in the remainder of this section.

### 13.6.3 Homogeneous Equilibrium Flow Model

The simplest of all two-phase (2-$\phi$) flow models is the *homogeneous equilibrium model* (HEM), where the liquid and vapor phases are assumed homogeneously mixed together at thermal equilibrium and flow with one velocity. The model is also known as the 1$V$1$T$ or *UVUT* model because the model uses a single, uniform velocity for a mixture at a single, uniform temperature. The HEM is obviously a crude representation of the separation that exists between the phases and the non-equilibrium thermodynamic conditions of two-phase flow. With judicious empirical correlations, the HEM may, however, provide approximate but useful analysis of complex two-phase flow problems.

In terms of the densities $\rho_f$ and $\rho_g$ and the mass flow rates $W_f$ and $W_g$ for liquid and vapor, respectively, we may determine the homogeneous density $\rho_h$

$$\frac{1}{\rho_h} = \left(\frac{W_g}{\rho_g} + \frac{W_f}{\rho_f}\right)\frac{1}{W} = \frac{x}{\rho_g} + \frac{1-x}{\rho_f}, \tag{13.116}$$

where the *flow quality* $x$ of vapor is defined as the vapor fraction of the total mass flow rate $W$

$$x = \frac{W_g}{W} = \frac{W_g}{W_f + W_g}. \tag{13.117}$$

In terms of flow quality $x$ and *saturated liquid and vapor enthalpies* $h_f$ and $h_g$, respectively, define the mixture enthalpy $h$

$$h = xh_g + (1 - x)h_f. \tag{13.118}$$

This is equivalent to calculating flow quality $x$ as

$$x = (h - h_f)/h_{fg}, \tag{13.119}$$

where $h_{fg} = h_g - h_f$ is the *latent heat of vaporization.*

In the HEM, conservation equations derived in Section 13.2 may be immediately used, provided one replaces density $\rho$ and specific enthalpy $h$ in the conservation equations with the homogeneous density $\rho_h$ of Eq. (13.116) and the mixture enthalpy of Eq. (13.118). Actual solution of the HEM equations will, of course, require regular supply of empirical correlations for the heat transfer coefficient, friction factor, and other relevant parameters.

### 13.6.4 Slip Flow Model

The traditional two-phase flow model used in reactor T/H analysis is the *slip flow model* and represents a significant improvement over the HEM. The slip flow model is a separated flow model, where each of the liquid and vapor phases is assigned a separate velocity with a fraction of flow area, and distinct properties. Interface effects between the two phases are, however, neglected, i.e. no friction or surface tension between the phases is considered. In addition, the vapor and liquid phases are assumed in thermal equilibrium at a uniform temperature. Hence, the slip flow model is designated as a $2V1T$ model. It is also known as a three-equation model, because two-phase flow is represented by three mixture conservation equations. Although it is still a highly idealized model, the slip flow model has proven to be quite useful in many applications including T/H reactor system analysis. One of the major limitations of the model, however, is its inherent inability to account for nonequilibrium phenomena including subcooled nucleate boiling.

With the subscripts $f$ and $g$ denoting the saturated liquid and vapor phases, respectively, we consider two separate regions in the flow cross section, with phase densities $\rho_f$ and $\rho_g$, phase velocities $v_f$ and $v_g$, and associated areas $A_f$ and $A_g$, as schematically shown in Figure 13.19. We then obtain mass flow rates for the separate phases:

$$W_f = \rho_f v_f A_f, \tag{13.120a}$$

$$W_g = \rho_g v_g A_g. \tag{13.120b}$$

For a channel with total cross-sectional area $A$, we may define *void fraction* $\alpha$ as the local volume fraction of vapor or equivalently as the fractional cross-sectional

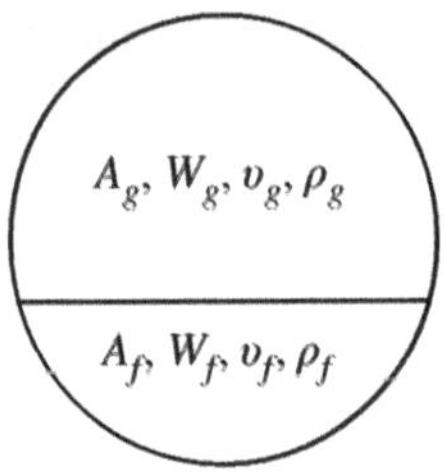


**Figure 13.19** Two separate phases in a flow channel.

area occupied by the vapor phase:

$$\alpha = \frac{A_g}{A} = \frac{A_g}{A_f + A_g}. \tag{13.121}$$

In terms of void fraction, we may define a *local (average) density* of the two-phase mixture, sometimes known as the *slip density*:

$$\rho = \alpha\rho_g + (1-\alpha)\rho_f = \rho_s. \tag{13.122}$$

Since the two-phase mixture density of Eq. (13.122) is given as a volume-weighted average density, it is a mixture density that can be measured in the laboratory and should be used as the density in all conservation equations for two-phase mixtures. In order to determine the local mixture density, however, a relationship between void fraction $\alpha$ and flow quality $x$ has to be empirically obtained. For this purpose, with Eqs. (13.117), (13.120), and (13.121), obtain

$$W_g = \rho_g v_g A_g = \alpha\rho_g v_g A = xW,$$

or

$$v_g = \frac{xGA}{\rho_g A_g} = \frac{xG}{\alpha\rho_g} \tag{13.123}$$

and similarly

$$v_f = \frac{(1-x)G}{(1-\alpha)\rho_f}. \tag{13.124}$$

With Eqs. (13.123) and (13.124), an expression for the *slip ratio* follows

$$S = \frac{v_g}{v_f} = \frac{x}{1-x}\frac{1-\alpha}{\alpha}\frac{\rho_f}{\rho_g}. \tag{13.125}$$

Solving Eq. (13.125) for void fraction yields

$$\alpha = \left[1 + \frac{1-x}{x}\frac{\rho_g}{\rho_f}S\right]^{-1},$$

or alternatively [Tho64]

$$\alpha = \frac{\gamma x}{1 + x(\gamma - 1)}, \tag{13.126}$$

in terms of Thom's *slip factor* $\gamma$ defined as

$$\gamma = \frac{v_f}{v_g}\frac{\rho_f}{\rho_g} = \frac{1}{S}\frac{\rho_f}{\rho_g}. \tag{13.127}$$

In Thom's slip flow model, parameter $\gamma$ is empirically correlated as a simple function of system pressure $p$ so that $\gamma = 1$ at the critical pressure $p = p_c$ and monotonically increases as $p$ decreases. For water, $p_c$ = 3,260 psia, and the void fraction-flow quality relationship of Eq. (13.126) is illustrated in Figure 13.20, where the sensitivity of $\alpha$ to $x$ for small values of $x$ and at low pressure is noteworthy. Through a simple tabulation of the two-phase parameter $\gamma(p)$, Eq. (13.126) provides a simple method of evaluating void fraction $\alpha$, given flow quality $x$ or equivalently through mixture enthalpy Eq. (13.118). One noteworthy feature of Thom's model is that Eq. (13.126) yields correct values of void fraction for two limiting cases: $\alpha = 0$ when $x = 0$ and $\alpha = 1$ when $x = 1$. In some other empirical correlations [Ton97], void fraction $\alpha$ is fitted directly as a function of flow quality $x$, which may not satisfy the necessary relationships for the two limiting cases, $x = 0$ and 1.

With the slip flow model, we may use the continuity equation obtained in Sections 13.2 and 13.4 directly to describe a two-phase flow, provided we use the local density defined in Eq. (13.122) as the mixture density. For a channel flow, for example, use Eq. (13.70) together with $G = \alpha\rho_g v_g + (1-\alpha)\rho_f v_f$. The equation of energy conservation has to be, however, modified to represent the phase separation explicitly, while the equation of motion for channel flow can take the same functional form of Eq. (13.72) provided the flow gradient term $R$ is modified to reflect two-phase phenomena. For the momentum conservation equation (13.72), the momentum flux term is rewritten

$$\left(\frac{dp}{dz}\right)_{mom} = \frac{d}{dz}\left(\frac{G^2}{\rho}\right) = \frac{d}{dz}\left[\alpha\rho_g v_g^2 + (1-\alpha)\rho_f v_f^2\right]. \tag{13.128}$$

**Figure 13.20** Void fraction vs. flow quality.

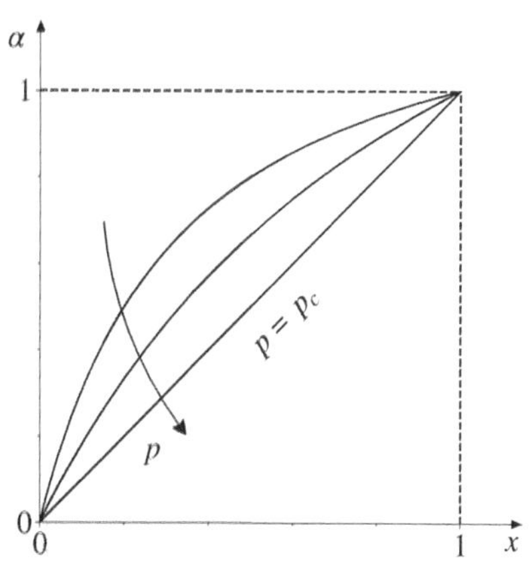

Substitution of Eqs. (13.123) and (13.124) into Eq. (13.128) yields a simple relationship

$$\left(\frac{dp}{dz}\right)_{mom} = \frac{d}{dz}\left(\frac{G^2}{\rho_m}\right), \tag{13.129}$$

provided we define the *momentum density* $\rho_m$

$$\frac{1}{\rho_m} = \frac{x^2}{\alpha\rho_g} + \frac{(1-x)^2}{(1-\alpha)\rho_f}. \tag{13.130}$$

The momentum density is introduced with the express purpose of casting the two-phase momentum flux term in the same functional form as the single-phase counterpart and is clearly not a density that can be measured readily in the laboratory. Note that the three two-phase densities introduced so far, $\rho_h$, $\rho_s$, and $\rho_m$, are equal to one another only in the HEM limit when the slip between phases is assumed negligible.

The frictional pressure gradient for two-phase flow is evaluated by multiplying the corresponding single-phase gradient by an empirical *two-phase friction multiplier* $\phi^2_{TP}$ usually defined as

$$\phi^2_{TP} = \frac{\left(\dfrac{dp}{dz}\right)_{FTP}}{\left(\dfrac{dp}{dz}\right)_{FLO}}, \tag{13.131}$$

where

$\left(\dfrac{dp}{dz}\right)_{FTP}$ = frictional pressure gradient in two-phase mixture,

$\left(\dfrac{dp}{dz}\right)_{FLO}$ = frictional pressure gradient due to liquid only.

In terms of the multiplier $\phi^2_{TP}$, determine the two-phase pressure gradient

$$\left(\frac{dp}{dz}\right)_{fric} = \left(\frac{dp}{dz}\right)_{FTP} = \left(\frac{dp}{dz}\right)_{FLO}\phi^2_{TP} = \frac{2G^2}{\rho}\frac{f_{TP}}{D_h} = \frac{2G^2}{\rho_f}\frac{f}{D_h}\phi^2_{TP}. \tag{13.132}$$

Several different two-phase friction multipliers have been introduced in the literature, and $\phi^2_{TP}$ in Eq. (13.131) is sometimes also written as $\phi^2_{LO}$, e.g. in the Martinelli–Nelson correlation [Ton97], which is one of the main empirical correlations for the two-phase friction multiplier. In Thom's slip flow model [Tho64], an integral of $\phi^2_{TP}$ over a boiling channel is empirically correlated, together with the momentum density $\rho_m$ of Eq. (13.130), to facilitate frictional pressure drop calculations. The two-phase friction multiplier often increases the frictional pressure drop of an equivalent single-channel flow by an order of magnitude or more for typical BWR configurations. Similar to the need to introduce the two-phase friction multiplier, the irreversible loss coefficients $K_i$ in Eq. (13.76) have to be replaced by empirical two-phase loss coefficients $K_{TP,i}$. Finally, the gravitational pressure gradient can simply be written in terms of local density $\rho_s$; Thom tabulates the gravitational pressure drop also as a function of the exit quality of a boiling channel. Collecting terms and modifying the single-phase equation (13.76) yields an expression for the pressure drop $\Delta p$ in a boiling channel:

$$\Delta p = G^2\Delta\left(\frac{1}{\rho_m}\right) + \frac{2G^2}{D_h}\int_0^L \frac{f}{\rho_f}\phi_{TP}^2 dz + \int_0^L \rho g dz + \sum_i \frac{G_i^2}{2\rho_{m,i}}(\sigma_{i+1}^2 - \sigma_i^2) + \sum_i K_{TP,i}\frac{G_i^2}{2\rho_{f,i}}. \tag{13.133}$$

With similar modification of various pressure gradients making up the term $R$ in Eq. (13.72), the same functional form of the time-dependent momentum conservation equation may also be used. For the channel energy conservation equation (13.78), volumetric enthalpy $\rho h$ is modified to account for the phase separation

$$\rho h = \alpha\rho_g h_g + (1-\alpha)\rho_f h_f,$$

which may be rewritten, with the $\alpha$–$x$ relationship of Eq. (13.126)

$$\rho h = \rho\cdot h - h_{fg}\left(\frac{\rho_f}{\gamma} - \rho_g\right)\alpha(1-x). \tag{13.134}$$

The convective enthalpy flow term is now written as

$$Gh = \alpha\rho_g v_g h_g + (1-\alpha)\rho_f v_f h_f = G\left[xh_g + (1-x)h_f\right] = G\cdot h, \tag{13.135}$$

where Eqs. (13.123) and (13.124) are used. Comparing Eqs. (13.134) and (13.135) indicates that the convective enthalpy flow term $Gh$ may be treated simply as a product of $G$ and $h$, just as in a single-phase flow, while the volumetric enthalpy $\rho h$ has to include a correction term to the product $\rho \cdot h$. This complexity in the energy equation is inherent in any two-phase formulation, regardless of which void fraction-flow quality correlation is used. Substituting Eqs. (13.134) and (13.135) into Eq. (13.78) finalizes the two-phase energy conservation equation for channel flow

$$\frac{\partial}{\partial t}(\rho h) = -\frac{\partial}{\partial z}(G\cdot h) + \frac{Mq_w}{A} + \frac{\partial p}{\partial t} + S, \tag{13.136}$$

where the term $\rho h$ in the time derivative is to be evaluated through Eq. (13.134).

A number of empirical correlations are necessary to solve the two-phase fluid conservation equations (13.70), (13.72), and (13.136). One such key empirical relationship is the void fraction-flow quality correlation of Eq. (13.126) or equivalent. With these empirical correlations, the fluid conservation equations may be solved together both for steady-state and transient two-phase problems. To illustrate the complexity of two-phase flow analysis in general, consider coolant density along the length of a BWR channel. Corresponding to the fluid enthalpy and temperature distributions of Eqs. (13.101) and (13.102) illustrated in Figure 13.15, the void fraction and coolant density are plotted in Figure 13.21. Due to the highly nonlinear nature of the $\alpha$–$x$ correlation for small values of flow quality $x$, a sharp variation may emerge in the coolant density near the inception of bulk boiling at $z_b$. As indicated by the dashed curves, however, the actual variations in the density and void fraction are not so sharp due to the presence of subcooled nucleate boiling. The density variation shown in Figure 13.21 has significant effects on the energy and momentum balance for the channel, and the pressure drop equation (13.133) has to duly reflect the density and void fraction variations along the channel together with the appropriate two-phase correction factors. In actual design calculations, especially in a BWR core design, neutron flux and hence the heat generation rate are closely related to the local moderator density. Hence, the pressure drop and density calculations have to be performed iteratively, coupled with the neutronic calculations, as discussed in the introduction to the chapter. Besides, one also has to consider flow mixing among different coolant channels for accurate T/H analysis, such as in subchannel analysis, where individual fuel rods and the associated coolant channels are explicitly and separately represented.

**Example 13.4** As a summary of the discussion on the single- and two-phase flow representations, evaluate the pressure drop across the core for two comparable LWR designs, AP600 [Wes00] and SBWR [GEN92] plants. The 600-MWe plants received design certification from the US Nuclear Regulatory Commission

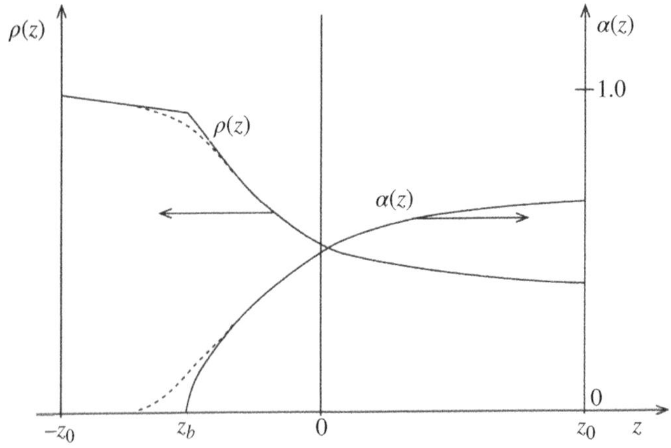


**Figure 13.21** Void fraction and coolant density variation in a BWR channel.

and served as the basis for the two representative Generation III+ designs, AP1000 [Hon12] and ESBWR [GEH14] plants. Use design parameters summarized in Table 13.3 together with empirical two-phase multipliers from Thom's slip flow model [Tho64] for the BWR configuration.

For the AP600 design, the single-phase model of Eq. (13.76) suffices, providing

- $Re = GD_h/\mu = 3.12 \times 10^5$, with viscosity $\mu = 9.02 \times 10^{-5}\,\text{kg} \cdot \text{m}^{-1}\,\text{s}^{-1}$,
- Fluid density $\rho = 0.726\,\text{Mg} \cdot \text{m}^{-3}$, $\Delta p_{acc} = G^2 \left( \frac{1}{\rho_{out}} - \frac{1}{\rho_{in}} \right) = 0.97\,\text{kPa}$, friction factor $f = 3.35 \times 10^{-3}$, $\Delta p_{fric} = f \frac{G^2}{D_h} \left( \frac{1}{\rho_{out}} + \frac{1}{\rho_{in}} \right) H = 16.4\,\text{kPa}$, $\Delta p_{elev} = 26.1\,\text{kPa}$, for a total pressure drop across the core $\Delta p_c = 43.4\,\text{kPa}$.

For the SBWR plant, Eq. (13.133) is used to represent the two-phase flow regime with Thom's empirical parameters [Tho64]

- $Re = GD_h/\mu = 1.79 \times 10^5, f = 3.85 \times 10^{-3}$, $\mu = 9.39 \times 10^{-5}\,\text{kg} \cdot \text{m}^{-1}\,\text{s}^{-1}$,
- Inlet enthalpy $h_{in} = 1,227\,\text{kJ} \cdot \text{kg}^{-1}$, $h_f = 1,273\,\text{kJ} \cdot \text{kg}^{-1}$, $h_{fg} = 1,496\,\text{kJ} \cdot \text{kg}^{-1}$, $h_{out} = 1,491\,\text{kJ} \cdot \text{kg}^{-1}$,
- Boiling boundary $\ell_f = H(h_f - h_{in})/\Delta h = 0.484\,\text{m}$, $\ell_g = 2.26\,\text{m}$,
- Single-phase: $\Delta p_{acc} = G^2 \left( \frac{1}{\rho_f} - \frac{1}{\rho_{in}} \right) = 0.0388\,\text{kPa}$, $\Delta p_{fric} = f \frac{G^2}{D_h} \left( \frac{1}{\rho_f} + \frac{1}{\rho_{in}} \right)$ $\ell_f = 0.402\,\text{kPa}$, $\Delta p_{elev} = 3.54\,\text{kPa}$, $\Delta p_{1-\phi} = 3.98\,\text{kPa}$,

**Table 13.3** Thermal-hydraulic parameters of AP600 and SBWR designs.

| **Design parameters** | **AP600** | **SBWR** |
|---|---|---|
| Core power $P$ (MWt) | 1,845 | 1,875 |
| Core flow rate $W$ ($\text{Mg} \cdot \text{s}^{-1}$) | 8.05 | 7.12 |
| Number of fuel rods | 38,280 | 43,920 |
| Fuel length $H$ (m) | 3.66 | 2.74 |
| Inlet coolant temperature $T_{in}$ (K) | 549.9 | 551.3 |
| Mass velocity $G$ ($\text{Mg} \cdot \text{m}^{-2}\,\text{s}^{-1}$) | 2.39 | 1.11 |
| Flow area per rod $A$ ($\text{mm}^2$) | 87.7 | 146 |
| Coolant enthalpy rise $\Delta h$ ($\text{kJ} \cdot \text{kg}^{-1}$) | 230 | 264 |
| Hydraulic diameter $D_h$ (mm) | 11.8 | 15.2 |
| Average flow speed $\langle v \rangle$ ($\text{m} \cdot \text{s}^{-1}$) | 3.26 | 0.914 |
| Pressure drop across core $\Delta p_c$ (kPa) | 121[a)] | 22.8 |

a) Includes losses at inlet and outlet plena.

- Two-phase: $\Delta p_{acc} = \dfrac{G^2}{\rho_f} r_2 = 3.48\,\text{kPa}$, $\Delta p_{fric} = 2\dfrac{G^2}{D_h}\dfrac{f}{\rho_f}\ell_g r_3 = 5.12\,\text{kPa}$, $\Delta p_{elev} = \rho_f g \ell_g r_4 = 9.31\,\text{kPa}$, $\Delta p_{2-\phi} = 17.9\,\text{kPa}$, for a total pressure drop $\Delta p_c = 21.9\,\text{kPa}$.

Note that the total pressure drop $\Delta p_c = 21.9\,\text{kPa}$ calculated is in good agreement with 22.8 kPa tabulated for the SBWR core, but the significant difference in $\Delta p_c$ for the AP600 core is due to the inability to represent the pressure losses due to inlet and outlet orifices. The Fanning friction factor $f$ is calculated via the Blasius formula of Eq. (13.66) although *Re* for both the AP600 and SBWR designs is somewhat outside the usual range of applicability. Thom's two-phase multipliers $r_2$, $r_3$, and $r_4$ also illustrate an alternate treatment for the momentum density $\rho_m$ in Eq. (13.133).

### 13.6.5 Drift Flux Model

Another two-phase flow model that has gained popularity in recent years is the *drift flux model* [Wal69, Tod12]. The drift flux model is usually formulated separately for the liquid and vapor phases, resulting in a *six-equation* or *two-fluid model*, but we limit our discussion mostly to a four-equation formulation, which accounts for subcooled nucleate boiling. The six-equation model is usually structured as a $2V2T$ model, representing the effects of unequal phase velocities and those of non-equilibrium thermodynamics. Our primary purpose here is to introduce the basic concepts behind the drift flux model and to illustrate the differences between the slip flow and drift flux models. Since we desire to represent non-equilibrium effects, we use the subscripts $v$ and $\ell$ to represent vapor and liquid phases, respectively, instead of the subscripts g and $f$ used in Sections 13.6.3 and 13.6.4 for saturated vapor and liquid, respectively.

The main distinguishing feature of the drift flux formulation is its use of the *drift velocity* $v_{DF}$, defined as the vapor phase speed relative to the *volumetric flux* $j$

$$v_{DF} = v_v - j, \tag{13.137}$$

where $j$ is given in terms of void fraction $\alpha$ as

$$j = \alpha v_v + (1-\alpha)v_\ell = (W_v/\rho_v + W_\ell/\rho_\ell)/A. \tag{13.138}$$

The volumetric flux $j$ is the volume flow rate per unit cross-sectional area and represents the speed of the center of volume of the two-phase mixture. Using the relations $G = \alpha\rho_v v_v + (1-\alpha)\rho_\ell v_\ell$ and $\rho = \alpha\rho_v + (1-\alpha)\rho_\ell$ yields an expression for the speed of the center of mass of the mixture

$$\frac{G}{\rho} = v_v - \frac{1-\alpha}{\rho}\rho_\ell(v_v - v_\ell) = v_v - \frac{\rho_\ell}{\rho} v_{DF}$$

and

$$v_v = \frac{G}{\rho} + \frac{\rho_\ell}{\rho} v_{DF}, \tag{13.139a}$$

$$v_\ell = \frac{G}{\rho} - \frac{\alpha}{1-\alpha} \frac{\rho_v}{\rho} v_{DF}. \tag{13.139b}$$

A relationship connecting the *relative speed* $v_r$, between the vapor and liquid phases, to drift velocity $v_{DF}$ also follows:

$$v_r = v_v - v_\ell = \frac{v_{DF}}{1-\alpha}. \tag{13.140}$$

The key empirical correlation of the drift flux model involves the representation of drift velocity $v_{DF}$ as a function of void fraction $\alpha$. This is one of several empirical correlations known usually as *constitutive relations* in the drift flux formulation. Given the mass velocity $G$ and void fraction $\alpha$, the drift flux model uses a drift velocity correlation and Eqs. (13.139) to determine the phase velocities $v_v$ and $v_\ell$. This is to be contrasted with the slip flow model, where flow quality $x$ and mass velocity $G$ are used to calculate phase velocities with Eqs. (13.123) and (13.124) together with a void fraction correlation, e.g. Eq. (13.126). Void fraction $\alpha$ plays the role of a primary system variable in the drift flux model, whereas void fraction $\alpha$ is determined from flow quality $x$, or mixture enthalpy $h$, in the slip flow model. In the slip flow model, slip ratio $s$ of Eq. (13.125) is defined to be positive so that the liquid and vapor phases should move in the same direction. In contrast, Eqs. (13.139) allow for phase velocities $v_v$ and $v_\ell$ in the opposite directions for the drift flux model. This feature is useful for modeling a counter-current flow of vapor and liquid phases as would occur during the refill phase of a large-break loss-of-coolant accident [Lee17], when the emergency core cooling water flows downward against the steam flowing upward in the core annulus.

Related to drift velocity $v_{DF}$ is the drift flux, which is defined as the volumetric flux of a component relative to a surface moving at the center-of-volume speed. Thus, for vapor and liquid phases, respectively, we define *drift flux*

$$j_v = \alpha(v_v - j)$$

and

$$j_\ell = (1-\alpha)(v_\ell - j).$$

The mixture enthalpy $h$ is normally defined in the drift flux model in terms of the *vapor mass concentration* $c = \alpha\rho_v/\rho$

$$h = ch_v + (1-c)h_\ell \tag{13.141}$$

and flow quality $x$ is related to $c$

$$x = c + D/G, \tag{13.142}$$

where the *mass diffusion rate* $D = cv_{DF}$ is introduced. Void fraction $\alpha$ is finally related to vapor mass concentration $c$:

$$\alpha = \frac{\gamma c}{1 + (\gamma - 1)c}, \tag{13.143}$$

with $\gamma = \rho_\ell / \rho_v$.

In the drift flux formulation, the mass and momentum conservation equations for the mixture take the same functional form as the slip flow equations, e.g. Eqs. (13.70) and (13.72) for a channel flow. The convective momentum flux term in Eq. (13.72) may be rewritten as

$$\left(\frac{dp}{dz}\right)_{mom} = \frac{\partial}{\partial z}\left[\alpha\rho_v v_v^2 + (1-\alpha)\rho_\ell v_\ell^2\right] = \frac{\partial}{\partial z}\left[\frac{G^2}{\rho} + \frac{Dv_{DF}}{1-\alpha}\right], \tag{13.144}$$

in contrast to the expression $G^2/\rho_m$ of Eq. (13.129) for the slip flow model. The mixture energy equation (13.136) may now be rewritten in the drift flux model

$$\frac{\partial}{\partial t}(\rho \cdot h) = -\frac{\partial}{\partial z}(G \cdot h) - \frac{\partial}{\partial z}D\left(h_v - h_\ell\right) + \frac{Mq_w}{A} + \frac{\partial p}{\partial t} + S. \tag{13.145}$$

The volumetric enthalpy term $\rho h$ is now written simply as a product of $\rho$ and $h$, with the definition for mixture enthalpy $h$ given in Eq. (13.141), but the convective energy flow term requires an additional term involving mass diffusion rate $D$. Thus, for both the drift flux and slip flow models, the energy conservation equation requires modifications to the single-phase counterpart, albeit in two different ways.

To account for subcooled nucleate boiling, one of two key nonequilibrium effects in boiling channels, one may consider the mass or energy conservation equation for the liquid phase separately. In the four-equation formulation implemented in the drift-flux version of the TRANSG code [Cru81], the liquid mass and energy conservation equations are coupled in a conservative form to yield a liquid conservation equation

$$(1-\alpha)\rho_\ell\frac{\partial h_\ell}{\partial t} = -\left[(1-c)G - D\right]\frac{\partial h_\ell}{\partial z} + \frac{Mq_\ell}{A} + \frac{\partial(1-\alpha)p}{\partial t}, \tag{13.146}$$

where $q_\ell$ is the wall heat flux directly deposited in the liquid phase, expressed through an empirical correlation. In general, the liquid conservation equation (13.146) has to be solved together with the mixture conservation equations (13.70), (13.72), and (13.145) in a fully coupled form. In detailed T/H codes, the presence of superheated vapor mixed with saturated liquid is represented through a vapor conservation equation, together with separate pressures for the liquid and vapor phases, resulting in a six-equation or two-fluid model discussed further in Section 13.8.1.

## 13.7 Thermal Hydraulic Limitations and Power Capability

We have discussed in the introduction to the chapter that core performance limits are usually imposed through the limiting fuel temperature and fuel clad heat flux. We have discussed in Section 13.5 how the fuel hot spot can be determined so that any fuel temperature limitations can be accounted for. Additional discussions included the design practice to limit the heat flux at hot spot or hot channel so that the $q_{DNB}$ or CHF is avoided. The figures of merit associated with the CHF limits are now evaluated first and then combined with the power peaking factor limits to determine the power capability of a given reactor core.

### 13.7.1 DNB Ratio and Number of Fuel Rods Reaching DNB

With the goal of avoiding the occurrence of CHF in a reactor core, effort is made both in the design and operation of the plant to ensure that, given the calculated heat flux $q(z)$, the DNB ratio

$$DNBR(z) = \frac{q_{DNB}^P}{q(z)} > 1.0,$$

everywhere in the core, with the best prediction available for the DNB heat flux $q_{DNB}^P$, which is subject to uncertainties. In the well-known W-3 DNB correlation [Ton96], $q_{DNB}$ is expressed as a function of the pressure $P$, flow quality $x$, mass velocity $G$, hydraulic diameter $D_h$, inlet enthalpy $h_{in}$, saturation liquid enthalpy $h_f$, and channel geometry. Another correlation [Kim97] is derived in terms of four dimensionless parameters. The correlations are typically obtained for coolant channels with uniform heat flux distributions, and there remain considerable disagreements between the correlations and measurements, even with corrections made for non-uniform heat flux distributions.

The differences between measured DNB heat flux $q_{DNB}^M$ and DNB heat flux $q_{DNB}^P$ predicted with the W-3 correlation lie within ±20% [Ton97] such that the probability for $\{q_{DNB}^P/q_{DNB}^M > 1.30\}$ is 5% with a 95% confidence for the standard deviation $\sigma$ for the DNB correlation. This may be rewritten equivalently for the probability $P(DNB)$ of reaching DNB, with the understanding that $q_{DNB}^M$ represents the actual DNB heat flux expected for the given flow channel with heat flux $q(z)$ calculated

$$\begin{aligned} P(DNB) &= P\{q_{DNB}^M < q(z)\} = P\left\{\frac{q_{DNB}^M}{q_{DNB}^P} < \frac{q(z)}{q_{DNB}^P}\right\} \\ &= P\left\{\frac{q_{DNB}^P}{q_{DNB}^M} > DNBR\right\} = 0.05 \text{ for } DNBR = 1.30, \end{aligned} \tag{13.147}$$

with a 95% confidence for $\sigma$ of the DNB correlation. Thus, $MDNBR = 1.30$ implies that the hot rod(s) have $DNBR = 1.30$ and the remaining fuel rods have $DNBR > 1.30$.

This may be restated that there is a 95% probability for the hot rods avoiding DNB with a 95% confidence level, or simply as a 95-95% probability [Lee17] for the hot rods not exceeding DNB. Following the same logic as Eq. (13.147) for other values of *DNBR* allows a plot of *P(DNB|DNBR)*, representing the probability of reaching DNB, as a function of *DNBR* in Figure 13.22a. Corresponding to the DNB probability curve, a *fuel census curve* may be generated to calculate the number $N$ of fuel rods with *DNBR* less than indicated in Figure 13.22b. The radial power peaking factor $F_{xy}$ or $F_{\Delta h}$ corresponding to *DNBR* is also indicated in the figure, together with the number $\Delta N_i$ of fuel rods in group $i$ with $DNBR = DNBR_i$, $i = 1, \dots, I$. The number $N(DNB)$ of fuel rods expected to reach DNB for the core as a whole may be calculated by convoluting the number of fuel rods $\Delta N_i$ of Figure 13.22b with the $P(DNB|DNBR_i)$ of Figure 13.22a

$$N(DNB) = \sum_{i=1}^{I} \Delta N_i \times P(DNB_i|DNBR_i) \leq 5 \sim 50 \text{ out of 50,000 rods,} \tag{13.148}$$

for a typical PWR core with $MDNBR = 1.30$.

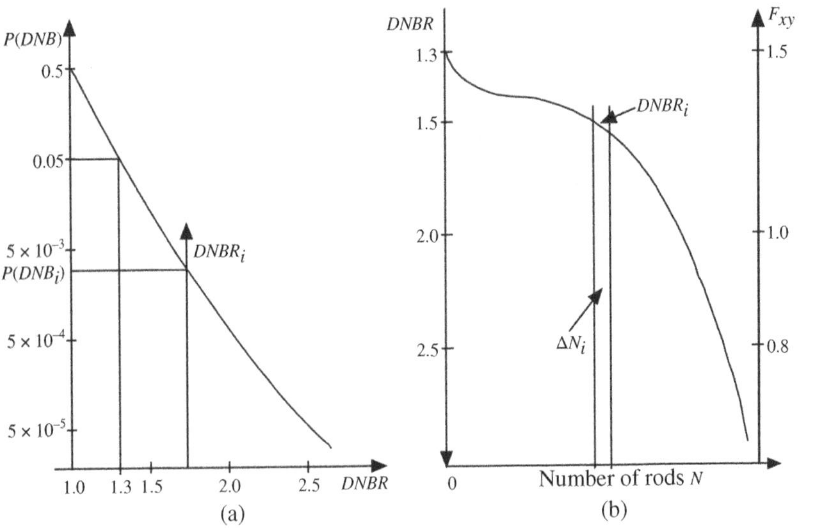


**Figure 13.22** Graphical construct to determine number of fuel rods reaching DNB. (a) DNB probability as a function of the DNB ratio and (b) fuel census curve representing the DNB ratio vs. the corresponding number of rods $N$ less than indicated. Radial peaking factor $F_{xy}$ corresponding to *DNBR* is also indicated.

### 13.7.2 Non-uniform Heat Flux Correction

Most of the channel flow data collected for DNB correlations have been obtained for uniform heat flux distributions, and a model has been developed [Ton96] to account for the effects of non-uniform heat flux distributions on $q_{DNB}$. The model is based on the observation that a superheated liquid layer and tiny bubbles essentially determine the CHF and that the bubble layer isolates the superheated liquid layer with tiny bubbles from the bulk stream, as illustrated in Figure 13.23. Furthermore, the boundary layer carries superheat and bubbles from upstream and hence provides a memory effect such that the CHF depends primarily on the enthalpy $H$ of the superheated liquid layer.

A 1-D energy balance is set up along the coolant channel with wetted perimeter or width $M$ at distance $z$ for the superheated liquid layer with enthalpy $H(z)$ and thickness $s$, density $\rho$, speed $\upsilon$, viscosity $\mu$, and heat capacity $C_p$

$$sM\rho\upsilon\frac{dH(z)}{dz} + hM[T(z) - T_b] = Mq(z), \tag{13.149}$$

where the total heat flux for the layer comprises the wall heat flux $q(z)$ and convective heat flux with heat transfer coefficient $h$ from the superheated liquid layer at temperature $T(z)$ to the bubble layer with enthalpy $H_b$ and temperature $T_b$. Equation (13.149) may be simplified to

$$\frac{d}{dz}[H(z) - H_b] + C^*[H(z) - H_b] = \frac{C^* C_p}{h}q(z), \; C^* \equiv \frac{h}{\rho C_p \upsilon s}. \tag{13.150}$$

Select $z$ such that $H(0) = H_b$, with $H_b$ set equal to the enthalpy at local boiling, i.e. $H_b \simeq h_f$, which allows integrating Eq. (13.150) to obtain

$$\begin{aligned} H(z) - H_b &= \frac{C^* C_p}{h}\int_0^z q(z')\exp[-C^*(z - z')]dz', \\ &= \frac{C_p q}{h}[1 - \exp(-C^* z)], \; q(z) = \text{constant}. \end{aligned} \tag{13.151}$$

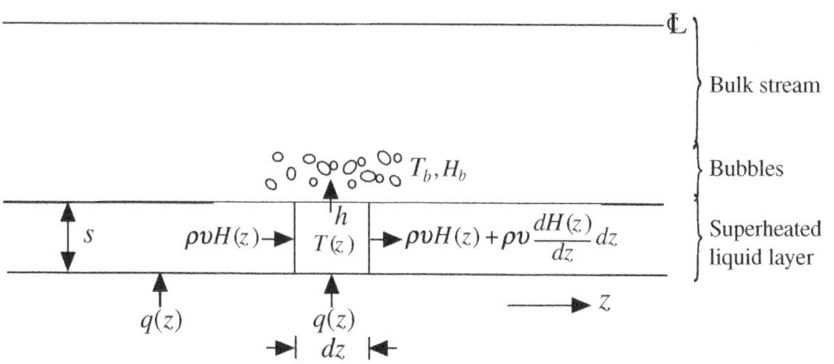


**Figure 13.23** Coolant channel illustrating superheated liquid layer with tiny bubbles between bubble layer and heated wall.

Assume that DNB occurs at the *boiling length* $z = \ell_{DNB}$, where $H(\ell_{DNB})$ of the superheated liquid layer represents some *characteristic enthalpy* for a given channel with mass flow rate $W$, hydraulic diameter $D_h$, wetted perimeter $M$, and other relevant parameters, such that $H(\ell_{DNB})$ is equal both for $\ell_{DNB,U}$ and $\ell_{DNB,N}$, corresponding to uniform and non-uniform heat flux distributions, respectively. With the specification for $H(0) \simeq h_f$, the boiling length $\ell_{DNB}$ represents the location of DNB or CHF measured from the inception of local boiling.

The uniform heat flux at $z = \ell_{DNB}$ is obtained from Eq. (13.151)

$$q_{DNB,U}(z) = \frac{C^*}{1 - \exp(-C^* z)} \int_0^z q(z') \exp[-C^*(z - z')]dz' \equiv F(z) q_{DNB,N}(z), \tag{13.152}$$

which provides the definition for the *correction factor* $F(z)$ for non-uniform heat flux:

$$F(z) = \frac{C^*}{q_{DNB,N}(z)[1 - \exp(-C^* z)]} \int_0^z q_{DNB,N}(z') \exp[-C^*(z - z')]dz'. \tag{13.153}$$

In actual determination of the *equivalent non-uniform DNB heat flux*

$$q_{DNB,N}(z) = \frac{q_{DNB,U}(z)}{F(z)}, \tag{13.154}$$

where $F(z)$ is evaluated with the calculated heat flux $q(z)$ substituted for $q_{DNB,N}(z)$ and the appropriate DNB correlation, e.g. the W-3 correlation, used for $q_{DNB,U}(z)$. In addition, an empirical formulation for the parameter $C^*$ as a function of the flow quality $x_{DNB}$ at DNB and mass velocity $G$ is used instead of the definition introduced in Eq. (13.150). We may develop the general trend by considering two representative cases:

(1) For subcooled or saturated boiling with $0 < x_{DNB} \leq 0.1$, heat transfer takes place mainly between the turbulent flow with small bubbles and heated surface representing primarily nucleate boiling such that $C^*$ is large or $C^* \ell_{DNB} \gg 1.0$, which then reduces Eq. (13.152) to

$$q_{DNB,U}(z) \simeq \frac{C^* q(z)}{1 - \exp(-C^* z)} \frac{1 - \exp(-C^* z)}{C^*} = q(z) \text{ for } z = \ell_{DNB}.$$

This indicates that $F(z) \simeq 1.0$ and the memory effect associated with the superheated liquid layer is small and local effects dominate in this primarily subcooled boiling regime.

(2) In the bulk boiling regime with $x_{DNB} > 0.1$, $C^* \ell_{DNB} \ll 1$ so that Eq. (13.152) is written in terms of the heat flux $\langle q \rangle$ averaged over the interval $[0, z]$

$$q_{DNB,U}(z) \simeq \frac{C^*}{1 - \exp(-C^* z)} \langle q \rangle \frac{1 - \exp(-C^* z)}{C^*} = \langle q \rangle = F(z) q_{DNB,N}(z),$$

indicating that the memory effect is significant in this flow regime with $F(z) < 1.0$.

### 13.7.3 Iterative Determination of DNB Ratio

With the implementation of the non-uniform correction factor $F(z)$ clarified, we turn to the task of accurately evaluating MDNBR in an iterative manner, as illustrated for core height $H$ in Figure 13.24. The procedure follows these steps:

(1) Using the appropriate DNB correlation, e.g. the W-3 correlation, find the DNB heat flux $q_{DNB,U}(z) = q_{DNB,U}[x(z)], C^*[x(z)]$, and $F(z)$ with the nominal $q(z)$ and $x(z)$ calculated for the channel.
(2) Determine $q_{DNB,N}(z) = q_{DNB,U}(z)/F(z)$ using Eq. (13.152) and calculate $MDNBR = min\{q_{DNB,N}(z)/q(z)\} \equiv MDNBR_A$.
(3) Update iteratively the flow quality $x(z)$ corresponding to $q_{DNB,N}$ at the location of MDNBR by increasing the magnitude of $q(z)$ while retaining the heat flux profile, and repeat steps (1) and (2) to arrive at the point marked ⊗ where the adjusted $q(z)$ equals $q_{DNB,N}$ and the converged $MDNBR = MDNBR_B$ is obtained.

Through the iteration process, effort is made to use the flow quality $x(z)$ that matches the actual $q_{DNB}$ expected for the channel, rather than the nominal $x(z)$

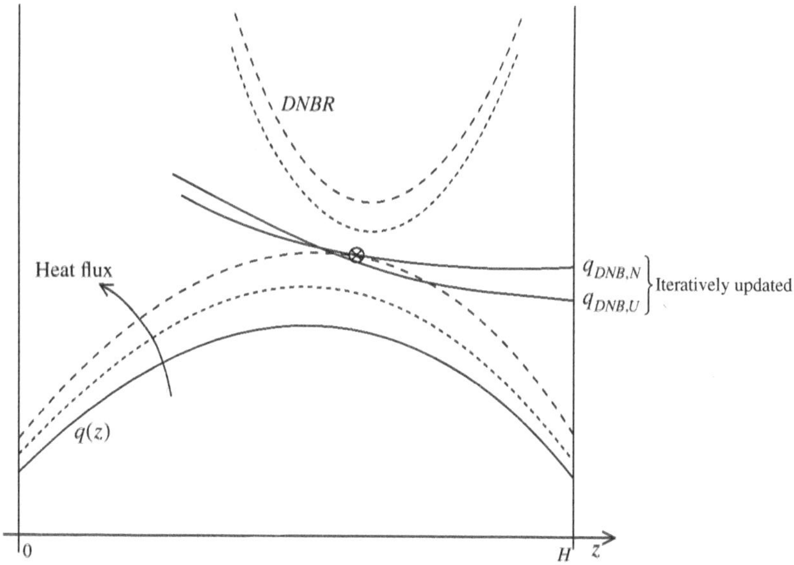


**Figure 13.24** Iterative determination of MDNBR with accurate estimate of flow quality.

that would be considerably smaller. The converged $MDNBR_B$ is usually smaller than $MDNBR_A$, because $q_{DNB}$ decreases as flow quality $x$ increases and represents the consistent limiting DNBR for the hot channel under consideration. Furthermore, as the heat flux $q(z)$ is skewed to the top of the core, where the quality $x$ corresponding to the *MDNBR* increases, thereby resulting in a smaller *MDNBR*. Hence, in general, the location of the hot spot as well as the axial power peaking factor $F_z$ itself is important in determining the *MDNBR* for the hot channel.

With increasing computation capability available for both steady-state and transient thermal-hydraulic calculations, it is often possible to perform subchannel analysis with the nominal power distributions and $MDNBR = 1.54$ for hot rods with the WRB-2M correlation [Hon12] while the engineering hot channel factor $F_q^E$ and uncertainty hot channel factor $F_q^U$ are represented statistically. This is the procedure taken with the VIPRE code [Sun99] for AP1000 design calculations.

CHF calculations for BWR cores are represented, similar to the DNBR for PWR cores, through the *critical power ratio* (CPR)

$$CPR(z) = \frac{P_c}{P(z)}, \tag{13.155}$$

where the *critical power* $P_c$ represents the power level reaching CHF and the axial power distribution $P(z)$ is determined through coupled 3-D nuclear-thermal-hydraulic calculations. The GEXL correlation [Eck73] provides *critical quality* $x_c$ as a function of channel parameters including the channel pressure $p$, mass velocity $G$, and hydraulic diameter $D_h$

$$x_c = f(p, G, D_h, \ell_B, \text{ power distribution}), \tag{13.156}$$

where the *boiling length* $\ell_B \simeq \ell_{DNB}$ is defined earlier as the distance between the onset of bulk boiling and CHF.

Figure 13.25 illustrates the iterative use of the GEXL correlation similar to the iterative approach estimating a realistic flow quality $x$ and associated DNB heat flux $q_{DNB}$ illustrated in Figure 13.24. With the GEXL correlation, power

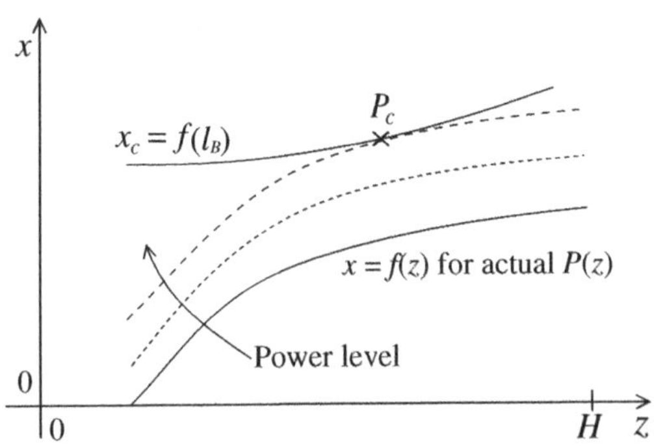


**Figure 13.25** Iterative determination of MCPR with accurate estimate of flow quality.

level for the hot channel is iteratively increased to arrive at the flow quality curve intersecting the critical quality curve, which provides the desired estimate for the critical power $P_c$ and yields the *CPR* through Eq. (13.155). With the GEXL correlation representing boiling regimes with high flow quality $x$, the BWR correlation has a smaller variance than the DNBR correlations covering subcooled or low-quality boiling regimes. This allows the use of the *minimum critical power ratio* $MCPR = 1.04 \sim 1.05$ as an indication that 99.9% of fuel rods in a BWR core are expected to avoid reaching CHF, equivalent to the use of $MDNBR = 1.30$ with the W-3 correlation for a PWR core.

### 13.7.4 Power Capability Determination

Setting the limiting fuel temperature $T_{fm}$ equal to the *fuel melting point* $T_{mp}$ provides us with an initial estimate for the *total power output* $P_t$ allowed in a reactor core. We may apply Eq. (13.100) to an idealized cylindrical core, with a uniform fission cross section $\Sigma_f$, containing $N$ fuel rods of length $H$ and wetted perimeter $M$

$$P_t = N\langle P\rangle_r = N\langle\psi\rangle_r P(0) = \frac{N}{F_{\Delta h}}\frac{WC_p(T_{mp} - T_1)}{\theta_{fm}}, \tag{13.157}$$

where $\theta_{fm}$ is the value of the dimensionless fuel temperature of Eq. (13.99) corresponding to $T_{mp}$. For operating reactor cores, the limiting rod power may not be equal to the power $P(0)$ produced in the center rod. Since $\theta_{fm}$ is approximately independent of the axial power shape, Eq. (13.157) suggests the limiting power output can be determined essentially as a function of $T_{mp}$ and the radial power peaking factor $F_{\Delta h}$. Thus, $P_t$ is inversely proportional to $F_{\Delta h}$, and the power output $P_t$ increases as the radial power distribution flattens. Note also that $\theta_{fm}$ is nearly equal to the axial peaking factor $F_z$ and hence Eq. (13.157) indicates that approximately the total power output will be inversely proportional to the overall hot channel factor $F_q$. The simple choice of $T_{mp}$ as the limiting fuel temperature can, of course, be altered if it is more prudent to choose some other temperature below $T_{mp}$.

The situation is somewhat different when we consider the CHF limitation in terms of the DNBR or CPR. Because CHF or $q_{DNB}$ decreases in general as flow quality $x$ increases, MDNBR is expected to occur somewhat downstream from the location of peak heat flux, as illustrated for a typical PWR configuration in Figure 13.24. For a given MDNBR limit, consider the power capability of a given reactor core:

$$P_t = N\langle P\rangle_r = NMH\langle q\rangle_{r,z} = NMH\frac{\langle q\rangle_{r,z}}{q_{max}}\frac{q_{max}}{q_{DNB}}q_{DNB} \simeq \frac{NMH}{F_q}\frac{q_{DNB}}{MDNBR}. \tag{13.158}$$

The allowable power output $P_t$ increases as the power distribution becomes flatter, similar to the observation of power capability subject to a fuel temperature limit, and as the limiting CHF increases. In a typical reactor core, the location of $T_{fm}$ and that of MDNBR do not coincide with one another, and the maximum power output has to be determined by comparing the maximum heat flux allowable under the limitations on both $T_{fm}$ and MDNBR.

The actual strategy to represent both constraints requires the determination of the maximum power peaking factor $F_q$ or $F_z$ allowed under operating conditions, illustrated with the actual operating data [Hon12] in Figure 13.26. The power peaking factor plot is incidentally known by its nickname, the *flyspec curve*. As part of the *constant axial offset control* (CAOC) discussed in Chapter 16, the limiting

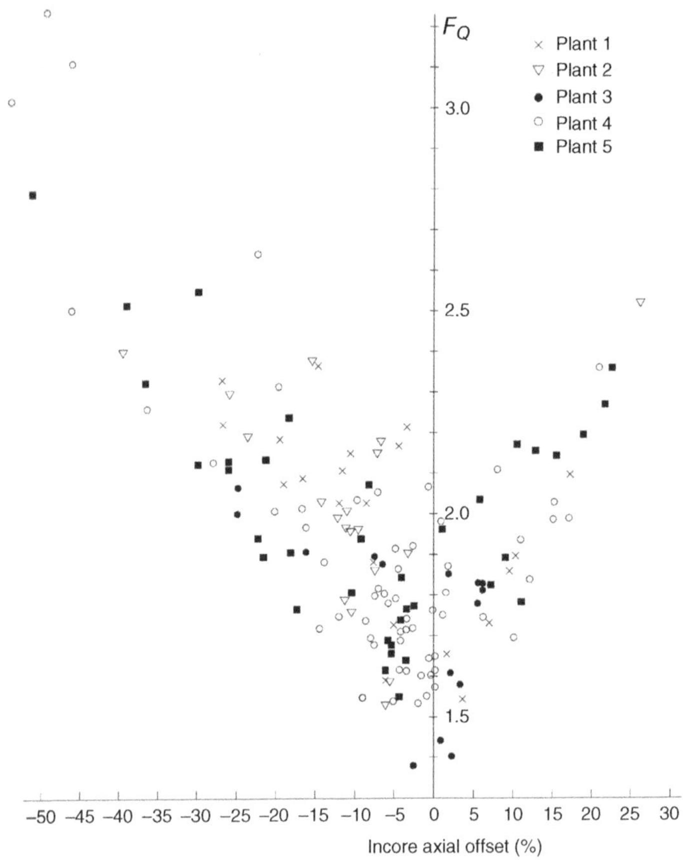


**Figure 13.26** Power peaking factor as a function of axial offset for PWR plants. *Source:* Hone [Hon12].

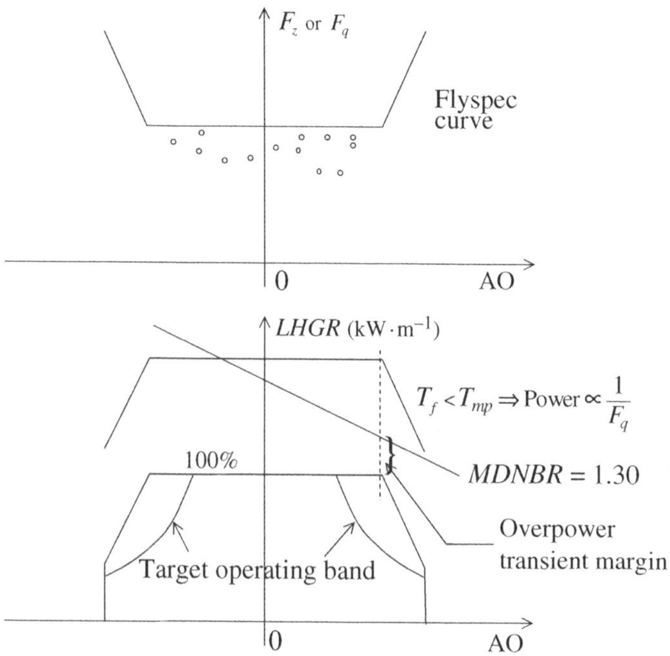


**Figure 13.27** Power capability determination accounting for limiting fuel temperature and heat flux.

power peaking factor $F_q$ may be determined for a particular plant subject to the operating band of axial offset (AO) of power.

As illustrated in Figure 13.27, the limiting $F_q$ is reflected into the maximum power allowed, in terms of the linear heat generation rate (LHGR) of Eq. (13.45), subject to the operating AO band. The tent-shape envelope is then compared with the LHGR allowed for nominal conditions under the MDNBR constraint, e.g. $MDNBR = 1.30$, which determines the maximum allowable LHGR. Finally, allowance has to be made for an overpower transient margin of typically ~ 9% including a 2% calorimetric error and a 2% decalibration error for the power distribution monitoring system to arrive at the target operating band in AO of power.

## 13.8 Thermal-Hydraulic Models for Nuclear Plant Analysis

Thermal-hydraulic models used in power plant simulation programs vary from the simple HEM to the full-blown six-equation, two-fluid models. The production simulation codes, e.g. RELAP5, RETRAN, and TRACE [NRC01, Pet81, NRC06], solve distributed-parameter representation of two-fluid, six-equation models governing the fluid flow and thermo-dynamics of the system, together with lumped-parameter models for the steam dome of nuclear steam generators, steam turbines, coolant pumps, and other systems. Basic formulation of T/H models in the NPP system codes is illustrated with the RELAP5 code as a primary example in this section, beginning with the governing two-fluid equations for fluid flows in various subsystems and components in the nuclear steam supply system (NSSS).

### 13.8.1 Light Water Reactor System Modeling Codes

The RELAP5 formulation begins with the conservation equations for mass, momentum, and energy summarized in Table 13.2, with the interactions between the vapor and liquid phases explicitly modeled. The fluid conservation equations are developed for 1-D flow channel with arbitrary orientation and area changes associated with expansions and contractions but with a number of simplifying assumptions:

(1) The fluid pressure is uniform across the vapor and liquid phases.
(2) The Reynolds dissipative heat flux is ignored in the energy equations for both phases.
(3) Transfer of mass, energy, and momentum takes place across the interface between the phases, which, however, does not have any volume associated with it. Any friction force at the interface is likewise ignored.

The phasic equations may be formally derived by averaging fluid conservation equations over a differential volume associated with each phase along the flow channel as well as over time, but simple cross-sectional areas may be assigned to each phase as discussed in Sections 13.6.4 and 13.6.5 for the slip flow and drift flux models, respectively. Assigning the void fraction $\alpha$ as the vapor fraction $\alpha_v$ and $1 - \alpha$ as the liquid fraction $\alpha_\ell$, together with the density $\rho_v$ and speed $v_v$ for the vapor phase and the density $\rho_\ell$ and speed $v_\ell$ for the liquid phase, generates the mass conservation equations

$$\begin{aligned} &\frac{\partial}{\partial t}(\alpha_v \rho_v) + \frac{\partial}{\partial z}(\alpha_v \rho_v v_v) = \Gamma, \\ &\frac{\partial}{\partial t}(\alpha_\ell \rho_\ell) + \frac{\partial}{\partial z}(\alpha_\ell \rho_\ell v_\ell) = -\Gamma, \end{aligned} \tag{13.159}$$

where $\Gamma$ represents the vapor generation rate comprising mass transfer at the vapor–liquid interface in the bulk fluid and near the channel wall. The momentum conservation is combined with the continuity equation for each phase to yield

$$
\begin{aligned}
\alpha_v\rho_v\frac{\partial v_v}{\partial t} + \frac{\alpha_v\rho_v}{2}\frac{\partial v_v^2}{\partial z} &= -\alpha_v\frac{\partial p}{\partial z} - \alpha_v\rho_v g - F_{wv} - F_{iv} + \Gamma(v_i - v_v),\\
\alpha_\ell\rho_\ell\frac{\partial v_\ell}{\partial t} + \frac{\alpha_\ell\rho_\ell}{2}\frac{\partial v_\ell^2}{\partial z} &= -\alpha_\ell\frac{\partial p}{\partial z} - \alpha_\ell\rho_\ell g - F_{w\ell} - F_{i\ell} - \Gamma(v_i - v_\ell),
\end{aligned}
\tag{13.160}
$$

where the force terms on the RHS represent

$F_{wv}, F_{w\ell}$ = wall friction forces on vapor and liquid, respectively,

$F_{iv}, F_{i\ell}$ = interface friction forces on vapor and liquid, respectively,

and the interfacial momentum transfers associated with vapor generation are explicitly written in terms of the interface speed $v_i$ and vapor generation rate $\Gamma$. The RELAP5 formulation includes additional terms associated virtual mass effects representing bubble motions in the channel, which are not included in Eqs. (13.160). Finally, the energy equations are written in terms of the internal energy $U$

$$
\begin{aligned}
\frac{\partial}{\partial t}(\alpha_v\rho_v U_v) + \frac{\partial(\alpha_v\rho_v v_v U_v)}{\partial z} &= -p\frac{\partial(\alpha_v v_v)}{\partial z} - p\frac{\partial\alpha_v}{\partial t} + Q_{wv} + Q_{fv} + Q_{iv} + \Gamma h_v,\\
\frac{\partial}{\partial t}(\alpha_\ell\rho_\ell U_\ell) + \frac{\partial(\alpha_\ell\rho_\ell v_\ell U_\ell)}{\partial z} &= -p\frac{\partial(\alpha_\ell v_\ell)}{\partial z} - p\frac{\partial\alpha_\ell}{\partial t} + Q_{w\ell} + Q_{f\ell} + Q_{i\ell} - \Gamma h_\ell,
\end{aligned}
\tag{13.161}
$$

where the energy source terms on the RHS represent

$Q_{wv}, Q_{w\ell}$ = wall heat fluxes for vapor and liquid, respectively,

$Q_{fv}, Q_{f\ell}$ = wall friction and pump effects for vapor and liquid, respectively,

$Q_{iv}, Q_{i\ell}$ = interface heat fluxes for vapor and liquid, respectively.

and the bulk energy transfers associated with the vapor generation are explicitly represented with the vapor enthalpy $h_v$, liquid enthalpy $h_\ell$, and $\Gamma$. In line with assumption (2), the Reynolds heat flux associated with viscous heating is neglected, but the second term on the RHS involving the time derivate of void fraction appears for both equations as the result of a specific time and space averaging process adopted [Ran94]. The time derivate terms are not present in some other two-fluid formulations [Ste79, Ish77]. The RELAP5 formulation includes additional terms associated with bulk interface mass transfer and thermal boundary layer, but those terms are not represented in Eqs. (13.161).

Since the interface between the phases is assumed to have no volume, the sum of the transfer rates of mass, momentum, and energy should vanish at the interface. The interface condition for the mass transfer rate $\Gamma$ has already been accounted for explicitly in deriving Eq. (13.159). For the momentum equations (13.160) and energy equations (13.161), the interface condition yields

$$\Gamma = -\frac{F_{iv} + F_{i\ell}}{v_v - v_\ell} = -\frac{Q_{iv} + Q_{i\ell}}{h_v - h_\ell}. \tag{13.162}$$

The interface conditions may be used effectively when the phasic equations are combined to generate mixture conservation equations. It should also be noted that the sum of the wall friction forces $F_{wv}$ and $F_{w\ell}$ equals the normal friction force $F_w$ for the channel. Likewise, the sum of wall heat fluxes $Q_{wv}$ and $Q_{w\ell}$ and of wall friction energy terms $Q_{fv}$ and $Q_{f\ell}$ should equal the total wall heat flux $Q_w$ and total friction energy $Q_f$, respectively.

The RELAP5 code tracks boric acid dissolved in the coolant for PWR plants as an additional component with density $\rho_b$ in the liquid mass conservation equation and non-condensable gases with mass fraction or quality $x_n$ in the vapor phase equations. Thus, the code solves the governing two-fluid equations (13.159) through (13.161) for six variables $[p, \alpha, v_v, v_\ell, U_v, U_\ell]$ together with additional field equations for $[\rho_b, x_n]$ for a total of eight variables describing the NSSS coolant system. The equations are combined and transformed into structures more convenient for numerical integration of the partial differential equations, together with empirical constitutive relationships for the void fraction $\alpha$, vapor generation rate $\Gamma$, and other force and energy terms introduced in Eqs. (13.160) and (13.161). The equation of state $\rho = f(h, p)$ connecting water density $\rho$ to enthalpy $h$ and pressure $p$ is obtained with the ASME steam table [Par00] to evaluate phasic densities $\rho_v$ and $\rho_\ell$ and other properties of water. Heat transfer from fluid to solid surfaces in various components is represented through the time-dependent heat conduction equation (13.30). The reactor core power output is represented by a point reactor kinetics equation discussed in Chapter 8.

The discretized field equations are solved with a staggered mesh structure for control volumes distributed over a coolant channel, where, as illustrated in Figure 13.28, extrinsic system variables, e.g. mass flow rate $W$ and flow speed, are evaluated at the cell boundary coupled to intrinsic variables, e.g. pressure $p$, density $\rho$, and enthalpy $h$, evaluated at the cell center. The temporal integration schemes allow semi-implicit and nearly-implicit formulations allowing efficient numerical analysis for complex NSSS systems. The RELAP5 numerical algorithms [NRC01] represent a significant enhancement over earlier versions of the code, which employed explicit integration routines and were often limited to fine time-steps subject to the Courant–Friedrichs–Lewy condition [Str89].

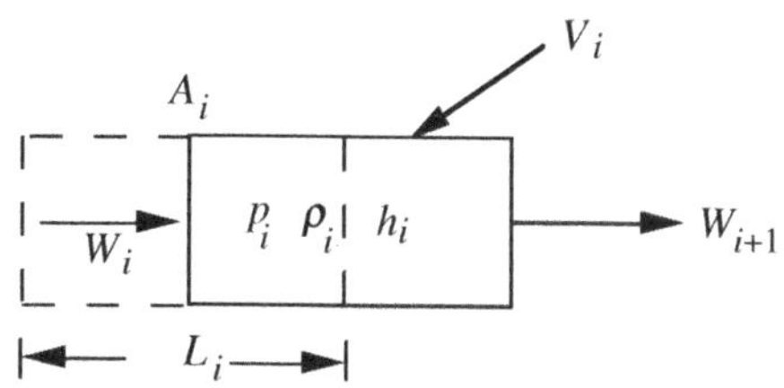


**Figure 13.28** Staggered mesh structure for control volume *i*.

The numerical algorithms seem to have benefited from the implicit continuous Eulerian (ICE) scheme developed by Harlow and Amsden [Har71] and somewhat similar to the extended ICE (EICE) formulation implemented in the TRANSG code [Cru81] for the transient analysis of nuclear steam generators. The RELAP5 code also provides a self-initialization capability for consistent steady-state formulation, at least for relatively simple systems; for realistic complex systems, however, the code still relies on a null-transient strategy whereby iterative adjustments of input parameters are made until the code output converges to a consistent set.

The RELAP5 code has been developed as a *best-estimate* NPP simulation code as part of the *Final Acceptance Criteria* [Lee17] for performing design and verification calculations for the *emergency core cooling system* (ECCS), which protects the plant in design basis accidents, especially the *loss of coolant accident* involving the rupture of primary coolant pipes. Thus, the code offers ability to model the details of the entire NSSS illustrated by the nodalization diagram for a two-loop PWR plant in Figure 13.29, coupled to the balance of plant (BOP) including the steam turbine generator system. The diagram illustrates a reactor core with coolant channels, two steam generators with two cold legs and one hot leg each, and a pressurizer, together with various ECCS systems. The designation and connection of control volumes and junctions are also illustrated in the diagram.

The RELAP5 code has the capability to represent any combination of coolant channels and heat structures, with input variables to various junctions determined as a function of time via flexible control logic. The code is limited to water steam table, but an extended version of the code developed as RELAP5-3D/ATHENA [INL14] provides access to several alternate coolants, including heavy water, helium, liquid metal, and molten salt, as well as 3-D fluid flow and point reactor kinetics formulations. An advanced system analysis code RELAP-7 [Ber14] has been under development within the consolidated software framework, Multi-Physics Object Oriented Simulation Environment (MOOSE) [Per20]. The code is written with the object oriented programming language C++ to cover seven-equation two-phase flow, reactor core heat transfer, and reactor kinetics models.

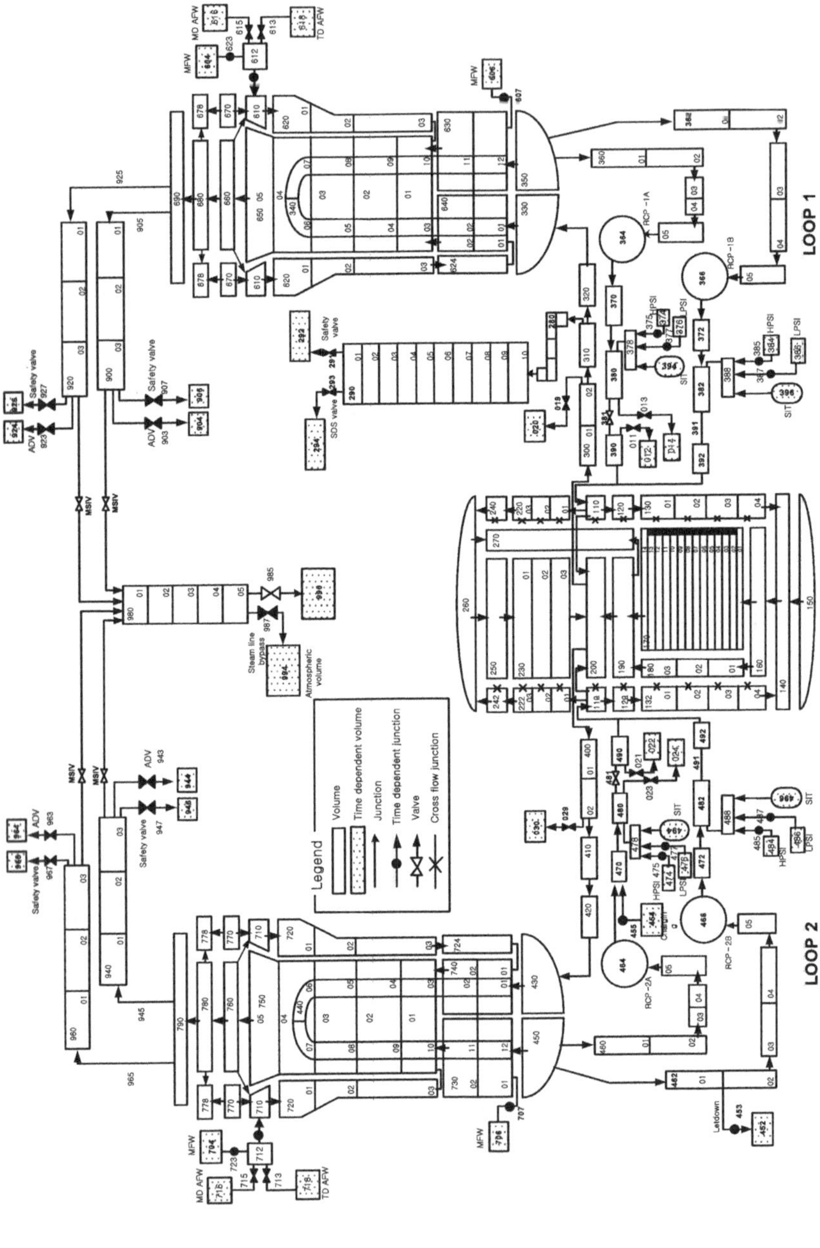


**Figure 13.29** NSSS nodalization diagram for PWR plant.

### 13.8.2 Subchannel Analysis Codes

A number of thermal-hydraulic codes have been developed starting from the THINC code [Che72] to represent the detailed fluid flow and heat transfer phenomena surrounding fuel rods in PWR and BWR fuel assemblies. The THINC code was used in two-step calculations in the early days of PWR development to obtain the overall power and flow distributions across the core in the first step, followed by detailed subchannel calculations in the limiting or hot assembly. The detailed subchannel analysis is necessary to account for the CHF in determining the power capability of a reactor core as discussed in Section 13.7. The VIPRE-01 code [Ste89] has been used for the subchannel analysis of the AP1000 design [Hon12].

The codes have evolved over the years to represent the two-phase flow in 3-D two-fluid formulations together with a separate field for the droplets in the COBRA-TF code [Avr14]. The CTF numerical formulation is based on a semi-implicit pressure-linked formulation [Pat80] where the 3-D fluid velocity field is iteratively updated with the system pressure as the fiducial variable. The code solves three momentum equations covering the vapor, liquid, and droplet fields and two energy equations for the vapor and liquid phases [Tod99], accounting for cross-channel flows coupled to axial flows in fuel assemblies. The continuous liquid phase is structured in thermal equilibrium with the droplet field. Entrainment and de-entrainment models provide the mass transfer between the two liquid fields. The exchange of mass, momentum, and energy through turbulent mixing or diffusion is modeled explicitly in addition to the normal molecular diffusion process.

### 13.8.3 Sodium-Cooled Fast Reactor Codes

Development of thermal-hydraulic and safety analysis codes for SFRs has been focused at the Argonne National Laboratory, and recently centered around the SASSYS-1/SAS4A package [Cah12]. The thermal-hydraulic package features single and two-phase sodium coolant dynamics for single and multiple-pin fuel assembly models. In light of the importance of reactivity feedback effects due to sodium voiding in SFRs, as discussed in Chapter 14, effort has been made to represent incipient bubble formation and growth, film evaporation and stripping, fuel-coolant interaction, and fuel pin disruption with the subchannel models.

The SASSYS-1 code performs multiple-pin subchannel thermal-hydraulic calculations, while the SAS4A code offers the ability to perform coupled nuclear-thermal-hydraulic calculations accounting for detailed reactivity feedback effects, including the fuel temperature Doppler feedback, fuel and cladding axial expansion, core radial expansion, fuel and cladding relocation, and control-rod driveline expansion. With the ability to represent sodium boiling and associated fuel reconfiguration in detail, the code package has been used for assessing safety margins in design basis accidents and anticipated transient without scram (ATWS) events. The SAS4A/SASSYS-1 package represents, in addition to sodium, several liquid-metal coolants including NaK, Pb, and Pb-Bi. Spatial kinetics calculation may be performed by coupling SAS4A with the DIF3D-K code [Tai92].

### 13.8.4 Containment Analysis Codes

Another class of thermal-hydraulic codes developed for NPP analysis is the containment analysis code, with the CONTAIN code [Mur97] as a primary example. The code solves a set of lumped-parameter fluid conservation equations together with the equation of state for a mixture of gases and liquids to predict containment behavior during reactor accidents. The system volume is divided into control volumes or cells, and each cell comprises an upper atmosphere region and, if necessary, a lower pool region. The exchange of mass, energy, and momentum between cells as well as between the upper and lower regions of a cell may be represented through a variety of flow paths and energy transfer processes. The code cannot represent detailed in-vessel phenomena but may accept time-dependent primary system mass and energy flow rates from other system codes, e.g. the RELAP5 code, as a boundary condition.

The basic representation of the CONTAIN control volumes is illustrated conveniently by the mass conservation equations in terms of mass inventory $M^u_{k,i}$ for component $k$ in the upper atmospheric region of cell $i$

$$\frac{dM^u_{k,i}}{dt} = \sum_j \left( f^u_{k,j} W^u_{j\to i} - f^u_{k,i} W^u_{i\to j} \right) + W^u_{k,i,source} - W^u_{k,i,sink}, \tag{13.163}$$

where $f^u_{k,j} = M^u_{k,j} / \sum_n M^u_{n,j}$ is the fraction of the upper region mass in component $k$ for cell $j$ and $W^u_{j\to i}$ the total upper region mass flow rate from cell $j$ to cell $i$ so that $f^u_{k,j} W^u_{j\to i}$ represents the upper region flow rate for component $k$ from cell $j$ to cell $i$. The last two terms in Eq. (13.163) represent the effective flow rates of component $k$ into and out of cell $i$, respectively, due to evaporation, condensation, or direct flows, e.g. the suppression chamber vent flow out of the cell. The lower pool comprises only liquid water and hence the continuity equation is written simply in terms of the mass $M^p_i$, flow rate $W^p_{j\to i}$ between cells, and direct flow rates into and out of cell $i$,

$$\frac{dM^p_i}{dt} = \sum_j \left( W^p_{j\to i} - W^p_{i\to j} \right) + W^p_{i,source} - W^p_{i,sink}. \tag{13.164}$$

The energy conservation equations for each cell are solved for the total energy inventories of the mixture and water for the upper and lower regions, respectively, without separating out contributions from each component. Contributions from convection, conduction, gravitational forces, and direct flows are represented in the energy conservation equations. Because containment analyses typically involve low flow rates, however, kinetic energy terms are neglected. The momentum conservation equation is written for a single pressure for each cell to represent normal pressure drops, including acceleration, gravitational, frictional, and form loss terms. For a cell consisting of multiple flow segments, the time derivative of the mass velocity is integrated in space over the entire flow path in the cell in a momentum integral approach [Mey61]. The equation of state is written separately for the upper and lower regions, with all gases represented via an ideal gas approximation.

### 13.8.5 Computational Fluid Dynamics Codes

With continuing improvement in the computational power, there is increasing interest in performing accurate modeling of turbulent phenomena in fluid flow and multi-physics simulations. One particular development is in CFD codes, with the STAR-CCM+ code [CDa16] as a prime example. In CFD codes, effort is made to represent fluctuations in fluid density, momentum, and energy as accurately as possible, with the *Reynolds Averaged Navier–Stokes (RANS) equation* used to represent the Reynolds stress due to turbulent velocity fluctuations.

Construct fluid velocity $\mathbf{v} = \overline{\mathbf{v}} + \mathbf{v}' \equiv \langle \mathbf{v} \rangle + \mathbf{v}'$ as in Eq. (13.62) to account explicitly for the turbulent component $\mathbf{v}'$ in addition to the time-averaged or ensemble-averaged velocity $\overline{\mathbf{v}} \equiv \langle \mathbf{v} \rangle$, together with $p = \overline{p} + p'$ and $\tau_{ij} = \overline{\tau}_{ij} + \tau'_{ij}$, and evaluate, for an incompressible fluid, an ensemble average of the convective flow of momentum in the Navier–Stokes equation (13.16)

$$\rho\nabla \cdot \langle \mathbf{v}\mathbf{v} \rangle = \rho\nabla \cdot \langle (\langle \mathbf{v} \rangle + \mathbf{v}')(\langle \mathbf{v} \rangle + \mathbf{v}') \rangle = \rho\langle \mathbf{v} \rangle \cdot \nabla \langle \mathbf{v} \rangle + \rho\nabla \cdot \langle \mathbf{v}'\mathbf{v}' \rangle, \tag{13.165}$$

with the term $\rho\langle \mathbf{v} \rangle \nabla \cdot \langle \mathbf{v} \rangle$ dropped via Eq. (13.11). Substituting Eq. (13.165) into the Navier–Stokes equation (13.13) for the $i$th velocity component $v_i$ yields the *RANS equation*

$$\rho\frac{\partial \overline{v}_i}{\partial t} + \rho\sum_{j=1}^{3}\overline{v}_j\frac{\partial \overline{v}_i}{\partial x_j} = -\rho\sum_{j=1}^{3}\frac{\partial \overline{v_i' v_j'}}{\partial x_j} - \sum_{j=1}^{3}\frac{\partial \overline{\tau}_{ij}}{\partial x_j} - \frac{\partial \overline{p}}{\partial x_i} + \rho g_i, i = 1, \dots, 3, \tag{13.166}$$

with the *Reynolds stress* $\rho\overline{v_i' v_j'}$ playing the same role as the shear stress $\tau_{ij}$. In simple applications of the RANS equation, the Reynolds stress, divided by density $\rho$, is expressed through the *eddy viscosity* $\nu$ and turbulent kinetic energy $k$

$$\overline{v_i'v_j'} = -\nu\left(\frac{\partial \overline{v}_i}{\partial x_j} + \frac{\partial \overline{v}_j}{\partial x_i}\right) + \frac{2}{3}k\delta_{ij} \tag{13.167}$$

and added to the shear stress $\tau_{ij}$ to yield the ensemble-averaged speed $\overline{v}_i, i = 1, \ldots, 3$. A transport equation may be obtained also for turbulent kinetic energy $k$ as a higher moment of the Navier–Stokes equation.

The CFD codes provide options for solving Eqs. (13.166) and (13.167) together with additional equations representing turbulent fluid flow. In the $k$–$\epsilon$ model, separate equations are introduced to represent turbulent kinetic energy $k$ and dissipation $\epsilon$ of kinetic energy, while in the $k$–$\omega$ mode, the dissipation is represented by the specific dissipation rate per turbulent kinetic energy $\omega = \epsilon/k$, both with the eddy viscosity model of Eq. (13.167). For many fluid flow problems, the $k$–$\omega$ formulation provides a better representation of boundary layer effects [Pop09, Dur10, CDa16]. A full-blown turbulent flow representation via the *direct numerical solution* (DNS) is also offered so that it may be combined with the $k$–$\epsilon$ or $k$–$\omega$ model for numerical efficiency in the *large eddy simulation* (LES) or *detached eddy simulation* (DES) option [Pop09, Dew11, CDa16].

## 13.9 Advances in Nuclear Plant Modeling System

Although we have developed general T/H models applicable to single- and two-phase flow problems in nuclear reactor analysis, our analytical formulations and examples have been limited to closed, single-channel formulations. Furthermore, we have decoupled the calculation of radial temperature distribution in a fuel rod, and the associated diametral gap and fuel clad, from the calculation of axial bulk coolant temperature distribution. Of course, in any numerical calculation of fuel and coolant temperature distributions, the radial heat conduction models of Sections 13.3.1 and 13.3.2 should be solved together with the axial convection models of Section 13.5 in a coupled manner.

In a coupled nuclear-T/H analysis, the heat flux distribution $q(z)$ is affected by the T/H feedback associated with the fuel temperature and moderator density distributions. The feedback effects are of crucial importance in BWR cores because of the rapid variation of coolant density shown in Figure 13.21. Furthermore, because of significant increases in the frictional pressure drop due to boiling, the channel flow rate itself has to be iteratively updated in a global 3-D calculation for BWR analysis. This is because the radial flow distribution will be affected by the flow resistance varying from channel to channel. In addition, in a detailed subchannel analysis, where individual coolant channels are explicitly represented, cross flow between channels should also be accounted for.

A large number of power plant simulation models with varying levels of sophistication in T/H formulations have been developed over the years, with RELAP5, RETRAN, RELAP5-3D, TRACE, and COBRA-TF as the representative software available in 2024. The complexity and sophistication of power plant simulation models can generally be categorized by the formulation to handle time-dependent two-phase flows in the primary and secondary sides of a nuclear power plant. Only a brief introduction has been provided for the representative T/H codes together with the CONTAIN code for containment system analysis and STAR-CCM+ as an example of CFD codes featuring multi-physics capability built around detailed CFD formulations [Dur10, Dew11]. Another widely used CFD software is the ANSYS Fluent code [Sto18], which provides an interface with structural computational tools. On the other hand, COMSOL Multiphysics [Com18] allows for accurate finite-element solutions of various multi-physics partial differential equations, including coupled T/H-structural analysis, for arbitrary geometries.

A number of reactor simulation tools have been developed in the Nuclear Energy Advanced Modeling and Simulation (NEAMS) program [Sta19] through the aegis of the US Department of Energy and may be accessed via the open-source Multiphysics Object-Oriented Simulation Environment (MOOSE) framework [She23, Per20]. The key NEAMS tools include: (a) Griffin [Lee21] providing diffusion and transport theory solvers for reactor physics analysis, (b) Sam [HuR21] supporting fast-running transient T/H analysis for nuclear plants, (c) Pronghorn [Nov21] providing coarse-mesh T/H analysis for non-LWR systems, and (d) Bison [Wil21] tailored for finite-element fuel performance analysis, as well as the COBRA-TF code mentioned earlier. All of the NEAMS software may conveniently be accessed via the MOOSE framework, which provides convenient tools for automatic differentiation, system interface, and mesh generation for system components, including fuel pin, fuel assembly, reactor core, reflector, control drum, and heat pipe, followed by automatic tagging tools for efficient post-processing of various data generated by the system. The NEAMS system covers, in fact, all areas of major reactor analysis tasks and a number of benchmark cases using the NEAMS suite of codes via the MOOSE framework for representative nuclear plant designs have also been made available through an online GitHub, Virtual Test Bed repository [Giu23].

## References

**[Avr14]** M. Avramova, "CTF - A Thermal-Hydraulic Subchannel Code for LWRs Transient Analyses, Users' Manual," Nuclear Engineering Program, Pennsylvania State University (2014).

**[Ber14]** R.A. Berry, J.W. Peterson, H. Zhang et al., "RELAP-7 Theory Manual," INL/EXT-14-31366, rev. 1, Idaho National Laboratory (2014).

**[Bir07]** R.B. Bird, W.E. Stewart, E.N. Lightfoot, *Transport Phenomena*, Wiley, 1st ed. (1960), 2nd ed. (2007).

**[Cah12]** J.E. Cahalan and T.H. Fanning, "The SAS4A/SASSYS-1 Safety analysis Code System" ANL/NE-12/4, Argonne National Laboratory (2012).

**[CDa16]** CD-adapco, "STAR-CCM+ Documentation," version 10.06, CD-adapco (2016).

**[Che72]** H. Chelemer, J. Weisman, and L.S. Tong, "Subchannel Thermal Analysis of Rod Bundle Cores," *Nucl. Eng. Des.* **21**, 35 (1972).

**[Col72]** J.G. Collier, *Convective Boiling and Condensation*, McGraw-Hill (1972).

**[Com18]** COMSOL.com, Comsol, Inc., "COMSOL Mutiphysics Modeling Software," vers. 5.4, COMSOL.com, Comsol, Inc. (2018).

**[Cru81]** M.W. Crump and J.C. Lee, "Nonlinear Transient Analysis of Light Water Reactor Steam Generators Using an Implicit Eulerian Method," *Nucl. Sci. Eng.* **77**, 192 (1981); also J.C. Lee et al., "Transient Modeling of Steam Generator Units in Nuclear Power Plants: Computer Code, TRANSG-01," EPRI NP-1368, Electric Power Research Institute (1980).

**[Dew11]** A. Dewan, *Tracking Turbulent Flow in Engineering*, Springer (2011).

**[Dur10]** P.A. Durbin and B.A. Pettersson Reif, *Statistical Theory and Modeling for Turbulent Flows*, Wiley (2010).

**[Eck73]** E.C. Eckert, J. Gonzalez, E. Janssen et al., "General Electric BWR Thermal Analysis Basis (GETAB): Data, Correlation and Design Application," NEDO-10958, General Electric Company (1973).

**[GEH14]** GE Hitachi Nuclear Energy, "ESBWR Design Control Document," 26A6642AP, rev. 10, GE Hitachi Nuclear Energy (2014).

**[GEN92]** GE Nuclear Energy, "SBWR Standard Safety Analysis Report," Report 25A5113, rev. A, GE Nuclear Energy (1992).

**[Ghi17]** S.M. Ghiaasiaan, *Two-Phase Flow, Boiling, and Condensation: In Conventional and Miniature Systems*, 2nd ed., Cambridge University Press (2017).

**[Giu23]** G.L. Giudicelli, A. Abou-Jaoude, A.J. Novak et al., "The Virtual Test Bed (VTB) Repository: A Library of Reference Reactor Models Using NEAMS Tools," *Nucl. Sci. Eng.* **197**, 2217 (2023).

**[Har71]** F.H. Harlow and A.A. Amsden,"A Numerical Fluid Dynamics Calculation Method for All Flow Speeds," *J. Comput. Phys.* **8**, 197 (1971).

**[Hon12]** M. Hone, S. Skidmore, M. Misvel, and E. Resch, "AP1000 Core Reference Report," WCAP-17524-NP, Westinghouse Electric Company (2012).

**[HuR21]** R. Hu, L. Zou, G. Hu et al., "SAM Theory Manual," ANL/NE-17/4 Rev. 1, Argonne National Laboratory (2021).

**[INL14]** Idaho National Laboratory, The RELAP5-3D Code Development Team, "RELAP5-3D Code Manual Volume I: Code Structure, System Models and Solution Methods," INL-EXT-98-00834-V1, rev. 4.2, Idaho National Laboratory (2014).

**[Ish77]** M. Ishii, "One-dimensional Drift-flux Model and Constitutive Equations for Relative Motion Between Phases in Various Two-phase Flow Regimes," ANL-77-47, Argonne National Laboratory (1977).

**[Kim97]** H.G. Kim and J.C. Lee, "Development of a Generalized Critical Heat Flux Correlation through the Alternating Conditional Expectation Algorithm," *Nucl. Sci. Eng.* **127**, 300 (1997).

**[Lee17]** J.C. Lee and N.J. McCormick, *Risk and Safety Analysis of Nuclear Systems*, Wiley (2011/2017).

**[Lee72]** J.C. Lee and D.H. Risher, "Power Peaking Factor due to Fuel Densification," ND-III-38, Westinghouse Electric Corporation (1972).

**[Lee21]** C. Lee, Y.S. Jung, H. Park et al., "Griffin Software Development Plan," ANL/NSE-21/23, Argonne National Laboratory (2021).

**[Mey61]** J.E. Meyer, "Hydrodynamic Models for the Treatment of Reactor Thermal Transients," *Nucl. Sci. Eng.* **10**, 269 (1961).

**[Mur97]** K.K. Murata, D.C. Williams, R.o. Griffith et al., "Code manual for CONTAIN 2.0: A computer code for nuclear reactor containment analysis," NUREG/CR-6533, US Nuclear Regulatory Commission (1997).

**[Nov21]** A. Novak, S. Schunert, R.W. Carlsen et al., "Pronghorn: A Multidimensional Coarse-Mesh Application for Advanced Reactor Thermal Hydraulics,"*Nucl. Technol.* **207**, 1015 (2021).

**[NRC01]** US Nuclear Regulatory Commission, "RELAP5/MOD3.3 Code Manual, Volume 1: Code Structure, Systems Models, and Solution Methods," NUREG/CR-5535, rev. 1, US Nuclear Regulatory Commission (2001).

**[NRC06]** US Nuclear Regulatory Commission, "TRACE V4.274 User's Manual," US Nuclear Regulatory Commission (2006).

**[Par00]** W.T. Parry, J.C. Bellows, J.S. Gallagher, and A.H. Harvey, *ASME International Steam Tables for Industrial Use*, CRTD-Vol. 58, American Society of Mechanical Engineers (2000).

**[Pat80]** S.V. Patankar, *Numerical Heat Transfer and Fluid Flow*, Hemisphere Publishing (1980).

**[Per20]** C.J. Permann, D.R. Gaston, D. Andrs et al., "MOOSE: Enabling Massively Parallel Multiphysics Simulation," *SoftwareX* **11**, 100430 (2020).

**[Pet81]** C.E. Peterson, J.H. McFadden, M.P. Paulsen, and G.C. Gose, "RETRAN - A Program for One-Dimensional Transient Thermal Hydraulic Analysis of Complex Fluid Flow Systems," EPRI NP-1850-CCM, Electric Power Research Institute (1981).

**[Pig65]** T.H. Pigford, "Temperature Distributions in Nuclear Reactor Cores," unpublished lecture notes, University of California (1965).

**[Pop09]** F. Pope, *Turbulent Flows*, Cambridge Press (2009).

**[Ran94]** V.J. Ransom, "The RELAP5 Two-fluid Model and Associated Numerical Methods," unpublished lecture notes, Purdue University (1994).

**[Moo73]** K.V. Moore and W.H. Rettig, "RELAP4 – A Computer Program for Transient Thermal-Hydraulic Analysis," ANCR-1127, Aerojet Nuclear Company (1973).

**[She23]** E. Shemon, Y. Miao, S. Kumar et al., "MOOSE Reactor Module: An Open-Source Capability for Meshing Nuclear Reactor Geometries," *Nucl. Sci. Eng.* **197**, 1656 (2023).

**[Sta19]** C. Stanek, "Overview of DOE-NE NEAMS Program," LA-UR-19-22247, Los Alamos National Laboratory (2019).

**[Ste79]** J.B. Stewart, "Calculation of Transient Boiling Flow in Channels," *J. Comput. Phys.* **30**, 61 (1979).

**[Ste89]** C.W. Stewart and J.M. Cuta, "VIPRE-01: A Thermal-Hydraulic Code for Reactor Core," NP-2511-CCM-A, Electric Power Research Institute (1989),

**[Sto18]** T. Stolarski, Y. Nakasone, and S. Yoshimoto, *Engineering Analysis with ANSYS Software*, 2nd ed., Elsevier (2018).

**[Str89]** K.C. Strikwerda, *Finite Difference Schemes and Partial Differential Equations*, Wadsworth and Brooks (1989).

**[Sun99]** Y.X. Sung, P. Schueren, and A. Meliksetian, "VIPRE-01 Modeling and Qualification for Pressurized Water Reactor Non-LOCA Thermal-Hydraulic Safety Analysis," WCAP-15306-NP-A, Westinghouse Electric Company (1999).

**[Tai92]** T.A. Taiwo and H.S. Khalil, "An Improved Quasistatic Optimization for the DIF3D Nodal Kinetics Code," *Proc. Topl. Mtg. Advances in Reactor Physics*, 2-469 (1992).

**[Tho64]** J.R.S. Thom, "Prediction of Pressure Drop During Forced Circulation Boiling of Water," *Int. J. Heat Mass Transfer* **7**, 709 (1964).

**[Tod99]** D.R. Todd, C. Frepoli, and L.E. Hochreiter, "Development of a COBRA-TF model for the Penn State University - Rod Bundle Heat Transfer Program," *Proc. 7th Int. Conf. Nucl. Eng.* (1999).

**[Tod12]** N.E. Todreas and M.S. Kazimi, *Nuclear Systems I: Thermal Hydraulic Fundamentals; II: Elements of Thermal Hydraulic Design*, 2nd ed., Hemisphere Publishing (2012).

**[Ton67]** L.S. Tong, "Heat Transfer in Water-Cooled Nuclear Reactors," *Nucl. Eng. Des.* **6**, 301 (1967).

**[Ton97]** L.S. Tong and Y.S. Tang, *Boiling Heat Transfer and Two-Phase Flow*, CRC Press (1997).

[Ton96] L.S. Tong and J. Weisman, *Thermal Analysis of Pressurized Water Reactors*, 3rd ed., American Nuclear Society (1996).
[Wal69] G.B. Wallis, *One-Dimensional Two-phase Flow*, McGraw-Hill (1969).
[Wes00] Westinghouse Electric Company, "AP600 Design Control Document," GW-GL-700, WCAP-15150, Westinghouse Electric Company (2000).
[Whi86] F.M. White, *Fluid Mechanics*, McGraw-Hill (1986).
[Wil21] R. Williamson, J.D. Hales, S.R. Novascone et al., "BISON: A Flexible Code for Advanced Simulation of the Performance of Multiple Nuclear Fuel Forms," *Nucl. Technol.* **207**, 954 (2021).

## Problems

**13.1** Show that the three definitions of two-phase densities given in Eqs. (13.116), (13.122), and (13.130) become identical in the limit of slip ratio $s = 1.0$.

**13.2** For the AP1000 core, use the total effective coolant flow rate of $13.5\,\text{Mg} \cdot \text{s}^{-1}$ at a net power output of 3,400 MWt and assume that the flow is uniformly distributed over 41,448 fuel rods in the core with fuel rod radius $a = 4.757$ mm and unit cell radius $b = 7.137$ mm. Assume also that the coolant is incompressible and that the friction force between the fuel rods and coolant water may be represented by the Blasius formula of Eq. (13.66) for the friction factor $f$, expressed in terms of an equivalent hydraulic diameter. You may use the formula, although the Reynolds number may lie slightly outside the accepted range. The form loss coefficient $K$ for the coolant channel is estimated to be 15. Determine the total pressure drop across the reactor core. Compare the result with that given in the design control document (DCD) and discuss.

**13.3** (a) Calculate the average coolant enthalpy rise and average coolant temperature rise across the core for $N = 41,448$ fuel rods, and the core-average coolant temperature for the AP1000 core. Compare the results with those given in DCD, together with an equivalent core diameter of 3.04 m. (b) Using the AP1000 pellet radius $a = 4.12$ mm, clad thickness $t_c = 0.572$ mm, and gap thickness $t_g = 0.056$ mm, calculate the core-average power density $P$ [$\text{kW} \cdot \text{m}^{-3}$] in terms of both core volume and fuel rod volume, and average fuel-rod heat flux $q$ [$\text{kW} \cdot \text{m}^{-2}$], and compare them with the corresponding DCD values. (c) Using Eq. (13.52), calculate the overall heat transfer coefficient and the core-average fuel centerline temperature. You may use the following data: (i) $UO_2$ thermal conductivity $k_f = 2.9\,\mathrm{W \cdot m^{-1}\,K^{-1}}$, (ii) Zircaloy clad thermal conductivity $k_c = 16\,\mathrm{W \cdot m^{-1}\,K^{-1}}$, (iii) gap conductance $h_g = 8.5\,\mathrm{kW \cdot m^{-2}\,K^{-1}}$, and (iv) convective heat transfer coefficient for water $h = 35\,\mathrm{kW \cdot m^{-2}\,K^{-1}}$.

**13.4** A bare cylindrical reactor of core height $H$ is cooled by water flowing axially through channels containing $UO_2$ fuel rods. The inlet temperature of the coolant water is $T_1$. The maximum allowable fuel temperature and the outlet coolant temperature in the central channel are given as $T_m$ and $T_2$, respectively, so that the maximum value of the dimensionless temperature profile is given as

$$\theta_{fm} = \frac{T_m - T_1}{T_2 - T_1} = \frac{1}{2}\left[1 + \sqrt{1 + \left(\frac{\pi W C_p}{MHU}\right)^2}\right] = 2.$$

If the coolant is replaced by another fluid with the same $T_1$ but with a very large heat capacity, what is the fractional increase in the maximum allowable power output with this alternate coolant? Assume that the mass flow rate $W$ and overall heat transfer coefficient $U$ will remain unchanged with the new coolant. Assume also that the axial and radial power distributions are not affected by the coolant change.

**13.5** The heat flux distribution $q(z)$ in a PWR coolant channel with core height $H$ is represented by $q(z) = q_0 \cos \pi z/H$. In an alternate design, the fissile loading is varied axially so that a uniform heat flux distribution is obtained with the same power output in the channel. Assume that the inlet coolant temperature $T_1$, overall heat transfer coefficient $U$ connecting the fuel centerline and bulk coolant temperatures, wetted perimeter $M$, and mass flow rate $W$ remain unchanged in the alternate design. (a) Obtain expressions for the coolant and fuel centerline temperature distributions for both designs. (b) Plot and compare the temperature distributions schematically, and discuss the advantages and disadvantages of the alternate design relative to the reference design.

**13.6** Combine each pair of the two-phase conservation equations (13.159) through (13.161), together with interface conditions including Eq. (13.162), and obtain the equations representing the conservation of mass, momentum, and energy for the two-phase mixture. Compare the equations with the single-phase conservation equations.

**13.7** Derive the overall heat transfer coefficient $U_c$ of Eq. (13.52) for a cylindrical fuel rod with a diametral gap and clad, together with coolant flow represented by a convective heat transfer coefficient.

**13.8** A common phenomenon in oxide fuel pellets is the formation of central voids caused by the migration of voids along the thermal gradient. When loaded into a fuel clad, $UO_2$ pellets have a uniformly distributed porosity of 5~10%. This porosity consists of small gas voids in the fuel matrix. In the hotter regions of the pellet, these voids migrate up the thermal gradient. In fact, the fuel sublimes from the hotter surface of the void and condenses on the cooler surface. The size of the central void depends upon fuel surface temperature, operating time, linear power density, and initial fuel porosity. (a) Derive an expression for the temperature rise in a hollow cylinder with uniform internal heat generation. (b) Assume that 50% of the porosity of a fuel pellet operating at LHGR = 35.1 kW · m$^{-1}$, corresponding to the AP600 peak power density, migrates to form a central void. Find the maximum fuel temperature before and after the void migration if the $UO_2$ thermal conductivity remains constant at $k = 2.9$ W · m$^{-1}$ K$^{-1}$. Other fuel data include (i) pellet diameter = 9.32 mm, (ii) pellet surface temperature = 890 K, (iii) maximum $UO_2$ density (zero porosity) = 10.95 Mg · m$^{-3}$, and (iv) initial porosity = 8%.

**13.9** An incompressible, Newtonian fluid is flowing in steady state in the annular region between two coaxial circular cylinders of radii $\kappa R$ and $R$, and length $L$. As in the Hagen–Poiseuille flow, assume that the axial fluid speed $v_z$ at distances far from the inlet and outlet can be considered a function only of radius $r$. Obtain expressions for $v_z(r)$, shear stress $\tau_{rz}(r)$, and mass flow rate $W$ through the coaxial cylinders.

**13.10** An incompressible, Newtonian fluid is flowing in steady state in a cylindrical channel of constant cross section. At the channel wall, axially uniform heat flux $q$ is established and the radial velocity profile is described by the Hagen–Poiseuille model. The axial heat conduction may be neglected. Determine the asymptotic radial temperature profile

$$\theta(r) = \frac{T(z,0) - T(z,r)}{T(z,0) - T_b(z)},$$

if a spatially uniform volumetric source $Q$ is added to the fluid volume, as a function of the ratio $\psi$ of the total rate at which heat is deposited in the fluid to the rate of heat generation through volumetric heating. What is the radial position of maximum fluid temperature if $\psi = 0.75$?

**13.11** An incompressible coolant flows axially along a cylindrical fuel element in a bare cylindrical reactor with a negligible neutron extrapolation distance. The power output is limited by the melting temperature $T_{mp}$ of the fuel element. For a smooth fuel surface with the overall heat transfer coefficient $U_s$, representing the heat transfer from the fuel centerline to the bulk coolant, this maximum centerline temperature occurs at a distance from the inlet equal to 2/3 of the total fuel element length. In an effort to increase the power output, it is proposed to increase the overall heat transfer coefficient to $U_r$ by roughening the surface over the entire length of the fuel element. The mass flow rate $W$ and inlet coolant temperature $T_1$ are to be held constant and assume $U_r/U_s = 3$. Obtain the maximum power level increase that can be achieved through this approach.

**13.12** Consider a diffuser in a circular pipe, where the cross-sectional areas upstream and downstream of the diffuser are $A_1$ and $A_2$, respectively. Assume the fluid is inviscid and incompressible, and neglect gravitational forces on the fluid. Starting from the equation of mechanical energy balance, derive Bernoulli's equation, and obtain the steady-state pressure differential across the diffuser with $A_1 < A_2$.

**13.13** In an experimental reactor design, the primary water flows downward in the shell side, while the feedwater flows upward inside the tubes in a counter-current once-through steam generator. In a simple thermal-hydraulic analysis, the heat transferred from the shell side into the tube side is represented by wetted perimeter $M$ and length $H$ of the tubes, effective heat transfer coefficient $U$ between the primary and secondary sides, average primary temperature $T_p$, and feedwater temperature $T_s(z)$ and enthalpy $h_s(z)$ at elevation $z$ of the tubes. Given flow rate $W$, heat capacity $C_p$, inlet temperature $T_{in}$, and outlet steam quality $x_e$ of the feedwater, obtain an expression for the elevation $z_0$ at the inception of boiling in the tubes and $T_s(z)$ for $z \le z_0$.

**13.14** Power production in a cylindrical reactor of core height $H$ is limited by the fuel surface temperature. Normally (Case A) the reactor is cooled by water flowing in a single-direction flow. It is possible, although mechanically complicated, to increase the power output by operating with split flow of coolant (Case B). The coolant is introduced at the axial midplane, where the heat flux is the highest, and splits to flow outward. The hot spot occurs at $z = H/6$ from the midplane in Case A. Assume that the fluid properties do not change with temperature and that the neutron flux distribution is not perturbed by the flow splitting in Case B. Compare the two cases having the same local coolant flow rate $W$ in a given coolant channel, the same inlet coolant temperature $T_1$, and the same allowable maximum surface temperature to determine the ratio of power outputs and coolant temperature rises for the two cases.

**13.15** In a thermal-hydraulic design study for a PWR core, it is suggested that the fuel centerline temperature distribution $T_c(z)$ be made uniform along the length of each fuel rod. Determine the axial power distribution $P(z)$ that would be required to yield such a temperature distribution for each rod.

**13.16** A radioisotope thermoelectric generator (RTG) design features a $^{238}PuO_2$ sphere that generates thermal energy at a rate of $S = 5.2\,MW \cdot m^{-3}$ with 5.50 MeV $\alpha$-particles and a half-life of 87.74 yr. Due to the short range of the $\alpha$ particles, the energy is deposited essentially at the decay site and hence $S$ is spatially uniform throughout the $^{238}PuO_2$ sphere. The surface of the sphere is covered with telluride thermoelectric elements, which should not be operated at temperatures above $T_s = 575$ K. The maximum temperature $T_m = 3,030$ K is chosen for the core to avoid melting of the $^{238}PuO_2$ fuel material with thermal conductivity $k = 4.0\,W \cdot m^{-1}\,K^{-1}$. The RTG is expected to have an energy conversion efficiency of 5%. Determine (a) the maximum allowable heat generation rate $P_{th}$, (b) maximum electrical power $P_e$, and (c) radius $R$ of the $^{238}PuO_2$ core.

**13.17** A flat-plate fuel element of thickness $a$ in the $x$-direction is cooled at $x = a$ by boiling fluid so that the surface temperature of the fuel element in contact with the fluid is maintained at $T_s$. At $x = 0$, the fuel element is insulated. The fuel plate is infinite in extent in the $y$- and $z$-directions. The volumetric heat generation rate is uniform in the $y$- and $z$-directions, but is distributed as $S(x) = Q \cosh \kappa x$ [kW $\cdot$ m$^{-3}$] in the $x$-direction. (a) Determine the heat generation rate $Q$ together with the temperature distribution $T(x)$ within the fuel plate so that the fuel temperature does not exceed $T_m$. (b) Obtain the power $P/A$ [kW $\cdot$ m$^{-2}$] produced per unit cross-sectional area of the plate.

# 14

# Power Coefficients of Reactivity

One of the key parameters affecting passive safety and inherent stability of nuclear reactors is the amount of reactivity change due to a power level variation, which is referred to as the *power coefficient of reactivity* (PCR). Since any power level variation in general involves changes in the temperature of core material, the PCR is related to, and in fact consists of, the temperature coefficients of reactivity. As temperature changes affect the core reactivity and the reactivity in turn determines the power and temperature variations, the PCR can also be considered a measure of the temperature feedback on reactivity. We discuss in this chapter how power and temperature variations affect neutron reaction rates and hence the reactivity, and study key parameters that control the PCR. Our discussion will be primarily focused on the behavior of LWRs, but due considerations will be given also for sodium-cooled fast spectrum reactors (SFRs).

We begin in Section 14.1 with a general discussion of physical mechanisms affecting the effective multiplication factor $k_{eff}$ of a reactor. To delineate various effects of a power level change on reactivity, definitions for several reactivity coefficients are introduced in Section 14.2. Section 14.3 presents a quick review of the two-group model of reactivity, developed in Section 7.3, which forms the basis for the physical representation of the temperature feedback on reactivity. A particular emphasis is placed on the effect of Doppler broadening of absorption resonances in the fuel material. Our analysis is based on a simple model of the core consisting of fuel and moderator regions. Section 14.4 discusses how reactivity coefficients depend on reactor physics parameters, e.g. fuel enrichment, soluble boron concentration in a PWR, void fraction in a BWR, and fuel burnup. Sections 14.5 and 14.6 discuss the reactivity coefficients of particular interest to SFRs and a quasi-static reactivity feedback model for SFR transient events, respectively.

*Nuclear Reactor Physics and Engineering*, Second Edition. John C. Lee.
© 2025 John Wiley & Sons, Inc. Published 2025 by John Wiley & Sons, Inc.
Companion website: www.wiley.com/go/LeeNRPE2

## 14.1 Physical Phenomena Affecting Core Reactivity

Although power level variations are directly felt in the temperatures of various materials making up the core, the resulting effects on the core reactivity are manifested primarily through changes in the physical densities and hence nuclear densities of the core materials, namely, fuel and moderator in our two-region unit-cell model. This is because the number density of a nuclide affects the macroscopic cross section and hence the reaction rates of the nuclide. In addition to these *indirect* effects, there are *direct* effects of temperature changes on reactivity, a primary example of which is Doppler broadening of resonances. Finally, temperature changes in general influence the rate of neutron leakage out of the core and hence the core reactivity. We examine the three effects in more detail.

(1) As the power level increases, there are in general increases $\delta T_F$ and $\delta T_M$ in the temperature of the fuel and moderator, respectively, which cause thermal expansion and hence decrease $\delta\rho_F$ and $\delta\rho_M$ in fuel and moderator densities. The density changes $\delta\rho_F$ and $\delta\rho_M$ immediately result in decreases in number densities $\delta N_F$ and $\delta N_M$ and hence in various cross sections $\delta\Sigma_F$ and $\delta\Sigma_M$ of the fuel and moderator, respectively. Although reactivity is affected by any one of the cross sections for both the fuel and the moderator, absorption cross sections often play a major role in determining the reactivity coefficients. We should also remember that the density changes $\delta\rho_F$ and $\delta\rho_M$ should not always be considered as the result of temperature variations $\delta T_F$ and $\delta T_M$. For example, in BWRs, the coolant/moderator density could change, due to variations in the vapor content, without any change in the temperature of the two-phase mixture of vapor and liquid water.

    In addition to the direct effects of thermal expansion on macroscopic cross sections, the neutron flux spectrum experiences a shift due to changes in neutron moderation, which is in turn due to cross section changes $\delta\Sigma_F$ and $\delta\Sigma_M$. This spectral change perturbs two-group microscopic cross sections and hence macroscopic cross sections indirectly. Both of these direct and indirect effects of thermal expansion should be accounted for in general, although the direct effect is obviously of primary importance.

(2) As the fuel material heats up due to a power level increase, the fuel nuclei gain thermal energy and move around faster. This broadens the resonance absorption cross section $\overline{\sigma}_a(v)$ for neutrons of speed $v$, averaged over the thermal motion of target nuclei, and increases neutron absorption in fuel resonances. This effect, known as *Doppler broadening* of resonances, causes $k_{eff}$ to decrease, as discussed in Section 9.5 and contributes toward a negative fuel temperature coefficient (FTC) of reactivity.

Another direct effect of temperature variations on reactivity is associated with $\delta T_M$. For neutrons in thermal equilibrium at moderator temperature $T_M$, the thermal neutron energy $E_0$ of Eq. (3.39), corresponding to the most probable neutron speed $v_0$, is rewritten in terms of $T_M$

$$E_0 = \frac{1}{2}mv_0^2 = kT_M, \tag{14.1}$$

where $k$ is the Boltzmann constant. Thus, changes in the moderator temperature perturb the thermal spectrum and thermal reaction rates. These direct temperature feedback effects on reactivity are distinct from and in addition to the thermal expansion effects discussed earlier.

(3) Macroscopic cross section changes due to power level variations also affect the diffusion length or migration area and could result in significant changes in the leakage rate of neutrons. This effect is important in small research reactors, where as much as 25% of neutrons produced in the fission process leak out of the core. Thus, in research reactors and very likely in submarine reactors, a primary contribution to the *moderator temperature coefficient* (MTC) of reactivity comes from changes in the leakage probability of neutrons. This leakage contribution to the PCR is also important in fast reactors, e.g. the SFR, where differential expansions of fuel rods or control rod drives relative to the sodium pool are significant. This is discussed further in Section 14.5.

Evaluation of the thermal-hydraulic feedback effects is necessary for the proper determination of flux and power distributions in any nuclear reactor core. In addition, the PCR is necessary for determining the *power defect of reactivity*, which represents the decrease in reactivity due to the power rise from the hot zero to hot full power (HFP) condition. This information in turn is essential for the determination of the excess reactivity required in any core design analysis. One important safety guideline for nuclear reactors is the stipulation in the General Design Criteria [NRC71] that the PCR be negative in the power operating range for inherent reactor protection.

In addition, reactivity feedback effects are used effectively in various control applications. One key example in this regard is the use of recirculation flow as an efficient control mechanism in BWRs through changes in the void fraction and core average coolant density. Another example is the use of negative MTCs in PWRs for stretching an operating cycle when an unplanned need arises for an extended plant operation. Such an extended cycle operation, however, could result in decreased plant efficiency and somewhat degraded fuel configurations for the subsequent fuel cycle.

## 14.2 Relationship Between Reactivity Coefficients

Changes in reactivity due to power level variations are the result of several related but distinct phenomena and cannot in general be separated. For convenience, however, the PCR is often broken down into separate effects, e.g. contributions from variations in fuel temperature and moderator or coolant temperature. Recognizing that the reactivity change $\Delta\rho = \Delta k/k = \Delta \ln k$, we may write the PCR $\alpha_p$ as

$$\alpha_p = \frac{\partial \ln k}{\partial p} \simeq \frac{\partial \ln k}{\partial T_F}\frac{\partial T_F}{\partial p} + \frac{\partial \ln k}{\partial T_M}\frac{\partial T_M}{\partial p} + \frac{\partial \ln k}{\partial \rho_M}\frac{\partial \rho_M}{\partial p}, \tag{14.2}$$

where we note that in general, the moderator density could change independently of the moderator temperature. In PWRs, the direct moderator temperature feedback term $\partial \ln k/\partial T_M$ is usually an order of magnitude smaller than the density feedback term $\partial \ln k/\partial \rho_M$ arising from the temperature change. Hence, Eq. (14.2) is usually written in terms of the *FTC* $\alpha_F$ and MTC $\alpha_M$:

$$\begin{aligned}\alpha_p &= \frac{\partial \ln k}{\partial T_F}\frac{\partial T_F}{\partial p} + \left(\frac{\partial \ln k}{\partial T_M} + \frac{\partial \ln k}{\partial \rho_M}\frac{\partial \rho_M}{\partial T_M}\right)\frac{\partial T_M}{\partial p},\\ &\simeq \frac{\partial \ln k}{\partial T_F}\frac{\partial T_F}{\partial p} + \frac{\partial \ln k}{\partial \rho_M}\frac{\partial \rho_M}{\partial T_M}\frac{\partial T_M}{\partial p},\\ &= \alpha_F\frac{\partial T_F}{\partial p} + \alpha_M\frac{\partial T_M}{\partial p}.\end{aligned} \tag{14.3}$$

In Eq. (14.3), the FTC includes the effects due to Doppler broadening of absorption resonances and thermal expansion of fuel itself, while the MTC represents, in practice, the moderator density feedback effects.

In BWRs, the direct moderator temperature feedback term is again much smaller than the reactivity feedback due to vapor or void fraction changes. To emphasize the role of the void fraction change, the moderator density feedback term in Eq. (14.2) is rewritten for BWRs in terms of the *void coefficient of reactivity* (VCR) $\alpha_V$:

$$\begin{aligned}\alpha_p &\simeq \frac{\partial \ln k}{\partial T_F}\frac{\partial T_F}{\partial p} + \frac{\partial \ln k}{\partial \ln V_M}\frac{\partial \ln V_M}{\partial p}\\ &= \alpha_F\frac{\partial T_F}{\partial p} + \alpha_V\frac{\partial \ln V_M}{\partial p}.\end{aligned} \tag{14.4}$$

The FTC and MTC are usually expressed in units of [(%$\Delta k/k$)/°F] or equivalently [pcm $\cdot$ $K^{-1}$], with 1 pcm (per cent mille) = $10^{-5}$ $\Delta k/k$. The PCR and VCR are usually written in units of [(%$\Delta k/k$)/(% power)] and [(%$\Delta k/k$)/(% void)], respectively.

Determination of reactivity coefficients in power reactors often begins with a measurement of the *isothermal temperature coefficient* (ITC) of reactivity, defined as

$$\frac{\partial \ln k}{\partial T} = \frac{\partial \ln k}{\partial T_F} + \frac{\partial \ln k}{\partial T_M} = \alpha_F + \alpha_M. \tag{14.5}$$

The ITC is determined by measuring at zero power the reactivity perturbation due to a uniform temperature perturbation $\Delta T = \Delta T_F = \Delta T_M$. Since the FTC is usually much smaller than the MTC, we may determine the MTC from a measurement of the ITC, together with a design estimate of $\alpha_F$. The FTC may then be determined experimentally in a PWR, once the PCR is measured at power and the temperature derivatives $\partial T_M/\partial p$ and $\partial T_F/\partial p$ are estimated from a combination of calculations and measurements. At-power reactivity measurements are, however, often subject to considerable uncertainties because of the complex interplay among various feedback terms. The basis for all reactivity measurements in nuclear reactors, including power reactors, is usually the control rod worth calibration performed at zero power. In power reactors, however, the rod worth changes significantly as power increases, but these changes cannot be directly measured at power.

Another term that needs introduction is the *power defect of reactivity*, which is defined as the total decrease in reactivity from *hot zero power* (HZP) to *HFP*. Since the PCR is usually a function of power level itself, the power defect should be considered an integral of PCR over power.

## 14.3 Two-Group Representation of Reactivity Feedback

To gain a clear physical understanding of the reactivity feedback effects, we return to the two-group model for the infinite multiplication factor $k_\infty$ developed in Eqs. (7.27) and (7.39)

$$k_\infty = \frac{\nu\Sigma_{f1}}{\Sigma_{a1} + \Sigma_r} + \frac{\nu\Sigma_{f2}}{\Sigma_{a2}}\frac{\Sigma_r}{\Sigma_{a1} + \Sigma_r} = k_1 + k_2 = k_1 + pf\eta, \tag{14.6}$$

where $k_1$ and $k_2$ represent the contributions from fast and thermal fissions, respectively. The thermal fission contribution $k_2$ is further broken down into the *resonance escape probability* $p$ of Eq. (7.36), *thermal utilization* $f$ of Eq. (7.37), and the *number* $\eta$ *of neutrons released per thermal neutron absorption in fuel* defined in Eq. (7.38). The resonance escape probability $p$ is rewritten in Eqs. (9.62) and (11.9) in terms of the *effective resonance integral* $I$

$$p = \exp\left[-\frac{N_F}{\xi\Sigma_s} I\right] = \exp\left[-\frac{N_F}{\xi\Sigma_s}\int_0^u du\sigma_a(u)\phi(u)\right], \tag{14.7}$$

where we recall the interpretation of $I$ as the *flux-weighted effective absorption cross section*. With this interpretation, Figure 9.7 illustrates how an increase in fuel temperature $T_F$ (equivalent to $T$ in Figure 9.7) would reduce the spatial self-shielding of flux $\phi(u)$, due to Doppler broadening, and thus increase $I$ and decrease $p$.

In LWRs utilizing ceramic fuel in the form of $UO_2$, thermal expansion of fuel is relatively insignificant and all other parameters in Eq. (14.6) are essentially unperturbed. Similarly, changes in the nonleakage probabilities may be ignored in large LWR cores. Thus, a fuel temperature increase in LWRs usually results in a decrease in reactivity due to Doppler broadening of absorption resonances. In contrast, the fuel expansion effects could be significant in fast reactors with metallic fuel and should be explicitly accounted for.

The effects of moderator temperature changes in a PWR may first be represented in terms of the thermal utilization written explicitly for a fuel-moderator mixture

$$f = \frac{\Sigma_{a2}^F}{\Sigma_{a2}^F + \Sigma_{a2}^M}. \tag{14.8}$$

Suppose we experience a moderator temperature increase during a power maneuver or due to an accident. Due to this temperature increase, we expect a decrease in the moderator density and hence a decrease in the number density of water and hydrogen. This decrease in the water number density results in a decrease in the thermal absorption cross section of moderator $\Sigma_{a2}^M$, without much change in the thermal absorption cross section of fuel $\Sigma_{a2}^F$. Thus, a decrease in the neutron moderation, due to a decrease in the hydrogen-to-uranium (H/U) atomic ratio, results in an increase in thermal utilization $f$.

The other parameter that is affected by an increase in moderator temperature $T_M$ is the resonance escape probability $p$. Since the scattering cross section $\Sigma_s$ in Eq. (14.7) is mostly associated with moderator scattering, an increase in $T_M$ results in a decrease in $\Sigma_s$. Although due to spectral hardening, the resonance integral $I$ may decrease slightly, this effect is smaller than the decrease in $\Sigma_s$, with the result that $p$ itself decreases as $T_M$ increases.

Returning to Eq. (14.6), note that a change in $T_M$ hardly affects the parameters $k_1$ and $\eta$, and hence $k_{eff}$ or $k_\infty$ will decrease or increase as a result of competing changes in $p$ and $f$. These competing trends in $p$ and $f$ are similar to the effect of fuel lumping illustrated in Figure 11.4 and are sketched in the left-hand plot of Figure 14.1 as a function of the $(H/^{235}U)$ and (H/U) atomic ratios, the moderator-to-fuel number density ratio $(N_M/N_F)$, and equivalently in terms of moderator density $\rho_M$ and the inverse of moderator temperature $T_M$. Due to the competition between $p$ and $f$, for some value of the (H/U) atomic ratio or moderator density $\rho_M$, the effective multiplication factor $k_\infty$ will reach a maximum, as illustrated by a bell-shaped curve on the right-hand plot of Figure 14.1. The MTC

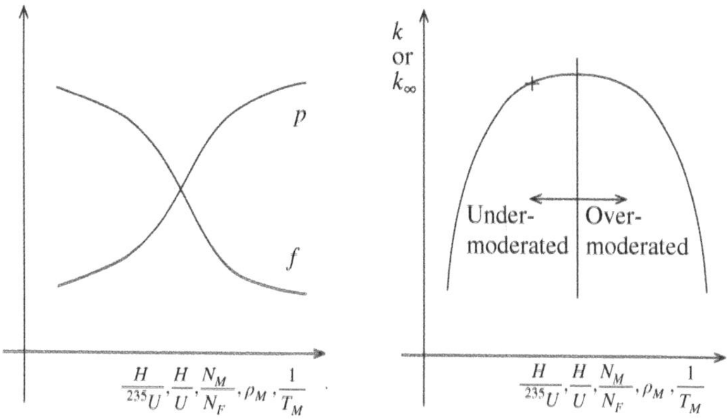


**Figure 14.1** Moderator temperature feedback effects on reactivity.

can be obtained as the slope of the $k_{eff}$ curve with respect to $T_M$. The left-hand half of the bell-shaped curve corresponds to an *under-moderated regime* so that any increase in $T_M$ or decrease in $\rho_M$ will result in sliding down the $k_{eff}$ curve, yielding a negative MTC. Furthermore, $\alpha_M$ itself becomes more negative as $T_M$ increases, since this corresponds to evaluating the slope further down the $k_{eff}$ curve.

LWR designs in the United States have always been chosen in the under-moderated regime marked by a plus sign to guarantee a negative $\alpha_M$ or $\alpha_V$. This key inherent safety feature was apparently violated in the ill-fated Chernobyl design, where a positive value of $\alpha_V$ was possible at low power with a small number of control rods inserted, and that is where the 1986 accident was initiated. The bell-shaped curve in Figure 14.1 is a succinct way of visualizing the moderator temperature feedback effects in LWRs and will prove to be quite useful in the next section.

## 14.4 Parametric Dependence of LWR Reactivity Coefficients

Having discussed how power level variations affect reactivity, we now examine how reactivity coefficients are influenced by key reactor physics parameters, e.g. fissile enrichment, soluble boron concentration, lumped neutron poison, and fuel burnup.

(1) The fissile enrichment of fuel has a direct effect on the neutron moderation and flux spectrum. In terms of the moderator temperature feedback effects illustrated in Figure 14.1, an increase in fissile enrichment is equivalent to decreasing the $(H/^{235}U)$ atomic ratio and hence making the system more

under-moderated and the flux spectrum harder. This means that the slope of the $k_{eff}$ curve becomes more negative, yielding a larger magnitude of the negative MTC. It should be noted that 0.1 wt% of $^{235}$U corresponds to approximately 1.0%$\Delta k/k$ of reactivity, worth about a month or so of full power operation in current LWR designs as discussed in Section 12.4.2.

(2) The concentration of $^{10}$B dissolved in coolant water as a chemical shim in PWRs influences the MTC through its effect primarily on thermal utilization $f$. Suppose an increase in moderator temperature takes place during a power maneuver. This results in a decrease in the moderator absorption cross section $\Sigma_{a2}^{M}$ due to a decrease in water number density $N_M$, as discussed in Section 14.3. When $^{10}$B atoms, with a large thermal absorption cross section, are homogeneously dissolved in water, $\Sigma_{a2}^{M}$ decreases further as the $^{10}$B number density decreases together with the water number density due to a moderator temperature rise. Hence, the increase in thermal utilization $f$, associated with an increase in $T_M$, is larger with $^{10}$B dissolved in water, and the MTC will become less negative. With a large $^{10}$B concentration, it is even possible to have a positive MTC, which is equivalent to operating the reactor in an over-moderated regime.

(3) The presence of lumped neutron absorbers, e.g. lumped burnable absorber (BA) rods or control rods, in LWRs affects the behavior of MTC in a manner distinct from that of soluble neutron absorbers. To study the effects of lumped absorbers, the definition of thermal utilization in Eq. (14.8) is extended to include the contribution $\Sigma_{a2}^{P}$ from the lumped absorbers to the thermal absorption cross section

$$f = \frac{\Sigma_{a2}^{F}}{\Sigma_{a2}^{F} + \Sigma_{a2}^{M} + \Sigma_{a2}^{P}}. \tag{14.9}$$

For an increase in moderator temperature $T_M$, $\Sigma_{a2}^{M}$ will decrease as usual. Due to the presence of $\Sigma_{a2}^{P}$, however, the soluble boron concentration is decreased in typical PWR designs and $\Sigma_{a2}^{M}$ itself is reduced, thereby decreasing the effect of any $T_M$ increase on MTC discussed in the preceding paragraph.

At the same time, the thermal diffusion length $L_2$, defined in Eq. (7.28), increases for the fuel and moderator regions, separated from the lumped absorbers, due to a reduction in the absorption cross section, $\Sigma_{a2}^{F} + \Sigma_{a2}^{M}$. Since $L_2$ is proportional to the distance that thermal neutrons travel between collisions on average, an increase in $L_2$ has the effect of increasing the likelihood that thermal neutrons encounter lumped absorbers during the migration. Thus, an increase in $T_M$ will increase *effectively* the parasitic absorption term $\Sigma_{a2}^{P}$ in Eq. (14.9), countering the decrease in $\Sigma_{a2}^{M}$. This is the second reason why the MTC of a PWR becomes more negative as lumped neutron absorbers are added. Because Eq. (14.9) does not fully reflect a heterogeneous lattice consisting of fuel, moderator, and parasitic absorbers, it cannot be used directly to explain why lumped absorbers act to cancel the increase in $f$ due to an increase in $T_M$; instead we have explained this effect in terms of an effective increase in $\Sigma_{a2}^P$.

(4) The VCR $\alpha_V$ for BWRs represents the coolant density feedback effect, as discussed in connection with Eq. (14.4). Hence, we may use the $k_{eff}$ curve of Figure 14.1 again to illustrate the dependence of VCR $\alpha_V$ on void fraction itself. As the void fraction increases, moderator density $\rho_M$ decreases, thus yielding a negative slope of the $k_{eff}$ curve and a negative value of $\alpha_V$ in an under-moderated regime. Furthermore, around a higher void fraction, i.e. further down the curve, the slope will become steeper, resulting in an increase in the magnitude of the negative $\alpha_V$ as a function of void fraction itself. Similar to the effect of lumped absorbers on the MTC of PWR cores, the insertion of cruciform control blades enhances the magnitude of the negative VCR in BWR cores.

(5) The fuel depletion in a reactor core also influences reactivity coefficients in a complex manner. In general, the evolution in fuel isotopics, especially the production of plutonium nuclides with low-lying resonances, could have a significant impact on reactivity coefficients. In LWRs, however, the primary fuel depletion effects on reactivity coefficients are those associated with control poisons. We illustrate the burnup dependence of $\alpha_M$ for PWRs and $\alpha_V$ for BWRs in Figure 14.2. Since PWRs operate with control rods essentially fully withdrawn, the MTC is influenced primarily by the soluble boron concentration, which decreases as the fuel burnup increases and excess reactivity decreases, and hence the MTC itself becomes more negative as a function of fuel burnup.

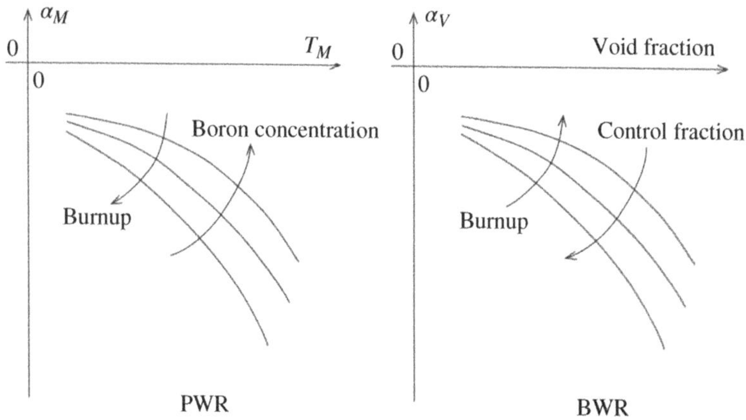


**Figure 14.2** Burnup dependence of reactivity coefficients in LWRs.

The situation is reversed for BWRs, because a BWR core typically operates with approximately 25% full-length equivalent of control blades inserted into the core at the beginning of cycle (BOC) and hence has the largest magnitude of the negative void coefficient at BOC. As fuel depletes and the excess reactivity decreases, the control blades are gradually withdrawn, making the void coefficient less negative.

## 14.5 Reactivity Coefficients in Sodium-Cooled Fast Reactor

In SFRs typically fueled with $^{239}$Pu and other fissile nuclides, and cooled with liquid sodium, coolant voiding could increase reactivity. This is primarily due to a peculiar cross section behavior of fissile nuclides around 100 ~ 200 keV, which is typically the mean energy of neutrons for these reactors. Around this energy, the capture-to-fission ratio $\alpha = \sigma_c/\sigma_f$ decreases as neutron energy increases, as illustrated in the ENDF/B-VIII plot [Bro18] of Figure 14.3 for $^{239}$Pu.

Thus, if sodium voiding were to take place and harden the flux spectrum, the parameter $\eta = \nu/(1+\alpha)$ and hence $k_\infty$ would increase. This tendency for a positive VCR is partly mitigated by an increase in diffusion constant $D$, which would increase neutron leakage. The net effect of sodium voiding is determined primarily by these two competing phenomena so that in most viable SFR designs, the sodium void coefficient is positive near the core center where the leakage effect is small but tends to become negative as the periphery of the core and the blanket regions

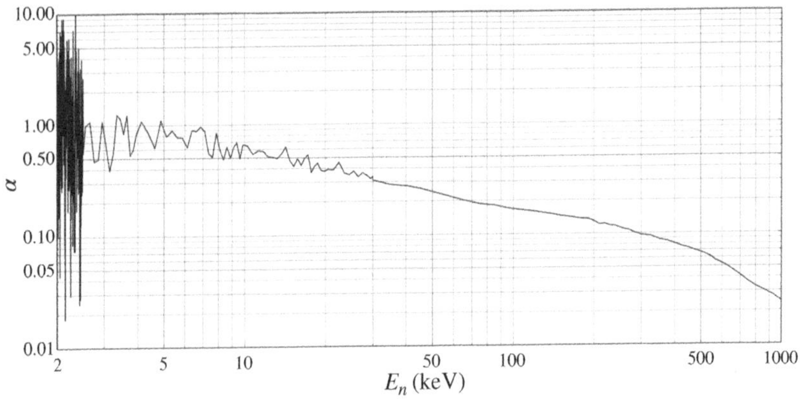


**Figure 14.3** Capture-to-fission cross section ratio for $^{239}$Pu. *Source*: Brown et al. [Bro18]/U.S. Department of Energy/Public domain.

are voided. Although the net PCR invariably would be sufficiently large negative in SFRs, this behavior of the sodium void coefficient has long been a concern in the development of this type of reactor.

We should of course recognize that the actual reactivity effects of sodium voiding in the SFR would involve a number of other phenomena including those associated with resonance absorptions, besides the two primary effects discussed here. In addition, in a pool-type SFR with metallic fuel, thermal expansion of structural components associated with the temperature increase of sodium in the pool could contribute significantly to the reactivity feedback. This is illustrated in the time-dependent feedback effects [Wig90] in Figure 14.4 for a metallic fuel SFR subject to a unprotected loss of flow (ULOF) event.

Both the axial expansion of metallic fuel and radial core expansion, together with the differential expansion of control rods into the core, generate substantial negative reactivity feedback. The control rod expansion effect is due to the thermal expansion of the control rod drive line resulting from the elevated sodium pool temperature. The rod line expansion effectively increases the insertion of control rod absorber into the core, thereby contributing a significant negative feedback. These structural and geometrical feedback effects, together with the leakage effects contributing to the sodium void coefficients, suggest that the SFR designs entailing large core sizes may require additional considerations to fully benefit from passive safety associated with negative feedback coefficients.

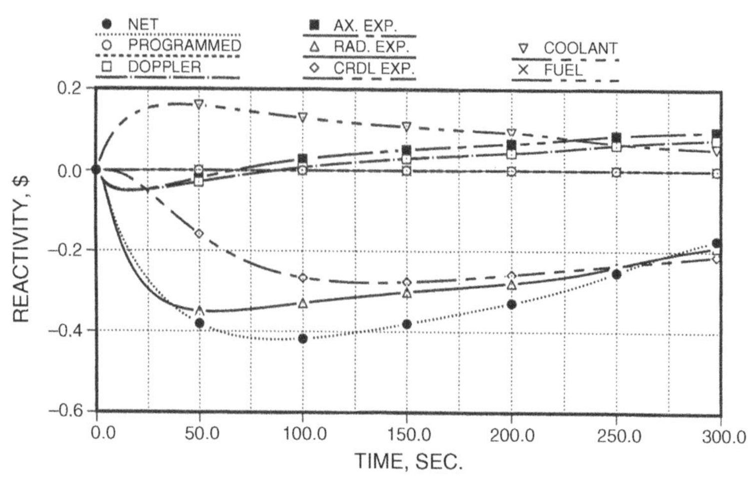


**Figure 14.4** Reactivity feedback for a ULOF transient in a metallic fuel SFR. *Source:* Wigeland et al. [Wig90]/U.S. Department of Energy/Public domain.

## 14.6 Reactivity Feedback Model for Sodium-Cooled Fast Reactor

For metal-fueled pool-type SFRs with passive safety characteristics, the net reactivity variation for the bulk of transient events of interest is small throughout the transient. This allows us to combine [Ott88,Pla87,Wad88] the temperature and flow feedback effects and formulate a convenient model for reactivity perturbation $\delta K(t)$ expressed in terms of relative power level $P(t)$ and relative flow rate $F(t)$

$$\delta K(t) = A[P(t) - 1] + B\left[\frac{P(t)}{F(t)} - 1\right] + C\delta T_{c,in}(t) \simeq 0, \tag{14.10}$$

where

$A$ = fuel temperature coefficient of reactivity,
$B$ = flow coefficient of reactivity, and
$C$ = inlet temperature coefficient of reactivity.

The sum of $A$ and $B$ is essentially the PCR representing power changes affecting both the fuel and coolant temperature distributions. For the EBR-II, the feedback coefficients $A$, $B$, and $C$ are all negative. Equation (14.10) represents the quasi-static formulation of the core reactivity behavior with $\delta K \simeq 0.0$.

For the EBR-II loss of flow without scram (LOFWS) test performed in 1986 [Pla87], the inlet coolant temperature $T_{c,in}$ remains nearly constant during the transient, and the associated feedback effect may be neglected. Equation (14.10) may then be solved for the time-dependent power-to-flow ratio

$$\frac{P(t)}{F(t)} = \frac{1 + A/B}{1 + (A/B)F(t)}, \tag{14.11}$$

while the change in the coolant temperature rise is given by

$$\delta[\Delta T_c(t)] = \Delta T_c(0)\left[\frac{P(t)}{F(t)} - 1\right], \tag{14.12}$$

in terms of the core-average coolant temperature rise $\Delta T_c(0)$ at rated power and flow. An asymptotic value of the power-to-flow ratio can be obtained approximately in the limit as $F(t)$ becomes vanishingly small:

$$\lim_{t\to\infty}\frac{P(t)}{F(t)} = 1 + \frac{A}{B} > 1.0. \tag{14.13}$$

Combining Eqs. (14.12) and (14.13) yields the change in the coolant outlet temperature

$$\delta T_{c,out}(\infty) = \frac{A}{B}\Delta T_c(0). \tag{14.14}$$

For the loss of heat sink without scram (LOHSWS) event [Pla87], the heat sink decreases, i.e. $P(\infty) = 0$, but the primary coolant flow remains nearly constant,

**Table 14.1** Representative feedback coefficients and temperature rises.

| Fuel type | $A$ (\$) | $B$ (\$) | $C$ (\$ · K$^{-1}$) | $\delta T_{c,out}(\infty)$ (K) LOFWS | $\delta T_{c,in}(\infty)$ (K) LOHSWS |
|---|---|---|---|---|---|
| Metal | −0.15 | −0.30 | −0.003 | 70 | 150 |
| Oxide | −1.70 | −0.40 | −0.004 | 595 | 525 |

allowing us to set $F(t) = 1$ in Eq. (14.10), which provides a simple expression for the asymptotic inlet coolant temperature rise:

$$\delta T_{c,in}(\infty) = \frac{A+B}{C} = \frac{\text{power coefficient of reactivity}}{\text{inlet temperature coefficient of reactivity}}. \tag{14.15}$$

In fact, in this transient, the coolant temperature rise $\Delta T_c$ decreases as the heat sink is lost and the inlet coolant temperature rise decreases the reactivity, thereby rendering the core to a high-temperature, low-power state.

With typical values [Ott88] estimated for the reactivity feedback coefficients $A$, $B$, and $C$ and $\Delta T_c(0) = 140$ K for the EBR-II system, Table 14.1 lists the asymptotic coolant temperature increases of Eqs. (14.14) and (14.15). The quasi-static formulations provide simple but valuable comparisons of the expected coolant temperature rises for the metal- and oxide-fueled SFR configurations. The temperature increases for both the postulated LOFWS and LOHSWS are much smaller for the metal-fueled core than those for the oxide-fueled core, indicating a greater potential for passive safety and relatively mild transients expected in metal-fueled pool-type SFRs.

As power level variations affect the reactivity of a reactor through thermal-hydraulic feedback, so do the reactivity coefficients affect transient behavior of the reactor. This was evident in the Chernobyl accident discussed earlier. Reactivity coefficients also play a key role in passive safety characteristics of nuclear power plants [Gol87,Pla87]. To illustrate the point, we discuss two types of transient events for metal-fueled SFR designs, which call for self-shutdown capability of the reactor even in the case of a primary sodium pump failure coupled with a scram failure.

In under-cooling events, exemplified by a ULOF or LOFWS event, the net reactivity remains vanishingly small during the transient represented by the quasi-static model of Eq. (14.10). Furthermore, the power transient primarily raises the fuel temperature, while the sodium coolant temperature is determined largely by the flow coastdown rate. This allows us to represent the reactivity balance in terms of a PCR $\alpha_p$ decoupled from a coolant coefficient of reactivity $\alpha_c$:

$$\delta K = \frac{\partial \ln k}{\partial T_F}\frac{\partial T_F}{\partial p}\delta P + \frac{\partial \ln k}{\partial T_c}\delta T_c = \alpha_p \delta P + \alpha_c \delta T_c \simeq 0. \tag{14.16}$$

Since both $\alpha_p$ and $\alpha_c$ are negative, an under-cooling event with $\delta T_c > 0$ can be terminated at a low power level corresponding to $\delta P < 0$, even in the case of a scram failure. To minimize the terminal power level, i.e. to have the largest possible reduction in power, however, we desire to make the power coefficient $\alpha_p$ as small negative as feasible.

This objective to reduce the magnitude of the negative PCR is rather contrary to the general concept behind inherent safety of nuclear reactors. In fact, consider a reactivity-induced transient initiated by the insertion of positive reactivity $\delta K_{ex}$ and again use a quasi-static approximation to obtain a reactivity balance:

$$\delta K = \delta K_{ex} + \alpha_p \delta P \simeq 0. \quad (14.17)$$

To minimize the power increase $\delta P$, it is necessary to maximize the magnitude of the negative power coefficient $\alpha_p$. This simple example illustrates rather succinctly that passive safety of nuclear power plants requires a careful balance between a number of conflicting objectives. This is merely one of the many challenges that lie ahead for nuclear engineers in the further development of Generation IV nuclear energy systems. Some tradeoff studies suggested by Eqs. (14.16) and (14.17) for SFR design have been presented in [Wad88].

## References

**[Bro18]** D.A. Brown, M.B. Chadwick, R. Capote et al., "ENDF/B-VIII.0: The 8th Major Release of the Nuclear Reaction Data Library with CIELO-project Cross Section, New Standards and Thermal Scattering Data," *Nucl. Data Sheets* **148**, 1 (2018).

**[Gol87]** G.H. Golden, H.P. Planchon, J.I. Sackett, and R.M. Singer, "Evolution of Thermal-Hydraulics Testing in EBR-II," *Nucl. Eng. Des.* **101**, 3 (1987).

**[NRC71]** US Nuclear Regulatory Commission, "General Design Criteria for Nuclear Power Plants," Title 10, *Code of Federal Regulations*, Part 50, Appendix A, US Nuclear Regulatory Commission (1971).

**[Ott88]** K.O. Ott, "Inherent Shutdown Capabilities of Metal-Fueled Liquid-Metal-Cooled Reactors During Unscrammed Loss-of-Flow and Loss-of-Heat Sink Incidents," *Nucl. Sci. Eng.* **99**, 13 (1988).

**[Pla87]** H.P. Planchon, J.I. Sackett, G.H. Golden, and R.H. Sevy, "Implications of the EBR-II Inherent Safety Demonstration Test," *Nucl. Eng. Des.* **101**, 75 (1987).

**[Wad88]** D.C. Wade and Y.I. Chang, "The Integral Fast Reactor Concept: Physics of Operation and Safety," *Nucl. Sci. Eng.* **100**, 507 (1988).

**[Wig90]** R.A. Wigeland, R.B. Turski, and R.K. Lo, "Relative Performance of Metallic Fuel and Oxide Fuel during Unprotected Accident Conditions in a 3500 MWth Pool-Type LMR," ANL-IFR-136, Argonne National Laboratory (1990).

## Problems

**14.1** A critical PWR configuration may be described by the following two-group model: (i) thermal utilization $f = 0.76$, resonance escape probability $p = 0.80$, number of neutrons released per thermal absorption in fuel $\eta = 1.25$, (ii) fast-fission contribution to the infinite multiplication factor $k_1 = 0.27$, (iii) nonleakage probability $P_{NL} = 1/1.03$, (iv) thermal neutron absorption in non-fuel material is evenly divided between moderator and lumped burnable absorbers (BAs), i.e. $\Sigma_{a2}^{P} = \Sigma_{a2}^{M}$ and the BA absorption cross section is independent of moderator density, and (v) scattering cross section $\Sigma_s$ for resonance neutrons can be assumed entirely due to that of water. If the water density is reduced by 3% due to a moderator temperature increase of 7 K, calculate the MTC for the reactor. Assume that changes in the microscopic cross sections may be ignored.

**14.2** In the two-group PWR model considered in Problem 14.1, obtain the maximum concentration of soluble boron that can be added to the coolant water without making the MTC of reactivity positive. Each addition of 100 ppm by weight of natural boron in water increases thermal absorption cross section $\Sigma_{a2}^{M}$ of water by 10%, which is to be compensated for by a 10% reduction in thermal absorption cross section $\Sigma_{a2}^{P}$ of lumped BAs to retain criticality at rated condition. Assume that boron is a pure thermal absorber, and consider a 3% decrease in the density of water due to a temperature increase of 7 K as in Problem 14.1.

**14.3** The two-group PWR parameters of Problem 14.1 may be used to represent a critical RBMK-1000 core loaded with low enrichment uranium (LEU) fuel assemblies, provided we set the control absorber cross section $\Sigma_{a2}^{P} = 0$ while keeping thermal utilization $f = 0.76$ and resonance escape probability $p = 0.80$. (a) If the water density is reduced by 3%, resulting in a 4% increase in the void fraction of the core, calculate the VCR $\alpha_V$ for the BWR core. (b) An alternate fuel design suggests increasing the fuel enrichment by 20% and adding burnable absorbers to restore criticality. For a preliminary scoping study, assume that the alternate design yields $\eta = 1.285$, $\Sigma_{a2}^{P}/\Sigma_{a2}^{M} = 0.2$, and $\Sigma_{a2}^{F}/\Sigma_{a2}^{M} = 3.403$. Repeat the $\alpha_V$ calculation for the alternate design proposed. Discuss the safety implications of the proposed design changes.

**14.4** Using the quasi-static reactivity feedback model for a LOFWS event in a SFR with metallic fuel with the observation that the sodium inlet coolant temperature rise $\delta T_{c,in}$ is negligible, (a) show that the power level $P(t)$ will follow the flow coastdown rate $F(t)$ but with a time lag. (b) With the feedback coefficients from Table 14.1 and $F(t) = \exp(-t/\tau)$, subject to time constant $\tau = 30$ s, determine the power level expected at $t = 90$ s.

**14.5** A molten salt reactor design features Li coolant flowing axially with $T_{in}$ = 600 K and $T_{out}$ = 900 K, providing a symmetric axial heat flux distribution $q(z) = q(0) \cos \pi z/H$. The Li coolant offers a volumetric expansion coefficient $\beta = -2.0 \times 10^{-4}\ \text{K}^{-1}$ and absorbs 10% of fission neutrons produced at nominal operating power. (a) Treating Li as a $1/v$-absorber and maintaining constant $T_{in}$ and mass flow rate $W$ during power maneuvers, obtain the coolant temperature coefficient of reactivity as a function of average coolant temperature $\langle T_b \rangle$. You may use the dimensionless temperature profile $\theta_b(z)$ of Eq. (13.90). (b) Determine the power coefficient of reactivity $\alpha_p$ associated with Li heating.

# 15

# Nuclear Energy Economics

As we envision further development of nuclear energy in the United States and across the world with the recent startup and construction of several AP1000, EPR, and APR-1400 plants, together with active developments of scores of advanced modular systems discussed in Chapter 1, there still remain a few issues that the nuclear community needs to address. They include the long-term disposal of irradiated nuclear fuel, proliferation risks associated with the reprocessing and recycling of used fuel, safety of a significantly larger fleet of nuclear power plants deployed, and economics of nuclear energy generation. These are the key issues certainly considered in the Generation IV Roadmap [DOE02] and the subsequent effort in the Nuclear Energy Research and Development Roadmap [DOE10]. The issues were also discussed in the reports of the two U.S. National Academy of Sciences committee published in 2023, with the first report [NAS23] focusing on the management and disposal of used nuclear fuel (UNF), while the second report emphasized the safety and economics of advanced reactor systems.

The Russian invasion of Ukraine in 2022 with the war continuing into the third year has raised significant questions for every nation of the world regarding the feasibility or the wisdom of relying on fossil fuel supplies from a few countries or regions of the world. Together with the need to reduce continued use of fossil fuel for the reduction of carbon emission for the world, there has emerged rational recognition for a multitude of nations to develop nuclear energy as soon and practical as possible. This is exemplified by the 2023 UN Framework Convention on Climate Change with the Conference of the Parties (COP) 28 that featured 24 nations pledging to triple nuclear plant generating capacity by 2050. Such an ambitious deployment of new NPPs depends ultimately on the economics of the Generation III+ and eventually the Generation IV power plant designs that are under active development. With this perspective in mind, we discuss in this chapter the economics of nuclear energy generation.

Our focus will be on the overall cost of electricity generation from NPPs including the cost associated with NPP construction, nuclear fuel cycle, and nuclear

*Nuclear Reactor Physics and Engineering*, Second Edition. John C. Lee.
© 2025 John Wiley & Sons, Inc. Published 2025 by John Wiley & Sons, Inc.
Companion website: www.wiley.com/go/LeeNRPE2

plant operation. We will review the basic features of engineering economics, limited to the extent necessary for determining the capital cost and fuel cycle components of nuclear electricity cost. We will then extend our discussion to compare the economics of nuclear electricity generation with those for other sources of electricity generation including the solar photovoltaics (PVs) and wind power generation. Section 15.1 begins with an overview of the electricity cost including a rough breakdown between the generation and transmission and distribution (T&D) costs and various components that should be considered for NPP economics. A brief overview of engineering economics is presented in Section 15.2, limited primarily to the concept of present worth and amortization of capital investment. We present in Section 15.3 the actual methodology and examples of nuclear electricity cost calculation, including the NPP capital cost, fuel cost, operation and maintenance expenses, and decommissioning expenses, finally providing the *levelized cost of electricity* (LCOE). This is followed by a review of issues associated with the increased capital and operating costs for nuclear power plants, together with perspectives on the economics of small modular reactor (SMR) systems in Section 15.4. Section 15.5 provides, for various energy sources, a comparison of (a) the electricity generation costs, accounting for actual availability of the generation systems, and (b) output-to-input energy ratios.

## 15.1 Electrical Energy Cost

The overall cost of electrical energy generated from any type of power plants is broken down into three components:

(1) Generation cost determined at the busbar connecting to the T&D system, and hence often called the *busbar cost*,
(2) Transmission cost, associated with the high-voltage transmission from the power plant to electric substations, and
(3) Distribution cost covering the distribution of electricity from substations to individual customers, commercial, and residential.

The demarcation between the transmission and distribution (T&D) systems is at the termination point of the transmission system, typically at 4.8 ~ 120 kV in the United States. For DTE circuits serving approximately 2.2 million customers with a generation capacity of 11 GWe, the demarcation is set at 13.2 kV. The transmission and distribution costs account for approximately 20% and 30%, respectively, of the total electricity cost, with the total T&D cost typically in the range of 9 ~ 11 cent/kWh, 90 ~ 110 mill/kWh, or equivalently $90 ~ 110/MWh of electricity. The remaining 45 ~ 50% of the electricity cost is generally attributed to the generation or busbar cost of 9 ~ 10 cent/kWh, with a total charge of

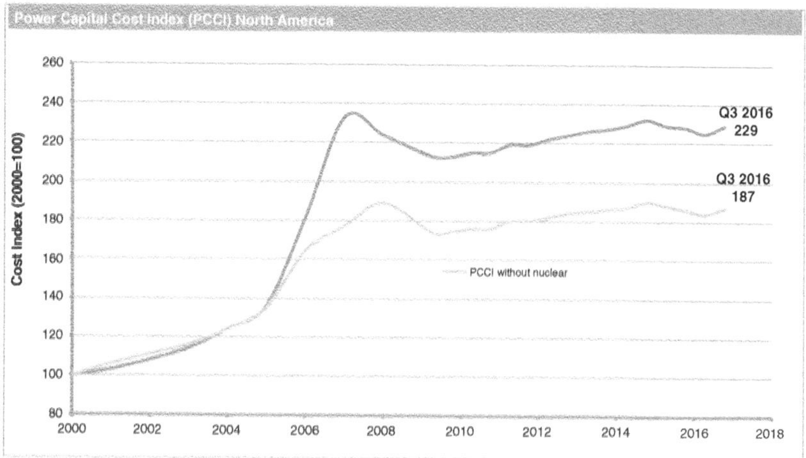


**Figure 15.1** Power capital cost index for North America, 2000–2016. *Source*: IHS website [IHS17].

18 ~ 21 cent/kWh for DTE residential customers in Michigan in 2024. State sales tax is also included as part of the T&D charges. The overall electricity charge for residential customers varies significantly from state to state in the United States.

The busbar cost for the reference Generation IV plant was estimated as \$34/MWh in the Generation IV Roadmap [DOE02] in 2002. This was based primarily on the projected economic data of the Westinghouse AP1000 design for the *N*th-of-a-kind (NOAK) power plants, not the first-of-a-kind (FOAK) plants, to be constructed. This estimate turned out to be quite optimistic, in light of large increases in the power capital costs index (PCCI) roughly over the past two decades [IHS17], especially during the boom years preceding the 2007 recession, as illustrated in Figure 15.1. Indeed, a PCCI increase of 129% is observed when nuclear plants are included, while all other construction costs excluding those of nuclear plants indicate a 87% increase.

The Generation IV estimate of \$34/MWh for the generation cost consists of three sub-costs:

(1) Capital cost of \$19/MWh, comprising the overnight construction cost (OCC) and interest during construction (IDC), levelized over the 40-year life of the plant, plus the plant decommissioning cost,
(2) Fuel cost of \$6/MWh, including the entire incore and excore fuel costs but excluding the plant decommissioning cost, and
(3) Operation and maintenance (O&M) cost of \$9/MWh.

It should be noted that the NPP capital cost is higher, even in this highly optimistic estimate, than that for either coal- or natural gas-fired power plants, but the NPP fuel cost is significantly lower. The O&M cost has also steadily increased over the past 20 yr or so largely due to additional security measures required, resulting in an estimate [Fes17] of $16/MWh in 2017. The large increases in the O&M cost have occurred despite high capacity factors (CFs) maintained at NPPs over the past three decades and has been a key factor in recent decisions to shut down and decommission several NPPs in deregulated merchant-fleet markets in the Unites States. We will discuss in detail in Section 15.3 how various generation costs may be estimated from the prevailing cost data.

Before moving into a brief review of engineering economics in Section 15.2, we discuss here the expenditures that should be included for determining the capital cost and the carrying charges for levelized cost calculations. The capital cost for a power plant consists of (a) the overnight capital cost (OCC) incurred if the plant were to be conceptually constructed instantaneously without any time delay and (b) *finance charges* or *IDC* that would accrue during the actual construction period. The OCC is typically broken down into two components:

(1) *Direct cost* including expenditures associated with construction labor, material, equipment, and land, and
(2) *Indirect cost* covering expenditures for contracting, design, engineering, inspections during construction, and working capital.

The *carrying charges*, also often called *financing charges*, expressed in terms of *discount rates*, include (a) the cost of the capital or money borrowed, (b) federal, state, and local taxes applicable, depending on the ownership of the plant, and (c) insurance premiums for the equipment and materials. The cost of the money borrowed depends on how the capital is raised. Traditionally, during the construction of the bulk of the NPPs currently in operation, the capital was divided approximately between 60% bonds and 40% stocks. At this point in 2024, it is, however, difficult to forecast how the capital will be raised for the construction of the NPPs currently on the drawing board. For the Generation IV Roadmap, a round figure of 10% per annum of carrying charges was adopted for the purpose of establishing a fiducial point to evaluate the merits of 100+ concepts submitted to the Generation IV roadmap team. It should be noted for our discussion here that the carrying charge rate of 10% per annum is considerably higher than the prevailing interest rate charged on commercial or residential loans.

The construction of the two AP1000 units each at the Vogtle Plant in Georgia and the V.C. Summer Plant in South Carolina began in 2013, with federal loan guarantees for the four Generation III+ plants in the United States. It should, of course, be noted that the loan guarantee program does not actually provide a loan but rather facilitates more favorable terms for the loan in return for a certain fee for the government. With significant construction delays and cost over-runs incurred for all four AP1000 units, two Summer units were eventually canceled and two

Vogtle units are now finally in commercial operation in 2024. Starting from the engineering procurement and construction (EPC) contracts of \$14B for the twin Vogtle units and \$10B for the twin Summer units, the best estimates for the construction projects have climbed to as high as \$30B for the twin-unit Vogtle project [Cho23]. The significant construction delays and cost over-runs resulted in the bankruptcy filing by the AP1000 supplier, Westinghouse Electric Company, and required restructuring of the company with Brookfield Business Partners.

Furthermore, as discussed in Section 1.5, the NuScale project that obtained the initial US Nuclear Regulatory Commission construction permit for the 12-module design, comprising 77-MWe units, has to cancel in early 2024 the Utah Associated Municipal Power Systems (UAMPS) consortium arrangement covering seven western states. The initial cost estimate of \$2.5B for the NOAK project, presented to the NAS committee in 2021, has turned out to be apparently an optimistic projection and the latest projection points to \$9.1B for a total plant output of 0.92 GWe [Cho23]. On a more positive note, however, it should be recognized that Korea Hydro and Nuclear Power recently constructed four 1.4-GWe APR-1400 units at the Barakah Plant providing 5.0-GWe to the United Arab Emirates within the initial budget of \$20B and a total construction period of 14 yr for the entire four units.

Thus, it is unclear how best to estimate the OCC of either a GWe-size plant or a SMR with the power output of 0.3 GWe or less, together with the effective carrying charges to apply for the purpose of estimating the total generation costs for any new nuclear plants. For the purpose of our engineering economics study, we have decided to follow the capital cost estimates of 2019 that the U.S. Energy Information Administration presented in 2020 [EIA20]. The 2019 cost data are also used in a number of energy economics studies, published in 2022 or 2023, comparing various electrical energy systems [Ide22], which allow us to present meaningful discussions on nuclear energy economics on a consistent basis albeit with the data not necessarily most up to date. The carrying charges are also selected so that our economic analysis may be compared with the recent studies.

## 15.2 Overview of Engineering Economics

The tasks at hand require the concept of present value of money accumulated over a period of time during which the interest charges on the money are compounded stepwise at regular intervals. If a sum of fund $S_n$ [\$] is accrued from an initial fund of $P$ [\$] at the $n$th compounding interval with an interest rate of $j$ per interval

$$S_n = P(1 + j)^n, \tag{15.1}$$

then we refer to $P$ as the *present value* or *present worth* of the sum $S_n$. The value at the plant construction time of fund $S_n$ needed for the next plant construction or

for a NOAK AP1000 plant. The OCC estimate in Table 15.1 should also be compared with a prediction of \$4.6 ~ 5.9 B/GWe for a 2.2-GWe advanced LWR plant in [Rot15] and recent tabulations covering 2020 estimates ranging over \$2.5 ~ 6.9 B/GWe [Rot22] and \$2.0 ~ 8.6 B/GWe [Ste22]. The lowest estimate of \$2.0 B/GWe reflects the recent successful AP1000 construction experience in China with low labor charges and the highest estimate of \$8.6 B/GWe corresponds to the difficult construction experience of the AP1000 system at the Vogtle Unit 3 Plant. A more recent estimate of the total construction cost for the twin Vogtle units connected finally to the grid in 2024, however, could be as high as \$30B [Cho23].

With the OCC estimate $C_{p0}$ = \$6.32/GWe, we may now calculate the interest during the construction (IDC) assuming (a) construction time of 4 yr, (b) carrying charge rate of 6.7% $\text{yr}^{-1}$, and (c) financing period of 30 yr, following the recent comparative energy cost study by Idel [Ide22] with one variation; a 4-yr construction period is assumed instead of 2 yr in Idel's study. To calculate the IDC, a simple assumption is made that the capital required for the plant construction is borrowed evenly over a period of the 4-yr construction time so that carrying charges may be incurred over one half of the OCC:

$$\text{IDC} = \$6.32\,\text{B} \times (6.7\%\,\text{yr}^{-1} \times 4\,\text{yr} \times 0.5) = \$0.84\text{B}. \tag{15.6}$$

Combining the OCC and IDC yields a *total capital cost* $C_p = C_{po} + IDC = \$7.16\text{B}$ for a 1.0-GWe plant. Note that the OCC $C_{po} = \$6.32\text{B}$ is equal to the present value $P$ of the plant, with the IDC calculation approximating Eq. (15.1)

$$S_n = P(1+j)^n \simeq P\left(1 + \frac{j \cdot n}{2}\right). \tag{15.7}$$

A CRF = 0.0782 $\text{yr}^{-1}$ from Eqs. (15.4) and (15.5) is used, corresponding to $j = 6.7\%\,\text{yr}^{-1}$ and a 30-yr financial arrangement, together with a capacity factor $F = 0.90$ for a plant with power rating $P_e = 1.0\,\text{GWe}$, to determine the *levelized capital cost*:

$$\begin{aligned} x_p\left(\frac{\$}{\text{MWh}}\right) &= \frac{\text{revenue required } (\$\,\text{yr}^{-1})}{\text{electricity generated } (\text{MWh}\cdot\text{yr}^{-1})} \\ &= \frac{C_p(\$)\cdot CRF\,(\text{yr}^{-1})}{P_e\,(\text{MWe})\cdot F\cdot(8,760\text{ hr}\cdot\text{yr}^{-1})} \\ &= \frac{\$7.16\times10^9\times0.0782\,\text{yr}^{-1}}{10^3\text{ MWe}\times0.90\times8,760\text{ hr}\cdot\text{yr}^{-1}} = \frac{\$71.0}{MWh}. \end{aligned} \tag{15.8}$$

Note here that the *capacity factor F* is defined as

$$F = \frac{\text{average power generated over plant life}}{\text{rated power}}. \tag{15.9}$$

In the Generation IV Roadmap, a reference OCC of \$1.25B per GWe was assumed with $CRF = 0.102$ corresponding to $j = 10\%\,\mathrm{yr}^{-1}$ and plant life of $n = 40\,\mathrm{yr}$, resulting in a levelized capital cost of \$19/MWh, which is lower by a factor of 3.7 than Eq. (15.8). In a comparative energy study [Ide22], with the same financing rate $j = 6.7\%\,\mathrm{yr}^{-1}$, $n = 30\,\mathrm{yr}$, and the assumption that the investment occurs over 2 yr, resulting in a slightly different cost recovery algorithm for the same $C_{po} = \$6.32\,\mathrm{B/GWe}$, a levelized capital cost of \$59/MWh is suggested.

### 15.3.2 Fuel Cost

The fuel component $x_c$ of the electricity generation cost for low enrichment uranium (LEU) fuel used currently in LWRs around the world accrues from the cost $C_c$ of the core, or the cost of fuel elements in the core, and consists of the cost associated with the uranium ore $C_u$, enrichment process $C_{enrich}$, fuel fabrication $C_{fab}$, spent fuel storage $C_{storage}$, and shipping and transportation $C_{s\&t}$ [Bun05]:

$$C_c = C_u + C_{enrich} + C_{fab} + C_{storage} + C_{s\&t}. \tag{15.10}$$

For our analysis, we ignore other miscellaneous expenses as well as the shipping and transportation cost $C_{s\&t}$, but include a generic cost associated with the *interim storage* of spent fuel, in lieu of the cost associated with a permanent repository for the irradiated fuel in the once-through LEU fuel cycle. For a closed fuel cycle, allowing for fuel reprocessing of UNF into mixed-oxide fuel (MOX) for recycling, the cost of reprocessed fuel should replace the uranium ore cost and enrichment cost in Eq. (15.10)

$$C_c = C_{rep} \times \frac{\text{wt\% TRU in MOX}}{\text{wt\% TRU in UNF}} + C_{fab} + C_{storage}, \tag{15.11}$$

where the reprocessing cost $C_{rep}$ is given in units of [\$/kgHM] and, for simplicity, we assume that all of the transuranics (TRUs) in the spent fuel would be processed into MOX fuel. This incurs a small error, because only about 90% of TRUs in the UNF are typically plutonium, with the remaining 10% comprising minor actinides, Am, Np, and Cm. Before discussing various components in Eqs. (15.10) and (15.11), consider the task to determine fuel cost $x_c$, given core cost $C_c$ in units of [\$/kgHM]. We should recall from our nuclear fuel cycle studies in Chapter 12 that the heavy metal refers to all nuclides with the atomic number $Z = 89$ or higher, thus including $^{232}_{89}\mathrm{Th}$, $^{235}_{92}\mathrm{U}$, $^{239}_{94}\mathrm{Pu}$ and other major and minor actinides.

Introduce the average *discharge fuel burnup B* (MWd/kgHM) for batches of fuel assemblies making up the core and thermal-to-electricity *conversion efficiency* $\eta$, which allow us to define the cost $x_c$ of assemblies per unit MWh of electricity generated

$$x_c\left(\frac{\$}{\mathrm{MWh}}\right) = \frac{C_c(\$/\mathrm{kgHM})}{B(\mathrm{MWh/kgHM})\cdot\eta}, \tag{15.12}$$

with discharge burnup $B$ expressed in units of MWd or MWh of *thermal* energy generated per kg of heavy metal. The discharge fuel burnup $B$ may be written in terms of the effective residence time of the fuel batch, with a nominal *cycle length* or *incore residence time* of $T_c$ years, converted into days, and rated thermal power density $P_t$

$$B = P_t\ (\text{MWt/kgHM}) \cdot F \cdot T_c\ (\text{d}), \tag{15.13}$$

where the product $F \cdot T_c$ is usually called the *equivalent full-power days*. Rewrite Eq. (15.12) in more convenient units

$$x_c\left(\frac{\$}{\text{MWh}}\right) = \frac{C_c\ (\$/\text{kgHM})}{\eta\left(\frac{\text{MWe}}{\text{MWt}}\right) \times B\left(\frac{\text{MWd}}{\text{kgHM}}\right) \times 24\left(\frac{\text{hr}}{\text{d}}\right)} = \frac{C_c}{24\eta \text{B}}, \tag{15.14}$$

with $B$ given now in conventional units of MWd/kgHM or GWD/MTU. We have ignored, for simplicity, the carrying charges associated with a short residence time of approximately 3 yr for fuel batches.

One key component of the cost of core $C_c$, or the cost of fuel assemblies, introduced in Eqs. (15.10) and (15.14) is the cost $C_{enrich}$ associated with the enrichment of natural uranium into slightly enriched uranium containing $3 \sim 5$ wt% of $^{235}$U for the once-through LEU cycle in the current generation of NPPs. To determine the enrichment cost, we consider an overall mass balance associated with the enrichment or separations process schematically illustrated in Figure 15.2.

We set up mass balance statements [Ben81,Gra79] for the feed flow rate $F$, product flow rate $P$, and waste flow rate $W$, expressed in units of [kg·s$^{-1}$]

$$F = P + W \tag{15.15}$$

and for the flow streams involving the desired nuclide $^{235}$U

$$x_F F = x_P P + x_W W, \tag{15.16}$$

where $\{x_i, i = F, P, W\}$ represent the *atom fraction* of the feed, product, and waste streams, respectively, also illustrated in Figure 15.2. For the purpose of fuel cycle cost calculations, assume that the enrichment $x_F = 0.007$ in the natural uranium feed, and use the waste atom fraction or *tails enrichment* $x_W = 0.002$ or 0.003.

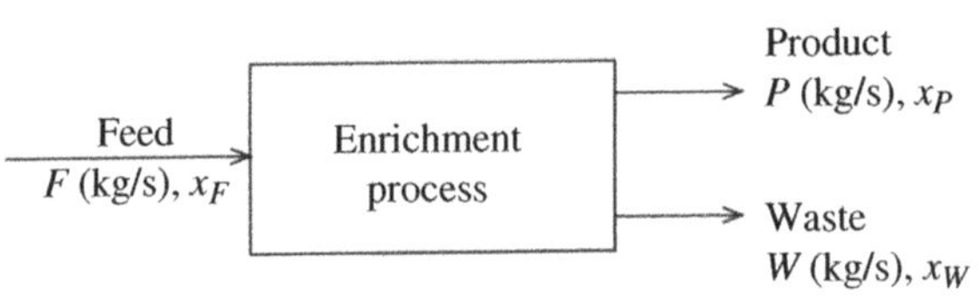


**Figure 15.2** Overall mass balance for the enrichment process.

Combining Eqs. (15.15) and (15.16), with $x_W = 0.002$ and $x_P = 0.047$ for the AP1000 design, obtain the ratios of the flow rates:

$$\frac{F}{P} = \frac{x_P - x_W}{x_F - x_W} = \frac{0.047 - 0.002}{0.007 - 0.002} = 9.0, \tag{15.17}$$

$$\frac{W}{P} = \frac{x_P - x_F}{x_F - x_W} = \frac{0.047 - 0.007}{0.007 - 0.002} = 8.0. \tag{15.18}$$

Using Eqs. (15.17) and (15.18) together with the *separative potential*

$$\phi(x_i) = (2x_i - 1)\ln\left(\frac{x_i}{1 - x_i}\right), \quad i = F, P, W, \tag{15.19}$$

provides the *separative work unit* (swu) associated with the entire separations process:

$$\text{Separative work unit} = \frac{\text{kg swu}}{\text{kg of product HM}} = \phi(x_P) + \frac{W}{P}\phi(x_W) - \frac{F}{P}\phi(x_F). \tag{15.20}$$

The mass flow rate $P$ [kg·s$^{-1}$] of the product stream essentially determines the kg swu in Eq. (15.20), and the enrichment cost is specified in terms of swu for a given enrichment facility. The simple box shown in Figure 15.2 actually consists of a large number of separation units in multiple stages of cascades [Ben81], but a simple view of the overall mass balance statements is taken with Eqs. (15.15) and (15.16) so that the separative work required in units of [kg swu/kgU] or simply [*kg swu*] may be calculated with Eq. (15.20).

Estimating costs for various fuel cycle parameters presents considerable challenges not unlike those for the capital cost, and two key references [Bun05,Rot15] are consulted to generate best estimates applicable for 2024 in Table 15.2. It should be recognized, in particular, that the uranium ore price has fluctuated, similar to that of crude oil in the spot market, over the past few decades. Note also that a minor approximation is made to leave out the cost of ~\$6/kgHM associated with the milling and conversion of uranium ore into $UF_6$ gas for the enrichment cascade. The MOX fuel reprocessing and fabrication costs updated [Rot15] in 2015 are nearly twice as high as those presented in a 2005 study [Bun05]. It is also worth noting that the Separation Technology and Transmutation Systems (STATS) Panel of the US National Academy of Sciences [Ras96] adopted, after a lengthy discussion on the economics of PUREX and pyroprocessing technologies, a round figure of \$1,000/kgHM for transuranics reprocessing in 1996.

In a September 2021 presentation to the US National Academy of Sciences (NAS) Committee on advanced reactors and UNF cycles, Orano USA suggested [Vex21] that an annual capacity of 60M kg swu is available worldwide, with Orano producing 7.5M kg swu·yr$^{-1}$ for a revenue of \$1.0 B/yr. The presentation essentially supports the enrichment cost \$120/kg swu listed in Table 15.2. For the same NAS

We may readily alter Eqs. (15.21) through (15.24) for different MOX loading fractions in LWRs or for sodium-cooled fast reactors (SFRs), where the entire core could be loaded with a larger TRU fraction. In addition, the cost basis for uranium cycles may be used for thorium fuel cycles for approximate initial estimates. For the high-assay low enrichment uranium (HALEU) fuel for several advanced reactor concepts discussed in Chapters 1 and 12, it remains to be seen if the same enrichment charges per kg swu could be considered with higher enrichments after the separative work requirements are evaluated with Eq. (15.20).

### 15.3.3 Operation and Maintenance Cost

As discussed in Section 15.1, NPPs have experienced in recent years significant increases in the O&M cost due to additional labor expenses for security personnel in the United States as well as increased maintenance tasks. Estimating O&M cost for NPPs is subject to large uncertainties, partly because details for the O&M costs at operating NPPs in the United States are generally considered proprietary information. Rothwell [Rot15] reports only general US Energy Information Agency statistics from 1995 for staffing requirements and suggests, for a 1.0-GWe plant, a total labor cost of \$50 M/yr for 630 employees, including 110 security personnel, together with miscellaneous maintenance expenses of \$25 M/yr. The 2017 DOE estimates [DOE17], reflecting updates from the 1987 analysis [Bow87], suggest a total O&M cost of \$179 M/yr comprising \$55 M/yr for labor and \$124 M/yr for other maintenance expenses.

We adopt the 2019 US EIA estimate [EIA20] of a fixed O&M cost $C_{om} = \$121\,\text{kW}^{-1}\,\text{yr}^{-1}$ and a variable O&M cost of \$2.4/MWh to obtain the total O&M cost:

$$x_{om} = \frac{\$121}{\text{kWe} \cdot (8,760\ \text{hr})} + \frac{\$2.4}{\text{MWh}} = \frac{\$16.2}{\text{MWh}}. \tag{15.25}$$

This estimate of \$16.2/MWh for the *levelized O&M cost* happens to be in good agreement with an estimate of \$16/MWh obtained [Fes17] for the Fermi II BWR plant. In the comprehensive energy economics study [Ide22], Idel modifies the 2019 US EIA estimate somewhat and combines O&M and fuel costs to arrive at the levelized cost of \$23/MWh, which is again in good agreement with our own estimate of \$23/MWh.

### 15.3.4 Decommissioning Cost

There is rather limited experience in decommissioning NPPs, and a recent World Nuclear Association report [Wor10] estimates the decommissioning cost as 9 ~ 15% of the initial capital cost. In contrast, a round figure of a third of the initial capital cost of \$1.5B was assumed, as a reference point, for the decommissioning

Combining Eqs. (15.15) and (15.16), with $x_W = 0.002$ and $x_P = 0.047$ for the AP1000 design, obtain the ratios of the flow rates:

$$\frac{F}{P} = \frac{x_P - x_W}{x_F - x_W} = \frac{0.047 - 0.002}{0.007 - 0.002} = 9.0, \tag{15.17}$$

$$\frac{W}{P} = \frac{x_P - x_F}{x_F - x_W} = \frac{0.047 - 0.007}{0.007 - 0.002} = 8.0. \tag{15.18}$$

Using Eqs. (15.17) and (15.18) together with the *separative potential*

$$\phi(x_i) = (2x_i - 1)\ln\left(\frac{x_i}{1 - x_i}\right), \quad i = F, P, W, \tag{15.19}$$

provides the *separative work unit* (swu) associated with the entire separations process:

$$\text{Separative work unit} = \frac{\text{kg swu}}{\text{kg of product HM}} = \phi(x_P) + \frac{W}{P}\phi(x_W) - \frac{F}{P}\phi(x_F). \tag{15.20}$$

The mass flow rate $P$ [kg·s$^{-1}$] of the product stream essentially determines the kg swu in Eq. (15.20), and the enrichment cost is specified in terms of swu for a given enrichment facility. The simple box shown in Figure 15.2 actually consists of a large number of separation units in multiple stages of cascades [Ben81], but a simple view of the overall mass balance statements is taken with Eqs. (15.15) and (15.16) so that the separative work required in units of [kg swu/kgU] or simply [*kg swu*] may be calculated with Eq. (15.20).

Estimating costs for various fuel cycle parameters presents considerable challenges not unlike those for the capital cost, and two key references [Bun05,Rot15] are consulted to generate best estimates applicable for 2024 in Table 15.2. It should be recognized, in particular, that the uranium ore price has fluctuated, similar to that of crude oil in the spot market, over the past few decades. Note also that a minor approximation is made to leave out the cost of ~\$6/kgHM associated with the milling and conversion of uranium ore into $UF_6$ gas for the enrichment cascade. The MOX fuel reprocessing and fabrication costs updated [Rot15] in 2015 are nearly twice as high as those presented in a 2005 study [Bun05]. It is also worth noting that the Separation Technology and Transmutation Systems (STATS) Panel of the US National Academy of Sciences [Ras96] adopted, after a lengthy discussion on the economics of PUREX and pyroprocessing technologies, a round figure of \$1,000/kgHM for transuranics reprocessing in 1996.

In a September 2021 presentation to the US National Academy of Sciences (NAS) Committee on advanced reactors and UNF cycles, Orano USA suggested [Vex21] that an annual capacity of 60M kg swu is available worldwide, with Orano producing 7.5M kg swu·yr$^{-1}$ for a revenue of \$1.0 B/yr. The presentation essentially supports the enrichment cost \$120/kg swu listed in Table 15.2. For the same NAS

**Table 15.2** Cost estimates for nuclear fuel cycle.

| Parameters | Cost ($) |
|---|---|
| Uranium ore price $C_u$ | 110/kgU |
| Enrichment cost $C_{enrich}$ | 120/kg swu |
| LEU fabrication cost $C_{fab}$ | 300/kgHM |
| MOX fuel reprocessing cost $C_{rep}$ | 2,500/kgHM |
| MOX fuel fabrication cost $C_{fab}$ | 2,300/kgHM |
| Interim storage cost $C_{storage}$ | 200/kgHM |

committee, M. Bunn suggested [Bun21] that the MOX fuel reprocessing cost could be $2,200 ~ 4,000/kgHM, allowing us to keep an earlier estimate of $2,500/kgHM in Table 15.2.

**Example 15.1** Determine the levelized fuel cost $x_c$, for the equilibrium fuel cycle data [Dru07] for AP1000 with the power rating of 3.4 GWt and 1.09 GWe.

(a) Power density $P_t = 40.3$ kWt/kgU,
(b) Thermal efficiency $\eta = 0.321$,
(c) Equilibrium feed enrichment $e = 4.7$ wt%,
(d) Equilibrium cycle length $T_c = 3 \times 18$ months,
(e) Capacity factor $F = 0.90$, and
(f) Discharge burnup $B = 49$ MWd/kgU.

Substituting the fuel cycle data (a), (d), and (e) into Eq. (15.13) yields $B = 59$ MWd/kgU, which is 20% larger than the actual Westinghouse calculation in (f). This is the result of a detailed fuel-shuffling scheme adopted that would load once- or twice-burned fuel elements to the outer regions of the core, where the power density is significantly lower than the core average. This is a simple reminder that the equilibrium cycling method of Section 12.4, under the assumption of equal power sharing among the three enrichment zones, may only yield approximate discharge burnup estimates.

For the product and waste enrichments, $x_W = 0.002$ and $x_P = 0.047$, Eqs. (15.19) and (15.20) yield a separative work of 8.2 kg swu. We substitute the cost data into Eq. (15.10) to obtain:

$$\begin{aligned} C_c &= C_u + C_{enrich} + C_{fab} + C_{storage}, \\ &= \frac{\$110}{\text{kgHM}} \times \left(\frac{F}{P} = 9.0\right) + \frac{\$120}{\text{kg swu}} \times 8.2 \left(\frac{\text{kg swu}}{\text{kgHM}}\right) + \frac{\$300}{\text{kgHM}} + \frac{\$200}{\text{kgHM}}, \\ &= \frac{\$2{,}474}{\text{kgHM}}, \end{aligned} \tag{15.21}$$

where every term is expressed in units of [$/kgHM], representing kg of *enriched uranium*. This explains why the uranium ore term contains the feed-to-product mass ratio $F/P$. The estimate of $2,474/kgHM translates approximately into $1.2M per PWR fuel assembly generally accepted in the industry in the early 2020s.

Substituting $C_c = \$2{,}474/\text{kgHM}$ into Eq. (15.12), with $B = 49$ MWd/kgHM for the AP1000 plant, yields the *levelized fuel cost*:

$$x_c = \frac{C_c}{24\eta B} = \frac{\$2{,}474/\text{kgHM}}{24 \times 0.321 \times 49\ \text{MWd/kgHM}} = \frac{\$6.6}{\text{MWh}}. \tag{15.22}$$

This estimate is in reasonable agreement with a reference value of $6.0/MWh adopted in the Generation IV Roadmap and a separate DTE estimate [Win14] of $7.0/MWh but somewhat less than an estimate of $8.4/MWh based on the 2019 EIA data [Ide22].

**Example 15.2** Repeat the levelized fuel cost calculation for a PWR with a quarter of the core loaded with MOX fuel assemblies containing 5.0 wt% TRU and the remaining three quarters loaded with LEU assemblies, currently 25–30% of the core is loaded with recycled MOX assemblies in many PWRs operating in France. We also recall from our fuel cycle studies in Chapter 12 that the UNF from the current generation of LWRs contains approximately 1.0 wt% of TRUs.

With the MOX fuel reprocessing and fabrication data from Table 15.2, we may than calculate first the cost for MOX assemblies:

$$\begin{aligned} C_c &= C_{rep} \times \frac{\text{wt\% TRU in MOX}}{\text{wt\% TRU in UNF}} + C_{fab} + C_{storage}, \\ &= \frac{\$2,500}{\text{kgHM}} \times \left(\frac{5\%}{1\%}\right) + \frac{\$2,300}{\text{kgHM}} + \frac{\$200}{\text{kgHM}} = \frac{\$15,000}{\text{kgHM}}. \end{aligned} \tag{15.23}$$

Combining the MOX fuel assembly cost with that for the LEU fuel assembly cost $2,474/kgHM from Eq. (15.21) yields

$$C_c \text{ for MOX core} = \frac{1}{4}C_c(\text{MOX}) + \frac{3}{4}C_c(\text{LEU}) = \frac{\$5,606}{\text{kgHM}}. \tag{15.24}$$

Comparing Eqs. (15.24) and (15.21) indicates a net increase of $3, 132/kgHM for the MOX core. Substituting the MOX core cost of $5,606/kgHM into Eq. (15.12) yields the fuel cost $x_c = \$14.8/\text{MWh}$, indicating a penalty of $8/MWh. This additional cost associated with fuel reprocessing and recycling is somewhat larger than 5% of the total generation cost, which the French suggested [Bou03] was readily acceptable in view of the long-term resource and waste management of nuclear fuel. This disparity may reflect substantially higher cost parameters assumed in Table 15.2 than in a previous recycling analysis [Bun05]. Another study by the Boston Consulting Group [Bos06] presents additional cost bases for fuel cycle analyses and advances an economic case in support of closed fuel cycles. The study was supported by AREVA Company.

We may readily alter Eqs. (15.21) through (15.24) for different MOX loading fractions in LWRs or for sodium-cooled fast reactors (SFRs), where the entire core could be loaded with a larger TRU fraction. In addition, the cost basis for uranium cycles may be used for thorium fuel cycles for approximate initial estimates. For the high-assay low enrichment uranium (HALEU) fuel for several advanced reactor concepts discussed in Chapters 1 and 12, it remains to be seen if the same enrichment charges per kg swu could be considered with higher enrichments after the separative work requirements are evaluated with Eq. (15.20).

### 15.3.3 Operation and Maintenance Cost

As discussed in Section 15.1, NPPs have experienced in recent years significant increases in the O&M cost due to additional labor expenses for security personnel in the United States as well as increased maintenance tasks. Estimating O&M cost for NPPs is subject to large uncertainties, partly because details for the O&M costs at operating NPPs in the United States are generally considered proprietary information. Rothwell [Rot15] reports only general US Energy Information Agency statistics from 1995 for staffing requirements and suggests, for a 1.0-GWe plant, a total labor cost of \$50 M/yr for 630 employees, including 110 security personnel, together with miscellaneous maintenance expenses of \$25 M/yr. The 2017 DOE estimates [DOE17], reflecting updates from the 1987 analysis [Bow87], suggest a total O&M cost of \$179 M/yr comprising \$55 M/yr for labor and \$124 M/yr for other maintenance expenses.

We adopt the 2019 US EIA estimate [EIA20] of a fixed O&M cost $C_{om}$ = \$121 kW$^{-1}$ yr$^{-1}$ and a variable O&M cost of \$2.4/MWh to obtain the total O&M cost:

$$x_{om} = \frac{\$121}{\text{kWe}\cdot(8,760\ \text{hr})} + \frac{\$2.4}{\text{MWh}} = \frac{\$16.2}{\text{MWh}}. \tag{15.25}$$

This estimate of \$16.2/MWh for the *levelized O&M cost* happens to be in good agreement with an estimate of \$16/MWh obtained [Fes17] for the Fermi II BWR plant. In the comprehensive energy economics study [Ide22], Idel modifies the 2019 US EIA estimate somewhat and combines O&M and fuel costs to arrive at the levelized cost of \$23/MWh, which is again in good agreement with our own estimate of \$23/MWh.

### 15.3.4 Decommissioning Cost

There is rather limited experience in decommissioning NPPs, and a recent World Nuclear Association report [Wor10] estimates the decommissioning cost as 9 ~ 15% of the initial capital cost. In contrast, a round figure of a third of the initial capital cost of \$1.5B was assumed, as a reference point, for the decommissioning cost of a 1.0-GWe plant in the Generation IV Roadmap. A recent experience with the decommissioning of the Big Rock Point boiling water reactor, however, suggests that this assumption may have been somewhat optimistic.

For our illustrative example, we will assume 20% of the initial capital cost of \$7.16B considered in Section 15.3.1 as the decommissioning cost. Following the steps taken to determine the capital cost $x_p$ in Eq. (15.8), and recognizing that we need to accumulate a sum of fund $C_{decom}$ over the plant life $T_p = 30\,\text{yr}$ assumed, we obtain the *decommissioning component* of the electric generation cost:

$$x_{decom} = \frac{C_{decom} \cdot (R/S_n)}{P_e(\text{kWe}) \cdot F \cdot (8,760\ \text{hr} \cdot \text{yr}^{-1})}. \tag{15.26}$$

Equation (15.3) is invoked to calculate the revenue $R$ to be collected per year so that we may accumulate $S_n$ over $T_p$.

For the purpose of the levelized cost calculation for $C_{decon}$, we may use a real interest rate $j = 5\%\,\text{yr}^{-1}$ rather than the total discount rate of $6.7\%\,\text{yr}^{-1}$ assumed in the plant capital cost calculation in Section 15.3.1:

$$\frac{R}{S_n} = \frac{j}{(1+j)^n - 1} = \frac{0.05}{(1.05)^{30} - 1} = \frac{0.05}{3.32} = 0.0151. \tag{15.27}$$

Substituting Eq. (15.27) into Eq. (15.26) finally yields

$$x_{decom} = \frac{\$1.43 \times 10^9 \times 0.0151}{1.0 \times 10^3(\text{MWe}) \times 0.9 \times 8,760(\text{hr} \cdot \text{yr}^{-1})} = \frac{\$2.7}{\text{MWh}}. \tag{15.28}$$

This indicates that the decommissioning cost, albeit significant compared with the construction cost itself, adds <4% to the capital cost estimated in Section 15.3.1, and is almost insignificant in the projection of the total cost of generating electricity in new nuclear power plants. Thus, the decommissioning cost may be subsumed into the capital cost, as was done in the Generation IV Roadmap.

### 15.3.5 Levelized Cost of Electricity

With the four components of the generation cost determined in Eqs. (15.8), (15.22), (15.25), and (15.28), we finally obtain the total electricity generation cost or busbar cost, often referred to as *LCOE*, of \$96.5/MWh or 9.7 cent/kWh for an AP1000 plant. This estimate for LCOE is 170% above the estimate of \$34/MWh used as a fiducial point to evaluate various alternate NPP designs in the Generation IV Roadmap [DOE02] and reflects a steep rise in the cost of construction materials due to rapidly increasing demands worldwide as well as a substantial increase in the O&M cost, as discussed in Section 15.1.

The LCOE estimate of \$97/MWh is somewhat higher than Rothwell's estimate [Rot22] of \$78/MWh and Idel's estimate [Ide22] of \$82/MWh, both of which are based mostly on the 2019 US EIA estimates of OCC, fuel cost, and O&M cost.

The difference of \$15/MWh between our LCOE estimate of \$97/MWh and Idel's estimate of \$82/MWh appears mostly due to differences in the CRF, IDC, and decommissioning cost. For example, Idel assumed a construction period of 2 yr, compared with 4 yr adopted in our study.

## 15.4 Impact of Increased Capital and O&M Costs

The LCOE of \$97/MWh presented in Section 15.3.5 is fairly similar to the current cost of \$90 ~ 100/MWh in most regions of the United States and the OCC estimate of \$6.32 B/GWe is fairly consistent with a recent benchmarking [Ste22] of large PWR construction experience indicating the total OCC of \$5.5 ~ 7.5 B/GWe for FOAK large PWR plants. The OCC estimates also present direct and indirect costs in similar categories and similar breakdown between direct and indirect costs as summarized in Table 15.1. One could also consult the cost estimate guidelines suggested for the Generation IV nuclear energy systems [Gif07], but the process would still require recent cost databases.

It should, however, be recognized that even the most recent cost estimates do not reflect the difficult experience with the four AP1000 construction projects at the Vogtle and Summer plants. Due to cost over-runs and delays in the construction projects, the Summer projects were canceled and Vogtle Units 3 and 4 were completed with construction costs possibly exceeding \$30B [Cho23]. A summary of construction experience for the FOAK AP1000 plants also suggests [Ste22] that the cost over-runs are partially related to new material and construction methods and design updates required to accommodate regulatory changes. For example, Westinghouse Electric Company introduced composite steel plates for the shield buildings. Similar construction delays and cost over-runs have been experienced also with the 1.6-GWe EPR project at the Olkiluoto Unit 3 Plant in Finland, requiring €11 or \$12B and 18 yr to complete the project [NEI24]. This reflects again a FOAK construction project, where Framatome was unable to utilize the excellent experience and know-how that the company had accumulated during the successful establishment of a nuclear infrastructure in France involving 30 PWRs of similar designs.

The difficult experience with both the Vogtle and Olkiluoto projects is reflected unfortunately in the recent decision to cancel the NuScale SMR project. As the first SMR project in the United States with the construction permit approved by the US Nuclear Regulatory Commission, the project was proposed in 2014 with a cost estimate of \$4.2B for 12 SMRs each producing 60 MWe with the support of Utah Associated Municipal Power System (UAMPS). The construction cost estimate apparently escalated to \$9.1B for a 12×77-MWe = 0.92 GWe system, with an estimated LCOE of \$89/MWh [Cho23]. Although this LCOE estimate is quite similar to our best estimate in early 2024, it was judged too expensive compared with alternate electricity sources. Additional comparisons between the LCOE of nuclear and alternate energy sources are discussed further in Section 15.5.

It is expected, however, that nuclear electricity generation would become competitive with other electricity generation plants, as we learn to trim the capital costs with multiple Generation III+ units and Generation IV SMR plants built over some period of time. In the current deregulated electricity market, it has also become a challenge to economically operate even fully amortized NPPs in a small merchant fleet in competition with well-subsidized natural gas plants, resulting in recent decisions to prematurely decommission several NPPs, including the Kewaunee and Vermont Yankee Plants. This is due to a large number of operations and support staff, together with security personnel, in every NPP in the United States that has increased the O&M cost significantly over the past two decades.. There is, however, some glimmer of hope that many communities recognize the importance of stable, sustained power generation capability of nuclear plants, exemplified by the effort being made in early 2024 by the State of Michigan to offer financial support to reapply for the operating license of the decommissioned Palisades Nuclear Plant in Covert, Michigan. This initiative also carries the loan guarantee of $1.5B from the US Department of Energy.

In addition to the concerns over the increased electricity generation costs, the investment risk associated with the capital cost of up to $6 ~ 10B for one large power plant has emerged as a serious concern for every utility company planning to build nuclear plants during this decade. This is because the total asset basis of many utility companies including the DTE Energy Company is not much larger than the projected cost of one new plant. This is obviously the reason for growing interest in SMRs that could offer an efficient construction of power modules comprising rail-shippable reactor vessels and associated equipment with incremental financial investment. It is noteworthy that, together with SMRs with Generation IV features employing different compositions and layouts of fuel and coolant systems, smaller designs, AP300 and BWRX-300, based on the established Generation III LWR plant structures, but with reduced total capital requirements are also proposed.

One study, although performed before the sharp increase in the PCCI shown in Figure 15.1, considered the economics of the 380-MWe S-PRISM design [Boa00] for modular SFRs. In this study, the busbar cost for a 1.5-GWt design would be $50/MWh, compared with $40/MWh for a similar 3.5-GWt unit. With the capital cost making up approximately three-quarters of the busbar cost, a capital cost penalty of approximately 20% may be attributed to the small modular design. Indeed, regardless of the size of a plant, certain costs including those associated with licensing, security, land and switchyard, and O&M may be nearly independent of the plant size. Thus, it may be prudent to assume a certain penalty in busbar cost per MWh for SMRs until the economy of multiples or economy of small scale is demonstrated through actual SMR deployments over the next decade or so.

As a continuing effort for the S-PRISM design, a study [Boa01] of the development and generation costs for the modular SFR design was published around the time when the Generation IV Roadmap was released. The Economics Modeling Working Group of the Generation IV International Forum also released a set of detailed guidelines [Gif07] for estimating costs for the design, construction, O&M, fuel cycle, and decommissioning for the FOAK through NOAK commercial nuclear plants. In addition, some guidelines are included for estimating costs for modular units, although no specific examples are given.

Another publication [IAE13] by the International Atomic Energy Agency provides specific guidelines and a test case for the capital cost of SMRs. In this report, according to the economy of scale, the OCC $C_{p0}$ [\$/GWe] is assumed inversely proportional to the 0.4-th power of the electrical power rating $Pe$ [GWe] introduced in Eq. (15.8). Some credits are, however, suggested to reflect (a) learning curve advantages for the $N$th unit, (b) safety and system enhancements achievable with smaller units, (c) an expedited construction schedule, and (d) distributing capital expenditures in a more timely manner. Thus, for the case where the power rating $P_e^*$ for a SMR is one-fourth of the rating $P_e$ for a regular NPP, i.e. $P_e^* = 0.25P_e$, the SMR OCC may initially be estimated as

$$C_{po}(P_e^*) = C_{po}(P_e)\left(\frac{P_e}{P_e^*}\right)^{0.4} = 1.74\ C_{po}(P_e), \tag{15.29}$$

reflecting a 74% penalty due to the small scale. Accounting, however, for a NOAK learning factor of 0.78, a system enhancement factor of 0.85, and a construction schedule factor of 0.94, the overall OCC penalty is reduced to 9%. This approach may be used for capital cost estimates, albeit somewhat optimistic, for SMR designs under development.

In an updated NPP economics study [Rot15], G. Rothwell provides alternate estimates for capital, O&M, and fuel costs for SMRs with different power ratings and fuel enrichments, suggesting that perhaps a 20 ~ 35% penalty may have to be assumed for SMRs both for FOAK and NOAK deployments. At this early stage of the AP300 and BWRX-300 SMR development, it is recognized that the 300-MWe LWR plants may cost as much as 50% more per GWe capacity than the full-size AP1000 and ESBWR plants discussed in Chapter 1 [Bra24]. Generation IV International Working Group has been working on various cost estimating guidelines [Gif07], but the guidelines will have to be revisited with the recent experience with Generation III+ plants in the United States, Europe, and Asia. As a crucial part of the rebirth of nuclear energy development, the US nuclear industry has to rebuild the infrastructure that supported the successful construction of approximately 100 nuclear plants. The effort will be required to successfully launch the Generation IV and SMR plants currently under development.

## 15.5 Energy Generation Efficiency of Various Technologies

### 15.5.1 Electric Energy Costs for Nuclear and Other Technologies

With the estimates and recent experience of electricity generation costs for nuclear plants reviewed in Section 15.4, we now turn to the task of comparing lifetime costs of alternate electricity generating technologies with that for nuclear power plants. For this task, we return to a comparative study [Ide22] that we have already consulted for estimating capital cost in Section 15.3. The study uses a consistent set of capital cost and O&M data mostly from the US Energy Information Administration forecast for 2020 [EIA20] covering conventional and renewable energy sources together with nuclear energy, as summarized in Table 15.3.

We note first that natural gas combined cycle (CC) systems with a high capacity factor would provide LCOE quite competitive with those of solar PV and wind generation systems even with low capacity factors (CFs). This observation simply represents that natural gas, solar PV, and wind electrical plants could be built with low overnight capital costs (OCCs). It is readily recognized that nuclear power plants with high OCCs are not as competitive as natural gas and renewable technologies for electricity generation on the basis of design-base capacity factors. This is a challenge that the nuclear energy community duly recognizes as effort is being made to

**Table 15.3** Cost estimates for various energy technologies.

| Technology | OCC ($B/GWe) | Fixed O&M cost ($/kWe-yr) | Variable O&M and fuel cost ($/MWh) | CF (%) | LCOE ($/MWh) |
|---|---|---|---|---|---|
| Nuclear | 6.317 | 121 | 8 | 90 | 82 |
| Biomass | 4.401 | 125 | 28 | 83 | 95 |
| Coal | 3.661 | 40 | 25 | 95 | 76 |
| Natural gas CC | 1.079 | 14 | 18 | 87 | 38 |
| Natural gas CT | 0.710 | 7 | 28 | 30 | 67 |
| Solar PV | 1.331 | 15 | 0 | 29 | 36 |
| Wind | 1.319 | 26 | 0 | 40 | 40 |

*Source*: Adapted from Idel [Ide22].

**Table 15.4** Effective levelized cost of electricity.

| Technology | ERCOT Capacity (GW) | ERCOT ECF (%) | ERCOT ELCOE ($/MWh) | Germany Capacity (GW) | Germany ECF (%) | Germany ELCOE ($/MWh) |
|---|---|---|---|---|---|---|
| Nuclear | 52 | 76.9 | 96 | 71 | 79.9 | 92 |
| Biomass | 53 | 76.0 | 104 | 75 | 76.0 | 104 |
| Coal | 58 | 68.9 | 94 | 77 | 73.7 | 88 |
| Natural gas CC | 69 | 58.2 | 57 | 82 | 69.9 | 47 |
| Natural gas CT | 69 | 58.1 | 35 | 83 | 69.6 | 29 |
| Solar PV | 438 | 10.4 | 100 | 3,564 | 1.5 | 696 |
| Wind | 517 | 7.8 | 205 | 906 | 5.9 | 271 |

*Source*: Adapted from Idel [Ide22].

develop SMRs with competitive capital cost as well as improve construction tasks for Generation III+ systems. It should be, however, recognized that the solar PV, wind, and natural gas combustion turbine (CT) plants operate with low CFs and hence have to be augmented with storage systems or other energy sources.

In the comparative energy technology study, Idel proposes [Ide22] to review and analyze two communities with fairly large electrical generation capabilities, Germany and the Electricity Reliability Council of Texas (ERCOT) market, as summarized in Table 15.4. For the purpose of studying nuclear energy economics, we note that the ERCOT market carries a nuclear generation capacity of 52 GW and Germany a slightly larger capacity of 71 GW. We also note that both markets rely significantly on large solar PV and wind powers. In particular, the German market relies on a nominal generation capacity of 3,564 GW solar PV plants, augmented by pumped hydro facilities. In other markets, solar PV generators are supported by large-scale battery systems [Ide22]. Thus, a suggestion is made to consider *effective capacity factors* (ECFs) to reflect the actual availability of various systems over a period of 8 yr covering 2012–2019 for the German and ERCOT markets, with the solar data for 2012 in Texas excluded for an unusually low capacity for the particular year.

The ECFs for seven electrical generation systems in both the ERCOT and Germany indicate less than the nominal CF values listed in Table 15.3, with the exception of the natural gas CT systems in both markets indicating higher utilization than designed for. In particular, the ECF for the solar PV plants in Germany is only 1.5% compared with the nominal or design CF = 29%. The ECFs for wind systems are equally very low compared with the nominal CF = 40% in both markets.

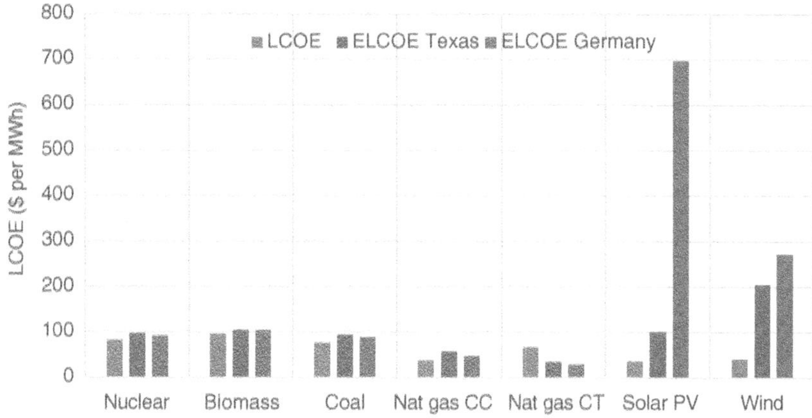


**Figure 15.3** Effective LCOE comparison for various technologies. *Source*: Adapted from Idel [Ide22].

Thus, the *effective levelized cost of electricity* (ELCOE) is calculated for each technology with the appropriate ECF value and compared in Table 15.4, which indicates that the ELCOE values are much higher for the solar PV and wind markets especially in Germany. A three-way comparison involving the nominal LCOE and ELCOE values, for Texas and Germany, is presented in Figure 15.3. It should, of course, be recognized that the ELCOE values should include the cost associated with storage systems, e.g. battery banks. With the storage system costs included, Idel indicates ELCOE = $1,380/MWh in Germany and $413/MWh in Texas for solar PV. The comparison, however, requires additional assumptions and details for the storage and dispatch systems, and hence is not presented here. Without the complications for the storage system details, Table 15.4 and Figure 15.3 simply illustrate the importance of stable nuclear electricity in the overall electrical energy market.

### 15.5.2 Energy Return on Investment for Various Technologies

Having analyzed and compared the effective cost of electricity generation from various technologies, represented in terms of effective LCOE in Section 15.5.1, we now discuss the efficiency of various technologies represented in terms of the lifetime energy production normalized by the energy required for establishing the electrical plant. The output-to-input ratio of energy production is designated as the *energy return on investment* (EROI) in a paper [Wei13] that discusses the economic efficiency and life cycle of key technologies available today. The data tabulated in Table 15.5 are selected from various sources to represent the relevant

**Table 15.5** Energy return on investment for electrical energy technologies.

| Technology | Nuclear | Nat gas CC | Solar PV | Wind | Coal |
|---|---|---|---|---|---|
| Installed capacity (kW) | $1.34 \times 10^6$ | $8.20 \times 10^5$ | 0.0928 | $1.50 \times 10^3$ | $5.09 \times 10^5$ |
| Full load (hr) | 8,000 | 7,500 | 1,000 | 2,000 | 7,500 |
| Lifetime (yr) | 60 | 35 | 25 | 20 | 50 |
| Energy output (TJ) | $2.32 \times 10^6$ | $7.75 \times 10^5$ | 0.00835 | 216 | $6.87 \times 10^5$ |
| Energy input (TJ) | $3.09 \times 10^4$ | $2.69 \times 10^4$ | 0.00217 | 13.2 | $2.40 \times 10^4$ |
| Construction (TJ) | $4.05 \times 10^3$ | 470 | 0.00177 | 12.9 | $1.97 \times 10^3$ |
| O&M (TJ) | $6.90 \times 10^3$ | 255 | 0.0004 | 0.3 | $7.40 \times 10^3$ |
| Fuel (TJ) | $1.88 \times 10^4$ | $2.61 \times 10^4$ | | | $1.45 \times 10^4$ |
| Decommissioning (TJ) | $1.15 \times 10^3$ | 30 | | | 91 |
| EROI | 75 | 29 | 3.8 | 16 | 29 |

*Source*: Weissbach et al. [Wei13]/U.S. Department of Energy.

technologies and are subject to minor variations depending on particular attributes for the energy input and output selected. The general trend, tabulated in Table 15.5 indicates, however, significantly different EROI values among the technologies, in particular, EROI of 75 for nuclear power plants compared with 3.8 for solar PV systems.

The lifetime of nuclear plants is considered 60 yr in Table 15.5, as is the case currently for most nuclear power plants in the United States, and the practice is expected in other countries as the technology matures. If a typical initial license period of 40 yr is selected for nuclear plants, the EROI value will decrease to close to 50, which still is significantly larger than the corresponding EROI values for other technologies. The significant energy input requirement for nuclear power plants in Table 15.5 reflects mostly the effort required for fuel enrichment tasks performed with the gaseous diffusion process. Thus, the fuel energy input could decrease substantially with the gas centrifuge process utilized, possibly providing EROI as high as 105. It should be also recognized that most of the advanced reactor systems utilizing the Generation IV technology presents a higher thermal efficiency, which is also expected to provide higher EROI values compared with the current generation of nuclear power plants.

## References

**[Ben81]** M. Benedict, T.H. Pigford, and H.W. Levi, *Nuclear Chemical Engineering*, McGraw-Hill (1981).

**[Boa00]** C.E. Boardman, A.E. Dubberley, and M. Hui, "Optimizing the Size of the S-PRISM Reactor," *Proc. 8th Int. Conf. Nucl. Engineering*, ICONE-8 (2000).

**[Boa01]** C.E. Boardman, M. Hui, D.G. Carroll, and A.E. Dubberley, "Economic Assessment of S-PRISM Including Development and Generating Costs," *Proc. 9th Int. Conf. Nucl. Engineering*, ICONE-9 (2001).

**[Bos06]** Boston Consulting Group, "Economic Assessment of Used Nuclear Fuel Management in the United States" (2006).

**[Bou03]** J. Bouchard, E. Proust, J.J. Gautrot, and B. Tinturier, "Economics of Nuclear Energy Production Systems: Reactors and Fuel Cycle," *Global 2003 Conference* (2003).

**[Bow87]** H.I. Bowers, L.C. Fuller, and M.L. Myers, "Cost Estimating Relationships for Nuclear Power Plant Operation and Maintenance," ORNL/TM-10563, Oak Ridge National Laboratory (1987).

**[Bra24]** J. Bradfute, Westinghouse Electric Company, private communication, January 2024.

**[Bun21]** M. Bunn, "The Economics of Reprocessing and Recycling vs. Direct Disposal of Spent Nuclear Fuel," presentation to the US National Academy of Sciences, 7 June 2021, https://doi.org/10.17226/26500 (2023).

**[Bun05]** M. Bunn, J.P. Holdren, S. Fetter, and B. Van der Zwaan, "The Economics of Reprocessing versus Direct Disposal of Spent Nuclear Fuel," *Nucl. Technol.* **150**, 209 (2005).

**[Cho23]** A. Cho, "Deal to Build Pint-Size Nuclear Reactors is Canceled," *Science* **382**, 749 (2023).

**[DOE02]** *A Technology Roadmap for the Generation IV Nuclear Energy Systems*, https://www.gen-4.org/gif/jcms/c40473/a-technology-roadmap-for-generation-iv-nuclear-energy-systems (2002).

**[DOE10]** US Department of Energy, "Nuclear Energy Research and Development Roadmap," US Department of Energy (2010).

**[DOE17]** US Department of Energy, "ARPA-E MEITNER Program Funding Opportunity Announcement," DOE FOA-0001798. US Department of Energy (2017).

**[DOL17]** US Department of Labor, "Consumer price index," https://www.bls.gov/cpi/, US Department of Labor (2017).

**[Dru07]** K. Drudy, Westinghouse Electric Company, private communication, March 2007.

**[EIA20]** U.S. Energy Information Administration, "The Electricity Market Module of the National Energy Modeling System: Model Documentation 2020," U.S. Energy Information Administration (2020).

**[Fes17]** P. Fessler, DTE Electric Company, private communication, April 2017.

**[Gif07]** Generation IV International Forum, "Cost Estimating Guidelines for Generation IV Nuclear Energy Systems," rev. 4.2, GIF/EMWG/2007/004, Generation IV International Forum (2007).

**[Gra79]** H.W. Graves Jr., *Nuclear Fuel Management*, Wiley (1979).

**[IAE13]** International Atomic Energy Agency, "Approaches for Assessing the Economic Competitiveness of Small and Medium Sized Reactors," IAEA Nuclear Energy Series No. NP-T-3.7, International Atomic Energy Agency (2013).

**[Ide22]** R. Idel, "Levelized Full System Costs of Electricity," *Energy* **259**, 124905 (2022).

**[IHS17]** IHS website, http://ihsindexes.com/pcci-graph.htm (2017).

**[NAS23]** J.D. Lee, P.A. Baisden, R.C. Ewing et al., "Merits and Viability of Different Nuclear Fuel Cycles and Technology Options and the Waste Aspects of Advanced Nuclear Reactors," US National Academy of Sciences, https://doi.org/10.17226/26500 (2023).

**[NEI24]** Nuclear Engineering International, "Olkiluoto 3 begins operation," Nuclear Engineering International, 18 March 2024.

**[Ras96]** N.C. Rasmussen, J. Buckham, T.J. Burke et al., *Nuclear Wastes: Technologies for Separations and Transmutation*, National Academy Press (1996).

**[Rot15]** G. Rothwell, *The Economics of Future Nuclear Power*, Routledge, Taylor & Francis Group (2015).

**[Rot22]** G. Rothwell, "Projected Electricity Costs in International Nuclear Power Markets," *Energy Policy* **164**, 112905 (2022).

**[Ste22]** W.R. Stewart and K. Shirvan, "Capital Cost Estimation for Advanced Nuclear Power Plants," *Renewable and Sustainable Energy Reviews* **155**, 111880 (2022).

**[Vex21]** A. Vexler, "HALEU," presentation to the US National Academy of Sciences committee, 15 September 2021, https://doi.org/10.17226/26500 (2023).

**[Wei13]** D. Weissbach, G. Ruprecht, A. Huke et al., "Energy Intensities, EROIs (Energy Returned on Invested), and Energy Payback Times of Electricity Generating Power Plants," *Energy* **52**, 210 (2013).

**[Win14]** J. Wines, DTE Electric Company, private communication, October 2014.

**[Wor10]** World Nuclear Association, "The Economics of Nuclear Power," World Nuclear Association, www.world-nuclear.org/info/inf02.html (2010).

## Problems

**15.1** Review the enrichment cascade theory in [Ben81] and derive the expression for the separative potential of Eq. (15.19).

**15.2** Using the reactivity-based instant cycling method from Section 12.5, compare the fuel cycle cost for two equilibrium feed enrichments of 3.0 wt% and 4.5 wt%. Assume the outage time required for preventive maintenance and refueling is independent of the cycle length.

**15.3** Perform parametric studies with various costs, including the uranium price and O&M cost, to assess the variability of the LCOE for nuclear power plants.

# 16

# Space-Time Kinetics and Reactor Control

Basic concepts and techniques for representing the dynamic behavior of nuclear reactors are discussed in Chapter 8 in terms of the point reactor kinetics equation. Because the point kinetics equation (PKE) is derived with the assumption that the spatial distributions of the flux and power do not change during a transient, its accuracy is limited in many transients of interest. This is due to the fact that any reactivity insertions, e.g. control rod movements, would entail substantial changes to the flux distribution in the reactor core. Furthermore, except in a zero-power reactor, any changes in the flux and power levels also induce changes in the flux distribution associated with thermal-hydraulic feedback effects. By definition, a zero-power reactor is a multiplying system that could allow self-sustaining chain reactions but produces no susceptible heat.

For mathematical simplicity and physical clarity, the criticality condition involving the material and geometrical buckling concepts in Chapter 5 and the derivation and applications of the PKE in Chapter 8 are presented in terms of energy-independent or one-group neutron diffusion equation. The changes in the spatial power distribution are typically accompanied by changes in the energy dependence of the neutron flux itself. These features of reactor transient phenomena require that the general space-time kinetics formulation should be based on the energy-dependent neutron diffusion equation. For notational convenience, the delayed neutron precursor concentrations are simplified into the one equivalent group notation introduced in Chapter 8.

Formulation and solution of the space-time kinetics equations are discussed in Section 16.1, followed by space-time kinetics problems of significant interest for nuclear reactor analysis in Section 16.2. Included are the space-time kinetics issues closely related to the buildup of the fission product $^{135}$Xe and the associated reactivity poisoning. Basic formulations of reactor control strategy and a few examples follow in Section 16.3, together with model-based control formulations in Section 16.4. Alternate control strategies are discussed in Section 16.5, with the system estimation algorithm via Kalman filtering in Section 16.6.

*Nuclear Reactor Physics and Engineering*, Second Edition. John C. Lee.
© 2025 John Wiley & Sons, Inc. Published 2025 by John Wiley & Sons, Inc.
Companion website: www.wiley.com/go/LeeNRPE2

## 16.1 Space-Time Reactor Kinetics

The time-dependent diffusion equation (4.39) is now modified to represent explicitly the delayed neutron fraction $\beta$ and fission spectrum $\chi_d(E)$, with the precursor concentration $C(\mathbf{r}, t)$, and prompt neutron fission spectrum $\chi_p(E)$

$$\begin{aligned}\frac{1}{v}\frac{\partial\phi(\mathbf{r},E,t)}{\partial t} &= (1-\beta)\chi_p(E)\int_0^\infty dE'\nu\Sigma_f(E')\phi(\mathbf{r},E',t) + Q(\mathbf{r},E,t)\\ &+ \chi_d(E)\lambda C(\mathbf{r},t) + \int_0^\infty dE'\,\Sigma_s(E'\to E)\phi(\mathbf{r},E',t)\\ &- \Sigma_t(E)\phi(\mathbf{r},E,t) + \nabla\cdot D(E)\nabla\phi(\mathbf{r},E,t),\end{aligned} \tag{16.1}$$

which is rewritten in terms of the production operator $L_2$ and the *net* loss operator $L_1$

$$\begin{aligned}\frac{1}{v}\frac{\partial\phi(\mathbf{r},E,t)}{\partial t} &= (1-\beta)\chi_p(E)L_2\phi(\mathbf{r},E,t) + Q(\mathbf{r},E,t)\\ &+ \chi_d(E)\lambda C(\mathbf{r},t) - L_1\phi(\mathbf{r},E,t).\end{aligned} \tag{16.2}$$

The operator notation is similar to that adopted in Chapter 10 but slightly modified to clarify the use of $\chi_p(E)$. The balance equation for the delayed neutron precursors is then set up

$$\frac{\partial C(\mathbf{r},t)}{\partial t} = -\lambda C(\mathbf{r},t) + \beta L_2\phi(\mathbf{r},E,t). \tag{16.3}$$

It should also be noted that the delayed neutron fraction $\beta$ should account for the differences in the fission spectra $\chi_p(E)$ and $\chi_d(E)$ of Figure 2.6 so that the actual value of $\beta$ should be obtained as an effective fraction $\beta_{eff}$.

### 16.1.1 Numerical Solution of Space-Time Kinetics Equation

The space-time kinetics equations are based on the steady-state form of the neutron diffusion equation, which also serves as the basis for numerical solution of the coupled equations (16.2) and (16.3). Our focus is on two representative solution techniques, one requiring a direct solution of the equations discretized in time and another requiring the factorization of the flux $\phi(\mathbf{r}, E, t)$ that makes an efficient use of the PKE solver. Other techniques for solving the space-time kinetics equations are discussed in a number of papers, in particular, a comprehensive review article [Sut96].

### 16.1.2 Direct Solution of Space-Time Kinetics Equation

The balance equation (16.2) for $\phi(\mathbf{r}, E, t)$ is rewritten, with $L = (1-\beta)\chi_p(E)L_2 - L_1$ and $Q(\mathbf{r}, E, t) = 0$, for the purpose of discretizing the equation in time

$$\frac{1}{v}\frac{\partial\phi(\mathbf{r},E,t)}{\partial t} \equiv L\phi(\mathbf{r},E,t) + \chi_d(E)\lambda C(\mathbf{r},t) \equiv R(\mathbf{r},E,t), \tag{16.4}$$

and a $\theta$-differencing technique [Dah74] is applied with the notation $\phi_n(\mathbf{r}, E) = \phi(\mathbf{r}, E, t_n), C_n(\mathbf{r}) = C(\mathbf{r}, t_n), R_n(\mathbf{r}, E) = R(\mathbf{r}, E, t_n)$, and time step $\Delta t = t_n - t_{n-1}$ to obtain

$$\begin{aligned}\frac{1}{v}\frac{\phi_n(\mathbf{r}, E) - \phi_{n-1}(\mathbf{r}, E)}{\Delta t} &= \theta R_n(\mathbf{r}, E) + (1-\theta)R_{n-1}(\mathbf{r}, E) \\ &= \theta[L\phi_n(\mathbf{r}, E) + \chi_d(E)\lambda C_n(\mathbf{r})] + (1-\theta)R_{n-1}(\mathbf{r}, E).\end{aligned} \tag{16.5}$$

This is followed by an implicit treatment with $\theta = 1.0$ to Eq. (16.3), with the recognition that the time evolution of the delayed neutron precursors, even for those with the shortest mean life, is usually slower than that for the neutron flux, which yields

$$\frac{C_n(\mathbf{r}) - C_{n-1}(\mathbf{r})}{\Delta t} = -\lambda C_n(\mathbf{r}) + \beta L_2\phi_n(\mathbf{r}, E). \tag{16.6}$$

Rewrite Eq. (16.6) for $C_n(\mathbf{r})$

$$\left(\lambda + \frac{1}{\Delta t}\right) C_n(\mathbf{r}) = \frac{C_{n-1}(\mathbf{r})}{\Delta t} + \beta L_2\phi_n(\mathbf{r}, E)$$

and obtain

$$\lambda C_n(\mathbf{r}) = \frac{\lambda[\beta L_2\phi_n(\mathbf{r}, E)\Delta t + C_{n-1}(\mathbf{r})]}{1 + \lambda\Delta t}. \tag{16.7}$$

Substituting Eq. (16.7) into Eq. (16.5) yields

$$\begin{aligned}&\left[L + \frac{\lambda\Delta t}{1+\lambda\Delta t}\beta\chi_d(E)L_2 - \frac{1}{\theta v\Delta t}\right]\phi_n(r.E) \\ &= \frac{\theta-1}{\theta}R_{n-1}(\mathbf{r}, E) - \frac{\chi_d(E)}{1+\lambda\Delta t}\lambda C_{n-1}(\mathbf{r}) - \frac{\phi_{n-1}(\mathbf{r}, E)}{\theta v\Delta t}.\end{aligned} \tag{16.8}$$

The neutron flux $\phi_n(\mathbf{r}, E)$ at time step $t_n$ can now be solved with the terms containing the neutron flux and delayed neutron precursors at time step $t_{n-1}$ collected on the RHS serving as a fixed source for Eq. (16.8). The actual solution involves discretizing various terms in both space and energy with slight modifications to the standard steady-state multigroup diffusion theory formulations. The choice of the differencing parameter $\theta$ as well as the time step size $\Delta t$ depends on the particular transients being simulated, with the implicit treatment $\theta = 1.0$ often allowing for efficient simulations. The choice of $\theta = 0.5$ representing the Crank–Nicolson scheme usually provides a stable solution.

Another approach for the direct solution of the space-dependent kinetics equations (16.2) and (16.3) involves an analytic integration of the precursor equation with a linear or quadratic relationship for the time dependence of the flux [Dow06]. A linear formulation entails

$$\phi(\mathbf{r}, E, t) = \phi_n(\mathbf{r}, E) + \frac{t_n - t}{\Delta t}[\phi_{n-1}(\mathbf{r}, E) - \phi_n(\mathbf{r}, E)],\ \Delta t = t_n - t_{n-1}, \tag{16.9}$$

which is substituted into the precursor equation (16.3) to yield

$$\begin{aligned}\lambda C_n(\mathbf{r}) &= \lambda C_{n-1}(\mathbf{r})e^{-\lambda\Delta t} + \beta\lambda L_2 \int_0^{\Delta t} \phi(\mathbf{r}, E, \tau)e^{-\lambda(t_n-\tau)}d\tau \\ &= \lambda C_{n-1}(\mathbf{r})e^{-\lambda\Delta t} + a_1 L_2\phi_n(\mathbf{r}, E) + a_2 L_2\phi_{n-1}(\mathbf{r}, E).\end{aligned} \tag{16.10}$$

Equation (16.10) is then used instead of Eq. (16.7) in Eq. (16.5) to obtain a slightly improved version of Eq. (16.8). The algebra for evaluating the coefficients $a_1$ and $a_2$ is left as an exercise and may be compared with the details given for a quadratic time-dependent treatment for the two-group flux evolution [Dow06].

### 16.1.3 Quasi-static Formulation of Kinetics Equation

A popular method for the solution of the space-time kinetics equations is the *quasi-static formulation* [Ott69], which involves representing the time evolution of the neutron flux primarily through the point-kinetics equations discussed in Chapter 8, but with the space- and energy-dependence of the flux allowed to change during a transient. The scalar flux is written as a product of the *shape function* $\psi(\mathbf{r}, E, t)$ and the *amplitude function* $n(t)$

$$\phi(\mathbf{r}, E, t) = \psi(\mathbf{r}, E, t)n(t), \tag{16.11}$$

where the bulk of the time dependence in $\phi(\mathbf{r}, E, t)$ is represented through $n(t)$, with corrections introduced through $\psi(\mathbf{r}, E, t)$. An extra degree of freedom introduced through the shape function is handled through a normalization condition introduced in the form of a weighting function $W(\mathbf{r}, E)$, which is usually obtained as the steady-state adjoint flux

$$\begin{aligned}&W(\mathbf{r}, E) = \phi^\dagger(\mathbf{r}, E), L_1^\dagger\phi^\dagger(\mathbf{r}, E) = \chi(E)L_2^\dagger\phi^\dagger(\mathbf{r}, E) \\ &\text{with } \chi(E) = \chi_p(E)(1-\beta) + \chi_d(E)\beta.\end{aligned} \tag{16.12}$$

The normalization condition is set up as

$$\left\langle \phi^\dagger(\mathbf{r}, E), \frac{1}{v}\psi(\mathbf{r}, E, t) \right\rangle = \text{constant} = \gamma, \tag{16.13}$$

with the inner product representing integrations both over the neutron energy $E$ and the reactor volume $V$.

Introducing the factorization equation (16.11) into Eq. (16.2) with the external source term suppressed yields

$$\begin{aligned}\frac{1}{v}\frac{\partial\psi(\mathbf{r}.E, t)}{\partial t}n(t) + \frac{\psi(\mathbf{r}, E, t)}{v}\frac{dn(t)}{dt} &= (1-\beta)\chi_p(E)L_2\psi(\mathbf{r}, E, t)n(t) \\ &\quad - L_1\psi(\mathbf{r}, E, t)n(t) + \chi_d(E)\lambda C(\mathbf{r}, t).\end{aligned} \tag{16.14}$$

Pre-multiply Eq. (16.14) by $\phi^\dagger(\mathbf{r},E)$ and perform integrations over $E$ and $V$ to obtain

$$\frac{d\gamma}{dt}n(t)+\gamma\frac{dn(t)}{dt}=(1-\beta)\langle\phi^\dagger,\chi_p(E)L_2\psi(\mathbf{r},E,t)\rangle n(t)$$
$$-\langle\phi^\dagger(\mathbf{r},E),L_1\psi(\mathbf{r},E,t)\rangle n(t)+\lambda\langle\phi^\dagger(\mathbf{r},E),\chi_d(E)C(\mathbf{r},t)\rangle. \tag{16.15}$$

Invoking the normalization condition of Eq. (16.13) with $\gamma=1$ converts Eq. (16.15) into the familiar PKE for the amplitude function $n(t)$

$$\frac{dn(t)}{dt}=\frac{K(t)-1}{\Lambda}n(t)+\lambda C(t), \tag{16.16}$$

with

$$\ell=\frac{1}{\langle\phi^\dagger(\mathbf{r},E),L_1\psi(\mathbf{r},E,t)\rangle},\ k(t)=\frac{\langle\phi^\dagger(\mathbf{r},E),\chi(E)L_2\psi(\mathbf{r},E,t)\rangle}{\langle\phi^\dagger(\mathbf{r},E),L_1\psi(\mathbf{r},E,t)\rangle},$$
$$\beta_{eff}=\frac{\langle\phi^\dagger(\mathbf{r},E),\beta\chi_d(E)L_2\psi(\mathbf{r},E,t)\rangle}{\langle\phi^\dagger(\mathbf{r},E),\chi(E)L_2\psi(\mathbf{r},E,t)\rangle},\ C(t)=\frac{\langle\phi^\dagger(\mathbf{r},E),\chi_d(E)C(\mathbf{r},t)\rangle}{\langle\phi^\dagger(\mathbf{r},E),L_1\psi(\mathbf{r},E,t)\rangle},$$
$$K(t)=\frac{k(t)-1}{k(t)\beta_{eff}},\ \Lambda=\frac{\ell}{k\beta_{eff}}.$$

Applying a similar inner-product step to the precursor equation (16.3) generates the familiar precursor balance equation in terms of the amplitude function $n(t)$ and the normalized neutron lifetime $\Lambda$:

$$\frac{dC(t)}{dt}=-\lambda C(t)+\frac{n(t)}{\Lambda}. \tag{16.17}$$

The neutron lifetime $\ell$ and $\Lambda$ now evolve, albeit slowly, over time, reflecting the time-dependent shape function $\psi(\mathbf{r},E,t)$. Accounting for the time dependence of $\ell$ and $\Lambda$ now requires obviously the solution of Eqs. (16.14) and (16.3) reorganized explicitly for $\psi(\mathbf{r},E,t)$

$$\frac{1}{v}\frac{\partial\psi(\mathbf{r}.E,t)}{\partial t}=\left[L-\frac{1}{v}\frac{\dot{n}(t)}{n(t)}\right]\psi(\mathbf{r},E,t)+\frac{\chi_d(E)}{n(t)}\lambda C(\mathbf{r},t), \tag{16.18a}$$

$$\frac{\partial C(\mathbf{r},t)}{\partial t}=-\lambda C(\mathbf{r},t)+\beta L_2\psi(\mathbf{r},E,t)n(t), \tag{16.18b}$$

which are similar to the original balance equations (16.2) and (16.3), but require the use of the amplitude function $n(t)$ and its derivative. Thus, Eqs. (16.16) and (16.17) are solved together with Eqs. (16.18) but with different time step sizes. The PKEs (16.16) and (16.17) are solved with fine time steps, while Eqs. (16.18) are solved with larger time steps with algorithms similar to the direct solution formulations from Section 16.1.2 and reflected back in updated evaluations of $K(t)$, $\Lambda$, and $C(t)$. With thermal-hydraulic feedback effects reflected in the neutron cross sections and hence in the diffusion theory operators $L_1$ and $L_2$, the shape-function time steps, as well as the amplitude-function time steps, are also adjusted to reflect the particular transients simulated. In practical implementations, the point kinetics parameters are updated to reflect the cross section variations for the solution of Eqs. (16.16) and (16.17) but without updating $\psi(\mathbf{r}, E, t)$ during one shape-function time step.

This is the algorithm used in the FX2-TH [Sho78], QUASAR [Lee72], and PARCS [Dow06] codes for the solution of the time-dependent diffusion equations in multigroup and multidimensional formulations. A similar idea has been implemented effectively in the time-dependent solution [Zhu16] of the neutron transport equation in the MPACT code, with an intermediate step added for the 3-D coarse-mesh diffusion solver between the method of characteristics (MOC) transport solver in 2-D planes and the PKE solver.

**Example 16.1** Compare the point kinetics solution using constant values of $\Lambda$ and $\beta_{eff}$ with the quasi-static solution using the PARCS code.

The PARCS code is exercised with six delayed neutron groups in a hot zero power (HZP) control rod ejection scenario [Bar99]. The transient is initiated at a relative power level of $1.0 \times 10^{-4}\%$ of full power, with the control rod ejection represented in a three-dimensional two-group diffusion theory formulation. The control rod ejection occurs over 0.3 s, resulting in a super-prompt critical transient with a maximum reactivity of 1.08 dollar, which is terminated by fuel temperature feedback. The reactivity $K(t)$ for the PKE solution is derived from the quasi-static formulation with the time-dependent shape function $\psi(\mathbf{r}, E, t)$ and illustrated in Figure 16.1. The PKE solution through a simple Crank–Nicolson scheme [Dah74] yields the peak power at 0.540 s in agreement with the quasi-static solution. The PKE peak power of 1.2384 also compares fairly well with the PARCS value of 1.2620, as illustrated in Figure 16.2, where an inset clarifies the differences around the power peak. This simple exercise indicates that as long as the reactivity is calculated to represent spatial effects accurately, the power level calculation would be also sufficiently accurate. This was observed also in an earlier study where the HZP rod ejection for a PWR core was simulated with a QUASAR 1-D axial quasi-static formulation [Lee72] with the reactivity insertion represented through a radial synthesis calculation. The 1-D rod ejection calculation, with the synthesized reactivity, compared fairly well with the 3-D TWINKLE calculation [Ris75].

### 16.1.4 Reactivity Determination from Multiple Detectors

As another application of the quasi-static formulation of Section 16.1.3, we develop a method to determine the reactivity from multiple detector signals. Recall first that a general method to determine the reactivity from flux or power

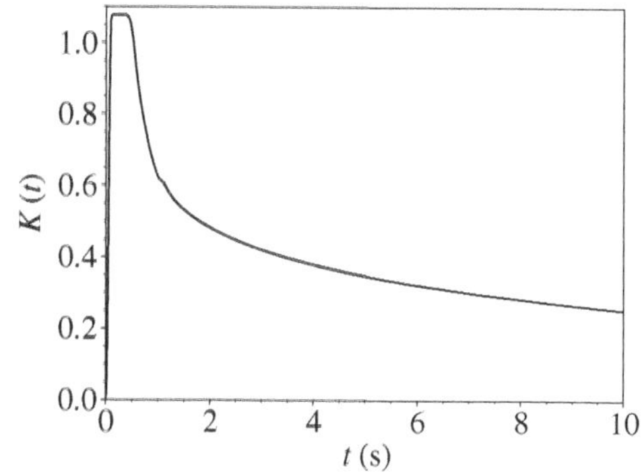


**Figure 16.1** Reactivity insertion $K(t)$ [dollar] arising from control rod ejection from a HZP configuration.

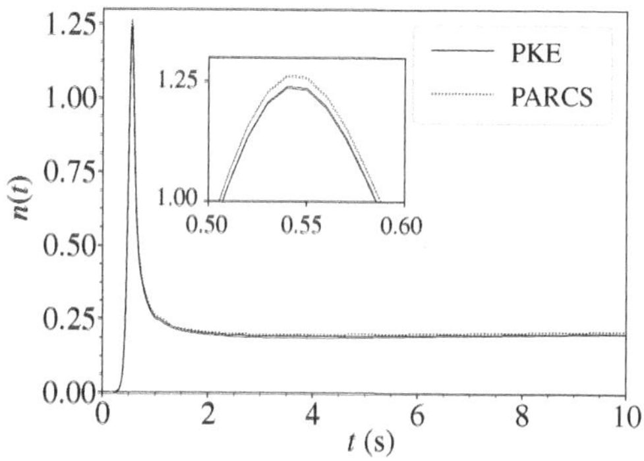


**Figure 16.2** Comparison of power transient $n(t)$ due to HZP control rod ejection between the PKE solution and quasi-static solution with the PARCS code. Reactivity $K(t)$ for the PKE calculation is derived from the PARCS calculation.

level measurements is the reactivity meter algorithm of Eq. (8.93). The reactivity $K(t)$ determined via the reactivity meter algorithm, however, depends on the detector location relative to the position where the reactivity perturbation is introduced, e.g. through a control rod movement in a critical reactor. In a source- or accelerator-driven subcritical system, the reactivity determined depends on the relationship between the location of the detector and that of a time-dependent source. Instead of taking an arithmetic average of either multiple detector signals or reactivity estimates obtained from the detectors, now consider a systematic application of the quasi-static formulation summarized in Eqs. (16.18) to determine the amplitude function $n(t)$ and reactivity $K(t)$.

The quasi-static reactivity meter algorithm [Mor84] requires representing the detector signal $R(\mathbf{r}_0, t)$ for a detector with an effective cross section $\Sigma_d(E)$ at position $\mathbf{r}_0$ and neutron flux $\phi(\mathbf{r}_0, E, t)$ in terms of the amplitude function

$$n(t) = \frac{R(\mathbf{r}_0, t)}{\langle \Sigma_d(E), \psi(\mathbf{r}_0, E, t)\rangle}, \tag{16.19}$$

where the inner product represents an integral over energy. With a sufficiently accurate evaluation of the shape function $\psi(\mathbf{r}_0, E, t)$, the detector signal $R(\mathbf{r}_0, t)$ can be corrected to yield $n(t)$ representing the whole core. For an efficient evaluation of the shape function, a combination of modal expansion and local function techniques was developed [Mor84] for the shape function,

$$\psi(\mathbf{r}, E, t) = h(\mathbf{r}, E, t) + f(\mathbf{r}, E, t), \tag{16.20}$$

where the substantial flux perturbations in the vicinity of variations in the reactor parameters may be represented by the local component $f(\mathbf{r}, E, t)$, while the overall smooth perturbations in flux can be represented by the modal component $h(\mathbf{r}, E, t)$. With the perturbation $\delta L$ in the operator $L = (1 - \beta)\chi_p(E)L_2 - L_1$, the shape function equation (16.18a) is separated into two parts:

$$\frac{1}{v}\frac{\partial f(\mathbf{r}, E, t)}{\partial t} + L_1 f(\mathbf{r}, E, t) = \delta L\psi(\mathbf{r}, E, t), \tag{16.21}$$

$$\begin{aligned} \frac{1}{v}\frac{\partial h(\mathbf{r}, E, t)}{\partial t} &= \left[L - \frac{1}{v}\frac{\dot{n}(t)}{n(t)}\right] h(\mathbf{r}, E, t) \\ &+ \left[(1 - \beta)\chi_p(E)L_2 - \frac{1}{v}\frac{\dot{n}(t)}{n(t)}\right] f(\mathbf{r}, E, t) + \frac{\chi_d(E)}{n(t)}\lambda C(\mathbf{r}, t). \end{aligned} \tag{16.22}$$

The local function $f(\mathbf{r}, E, t)$ in Eq. (16.21) is written explicitly for perturbations involving neutron absorptions in the operator $L_1$ and may be obtained with the time derivative term and the second-order term $\delta L\delta\psi(\mathbf{r}, E, t)$ neglected:

$$L_1 f(\mathbf{r}, E, t) = \delta L\psi(\mathbf{r}, E, 0), \tag{16.23}$$

accounting primarily for instantaneous local variations. The solution is substituted into Eq. (16.22), and the modal function $h(\mathbf{r}, E, t)$ is expanded through lambda modes $\phi_n(\mathbf{r}, E)$:

$$h(\mathbf{r}, E, t) = \sum_{n=1}^{N} a_n(t)\phi_n(\mathbf{r}, E), \tag{16.24}$$

The lambda modes are introduced in Eq. (10.48), with explicit accounting given here for the combined neutron spectrum $\chi(E)$:

$$L_1\phi_n(\mathbf{r}, E) = \frac{\chi(E)}{\lambda_n} L_2\phi_n(\mathbf{r}, E), \chi(E) = (1 - \beta)\chi_p(E) + \beta\chi_d(E). \tag{16.25}$$

With the lambda mode expansion of Eq. (16.24) substituted into Eq. (16.22), pre-multiply by the adjoint lambda mode $\phi_m^{\dagger}(\mathbf{r}, E)$ and integrate over neutron energy $E$ and reactor volume $V$. A similar step is taken for the delayed neutron precursor equation (16.18b) and the orthonormality properties for the lambda modes are invoked, with the cross-product terms involving lambda modes of different orders recognized as the second order. This yields a set of ODEs for the expansion coefficients $a_n(t)$ for the modal function and those for the precursor concentration $C(\mathbf{r}, t)$. With the local function equation (16.23) obtained efficiently

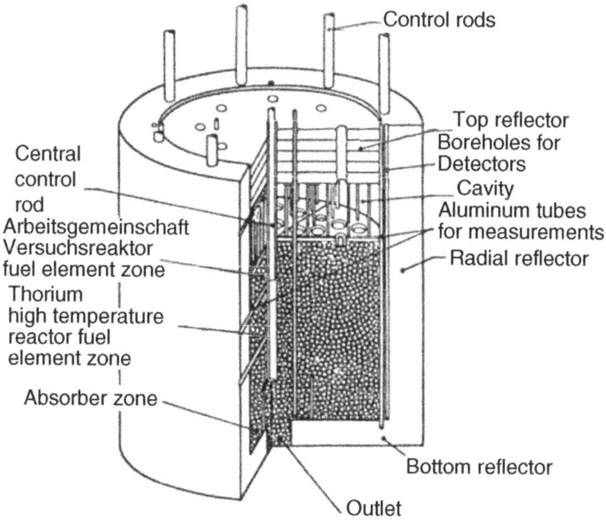


**Figure 16.3** Overall layout of the KAHTER critical assembly for the pebble-bed reactor. *Source*: Moreira and Lee [Mor84].

via a fixed-source multigroup diffusion equation solver, the set of ODEs for the modal function $h(\mathbf{r}, E, t)$ and delayed neutron precursor concentration $C(\mathbf{r}, t)$ is solved to generate the shape function $\psi(\mathbf{r}, E, t)$ and the amplitude function $n(t)$ via Eq. (16.19). The details for extracting the lambda mode expansion coefficient $a_n(t)$ are clarified further in Section 16.2 for space-time xenon oscillation analysis.

The modal-local approach for evaluating the shape function $\psi(\mathbf{r}, E, t)$ was applied to the space-time analysis [Mor84] of control-rod worth measurements at the KAHTER pebble-bed gas-cooled reactor test facility [Ker80] in Jülich, Germany. Figure 16.3 illustrates the cylindrical critical test facility with a core diameter of 2.16 m and a height of 3.0 m, surrounded by a radial graphite reflector 0.6 m in thickness and graphite reflectors 0.6 and 0.5 m in thickness at the top and the bottom, respectively. Illustrated also are the control rod configurations and the upper plenum gas region. Figure 16.4 presents a top view of the facility indicating both the eight top reflector and eight radial reflector absorber rods as well as the eight neutron instrumentation channels. Figure 16.3 also shows the 6-cm pebbles loaded into the critical assembly.

The integral rod worth measurements were conducted with the top reflector rods inserted into the central region of the cylindrical assembly, and count rates are plotted in Figure 16.5 for 10 s following the initiation of rod insertion. The actual rod insertion took place over ~4.0 s. Significant position dependence of the raw count rates is noted for the detectors 1 through 4 located in channel A4 at a

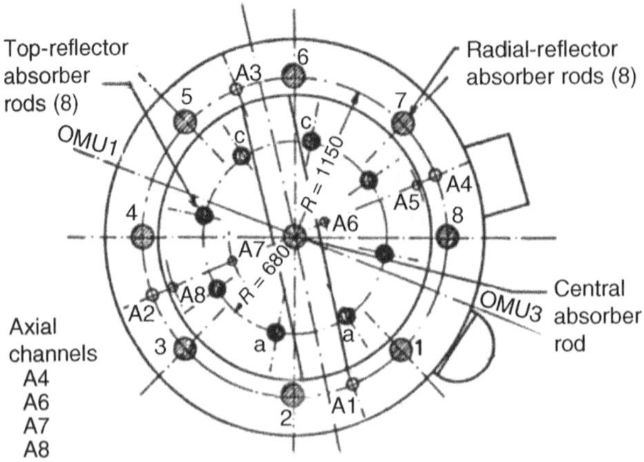


**Figure 16.4** Top view of the instrumentation channel layout for the KAHTER facility. The radii are expressed in mm. *Source*: Moreira and Lee [Mor84].

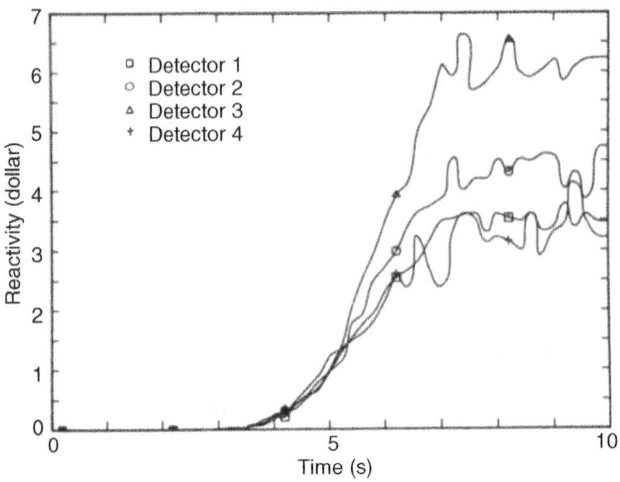


**Figure 16.5** Integral rod worth as a function of time after rod insertion without space-time correction. *Source*: Moreira and Lee [Mor84]/Taylor & Francis Group.

height of 1.2 m, in channel A6 at 1.8 m, in channel A7 at 2.4 m, and in channel A8 at 0.4 m from the bottom of the core, respectively. The shape function $\psi(\mathbf{r}, E, t)$ was calculated with the modal-local quasi-static formulation of Eqs. (16.21) and (16.22) with the CITATION code [Fow71] in $(r–\theta–z)$ three-dimensional geometry with 6 energy groups, 27 radial meshes, 6 azimuthal angular meshes, and 41 axial meshes. The lambda mode calculation was limited to the first harmonic in Eq. (16.24). The

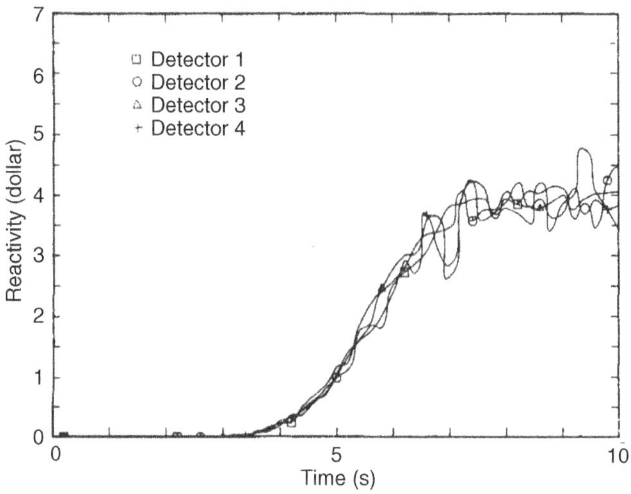


**Figure 16.6** Integral rod worth as a function of time after rod insertion corrected with the modal-local method. *Source*: Moreira and Lee [Mor84]/Taylor & Francis Group.

KAHTER rod worth measurements were corrected via the amplitude function $n(t)$ from Eq. (16.19) for each of the four detectors, and the corrected rod worth plots in Figure 16.6 show significantly reduced dependence on the location of the detector, despite the residual presence of statistical fluctuations. The corrected integral rod worth of $(3.87 \pm 0.23)$ dollar averaged over the four instrumentation channels indicates a good agreement with the rod worth of $(3.91 \pm 0.12)$ dollar obtained [Ker80] with an adiabatic treatment [Ott69] requiring static eigenvalue calculations for several intermediate reactor configurations.

## 16.2 Space-Time Power Oscillations due to Xenon Poisoning

The issues associated with the buildup of the fission product $^{135}$Xe in light water reactor (LWR) cores are studied as part of the fuel cycle analysis in Chapter 12. In addition to the considerations required to account for the reactivity poisoning in power operation due to the large thermal absorption cross section of $^{135}$Xe, an increased buildup of $^{135}$Xe following reactor shutdown from an equilibrium xenon concentration requires additional provisions in reactor operations in general. Furthermore, in large LWR cores, the flux and power distributions will undergo space-time oscillations even with a relatively constant power output of the core. We investigate the space-time oscillations due to $^{135}$Xe poisoning in

PWR cores and formulate optimal control strategies both for the post-shutdown xenon buildup and space-time oscillations. Effort is made first to formulate the space-time kinetics framework using perturbation theory, discussed in Chapter 10, and we move next to study control strategies.

### 16.2.1 Modal Analysis of Space-Time Xenon-Power Oscillations

We begin our space-time power oscillation study rewriting the balance equations (12.40) and (12.41) for $I(z,t)$ and $X(z,t)$ representing the $^{135}$I and $^{135}$Xe distributions, respectively, in a one-dimensional axial formulation

$$\frac{\partial X(z,t)}{\partial t} = \gamma_X \Sigma_f \phi(z,t) + \lambda_I I(z,t) - \lambda^*(z,t) X(z,t), \tag{16.26}$$

$$\frac{\partial I(z,t)}{\partial t} = \gamma_I \Sigma_f \phi(z,t) - \lambda_I I(z,t), \tag{16.27}$$

where $\lambda^*(z,t) = \lambda_X + \sigma_X \phi(z,t)$ representing the effective destruction rate of $^{135}$Xe and the $^{135}$I destruction due to neutron absorption is suppressed. Movement of control rods in and out of the core in a PWR plant could perturb the axial flux distribution $\phi(z,t)$ in the core significantly, resulting in perturbations for the destruction rate of $^{135}$Xe and the production rate of $^{135}$I. The perturbed $^{135}$I production rate would in turn result, with a time delay, in perturbations in $X(z,t)$, reversing the initial perturbations, and causing the transient in the reverse direction. The sequence of events is illustrated schematically in Figure 16.7, starting with the axial flux perturbation $\delta\phi(z,t)$ associated with the flux shifting toward the bottom of the core.

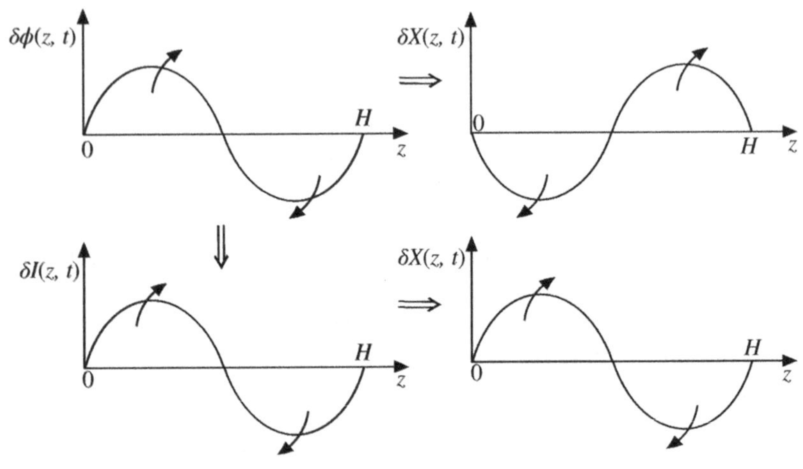


**Figure 16.7** Evolution of flux perturbations into perturbations for xenon and iodine concentrations.

This results in shifting the $^{135}$Xe distribution toward the top of the core, while the $^{135}$I distribution is gradually shifted toward the bottom of the core. With a time delay characteristic of the $^{135}$I decay constant $\lambda_I$, the $^{135}$Xe distribution will be shifted toward the bottom, causing a $\phi(z,t)$ shift toward the top of the core, thereby reversing the initial oscillations.

An analytical formulation for the space-time xenon oscillations is set up in terms of perturbation theory introduced in Chapter 10, where flux perturbation $\delta\phi(z,t)$ and associated perturbations $\delta X(z,t)$ and $\delta I(z,t)$ are expressed in terms of the lambda modes with the fundamental mode representing the steady-state one-group neutron diffusion equation

$$L_0\phi_0(z) = \left(L_1 - \frac{L_2}{\lambda_0}\right)\phi_0(z) = 0,\ \lambda_0 = k_{eff} = 1.0, \tag{16.28}$$

together with the corresponding steady-state distributions $X_0(z)$ and $I_0(z)$. The perturbed flux and perturbed $^{135}$Xe, and $^{135}$I distributions may then be represented in terms of the $n$th order lambda-mode eigenfunctions:

$$\phi(z,t) = \phi_0(z) + \delta\phi(z,t) \simeq \phi_0(z) + \sum_{i=1}^{n} a_i(t)\phi_i(z), \tag{16.29a}$$

$$X(z,t) = X_0(z) + \delta X(z,t) \simeq X_0(z) + \sum_{i=1}^{n} b_i(t)\phi_i(z), \tag{16.29b}$$

$$I(z,t) = I_0(z) + \delta I(z,t) \simeq I_0(z) + \sum_{i=1}^{n} c_i(t)\phi_i(z). \tag{16.29c}$$

The perturbations $\delta\phi(z,t), \delta X(z,t)$, and $\delta I(z,t)$ illustrated in Figure 16.7 may in fact be visualized as the first-order lambda mode $\phi_1(z)$. The remaining task now requires representing the perturbations in the diffusion theory operator $L_0$ together with Eqs. (16.26) and (16.27) to solve for the time-dependent expansion coefficients $a_i(t), b_i(t)$, and $c_i(t)$.

The perturbation $\delta L$ in the diffusion theory operator now includes control maneuvers $\delta L_c$ that may be required to control space-time flux-xenon oscillations, changes $\sigma_X \delta X(z,t)$ in the absorption cross section, and thermal-hydraulic feedback effects $\alpha_T \delta\phi(z,t)$

$$\delta L = \delta L_c + \sigma_X \delta X(z,t) + \alpha_T \delta\phi(z,t), \tag{16.30}$$

so that perturbation theory yields an expression connecting $\delta L$ to $\delta\phi(z,t)$

$$\delta L\phi_0(z) + L_0\delta\phi(z,t) = 0. \tag{16.31}$$

Likewise, substituting Eqs. (16.29) into Eqs. (16.26) and (16.27) yields

$$\frac{\partial \delta X(z,t)}{\partial t} = [\gamma_X\Sigma_f - \sigma_X X_0(z)]\delta\phi(z,t) - \lambda_X^*(z,t)\delta X(z,t) + \lambda_I \delta I(z,t), \tag{16.32}$$

$$\frac{\partial \delta I(z,t)}{\partial t} = \gamma_I \Sigma_f \delta\phi(z,t) - \lambda_I \delta I(z,t). \tag{16.33}$$

Now pre-multiply Eq. (16.31) by the adjoint lambda mode $\phi_i^\dagger(z)$ and perform an integral over the reactor volume or equivalently core height $H$ to obtain

$$\langle \phi_i^\dagger, \delta L \phi_0 \rangle + \langle \phi_i^\dagger, L_0 \delta\phi \rangle = 0$$

or equivalently,

$$\begin{aligned}\langle \phi_i^\dagger, \delta L_c \phi_0 \rangle + \left\langle \phi_i^\dagger, \sigma_X \sum_j b_j \phi_j \phi_0 \right\rangle + \left\langle \phi_i^\dagger, \alpha_T \sum_j a_j \phi_j \phi_0 \right\rangle \\ + \left\langle \phi_i^\dagger, L_0 \sum_j a_j \phi_j \right\rangle = 0.\end{aligned} \tag{16.34}$$

Recalling the orthonormality properties of the lambda modes in Eq. (10.48)

$$\left\langle \phi_i^\dagger, \frac{L_2}{\lambda_j} \phi_j \right\rangle = \delta_{ij} = \left\langle \frac{L_2^\dagger}{\lambda_i} \phi_i^\dagger, \phi_j \right\rangle = \langle L_1^\dagger \phi_i^\dagger, \phi_j \rangle, \tag{16.35}$$

introduce an approximation to drop cross-product terms in the second and third terms in Eq. (16.34). Furthermore, Eq. (10.49c) is invoked to simplify the last term representing the perturbed flux:

$$\langle \phi_i^\dagger, L_0 \delta\phi \rangle = \sum_j^n a_j \left(1 - \frac{\lambda_j}{\lambda_0}\right) \delta_{ij} = a_i(1 - \lambda_i), \ \forall i. \tag{16.36}$$

Effort is also made now to focus on the first harmonic mode $\phi_1(z)$, which allows us to introduce a simplified notation $a(t) = a_1(t)$, $b(t) = b_1(t)$ and $c(t) = c_1(t)$, and rewrite Eq. (16.34)

$$\langle \phi_1^\dagger, \delta L_c \phi_0 \rangle + \langle \phi_1^\dagger, \sigma_X \phi_1 \phi_0 \rangle b(t) + \langle \phi_1^\dagger, \alpha_T \phi_1 \phi_0 \rangle a(t) + (1 - \lambda_1) a(t) = 0. \tag{16.37}$$

Equation (16.37) is solved for $a(t)$

$$a(t) = g_1 b(t) + u(t) \equiv x_1(t) + u(t), \tag{16.38}$$

by defining

$$g_1 = -\frac{\langle \phi_1^\dagger, \sigma_X \phi_1 \phi_0 \rangle}{1 - \lambda_1 + \langle \phi_1^\dagger, \alpha_T \phi_1 \phi_0 \rangle}, \quad u(t) = -\frac{\langle \phi_1^\dagger, \delta L_c \phi_0 \rangle}{1 - \lambda_1 + \langle \phi_1^\dagger, \alpha_T \phi_1 \phi_0 \rangle}.$$

Note that $a(t)$ is the expansion coefficient or the amplitude function for the first harmonic mode visualized in Figure 16.7 and may be represented as the *axial offset (AO) of power* or equivalently as the AO of flux introduced in Eq. (12.65)

$$\text{AO} = \frac{P_T - P_B}{P_T + P_B}, \tag{16.39}$$

where $P_T$ and $P_B$ represent the power produced in the top and bottom halves of the core, respectively. With this interpretation for $a(t)$, the parameters $x_1(t)$ and $u(t)$ now represent the xenon-induced flux perturbations and control actions, respectively, both expressed in units of AO of power.

Applying a similar inner-product operation with the adjoint harmonics to Eqs. (16.32) and (16.33), and limiting the analysis to the first harmonics provide expressions required for the xenon and iodine expansion coefficients $b(t)$ and $c(t)$,

$$\frac{db(t)}{dt} = \left(\gamma_X\overline{\Sigma_f} - \overline{\sigma_X X_0}\right) a(t) - \left(\lambda_X + \overline{\sigma_X \phi_0}\right) b(t) + \lambda_I c(t), \tag{16.40}$$

$$\frac{dc(t)}{dt} = \gamma_I \overline{\Sigma_f} a(t) - \lambda_I c(t), \tag{16.41}$$

with definitions for the weighted-average parameters

$$\overline{\Sigma_f} = \langle \phi_1^\dagger, \Sigma_f L_1 \phi_1 \rangle,\ \overline{\sigma_X X_0} = \langle \phi_1^\dagger, \sigma_X X_0 L_1 \phi_1 \rangle,\ \overline{\sigma_X \phi_0} = \langle \phi_1^\dagger, \sigma_X \phi_0 L_1 \phi_1 \rangle.$$

Now introduce the parameter $g_1$ of Eq. (16.38) into Eqs. (16.40) and (16.41) and rearrange them to obtain the final form of the expressions for the xenon and iodine expansion coefficients

$$\begin{aligned}\frac{dx_1(t)}{dt} &= f_3 a(t) - \left(\lambda_X + \overline{\sigma_X \phi_0}\right) x_1(t) - f_2 x_2(t), \\ &= f_1 x_1(t) - f_2 x_2(t) + f_3 u(t),\end{aligned} \tag{16.42}$$

$$\frac{dx_2(t)}{dt} = \lambda_I [a(t) - x_2(t)] = \lambda_I x_1(t) + \lambda_I u(t) - \lambda_I x_2(t), \tag{16.43}$$

where

$$f_1 = -g_1\left(\overline{\sigma_X X_0} - \gamma_X \overline{\Sigma_f}\right) - \left(\lambda_X + \overline{\sigma_X \phi_0}\right),$$

$$f_2 = -g_1 \gamma_I \overline{\Sigma_f}, f_3 = -g_1\left(\overline{\sigma_X X_0} - \gamma_X \overline{\Sigma_f}\right).$$

Equations (16.42) and (16.43) may now be rewritten in terms of a two-dimensional vector $x(t) = [x_1(t), x_2(t)]^T$ describing the time evolution of the xenon–iodine first harmonics

$$\frac{dx(t)}{dt} = \frac{d}{dt}\begin{bmatrix} x_1(t) \\ x_2(t) \end{bmatrix} = \begin{bmatrix} f_1 & -f_2 \\ \lambda_I & -\lambda_I \end{bmatrix}\begin{bmatrix} x_1(t) \\ x_2(t) \end{bmatrix} + u(t)\begin{bmatrix} f_3 \\ \lambda_I \end{bmatrix} \equiv Ax(t) + u(t)B, \tag{16.44}$$

together with Eq. (16.38) for the AO of power $a(t) = x_1(t) + u(t)$. To be consistent with control theory literature, the vector $x(t)$ is not boldfaced in Eq. (16.44) and for the rest of the chapter.

The weighted-average parameters introduced in Eqs. (16.38), (16.40), and (16.41) may be determined with any numerical diffusion equation solvers with the proper xenon–iodine dynamics represented. It is worth noting that a steady-state form of the neutron diffusion equation is used in the formulation of xenon–iodine dynamics, because the half-lives of $^{135}$Xe and $^{135}$I are on the order of several hours and hence the neutron population assumes a quasi-static state as the $^{135}$Xe and $^{135}$I distributions in the core evolve in time. In practice, however, it is more convenient to perform simple numerical experiments with the diffusion equation solver and determine the parameters $f_1$,$f_2$, and $f_3$ directly and represent $u(t)$ in units of AO of power as control rods are exercised [Sch80].

### 16.2.2 Stability of Space-Time Xenon-Power Oscillations

Given the governing equation (16.44) for the xenon–iodine dynamics, the stability of the core against spatial power oscillations can be studied for free-running oscillations by setting the control rod movement $u(t) = 0$ and evaluating the eigenvalues $\xi$ of the $(2 \times 2)$ matrix $A$ by setting up

$$Ax = \xi x \implies x(t) = x(0)e^{\xi t},\ \xi = \alpha \pm i\omega. \tag{16.45}$$

For the $(2 \times 2)$ matrix $A$, the eigenvalues satisfy the quadratic equation

$$\xi^2 + (\lambda_I - f_1)\xi + \lambda_I(f_2 - f_1) = 0, \tag{16.46}$$

providing

$$\alpha = \frac{f_1 - \lambda_I}{2}, \omega^2 = -(\lambda_I + \alpha)^2 + \lambda_I f_2. \tag{16.47}$$

The eigenvalue $\xi$ can be used to rewrite

$$f_1 = \lambda_I + 2\alpha, f_2 = \frac{(\lambda_I + \alpha)^2 + \omega^2}{\lambda_I}. \tag{16.48}$$

A free-running oscillation grows in magnitude if the *stability index* $\alpha > 0$, while the period of oscillations is given by $T = 2\pi/\omega$. The factor $\exp(\alpha T)$ represents the ratio of the amplitudes of the AO oscillations over two successive cycles and is known as the *decay ratio*; the oscillation is unstable if $\alpha > 0$ or $\exp(\alpha T) > 1.0$. By performing numerical experiments simulating free-running oscillations with a diffusion equation solver with the xenon–iodine dynamics represented, the period of oscillations $T$ and stability index $\alpha$ may be determined, which then provide numerical values of the parameters $f_1$ and $f_2$ via Eq. (16.48) without performing the complicated integrals introduced in Eq. (16.42). The parameter $f_3$ likewise may be determined by simulating a pulse-insertion of the control rod bank of interest illustrated in Figure 16.8. With Eq. (16.42) evaluated right after the control bank insertion into an equilibrium state, we obtain $a(0+) = u_0$ and

$$\frac{da(0+)}{dt} = \frac{dx_1(0+)}{dt} = f_1x_1(0+) - f_2x_2(0+) + f_3u_0 = \frac{dx_1(0-)}{dt} + f_3u_0 = f_3u_0. \tag{16.49}$$

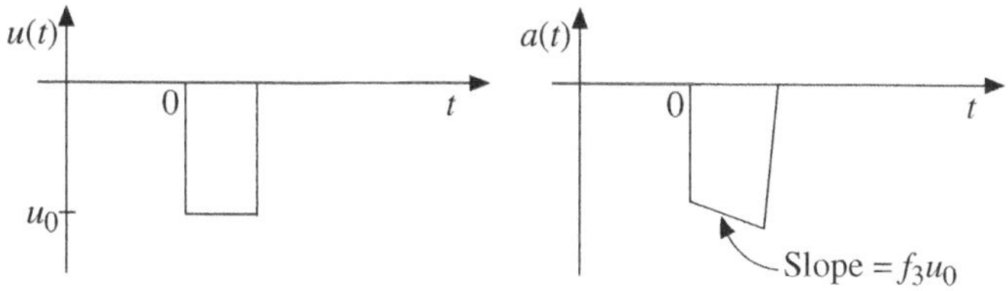


**Figure 16.8** Determination of system parameter $f_3$ via simulating a control rod bank insertion.

Thus, measuring the slope of the AO transient associated with the control rod bank insertion over a short rectangular pulse conveniently yields $f_3$. For numerical simulations of space-time xenon transients, care should be taken to use relatively fine time steps. If computational resources are limited, a $\theta$-weighted scheme similar to the space-time kinetics algorithm [Lee74] of Section 16.1.2 may be adopted. For more realistic analysis and accurate determination of the system parameters, actual tests should of course be performed for the reactor core of interest whenever possible.

Having obtained an expression for the stability index $\alpha$, we may now evaluate the condition that could result in unstable oscillations, as studied originally by Randall and St. John [Ran58] and somewhat later by Ash [Ash79] and others. For this purpose, return to the definition for the parameter $f_1$ in Eqs. (16.42) and (16.48) and approximate the three weighted averages in terms of simple volume-average $\Sigma_f$, volume-average xenon concentration $X_0$, and volume-average flux $\phi_0$, respectively

$$\overline{\Sigma_f} \simeq \Sigma_f, \overline{\sigma_X X_0} \simeq \sigma_X X_0, \overline{\sigma_X \phi_0} \simeq \sigma_X \phi_0,$$

together with the feedback coefficient $\alpha_T = 0$. We may then translate the instability condition $(f_1 - \lambda_I) > 0$ to

$$-(\lambda_I + \lambda_X + \sigma_x \phi_0) + \frac{\langle \phi_1^\dagger, \sigma_X \phi_0 \phi_1 \rangle (\sigma_X X_0 - \gamma_X \Sigma_f)}{1 - \lambda_1} > 0. \tag{16.50}$$

With the orthonormality condition of the lambda modes, Eq. (16.35), the instability condition is rewritten as

$$\frac{\sigma_X}{\nu \Sigma_f} \phi_0 \left( \frac{\sigma_X X_0 - \gamma_X \Sigma_f}{\lambda_I + \lambda_X + \sigma_X \phi_0} \right) > \frac{1}{\lambda_1} - 1, \tag{16.51}$$

which is equivalent to

$$\frac{\sigma_X}{\nu \Sigma_f} F(\phi_0) = \frac{\gamma}{\nu} \left( \frac{1}{1 + \lambda_X/(\sigma_X \phi_0)} - \frac{\gamma_X}{\gamma} \right) \left( 1 + \frac{\lambda_I + \lambda_X}{\sigma_X \phi_0} \right)^{-1} > \frac{1}{\lambda_1} - 1, \tag{16.52}$$

in terms of the stability function $F(\phi_0)$ [Ash79] with a missing term duly added.

**Example 16.2** Evaluate and discuss the instability condition of Eq. (16.52) for typical PWR parameters similar to those of the AP1000 design.

Recall the $^{135}$Xe and $^{135}$I parameters considered in Section 12.5:

$$\gamma_I = 0.064, \gamma_X = 0.002, \sigma_X = 1.5 \text{ Mb}, \lambda_I = 2.89 \times 10^{-5} \text{ s}^{-1}, \lambda_X = 2.08 \times 10^{-5} \text{ s}^{-1},$$

together with $\Sigma_f = 0.065$ cm$^{-1}$, $\nu\Sigma_f = 0.0159$ cm$^{-1}$, and $\phi_0 = 5.28 \times 10^{13}$ cm$^{-2}$ s$^{-1}$. The parameters yield equilibrium concentrations

$$I_0 = \gamma_I \Sigma_f \phi_0 / \lambda_I = 7.60 \times 10^{-9} [\text{nucleus b}^{-1} \text{ cm}^{-1}],$$

$$X_0 = \gamma \Sigma_f \phi_0 / (\lambda_X + \sigma_X \phi_0) = 2.26 \times 10^{-9} [\text{nucleus b}^{-1} \text{ cm}^{-1}].$$

Corresponding to these equilibrium concentrations and average thermal flux $\phi_0$, Eq. (16.51) yields

$$\frac{\sigma_X}{\nu\Sigma_f} F(\phi_0) = 0.027 \left( \frac{1}{1.26} - 0.03 \right) \frac{1}{1.63} = 0.013.$$

In terms of the migration area $M^2$ and geometrical buckling $B^2$, we introduce an approximate estimate of the nonleakage probability $P_{NL} = 1/(1 + M^2B^2) = 0.975$, and normal criticality for the core $\lambda_0 = k_0 = 1.0$ to obtain an estimate for the eigenvalue for the first lambda mode $\lambda_1 = (1 + M^2B^2)/(1 + 4M^2B^2) = 1/1.073$. With this estimate of $\lambda_1$, Eq. (16.51) yields $\sigma_X F(\phi_0)/\nu\Sigma_f < 1/\lambda_1 - 1$, suggesting that AP1000 will be stable against axial xenon oscillations. On the other hand, if we consider just the axial buckling $B^2 = 5.42 \times 10^{-5}$ cm$^{-2}$ for a core height of 4.27 m and a typical value of the migration area $M^2 = 50$ cm$^2$, an alternate estimate of $\lambda_1 = 1/1.0081$ is obtained, indicating the possibility of unstable first-harmonic axial xenon oscillations for the 14-foot AP1000 core.

This analysis could be also compared with an earlier analysis by Randall and St. John [Ran58], where for sufficiently high thermal flux, a stability criterion of $(H/M)^2 < 1,000$ is suggested for core height $H$ and migration area $M^2$. This simple criterion would suggest that for the AP1000 core $(H/M)^2 = 3,600$, rendering the core unstable. In the overall stability analysis, however, effects of temperature feedback represented by the coefficient $\alpha_T$ in Eq. (16.30) and flat flux approximations introduced in Eq. (16.50) should also be considered.

The stability of PWR cores against axial xenon oscillations was actually verified in 1970 through performing two free-running oscillation tests [Lee71a, Lee71b] at the Rochester Gas and Electric (RGE) Ginna Unit 1 plant, which was one of the first 12-foot Westinghouse PWR plants that went into operation in the early days of PWR development. The tests were performed fairly early in the first cycle but corresponding to two different core-average burnup states: 1.55 and 7.70 MWd/kgHM

**Table 16.1** Stability index and oscillation period measurements at the RGE Plant.

| Core average burnup (MWd/kgHM) | Stability index ($hr^{-1}$) | Oscillation period (hr) | Critical boron concentration (ppm) |
|---|---|---|---|
| 1.55 | −0.041 | 32.4 | 1,065 |
| 7.70 | −0.014 | 27.2 | 700 |

[Lee71a, Hon12]. The transients were induced by a pulse-insertion of control rod bank D into the core, thereby shifting the power distribution toward the bottom of the core. This is indicated by the AO of power becoming sharply negative. The core power output was kept constant during the subsequent axial power oscillations monitored by the incore and excore neutron instrumentation systems. In both tests, the AO of power was determined by a set of excore neutron detectors monitoring the power distribution in the bottom and top halves of the core in each of the four quadrants. The implementation of the excore detectors is discussed in connection with the detector response function in Section 10.6, with the typical locations illustrated in Figure 10.3.

Stability index $\alpha$ and oscillation period $T = 2\pi/\omega$ were determined via Eq. (16.45) from the AO oscillation measurements and summarized in Table 16.1. Comparison of the stability indices indicates significant burnup dependence, which required some discussion and analysis, especially in view of the increasing magnitude of the negative moderator temperature coefficient (MTC) of reactivity as a function of fuel burnup. The effect of fuel burnup on the MTC was discussed in more detail in Chapter 14. The feedback coefficient $\alpha_T$ in Eq. (16.30) could conceptually represent both the moderator and fuel temperature feedback effects, with the fuel temperature feedback effects remaining small and nearly constant during the fuel cycle. The effects of MTC on first-harmonic xenon oscillations required a bit more detailed analysis than that represented in Eq. (16.30) and subsequent definitions of $x_1(t)$ and $u(t)$ in Eq. (16.38). The oscillation period of 24~30 hr is closely related to the half-lives of $^{135}I$ and $^{135}Xe$.

The dependence of core stability on fuel burnup was analyzed with a one-dimensional two-group diffusion theory code featuring formulations for transient xenon–iodine dynamics and moderator and fuel temperature feedback effects. Numerical simulations of the test conditions and free-running oscillations indicate that the burnup effects on core stability are due to a combination of the axial power distribution and moderator temperature feedback evolving as a function of fuel burnup. The axial power distribution becomes flatter as the fuel burnup increases, which in turn increases the weighted average parameters introduced to determine the parameters $f_1$, $f_2$, and $f_3$ in Eq. (16.49). This tends to make the core more susceptible to axial xenon oscillations [Ran58]. As indicated in Chapter 14, the MTC becomes more negative as a function of moderator temperature and as a function of fuel burnup. The effect of fuel burnup on MTC is mostly due to a decrease in the soluble boron concentration such that the derivative of MTC with respect to moderator temperature $T_M$ becomes more negative and increases in magnitude as fuel burnup increases. The AP1000 Design Control Document [Wes03] indicates that the stability index for axial xenon oscillations for PWR cores becomes zero at a core average burnup of 12.0 MWd/kgHM, but the control of xenon oscillations would be possible with control rod movements.

The effects of the burnup dependence of MTC on core stability may be illustrated by considering the first harmonic flux distribution $\delta\phi(z, t)$ from Figures 16.7 and 16.9a. The effects of perturbations in the flux and power distributions on the moderator temperature distribution may be evaluated in turn via the moderator temperature distributions $T_M(z, t)$ in Figure 16.9b, where the dotted curve shows the perturbed distribution compared with the solid curve for the nominal power distribution. The perturbation $\delta T_M(z, t)$ in the moderator temperature distribution is

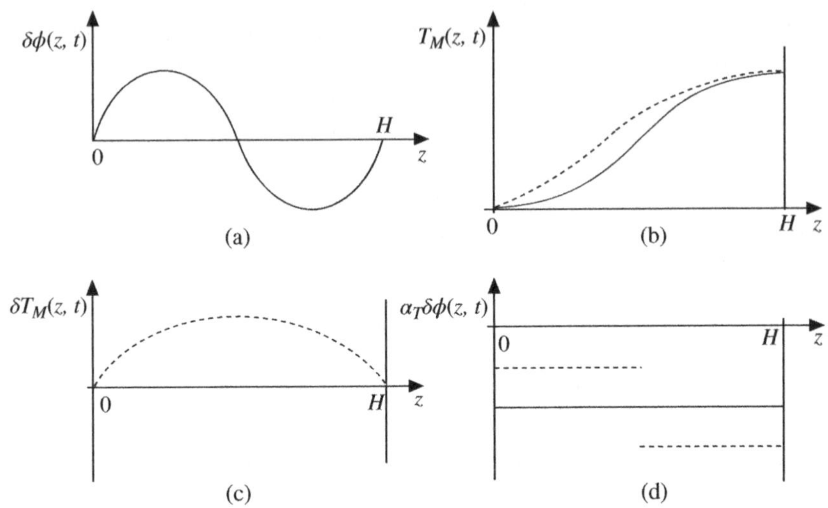


**Figure 16.9** Effect of flux perturbations on moderator temperature and reactivity feedback. Plot (a) indicates the flux distribution shifted toward the bottom half of the core, (b) effect of flux perturbation on the moderator temperature (dotted curve) compared with the nominal distribution (solid curve), (c) perturbations in the moderator temperature distribution, and (d) moderator temperature feedback averaged over the top and bottom halves of the core (dotted lines) relative to the core average feedback (solid line).

plotted as a dotted curve in Figure 16.9c. Figure 16.9d plots the resulting feedback effects $\alpha_T \delta\phi(z,t)$ of Eq. (16.30), averaged over the bottom and top halves as dotted lines, relative to the core average feedback effect illustrated as a solid line. The perturbation $\delta T_M(z,t)$ manifests as a larger negative temperature feedback in the top half of the core, due to a larger magnitude of the negative MTC, than that in the bottom half, thereby enhancing the imbalance in the perturbation $\delta L$ of the diffusion operator between the top and bottom halves of the core.

The enhanced imbalance in the negative feedback tends to increase the imbalance in the xenon concentration $X(z,t)$ and decrease the core stability. As the core average fuel burnup increases, the derivative of the negative MTC with respect to $T_M$ increases in magnitude and the imbalance in the negative MTC between the top and bottom halves of core increases, thereby increasing the destabilizing effect of the negative MTC derivative as a function of $T_M$. Comparison of the baseline AO values in the two Ginna tests also indicates that the axial power distribution shifted from the bottom to the top of the core gradually as a function of fuel burnup, making the core less stable.

### 16.2.3 Space-Time Xenon-Power Oscillations in $X-Y$ Plane

In addition to the xenon-induced axial power oscillations discussed so far in this section, the possibility of xenon-induced power oscillation in the $X$–$Y$ plane should be considered as the core size increases. For this purpose, one could approximate the reactor core as an equivalent cylindrical core with core height $H$ and radius $R$ and consider the magnitude of the flux harmonics in the radial plane of a cylindrical reactor considered in analogy to an oscillating circular membrane. The space-time analysis of axial xenon oscillations was presented in terms of the first axial lambda mode $\phi_1(z)$ with eigenvalue $\lambda_1$. This reflects the recognition that the least stable axial harmonic mode of the flux distribution is the first harmonic mode. The concept of the dynamic eigenvalue $\alpha_n$ is considered here together with that of geometric buckling $B_g^2 = B_0^2$ introduced in Eq. (5.62).

For flux oscillations in the radial plane of a cylinder with radius $R$, the general solution $\phi(r,\theta,t)$ at radius $r$ and azimuthal angle $\theta$ may be written as

$$\phi(r,\theta,t) = \sum_{n,k} T_{nk}(t)\phi_{nk}(r,\theta) = \sum_{n,k} T_{nk}(0)\exp(-\alpha_{nk} v D t)\phi_{nk}(r,\theta), \tag{16.53}$$

where $\phi_{nk}(r,\theta) = J_n(B_{nk}r)\sin n\theta$ is the solution [Kap66] to the one-group neutron diffusion equation

$$\nabla^2\phi_{nk}(r,\theta) + B_{nk}^2\phi_{nk}(r,\theta) = 0, \text{subject to } \phi_{nk}(R,\theta) = 0 \ \forall \ \theta, \tag{16.54}$$

with the eigenvalue $B_{nk}^2 = B_0^2 + \alpha_{nk}$. It should be noted that the dynamic eigenvalue $\alpha_n$ of Eq. (5.62) is generalized here to $\alpha_{nk}$ to represent the eigenfunctions in the

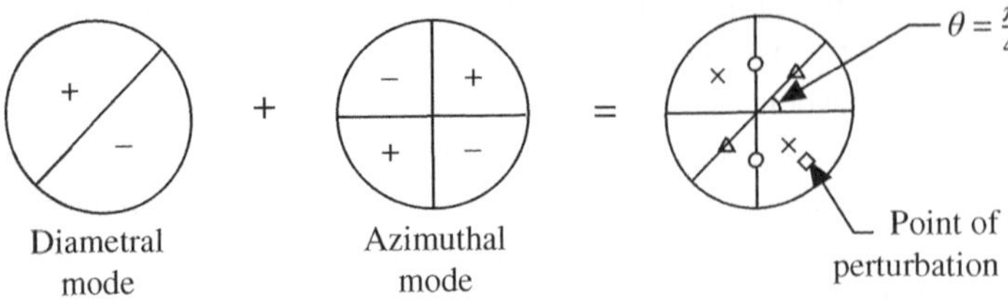


**Figure 16.10** Composite mode $\phi_1(r, \theta, t)$ representing a combination of the diametral and first azimuthal modes.

two-dimensional space $(r, \theta)$, and $J_n$ is the $n$th order Bessel function of the first kind. Corresponding to the fundamental mode, $\alpha_{01} = 0$ and $B_0^2 = B_m^2 = (2.405/R)^2$, with the first zero $\nu_0 = 2.405$ of $J_0$.

The first few harmonic modes correspond to the eigenvalues

$$\alpha_{02} = 4.27B_0^2,\ \alpha_{11} = 1.54B_0^2,\ \alpha_{12} = 7.51B_0^2,\ \alpha_{21} = 3.56B_0^2, \tag{16.55}$$

indicating that the first two overtones that may be excited are the diametral mode

$$\phi_{11}(r, \theta) = J_1(B_{11}r)\sin\theta,\ B_{11}^2 = 2.54B_0^2, \tag{16.56}$$

and first azimuthal mode

$$\phi_{21}(r, \theta) = J_2(B_{21}r)\sin 2\theta,\ B_{21}^2 = 4.56B_0^2, \tag{16.57}$$

together with the nodal lines, illustrated in Figure 16.10. The nodal lines are the lines along which the eigenfunction $\phi_{nk}(r, \theta)$ remains zero. The two low-order harmonics are combined to represent xenon-induced power oscillation tests [Fra72, Hon12] performed in 1971 at one of the first three-loop Westinghouse PWR plants that went into operation. For this purpose, the phase for the diametral mode is set so that $\theta = -\pi/4$ at the location of the flux perturbation that induced the oscillations and the phase for the azimuthal mode is selected to reflect the observation that a half-core symmetry was maintained during the oscillations, yielding the dominant composite harmonic mode:

$$\phi_1(r, \theta, t) = T_{11}(t)J_1(B_{11}r)\sin(\theta - \pi/4) + T_{21}(t)J_2(B_{21}r)\sin 2\theta. \tag{16.58}$$

Key observations can be made from Eq. (16.58) and Figure 16.10:

(1) Measurements at points marked [o] would yield stability characteristics of the diametral modes, in contrast to the contributions from the azimuthal mode at points marked [Δ].
(2) Points marked [×] would indicate contributions from both the diametral and azimuthal modes.

The stability characteristics of the diametral modes with the lowest eigenvalue $\alpha_{11} = 1.54B_0^2$ may be obtained by measuring the quadrant-average power fractions, usually known as the *quadrant tilts of power*, in two symmetrically opposite quadrants, equivalent to points marked [×], and taking the difference to subtract out the azimuthal mode contributions. In a $X$–$Y$ xenon oscillation test performed at a three-loop PWR plant [Hon12], an impulse insertion of a rod cluster control (RCC) unit in the lower-right quadrant, corresponding to the point [◇] in Figure 16.10, induced flux and power oscillations at a core average burnup of 1.54 MWd/kgHM. The power oscillations were monitored by both incore neutron detectors and thermocouples over a period of approximately 63 hr. The thermocouple plot of the quadrant tilt difference of power in Figure 16.11 yielded a stability index $\alpha = -0.076\,\text{hr}^{-1}$ and oscillation period $T = 29.6$ hr, indicating highly convergent xenon-induced power oscillations in the $X$–$Y$ plane for a three-loop Westinghouse PWR plant with 157 (15 × 15) fuel assemblies. The impulse motion of the RCC unit induced small axial power oscillations, which did not impact the oscillations in the $X$–$Y$ plane. Computer simulations also indicated that trace contributions from the azimuthal mode did not significantly affect the measured stability of the diametral mode oscillations for the three-loop PWR plant. A subsequent test at a core-average burnup of 12.9 MWd/kgHM indicates increased stability [Hon12] against $X$–$Y$ xenon oscillations in the three-loop PWR core.

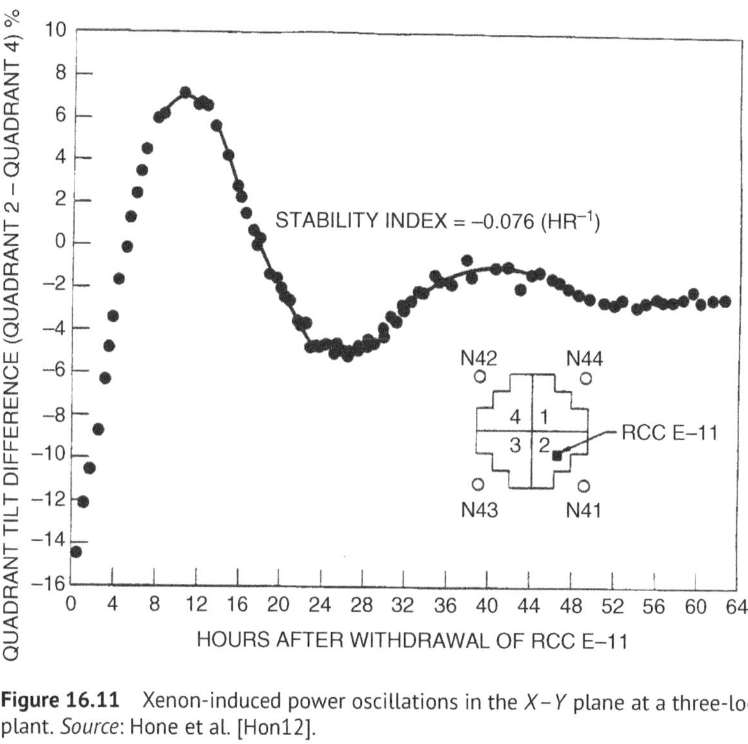


**Figure 16.11** Xenon-induced power oscillations in the $X$–$Y$ plane at a three-loop PWR plant. *Source*: Hone et al. [Hon12].

## 16.3 Time-Optimal Reactor Control

Building upon the dynamic reactor characteristics studied in Chapter 8, the xenon poisoning issues introduced in Chapter 12, and the associated space-time stability analysis covered in Section 16.2, we now turn our attention to control strategies relevant to safe and efficient operation of the reactor core. Our discussion is directed to optimal control through classical variational principle and through model-based control formulations that have experienced in recent years significant applications in many branches of engineering. Time-optimal control of post-shutdown xenon transient and axial space-time oscillations is studied in Sections 16.3.1 and 16.3.2, respectively.

### 16.3.1 Optimal Control of Xenon-Induced Transients

Recall that the $^{135}$Xe concentration tends to grow after the reactor is shut down from power operation due to a competition between the decay of $^{135}$Xe nuclei and decay of $^{135}$I nuclei, before $^{135}$Xe nuclei decay away. The overall transient phenomenon and the typical time of 11.3 hr at which the post-shutdown $^{135}$Xe concentration would reach a maximum are illustrated through Eq. (12.59) and Figure 12.12. We propose to develop a shutdown power program so that the reactor operation is terminated in minimal time without the xenon reactivity penalty exceeding the reserve reactivity. The minimum-time optimal control task is formulated via a variational principle proposed by Pontryagin [Pon62].

For this task, recast [Rob65] the $^{135}$I–$^{135}$Xe balance equations (12.42) and (12.43) by defining $x_1(t) = I(t)/\Sigma_f$ [cm$^{-2}$] and $x_2(t) = X(t)/\Sigma_f$ [cm$^{-2}$], with $x = (x_1, x_2)^T$ and $f = (f_1, f_2)^T$:

$$\frac{dx_1(t)}{dt} = \gamma_I \phi(t) - \lambda_I x_1(t) \equiv f_1(x, \phi), \tag{16.59}$$

$$\begin{aligned} \frac{dx_2(t)}{dt} &= \gamma_X \phi(t) + \lambda_I x_1(t) - \lambda^*(t) x_2(t) \equiv f_2(x, \phi), \\ \lambda^*(t) &= \lambda_X + \sigma_X \phi(t). \end{aligned} \tag{16.60}$$

A Hamiltonian $H$ representing the motion of the system in the phase plane $(x_1, x_2)$ is constructed by introducing adjoint functions $p_1(t)$ and $p_2(t)$, with $p = (p_1, p_2)^T$, and a Lagrangian multiplier $\eta$ for the constrained boundary $S(x) = x_2 - x_{2m}$

$$\begin{aligned} H &= p_1(t) f_1(x, \phi) + p_2(t) f_2(x, \phi) + \eta^T S(x) = p^T f + \eta^T S \\ &= (\gamma_I \phi - \lambda_I x_1) p_1 + (\gamma_X \phi + \lambda_I x_1 - \lambda^* x_2) p_2 + \eta (x_2 - x_{2m}), \end{aligned} \tag{16.61}$$

where

$$\eta = \begin{cases} = 0, & S < 0, \\ \geq 0, & S = 0. \end{cases}$$

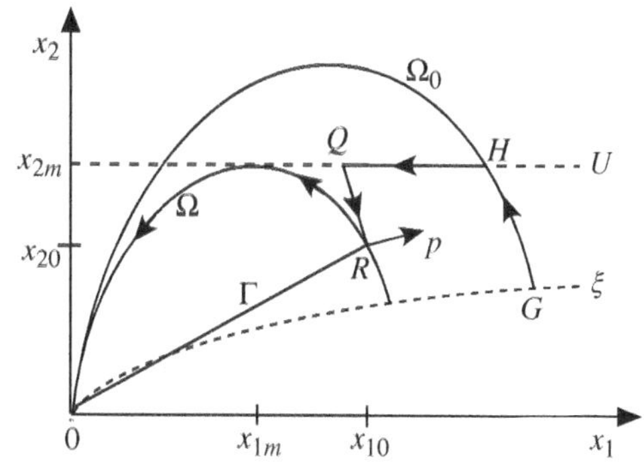


**Figure 16.12** Phase-plane trajectory for the time-optimal shutdown strategy.

The optimal shutdown flux program entails obtaining $\phi(t)$, given the maximum allowable xenon concentration $x_{2m}$ such that the total shutdown time $t_f$ is minimized, with $x_2(t_f) = 0$ and $x_2(t) \leq x_{2m}$, $t \in [0, t_f]$. The optimal shutdown trajectory in the phase plane $(x_1, x_2)$ is determined by minimizing the Hamiltonian $H$, which is conceptually equivalent to the total energy comprising the kinetic and potential energy. The optimal trajectory, illustrated in Figure 16.12, intuitively involves trajectory $\overline{GH}$ along the shutdown curve $\Omega_0$ starting with the operating flux level $\phi_0 = \phi(0)$ at point $G$ on the equilibrium xenon–iodine curve $\xi$ and continuing along the constrained boundary $U$ where $S(x) = 0$ or $x_2 = x_{2m}$. This is in contrast to the simple trajectory $\Omega_0$ all the way to the origin $(0, 0)$ for the case of a sufficient reactivity margin such that $x_{2m}$ lies completely above $\Omega_0$. Although a bit more formal treatment is presented in the original optimal time study [Rob65], we focus our analysis for the trajectory from point $Q$ on the constrained boundary $U$ to point $R$ on the final shutdown curve $\Omega$ that can take the trajectory to the origin without violating the constraint $x_2 \leq x_{2m}$.

The optimality condition for the control variable $\phi(t)$ requires minimizing the Hamiltonian

$$\frac{\partial H}{\partial \phi} = \left(\frac{\partial f}{\partial \phi}\right)^T p = 0 \tag{16.62}$$

together with the requirement that the adjoint function $p(t)$ satisfy an equation equivalent to the Euler–Lagrange equation in mechanics:

$$\frac{dp}{dt} = -\frac{\partial H}{\partial x} = -\left(\frac{\partial f}{\partial x}\right)^T p - \left(\frac{dS}{dx}\right)^T \eta. \tag{16.63}$$

When $H$ is a linear function of the control variable, Eq. (16.62) does not provide a solution and has to be modified to yield a bang-bang solution:

$$\phi = \begin{cases} 0, & \dfrac{\partial H}{\partial \phi} > 0, \\ \phi_{max}, & \dfrac{\partial H}{\partial \phi} < 0. \end{cases} \tag{16.64}$$

In addition, for the time-optimality problem under consideration, the time $t_3$ at which the trajectory reaches the shutdown trajectory $\Omega$ is not fixed, which then requires the use the transversality condition at $t = t_3$

$$p^T(t_3)\dot{x}(t_3) = 0 \Longrightarrow \sum_{1=1}^{2} p_i(t_3)dx_i(t_3) = 0, \tag{16.65}$$

which implies that the adjoint vector $p(t_3) = [p_1(t_3), p_2(t_3)]^T$ is perpendicular to the system trajectory at point $R$ on $\Omega$. From the shutdown equations (12.58) and (12.59) rewritten

$$x_1(t) = x_{10}\exp(-\lambda_I t), \tag{16.66a}$$

$$x_2(t) = \left(x_{20} + \frac{x_{10}}{1-\omega}\right)\exp(-\lambda_X t) - \frac{x_{10}}{1-\omega}\exp(-\lambda_I t),\ \omega = \frac{\lambda_X}{\lambda_I}, \tag{16.66b}$$

with $(x_{10}, x_{20})$ corresponding to the point $R$ yet to be determined, a phase-plane representation for the shutdown trajectory $\Omega$ is obtained in terms of the peak xenon concentration $x_{2m}$:

$$x_2 \equiv g(x_1) = \left[\frac{x_{2m}}{(\omega x_{2m})^{\omega}} + \frac{(\omega x_{2m})^{(1-\omega)}}{1-\omega}\right] x_1^{(\omega-1)} - \frac{1}{1-\omega}. \tag{16.67}$$

The bang-bang solution of Eq. (16.64) requires that $\partial H/\partial\phi < 0$ with $\phi(t_3-) = \phi_{max}$ on the shutdown curve $\Omega$ arriving at point $R$ where $\partial H/\partial\phi > 0$ with $\phi(t_3+) = 0$. The terminal point $R$ is thus defined by

$$\frac{\partial H}{\partial\phi} = \gamma_I p_1 + (\gamma_X - \sigma_X x_2)p_2 = 0, \tag{16.68a}$$

$$\frac{d}{dt}\left(\frac{\partial H}{\partial\phi}\right) = \gamma_I\lambda_I p_2\left[\frac{p_1}{p_2} - 1 + \frac{\gamma_x\lambda_x}{\gamma_I\lambda_I} - \frac{\sigma_X x_1}{\gamma_I}\right] \geq 0. \tag{16.68b}$$

The adjoint equation (16.63) on the shutdown curve $\Omega$ yields

$$\frac{dp_1}{dt} = -\frac{\partial H}{\partial x_1} = p_1\lambda_I - p_2\lambda_I, \tag{16.69a}$$

$$\frac{dp_2}{dt} = -\frac{\partial H}{\partial x_2} = p_2\lambda^*. \tag{16.69b}$$

The transversality condition of Eq. (16.65) with the slope $dx_2/dx_1 = dg(x_1)/dx_1 < 0$ for Eq. (16.67) and the solution for $p_2(t) > 0$ implies that $p_1(t) > 0$, i.e. the adjoint vector $p = (p_1, p_2)^T$ is an outward normal vector to $\Omega$ at the terminal point $t = t_3$ landing on $\Omega$, as illustrated in Figure 16.12. Substituting Eq. (16.68a) into Eq. (16.68b) provides the line $\Gamma$ that intersects $\Omega$ at $R$:

$$x_2 = x_1 + \frac{1}{\sigma_X}\left(\gamma - \frac{\gamma_X\lambda_X}{\lambda_I}\right) = x_1 + C_0,\ C_0 = 4.3\times10^{-8}\ \text{b}^{-1}. \tag{16.70}$$

With Eq. (16.70) for the line $\Gamma$ and hence the arrival point $R = (x_{10}, x_{20})$ on $\Omega$ determined, the junction point $Q = [x_1(t_2), x_{2m}]$ and switching time $t_2$ on the constraint boundary $U$ may be obtained by requiring that the xenon–iodine trajectory be perpendicular to $p = (p_1, p_2)^T$ with $\phi(t) = \phi_{max}$. When the switching time $t_2$ and $Q$ are determined, the balance equations are solved together with $dx_2(t)/dt = 0$ and $x_2(t) = x_{2m}$ on the constraint boundary $U$ to determine the time-dependent flux $\phi(t)$ that will land the system at $Q$ from point $H$. It may be noted that the adjoint equation (16.69b) should be modified on $U$ to account for the state-space constraint $S(x) = x_2 - x_{2m} = 0$ as in Eq. (16.63), allowing for a convenient selection of the Lagrange multiplier $\eta$ so that the continuity of the Hamiltonian $H$ is maintained at the junction point $Q$. Once the system arrives at $R$ on the shutdown trajectory $\Omega$, the system will coast down to the origin $(0, 0)$ with $\phi = 0$.

The overall optimal trajectory is schematically illustrated in Figure 16.13 with the arrival point $H$ at time $t_1$ on the constraint boundary $U$, junction point $Q$ at time $t_2$, and terminal point $R$ on the shutdown curve $\Omega$ at time $t_3$ indicated. The time-optimal shutdown program was experimentally tested [Rob67] at the 2.0-MW Livermore Pool Type Reactor (LPTR) in 1965, where the travel time $(t_2 - t_1) = 2.5$ hr on the boundary is followed by full-power operation at $\phi(t) = \phi_{max}$ for $(t_3 - t_2) = 1.8$ hr. Approximately 7 hr of additional reactor operation was allowed, avoiding a 16-hr period during which the $^{135}Xe$ buildup would have prevented reactor operation.

The time-optimal shutdown program discussed here may be contrasted to other optimal xenon shutdown programs [Ash59, Ros64], where the maximum $^{135}Xe$ is to be minimized during the shutdown period. The xenon shutdown problem studied via Pontryagin's maximum principle is not of practical concern in power reactors where a sufficient reactivity margin exists throughout the reactor cycle. The problem could present some operational limitations in research reactors and

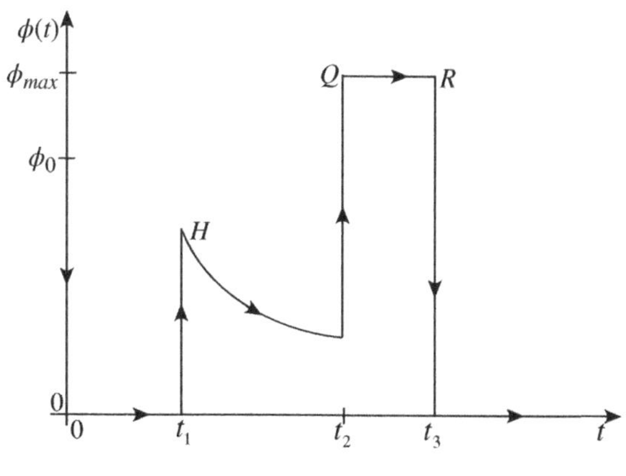


**Figure 16.13** Time-domain flux program for time-optimal shutdown strategy.

submarine reactors, where practical shutdown programs have apparently been developed over the years.

### 16.3.2 Control of Spatial Xenon Oscillations

With the time-domain optimal control for efficient handling of the reactivity constraints associated with post-shutdown xenon buildup discussed in Section 16.3.1, another issue more applicable to power reactors is now presented. This involves the space-time xenon oscillations discussed via a modal analysis framework in Section 16.2.1 and an application of Pontryagin's maximum principle for time-optimal control of the oscillations [Sch80].

Given the xenon–iodine phase-plane equation (16.44), an equilibrium solution for the system state $x = (x_1, x_2)^T = -uA^{-1}B = uz$ with some control $u$ should lie on a target line $\mathscr{L}_1$ displayed in Figure 16.14

$$\mathscr{L}_1 = \{x | x = \gamma z,\ U^- \leq \gamma \leq U^+\}, \tag{16.71}$$

where

$$z = \begin{pmatrix} K \\ L \end{pmatrix},\ K = \frac{f_3 - f_2}{f_2 - f_1}, L = \frac{f_3 - f_1}{f_2 - f_1}, f_3 - f_1 = \lambda^*. \tag{16.72}$$

The diverging spiral shown in Figure 16.14 represents [Sch80] the phase-plane trajectory for an unstable space-time xenon oscillation corresponding to the stability index $\alpha > 0$ and centered at the origin. The optimal solution of the space-time xenon oscillation consists of a minimal-time arrival at the target line

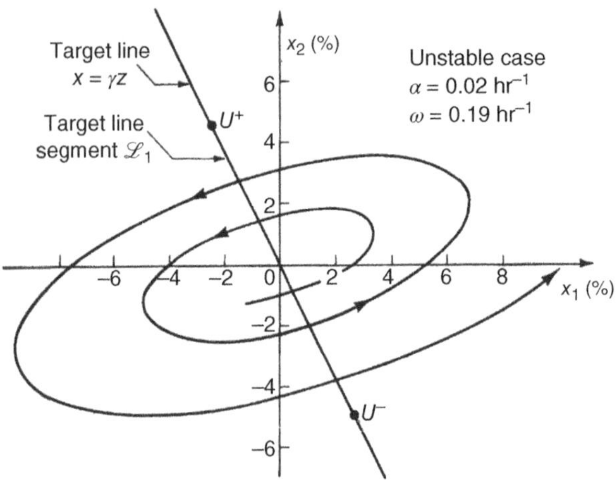


**Figure 16.14** Phase-plane trajectory for free-running oscillation and target line for control. *Source*: Schulz and Lee [Sch80]/Taylor & Francis Group.

and relaxation of the control $u$ within the bound $[U^-, U^+]$ to kill the ongoing oscillation. Application of the maximum principle results in a bang-bang motion of control rods in a PWR plant, which may, however, entail control rod movements outside the normal allowable range.

A modified approach is taken that accounts effectively for operational constraints representing acceptable power peaking factors. In this approach, a time-optimal control strategy is developed that can account for constraints [Sip76] on the AO of power $a(t)$ defined in Eq. (16.38) by recasting the phase-plane equation (16.44) into

$$\frac{dx(t)}{dt} = \frac{d}{dt}\begin{bmatrix} x_1(t) \\ x_2(t) \end{bmatrix} = \begin{bmatrix} -\lambda^* x_1(t) - f_2 x_2(t) + f_3 a(t) \\ -\lambda_I x_2(t) + \lambda_I a(t) \end{bmatrix}, \tag{16.73}$$

or equivalently

$$\frac{dx(t)}{dt} = Cx(t) + a(t)B \tag{16.74}$$

in terms of

$$C = \begin{pmatrix} -\lambda^* & -f_2 \\ 0 & -\lambda_I \end{pmatrix}, B = \begin{pmatrix} f_3 \\ \lambda_I \end{pmatrix}, \tag{16.75}$$

where the AO $a(t)$ plays the role of control. It should be noted that the variation of $a(t)$ requires the motion of control rods in a PWR plant.

Set up a Hamiltonian with a costate or adjoint vector $p(t)$

$$H = p^T \dot{x} = p^T(Cx + aB), \tag{16.76}$$

and apply Pontryagin's maximum principle for a time-optimal control to a generalized target $\mathscr{L}_2$ corresponding to an equilibrium condition

$$\mathscr{L}_2 = \{x | x = \gamma y,\ A^- \leq \gamma \leq A^+\}, \tag{16.77}$$

where

$$y = -C^{-1}B = \begin{pmatrix} K/L \\ 1 \end{pmatrix} = \frac{1}{L}z, z = \begin{pmatrix} K \\ L \end{pmatrix}. \tag{16.78}$$

The target line $\mathscr{L}_2$ is displayed in Figure 16.15, where the abscissa is compressed by an order of magnitude compared with that of Figure 16.14 to illustrate general control strategies corresponding to the AO control bound $a(t) \in [A^{-1}, A^+]$. Since the control $a$ appears as a linear relationship in the Hamiltonian, as is the case for the post-shutdown xenon control discussion in Section 16.3.1, $\partial H/\partial a$ does not yield a meaningful relationship for the optimal control, requiring a bang-bang motion of the control to minimize the Hamiltonian

$$a(t) = \begin{cases} A^-, & \dfrac{\partial H}{\partial a} > 0, \\[2ex] A^+, & \dfrac{\partial H}{\partial a} < 0. \end{cases} \tag{16.79}$$

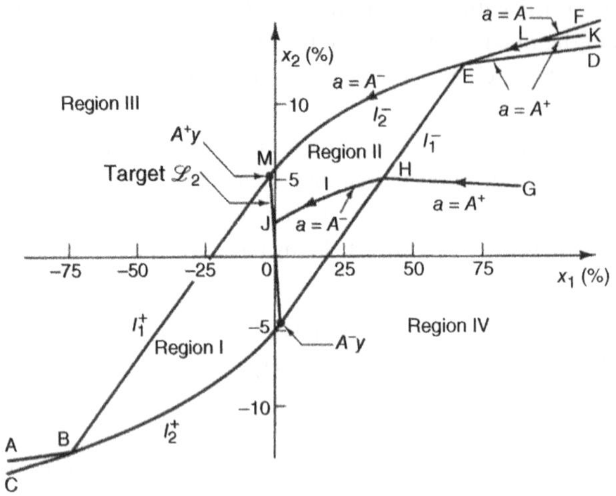


**Figure 16.15** Time-optimal trajectories for constrained AO control. *Source*: Schulz and Lee [Sch80]/Taylor & Francis Group.

The costate vector $p(t)$ satisfies

$$\frac{dp(t)}{dt} = -\frac{\partial H}{\partial x} = -C^T p(t), \tag{16.80}$$

yielding

$$p(t) = \exp(-C^T t)p(0). \tag{16.81}$$

The bang-bang variation of $a(t)$ in Eq. (16.79) takes place with the switching function $S(t)$

$$\begin{aligned} S(t) &= p^T(t)B \\ &= \left(f_3 - \frac{\lambda_i f_2}{\lambda_I - \lambda^*}\right) p_1(0)\exp(\lambda^* t) + \left[\lambda_I p_2(0) + \frac{\lambda_I f_2 p_1(0)}{\lambda_i - \lambda^*}\right] \exp(\lambda_I t), \end{aligned} \tag{16.82}$$

which indicates at most one switching possible along any optimal trajectory. Corresponding to a constant control $a(t) = A^+$ or $A^-$, obtain the solution to Eq. (16.74), with the initial condition $x_0 = x(0)$

$$x(t) = ay + \exp(Ct)(x_0 - ay), \tag{16.83}$$

where $y$ is defined for the target line $\mathscr{L}_2$ in Eq. (16.78).

In addition to the bang-bang control, a transversality condition has to be met due to the desire to control to a generalized target line $\mathscr{L}_2$

$$p^T(t_1)y = 0, \tag{16.84}$$

when the optimal trajectory arrives at the target line at time $t_1$. Resetting the arrival time $t_1 = 0$ and the switching time to $-t_s$ such that the switching function $S(-t_s) = 0$ provides a useful expression for the optimal switching trajectory $x(t)$:

$$\exp[(\lambda^* - \lambda_I)t_s] = \frac{\lambda^*}{\lambda_I}, \tag{16.85a}$$

$$x(-t_s) = ay + \frac{\exp(\lambda_I t_s)}{\lambda_I}(\gamma - a)B. \tag{16.85b}$$

Figure 16.15 illustrates how the phase plane is divided into four regions by switching curves $\ell_1^+$ and $\ell_1^-$ and boundary curves $\ell_2^+$ and $\ell_2^-$. At switching curve $\ell_1^+$ (or $\ell_1^-$), the control $a(t)$ switches from $a = A^-$ to $A^+$(or from $A^+$ to $A^-$), while the boundary trajectories $\ell_2^+$ (or $\ell_2^-$) is obtained from Eq. (16.83) with $a(t) = A^+$ (or $A^-$) with the initial conditions $x_0 = A^- y$ (or $A^+ y$). Depending on the initial point in the ($x_1$–$x_2$) phase plane, the optimal control $a(t)$ to the target line $\mathscr{L}_2$ lies in one of the four regions I, II, III, and IV indicated. Upon completing one of the appropriate controls, illustrated by the curve $\overline{GHIJ}$, the trajectory would lie on the target line at $x = \gamma y$, with $\gamma \in [A^-, A^+]$, and AO can then be finally relaxed to $a = \gamma$, reaching a new equilibrium ($x_1$–$x_2$) configuration.

There is, in practice, a region around the origin of the ($x_1$–$x_2$) phase space, i.e. regions I and II, for which the optimal control involves a single-pulse variation of $a(t)$, e.g. the trajectory $\overline{IJ}$. Thus, xenon-induced spatial oscillations could normally be controlled from region I or II and involve single-pulse controls $a(t) = A^+$ or $a(t) = A^-$ to $\mathscr{L}_2$, followed by relaxation to a steady-state condition.

A time-domain simulation of the phase-plane optimal trajectory was performed with the MID2 code, which is based on the two-group neutron diffusion equation solver ONED discussed in Chapter 6 and coupled with the xenon–iodine dynamics and one-dimensional axial thermal-hydraulic feedback representations. For this study, a PWR core with an active fuel length of 3.66 m operating near the end of a typical first cycle was simulated, yielding a configuration with a stability index $\alpha = 0.02\ \text{hr}^{-1}$ and oscillation period $T = 2\pi/\omega = 33.1$ hr. A simple one-pulse control was initiated with an AO constraint $a(t) \in [-15\%, 15\%]$ at $t = 5$ hr into a free-running oscillation illustrated in Figure 16.16. The optimal control strategy corresponds to the trajectory $\overline{IJ}$ of Figure 16.15, where the control rods are immediately moved so that $a = -15\%$. The control rods are subsequently exercised at 6-min intervals to maintain the AO constant at −15%. After approximately 1.25 hr on the bounding trajectory $a(t) = -15\%$, monitoring the ratio $R = x_1/x_2$ indicates that the generalized target $\mathscr{L}_2$ corresponding to $R = K/L$ is reached. This allows the control to relax so that $a(t) = x_2(t) \simeq -5\%$ is retained and no further control actions are required. Additional suboptimal

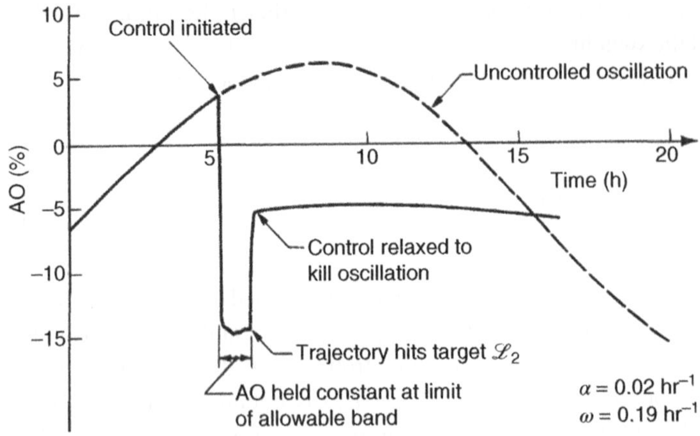


**Figure 16.16** Time-domain simulation of optimal control with AO constraints. *Source*: Schulz and Lee [Sch80]/Taylor & Francis Group.

control strategies following the time-optimal formulation are also reported [Sch80].

Two applications of the Pontryagin's maximum principle in Section 16.3 illustrate useful approaches for optimal control of two operational transients of interest in nuclear plants. The time-optimal control strategy of axial xenon oscillations could be integrated as part of load-follow maneuvers for nuclear plants that would play a more important role in the overall electric market around the world. This will be discussed further in the later part of this chapter. We now turn our attention to alternative control techniques that have been developed in the control theory community involved primarily with other engineering disciplines; the techniques rely heavily on modern development of the MATLAB technology.

## 16.4 Model-Based Reactor Control

We present model-based control formulations that make efficient use of the canonical system model and linear fractional transformation (LFT) introduced in Chapter 8. The discussion begins with a traditional feedback controller without explicit accounting for disturbance or measurement and moves to the $H_2$ and $H_\infty$ formulations employed in an augmented plant representation. The derivation of various optimization algorithms is presented via the variational calculus approach of Pontryagin's maximum principle in Appendix E, rather than formal state-space representations [Doy89].

### 16.4.1 Linear Quadratic Regulator

We return to a simple feedback controller comprising a system model for state $x$ with output $z$ in Eqs. (8.60) and (8.64)

$$z = \left( \begin{array}{c|c} A & B \\ \hline C & D \end{array} \right) u \equiv T_{zu}u, \; T_{zu} = T(u \to z),$$

or

$$\begin{aligned} \dot{x} &= Ax + Bu = f(x, u), \\ z &= Cx + Du, \end{aligned} \tag{16.86}$$

with the objective to minimize the output $z$ by optimizing the transfer function $T_{zu}$. With this purpose in mind, construct an objective function $J$ that represents the magnitude of output in the Euclidean or $H_2$ norm, for control duration $t_f$

$$\begin{aligned} J &= \frac{1}{2}\int_0^{t_f} |z(t)|^2 dt = \frac{1}{2}\int_0^{t_f} z^T(t)z(t)dt \\ &= \frac{1}{2}\int_0^{t_f} (x^T C^T Cx + u^T D^T Du)dt = -\int_0^{t_f} L(x, u)dt, \end{aligned} \tag{16.87}$$

with the assumption that the parameters $C$ and $D$ affect the output $z$ independently, i.e. $C^T D = 0$. Introducing a Hamiltonian with the normalization $D^T D = I$ and a costate or adjoint vector $p(t)$

$$H = -L + p^T f = \frac{1}{2}(x^T C^T Cx + u^T u) + p^T(Ax + Bu), \tag{16.88}$$

converts the objective function $J$ to an augmented objective function

$$J^* = -\int_0^{t_f} [L + p^T(\dot{x} - f)]dt = \int_0^{t_f} (H - p^T\dot{x})dt. \tag{16.89}$$

Minimizing $J^*$ via the steps taken in Appendix E yields the optimal control

$$\frac{\partial H}{\partial u} = p^T B + u^T = 0, \text{ yielding } u = -B^T p, \tag{16.90}$$

together with Hamilton's equations:

$$\begin{aligned} \frac{\partial H}{\partial p} &= f = Ax + Bu = Ax - BB^T p = \dot{x}, \\ \frac{dp}{dt} &= -\frac{\partial H}{\partial x} = -(A^T p + C^T Cx). \end{aligned} \tag{16.91}$$

Recast Eqs. (16.91) into a matrix form with the Hamiltonian matrix $\tilde{H}$:

$$\frac{d}{dt}\begin{pmatrix} x \\ p \end{pmatrix} = \begin{pmatrix} A & -BB^T \\ -CC^T & -A^T \end{pmatrix}\begin{pmatrix} x \\ p \end{pmatrix} = \begin{pmatrix} A & -R \\ -Q & -A^T \end{pmatrix}\begin{pmatrix} x \\ p \end{pmatrix} \equiv \tilde{H}\begin{pmatrix} x \\ p \end{pmatrix}. \tag{16.92}$$

Substituting a trial solution $p(t) = X(t)x(t)$ into Eq. (16.92) yields

$$-\dot{X} = XA + A^T X - XRX + Q, \tag{16.93}$$

where the auxiliary function $X(t)$ plays the role of state transition matrix $\Phi$ in the Kalman filter formulation discussed in Appendix F and illustrated further in Problem 16.13. For an infinite horizon problem with the final control time $t_f \rightarrow \infty$, set $\dot{X} = 0$ to obtain an *algebraic Ricatti equation*:

$$XA + A^T X - XRX + Q = 0. \tag{16.94}$$

Solution $X$ to Eq. (16.94) may be obtained via the MATLAB function *are(A, R, Q)*, *care(A, B, Q)*, or *ric($\tilde{H}$)*, and the final optimal control is provided for a *linear quadratic regulator (LQR)* or *linear quadratic Gaussian (LQG) controller*:

$$\dot{x} = (A + BK_c)x \equiv \hat{A}x, \text{ with } K_c = -B^T X \text{ and } u = K_c x. \tag{16.95}$$

A number of studies [Men92, Oka13] have been reported on LQG/LQR applications for nuclear reactor control, in particular, for load follow maneuvers, often involving a set of linearized models covering the entire range of power level variations.

**Example 16.3** Obtain the LQR controller for a system specified by

$$A = \begin{pmatrix} 0 & 1 \\ 0 & -3 \end{pmatrix}, B = \begin{pmatrix} 0 \\ 1 \end{pmatrix}, C = \begin{pmatrix} \sqrt{2} & 0 \\ 0 & 1 \end{pmatrix}.$$

Using *are(A, R, Q)* with $R = BB^T$ and $Q = CC^T$ yields

$$X = \begin{pmatrix} 5.0653 & 1.4142 \\ 1.4142 & 0.5817 \end{pmatrix}, K_c = -B^T X = (-1.4142 \;\; -0.5817),$$

and the controller solution

$$\frac{dx(t)}{dt} = \hat{A}x(t) = \begin{pmatrix} 0 & 1.0 \\ -1.4142 & -3.5817 \end{pmatrix} x(t).$$

The eigenvalues of the controller equation are $\{-0.4518, -3.1299\}$, indicating the controller is stable.

### 16.4.2 $H_2$ Controller

We extend the LQG controller to include a measurement vector $y$ and the associated disturbance or noise $w$ to represent a combined system model

$$\begin{pmatrix} \dot{x} \\ z \\ y \end{pmatrix} = \begin{pmatrix} A & B_1 & B_2 \\ C_1 & 0 & D_{12} \\ C_2 & D_{21} & 0 \end{pmatrix} \begin{pmatrix} x \\ w \\ u \end{pmatrix}. \tag{16.96}$$

The solution to the combined system model may be conveniently obtained by separate steps for the optimal control and the filter that minimizes the disturbance. For the control, return to the LQR formulation in terms of the transfer function $T_{zu} = T(u \to z)$

$$\begin{pmatrix} \dot{x} \\ z \end{pmatrix} = \begin{pmatrix} A & B_2 \\ C_1 & D_{12} \end{pmatrix} \begin{pmatrix} x \\ u \end{pmatrix} = T_{zu} \begin{pmatrix} x \\ u \end{pmatrix}. \tag{16.97}$$

The resulting LQR control is written with $u = K_c x = -B_2^T X x$ as

$$\begin{pmatrix} \dot{x} \\ u \end{pmatrix} = \begin{pmatrix} A - B_2 B_2^T X & 0 \\ -B_2^T X & 0 \end{pmatrix} \begin{pmatrix} x \\ z \end{pmatrix} = K \begin{pmatrix} x \\ z \end{pmatrix}, \tag{16.98}$$

with $X = ric(\tilde{H})$ for $\tilde{H} = \begin{pmatrix} A & -B_2 B_2^T \\ -C_1^T C_1 & -A^T \end{pmatrix}$.

The optimal filter for measurement $y$ and disturbance $w$ is obtained for the system model

$$\begin{pmatrix} \dot{x} \\ y \end{pmatrix} = \begin{pmatrix} A & B_1 \\ C_2 & D_{21} \end{pmatrix} \begin{pmatrix} x \\ w \end{pmatrix} = T_{yw} \begin{pmatrix} x \\ w \end{pmatrix}, \tag{16.99}$$

where the transfer function $T_{yw}$ represents the effect of disturbance or noise $w$ on the measurement error $(y - \hat{y})$ given the optimal observation $\hat{y}$. The term $w$ is further separated into disturbance $d$ and noise $n$, $w = (d \;\; n)^T$, to yield the measurement portion of Eq. (16.96):

$$\begin{aligned} \dot{x} &= Ax + B_1 d, \\ y &= C_2 x + D_{21} n, \;\; D_{21}^T D_{21} = I. \end{aligned} \tag{16.100}$$

Similar to the steps taken for the LQR controller, minimize the Euclidean or $H_2$ norm of the combined disturbance $w$

$$J = \frac{1}{2} \int_0^{t_f} |w(t)|^2 dt = \frac{1}{2} \int_0^{t_f} (d^T d + n^T n)^2 dt, \tag{16.101}$$

which can be obtained via minimizing the Hamiltonian with costate vector $p(t)$:

$$H = \frac{1}{2}(w^T w + p^T f) = \frac{1}{2}[d^T d + (y - C_2 x)^T (y - C_2 x)] + p^T f, f = \dot{x}. \tag{16.102}$$

The *worst disturbance* is obtained as

$$\frac{\partial H}{\partial d} = d^T + p^T B_1 = 0, \;\text{ or } d = -B_1^T p, \tag{16.103}$$

while Hamilton's equation yields

$$\frac{d}{dt} \begin{pmatrix} x \\ p \end{pmatrix} = \begin{pmatrix} A - B_1 B_1^T & \\ -C_2^T C_2 & -A^T \end{pmatrix} \begin{pmatrix} x \\ p \end{pmatrix} + \begin{pmatrix} 0 \\ C_2^T \end{pmatrix} y. \tag{16.104}$$

Setting $\hat{x} = x + Yp$ generates

$$\begin{aligned}\frac{d\hat{x}}{dt} &= \frac{dx}{dt} + \dot{Y}p + Y\dot{p}, \\ &= A\hat{x} + YC_2^T(y - C_2\hat{x}) + (\dot{Y} - AY - B_1B_1^T + YC_2^TC_2Y - YA^T)p.\end{aligned} \tag{16.105}$$

The infinite-horizon solution for the function

$$Y = ric(\tilde{H}) \text{ with } \tilde{H} = \begin{pmatrix} A^T & -C_2^TC_2 \\ -B_1B_1^T & -A \end{pmatrix} \tag{16.106}$$

allows combining Eqs. (16.98) and (16.105) to generate the final $H_2$ *controller*

$$\begin{aligned}\frac{d\hat{x}}{dt} &= A\hat{x} + B_2u + K_f(y - \hat{y}), \\ &= (A + B_2K_c - K_fC_2)\hat{x} + K_fy, \\ &\equiv \hat{A}\hat{x} + K_fy,\end{aligned} \tag{16.107}$$

where $K_f = YC_2^T$ is the Kalman gain and $\hat{y} = C_2\hat{x}$ is the optimal observation corresponding to the optimal system estimate $\hat{x}$, together with the optimal control $u = K_c\hat{x} = -B_2^TXx$ from Eq. (16.98).

The auxiliary function $Y$ again plays the role of state transition matrix in Kalman filtering. The control term $B_2u = B_2K_cx$ is set to $B_2K_c\hat{x}$ with the observation that the filtering covariance is unrelated to the control action. The $H_2$ controller is recast into a LFT structure

$$u = \left[\begin{array}{c|c} \hat{A} & K_f \\ \hline K_c & 0 \end{array}\right] y \equiv Ky, \tag{16.108}$$

with the overall transfer function $K$

$$K = K_c(sI - \hat{A})^{-1}K_f = K_c\Phi K_f, \ \Phi = (sI - \hat{A})^{-1}, \tag{16.109}$$

combining the control gain $K_c$ and filter gain $K_f$. The role of the $H_2$ controller serving as a *model-based controller* (MBC) for the actual plant $G$ is illustrated in Figure 16.17, where the control $u$ with gain $K_c$ is provided to $G$ and measurement $y$ is returned to the controller with filter gain $K_f$. The structure is written in terms of an augmented system variable $\hat{x}$ that includes an optimal control and a correction for the filter. With Eqs. (16.107) and (16.108), the MBC structure may be rewritten

$$\hat{x} = \Phi[K_f(y - C_2\hat{x}) + B_2K_c\hat{x}], \tag{16.110}$$

to illustrate the overall MBC structure of Figure 16.17.

### 16.4.3 $H_\infty$ Controller

With the $H_\infty$ controller, we take up the task of minimizing the output vector $z$ with full accounting given for the disturbance $w$ in the system equation (16.96)

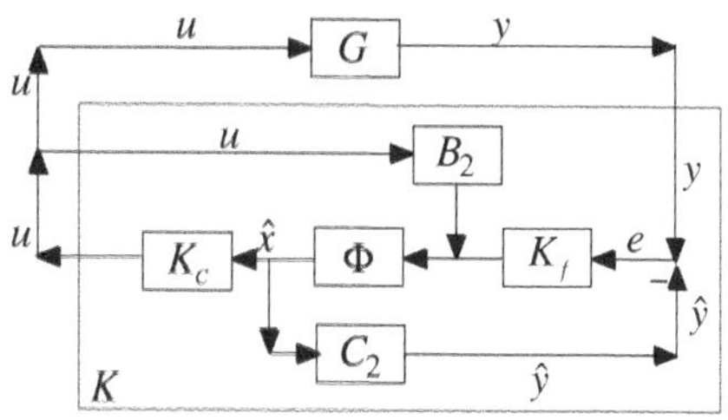


**Figure 16.17** Model-based controller $K$ with control gain $K_c$ and filter gain $K_f$ connected to the system transfer function $\Phi$ providing optimal control $u$ to the plant $G$, while measurement $y$ is fed back to the controller $K$.

and with the transfer function $T_{zw} = T(w \to z)$ represented in a $H_\infty$ *norm* via the *singular value* $\sigma_i[T_{zw}(j\omega)]$

$$\begin{aligned} ||T_{zw}||_\infty &= \sup_\omega ||T_{zw}(j\omega)||_\infty = \sup_\omega \max_i \sigma_i[T_{zw}(j\omega)], \\ \sigma_i(T_{zw}) &= [\lambda_i(T^*_{zw}T_{zw})]^{1/2}, \end{aligned} \tag{16.111}$$

with $\lambda_i$ representing the eigenvalue. The norm $||T_{zw}||_\infty$ represents the maximum magnitude of the transfer function in the Bode diagram or Nyquist plot for a simple transfer function. Thus, the objective for $H_\infty$ control is to update the control $K$ in Eq. (16.108) so that it could account for the *worst-case disturbance* to the systems with $||T_{zw}||_\infty \leq \gamma$ for the smallest $\gamma$. The control would then satisfy

$$\begin{aligned} ||z||_2^2 &= ||T_{zw}w||_2^2 = \frac{1}{2\pi}\int_{-\infty}^{\infty} |T_{zw}(j\omega)w(j\omega)|_2^2 d\omega \\ &\leq ||T_{zw}||_\infty^2 \frac{1}{2\pi}\int_{-\infty}^{\infty} |w(j\omega)|^2 d\omega = ||T_{zw}||_\infty^2 ||w||_2^2 \end{aligned} \tag{16.112}$$

or

$$||z||_2^2 - \gamma^2 ||w||_2^2 \leq 0,$$

with the last step in Eq. (16.112) resulting from Parseval's theorem [Arf13]. Equation (16.112) leads to the task of minimizing the objective function

$$J = \frac{1}{2}\int_0^{t_f} (|z|^2 - \gamma^2|w|^2)dt \tag{16.113}$$

and the corresponding Hamiltonian with costate vector $p(t)$

$$H = \frac{1}{2}(x^T C_1^T C_1 x + u^T u - \gamma^2 w^T w) + p^T(Ax + B_1 w + B_2 u), \tag{16.114}$$

with $D_{12}^T D_{12} = I$ and $C_1^T D_{12} = 0$. Minimizing $H$ yields

$$\begin{aligned} \text{Best control: } u &= -B_2^T p, \\ \text{Worst disturbance: } w &= -\gamma^{-2} B_1^T p. \end{aligned} \tag{16.115}$$

Taking steps similar to those for $H_2$ control requires solving two Ricatti equations

$$X_\infty = ric\begin{pmatrix} A & \gamma^{-2}B_1B_1^T - B_2B_2^T \\ -C_1^TC_1 & -A^T \end{pmatrix},$$
$$Y_\infty = ric\begin{pmatrix} A^T & \gamma^{-2}C_1^TC_1 - C_2^TC_2 \\ -B_1B_1^T & -A \end{pmatrix}, \tag{16.116}$$

which yield the optimal control and filter gain

$$K_c = -B_2^TX_\infty,$$
$$K_f = Y_\infty C_2^T. \tag{16.117}$$

Finally, the combined augmented system model is obtained as

$$\hat{A}_\infty = A + B_2K_c - Z_\infty K_fC_2 + \gamma^{-2}B_1B_1^TX_\infty, \tag{16.118}$$

with

$$Z_\infty = (I - \gamma^{-2}Y_\infty X_\infty)^{-1}. \tag{16.119}$$

The solution of the Ricatti equations (16.116) requires iteratively updating the parameter $\gamma$ until a feasible solution with a minimum value of $\gamma$ is reached. Implementing Eq. (16.118) results in modifying the filter gain $K_f$ in Eq. (16.107) by $Z_\infty K_f$ and adding another feedback loop representing $\gamma^{-2}B_1B_1^TX_\infty$. The $H_\infty$ formulation presented is one of several possible representations, which is better developed than other models [Doy89].

### 16.4.4 Augmented Plant Representation

The model-based controller $K$ illustrated in Figure 16.17 for $H_2$ control or one slightly modified for $H_\infty$ control may now be combined into an augmented plant structure where frequency weights are introduced to enhance the overall stability of the model-based control system both for low and high frequency responses. With the noise or disturbance $w$ input to the augmented plant $P$, effort is now made to shape the output $y$ through the weights $W_1$ and $W_2$ for *loop shaping*. In this approach, with the output $z$ representing deviations in the system state from nominal or equilibrium condition, the overall objective of the controller is to minimize the deviation $z$ both for low- and high-frequency regions through a judicious selection of the weights $W_1$ and $W_2$.

The augmented plant $P$ comprising the nominal plant $G$ and weights $W_1$ and $W_2$ in Figure 16.18 is driven by the input disturbance $w$ and control $u$ and generates the output signal $z$ and measurement signal $y$, while the controller $K$ generates

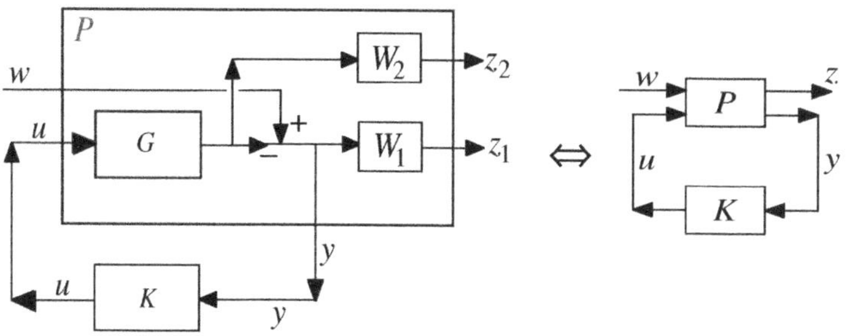


**Figure 16.18** Augmented plant $P$ comprising nominal plant $G$ and loop shaping weights $W_1$ and $W_2$ coupled with model based controller $K$ (LHS diagram), with $P$ indicated in a summary layout (RHS diagram).

control $u$ from measurement $y$, which is provided to the plant model

$$\begin{pmatrix} z \\ y \end{pmatrix} = P \begin{pmatrix} w \\ u \end{pmatrix} = \begin{pmatrix} z_1 \\ z_2 \\ y \end{pmatrix}, \tag{16.120}$$

with the transfer function $P$

$$P = \begin{pmatrix} W_1 & -W_1G \\ 0 & W_2G \\ I & -G \end{pmatrix} \equiv \left(\begin{array}{c|c} P_{zw} & P_{zu} \\ \hline P_{yw} & P_{yu} \end{array}\right),$$
$$P_{zw} = \begin{pmatrix} W_1 \\ 0 \end{pmatrix}, P_{zu} = \begin{pmatrix} -W_1G \\ W_2G \end{pmatrix}, P_{yw} = I, P_{yu} = -G. \tag{16.121}$$

With $u = Ky = K(I + GK)^{-1}w$, obtain the overall transfer function $T_{zw}$ connecting output $z$ to disturbance $w$

$$z = T_{zw}w = \begin{pmatrix} W_1S \\ W_2T \end{pmatrix}, \tag{16.122}$$

in terms of the *sensitivity transfer function* $S = (I + GK)^{-1}$ and *complementary sensitivity transfer function* $T = GK(I + GK)^{-1}$. Together with the obvious relationship $S + T = I$, the objective behind loop shaping is to choose frequency weights $W_1$ and $W_2$ such that the overall transfer function $T_{zw}$ is minimized for both low- and high-frequency ranges, by requiring

$$\begin{aligned} ||W_1(\omega)S(\omega)|| &\leq \gamma \text{ for low } \omega, \\ ||W_2(\omega)T(\omega)|| &\leq \gamma \text{ for high } \omega, \end{aligned} \tag{16.123}$$

where the weight selection is made via $H_\infty$ norm for a $H_\infty$ controller.

An early application of a $H_\infty$ controller to reactor control problems was made [Suz93] for the control of nuclear-coupled density wave oscillations (NCDWOs) experienced at the LaSalle Nuclear Plant in 1988 [Lee17]. The oscillations involved unstable oscillations of the two-phase boundary that excited power level oscillations with a frequency of 0.5 Hz at natural circulation flow. The oscillations were eventually terminated through a high flux scram without incurring damage to the plant. A phenomenological NCDWO model representing the water density and temperature in a BWR core is coupled to a PKE with one equivalent group of delayed neutrons. The BWR system model is connected to a power monitoring system and reactivity control system representing measurement $y$ and control $u$ providing a seventh-order system model. Low-order frequency weights are selected for the frequency weights to satisfy the loop shaping requirements:

$$W_1(s) = \frac{0.1s^3 + s^2}{325}, W_2(s) = \frac{200}{400s + 1}. \tag{16.124}$$

Application of the $H_\infty$ control algorithm shows that the system is stable for relatively large disturbances illustrated in Figure 16.19, when a multiplier $\epsilon$ representing the void reactivity feedback coefficient is varied parametrically. In contrast, the LQG control formulation indicates a tendency for unattenuated power oscillations, displayed in Figure 16.20. More recent applications include $H_\infty$ controller for load follow operations [Chi02] in a PWR and a combined use of a LQG algorithm and loop transfer recovery (LTR) structure [Lig14], equivalent to the

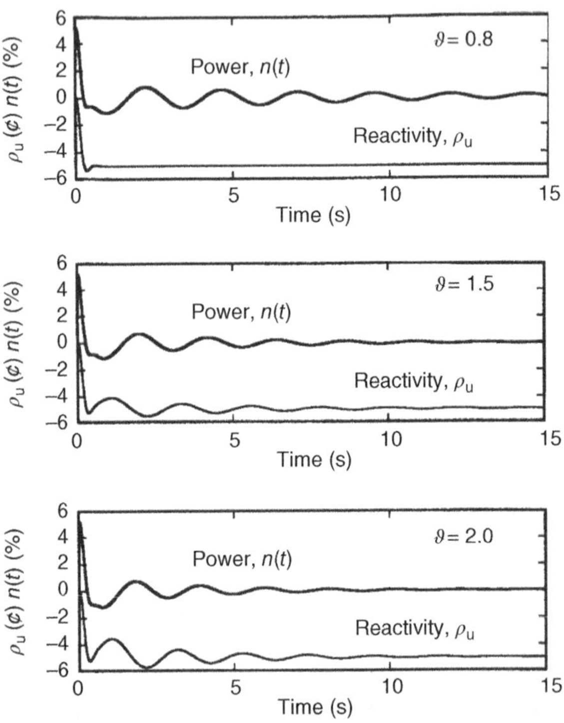


**Figure 16.19** $H_\infty$ controller indicating stable control of reactivity and power associated with NCDWOs in a BWR core. *Source*: Suzuki et al. [Suz93]/Taylor & Francis Group.

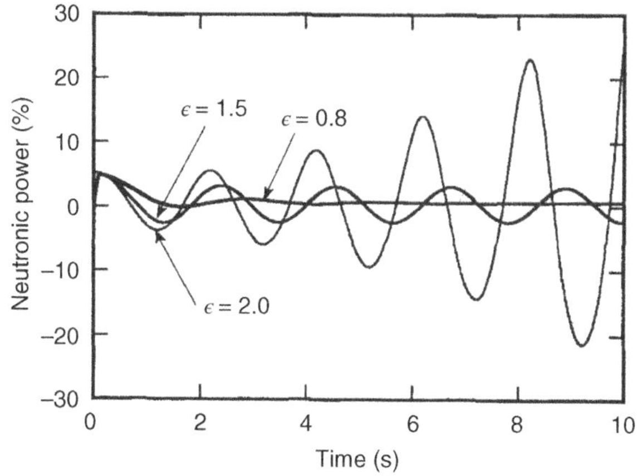


**Figure 16.20** LQG regulator for a BWR core indicating tendency of unstable power oscillations.
*Source*: Suzuki et al. [Suz93]/Taylor & Francis Group.

$H_\infty$ controller, for power and axial power distribution control of a PWR core represented by a two-point model. The phase and gain margins for $H_\infty$ control are usually larger than those for LQG control, indicating that $H_\infty$ control offers better stability against external disturbances and model parameter variations. It should, however, be noted that the $H_2$ and $H_\infty$ formulations presented in this chapter are all based on linear system models, which require successive linearization of nonlinear models in general. Additional developments will be required to fully accommodate nonlinear $H_2$ and $H_\infty$ formulations.

## 16.5 Alternate Reactor Control Techniques

In addition to the time-optimal and model-based control techniques discussed in this chapter, a number of other techniques have been successfully developed and demonstrated for nuclear reactor control. Examples of reactor control studies for different reactor types are also discussed in [Oka13].

Originally developed by N. Metropolis [Met53] in an effort to obtain the equation of state for various materials, the *simulated annealing* algorithm models the process of arriving at a thermodynamic equilibrium via repeated heating and cooling processes. In statistical mechanics, the equation of state may be obtained by calculating the weighted average of properties, e.g. energy $E$, for a large number of random configurations of the system with the Boltzmann factor as the weight. The Boltzmann factor $M(E) = \exp(-E/kT)$ represents the thermodynamic equilibrium state, where $k$ is the Boltzmann constant and $T$ is the temperature of the system. In a simulated annealing process, the feasible configurations are selected with a probability $M(E)$ and then simply averaged. Together with the development of the weighted or biased Monte Carlo algorithm, N. Metropolis played a pioneering role in developing digital computers.

For optimal control or optimization of engineering or physical systems, various feasible configurations with the objective function $J$, given as a function of control parameter $u$, are evaluated with a probability represented by the Boltzmann factor $M(J) = \exp(-J/kT)$ for high $T$, i.e. with a nearly uniform probability distribution, and a number of promising configurations are selected with the acceptance probability

$$p(\text{accept}) = \begin{cases} 1, & \Delta J < 0, \\ \exp(-\Delta J/kT), & \Delta J \geq 0. \end{cases} \tag{16.125}$$

The process is repeated at a lower $T$ with the configurations now focused around the promising regions until a converged optimal configuration is attained. If the cooling process results in either infeasible or inferior configurations, the temperature $T$ is raised by some increment and the process is repeated, simulating a physical annealing process of metals. The biased Monte Carlo algorithm using the Boltzmann factor $M(J)$ is compared in Figure 16.21 with the process for statistical mechanics where the potential function $V(r)$ given as a function of radius $r$ in a central force field is visualized. As the total energy $E = K + V$ including the kinetic energy $K$ is minimized at $r = r_0$, the Boltzmann factor $M(E)$ is maximized to attain thermal equilibrium. The process is analogous for optimization problems where the corresponding Boltzmann factor $M(J)$ is maximized for optimal configuration with $u = u_0$ when the objective function $J$ is minimized.

The simulated annealing algorithm is able to account for various control problems with complex constrains, but the process often requires some heuristic guidance in selecting the initial configurations or in the annealing schedule. Through the biased Monte Carlo algorithm, a number of optimal control and optimization algorithms have been developed, including those for optimal fuel loading pattern software, e.g. FORMOSA [Moo99] and ALPS [Joh92]. The stochastic algorithms, by their nature, often require a large number of evaluations of the system state and may end up in a local optimal state, which may then require some heuristic guidance. For example, the ALPS code may require as many as $10^5$ 3-D core calculations to perform a fuel assembly loading pattern search [Oui16].

Another popular stochastic algorithm is the *genetic algorithm* [Gol89] that mimics biological evolution processes by constructing bit strings that encode an objective function and uses stochastic algorithms to manipulate bit positions of the strings via crossover and mutation operators. Through multiple generations involving parent and progeny strings, a string with the highest objective function emerges, which then provides the desired optimal configuration. As a stochastic

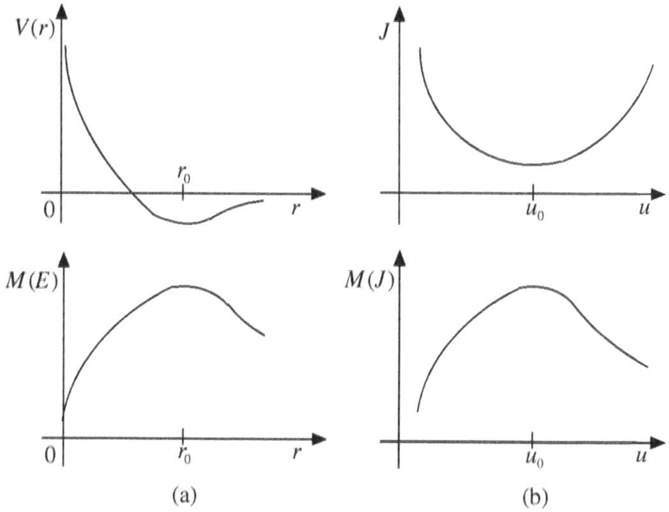


**Figure 16.21** Simulated annealing process for (a) statistical mechanics and (b) stochastic optimization.

algorithm, the genetic algorithm offers the strength and weakness similar to those of the simulated annealing algorithm.

Many of the early control studies for NPP systems were performed with long-standing proportional-integral-derivative (PID) techniques. Various examples for steam cycle controls are discussed in a Babcock & Wilcox handbook on steam technology [B&Wil72] including several PID combinations in feedforward and feedback control structures. Among the important control tasks for NPP systems are the three-element control of steam generators involving the feed water flow rate, steam flow rate, and water level, and the turbine- and reactor-based control of load variations in actual plant operations. A PID controller for feedwater heater level control [Mow98] for the Browns Ferry BWR plant model was studied with the RELAP5/MOD3 code [INL14]. The reactor pressure vessel and main steam modules are modeled as well as the recirculation pumps and feedwater heater modules, with the system interfaces represented nicely with the LabVIEW graphical software to demonstrate the efficiency of the proportional-integral (PI) and PID control techniques.

For load follow and operational transient maneuvers of PWR plants, a constant axial offset control (CAOC) procedure [Mor74, Sip76] has been developed primarily through iterative feedback control procedures. The primary purpose behind maintaining the AO within a band reflects the relationship between the power peaking factor and AO illustrated in the *flyspec curve* of Figure 13.26. Based on a number of 1-D and 3-D power distribution calculations reflecting the motion of both full- and part-length control rods in PWR cores, load follow capability for Westinghouse PWR plants was demonstrated at the Indian Point Unit 2 plant in 1974. With the *mechanical shim* (MSHIM) control strategy introduced for the new AP1000 design, two banks of traditional *black* control rods are exercised together with four banks of *gray* control rods of low reactivity worth in an overlapping strategy to provide reactivity variations required for load follow maneuvers. This is in contrast to the primary use of *chemical shim* in the form of soluble boron concentration for reactivity control in the current generation of PWR plants in operation. The MSHIM strategy is augmented by the traditional CAOC control maneuvers via full-length AO control rod banks. The MSHIM control strategy comprising overlapped maneuvers of the gray and regular control rod banks is demonstrated [Dru09] together with the BEACON [Bea94] or COLSS [Com90] online core monitoring system. A more systemic version of the MSHIM control maneuver was developed recently [Wan14] with a PI control strategy implemented with the MATLAB Simulink toolbox.

In contrast to the PWR load-following technology, BWR systems are able to perform load follow maneuvers conveniently by varying the recirculation flow rate without requiring control rod movements. This technique relies on the negative void coefficient of reactivity discussed in Section 14.1. An increase (or decrease) in recirculation flow temporarily reduces (or increases) the void fraction, which increases (or decreases) the reactivity, thereby increasing (or decreasing) the power level. Recirculation flow maneuvers could accommodate load variations typically up to 25% of rated power and at a rate of 1% per second.

Load follow or scheduling maneuvers for the sodium-cooled fast reactor (SFR) system have been also studied [Pas15, Pas17], with particular attention given to the passive safety characteristics of the system. The passive safety features are modeled with the power coefficient model from Eq. (14.10) involving the coolant inlet temperature $T_{in}$, relative power $P$, and power-to-flow ratio $P/F$, and control strategies are developed that could allow active control of the system without overriding the passive control characteristics. Multiple-input multiple-output (MIMO) control of the plant is effectively represented by multiple single-input single-output (SISO) control maneuvers that account for the relative gain or importance via characteristic time constants.

Associated with load-follow maneuvers for NPPs is the need to control the frequency of generated electricity supplied to the grid [IAE18]. Controlling the frequency on a continuous basis requires an increase in the electrical output when the system frequency decreases, since the power increase is proportional to the frequency increase via a proportionality constant known as the *droop* [Lok12]. In France, with close to 75% of electricity supplied by nuclear power plants, load follow maneuvers are effectively used, and a 20 mHz frequency change would require ~1% change in power output. In the United States, all of the 98 NPPs currently operate in a base load mode, usually operating at the maximum possible output of each plant.

## 16.6 Kalman Filtering for Optimal System Estimation

One of the techniques used as part of the model-based control formulation is Kalman filtering, which provides an optimal estimate of a dynamical system subject to modeling uncertainties and inherent observation errors. The linear Kalman filter formulation presented in Appendix F is demonstrated here first for a simple analytical model borrowed from [Jaz70] and then for system diagnosis applications. The basic formulation for a discretized linear Kalman filter is summarized for system state $x(k)$ and state transition matrix $\Phi(k|k-1)$ with variance $Q$ and observation $y(k)$ with variance $R$ at time step $k$:

$$x(k) = \Phi(k|k-1)x(k-1) + w(k), \ \left\langle w^T(k)w(k)\right\rangle = Q, \tag{16.126a}$$

$$y(k) = M(k)x(k) + v(k), \ \left\langle v^T(k)v(k)\right\rangle = R. \tag{16.126b}$$

Before the measurements at time step $k$, optimal state and variance are estimated

$$\hat{x}^-(k) = \Phi\hat{x}(k-1), \tag{16.127a}$$

$$P^-(k) = \Phi P(k-1)\Phi^T + Q(k). \tag{16.127b}$$

After measurements are taken, optimal estimates are obtained

$$\hat{x}(k) = \hat{x}^-(k) + K(k)\left[y(k) - M\hat{x}^-(k)\right], \tag{16.128a}$$

$$P(k) = \left[I - K(k)M\right]P^-(k). \tag{16.128b}$$

with the Kalman gain matrix $K(k) = P^-(k)M^T\left[MP^-(k)M^T + R\right]^{-1}$.

**Example 16.4** For an object falling down subject to gravity without friction, determine the optimal position and velocity along the trajectory given discrete observations for the position $x_1$ [Jaz70]. The system model for the falling object may be assumed exact but the initial estimates $\hat{x}_1(0) = 95$ for the position and $\hat{x}_2(0) = 1$ for the speed of the object are in error compared with the correct state $x(0) = (100 \ \ 0)^T$.

Set up the equation of motion

$$\frac{dx(t)}{dt} = \begin{pmatrix} \dot{x}_1 \\ \dot{x}_2 \end{pmatrix} = \begin{pmatrix} 0 & 1 \\ 0 & 0 \end{pmatrix} x(t) + \begin{pmatrix} 0 \\ -g \end{pmatrix} = F(t)x(t) + u(t),$$

where $u(t)$ is the control vector representing the acceleration of gravity $g$. Integrating the equation results in a discretized form of the system equation

$$x(k) = \begin{pmatrix} 1 & 1 \\ 0 & 1 \end{pmatrix} x(k-1) - \begin{pmatrix} 0.5 \\ 1.0 \end{pmatrix} g = \Phi x(k-1) + u, Q = \begin{pmatrix} 0 & 0 \\ 0 & 0 \end{pmatrix},$$

together with the observation equation for the position $x_1(k)$ with white Gaussian noise vector $v(k) = N(0, 1)$

$$y(k) = (1 \quad 0) \begin{bmatrix} x_1(k) \\ x_2(k) \end{bmatrix} + v(k) = M(k)x(k) + v(k), R = 1.$$

Perform filtering with initial estimates $\hat{x}(0) = \begin{pmatrix} 95 \\ 1 \end{pmatrix}, P(0) = \begin{pmatrix} 10 & 1 \\ 0 & 1 \end{pmatrix}$, and $g = 1$. Before measurement:

$$\hat{x}^-(1) = \Phi\hat{x}(0) + u(1) = \begin{pmatrix} 1 & 1 \\ 0 & 1 \end{pmatrix} \begin{pmatrix} 95 \\ 1 \end{pmatrix} - \begin{pmatrix} 0.5 \\ 1 \end{pmatrix} = \begin{pmatrix} 95.5 \\ 0 \end{pmatrix},$$

$$P^-(1) = \Phi P(0)\Phi^T + Q = \begin{pmatrix} 1 & 1 \\ 0 & 1 \end{pmatrix} \begin{pmatrix} 10 & 0 \\ 0 & 1 \end{pmatrix} \begin{pmatrix} 1 & 0 \\ 1 & 1 \end{pmatrix} = \begin{pmatrix} 11 & 1 \\ 1 & 1 \end{pmatrix}.$$

After measurement:

$$K(1) = P^-(1)M^T[MP^-(1)M^T + R]^{-1} = \begin{pmatrix} 0.9167 \\ 0.0833 \end{pmatrix},$$

$$\hat{x}^+(1) = \hat{x}^-(1) + K(1)[y(1) - M\hat{x}^-(1)] = \begin{pmatrix} 99.64 \\ 0.375 \end{pmatrix}, \; y(1) = 100.0,$$

$$P^+(1) = [I - K(1)M]P^-(1) = \begin{pmatrix} 0.9163 & 0.0833 \\ 0.0837 & 0.9167 \end{pmatrix}.$$

Filtering is continued for four additional steps, and the optimal estimates for the position of the falling object are plotted together with the measurement and exact solution in Figure 16.22. Note that despite a rather poor estimate of the initial position $x_1(0) = 95$, a sufficiently accurate measurement provides a good estimate at time step $k = 1$. It is moreover noteworthy that, despite a poor measurement, an accurate estimate is retained at $k = 3$. It should be mentioned that, due to the presence of control $u(t)$ in this simple illustration of Kalman filtering, the basic formulation has to be slightly adapted.

Another more realistic example is borrowed from a system diagnostics study [Has89] simulating the pressurizer dynamics of the Three Mile Island unit 2 (TMI-2) plant. In the 1979 TMI-2 accident, misdiagnosing the coolant flow leaking out of a stuck-open power-operated relief valve (PORV) played a significant role

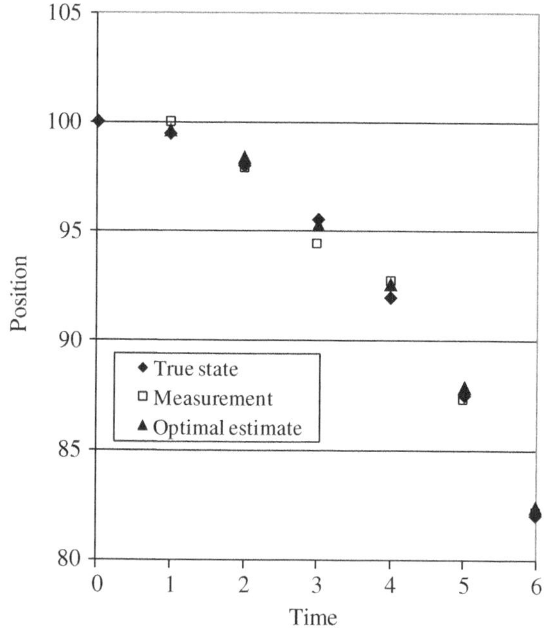


**Figure 16.22** Kalman filter estimates for falling body position compared with measurements and correct position.

in the severity of the unfortunate accident [Lee17]. In this study, Kalman filtering is applied to the pressurizer model comprising the state vector

$$x = \begin{pmatrix} P \\ M_1 \\ M_2 \\ h_1 \\ h_2 \\ W_{re} \\ W_{su} \end{pmatrix} = \begin{pmatrix} \text{pressurizer pressure} \\ \text{vapor mass inventory} \\ \text{liquid mass inventory} \\ \text{vapor enthalpy} \\ \text{liquid enthalpy} \\ \text{relief flowrate} \\ \text{surge flowrate} \end{pmatrix}$$

as part of a moveable boundary version of the TRANSG code [Cru81] representing the primary loop and steam generator. The measurement vector set to $y = (P \;\; L)^T$, with $L$ representing the liquid level in the pressurizer, a nominal simulation is performed with the assumption that the PORV failed fully open. Kalman filtering is then applied to obtain optimal estimates of the PORV relief flowrate $W_{re}$ and surge flowrate $W_{su}$. The pressurizer pressure and level obtained through Kalman filtering are compared with the nominal simulation results in Figure 16.23, where the effects of partially open PORV are systematically represented to show improved agreement with the observed data for both $P$ and $L$. Other examples of Kalman filtering for system diagnostics are discussed further in dynamic event tree applications [Lee17].

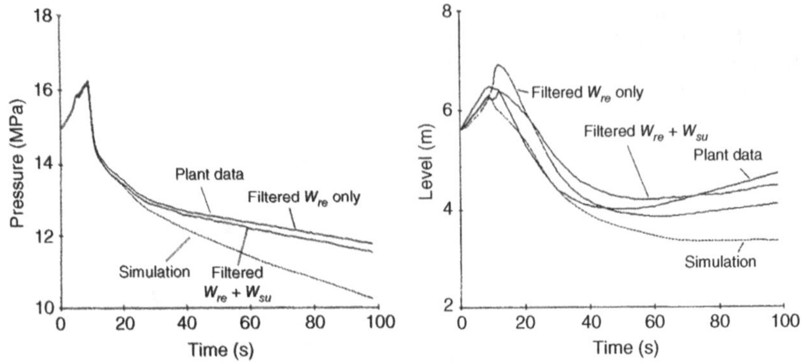


**Figure 16.23** Pressurizer pressure and level filtered for the TMI-2 accident scenario compared with the plant data. The nominal simulation represents the PORV in fully open state. *Source*: Hassberger and Lee [Has89]/Taylor & Francis Group.

## References


**[Arf13]** G.B. Arfken, H.J. Weber, and F.E. Harris, *Mathematical Methods for Physicist: A Comprehensive Guide*, 7th ed., Academic Press (2013).

**[Ash59]** M. Ash, R. Bellman, and R. Kalaba, "On Control of Reactor Shut-down Involving Minimal Xenon Poisoning," *Nucl. Sci. Eng.* **6**, 152 (1959)

**[Ash79]** M. Ash, *Nuclear Reactor Kinetics*, Wiley (1979).

**[Bar99]** M. Barber, R. Miller, H. Joo et al., "Coupled 3D Reactor Kinetics and Thermal-Hydraulic Code Activities at the US. Nuclear Regulatory Commission," M&C 99 (1999).

**[B&Wil72]** Babcock & Wilcox, *Steam – Its Generation and Use*, Babcock & Wilcox (1974).

**[Bea94]** C.L. Beard and T. Morita, "BEACON - Core Monitoring and Operations Support System," WCAP-12472-P-A, Westinghouse Electric Company (1994).

**[INL14]** The RELAP5-3D Code Development Team, "RELAP5-3D Code Manual, Volume I; Code Structure, System Models and Solution Methods," INL-EXT-98-00834-V1. rev. 4.2, Idaho National Laboratory (2014).

**[Chi02]** S.G. Chi and N.Z. Cho, "$H_\infty$ Control Theory Applied to Xenon Control for Load-Following Operation of a Nuclear Reactor," *Nucl. Technol.* **137**, 127 (2002).

**[Com90]** Combustion Engineering, "Overview Description of the Core Operating Limit Supervisory System (COLSS)," CEN312-P Revision 02-P, Combustion Engineering (1990).

**[Cru81]** M.W. Crump and J.C. Lee, "Nonlinear Transient Analysis of Light Water Reactor Steam Generators Using an Implicit Eulerian Method," *Nucl. Sci. Eng.* **77**, 192 (1981); also J.C. Lee et al., "Transient Modeling of Steam Generator Units in Nuclear Power Plants: Computer Code, TRANSG-01," EPRI NP-1368, Electric Power Research Institute (1980).

**[Dah74]** G. Dahlquist and A. Björck, *Numerical Methods*, Prentice-Hall (1974).

**[Dow06]** T. Downar, Y. Xu, T. Kozlowski, and D. Carlson, "PARCS v2.7 US NRC Core Neutronics Simulator," Purdue University (2006).

**[Doy89]** J.C. Doyle, K. Glover, P.P. Khargonekar, and B.A. Francis, "State-Space Solution to Standard $H_2$ and $H_\infty$ Control Problems," *IEEE Trans. Autom. Control* **34**, 831 (1989).

**[Dru09]** K.J. Drudy, T. Morita, and B.T. Connelley, "Robustness of the MSHIM Operation and Control Strategy in the AP1000 Design," *Proc. 17th Int. Conf. Nucl. Eng. (ICONE17)* (2009).

**[Fow71]** T.B. Fowler, D.R. Vondy, and G.W. Cunningham, "Nuclear Reactor Core Analysis Code: CITATION," ORNL-TM-2496, rev. 2, Oak Ridge National Laboratory (1971).

**[Fra72]** F.J. Frank, D.P. Felentzer, and J.C. Lee, "Measured Characteristics of Xenon-Induced Spatial Oscillations in the H. B. Robinson Unit 2 Nuclear Power Plant," WCAP-7936, Westinghouse Electric Corporation (1972).

**[Gol89]** D.E. Goldberg, *Genetic Algorithms in Search, Optimization, and Machine Learning*, Addison-Wesley (1989).

**[Has89]** J.A. Hassberger and J.C. Lee, "A Simulation-Based Expert System for Nuclear Power Plant Diagnostics," *Nucl. Sci. Eng.* **102**, 153 (1989).

**[Hon12]** M. Hone, S. Skidmore, M. Misvel, and E. Resch, "AP1000 Core Reference Report," WCAP-17524-NP, Westinghouse Electric Company (2012).

**[IAE18]** International Atomic Energy Agency, "Non-baseload Operation in Nuclear Power Plants: Load Following and Frequency Control Modes of Flexible Operation," NP-T-2.23, International Atomic Energy Agency (2018).

**[Jaz70]** A.H. Jazwinski, *Stochastic Processes and Filtering Theory*, Academic Press (1970).

**[Joh92]** B. Johansen, Y.A. Chao, and A.L. Casadei, "ALPS: an advanced loading pattern search program," *Proc. Tech. Meeting and Workshop on In-core Fuel Management: Reloading Techniques*, Vienna, Austria (1992).

**[Kap66]** S. Kaplan and J.B. Yasinsky, "Natural Modes of the Xenon Problem With Feedback – An Example," *Nucl. Sci. Eng.* **25**, 430 (1966).

**[Ker80]** Kernforschungsanlage Jülich, "Kritisches Experiment KAHTER OTTO," Statusbericht No. 6, Kernforschungsanlage Jülich (1980).

**[Lee17]** J.C. Lee and N.J. McCormick, *Risk and Safety Analysis of Nuclear Systems*, Wiley (2011/2017).

**[Lee71a]** J.C. Lee, K.A. Jones, D.P. Felentzer, and D. Rawle, "Axial Xenon Transient Tests at the Rochester Gas and Electric Reactor," WCAP-7964, Westinghouse Electric Corporation (1971).

**[Lee71b]** J.C. Lee, K.A. Jones, and W.I. McCoy, "Axial Stability Measurements at the Rochester Gas and Electric Reactor," *Trans. Am. Nucl. Soc.* **14**, *Suppl. 2*, 15 (1971).

**[Lee72]** J.C. Lee, C.G. Poncelet, and L. Eisenhart, "Quasi-static Code," Internal report, Westinghouse Electric Corporation (1972).

**[Lee74]** J.C. Lee, "Application of an Improved Time-Difference Method to Spatial Xenon Transient Analysis," *Nucl. Sci. Eng.* **54**, 206 (1974).

**[Lig14]** G. Li, "Modeling and LQG/LTR Control for Power and Axial Power Difference of Load-Follow PWR Core," *Ann. Nucl. Energy* **68**, 193 (2014).

**[Lok12]** A. Lokhov, "Load-Following Capabilities of NPPs," *Nucl. Eng. Int.* **57–694**, 34 (2012).

**[Men92]** S.K. Menon and A.G. Parlos, "Gain-Scheduled Nonlinear Control of U-Tube Steam Generator Water Level," *Nucl. Sci. Eng.* **111**, 294 (1992).

**[Met53]** N. Metropolis, A.W. Rosenbluth, M.N. Rosenbluth, and A.H. Teller, "Equation of State Calculations by Fast Computing Machine," *J. Chem. Phys.* **21**, 1087 (1953).

**[Moo99]** B.R. Moore, P.J. Turinsky, A.A. Karve,"FORMOSA-B: A Boiling Water Reactor In-core Fuel Management Optimization Package," *Nucl. Technol.* **126**, 153 (1999).

**[Mor74]** T. Morita, L.R. Scherpereel, K.I. Dzikowski et al., "Power Distribution Control and Load Following Procedures," WCAP-8403, Westinghouse Electric Corporation (1974).

**[Mor84]** J. Moreira and J.C. Lee, "Space-Time Analysis Reactor-Control-Rod-Worth Measurements," *Nucl. Sci. Eng.* **86**, 91 (1984).

**[Mow98]** J.A. Mowrey, P.R. Boylan, and S.I. Abdel-Khalik, "Modeling of Boiling Water Reactor Feedwater Control Algorithms," *Nucl. Technol.* **124**, 223 (1998).

**[Oka13]** Y. Oka and K. Suzuki, eds., *Nuclear Reactor Kinetics and Reactor Control*, Springer Japan (2013).

**[Ott69]** K.O. Ott and D.A. Meneley, "Accuracy of the Quasistatic Treatment of Spatial Reactor Kinetics," *Nucl. Sci. Eng.* **36**, 402 (1969).

**[Oui16]** M. Ouisloumen, "Westinghouse Electric Company LLC - Overview," Presentation at Nuclear Engineering Radiological Sciences Colloquium, University of Michigan (2016) and private communication, February 2017.

**[Pas15]** S. Passerini and R.B. Vilim, "Designing for Inherent Control in Liquid-Metal Advanced Small Modular Reactors," *Nucl. Technol.* **191**, 254 (2015).

[Pas17] S. Passerini, R. Ponciroli, and R.B. Vilim, "Impact of Active Control on Passive Safety Response Characteristics of Sodium-Cooled Fast Reactors: I – Theoretical Background," *Nucl. Technol.* **199**, 1 (2017).

[Pon62] L.S. Pontryagin, V.G. Boltyanskii, R.G. Gamkrelidze, and E.F. Mishchenko, *The Mathematical Theory of Optimal Processes*, Interscience Publishers (1962).

[Ran58] D. Randall and D.S. St. John, "Xenon Spatial Oscillations," *Nucleonics* **16**, 82 (1958).

[Ris75] D.H. Risher Jr. and R.F. Barry, "TWINKLE – A Multi-Dimensional Neutron Kinetics Computer Code," WCAP-8028-A, Westinghouse Electric Corporation (1975).

[Rob65] J.J. Roberts and H.P. Smith Jr., "Time Optimal Solution to the Reactivity-Xenon Shutdown Problem," *Nucl. Sci. Eng.* **22**, 470 (1965).

[Rob67] J.J. Roberts, R.F. Fleming, and H.P. Smith Jr., "Experimental Application of the Time-Optimal Xenon Shutdown Program," *Nucl. Sci. Eng.* **27**, 573 (1967).

[Ros64] Z.R. Rosztoczy and L.E. Weaver, "Optimum Reactor Shutdown Program for Minimum Xenon Buildup," *Nucl. Sci. Eng.* **20**, 318 (1964).

[Sch80] E.J. Schulz and J.C. Lee, "Time-Optimal Control of Spatial Xenon Oscillations to a Generalized Target," *Nucl. Sci. Eng.* **73**, 140 (1980).

[Sho78] R.A. Shober, D.R. Ferguson, and T.A. Daly, "FX2: A Two-dimensional Nuclear Reactor Kinetics Code with Thermal and Hydraulics Feedback," ANL-78-97, Argonne National Laboratory (1978).

[Sip76] P.J. Sipush, R.A. Kerr, A.P. Ginsberg et al., "Load Follow Demonstrations Employing Constant Axial Offset Power-Distribution Control Procedures," *Nucl. Technol.* **31**, 12 (1976).

[Sut96] T.M. Sutton and B.N. Aviles, "Diffusion Theory Methods for Spatial Kinetics Calculations," *Prog. Nucl. Energy* **30**, 119 (1996).

[Suz93] K. Suzuki, J. Shimazaki, and Y. Shinohara, "Application of $H_\infty$ Control Theory to Power Control of a Nonlinear Reactor Model," *Nucl. Sci. Eng.* **115**, 142 (1993).

[Wan14] P. Wang, J. Wan, Z. Chen et al., "Dynamic Simulation and Study of Mechanical Shim (MSHIM) Core Control Strategy for AP1000 Reactor," *Ann. Nucl. Energy* **72**, 49 (2014).

[Wes03] Westinghouse Electric Company, "AP1000 Design Control Document," APR-GW-GI-700, rev. 3, Westinghouse Electric Company (2003).

[Zhu16] A. Zhu, Y. Xu, and T. Downar, "A Multilevel Quasi-Static Kinetics Method for Pin-Resolved Transport Transient Reactor Analysis," *Nucl. Sci. Eng.* **182**, 435 (2016).

## Problems

**16.1** Derive the coefficients $a_1$ and $a_2$ in Eq. (16.10) for the linear flux evolution of Eq. (16.9).

**16.2** Show that $\phi_{nk}(r,\theta) = J_n(B_{nk}r)\sin n\theta$ satisfies Eq. (16.54).

**16.3** Obtain an expression for the costate vector $p(t)$ of Eq. (16.81) and the associated switching function $S(t)$ of Eq. (16.82).

**16.4** Using one-group neutron diffusion theory and introducing a Lagrangian multiplier, obtain the optimal radius-to-height ratio for a cylindrical reactor so that the neutron leakage probability is minimized.

**16.5** Through direct evaluation, show that the Hamiltonian is conserved throughout the system trajectory for Example E.1.

**16.6** A rocket ship is descending to earth vertically from an initial position $h$ within the atmosphere. The upward thrust of the rocket engine has maximum value $u_m$ so that $0 \le u \le u_m$, with $u_m > g$, the acceleration of gravity. Assume that the mass $m$ of the rocket ship remains constant during the descent so that we may set $m = 1.0$ in some unit. (a) Determine the optimal maneuver of the rocket and the trajectory, in the phase plane of height $x_1$ and speed $x_2$, for a minimum-time descent to the earth surface, $x_1 = 0$, with the requirements that both the initial and final speeds equal zero. (b) Plot the optimal trajectory in the phase plane and determine an expression for the switching time.

**16.7** With the optimal rocket descent solution of Problem 16.6, determine the rocket control that minimizes the energy expended during the descent. For this formulation, construct a cost or objective function $x_0$ as the rate of energy expended as part of the Hamiltonian $H$. (a) How many switchings would be allowed for the optimal maneuver so that $H$ may be minimized? (b) Based on your answer for part (a), compare the total descent time with that obtained in Problem 16.6, without explicitly solving for the system trajectory.

**16.8** Evaluate the stability of the LQG controller of Example 16.3 with the system matrix modified to

$$A = \begin{pmatrix} 0 & 1 \\ 0 & -1 \end{pmatrix}.$$

**16.9** Formulate a simulated annealing algorithm for the time-optimal solution with constrained speed considered in Example E.2. An objective function may be set up to represent the travel time combined with a penalty function for violating the speed limit.

**16.10** For a system modeled by state vector $x$ and measurement vector $y$, assume that the state is completely measurable, i.e. $x = y$. Furthermore, if the covariance matrix $P^-$ for state $x$ before measurements at time step $t_k$ is approximated by a diagonal matrix whose elements are uniformly equal to $\sigma_x^2$ and the measurement covariance matrix $R$ is represented by a diagonal matrix of elements $\sigma_y^2$, obtain a simplified expression for the Kalman gain matrix $K$ for the system at $t = t_k$. Provide a physical interpretation of the result. Show also that the optimal posterior system estimate $x^+$ given the measurement $y$ represents the variance-weighted average of measurement $y$ and prior system estimate $x^-$.

**16.11** For the critical assembly described by the Nordheim–Fuchs model from Section 8.4.2, determine the reactivity control starting at the peak of the power pulse so that the power can be kept constant at $n_{max}$. Assume that the reactivity control will take effect over a short period of time so that the heat transfer and delayed neutron effects may be ignored.

**16.12** An airline pilot is at position $x_0$ with speed $v_0$ heading for an airport at position $x_1$ that must be reached with speed $v_1$. The thrust (forward force) is limited to a range of $0 \leq u \leq u_m$, which is opposed by a drag force of the form $-C_0 v$, with $C_0 > 0$. Assume that the mass of the plane $m = 1$ in a suitable unit. (a) Prove that the time-optimal flight strategy involves a bang-bang control of thrust with at most one switching. (b) If the landing speed $v_1$ is not fixed, show that the time-optimal solution still involves a bang-bang control of thrust but without any switching during the flight.

**16.13** Given the state-space representation of Eq. (16.86), show that the system state $x(t)$ can be written in terms of the state transition matrix $\phi(t - t_0) = \phi(t, t_0)$

$$x(t) = \phi(t, t_0)x(t_0) + \int_{t_0}^{t} \phi(t, \tau)Bu(\tau)d\tau.$$

**16.14** Derive the phase-plane representation of the xenon shutdown trajectory $\Omega$ of Eq. (16.67).

**16.15** Develop a computer program to simulate the time-optimal shutdown trajectory derived in Section 16.3.1.

**16.16** Derive the switching function of Eq. (16.82) for the optimal control of space-time xenon oscillations.

**16.17** Incorporate the optimal control strategy for space-time xenon oscillations derived in Section 16.3.2 into a computer program that simulates the load follow maneuvers in PWR plant.

# 17

# Elements of Neutron Transport Theory

Various techniques for solving the neutron transport equation are discussed in this final chapter, for the purpose of providing fundamental concepts and formulations augmenting simplified treatments for various reactor physics topics covered in earlier chapters. Formulations for collision probability (CP), escape probability, and blackness related to lattice physics analysis of Chapter 11 are discussed in Sections 17.1 through 17.3. Section 17.4 finally presents an overview of recent numerical techniques developed for solving the neutron transport equation in general.

## 17.1 Collision Probability Method

Mathematical formulations for calculating the probability of neutrons interacting with lumped absorbing media are presented in this section. Through the study of collision probability concepts, we will obtain practical algorithms to solve the neutron transport equation in complex geometry. Collision probability relationships of importance in heterogeneous core analysis will also be derived. For notational convenience, we will use energy-independent formulations throughout the section.

We first derive in Section 17.1.1 an integral form of the neutron transport equation and indicate how the transport kernel is related to the probability of neutrons escaping from one region to another. This then introduces the reciprocity relationships for both angular and scalar fluxes in Section 17.1.2, leading to the collision probability concept in Section 17.1.3.

### 17.1.1 Integral Transport Equation

We begin with the steady-state, one-group neutron transport equation

$$\mathbf{\Omega} \cdot \nabla\psi(\mathbf{r}, \mathbf{\Omega}) + \Sigma(\mathbf{r})\psi(\mathbf{r}, \mathbf{\Omega}) = S(\mathbf{r}, \mathbf{\Omega}), \tag{17.1}$$

*Nuclear Reactor Physics and Engineering*, Second Edition. John C. Lee.
© 2025 John Wiley & Sons, Inc. Published 2025 by John Wiley & Sons, Inc.
Companion website: www.wiley.com/go/LeeNRPE2

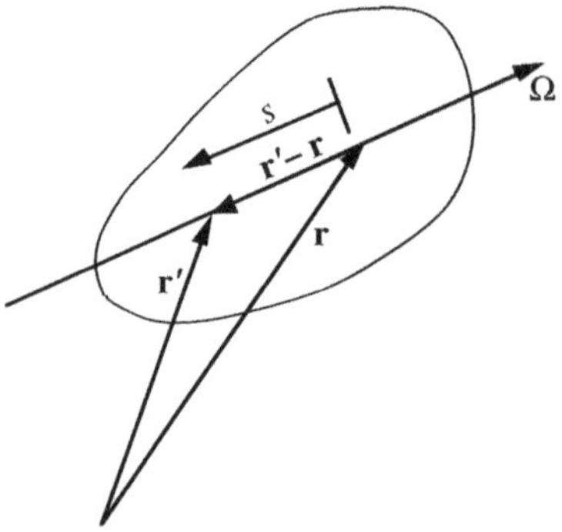


**Figure 17.1** Geometry for a diffusing lump.

where the source term $S(\mathbf{r}, \mathbf{\Omega})$ includes the slowing-down source, fission source, and external sources, if present. Now introduce an isolated lump of diffusing material of volume $V$ and surface area $A$, and define a path length $s$ that increases in the direction opposite that of the neutron motion along $\mathbf{\Omega}$, as illustrated in Figure 17.1.

Write the leakage term in Eq. (17.1) at

$$\mathbf{r}' = \mathbf{r} - s\mathbf{\Omega} \tag{17.2}$$

as a directional derivative in terms of $s$, and obtain

$$-\frac{\partial}{\partial s}\psi(\mathbf{r} - s\mathbf{\Omega}, \mathbf{\Omega}) + \Sigma(\mathbf{r} - s\mathbf{\Omega})\psi(\mathbf{r} - s\mathbf{\Omega}, \mathbf{\Omega}) = S(\mathbf{r} - s\mathbf{\Omega}, \mathbf{\Omega}). \tag{17.3}$$

Equation (17.3) is an ordinary differential equation in path length $s$ and can be readily integrated:

$$\begin{aligned}\psi(\mathbf{r} - s\mathbf{\Omega}, \mathbf{\Omega}) = \exp\left[\int_{s_0}^{s} ds'\Sigma(\mathbf{r} - s'\mathbf{\Omega})\right]\psi(\mathbf{r} - s_0\mathbf{\Omega}, \mathbf{\Omega}) \\ - \int_{s_0}^{s} S(\mathbf{r} - s'\mathbf{\Omega}, \mathbf{\Omega}) \exp\left[\int_{s'}^{s} ds''\Sigma(\mathbf{r} - s''\mathbf{\Omega})\right] ds'.\end{aligned} \tag{17.4}$$

Now suppose the lump of volume $V$ is convex and surrounded by vacuum, and place $s_0$ on the surface, with $\mathbf{\Omega}$ directed inward to the lump. With the vacuum boundary condition, $\psi(\mathbf{r} - s_0\mathbf{\Omega}, \mathbf{\Omega}) = 0$, the first term on the RHS of Eq. (17.4) vanishes. Since $S(\mathbf{r}, \mathbf{\Omega}) = 0$ outside $V$, extend the limit of integral to $s_0 = \infty$ to rewrite Eq. (17.4):

$$\psi(\mathbf{r} - s\mathbf{\Omega}, \mathbf{\Omega}) = \int_{s}^{\infty} S(\mathbf{r} - s'\mathbf{\Omega}, \mathbf{\Omega}) \exp\left[\int_{s'}^{s} ds''\Sigma(\mathbf{r} - s''\mathbf{\Omega})\right] ds'.$$

Set $s = 0$, i.e. move from location $\mathbf{r}'$ to $\mathbf{r}$, and let $s' \to s$ to obtain

$$\psi(\mathbf{r}, \mathbf{\Omega}) = \int_{0}^{\infty} S(\mathbf{r} - s\mathbf{\Omega}, \mathbf{\Omega}) e^{-\tau(\mathbf{r}, \mathbf{r} - s\mathbf{\Omega})} ds,$$

with the optical path length defined as

$$\tau(\mathbf{r}, \mathbf{r} - s\mathbf{\Omega}) = \int_{0}^{s} \Sigma(\mathbf{r} - s'\mathbf{\Omega}) ds'. \tag{17.5}$$

Using Eq. (17.2) results in a final expression for the angular flux

$$\psi(\mathbf{r}, \mathbf{\Omega}) = \int_0^\infty S(\mathbf{r}', \mathbf{\Omega}) e^{-\tau(\mathbf{r},\mathbf{r}')} ds, \tag{17.6}$$

where Eq. (17.5) now represents the *optical path length* between $\mathbf{r}$ and $\mathbf{r}'$

$$\tau(\mathbf{r}, \mathbf{r}') = \int_0^s \Sigma(\mathbf{r} - s'\mathbf{\Omega}) ds', \tag{17.7}$$

or simply $s\Sigma$ for constant $\Sigma$.

Equation (17.6) essentially represents the simple observation that the angular flux $\psi(\mathbf{r}, \mathbf{\Omega})$ is the sum of the contributions from sources at $\mathbf{r}'$ of neutrons moving in direction $\mathbf{\Omega}$ that are attenuated by the factor $\exp[-\tau(\mathbf{r}, \mathbf{r}')]$ in traveling from $\mathbf{r}'$ to $\mathbf{r}$ along the direction $\mathbf{\Omega}$. Equation (17.6) can further be integrated over solid angle $\mathbf{\Omega}$ to yield

$$\phi(\mathbf{r}) = 4\pi \int_V S\left(\mathbf{r}', \frac{\mathbf{r} - \mathbf{r}'}{|\mathbf{r} - \mathbf{r}'|}\right) T(\mathbf{r}, \mathbf{r}') d\mathbf{r}', \tag{17.8}$$

where

$$T(\mathbf{r}, \mathbf{r}') = \frac{e^{-\tau(\mathbf{r},\mathbf{r}')}}{4\pi|\mathbf{r} - \mathbf{r}'|^2} \tag{17.9}$$

is the *point transport kernel*. Note that the differential solid angle $d\mathbf{\Omega}$ subtended by a differential area $dA$ at distance $s$ is defined as $d\mathbf{\Omega} = dA/s^2$ and hence

$$ds d\mathbf{\Omega} = \frac{d\mathbf{r}'}{|\mathbf{r} - \mathbf{r}'|^2}. \tag{17.10}$$

Equation (17.8) is known as Peierls' equation, and the kernel $T(\mathbf{r}, \mathbf{r}')$ physically represents the scalar flux at $\mathbf{r}$ due to a unit isotropic point source of neutrons at $\mathbf{r}'$ moving in the direction $\mathbf{\Omega} = (\mathbf{r} - \mathbf{r}')/|\mathbf{r} - \mathbf{r}'|$. For isotropic sources distributed in a homogeneous medium, Eq. (17.8) reduces to

$$\phi(\mathbf{r}) = \int_V S(\mathbf{r}') T(\mathbf{r}, \mathbf{r}') d\mathbf{r}' = \frac{1}{4\pi} \int_V S(\mathbf{r}') e^{-\Sigma|\mathbf{r}-\mathbf{r}'|} \frac{d\mathbf{r}'}{|\mathbf{r} - \mathbf{r}'|^2}. \tag{17.11}$$

This is an integral form of the neutron transport equation that serves as the starting point for the collision probability (CP) methods used for heterogeneous lattice analysis.

### 17.1.2 Reciprocity Relationship

For a point neutron source of strength $Q$ [neutron $\cdot$ s$^{-1}$] released in direction $\mathbf{\Omega}_0$ at position $\mathbf{r}_0$

$$S(\mathbf{r}', \mathbf{\Omega}) = Q\delta(\mathbf{r}' - \mathbf{r}_0)\delta(\mathbf{\Omega} - \mathbf{\Omega}_0), \tag{17.12}$$

an expression for the angular flux is obtained [Cas53] from Eq. (17.6), together with Eq. (17.10):

$$\begin{aligned}\psi(\mathbf{r},\boldsymbol{\Omega}) &= \int_0^\infty S(\mathbf{r}',\boldsymbol{\Omega})e^{-\tau(\mathbf{r},\mathbf{r}')}ds\int_{\boldsymbol{\Omega}}\delta(\boldsymbol{\Omega}-\boldsymbol{\Omega}')d\boldsymbol{\Omega}' \\ &= \int_V Q\delta(\mathbf{r}'-\mathbf{r}_0)\delta(\boldsymbol{\Omega}-\boldsymbol{\Omega}_0)e^{-\tau(\mathbf{r},\mathbf{r}')}\delta(\boldsymbol{\Omega}-\boldsymbol{\Omega}')\frac{d\mathbf{r}'}{|\mathbf{r}-\mathbf{r}'|^2} \\ &= \frac{Qe^{-\tau(\mathbf{r},\mathbf{r}_0)}}{|\mathbf{r}-\mathbf{r}_0|^2}\delta\left(\frac{\mathbf{r}-\mathbf{r}_0}{|\mathbf{r}-\mathbf{r}_0|}-\boldsymbol{\Omega}\right)\delta(\boldsymbol{\Omega}-\boldsymbol{\Omega}_0) = \psi(\mathbf{r},\boldsymbol{\Omega}|\mathbf{r}_0,\boldsymbol{\Omega}_0).\end{aligned} \tag{17.13}$$

The symmetry in Eq. (17.13) establishes a *reciprocity relationship for angular flux*

$$\psi(\mathbf{r},\boldsymbol{\Omega}|\mathbf{r}_0,\boldsymbol{\Omega}_0) = \psi(\mathbf{r}_0,-\boldsymbol{\Omega}_0|\mathbf{r},-\boldsymbol{\Omega}), \tag{17.14}$$

representing the equivalence between the angular flux $\psi(\mathbf{r},\boldsymbol{\Omega})$ due to a point source at $\mathbf{r}_0$ of neutrons moving in direction $\boldsymbol{\Omega}_0$ and $\psi(\mathbf{r}_0,-\boldsymbol{\Omega}_0)$ due to a point source at $\mathbf{r}$ of neutrons moving in direction $-\boldsymbol{\Omega}$. For a general case where neutron scattering is explicitly represented, a similar approach may be taken to support the applicability of Eq. (17.14) provided the scattering is isotropic in the laboratory system.

Likewise, from Eq. (17.11), obtain scalar flux $\phi(\mathbf{r}|\mathbf{r}_0)$ at $\mathbf{r}$ in a homogeneous medium due to a unit point source at $\mathbf{r}_0$:

$$\begin{aligned}\phi(\mathbf{r}|\mathbf{r}_0) &= \frac{1}{4\pi}\int_V \delta(\mathbf{r}'-\mathbf{r}_0)e^{-\Sigma|\mathbf{r}-\mathbf{r}'|}\frac{d\mathbf{r}'}{|\mathbf{r}-\mathbf{r}'|^2} \\ &= \frac{e^{-\Sigma|\mathbf{r}-\mathbf{r}_0|}}{4\pi|\mathbf{r}-\mathbf{r}_0|^2}.\end{aligned} \tag{17.15}$$

Hence, we readily note a *reciprocity relationship for scalar flux*

$$\phi(\mathbf{r}|\mathbf{r}_0) = \phi(\mathbf{r}_0|\mathbf{r}), \tag{17.16}$$

equivalent to Eq. (17.14).

### 17.1.3 Transport Kernel and Collision Probability

The transport kernel $T(\mathbf{r},\mathbf{r}')$ given in Eq. (17.9) can also be interpreted as the probability that neutrons, isotropically distributed in a unit volume element at $\mathbf{r}'$, move into a unit volume element at $\mathbf{r}$ without suffering collisions, i.e. a first-flight escape probability. With this interpretation for the transport kernel $T(\mathbf{r},\mathbf{r}')$, we may now derive the CP formulation of the neutron transport equation.

Consider the volume $V$ of our interest divided into $N$ homogeneous subregions of volume $V_n, n = 1,\dots,N$, with $\Sigma = \Sigma_t$ assumed spatially uniform within each region, i.e.

$$\Sigma(\mathbf{r}) = \Sigma_n, \forall\mathbf{r}\in V_n, n = 1,\dots,N.$$

The probability $P_{mn}$ that a neutron born isotropically in $V_m$ has its next collision in $V_n$ is written as

$$P_{mn} = \frac{\int_{4\pi} d\Omega \int_{V_n} d\mathbf{r}\Sigma_n \int_{V_m} d\mathbf{r}' S\left(\mathbf{r}', \frac{\mathbf{r}-\mathbf{r}'}{|\mathbf{r}-\mathbf{r}'|}\right) T(\mathbf{r},\mathbf{r}')}{\int_{V_m} d\mathbf{r} S(\mathbf{r})}, \tag{17.17}$$

which yields, together with a discretized form of Eq. (17.8),

$$\int_{V_n} d\mathbf{r}\Sigma_n \phi(\mathbf{r}) = \sum_m \int_{4\pi} d\Omega \int_{V_n} d\mathbf{r}\Sigma_n \int_{V_m} d\mathbf{r}' S\left(\mathbf{r}', \frac{\mathbf{r}-\mathbf{r}'}{|\mathbf{r}-\mathbf{r}'|}\right) T(\mathbf{r},\mathbf{r}'),$$

or equivalently

$$\Sigma_n \overline{\phi}_n V_n = \sum_{m=1}^{N} P_{mn} \overline{S}_m V_m, \quad n = 1, \ldots, N, \tag{17.18}$$

where the region-average flux and source are defined as

$$\overline{\phi}_n = \frac{1}{V_n}\int_{V_n} d\mathbf{r}\phi(\mathbf{r}) = \frac{1}{V_n}\int_{V_n} d\mathbf{r} \int_{4\pi} d\Omega \psi(\mathbf{r},\Omega),$$
$$\overline{S}_n = \frac{1}{V_n}\int_{V_n} d\mathbf{r} S(\mathbf{r}) = \frac{1}{V_n}\int_{V_n} d\mathbf{r} \int_{4\pi} d\Omega S(\mathbf{r},\Omega).$$

Except when $S(\mathbf{r})$ is limited to an external neutron source, Eq. (17.18) represents a set of $N$ coupled linear algebraic equations for the region fluxes $\overline{\phi}_n$ and has to be solved iteratively.

Some savings in the effort to calculate the collision probabilities $P_{mn}$ can be achieved if we assume the source $S(\mathbf{r},\Omega)$ is isotropic and uniform over each region $V_n$. A reciprocity relationship is then obtained from Eq. (17.17):

$$\Sigma_m P_{mn} V_m = \frac{\Sigma_n}{\overline{S}_n} \int_{V_m} d\mathbf{r}\Sigma_m \int_{V_n} d\mathbf{r}' \overline{S}_n T(\mathbf{r},\mathbf{r}') = \Sigma_n P_{nm} V_n. \tag{17.19}$$

For example, if we consider a two-region unit cell, consisting of a fuel region of volume $V_F$ surrounded by a moderator region of volume $V_M$ as illustrated in Figure 17.2, Eq. (17.19) implies

$$\Sigma_M P_{MF} V_M = \Sigma_F P_{FM} V_F. \tag{17.20}$$

From the neutron balance for a unit cell, we also obtain

$$P_{FM} = 1 - P_{FF}, \tag{17.21}$$

$$P_{MM} = 1 - P_{MF}. \tag{17.22}$$

We may start with the self-collision probability $P_{FF}$, which is equal to the first-flight collision probability $P_c$ to be discussed in Section 17.2. Then, using Eqs. (17.21), (17.20), and (17.22), in that order, we can evaluate all other collision probabilities $P_{FM}$, $P_{MF}$, and $P_{MM}$ for the unit cell.

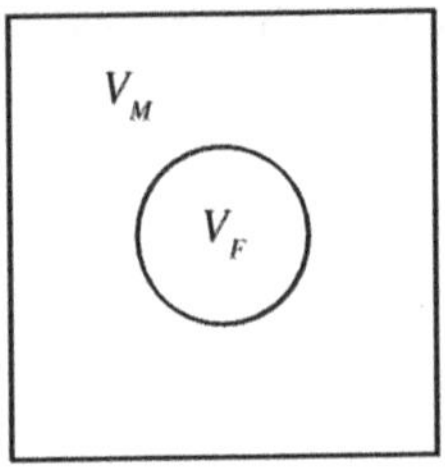


**Figure 17.2** Two-region unit cell.

## 17.2 First-Flight Escape Probability and Dirac Chord Method

The basic concept of the first-flight escape probability and Dirac's chord method for evaluating the probability are introduced in this section. We then study how a measure of flux depression in an absorbing lump can be obtained through the escape probability and derive various relationships among collision and escape probabilities.

Defining $P(\mathbf{r})$ as the probability that a neutron born at $\mathbf{r}$ within volume $V$ and surface area $A$ escapes without suffering a collision in $V$, we obtain the average probability $P_0$ that neutrons of source strength $S(\mathbf{r})$ isotropically distributed in $V$ escape without collisions:

$$P_0 = \frac{\int_V d\mathbf{r} S(\mathbf{r}) P(\mathbf{r})}{\int_V d\mathbf{r} S(r)}. \tag{17.23}$$

With $d\boldsymbol{\Omega}$ representing the solid angle subtended by the area $dA$ at distance $s$ from point $\mathbf{r}$, as illustrated in Figure 17.3, the first-flight escape probability $P(\mathbf{r})$ is written in terms of the transport kernel of Eq. (17.9)

$$P(\mathbf{r}) = \int_A \frac{e^{-\Sigma s}}{4\pi s^2} (\mathbf{n} \cdot \boldsymbol{\Omega})\, dA = \int_\Omega e^{-\Sigma s} \frac{d\Omega}{4\pi}, \tag{17.24}$$

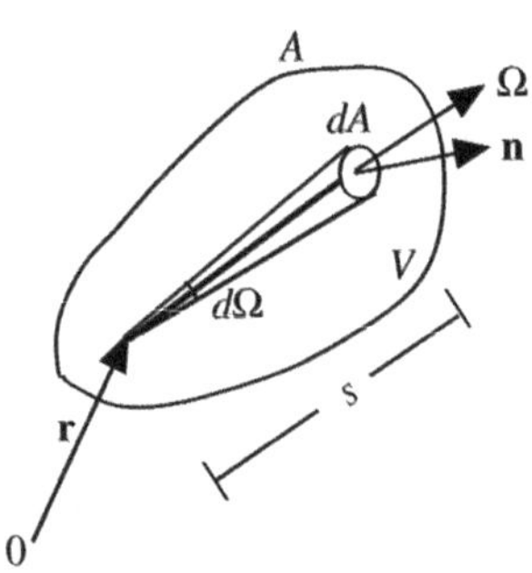


**Figure 17.3** Geometry for escape probability calculation.

where $d\Omega = (\mathbf{n} \cdot \boldsymbol{\Omega})dA/s^2$ for an outward normal vector $\mathbf{n}$. For a spatially uniform source distribution, substituting Eq. (17.24) into Eq. (17.23) yields

$$P_0 = \frac{1}{4\pi V}\int_V d\mathbf{r}\int_\Omega d\Omega e^{-\Sigma s}. \tag{17.25}$$

Since the derivation of Eqs. (17.24) and (17.25) has been limited to the probability that neutrons escape without suffering collisions in the lump, $P_0$ represents a *first-flight escape probability*. For a purely absorbing lump, i.e. $\Sigma_a = \Sigma = \Sigma_t$, $P_0$ is equal to the net escape probability. For a scattering medium, the net escape probability should take into account multiple-collision events, as discussed further in Section 17.3. Note from Eq. (17.25) that $P_0$ is a function of $\Sigma$ and geometry of the lump. In terms of the first-flight escape probability, define the *first-flight collision probability* $P_c$

$$P_c = 1 - P_0, \tag{17.26}$$

which represents the average probability that neutrons born isotropically and uniformly in a lump will have first collisions in the lump. Similar to the observation for $P_0$, the first-flight collision probability $P_c$ becomes, for a purely absorbing medium, the net absorption probability for neutrons born isotropically and uniformly in the medium.

Actual calculation of the escape probability $P_0$ is often conveniently performed through the chord length method introduced by Dirac [Dir43]. If we picture the volume $V$ consisting of a collection of tubes of cross-sectional area $(\mathbf{n}_i \cdot \boldsymbol{\Omega})dA$ and length $s$, with differential length $d\ell$, volume element $d\mathbf{r}$ is given as

$$d\mathbf{r} = (\mathbf{n}_i \cdot \boldsymbol{\Omega})dAd\ell. \tag{17.27}$$

Here, $\mathbf{n}$ and $\mathbf{n}_i$ are the outward and inward unit normal vectors, respectively, at $dA$, and $s(\boldsymbol{\Omega})$ is equal to the length of a chord drawn in direction $\boldsymbol{\Omega}$ from the surface element $dA$, as illustrated in Figure 17.4. Substituting Eq. (17.27) into Eq. (17.25) yields

$$\begin{aligned} P_0 &= \frac{1}{4\pi V}\int_A dA\int_\Omega d\Omega\int_0^s d\ell\, e^{-\Sigma\ell}\,(\mathbf{n}_i \cdot \boldsymbol{\Omega}) \\ &= \frac{1}{4\pi V\Sigma}\int_A dA\int_\Omega d\Omega(1 - e^{-\Sigma s})(\mathbf{n}_i \cdot \boldsymbol{\Omega}), \end{aligned} \tag{17.28}$$

where the integral is performed such that $n_i \cdot \boldsymbol{\Omega} > 0$. In terms of the volume element of Eq. (17.27), we may represent the volume of the lump as a sum of volumes of tubes of length $s(\boldsymbol{\Omega})$ and area of the base equal to $dA(\mathbf{n}_i \cdot \boldsymbol{\Omega})$ so that

$$V = \int_A dA(\mathbf{n}_i \cdot \boldsymbol{\Omega})\, s(\boldsymbol{\Omega}) \tag{17.29}$$

and determine the number of chords of length between $s$ and $s + ds$ in the solid angle $d\Omega$ to be proportional to $\mathbf{n}_i \cdot \boldsymbol{\Omega}$, the cosine of the angle between the inward

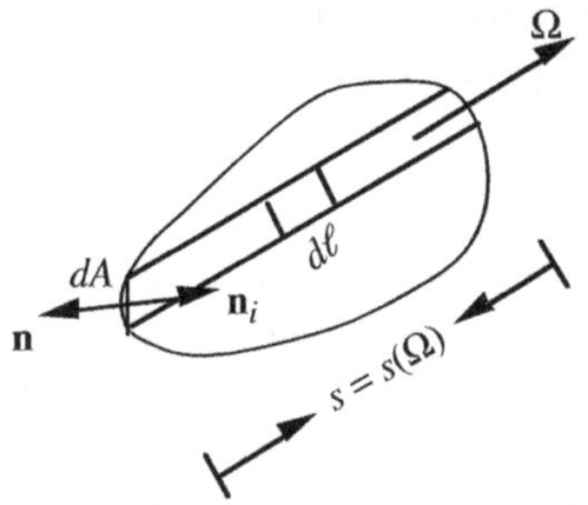


**Figure 17.4** Chord length in a lump of volume $V$.

normal vector $\mathbf{n}_i$ and direction $\boldsymbol{\Omega}$ at $dA$. The fraction $h(s)ds$ that chords have lengths between $s$ and $s + ds$ in $V$ may then be obtained as

$$h(s)\,ds = \frac{\int_A dA \int_{\Omega=\Omega(s)} d\Omega\, \mathbf{n}_i \cdot \boldsymbol{\Omega}}{\int_A dA \int_{\mathbf{n}_i\cdot\Omega>0} d\Omega\, \mathbf{n}_i \cdot \boldsymbol{\Omega}}. \tag{17.30}$$

The numerator is proportional to the number of chords of length $s \sim s + ds$ summed over all possible values of $\boldsymbol{\Omega}$ and surface area $A$. The denominator is proportional to the total number of chords of all lengths that lie within $V$ such that $\mathbf{n}_i \cdot \boldsymbol{\Omega} > 0$ and can be readily evaluated

$$\int_A dA \int_{\mathbf{n}_i\cdot\Omega>0} d\Omega\, \mathbf{n}_i \cdot \boldsymbol{\Omega} = \int_A dA \int_0^1 d\mu\mu \int_0^{2\pi} d\varphi = \pi A. \tag{17.31}$$

The average length of chords in $V$ can then be determined with Eqs. (17.29) through (17.31)

$$\bar{s} = \int_s ds\, s\, h(s) = \frac{1}{\pi A}\int_\Omega d\Omega \int_A dA\ \ s(\boldsymbol{\Omega})(\mathbf{n}_i \cdot \boldsymbol{\Omega}) = \frac{V}{\pi A}\int_{4\pi} d\Omega = \frac{4V}{A}, \tag{17.32}$$

with all possible directions now considered from $dA$ for a convex body. This expression for the *mean chord length* is usually attributed to P. Dirac [Dir43,Cas53] but was actually first obtained by A. Cauchy in 1908 [Wei58]. Cauchy proved that the projected area of a convex body with surface area $A$ is $A/4$, as is the case for a sphere, and obtained the average length of the chords making up the body as the volume divided by the projected area. This derivation provides a simple physical interpretation for the mean chord length.

We now perform the integration over $\boldsymbol{\Omega}$ in Eq. (17.28) for the first-flight escape probability $P_0$ by invoking the chord length distribution function from Eq. (17.30) for the lump with the chord lengths covering the interval $[s_{min}, s_{max}]$:

$$P_0 = \frac{\pi A}{4\pi V\Sigma}\int_{s_{min}}^{s_{max}} ds\, h(s)(1 - e^{-\Sigma s}) = \frac{1}{\bar{s}\Sigma}\int_{s_{min}}^{s_{max}} ds\, h(s)(1 - e^{-\Sigma s}). \tag{17.33}$$

**Figure 17.5** Chord length for a semi-infinite slab of thickness *a*.

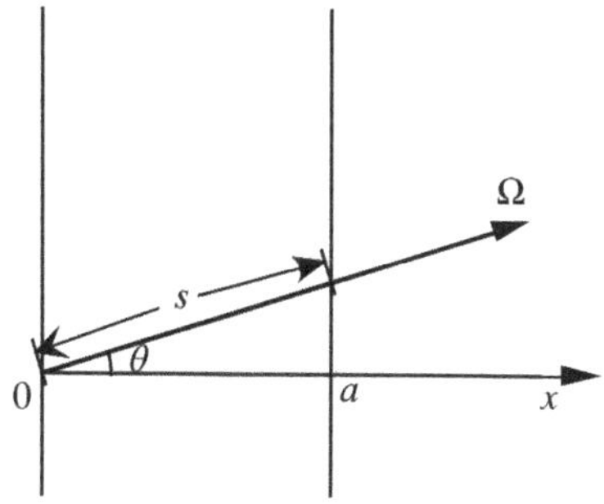


Since the mean chord length from Eq. (17.32) is a simple function of the volume and surface area of the lump, we expect that the escape probability can be characterized primarily in terms of the mean chord length regardless of the particular geometry and shape of the lump involved. The mean chord length is, in fact, very similar to the definition of the effective hydraulic diameter in fluid mechanics. Although empirical in nature, the effective hydraulic diameter can be used to represent characteristics of turbulent fluid flow nearly independent of the flow geometry.

**Example 17.1** Consider a semi-infinite slab of thickness $a$, illustrated in Figure 17.5, and obtain the first-flight escape probability $P_0$.

The chord length distribution function from Eq. (17.30) may be written in terms of the cosine of the polar angle $\mu = \cos\theta$

$$h(\mu)d\mu = \frac{1}{\pi A}\int_A dA \int_\mu^{\mu+d\mu} d\mu'\mu' \int_0^{2\pi} d\varphi = 2\mu d\mu = -h(s)ds,$$

which shows $h(s) = 2a^2/s^3$. Substitution of $h(s)$ into Eq. (17.33) yields

$$P_0 = \frac{a}{\Sigma}\int_a^\infty (1 - e^{-\Sigma s})\frac{ds}{s^3} = \frac{1}{a\Sigma}\left[\frac{1}{2} - E_3(\Sigma a)\right]. \tag{17.34}$$

The first-flight escape probability for slab geometry may be obtained also from Eq. (17.24) simplified for the slab

$$P(x) = \frac{1}{2}\left[\int_0^1 e^{-\Sigma x/\mu}d\mu + \int_0^1 e^{\Sigma(a-x)/\mu}d\mu\right] = \frac{1}{2}\left[E_2(\Sigma x) + E_2\{\Sigma(a-x)\}\right],$$

which is averaged over the slab thickness to yield Eq. (17.34). For the semi-infinite slab of thickness $a$, the mean chord length $\bar{s} = 2a$, representing the left- and right-hand surfaces through which neutrons may leak out of the slab.

**Example 17.2** Obtain the first-flight escape probability for a sphere of radius $a$, illustrated in Figure 17.6.

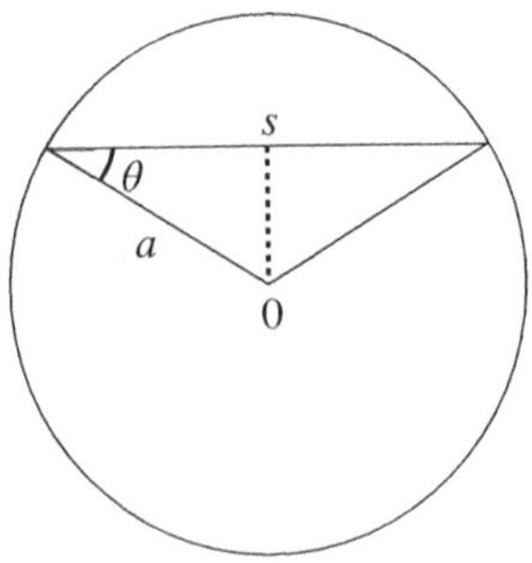


**Figure 17.6** Chord length for a sphere of radius $a$.

With the chord length $s$ chosen in Figure 17.6, the distribution function $h(s)$ is determined via $\mu = \cos\theta$ such that

$$h(s)ds = \frac{1}{\pi A}\int_A dA \int_{\mu}^{\mu+d\mu} d\mu'\mu' \int_0^{2\pi} d\varphi = 2\mu d\mu = \frac{s}{2a^2}ds.$$

Substitution of $h(s)$ into Eq. (17.33) yields

$$P_0 = \frac{3}{4\Sigma a}\int_0^{2a}(1 - e^{-\Sigma s})\frac{sds}{2a^2} = \frac{3}{8(\Sigma a)^3}\left[2(\Sigma a)^2 - 1 + (1 + 2\Sigma a)e^{-2\Sigma a}\right], \tag{17.35}$$

with the mean chord length $\bar{s} = \dfrac{4a}{3}$.

The first-flight collision probability $P_c$ defined in Eq. (17.26) as the complement of $P_0$ has been evaluated for a number of standard geometries and is tabulated [Cas53]. As noted in connection with Eq. (17.33), collision and escape probabilities are characterized primarily by the mean chord length. Hence, the slab geometry result of Eq. (17.34) may be used quite accurately for other geometries, provided an effective slab thickness $a$ is determined to preserve the mean chord length. Since we have limited ourselves, however, to a convex volume, with no re-entrant surface allowed, the surface area $A$ for calculating the mean chord length should also exclude surface areas over which there is no net leakage or current of neutrons. This point is important, for example, for a unit cell in Figure 17.2, where the mean chord length for the moderator region should exclude the unit-cell outer boundary; by the definition of a unit cell, there is zero net leakage at the cell boundary. This is illustrated for the simple unit-cell model discussed in Chapter 11. Note also that the self-collision probabilities $P_{FF}$ and $P_{MM}$ from Eqs. (17.21) and (17.22) represent the first-flight collision probabilities for the fuel and moderator regions, respectively.

The first-flight escape probability $P_0$ for a sphere and infinitely long cylinder are compared with that for a semi-infinite slab of equal mean chord length in

**Table 17.1** Escape probability as a function of mean chord length.

| $\bar{s}\Sigma$ | Slab | Sphere | Cylinder | Wigner rational approximation | Carlvik two-term approximation |
|---|---|---|---|---|---|
| 0.04 | 0.952 | 0.978 | 0.974 | 0.962 | 0.974 |
| 0.1 | 0.902 | 0.946 | 0.939 | 0.909 | 0.937 |
| 5.0 | 0.193 | 0.193 | 0.193 | 0.167 | 0.196 |

Table 17.1. Included also are Wigner's rational approximation and Carlvik's two-term approximation introduced in Chapter 11. We note that for sufficiently large values of $\bar{s}\Sigma$, $P_0$ for the sphere and infinitely long cylinder may be represented by the simple slab-geometry result with the equivalent $\bar{s}\Sigma$ to a reasonable approximation. In addition, Table 17.1 indicates the degree of accuracy involved with the traditional Wigner's rational approximation and Carlvik's two-term rational approximation of Eq. (11.68) developed for cylindrical fuel rods. The fitting parameters chosen for Carlvik's approximation [Sta83] are

$$\alpha_1 = 2, \quad \alpha_2 = 3, \quad \beta_1 = 2, \quad \beta_2 = -1$$

for

$$P_0 = \frac{\beta\alpha_1}{x+\alpha_1} + \frac{(1-\beta)\alpha_2}{x+\alpha_2} \equiv \sum_{i=1}^{2} \frac{\beta_i\,\alpha_i}{x+\alpha_i}, \quad x = \bar{s}\Sigma. \tag{17.36}$$

## 17.3 Flux Depression Calculation and Blackness

We discuss in this section how some of the concepts related to the escape probability of Section 17.2 could be used to derive alternate expressions useful in the collision probability solution of the neutron balance equations. The basic concepts are also extended to diffusing media with finite scattering cross sections.

### 17.3.1 Escape Probability and Flux Depression Factor

We begin with the derivation of an expression for the depression of the flux inside an absorbing lump and illustrate how the first-flight escape probability $P_0$ for outgoing neutrons is related to the first-flight collision probability for incoming neutrons. Consider a lump of volume $V$ immersed in an infinite bath of mono-energetic neutrons, and assume that the neutrons are incident isotropically on the lump. Given the asymptotic scalar flux $\phi_\infty$ of neutrons in the infinite medium, the number $N_i$ of neutrons incident on the lump per unit time from the surrounding medium is obtained as

$$N_i = \int_A dA \int_{\mathbf{n}_i \cdot \mathbf{\Omega} > 0} d\mathbf{\Omega}\, (\mathbf{n}_i \cdot \mathbf{\Omega}) \psi(\mathbf{r}, \mathbf{\Omega}) = \int_A dA \int_{\mathbf{n}_i \cdot \mathbf{\Omega} > 0} d\mathbf{\Omega} \frac{\phi_\infty}{4\pi} (\mathbf{n}_i \cdot \mathbf{\Omega}) = \frac{\phi_\infty A}{4}, \tag{17.37}$$

where the isotropic angular flux $\psi(\mathbf{r}, \mathbf{\Omega})$ of incident neutrons is duly recognized, together with Eq. (17.31). Likewise, the number $N_0$ of neutrons leaving the lump per unit time without suffering any collision is

$$N_0 = \int_A dA \int_{\mathbf{n}_i \cdot \mathbf{\Omega} > 0} d\mathbf{\Omega} \frac{\phi_\infty}{4\pi} (\mathbf{n}_i \cdot \mathbf{\Omega}) e^{-\Sigma s}, \tag{17.38}$$

where the exponential term accounts for the attenuation of the incident beam of neutrons traversing a distance $s$ along $\mathbf{\Omega}$.

Combining Eqs. (17.37) and (17.38) yields an expression for the number $N_a$ of neutrons suffering first collisions per unit time in the lump

$$N_a = N_i - N_0 = \frac{\phi_\infty}{4\pi} \int_A dA \int_{\mathbf{n}_i \cdot \mathbf{\Omega} > 0} d\mathbf{\Omega} (1 - e^{-\Sigma s}) (\mathbf{n}_i \cdot \mathbf{\Omega}), \tag{17.39}$$

which can be also written in terms of the average flux $\overline{\phi}$ within the lump

$$N_a = \Sigma \overline{\phi} V.$$

Comparing Eqs. (17.28) and (17.39) indicates that

$$N_a = \Sigma \phi_\infty V P_0.$$

Hence, we obtain

$$P_0 = \frac{\overline{\phi}}{\phi_\infty}, \tag{17.40}$$

as a simple measure of the *flux depression* in a purely absorbing lump with $\Sigma_t = \Sigma$ immersed in an infinite bath of mono-energetic neutrons that are isotropically distributed. This result is of considerable importance in accounting for flux depression in the fuel region in unit-cell analysis.

In terms of the collision rate $N_a$ of Eq. (17.39) and the rate $N_i$ of incident neutrons given in Eq. (17.37), determine an expression for the probability $\beta$ that neutrons isotropically incident on the lump will suffer first collisions in the lump

$$\beta = \frac{N_a}{N_i} = \bar{s} \Sigma P_0 = \bar{s} \Sigma (1 - P_c), \tag{17.41}$$

where the first-flight collision probability $P_c$ for outgoing neutrons from Eq. (17.26) is recalled. The *first-flight collision probability $\beta$ for incoming neutrons* is commonly known as the *first-flight blackness*. We observed in Section 17.2 that the first-flight probabilities $P_0$ and $P_c$ represent, for a purely absorbing medium, the net escape and absorption probabilities, respectively, for outgoing neutrons. For a purely absorbing medium, the first-flight blackness $\beta$ likewise represents the net absorption probability for incoming neutrons. Section 17.3.2 discusses how these first-flight probabilities can be used to obtain the corresponding net probabilities.

### 17.3.2 Net Escape Probability and Collision Probability

To obtain the net escape and collision probabilities, we assume, for simplicity, that neutrons are distributed uniformly across the lump in each generation of neutron collision. A correction for nonuniform distributions of neutrons is included in the Amouyal–Benoist–Horowitz model for thermal disadvantage factor calculation [Amo57] discussed in Section 11.4.2. For uniform neutron distributions, the net probabilities are obtained through a summation of contributions from successive generations. With this purpose in mind, clarify the physical interpretation of first-flight probabilities for a medium of volume $V$ and surface area $A$ with cross sections, $\Sigma_a$, $\Sigma_s$, and $\Sigma_t = \Sigma_a + \Sigma_s$, as illustrated in Figure 17.7:

$P_0$ = average probability that neutrons distributed uniformly and isotropically within $V$ escape without suffering an absorption or scattering collision in $V$,
$P_c$ = average probability that neutrons distributed uniformly and isotropically within $V$ suffer the first (or next) collisions in $V$,

with Eq. (17.41) rewritten as

$$\beta = \bar{s}\Sigma_t P_0 = \bar{s}\Sigma_t(1 - P_c).$$

In terms of the first-flight collision probability $P_c$, now define the *net escape probability*:

$P_0^*$ = average probability that neutrons distributed uniformly and isotropically within $V$ escape after any number of collisions in $V$,

Figure 17.7 Escape probability $P_0$ and average flux $\bar{\phi}$ in a volume immersed in $\phi_\infty$.

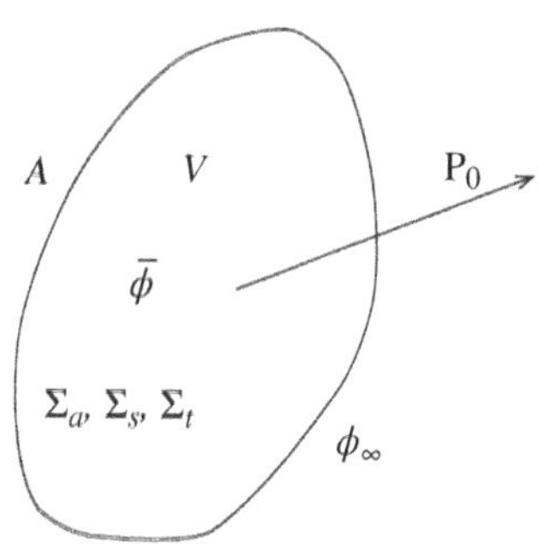

which may readily be obtained as

$$P_0^* = (1 - P_c) + P_c\gamma(1 - P_c) + P_c\gamma P_c\gamma(1 - P_c) + \cdots = \frac{1 - P_c}{1 - \gamma P_c}, \tag{17.42}$$

with $\gamma = \Sigma_s/\Sigma_t$. Likewise, define the net blackness or simply *blackness* as

$\beta^*$= average probability that neutrons uniformly and isotropically incident on the lump are absorbed in $V$ after any number of collisions in $V$,

to obtain

$$\beta^* = \beta[(1 - \gamma) + \gamma P_c(1 - \gamma) + \gamma P_c\gamma P_c(1 - \gamma) + \cdots]$$

or

$$\beta^* = \frac{\beta(1 - \gamma)}{1 - \gamma P_c} = \bar{s}\Sigma_a\left(\frac{1 - P_c}{1 - \gamma P_c}\right) = \bar{s}\Sigma_a P_0^*. \tag{17.43}$$

The expression for the blackness may also be obtained by using the net escape probability of Eq. (17.42):

$$\beta^* = \beta[(1 - \gamma) + \gamma(1 - P_0^*)] = \beta\left(1 - \gamma P_0^*\right). \tag{17.44}$$

In general, the probabilities $P_0$, $\beta$, $P_0^*$, and $\beta^*$ can be calculated via the first-flight collision probability $P_c$, tabulated in [Cas53] or some approximations discussed with Table 17.1. For a purely absorbing medium, simply note that $\beta^*$ and $P_0^*$ reduce to $\beta$ and $P_0$, respectively, as they should. We have obtained a number of relations among the escape and collision probabilities and illustrated the concept of Dirac's chord method as a practical method to obtain the first-flight probabilities. That the first-flight escape probability provides a measure of flux depression in an absorbing lump is also a result of some importance in heterogeneous core physics analysis, as illustrated in Section 11.5.1. Although our discussion has been based on an energy-independent model, an extension to energy-dependent models can be made in a straightforward manner.

### 17.3.3 Dancoff Factor for Fuel Lattice

For lattice physics analysis of fuel assemblies in nuclear reactors, the basic representation is often built around a unit cell geometry comprising a fuel element or rod surrounded by a cladding and a coolant or moderator region. Accurate evaluation of the effective resonance integral representing absorption rates around sharp resonances in a fuel rod also requires proper accounting for the presence of neighboring rods. This is necessary because some of the resonance energy neutrons moving in the moderator region toward a fuel rod under consideration suffer collisions in neighboring fuel rods. This *rod shadowing* phenomenon discussed in Chapter 11 is usually represented with the *Dancoff factor*,

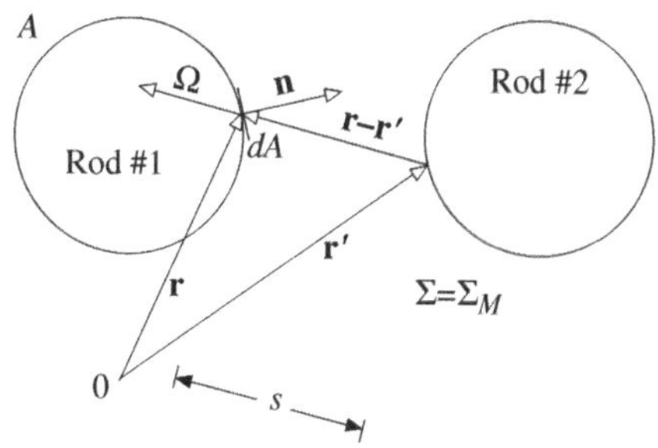


**Figure 17.8** Lattice arrangement for Dancoff factor evaluation.

which is derived through the collision probability concept introduced in this section.

Consider, as illustrated in Figure 17.8, fuel element #1 under consideration with surface area $A$ and neighboring element #2 immersed in a moderator of volume $V$ and total cross section $\Sigma = \Sigma_M$ with resonance neutrons of strength of $Q$ [neutron $\cdot$ cm$^{-3}$ s$^{-1}$] isotropically and uniformly distributed within $V$. The first-flight escape probability $P(\mathbf{r})$ from Eq. (17.24) provides an expression for the incoming current $J_-$ of neutrons for element #1 with surface area $A$ due to neutrons originating at all positions $\mathbf{r}'$ on fuel element #2 and arriving at the surface of element #1 without suffering any collision in the moderator:

$$J\_A = Q\int_A dA \int d\mathbf{r}' \frac{e^{-\Sigma s}}{4\pi s^2}(\mathbf{n}\cdot\boldsymbol{\Omega}), \quad s = |\mathbf{r}-\mathbf{r}'|. \tag{17.45}$$

Setting $d\mathbf{r}' = s^2 ds d\boldsymbol{\Omega}$ anchored at $\mathbf{r}$ on element #1 and integrating over the track length $s$, landing on element #2, yields a more useful expression:

$$J_- = \frac{Q}{4\pi\Sigma A}\int_A dA \int_{\mathbf{n}\cdot\boldsymbol{\Omega}<0} d\boldsymbol{\Omega}(1-e^{-\Sigma s})(\mathbf{n}\cdot\boldsymbol{\Omega}). \tag{17.46}$$

Recalling Eq. (17.31) allows rewriting Eq. (17.46) and introduces the definition of the *Dancoff factor C*

$$J_- = \frac{Q}{4\Sigma}\left[1 - \frac{1}{\pi A}\int_A dA \int_{\mathbf{n}\cdot\boldsymbol{\Omega}<0} d\boldsymbol{\Omega} e^{-\Sigma s}(\mathbf{n}\cdot\boldsymbol{\Omega})\right] = J_-^{\infty}(1-C), \tag{17.47}$$

with $J_-^{\infty} = Q/(4\Sigma)$ representing the incoming current due to an isotropic neutron source of strength $Q$ uniformly distributed in an infinite medium, without the presence of fuel element #2. It should be noted that the integral over $dA$ is performed over the surface of element #1 while the integral over $d\boldsymbol{\Omega}$ is carried out over the surface of element #2, thereby properly accounting for the transport of neutrons between the two fuel elements.

Comparing Eq. (17.46) with Eq. (17.28) shows that $J_-$ may also be rewritten as

$$J_- = \frac{QV_M}{A}P_{0M} = J_-^{\infty}\bar{s}_M\Sigma_M P_{0M} = J_-^{\infty}\beta_M, \tag{17.48}$$

where the subscript $M$ is introduced to clarify that all of the parameters now refer to the moderator region. It is also noted that the Dancoff factor may simply be written in terms of the first-flight blackness of the moderator, $C = 1 - \beta_M$, duly representing a reduction in $J_-^\infty$ due to resonance energy neutrons suffering first collisions in the moderator region. Equation (17.47) allows us to interpret the Dancoff factor $C$ also as the probability that resonance neutrons escaping a fuel element reach another fuel element without suffering collisions in the moderator region.

Dancoff and Ginsberg [Dan44] analytically performed the integration of Eq. (17.45) over two cylindrical fuel rods involving the Bickley–Nayler functions [Bic35]. In recent years, considerable effort has been made to obtain the Dancoff factor for various fuel lattices through analytical approaches [Tal07], including the chord length method [Jiw11] and Monte Carlo algorithms. Equation (17.47) also indicates that the Dancoff factor for a fuel rod amounts to reducing the surface area $A$ of the fuel element by the factor $1 - C$, hence the *rod shadowing effect*. The recognition that $C = 1 - \beta_M$ also allows a number of approximate representations [Arg63]. The simplest representation is Wigner's rational approximation for $P_{0M}$:

$$1 - C \simeq \frac{\bar{s}_M \Sigma_M}{1 + \bar{s}_M \Sigma_M} \simeq \beta_M. \tag{17.49}$$

Bell and Glasstone [Bel70] also suggested a complex expression allowing for both the use of first-flight blackness $\beta$ and multiple traverses of neutrons between the fuel element and surrounding moderator region. In this approach, the effective escape probability for the fuel element is equated to the total collision probability in the moderator

$$\begin{aligned} P_{0F}(1 - C_{eff}) &= P_{0F}\beta_M + P_{0F}(1 - \beta_M)(1 - \beta_F)\beta_M + \cdots \\ &= \frac{P_{0F}(1 - C)}{1 - C(1 - \beta_F)}, \end{aligned} \tag{17.50}$$

with $C = 1 - \beta_M$ and $\beta_F$ representing the first-flight blackness of the fuel region. Approaches similar to Eq. (17.50) are also suggested in [Dud76,Hen75,Tal07]. It should be finally noted that the Dancoff factor for a lattice with multiple fuel elements should, in principle, be summed over the immediately surrounding fuel elements.

## 17.4 Numerical Solution of Neutron Transport Equation

We now discuss numerical algorithms to evaluate the collision probability (CP) between various regions in a nuclear reactor core starting with the analytic expressions for escape and collision probabilities obtained in Sections 17.2 and 17.3.

This is followed by a brief discussion of general numerical algorithms for solving the neutron transport equations.

### 17.4.1 Collision Probability Calculation for Annular Geometry

For the purpose of developing CP relationships for a unit-cell geometry, apply the first-flight escape probabilities of Eqs. (17.24) and (17.25) to the case of an infinitely long cylinder of radius $R$ in Figure 17.9 and calculate the probability $P(0 \to R)$ that neutrons born uniformly and isotropically along the axis of the cylinder escape the cylinder without suffering collisions. The probability $P(0 \to R)$ may be considered the transmission probability for neutrons traversing an optical distance $\tau(R)$ defined as

$$\tau(R) = \int_0^R \Sigma(s)ds, \quad \Sigma = \Sigma_t. \tag{17.51}$$

Equation (17.24) is used to obtain $P(0 \to R)$ or written equivalently as $P(\tau)$

$$\begin{aligned} P(\tau) = P(0 \to R) &= \frac{1}{2}\int_0^\pi \exp(-\tau/\sin\theta)\sin\theta d\theta \\ &= \int_0^{\pi/2} \exp(-\tau/\sin\theta)\sin\theta d\theta. \end{aligned} \tag{17.52}$$

A simple coordinate transformation

$$\cosh u = 1/\sin\theta, du = -\cosh u d\theta$$

yields

$$P(\tau) = \int_0^\infty du \frac{\exp(-\tau\cosh u)}{\cosh^2 u} = Ki_2(\tau), \tag{17.53}$$

Figure 17.9 Infinitely long cylinder for transmission probability calculation.

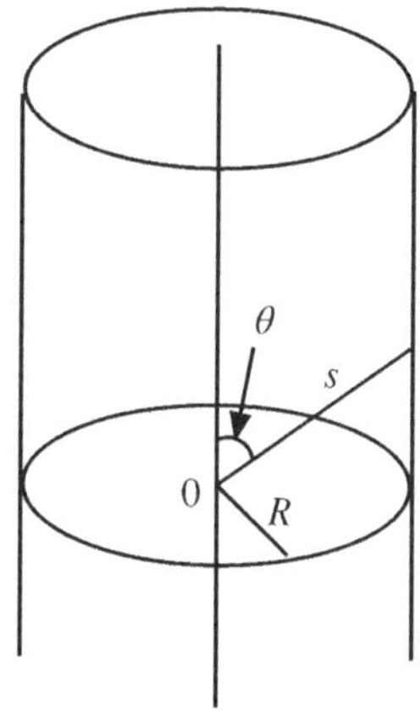

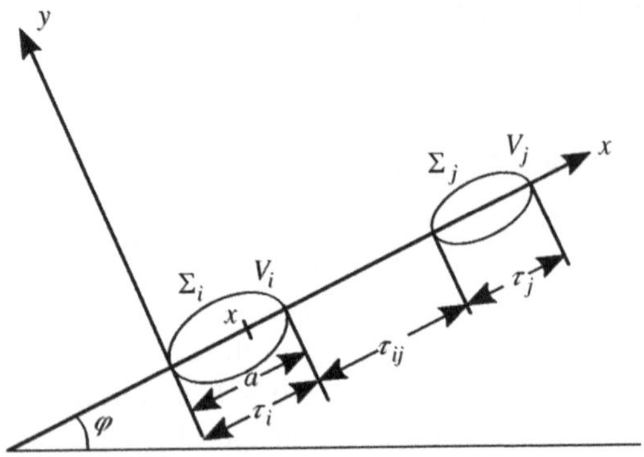


**Figure 17.10** Geometry for collision probability $P_{i\to j}$.

with the Bickley–Nayler function [Bic35] of order $n$ defined as

$$Ki_n(\tau) = \int_0^\infty du \frac{\exp(-\tau \cosh u)}{\cosh^n u}. \tag{17.54}$$

In terms of Eq. (17.53) for the transmission probability $P(\tau)$ for neutrons traveling an optical distance $\tau$ without suffering any collision, an expression is derived for the two-dimensional collision probability $P_{i\to j}$ that neutrons born uniformly and isotropically in volume $V_i$ will have next collisions in volume $V_j$. Generalize $P(\tau)$, as illustrated in Figure 17.10, to determine the probability $P(x, y, \varphi)$ that neutrons born isotropically at position $(x, y)$ traveling along direction $\varphi$ within volume $V_i$ will suffer next collisions in $V_j$. The probability can be written in terms of $P(\tau)$ of Eq. (17.53) as a product of the probability of traveling from position $(x, y)$ to volume $V_j$ and the probability of suffering collisions within $V_j$

$$\begin{aligned} P(x, y, \varphi) &= Ki_2[\Sigma_i(a - x) + \tau_{ij}][1 - Ki_2(\tau_j)] \\ &= Ki_2[\Sigma_i(a - x) + \tau_{ij}] - Ki_2[\Sigma_i(a - x) + \tau_{ij} + \tau_j], \end{aligned} \tag{17.55}$$

where $\tau_j$ and $\tau_{ij}$ are optical distances within volume $V_j$ and between the volumes $V_i$ and $V_j$, respectively. The probability $P_{i\to j}$ that neutrons born uniformly and isotropically in $V_i$ suffer next collisions in $V_j$ can be obtained as an integral of $P(x, y, \varphi)$ over the cross-sectional area of source volume $V_i$ and all possible directions $\varphi$ of neutron travel

$$\begin{aligned} P_{i\to j} &= \frac{1}{2\pi V_i} \int_0^{2\pi} d\varphi \int_{y_{min}(\varphi)}^{y_{max}(\varphi)} dy \int_0^a dx P(x, y, \varphi), \\ &= \frac{1}{2\pi V_i \Sigma_i} \int_0^{2\pi} d\varphi \int_{y_{min}(\varphi)}^{y_{max}(\varphi)} dy [Ki_3(\tau_{ij}) - Ki_3(\tau_{ij} + \tau_i) \\ &\quad - Ki_3(\tau_{ij} + \tau_j) + Ki_3(\tau_{ij} + \tau_i + \tau_j)], \end{aligned} \tag{17.56}$$

with $\tau_i$ representing the optical distance in $V_i$. The arguments of the Bickley–Nayler function $Ki_3$ in Eq. (17.56) indicate that the terms involving the path length between $V_i$ and $V_j$ and over the entire distance from the left edge of $V_i$ to the right edge of $V_j$ are summed, from which the terms covering the distance from the left edge of $V_i$ to the left edge of $V_j$ and the distance from the right edge of $V_i$ to the right edge of $V_j$ are subtracted.

For a cylindrical geometry, the integration over the azimuthal angle $\varphi$ cancels the term $2\pi$ in the denominator, yielding [Sta83]

$$\Sigma_i V_i P_{i\to j} = \int_{y_{i,min}}^{y_{i,max}} dy[Ki_3(\tau_{ij}) + Ki_3(\tau_i + \tau_j + \tau_{ij}) - Ki_3(\tau_i + \tau_{ij}) - Ki_3(\tau_j + \tau_{ij})]. \tag{17.57}$$

Integrating Eq. (17.57) over the right half of the cylinder for the neutrons traveling to the right half of the annular ring $V_j$ yields

$$\frac{1}{2}\Sigma_i V_i P_{i\to j} = \int_0^{r_i} dy[Ki_3(\tau^-_{i,j-1}) + Ki_3(\tau^-_{i-1,j}) - Ki_3(\tau^-_{i-1,j-1}) - Ki_3(\tau^-_{i,j})], \tag{17.58}$$

with the convention for optical distances between the boundaries of volume $V_i$ and $V_j$

$$\tau^{\pm}_{ij} = \tau_j \pm \tau_i = \sqrt{r_j^2 - y^2} \pm \sqrt{r_i^2 - y^2}, \quad i < j, \tag{17.59}$$

as illustrated in Figure 17.11. Adding the contributions of neutrons traveling to the left and accounting for the left half of the cylinder yields

$$\Sigma_i V_i P_{i\to j} = 2(S_{i-1,j-1} + S_{ij} - S_{i-1,j} - S_{i,j-1}), \tag{17.60}$$

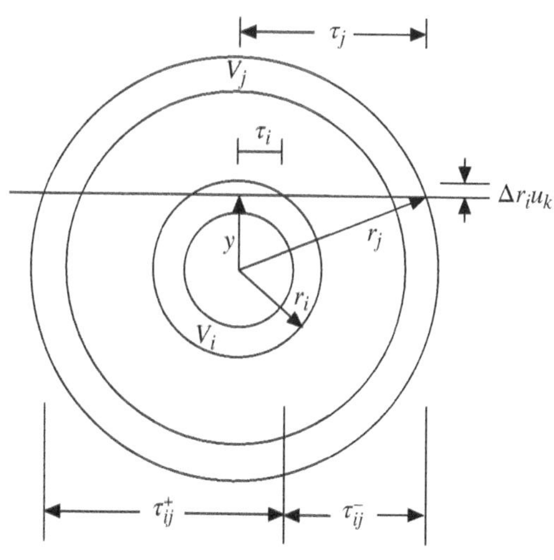


**Figure 17.11** Unit-cell geometry for collision probability calculation.

with

$$S_{ij} = \int_0^{r_i} dy[Ki_3(\tau_{ij}^+) - Ki_3(\tau_{ij}^-)]. \tag{17.61}$$

Finally, the self-collision probability $P_{i\to i}$ is evaluated via the escape probability obtained separately from Eq. (17.56)

$$P_{0,i} = \frac{1}{\Sigma_i V_i} \int_{y_{i,min}}^{y_{i,max}} dy[Ki_3(0) - Ki_3(\tau_i)] \tag{17.62}$$

and accounting properly for the neutrons traversing in both directions within $V_i$ yields the general relationship for the collision probability:

$$\Sigma_i V_i P_{i\to j} = \Sigma_i V_i \delta_{ij} + 2(S_{i-1,j-1} + S_{ij} - S_{i-1,j} - S_{i,j-1}), \quad i \le j. \tag{17.63}$$

**Example 17.3** Implement the collision probability formulation of Eq. (17.63) for a unit cell with a fuel rod radius of 4.75 mm, pitch of 12.60 mm, total cross sections $\Sigma_F = \Sigma_1 = 0.611\,\text{cm}^{-1}$, and $\Sigma_M = \Sigma_2 = 1.09\,\text{cm}^{-1}$. Compare the collision probability $P_{1\to 1}$ for the fuel rod with the self-collision probability $P_{FF} = P_c$ of Eq. (17.21).

For the evaluation of $S_{ij}$ involving the integral of the Bickley–Nayler function $Ki_3(\tau_{ij}^{\pm})$ over $y$, adopt the Gauss quadrature [Abr64] implemented in the FLURIG subroutine of the CPM-2 code [Jon87], beginning with

$$\int_0^{r_i} dyKi_3[\tau_{ij}^{\pm}(y)] = \sum_{\ell=1}^{i} \int_{r_{\ell-1}}^{r_\ell} dyKi_3[\tau_{ij}^{\pm}(y)] = \sum_{\ell=1}^{i} \Delta r_\ell \int_0^1 duKi_3[\tau_{ij}^{\pm}(u)],$$

$$\Delta r_\ell = r_\ell - r_{\ell-1}.$$

The algorithm closely follows the COPRAN subroutine [Sta83]. The integral over the normalized variable $u$ is then performed with $u = x^2$ via the Gauss quadrature of order $n$

$$\int_0^{r_i} dyKi_3[\tau_{ij}^{\pm}(y)] = \sum_{\ell=1}^{i} \Delta r_\ell \int_0^1 2Ki_3[\tau_{ij}^{\pm}(x^2)]xdx = \sum_{\ell=1}^{i} \Delta r_\ell \sum_{k=1}^{n} 2w_k Ki_3[\tau_{ij}^{\pm}(x_k^2)],$$

where the quadrature points $x_k$ and weights $w_k$ for order $n = 2$ are

$$(x_1, x_2) = (0.35505, 0.84494) \quad \text{and} \quad (w_1, w_2) = (0.18195, 0.31804).$$

With the terms $S_{ij}$ of Eq. (17.61) evaluated for $r_1 = 4.75$ mm and $r_2 = 7.108$ mm through the Gauss quadrature and substituted into Eq. (17.63), the first-flight collision probability for the two-region unit cell is obtained:

$$P(i \to j) = \begin{bmatrix} 0.27582 & 0.23099 \\ 0.10447 & 0.32719 \end{bmatrix}.$$

The self-collision probability $P(1 \to 1) = 0.27582$ for the fuel rod with the second-order Gauss quadrature compares favorably with $P_c = 0.2765$ interpolated from the analytical result for an infinitely long cylinder [Cas53] . A fourth-order Gauss quadrature yields $P(1 \rightarrow 1) = 0.27662$ in essentially perfect agreement with the analytical result. It should be noted that a 50-mesh trapezoidal integration yields $P(1 \rightarrow 1) = 0.27633$, indicating the accuracy and efficiency of the Gauss quadrature algorithm.

In terms of the two-dimensional collision probability $P_{i \rightarrow j}$, the transport probability $T(\mathbf{r}, \mathbf{r}')$ from Eq. (17.11) can finally be written in a discretized form

$$T_{i \rightarrow j,g} = P_{i \rightarrow j,g} / (\Sigma_{j,g} V_j), \tag{17.64}$$

where the energy dependence is now explicitly shown for group $g$. A sequential combination of one- and two-dimensional CP formulations forms the basis of the CPM-2 code for both fast and thermal spectrum calculations at the assembly level. The first step involves a one-dimensional fine-mesh, micro-group calculation, via the FLURIG routine, for each of the distinct fuel and absorber rod types. Fine-group fluxes from the micro-group calculations are then used to generate macro-group unit-cell average cross sections for each rod in the assembly. The code then performs a two-dimensional CP calculation using these coarse-mesh, coarse-group constants, representing the actual location of fuel rods and non-lattice regions of the assembly in $x$–$y$ geometry. This algorithm for the CPM-2 code [Jon87] and similar approaches taken for other lattice physics codes including CASMO [Ede92] represented an important step in the adoption of CP algorithms for LWR lattice physics analysis.

### 17.4.2 Discrete Ordinates Method

The main focus of lattice physics analysis at the unit cell or unit assembly level clearly involves the solution of the neutron transport equation (17.1) with the energy dependence and scattering collisions fully represented. One traditional method for solving the transport equation is the $P_n$ or spherical harmonics expansion discussed as part of the derivation for the neutron diffusion equation in Chapter 4. With the one-dimensional, one-group form the transport equation (4.44) rewritten here

$$\Sigma_t \psi(z, \mu) + \mu \frac{\partial \psi(z, \mu)}{\partial z} = S(z, \mu) + \rho(z, \mu), \tag{17.65}$$

expanding the angular flux $\psi(z, \mu)$ in terms of the Legendre polynomials $P_\ell(\mu)$, converts the transport equation into a set of differential equations connecting the Legendre moments $\phi_\ell(z)$ of the angular flux. The coupled equations for $\phi_\ell(z)$ are discretized in space, similar to the approach taken for the finite-difference solution of the neutron diffusion equation in Chapter 6 to yield $\psi(z, \mu)$.

Another popular approach to solve Eq. (17.65) is the $S_n$ or *discrete ordinates method* [Lar84,Lew84], which calculates the angular flux $\psi(z, \mu)$ for a few discrete values of direction $\mu$ and approximates the integral

$$\phi(z) = \int_{-1}^{1} \psi(z, \mu) d\mu = \sum_{n} \omega_n \psi(z, \mu_n), \tag{17.66}$$

in terms of a suitable set of quadrature weights $\omega_n$. For a discrete direction $\mu_n$, approximating the derivative by a first-order difference over a mesh interval $\Delta z_j = z_{j+1/2} - z_{j-1/2}$ yields

$$\Sigma_t \psi(z_j, \mu_n) + \mu_n \frac{\psi(z_{j+1/2}, \mu_n) - \psi(z_{j-1/2}, \mu_n)}{\Delta z_j} = S(z_j, \mu_n) + \rho(z_j, \mu_n), \tag{17.67}$$

where the cell-center flux $\psi(z_j, \mu_n)$ is obtained as a function of mesh-boundary fluxes $\psi(z_{j-1/2}, \mu_n)$ and $\psi(z_{j+1/2}, \mu_n)$. In the *diamond differencing (DD) scheme*, a simple arithmetic averaging is used:

$$\psi(z_j, \mu_n) = \frac{1}{2}[\psi(z_{j-1/2}, \mu_n) + \psi(z_{j+1/2}, \mu_n)]. \tag{17.68}$$

Given the source term $S(z_j, \mu_n)$ and scattering term $\rho(z_j, \mu_n)$, Eq. (17.67) is solved for mesh-boundary fluxes, following the direction of neutron travel for each $\mu_n$. To avoid numerical difficulties, including negative values of flux that may be encountered in the diamond differencing scheme, a number of alternate high-order schemes have been developed. One popular scheme, called the *linear discontinuous scheme*, approximates $\psi(z, \mu_n)$ for each $\mu_n$ by a linear function that is discontinuous at the mesh boundaries. In this scheme, two difference equations, similar to Eq. (17.67), are solved for each spatial cell and for each discrete direction.

Once the angular flux $\psi(z, \mu_n)$ is solved through the diamond differencing or alternative approaches, the source term $S(z, \mu_n)$ and scattering term $\rho(z, \mu_n)$ may be updated by using the latest estimate of $\psi(z, \mu_n)$, and the process is repeated until convergence is reached. In this traditional source iteration method, the convergence can be slow, since the *spectral radius* $\gamma$ representing the largest value of the magnitude of eigenvalues of the governing iteration matrix is equal to the ratio $c = \Sigma_s/\Sigma_t$. To overcome this difficulty, a number of alternate iteration schemes have been developed. In the *diffusion synthetic acceleration scheme*, the source iteration is accelerated by combining the discretized solution for $\psi(z, \mu_n)$ with a consistently discretized solution for a low-order approximation, usually diffusion theory or low-order $P_n$ formulation [Lar84,Lew84]. Significant accelerations can be attained in this synthetic approach, with the spectral radius reduced to $\gamma = 0.23c$, for slab-geometry transport problems. A number of discrete coordinates codes, including Denovo [Eva10] and PARTISN [Alc08], with various acceleration schemes are available for both lattice physics and global reactor physics analysis.

One acceleration algorithm for the solution of the transport equation coupled with low-order solution may be illustrated with the concept of the Eddington factor [Wil15]. With the $P_0$ component of Eq. (17.65) written for the scalar flux $\phi_0(z)$

$$\Sigma_t\phi_0(z) + \frac{d\phi_1(z)}{dz} = S_0(z) + \Sigma_s\phi_0(z), \tag{17.69}$$

consider the $P_1$ component written explicitly in terms of the second moment of $\psi(z,\mu)$

$$\Sigma_t\phi_1(z) + \frac{d}{dz}[E\phi_0(z)] = \Sigma_{s1}\phi_1(z),\ E\phi_0(z) = \int_{-1}^{1}\mu^2\psi(z,\mu)d\mu, \tag{17.70}$$

where the $P_1$ component of the source $S_1(z)$ is assumed negligible. The *Eddington tensor* $E\phi_0(z)$ represents the total contribution of the high-order terms of $\psi(z,\mu)$ without any truncation of the angular flux and may be written to provide a correction factor $\hat{D}$ for the low-order diffusion equation solver:

$$\frac{d}{dz}[E\phi_0(z)] = \frac{1}{3}\frac{d\phi_0(z)}{dz} - \Sigma_{tr}\hat{D}\phi_0(z),\ \Sigma_{tr} = \Sigma_t - \Sigma_{s1}. \tag{17.71}$$

This allows rewriting the $P_1$ equation (17.70) in terms of $\hat{D}$

$$\frac{1}{3}\frac{d\phi_0(z)}{dz} + \Sigma_t\phi_1(z) - \Sigma_{s1}\phi_1(z) = \Sigma_{tr}\hat{D}\phi_0(z), \tag{17.72}$$

with the correction factor $\hat{D}$ introduced

$$\hat{D} = \frac{1}{\hat{\phi}_0(z)}\left[\frac{1}{3\Sigma_{tr}}\frac{d\hat{\phi}_0(z)}{dz} + \hat{\phi}_1(z)\right], \tag{17.73}$$

where the scalar flux $\hat{\phi}_0(z)$ and current $\hat{\phi}_1(z)$ are evaluated with the *high-order transport solution*. Utilizing the concept of the Eddington factor allows the low-order $P_1$ equation or the diffusion equation to account consistently for the accurate transport solution in the coupled acceleration scheme.

### 17.4.3 Method of Characteristics

In the CPM-3 code [Jon02], the one-dimensional unit-cell calculations are replaced by two-dimensional CP calculations for the entire fuel assembly of arbitrary geometry, which makes the code much more versatile than the earlier version of the code. The algorithm entails shooting a number of rays along user-specified directions, representing the integration over angle $\varphi$, and subdividing the integration over $y$ into a number of intervals for numerical evaluation of Eq. (17.56). In addition, neutron reflections at assembly boundaries are represented by keeping track of the Bickley–Nayler functions in Eq. (17.56) in terms of optical distances traversed by the reflected neutron tracks, with a cutoff specified by the user.

Another deterministic transport algorithm gaining popularity in recent years is the *method of characteristics* (MOC), which could provide efficient solutions to 2-D or 3-D transport equations using essentially an integral transport equation (17.6). In this approach, the leakage term $\Omega \cdot \nabla\psi$ in Eq. (17.1) is represented along the direction of neutron travel $s$ in Eq. (17.3). This is equivalent to writing the leakage term in the 1-D 1-group transport equation (17.65) along the characteristic $s = z/\mu$, with $\mu = \cos\theta$, to yield

$$\mu\frac{\partial\psi(z,\mu)}{\partial z} = \frac{d\psi(s)}{ds} = -\Sigma_t(s)\psi(s) + S(s) + \rho(s), \tag{17.74}$$

which may conveniently be integrated for a spatially uniform $\Sigma_t$ to yield

$$\psi(s) = \exp(-\Sigma_t s)\left[\int_0^s ds'[S(s') + \rho(s')]\exp(\Sigma_t s') + \psi(0)\right]. \tag{17.75}$$

With a judicious selection of the directions for the paths or characteristics equivalent to those in the CP method, and duly accounting for the energy- and space-dependent cross sections, the MOC formulation is able to provide solutions to the general 3-D transport equation, where the characteristics may represent paths or tracks in the space-time domain. Recent developments include the DeCART code [Joo04] and the MPACT code [Koc13], which feature efficient 2-D steady-state and transient MOC formulations in the $(x$–$y)$ domain. The MPACT code couples 2-D MOC solver with the coarse mesh finite-difference (CMFD) diffusion solver in the $z$-direction, thereby providing accurate angular representations of the time-dependent neutron flux for pin-resolved 3-D assembly and global power distributions. The coupling of the 2-D MOC solver with the 3-D CMFD flux solver utilizes the Eddington factor construct introduced in Eqs. (17.72) and (17.73). Most of the lattice physics codes under active development, including CASMO5 [Fer17], LANCR02 [Glo09], POLARIS [Rea18], and APOLLO2.8 [San09], have implemented the MOC formulations either at the unit- or multi-assembly level. Time-dependent 2-D MOC formulations have been also developed that could provide accurate transient angular flux solutions with minimum storage requirements [Hof16].

### 17.4.4 Monte Carlo Algorithm

By selecting a host of pseudo-random numbers, Monte Carlo algorithms simulate the life or history of individual particles that follow physical laws of particle interaction and transport, as represented by the general neutron transport equation, without the need to discretize any of the spatial, energy, or directional variables. Monte Carlo algorithms offer the potential to provide accurate solutions for transport problems with complex geometries and material heterogeneities, with the solution accuracy limited only by the computing resources available. With rapid

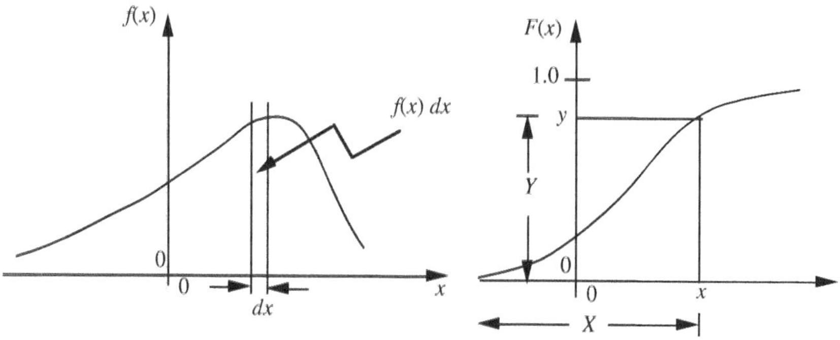


**Figure 17.12** Relationship between PDF and CDF.

advances made in computer hardware, there has been a significant increase in the popularity of Monte Carlo algorithms in both neutron and photon transport applications. This increased popularity owes in no small measure to the versatility that the Monte Carlo codes, including MCNP6 [Goo12], Keno [Rea18], and Serpent [Fri11], offer: (a) simple description of complex geometries using well-defined surfaces, (b) separate or coupled neutron and gamma transport calculations, and (c) cross-section libraries in a continuous energy structure, rather than in discrete group formulation.

The basic principle involved in a Monte Carlo representation of neutron transport and reactions may be illustrated by considering the probability density function (PDF) $f(x)$, as illustrated in Figure 17.12, for a random variable $X$ [Spi08,Lee17]. The PDF is defined such that the probability that a continuous random variable $X$ lies between $x$ and $x + dx$ is given by

$$P(x \leq X \leq x + dx) = f(x)dx, \tag{17.76}$$

which then provides the cumulative distribution function (CDF) $F(x)$:

$$P(X \leq x) = \int_{-\infty}^{x} f(x')dx' = F(x). \tag{17.77}$$

Choose another random variable $Y$, illustrated in Figure 17.12, and define $y = F(x)$ so that

$$\begin{aligned} P(Y \leq y) &= P(X \leq x) = P[X \leq F^{-1}(y)] = \int_{-\infty}^{F^{-1}(y)} f(x')dx' \\ &= F[F^{-1}(y)] = y. \end{aligned} \tag{17.78}$$

In analogy to the relationship between Eqs. (17.76) and (17.77) for the random variable $X$, Eq. (17.78) suggests [Ash65,Kal86] another PDF $g(y)$ corresponding to the random variable $Y$ such that

$$g(y)dy = P(y \leq Y \leq y + dy), \tag{17.79}$$

which then yields

$$g(y) = \begin{cases} 1, & 0 \le y \le 1, \\ 0, & \text{otherwise.} \end{cases} \tag{17.80}$$

Equation (17.80) simply states that the random variable $Y$ is uniformly distributed over the interval [0,1]. Thus, $x$ may be sampled as $F^{-1}(y)$ for each $y$ uniformly selected over [0,1].

For sampling the distance to the next collision site of a particle moving randomly in a medium with total cross section $\Sigma$, choose as the random event $X$ the distance at which the particle makes the next (or first) collision along its path. Recognizing that the macroscopic cross section $\Sigma$ represents the probability of collision per unit distance of travel yields the probability of collisions taking place between $x$ and $x + dx$

$$P(x \le X \le x + dx) = f(x)dx = \Sigma \exp(-\Sigma x)dx, \tag{17.81}$$

from which the CDF can be readily constructed:

$$F(x) = \int_{-\infty}^{x} f(x')dx' = \int_{0}^{x} f(x')dx' = 1 - \exp(-\Sigma x) = \xi. \tag{17.82}$$

For each random number $\xi$ selected uniformly over [0, 1], the CDF may be inverted to generate the random variable $x$ representing the distance to the next collision site:

$$x = -\frac{1}{\Sigma}\ln(1-\xi) = -\frac{1}{\Sigma}\ln\xi. \tag{17.83}$$

In the last expression, it is noted that if $\xi$ is randomly and uniformly distributed over [0,1], the variable $1 - \xi$ is equally distributed randomly and uniformly over the same interval. Once samples $x_i$, $i = 1, \ldots, N$, are obtained via Eq. (17.83), evaluate the mean free path of the particles as the sample mean $\langle x \rangle = \sum_{n=1}^{N} x_n / N$, which will approach the exact answer $1/\Sigma$ for a sufficiently large value $N$. A similar approach may be taken to account for the energy and angular dependence of reaction cross sections in a Monte Carlo solution of the neutron transport equation.

For a fixed source problem, Monte Carlo sampling naturally starts from the source location. A fission source problem begins with seeds of sources judiciously distributed and sampled until the source distribution is sufficiently converged. This requires discarding a number of initial generations before actual tallies are to be accumulated. Although the accuracy of Monte Carlo calculations is limited by the number of particle histories simulated, such calculations performed on workstations provide acceptable accuracies for many practical calculations, especially criticality calculations, where the eigenvalue is determined as a sum total of particle histories. Local flux or reaction rate calculations may, however, suffer from statistical fluctuations inherent in Monte Carlo calculations, especially in deep-penetration shielding problems.

The Serpent code [Fri11] has implemented the $B_1$ formulation to account for neutron leakage in both the unit-cell and unit-assembly lattice physics analysis for essentially arbitrary geometries and with accurate cross section libraries. For many practical lattice physics analyses, the code provides sufficiently accurate cross section libraries that could be used efficiently for subsequent whole-core multigroup diffusion theory and fuel depletion calculations. In fact, the code has been used for whole-core Monte Carlo analysis on its own with pin-cell resolutions and was used to generate two-group constants for a unit-cell PWR configuration for Examples 5.2 and 7.1.

## References

**[Abr64]** M. Abramowitz and I. Stegun, eds. *Handbook of Mathematical Functions with Formulas, Graphs, and Mathematical Tables*, US Government Printing Office (1964); reprinted by Dover (1970).

**[Alc08]** R.E. Alcouffe, R.S. Baker, J.A. Dahl et al., "PARTISN: A Time-Dependent, Parallel Neutral Particle Transport Code System," LA-UR-08-07258, Los Alamos National Laboratory (2008).

**[Amo57]** A. Amouyal, P. Benoist, and J. Horowitz, "Nouvelle Methode de Determination du Facteur d'Utilisation Thermique d'une Cellule," *J. Nucl. Energy* **6**, 79 (1957).

**[Arg63]** Argonne National Laboratory, *Reactor Physics Constants*, ANL-5800, 2nd ed., Argonne National Laboratory (1963).

**[Ash65]** M. Ash *Nuclear Reactor Kinetics*, McGraw-Hill (1965).

**[Bel70]** G.I. Bell and S. Glasstone, *Nuclear Reactor Theory*, Van Nostrand Reinhold (1970).

**[Bic35]** W.G. Bickley and J. Nayler, "A Short Table of the Functions $Ki_n(x)$ from $n = 1$ to $n = 16$," *Philos. Mag.* **20**, 343 (1935).

**[Cas53]** K.M. Case, F. de Hoffmann, and G. Placzek, *Introduction to the Theory of Neutron Diffusion*, Los Alamos Scientific Laboratory (1953).

**[Dan44]** S.M. Dancoff and M. Ginsburg, "Surface Resonance Absorption in a Close-Packed Lattice," CP-2157, US Atomic Energy Commission (1944).

**[Dir43]** P.A.M. Dirac, "Approximate Rate of Neutron Multiplication for a Solid of Arbitrary Shape and Uniform Density," MSD 5, part I, British Declassified Document Association (1943).

**[Dud76]** J.J. Duderstadt and L.J. Hamilton, *Nuclear Reactor Analysis*, Wiley (1976).

**[Ede92]** M. Edenius and F. Forssen, "CASMO-3, A Fuel Assembly Burnup Program," Studsvik/NFA-89/3, Rev. 2, Studsvik AB (1992).

**[Eva10]** T.M. Evans, A.S. Stafford, R.N. Slaybaugh, and K.T. Clarno, "Denovo: A New Three-Dimensional Parallel Discrete Ordinates Code in SCALE," *Nucl. Technol.* **171**, 171 (2010).

**[Fer17]** R. Ferrer, J. Hykes, and J. Rhodes, "CASMO5, A Fuel Assembly Burn-up Program: Methodology Manual," SSP-08/405, rev. 7, Studsvik Scandpower (2017).

**[Fri11]** E. Fridman and J. Leppänen, "On the Use of the Serpent Monte Carlo Code for Few-Group Cross Section Generation," *Ann. Nucl. Energy* **38**, 1399 (2011).

**[Glo09]** Global Nuclear Fuel, "LANCR02 Lattice Physics Model Description," NED-33376-A, rev. 3, Global Nuclear Fuel (2009).

**[Goo12]** T. Goorley, M. James, T. Booth et al., "Initial MCNP6 Release Overview," *Nucl. Technol.* **180**, 298 (2012).

**[Hen75]** A.F. Henry, *Nuclear-Reactor Analysis*, The MIT Press (1975).

**[Hof16]** A.J. Hoffman and J.C. Lee, "A Time-Dependent Neutron Transport Method of Characteristics Formulation With Time Derivative Propagation," *J. Comput. Phys.* **307**, 696 (2016).

**[Jiw11]** W. Ji, "Application of the Chord Method to Obtain Analytical Expressions for Dancoff Factors in Stochastic Media," *Nucl. Sci. Eng.* **169**, 19 (2011).

**[Jon87]** D.B. Jones, "ARMP-02 Documentation, Part II, CPM-2 Computer Code Manual," EPRI NP-4574-CCM, Electric Power Research Institute (1987).

**[Jon02]** D.B. Jones, K.E. Watkins, and M.L. Williams, "CPM-3 Computer Code Manual," EPRI-CPM-001-M-001, rev. A, Electric Power Research Institute (2002).

**[Joo04]** H.G. Joo, J.Y. Cho, K.S. Kim et al., "Methods and Performance of a Three-Dimensional Whole-Core Transport Code DeCART," *Proc. PHYSOR 2004* (2004).

**[Kal86]** M.H. Kalos and P.A. Whitlock, *Monte Carlo Methods*, vol. **I**, Wiley (1986).

**[Koc13]** B. Kochunas, B. Collins, D. Jabaay et al., "Overview of Development and Design of MPACT: Michigan Parallel Characteristics Transport Code," *Proc. M&C Sun Valley* **1**, 42 (2013).

**[Lar84]** E.W. Larsen, "Diffusion-Synthetic Acceleration Methods for Discrete Ordinates Problems," *Transport Theor. Stat. Phys.* **13**, 107 (1984).

**[Lee17]** J.C. Lee and N.J. McCormick, *Risk and Safety Analysis of Nuclear Systems*, Wiley (2011/2017).

**[Lew84]** E.E. Lewis and W.F. Miller Jr., *Computational Methods of Neutron Transport*, Wiley (1984).

**[Rea18]** B.T. Rearden and M.A. Jessee, eds., "SCALE Code System," vers. 6.2.3, ORNL/TM-2005/39, Oak Ridge National Laboratory (2018).

**[San09]** A. Santamarina, D. Bernard, P. Plaise et al., "APOLLO2.8; A validated code package for PWR neutronics," *Proc. ANFM IV* (2009).

**[Spi08]** M.R. Spiegel and L.J. Stephens, *Theory and Problems of Statistics*, 4th ed., McGraw-Hill (2008).

**[Sta83]** R.J.J. Stamm'ler and M.J. Abbate, *Methods of Steady-State Reactor Physics in Nuclear Design*, Academic Press (1983).

**[Tal07]** A. Talamo, "Analytical Calculation of the Average Dancoff Factor for Prismatic High-Temperature Reactors," *Nucl. Sci. Eng.* **156**, 343 (2007).

**[Wei58]** A.M. Weinberg and E.P. Wigner, *The Physical Theory of Neutron Chain Reactors*, University of Chicago Press (1958).

**[Wil15]** J. Wilert, H. Park, and W. Taitano, "Using Anderson Acceleration to Accelerate the Convergence of Neutron Transport Calculations with Anisotropic Scattering," *Nucl. Sci. Eng.* **181**, 342 (2015).

## Problems

**17.1** Repeat the collision probability calculation for the fuel rod in Example 17.2 for a rod radius 25% larger using both the trapezoidal rule and Gauss quadrature, and compare the results with that given in [Cas53].

**17.2** Determine an expression for the collision probability between two slabs using diffusion theory.

**17.3** Can the $P_1$ approximation for the angular flux satisfy the rigorous vacuum boundary condition $\psi(0, \mu) = 0, \mu < 0$, for a half space $z < 0$ in contact with vacuum at $z = 0$? What can you suggest to remedy the situation?

**17.4** Obtain the condition imposed on the $P_1$ approximation for the angular flux $\psi(0, \mu)$ at a vacuum boundary if the mean square error for $\psi(0, \mu)$ is minimized.

**17.5** (a) Obtain the scalar flux $\phi(x)$ in an infinite homogeneous absorbing medium at distance $x$ from an isotropic plane source of strength $S$ [neutron$\cdot$ cm$^{-2}$ s$^{-1}$] by solving the neutron transport equation. (b) Determine the mean square distance of neutron travel, and compare the result with the corresponding expression obtained via diffusion theory.

**17.6** In the first-flight transport model for a highly absorbing fuel rod, any neutron suffering either a scattering or absorption collision is assumed lost from the system, with the fuel region treated as a purely absorbing medium with an effective absorption cross section equal to the total cross section $\Sigma_{tF}$. (a) Set up a neutron balance statement for a fuel plate of half-thickness $a$, and obtain an expression for the lattice parameter $F = \phi_F(a)/\overline{\phi}_F$ in terms of the current-to-flux ratio $\alpha = -J(a)/\phi(a)$. (b) Obtain an expression for $\alpha$ with the one-group neutron transport equation for a source-free purely absorbing slab with the $P_1$ approximation for the angular flux $\psi(a, \mu)$ and $\psi(-a, \mu)$ of neutrons incident from the moderator region.

**17.7** A half-space $x < 0$ consisting of purely absorbing material with cross section $\Sigma_a$ is surrounded by vacuum, and neutrons of strength $S_0$ [neutron$\cdot$cm$^{-2}$s$^{-1}$] are distributed uniformly and isotropically in the medium. A homogeneous purely absorbing sphere of radius $a$ and cross section $\Sigma_0$ is placed at distance $b$ from the vacuum interface. Obtain the scalar flux $\phi(a + b)$ on the sphere.

**17.8** Obtain the one-dimensional transport kernel by solving for the scalar flux $\phi(x)$ at position $x$ due to a unit planar isotropic source of neutrons at position $x_0$ in a purely absorbing infinite medium with cross section $\Sigma$.

**17.9** The blackness of an isolated fuel rod of radius $a$ is $\beta_F^*$. Determine the blackness $\beta_{FC}^*$ of a unit cell of radius $b$ comprising the fuel rod and a surrounding coolant channel filled with helium gas at 10 MPa.

**17.10** Obtain the collision probability $P_{i\rightarrow j}$ that neutrons born uniformly and isotropically in a slab of thickness $h_i$ at $x_i$ will suffer next collisions in a slab of thickness $h_j$ at $x_j$ in an infinite medium with total cross section $\Sigma$ and scattering cross section $\Sigma_s$.

**17.11** For the Milne problem representing neutron streaming into vacuum from a half space $z > 0$, write down the $S_2$ discrete ordinates equations in terms of $\psi_1(z) = \psi(z, \mu_1)$ and $\psi_2(z) = \psi(z, \mu_2)$, with the symmetric Gauss quadrature set $\mu_1 = -\mu_2 = 1/\sqrt{3}, w_1 = w_2 = 1$. Combine the resulting ODEs into a second-order ODE for $\psi^+(z) = \psi_1(z) + \psi_2(z)$ and obtain an expression for the extrapolated end-point $z_0$. Assume the scattering is isotropic in the half space.

**17.12** In the step characteristic (SC) formulation of the discrete ordinates method, the diamond differencing (DD) scheme is modified to perform an integration of the source term $q_i$, assumed constant, over a spatial mesh interval $\Delta$ at $x_i$ to arrive at a form $\psi_{i+1/2} = \psi_{i-1/2} f_i + q_i(1 - f_i)/\Sigma_i$, with $\psi_{i\pm1/2} = \psi(x_{i\pm1/2}, \mu_j), \forall \mu_j$. (a) Obtain expressions for $f_i$ in terms $\epsilon_i = \Sigma_i \Delta/\mu_j$ and the angular flux $\overline{\psi}_i$ averaged over $\Delta$ at $x_i$. (b) Show that the SC method preserves the positivity of the angular flux.

**17.13** Show that the SC method in Problem 17.12 becomes identical to the standard DD scheme if the function $f_i$ is approximated by $(1 - 0.5\epsilon_i)/(1 + 0.5\epsilon_i)$.

# Appendix A

# Key Physical Constants

**Table A.1** Key physical constants.

| Parameter | Symbol | Definition/value |
|---|---|---|
| Atomic mass unit | amu | $1.660\,539 \times 10^{-27}$ kg |
| | | $9.314\,941 \times 10^{2}$ MeV |
| Avogadro number | $N_A$ | $6.022\,141 \times 10^{23}$ mol$^{-1}$ |
| Becquerel | Bq | 1 disintegration s$^{-1}$ |
| Boltzmann constant | $k$ | $1.380\,649 \times 10^{-23}$ J · K$^{-1}$ |
| Compton wave length | $\lambda_c$ | $2.426\,310 \times 10^{-12}$ m |
| Curie | Ci | $3.7 \times 10^{10}$ Bq |
| Electron charge | $q_e$ | $1.602\,177 \times 10^{-19}$ C |
| Electron rest mass | $m_e$ | $9.109\,384 \times 10^{-31}$ kg |
| | | 0.510 999 MeV |
| Electron volt | eV | $1.602\,177 \times 10^{-19}$ J |
| Molar gas constant | $R$ | 8.314 460 J · K$^{-1}$ · mol$^{-1}$ |
| Neutron rest mass | $m_n$ | $1.674\,927 \times 10^{-27}$ kg |
| Planck constant | $h$ | $6.626\,070 \times 10^{-34}$ J · s |
| | | $4.135\,668 \times 10^{-15}$ eV · s |
| Proton–electron mass ratio | $m_p/m_e$ | $1.836\,153 \times 10^{3}$ |
| Proton rest mass | $m_p$ | $1.672\,622 \times 10^{-27}$ kg |
| | | 1.007 276 amu |
| Speed of light in vacuum | $c$ | $2.999\,792 \times 10^{8}$ m · s$^{-1}$ |

*Source*: https://physics.nist.gov/cuu/Constants.

*Nuclear Reactor Physics and Engineering*, Second Edition. John C. Lee.
© 2025 John Wiley & Sons, Inc. Published 2025 by John Wiley & Sons, Inc.
Companion website: www.wiley.com/go/LeeNRPE2

# Appendix B

## Comparison of Major Reactor Types

A summary comparison of three major nuclear power plants under active deployment or development is presented in this appendix. The comparison is made for reactor parameters in four major categories: (a) core physics and fuel design, (b) reactor physics characteristics including reactivity feedback parameters, (c) thermal hydraulic characteristics, and (d) safety and control systems. The AP1000 design that forms the basis for several Generation III+ plants that began operation in 2018 and 2019 serves as the reference light water reactor (LWR) system for the comparison. Two Generation IV designs, sodium-cooled fast reactor (SFR) and gas-cooled very high temperature reactor (VHTR), under active development in the United States and elsewhere are chosen as two alternate designs primarily for their coolant types and design features, which are sufficiently different from the current generation of pressurized water reactor (PWR) and boiling water reactor (BWR) plants. For the purpose of simplified comparisons of key system parameters, the AP1000 design is selected to represent both PWR and BWR plants, but with key differences noted as appropriate (Table B.1).

*Nuclear Reactor Physics and Engineering*, Second Edition. John C. Lee.
© 2025 John Wiley & Sons, Inc. Published 2025 by John Wiley & Sons, Inc.
Companion website: www.wiley.com/go/LeeNRPE2

**Table B.1** Key features of major reactor types.

| **Reference design** | **Sodium fast reactor S-PRISM [Boa01]** | **High temperature reactor GT-MHR [Bal08]** | **Light water reactor AP1000 [Hon12]** |
|---|---|---|---|
| 1. Core and fuel design | | | |
| Power output (MWt/MWe) | 2 × 1,000/760 | 600/285 | 3,400/1,090 |
| Core height (m) | 1.02 | 7.9 | 4.27 |
| Core diameter (m) | 2.7 | 2.96 ID/4.84 OD | 3.04 |
| Assembly geometry | Hexagonal | Hexagonal | Square |
| Core pressure drop (kPa) | 410 | 7 | 280 |
| Coolant/moderator (vol%) | 37 | 20/50 | 55 |
| Fuel type | (U-Pu-Zr) + Na bonding | UCO 0.8 ~ 0.9 mm coated particle | $UO_2$ + gap |
| Fuel element | Rod 7 mm OD | Compact 13 mm OD × 50 mm H | Pellet 10 mm OD |
| Fission product retention/treatment | Plenum, [venting]$^+$ | Porous pyrolytic C layer | Plenum |
| Power density ($MW{\cdot}m^{-3}$) | 170 (driver + fuel blanket) | 6.5 | 109 |
| (MW/kg fissile) | 0.43 | 1.0 | 1.2 |
| (kW/kgHM) | 38 | 113 | 33 |
| Fissile inventory (Mg/GWe) | 6.2 | 1.9 | 2.6 |
| Fissile enrichment | 21.0 wt% Pu in U-Pu | 10 wt% $^{235}U$ | 2.3 ~ 4.5 wt%$^{235}U$ |
| Discharge burnup (MWd/kgHM) | 100 ~ 150 | 100 | 49 |
| Reactivity swing over cycle (%$\Delta k/k$) | 1 ~ 3 | <10 | 15 |
| Breeding/conversion ratio | 1.05 | 0.8 | 0.6 |

| | | | |
|---|---|---|---|
| **2. Physics characteristics** | | | |
| Average neutron energy | 200 keV | >eV | 0.2 eV |
| Neutron mean free path | 0.15 ~ 0.30 m (nearly homogeneous) | ~0.1 m (core average) <0.1 mm (TRISO particle) | ~10 mm (thermal) (heterogeneous) |
| Neutron leakage fraction | 0.20 ~ 0.30 | ~0.10 | 0.03 |
| Radial leakage | 0.12 ~ 0.18 | | 0.025 |
| Axial leakage | 0.08 ~ 0.12 | | 0.005 |
| Xenon and samarium poisoning | Unimportant | Important | Important |
| Void reactivity worth | −150 ~ 2,000 pcm core void | −10 pcm due to He loss | [BWR: −150 pcm/%void][+] |
| Fuel temperature coefficient of reactivity ($pcm \cdot K^{-1}$) | −(0.1 ~ 0.5) | −(3 ~ 5) | −(1 ~ 2) |
| Moderator temperature coefficient of reactivity ($pcm \cdot K^{-1}$) | ±0.4 ~ 1.0 | −(5 ~ 10) | −(10 ~ 80) |
| Overall hot channel factor $F_q$ | 1.5 (with blanket) | 2.2 ~ 3.0 | 2.5 |
| Neutronic analysis | Spectral-spatial coupling | Double heterogeneity | Spatial self-shielding |
| **3. Thermal-hydraulic parameters** | | | |
| Coolant inlet temperature (K) | 636 | 491 | 552 |
| Coolant temperature rise (K) | 147 | 362 | 43 |
| Primary system pressure (MPa) | 0.7 | 7.12 | 15.5 |
| Steam condition | 17.3 MPa, 741 K | 7.0 MPa, 1,120 K | 5.8 MPa, 547 K |
| Coolant system | Pool | Loop | Loop |
| Thermal efficiency (%) | 38 | 48 | 32 |

(Continued)

**Table B.1** (Continued)

| Reference design | Sodium fast reactor S-PRISM [Boa01] | High temperature reactor GT-MHR [Bal08] | Light water reactor AP1000 [Hon12] |
|---|---|---|---|
| 4. Safety and control | | | |
| Design basis accident (DBA) | Unprotected transient overpower leading to core disruptive accident | He depressurization, air/water ingress | loss of coolant accident, [class 9 accidents]+ |
| Time scale for DBA | >min (Na subcooling = 400 K) | ~hr (graphite heat sink) | s ~ min |
| Recriticality | Possible | Not likely | Nearly impossible |
| Passive safety features | Large sodium pool, low fuel temperature, air cooling of reactor vessel | Large graphite heat sink, passive shutdown, reactor cavity cooling | Gravity-driven water in IRWST, pressurized accumulator |
| Containment | Yes | No | Yes |
| Note | Pancake core reduces pumping power and positive void coefficient | TRISO = tristructural-isotropic UCO = uranium oxycarbide | IRWST = in-containment refueling water storage tank |

+ indicates alternate or possible variation.

## References

**[Bal08]** S.J. Ball and S.E. Fisher, "Next Generation Nuclear Plant Phenomena Identification and Ranking Tables (PIRTs)," NUREG/CR-6944, vol. **1**, US Nuclear Regulatory Commission (2008).

**[Boa01]** C.E. Boardman, M. Hui, D.G. Carroll, and A.E. Dubberley, "Economic Assessment of S-PRISM Including Development and Generating Costs," ICONE-9 (2001).

**[Hon12]** M. Hone, S. Skidmore, M. Misvel, and E. Resch, "AP1000 Core Reference Report," WCAP-17524-NP, Westinghouse Electric Company (2012).

# Appendix C

# Special Mathematical Functions

This appendix introduces some of the advanced mathematical functions used in reactor physics and engineering formulations. They include (a) gamma function introduced to efficiently handle integrals involving the exponential function, (b) Legendre polynomials and spherical harmonics invoked to represent the angular motion of neutrons, (c) Bessel functions introduced as part of the Laplacian for cylindrical geometry, and (d) Dirac delta function that simplifies integrals involving localized sources.

## C.1 Gamma Function

The gamma function may be defined in a number of different ways including

$$\Gamma(z) = \int_0^\infty e^{-t} t^{z-1} dt, \quad \Re e(z) > 0, \tag{C.1a}$$

$$= \lim_{n\to\infty} \frac{1 \cdot 2 \cdot 3 \cdots n}{z(z+1)(z+2)\cdots(z+n)} n^z, \quad z \neq 0, -1, -2, -3, \dots, \tag{C.1b}$$

$$\frac{1}{\Gamma(z)} = z e^{\gamma z} \prod_{n=1}^{\infty} \left(1 + \frac{z}{n}\right) e^{-z/n}, \tag{C.1c}$$

where $\gamma = 0.5772156619\cdots$ is the Euler–Mascheroni constant. Some of the useful properties include

$$\Gamma(z+1) = z\Gamma(z), \tag{C.2a}$$

$$\Gamma(n+1) = n!, \; n = \text{integer}, \tag{C.2b}$$

$$\Gamma(z)\Gamma(1-z) = \frac{\pi}{\sin z\pi}, \tag{C.2c}$$

$$\Gamma\left(\frac{1}{2}\right) = \sqrt{\pi}. \tag{C.2d}$$

*Nuclear Reactor Physics and Engineering*, Second Edition. John C. Lee.
© 2025 John Wiley & Sons, Inc. Published 2025 by John Wiley & Sons, Inc.
Companion website: www.wiley.com/go/LeeNRPE2

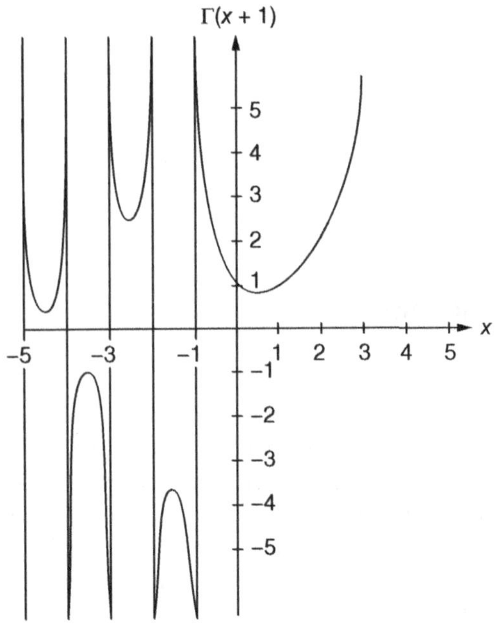


**Figure C.1** Gamma function $\Gamma(x+1)$ for real $x$. *Source*: Arfken et al. [Arf13]/with permission of Elsevier.

Figure C.1 illustrates the behavior of $\Gamma(x)$ for real values of $x$. Related to the gamma function is a two-parameter *incomplete gamma function*:

$$\gamma(z,x) = \int_0^x e^{-t} t^{z-1} dt, \tag{C.3a}$$

$$\gamma(z+1,x) = z\gamma(z,x) - x^z \exp(-x). \tag{C.3b}$$

The function $\gamma(z,x)$ becomes $\Gamma(z)$ in the limit of $x \to \infty$. A function related to the incomplete gamma function is the *exponential integral function* [Abr64]:

$$E_n(z) = \int_1^\infty x^{-n} \exp(-zx)dx, \tag{C.4a}$$

$$= z^{n-1} \int_z^\infty x^{-n} \exp(-x)dx, \quad n = 1, 2, \ldots, \tag{C.4b}$$

$$= \int_0^1 x^{n-2} \exp(-z/x)dx, \tag{C.4c}$$

$$= \frac{1}{n-1}[e^{-z} - zE_{n-1}(z)], \; n > 1. \tag{C.4d}$$

Useful properties of $E_n(z)$ include

$$E_0(z) = \frac{e^{-z}}{z}, \tag{C.5a}$$

$$\frac{dE_n(z)}{dz} = -E_{n-1}(z). \tag{C.5b}$$

The exponential integral function is tabulated in [Abr64, Gol59]. The *error function* is closely related to the Gaussian distribution for probability density functions and defined by

$$\mathrm{erf}(z) = \frac{2}{\sqrt{\pi}} \int_0^z \exp(-u^2)du = \frac{\gamma(0.5, z^2)}{\sqrt{\pi}}. \tag{C.6}$$

## C.2 Legendre Polynomial and Spherical Harmonics

When the Helmholtz equation or 3-D neutron diffusion equation is written in spherical polar coordinates, the component for the polar angle $\theta$ takes the form of the Legendre equation

$$(1 - x^2)\frac{d^2P_\ell(x)}{dx^2} - 2x\frac{dP_\ell(x)}{dx} + \ell(\ell + 1)P_\ell(x) = 0, \tag{C.7}$$

with the solution $P_\ell(x)$ obtained for $|x = \cos\theta| \leq 1.0$ as the *Legendre polynomial.* The polynomial may be generated by

$$P_\ell(x) = \frac{1}{2^\ell \ell!}\left(\frac{d}{dx}\right)^\ell (x^2 - 1)^\ell \tag{C.8}$$

and is listed in Table C.1 for a few low values of $\ell$.

Useful recursive relations for the Legendre polynomials include

$$P_n(-x) = (-1)^n P_n(x), \tag{C.9a}$$

$$(2n + 1)xP_n(x) = (n + 1)P_{n+1}(x) + nP_{n-1}(x), \quad n > 1, \tag{C.9b}$$

$$\frac{dP_{n+1}(x)}{dx} + \frac{dP_{n-1}(x)}{dx} = 2x\frac{dP_n(x)}{dx} + P_n(x), \tag{C.9c}$$

$$\frac{dP_{n+1}(x)}{dx} - \frac{dP_{n-1}(x)}{dx} = (2n + 1)P_n(x). \tag{C.9d}$$

**Table C.1** Legendre polynomials.

| |
|---|
| $P_0(x) = 1$ |
| $P_1(x) = x$ |
| $P_2(x) = \frac{1}{2}(3x^2 - 1)$ |
| $P_3(x) = \frac{1}{2}(5x^3 - 3x)$ |
| $P_4(x) = \frac{1}{8}(35x^4 - 30x^2 + 3)$ |
| $P_5(x) = \frac{1}{8}(63x^5 - 70x^3 + 15x)$ |

One important property of the Legendre polynomial is the orthonormality condition

$$\int_{-1}^{1} P_n(x)P_m(x)dx = \frac{2}{2n+1}\delta_{nm}, \tag{C.10}$$

where the *Kronecker delta* represents

$$\delta_{nm} = \begin{cases} 1, & n = m, \\ 0, & n \neq m. \end{cases} \tag{C.11}$$

The *associated Legendre polynomial* $P_\ell^m(x)$ is defined as the solution to the polar angle component of the Helmholtz equation where the azimuthal component of the solution is coupled explicitly through another parameter $m$

$$(1-x^2)\frac{d^2P_\ell^m(x)}{dx^2} - 2x\frac{dP_\ell^m(x)}{dx} + \left[\ell(\ell+1) - \frac{m^2}{1-x^2}\right]P_\ell^m(x) = 0. \tag{C.12}$$

The solution $P_\ell^m(x)$ may be obtained from the Legendre polynomial

$$P_\ell^m(x) = (-1)^m(1-x^2)^{m/2}\frac{d^m}{dx^m}P_\ell(x), \tag{C.13}$$

together with a key relation

$$P_\ell^{-m}(x) = (-1)^m\frac{(\ell-m)!}{(\ell+m)!}P_\ell^m(x). \tag{C.14}$$

The orthonormality condition for the associated Legendre polynomial is given as

$$\int_{-1}^{1} P_\ell^m(x)P_n^m(x)dx = \frac{2}{2\ell+1}\frac{(\ell+m)!}{(\ell-m)!}\delta_{\ell n}. \tag{C.15}$$

A normalized form of the associate Legendre polynomial is defined as the *spherical harmonics*

$$Y_\ell^m(\boldsymbol{\Omega}) = Y_\ell^m(\theta,\varphi) = \sqrt{\frac{2\ell+1}{4\pi}\frac{(\ell-m)!}{(\ell+m)!}}P_\ell^m(\mu)e^{im\varphi}, \quad \mu = \cos\theta, \tag{C.16}$$

with the orthonormality condition

$$\int_{4\pi} d\boldsymbol{\Omega} Y_\ell^m(\boldsymbol{\Omega})Y_\alpha^{\beta*}(\boldsymbol{\Omega}) = \int_0^{2\pi} d\varphi \int_{-1}^{1} d\mu Y_\ell^m(\mu,\varphi)Y_\alpha^{\beta*}(\mu,\varphi) = \delta_{m\beta}\delta_{\ell\alpha}. \tag{C.17}$$

The spherical harmonics provide a useful addition theorem [Mar76] for the Legendre polynomial

$$P_\ell(\boldsymbol{\Omega}\cdot\boldsymbol{\Omega}') = P_\ell(\mu_0) = \frac{4\pi}{2\ell+1}\sum_{m=-\ell}^{\ell} Y_\ell^m(\boldsymbol{\Omega})Y_\ell^{m*}(\boldsymbol{\Omega}'), \tag{C.18}$$

in terms of cosine of the angle $\theta_0$ between two unit directional vectors $\boldsymbol{\Omega}$ and $\boldsymbol{\Omega}'$

$$\mu_0 = \cos\theta_0 = \cos\theta\cos\theta' + \sin\theta\sin\theta'\cos(\varphi-\varphi'). \tag{C.19}$$

With an azimuthal symmetry, $m = 0$ and $e^{im\varphi} = 1.0$, the addition theorem simplifies to

$$P_\ell(\mu_0) = \frac{4\pi}{2\ell + 1} Y_\ell^0(\boldsymbol{\Omega}) Y_\ell^{0*}(\boldsymbol{\Omega}') = P_\ell(\mu) P_\ell(\mu'). \tag{C.20}$$

## C.3 Bessel Function

When the Helmholtz or wave equation is written in circular cylindrical coordinates, the angular part of the equation produces the Bessel equation

$$x^2 \frac{d^2 f(x)}{dx^2} + x \frac{df(x)}{dx} + (x^2 - n^2) f(x) = 0, \tag{C.21}$$

with the solution $J_n(x)$ defined as the *Bessel function of the first kind*:

$$J_n(x) = \sum_{m=0}^{\infty} \frac{(-1)^m}{m!(n+m)!} \left(\frac{x}{2}\right)^{n+2m} = (-1)^n J_{-n}(x). \tag{C.22}$$

Key recurrence formulas include

$$J_{n-1}(x) + J_{n+1}(x) = \frac{2n}{x} J_n(x), \tag{C.23a}$$

$$J_{n-1}(x) - J_{n+1}(x) = 2\frac{dJ_n(x)}{dx}, \tag{C.23b}$$

$$\frac{dJ_0(x)}{dx} = -J_1(x), \tag{C.23c}$$

$$\frac{d}{dx}[x^n J_n(x)] = x^n J_{n-1}(x), \tag{C.23d}$$

$$\frac{d}{dx}[x^{-n} J_n(x)] = -x^{-n} J_{n+1}(x), \tag{C.23e}$$

$$J_n(x) = \pm \frac{dJ_{n\pm 1}(x)}{dx} + \frac{n \pm 1}{x} J_{n\pm 1}(x). \tag{C.23f}$$

The Bessel functions $J_n(x)$ are plotted in Figure C.2, and the zeros of the functions are listed in Table C.2 for a few low orders. Additional roots, as well as numerical values, of the Bessel functions may be obtained via *Mathematica*, *Maple*, and other symbolic mathematics software.

The *Bessel functions of the second kind*, or Neumann functions, are defined as

$$Y_n(x) = \frac{\cos n\pi J_n(x) - J_{-n}(x)}{\sin n\pi}, \tag{C.24}$$

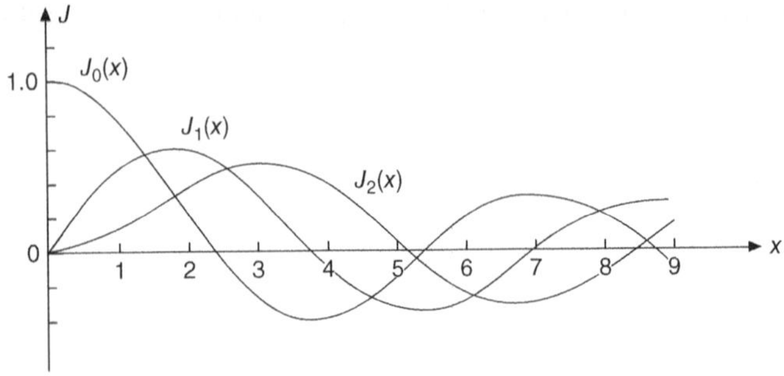


**Figure C.2** Bessel functions of the first kind $J_0(x), J_1(x)$, and $J_2(x)$. *Source*: Arfken et al. [Arf13]/with permission of Elsevier.

**Table C.2** Zeros of the Bessel functions of the first kind.

| Order of zeros | $J_0(x)$ | $J_1(x)$ | $J_2(x)$ |
|---|---|---|---|
| 1 | 2.4048 | 3.8317 | 5.1356 |
| 2 | 5.5201 | 7.0156 | 8.4172 |
| 3 | 8.6537 | 10.1735 | 11.6198 |
| 4 | 11.7915 | 13.3237 | 14.7960 |

with key recursive relationships

$$Y_{n-1}(x) + Y_{n+1}(x) = \frac{2n}{x} Y_n(x), \tag{C.25a}$$

$$Y_{n-1}(x) - Y_{n+1}(x) = 2\frac{dY_n(x)}{dx}, \tag{C.25b}$$

$$Y_{-n}(x) = (-1)^n Y_n(x). \tag{C.25c}$$

Three lowest orders of Bessel functions of the second kind $Y_n(x)$ are plotted in Figure C.3.

The *modified Bessel functions* satisfy

$$x^2\frac{d^2f(x)}{dx^2} + x\frac{df(x)}{dx} - (x^2 + n^2)f(x) = 0, \tag{C.26}$$

with the *modified Bessel functions of the first kind* obtained from $J_n(x)$ with a complex argument

$$I_n(x) = i^{-n}J_n(ix) = I_{-n}(x). \tag{C.27}$$

Similarly, the *modified Bessel functions of the second kind* are obtained as

$$K_n(x) = \frac{\pi}{2}i^{n+1}[J_n(ix) + iY_n(ix)]. \tag{C.28}$$

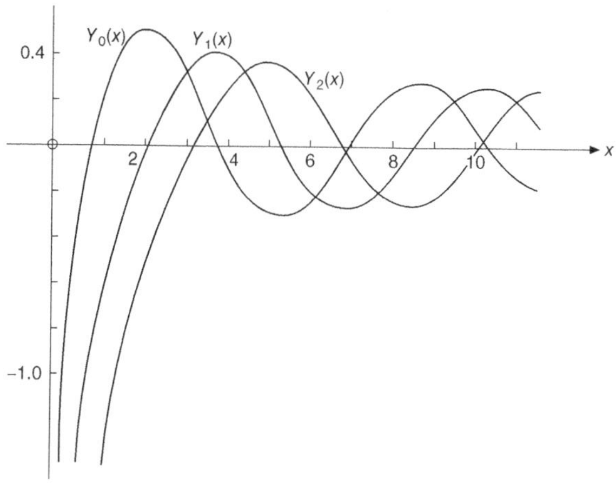


**Figure C.3** Bessel functions of the second kind $Y_n(x)$. *Source*: Arfken et al. [Arf13]/with permission of Elsevier.

Key recurrence relationships for $I_n(x)$ and $K_n(x)$ include

$$I_{n-1}(x) - I_{n+1}(x) = \frac{2n}{x} I_n(x), \tag{C.29a}$$

$$I_{n-1}(x) + I_{n+1}(x) = 2\frac{dI_n(x)}{dx}, \tag{C.29b}$$

$$K_{n-1}(x) - K_{n+1}(x) = -\frac{2n}{x} K_n(x), \tag{C.29c}$$

$$K_{n-1}(x) + K_{n+1}(x) = -2\frac{dK_n(x)}{dx}. \tag{C.29d}$$

Two lowest orders of the modified Bessel functions $I_n(x)$ and $K_n(x)$ are plotted in Figure C.4.

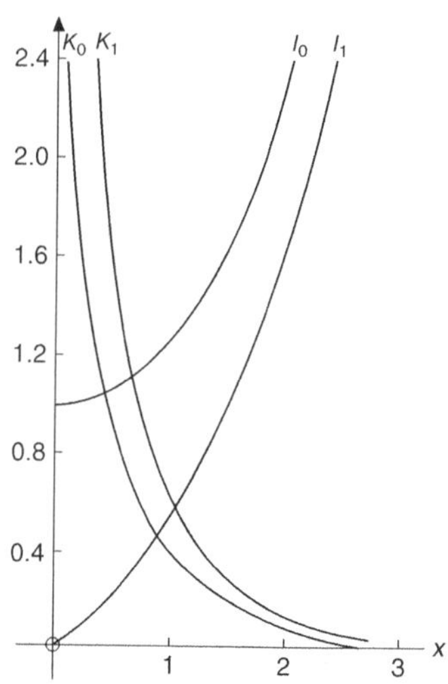


**Figure C.4** Modified Bessel functions $I_n(x)$ and $K_n(x)$. *Source*: Arfken et al. [Arf13]/with permission of Elsevier.

## C.4 Dirac Delta Function

The Dirac delta function may be obtained as a limiting form of the Gaussian function

$$\delta(x-x_0) = \lim_{a\to 0}\frac{1}{a\sqrt{\pi}}\exp\left[-\frac{(x-x_0)^2}{a^2}\right] = \frac{1}{\pi}\lim_{a\to 0+}\frac{a}{(x-x_0)^2+a^2}, \tag{C.30}$$

yielding the properties

$$\delta(x-x_0) = 0, \quad x \neq x_0, \tag{C.31a}$$

$$\int_{-\infty}^{\infty}\delta(x-x_0)dx = 1.0, \tag{C.31b}$$

$$\int_{-\infty}^{\infty}f(x)\delta(x-x_0)dx = f(x_0). \tag{C.31c}$$

Additional properties include [Fri56]

$$\delta(x) = \delta(-x) \tag{C.32a}$$

$$\delta(ax) = \frac{1}{|a|}\delta(x), \quad a \neq 0, \tag{C.32b}$$

$$\delta[f(x)] = \frac{\delta(x-a)}{|f'(a)|}, \quad f'(a) \neq 0, \tag{C.32c}$$

$$\int_{-\infty}^{\infty}\frac{d\delta(x-a)}{dx}f(x)dx = -\frac{df(a)}{dx}. \tag{C.32d}$$

## References

**[Abr64]** M. Abramowitz and I. Stegun, eds., *Handbook of Mathematical Functions with Formulas, Graphs, and Mathematical Tables*, US Government Printing Office (1964); reprinted by Dover (1970).

**[Arf13]** G.B. Arfken, H.J. Weber, and F.E. Harris, *Mathematical Methods for Physicist: A Comprehensive Guide*, 7th ed., Academic Press (2013).

**[Fri56]** B. Friedman, *Principles and Techniques of Applied Mathematics*, Wiley (1956).

**[Gol59]** H. Goldstein, *Fundamental Aspects of Reactor Shielding*, Addison-Wesley (1959).

**[Mar76]** H. Margenau and G.M. Murphy, *The Mathematics of Physics and Chemistry*, Krieger (1976).

# Appendix D

# Integral Transforms

Two examples of integral transforms are discussed in this appendix, primarily to present short tables of transforms useful for efficient space and time domain solutions of differential equations. Basic approaches for integral transforms are illustrated in Appendix D.1 for the Laplace transform, together with a brief discussion of the inverse transform techniques and a short table of Laplace transforms. A similar summary of Fourier transform techniques is presented in Appendix D.2. For the evaluation of inverse Laplace or Fourier transforms, the integral path in the complex plane is usually extended to generate a closed contour and invoke the properties related to the residue theorem of complex variables and theorems on limits of the contour. Appendix D.3 summarizes *Jordan's lemma*, which governs the evaluation of limiting contours.

## D.1 Laplace Transform

As a primary example of general integral transforms, a Laplace transform converts ordinary differential equations (ODEs), together with the appropriate initial conditions, into a simpler form of the equations in the transform domain. The simpler equations are then solved in the transform domain, and the resulting solution is inverted back into the physical domain where the desired solution is sought in the first place. This process is summarized in Figure D.1. For first-order ODEs of the type considered for the point kinetics equation in Chapter 8, the transformed equations take the form of coupled algebraic equations, which may be solved and then inverted back into the time domain. When the original equations are either of a higher order or in the form of partial differential equations, we may have to take either successive Laplace transforms or yet another kind of integral transform, e.g. Fourier transform following the Laplace transform, and follow through the inverse transforms in sequence.

*Nuclear Reactor Physics and Engineering*, Second Edition. John C. Lee.
© 2025 John Wiley & Sons, Inc. Published 2025 by John Wiley & Sons, Inc.
Companion website: www.wiley.com/go/LeeNRPE2

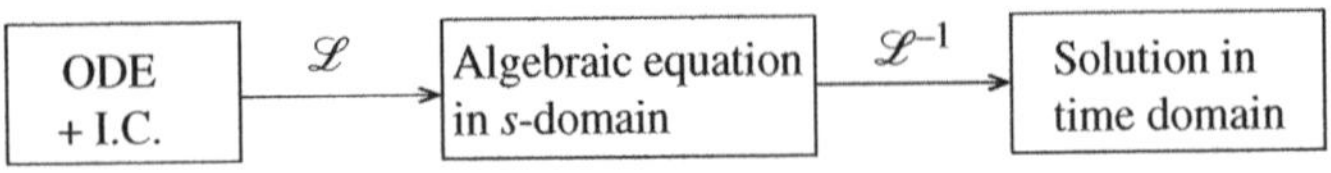


**Figure D.1** Application of Laplace transform to the ODE solution.

**Table D.1** A short table of Laplace transforms.

| Formula | $f(t)$ | $\bar{f}(s)$ |
|---|---|---|
| 1 | $f(at), a > 0$ | $\frac{1}{a}\bar{f}\left(\frac{s}{a}\right)$ |
| 2 | $f(t-a), a > 0$ | $e^{-as}\bar{f}(s)$ |
| 3 | $e^{-at}f(t)$ | $\bar{f}(s+a)$ |
| 4 | $\int_0^t f(\tau)d\tau$ | $\frac{\bar{f}(s)}{s}$ |
| 5 | $\frac{df(t)}{dt}$ | $s\bar{f}(s) - f(0+)$ |
| 6 | $(-1)^n t^n f(t)$ | $\frac{d^n\bar{f}(s)}{ds^n}$ |
| 7 | $t^n/\Gamma(n+1), n > -1$ | $s^{-(n+1)}$ |
| 8 | $\int_0^t f(\tau)g(t-\tau)d\tau$ | $\bar{f}(s)\bar{g}(s)$ |
| 9 | $e^{-at}$ | $\frac{1}{s+a}$ |
| 10 | $u(t-a), a > 0$ | $\frac{e^{-as}}{s}$ |
| 11 | $\delta(t-a)$ | $e^{-as}$ |
| 12 | $\cos at$ | $\frac{s}{s^2+a^2}$ |
| 13 | $\sin at$ | $\frac{a}{s^2+a^2}$ |

Note:
a) The integral in formula 8 is the convolution integral.
b) The function $u(t)$ in formula 10 is the Heaviside step function.
c) The function $\delta(t)$ in formula 11 is the Dirac delta function.

We begin with the definition of the Laplace transform of a general function $f(t)$

$$\bar{f}(s) = \int_0^{\infty} f(t)e^{-st}dt = \mathfrak{L}\{f(t)\}, \tag{D.1}$$

coupled to the inverse transform

$$f(t) = \frac{1}{2\pi i}\int_{\gamma-i\infty}^{\gamma+i\infty} \bar{f}(s)e^{st}ds = \mathfrak{L}^{-1}\left\{\bar{f}(s)\right\}. \tag{D.2}$$

The Laplace transform $\bar{f}(s)$ of function $f(t)$ is a function of complex variable $s = \sigma + i\omega$, and the inverse transform integral of Eq. (D.2) has to be performed in the complex domain, along an axis parallel to the imaginary axis and located at $\sigma = \gamma$, where $\gamma >$ real parts of singularities of $\bar{f}(s)$. The integral path is known as the *Bromwich path* [Sne51], and it becomes necessary often to distort and augment it, invoking the residue theorem and other properties of complex variables. Table D.1 presents key Laplace transform pairs.

## D.2 Fourier Transform

The Fourier transform is applicable to the solution of space-dependent differential equations, as a counterpart to the Laplace transform for time-dependent differential equations. We define the complex Fourier transform [Sne51] useful for infinite medium problems

$$\bar{f}(\xi) = \frac{1}{\sqrt{2\pi}}\int_{-\infty}^{\infty} f(x)e^{i\xi x}dx = \mathfrak{F}\{f(x)\}, \tag{D.3}$$

coupled to the inverse transform

$$f(x) = \frac{1}{\sqrt{2\pi}}\int_{-\infty}^{\infty} \bar{f}(\xi)e^{-i\xi x}d\xi = \mathfrak{F}^{-1}\left\{\bar{f}(\xi)\right\}. \tag{D.4}$$

Table D.2 presents a few key Fourier transform pairs. The Fourier sine and cosine transforms [Sne51] may be used for finite medium problems.

## D.3 Jordan's Lemma

For the evaluation of an inverse Laplace transform along the Bromwich path, the line integral is extended to cover a semicircular arc of radius $R$ enclosing the singularities of the transform $\bar{f}(s)$ so that the residue theorem of complex variables may be invoked. This process is meaningful only if it can be proven that the integral along the distorted contour vanishes in the limit of $R \to \infty$. Likewise, for the evaluation of an inverse Fourier transform involving $\bar{f}(\xi)$, the integral path $[-\infty, +\infty]$

**Table D.2** A short table of Fourier transforms.

| Formula | $f(x)$ | $\bar{f}(\xi)$ |
|---|---|---|
| 1 | $\dfrac{\sin ax}{x}$ | $\left(\dfrac{\pi}{2}\right)^{\frac{1}{2}}, \vert\xi\vert < a; 0, \vert\xi\vert > a$ |
| 2 | $e^{i\omega x}, p < x < q; 0,$ otherwise | $\dfrac{i}{\sqrt{2\pi}} \dfrac{e^{ip(\omega+\xi)} - e^{iq(\omega+\xi)}}{\xi}$ |
| 3 | $e^{-cx+i\omega x}, x > 0; 0,$ otherwise | $\dfrac{i}{\sqrt{2\pi}} \dfrac{1}{(\omega+\xi+ic)}$ |
| 4 | $e^{-px^2}, Re(p) > 0$ | $\dfrac{1}{\sqrt{2p}} e^{-\xi^2/4p}$ |
| 5 | $\cos px^2$ | $\dfrac{1}{\sqrt{2p}} \cos\left(\dfrac{\xi^2}{4p} - \dfrac{\pi}{4}\right)$ |
| 6 | $\sin px^2$ | $\dfrac{1}{\sqrt{2p}} \sin\left(\dfrac{\xi^2}{4p} + \dfrac{\pi}{4}\right)$ |
| 7 | $\dfrac{1}{\vert x\vert}$ | $\dfrac{1}{\vert\xi\vert}$ |
| 8 | $\dfrac{\cosh ax}{\cosh \pi x}, -\pi < a < \pi$ | $\sqrt{\dfrac{2}{\pi}} \dfrac{\cos(a/2)\cosh(\xi/2)}{\cos a + \cosh \xi}$ |
| 9 | $\dfrac{\sinh ax}{\sinh \pi x}, -\pi < a < \pi$ | $\sqrt{\dfrac{1}{2\pi}} \dfrac{\sin a}{\cos a + \cosh \xi}$ |
| 10 | $P_n(x), \vert x\vert < 1; 0, \vert x\vert > 1$ | $\dfrac{i^n}{\sqrt{\pi}} J_{n+\frac{1}{2}}(\xi)$ |

may be extended to cover a semicircular arc in the top or bottom half of the plane, with the requirement that the integral along the distorted contour vanishes. Evaluating the limiting integrals in the complex plane is governed by a theorem known as *Jordan's lemma*, which is summarized here [Arf13].

For a function $f(z) \to 0$ on an upper half circle $C_R$ with radius $R$ and center at the origin, consider an integral

$$I = \lim_{R\to\infty} \int_{C_R} e^{i\xi z} f(z) dz, \ \xi > 0.$$

With $|f(z)| < \epsilon, z = Re^{i\theta}$, and $dz = iRe^{i\theta} d\theta$, we obtain

$$|I| \leq \epsilon R \int_0^{\pi} e^{-\xi R \sin\theta} d\theta \leq 2\epsilon R \int_0^{\pi/2} e^{-2\xi R\theta/\pi} d\theta < \frac{\pi}{\xi}\epsilon \to 0. \tag{D.5}$$

This forms Jordan's lemma for $C_R$ in the first and/or second quadrants, with similar expressions for the other three quadrant combinations.

## References

[Arf13] G.B. Arfken, H.J. Weber, and F.E. Harris, *Mathematical Methods for Physicist: A Comprehensive Guide*, 7th ed., Academic Press (2013).

[Sne51] I.N. Sneddon, *Fourier Transforms*, McGraw-Hill (1951).

# Appendix E

# Calculus of Variation for Optimal Control Formulation

A brief introduction to the calculus of variation is presented for the derivation of optimal control and optimization formulations, starting with classical applications in mechanics. Variational calculus is first introduced in Hamilton's principle [Gol80], which states that the motion of a particle follows a path minimizing a time-integral of the Lagrangian of motion. The basic approach is extended to the general case of system evolution, providing optimal control and optimization strategies that form the basis for Pontryagin's maximum principle [Pon63] and model-based control algorithms [Doy89].

## E.1 Euler–Lagrange and Hamilton Equations

According to Hamilton's principle, the trajectory of a particle with kinetic energy $T$ and potential energy $V$ may be described by Lagrangian $L = T - V$ such that the time integral $J$ of $L$ over the travel time $[0, t_f]$ may be minimized

$$\delta J(\delta x) = 0 \quad \text{for} \quad J = \int_0^{t_f} L(x, \dot{x}) dt. \tag{E.1}$$

The Lagrangian is written in terms of position $x$ and speed $\dot{x}$, and the variation is taken over virtual displacement $\delta x$, as illustrated in Figure E.1:

$$\delta J(\delta x) = \frac{\partial J}{\partial x}\delta x = \int_0^{t_f} \left( \frac{\partial L}{\partial x}\delta x + \frac{\partial L}{\partial \dot{x}}\delta \dot{x} \right) dt. \tag{E.2}$$

Evaluating the integral over $\delta \dot{x} dt$ by parts yields

$$\delta J(\delta x) = \int_0^{t_f} \left[ \frac{\partial L}{\partial x} - \frac{d}{dt}\left( \frac{\partial L}{\partial \dot{x}} \right) \right] \delta x dt + \left. \frac{\partial L}{\partial \dot{x}}\delta x \right|_0^{t_f}. \tag{E.3}$$

*Nuclear Reactor Physics and Engineering*, Second Edition. John C. Lee.
© 2025 John Wiley & Sons, Inc. Published 2025 by John Wiley & Sons, Inc.
Companion website: www.wiley.com/go/LeeNRPE2

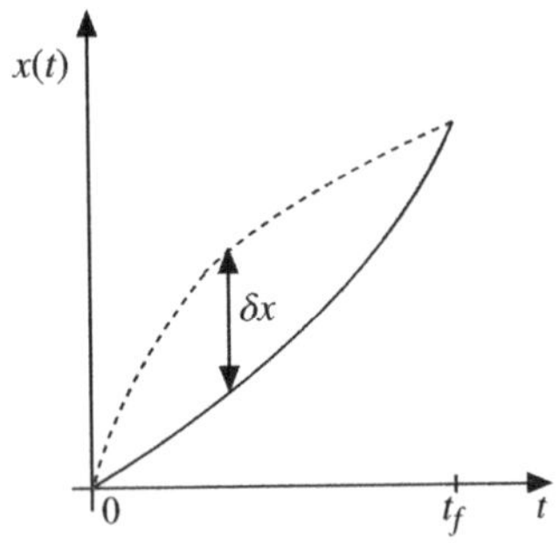


**Figure E.1** Virtual displacement $\delta x$ separating the actual particle trajectory (solid curve) and non-physical trajectory (dotted curve).

With the constraints $\delta x(0) = \delta x(t_f) = 0$, $\delta J(\delta x) = 0$ yields the *Euler–Lagrange equation*

$$\frac{\partial L}{\partial x} - \frac{d}{dt}\left(\frac{\partial L}{\partial \dot{x}}\right) = 0, \tag{E.4}$$

representing the motion of an object in general.

Recasting the Lagrangian via the Hamiltonian or total energy

$$H = T + V = p\dot{x} - L(x, \dot{x}) = H(x, p), \quad T = \frac{m\dot{x}^2}{2}, \tag{E.5}$$

in terms of linear momentum $p = m\dot{x}$ yields *Hamilton's equations*

$$\frac{\partial H}{\partial p} = \dot{x} = \frac{dx}{dt}, \tag{E.6}$$

$$\frac{\partial H}{\partial x} = -\frac{\partial L}{\partial x} = -\frac{dp}{dt}, \tag{E.7}$$

where Eq. (E.4) is used together with $p = \partial L/\partial \dot{x}$. For optimal control and optimization formulations, extend the concept of linear momentum $p$ to represent a costate or adjoint vector in the form of a Lagrangian multiplier, with control variable $u$ added to the system equation:

$$\dot{x} = f(x, u). \tag{E.8}$$

## E.2 Pontryagin's Maximum Principle

Consider a time-dependent optimization problem where an objective function $J$ is to be minimized over a time interval $[0, t_f]$ with the system equation

$$J = \phi[x(t_f)]. \tag{E.9}$$

Augment the objective function with Lagrange multipliers $p$ and $\eta$, both of which are piecewise continuous, and construct a Hamiltonian

$$J^* = \phi[x(t_f)] + \int_0^{t_f} dt[p^T(f - \dot{x}) + \eta^T S], \tag{E.10}$$

where a state-space constraint is introduced as $S(x) = 0$, and $\eta$ is to be selected such that

$$\begin{cases} \eta = 0 & \text{for} \quad S < 0, \\ \eta \geq 0 & \text{for} \quad S = 0. \end{cases} \tag{E.11}$$

It should be noted that the introduction of two Lagrange multipliers $p$ and $\eta$ does not perturb the objective function $J$ so that minimizing the augmented objective function $J^*$ accomplishes the same objective as minimizing the original objective function $J$. Introducing a Hamiltonian

$$H = p^T f + \eta^T S, \tag{E.12}$$

recasts $J^*$

$$J^* = \phi[x(t_f)] + \int_0^{t_f} dt(H - p^T \dot{x}). \tag{E.13}$$

The necessary condition for the optimal control $u$ then translates to setting $\delta J^* = 0$

$$\begin{aligned} \delta J^* &= \left.\frac{\partial \phi}{\partial x}\right|_{t_f} \delta x(t_f) + \int_0^{t_f} dt \left[ p^T \left( \frac{\partial f}{\partial x}\delta x + \frac{\partial f}{\partial u}\delta u - \frac{d}{dt}\delta x \right) + \eta^T \frac{dS}{dx}\delta x \right] = 0 \\ &= \left.\frac{\partial \phi}{\partial x}\right|_{t_f} \delta x(t_f) - \left. p^T \delta x \right|_0^{t_f} \\ &\quad + \int_0^{t_f} dt \left[ p^T \left( \frac{\partial f}{\partial x}\delta x + \frac{\partial f}{\partial u}\delta u \right) + \frac{dp^T}{dt}\delta x + \eta^T \frac{dS}{dx}\delta x \right], \end{aligned} \tag{E.14}$$

which yields, with the initial condition $\delta x(0) = 0$, a set of equations to be satisfied for the optimal control:

$$\frac{dp}{dt} = -\left(\frac{\partial f}{\partial x}\right)^T p - \left(\frac{dS}{dx}\right)^T \eta = -\frac{\partial H}{\partial x}, \quad \text{Hamilton's equation,} \tag{E.15}$$

$$\frac{\partial H}{\partial u} = \left(\frac{\partial f}{\partial u}\right)^T p = 0, \quad \text{optimality condition for } u, \tag{E.16}$$

$$p^T(t_f) = \left.\frac{\partial \phi}{\partial x}\right|_{t_f}, \quad \text{terminal condition.} \tag{E.17}$$

When the terminal time $t_f$ is not fixed, the *terminal condition* from Eq. (E.17) has to be modified to the transversality condition with the variation $\delta t_f$ of the terminal time reflected in the total variation $\Delta x(t_f)$ in the system trajectory in Figure E.2:

$$\Delta x(t_f) = \delta x(t_f) + \dot{x}(t_f)\delta t_f.$$

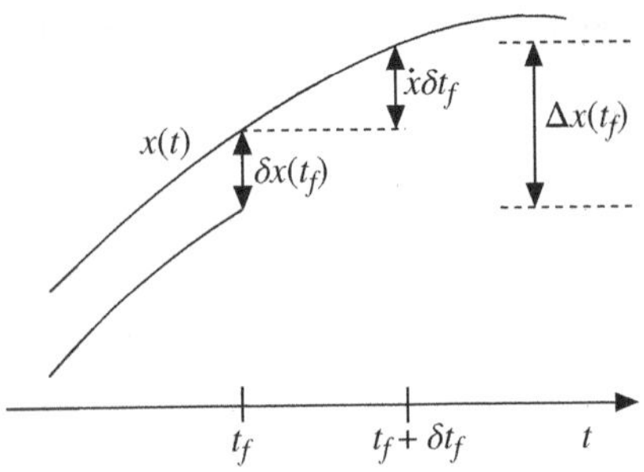


**Figure E.2** Total variation at the target trajectory with variable terminal time $t_f$.

With $\Delta x(t_f)$ replacing $\delta x(t_f)$ in the first RHS term of Eq. (E.14), the terminal terms are rewritten

$$\left.\frac{\partial \phi}{\partial x}\right|_{t_f} \Delta x(t_f) - p^T(t_f)\delta x(t_f) = \left[\left.\frac{\partial \phi}{\partial x}\right|_{t_f} - p^T(t_f)\right] \Delta x(t_f) + p^T(t_f)\dot{x}(t_f)\delta t_f = 0,$$

which requires

$$\sum_{i=1}^{n} p_i(t_f)dx_i(t_f) = 0,$$

for an $n$-dimensional trajectory. This requirement is the *transversality condition* [Bry63] that, for a trajectory where $t_f$ is variable, the adjoint vector is perpendicular to the trajectory at the time of arrival on the target

$$p^T(t_f)x(t_f) = 0. \tag{E.18}$$

For minimization of the objective function $J$ of Eq. (E.9) with the system equation (E.8) and Hamiltonian of Eq. (E.12), *Pontryagin's maximum principle* [Pon63] provides the optimal control $u$ by Eq. (E.16), with the help of adjoint function $p(t)$ specified by Eq. (E.15). Equations (E.17) and (E.18) provide the terminal condition and transversality conditions, respectively. Pontryagin's original formulation was augmented by a number of subsequent studies [Bry63,Jac71] especially for handling state-space constraints. The optimality condition in Eq. (E.16) may not provide a meaningful solution when $H$ displays a linear relationship in $u$, which may then require an alternate control strategy that minimizes $H$. This often entails a *bang–bang control* strategy involving the use of limiting values of $u \in [U_{min}, U_{max}]$ such that

$$u = \begin{cases} U_{min}, & \dfrac{\partial H}{\partial \phi} > 0, \\[2ex] U_{max}, & \dfrac{\partial H}{\partial \phi} < 0. \end{cases} \tag{E.19}$$

Subject to the state-space constraint $S(x) = 0$ with a non-negative piecewise continuous Lagrangian multiplier $\eta$, $p(t)$ may be discontinuous at junction points, i.e.

entrance and exit points for the constraint boundary, and a jump condition [Jac71] has to be established. Consider a time integral of Eq (E.15) across an entrance point at $t_1$

$$\int_{t_1^-}^{t_1^+} \frac{dp(t)}{dt} dt = -\int_{t_1^-}^{t_1^+} \left(\frac{\partial f}{\partial x}\right)^T p(t)dt - \int_{t_1^-}^{t_1^+} \left(\frac{dS}{dx}\right)^T \eta dt, \tag{E.20}$$

where the first integral on the RHS vanishes because $(\partial f/\partial x)^T$ is continuous across the junction and $p(t)$ would have at most a finite jump. Recognizing that $dS/dx$ is also continuous and $\eta$ is non-decreasing across the junction yields the desired *junction condition*

$$p(t_1^+) - p(t_1^-) = -\left(\frac{dS}{dx}\right)^T \mu(t_1), \tag{E.21}$$

with $\mu(t_1) \geq 0$. A similar junction condition has to be considered at the exit point on the constraint boundary.

**Example E.1** Obtain the minimum time solution and trajectory $x(t)$ for a vehicle with unit mass and inertialess control

$$\frac{d^2x(t)}{dt^2} = u(t), \quad |u| = 1,$$

to reach the final point $[x = 0, \dot{x} = 0]$ starting at point $[x(0) = x_{10}, \dot{x}(0) = 0]$.

Recast the equation of motion for a phase plane $(x_1, x_2) = (x, v = \dot{x})$

$$\frac{d}{dt}\begin{bmatrix} x_1 \\ x_2 \end{bmatrix} = \begin{bmatrix} x_2 \\ u \end{bmatrix} = \begin{bmatrix} f_1 \\ f_2 \end{bmatrix},$$

with the terminal time as the objective function $J = t_f$ to be minimized. The terminal condition of Eq. (E.17) becomes inactive for this time-optimal control problem and construct the Hamiltonian

$$H = p_1 f_1 + p_2 f_2.$$

Hamilton's equation (E.15) for the costate vector takes the form

$$\frac{d}{dt}\begin{bmatrix} p_1 \\ p_2 \end{bmatrix} = \begin{bmatrix} -\partial H/\partial x_1 \\ -\partial H/\partial x_2 \end{bmatrix} = \begin{bmatrix} 0 \\ -p_1 \end{bmatrix},$$

which yields

$p_1 = C_1, p_2 = -C_1 t + C_2$, with two constants $C_1$ and $C_2$, and

$$H = C_1 x_2 + (C_2 - C_1 t)u.$$

The optimal control requires minimizing $H$ via a bang–bang control

$$u(t) = \pm 1 \quad \text{for} \quad \partial H/\partial u = p_2(t) \lessgtr 0,$$

**Table E.1** Phase plane solution with constraints for Example E.1.

| $u(t)$ | $x_2(t)$ | $x_1(t)$ |
|---|---|---|
| $-1$ | $-t + s_2$ | $-(x_2)^2/2 + s_1$ |
| $+1$ | $t + s_2'$ | $(x_2)^2/2 + s_1'$ |

with $p_2(t)$ permitting only one switching, which also requires that $C_1$ and $C_2$ both have the same sign to allow the switching time $t_s > 0$. Integration of the system equation with $u(t) = \pm 1$ yields the solution summarized in Table E.1 with constants $s_1, s_1', s_2$, and $s_2'$ yet to be determined.

We claim that $C_2 > 0$; otherwise $u = 1$ at $t = 0$, which does not provide a feasible solution to the origin $(0, 0)$. Hence, the feasible optimal solution starts with $u = -1$ at $(x_{10}, 0)$ and switches to $u = +1$ at $t = t_s$. This then determines $s_2 = 0$, $s_1 = x_{10}, s_2' = -2t_s, s_1' = 0$, together with $x_2^2(t_s) = x_{10}, x_1(t_s) = x_{10}/2$, and $t_f = 2t_s$. Feasible solutions in the phase plane corresponding to that summarized in Table E.1 are illustrated in Figure E.3 for both $u = 1$ and $u = -1$. The time-optimal solution that reflects the initial and final conditions is clarified in Figure E.4, together with the bang-bang control motion.

The optimal solution confirms the intuitive nature of the solution, with the costate or adjoint function $p(t)$ facilitating the solution for the system trajectory, in this example, as a switching function for acceleration and deceleration. It is also instructive to calculate the Hamiltonian

$$
\begin{aligned}
H &= C_1 x_2(t) + C_1(t_s - t)u(t) \\
&= \begin{cases} -C_1 t - C_1(t_s - t) = -C_1 t_s, & u(t) = -1, \\ C_1(t - 2t_s) + C_1(t_s - t) = -C_1 t_s, & u(t) = +1, \end{cases}
\end{aligned}
$$

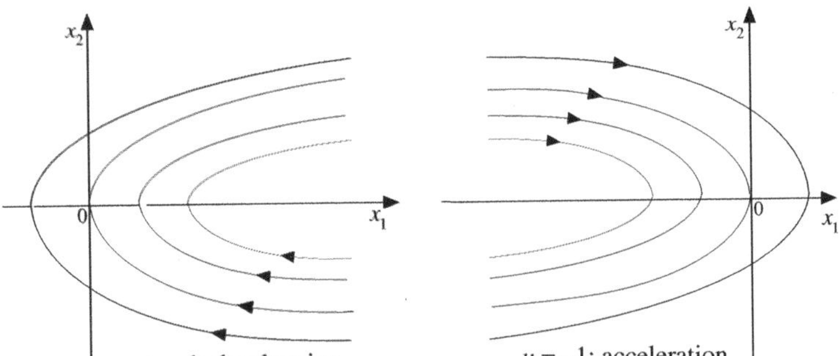


**Figure E.3** Feasible trajectories corresponding to $u = \pm 1$.

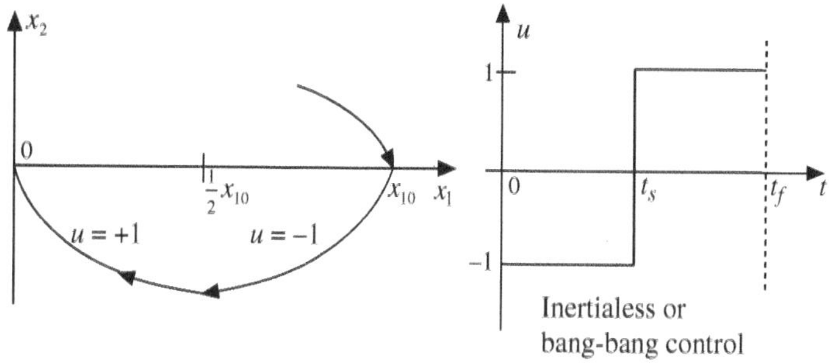


**Figure E.4** Optimal system trajectory and corresponding control for Example E.1.

which indicates that $H$ is constant through the entire trajectory. This is generally the case for *autonomous systems*, where the system equation $dx/dt$, as exemplified by Eq. (E.8), does not show an explicit dependence on time. The conservation of the Hamiltonian is equivalent physically to the conservation of total energy in classical mechanics without the presence of frictional or other dissipative forces.

**Example E.2** Repeat the time-optimal control problem from Example E.1 with the additional constraint that the speed of the vehicle is limited to $|x_2| \leq \xi$.

With the state-space constraint written as $S = -x_2 - \xi \leq 0$, the Hamiltonian is obtained as

$$H = p_1x_2 + p_2u + \eta(-x_2 - \xi)$$

and Hamilton's equation (E.15) for the costate vector is modified to

$$\frac{d}{dt}\begin{bmatrix} p_1 \\ p_2 \end{bmatrix} = \begin{bmatrix} -\partial H/\partial x_1 \\ -\partial H/\partial x_2 \end{bmatrix} = \begin{bmatrix} 0 \\ -p_1 + \eta \end{bmatrix},$$

Since the Lagrangian multiplier $\eta$ is active only on the constraint boundary for speed $x_2(t)$, it is necessary to consider three different segments for the trajectory and adjoint function $p_2(t)$ delineated by the entry time $t_1$ and exit time $t_2$ for the boundary, with the switching of control $u(t) = \mp 1$ determined by $p_2(t)$ via a bang-bang control strategy away from the boundary. The adjoint function $p_1(t) = C_1$, for some constant $C_1$, throughout the trajectory as in Example E.1. On the constraint boundary, minimizing $H$ through $\partial H/\partial u = 0$ indicates $p_2(t) = 0$ is a feasible solution for control $u(t) = 0$, allowing for the jump parameter $\mu = 0$ in Eq. (E.21) and accommodating $p_2(t)$ to be continuous at both $t_1$ and $t_2$. With the travel time $t_2 - t_1 = t_f - 2\xi$ on the boundary and feasibility condition $x_{10} = \xi(t_f - \xi)$, the overall trajectory is summarized in Table E.2 and the phase plane trajectory and the costate vector are illustrated in Figure E.5. The dotted curve indicates the feasible trajectory without the constraint on the speed as in Example E.1.

**Table E.2** Phase plane solution for Example E.2.

| Time interval | Constraint | $p_2(t)$ | $u(t)$ | $x_2(t)$ | $x_1(t)$ |
|---|---|---|---|---|---|
| (I) $0 \le t \le t_1$ | $S < 0, \eta = 0$ | $C_1(\xi - t)$ | $-1$ | $-t$ | $-x_2^2/2 + x_{10}$ |
| (II) $t_1 \le t \le t_2$ | $S = 0, \eta \ge 0$ | 0 | 0 | $-\xi$ | $\xi(\xi/2 - t) + x_{10}$ |
| (III) $t_2 \le t \le t_f$ | $S < 0, \eta = 0$ | $C_1(t_f - \xi - t)$ | $+1$ | $t - t_f$ | $x_2^2/2$ |

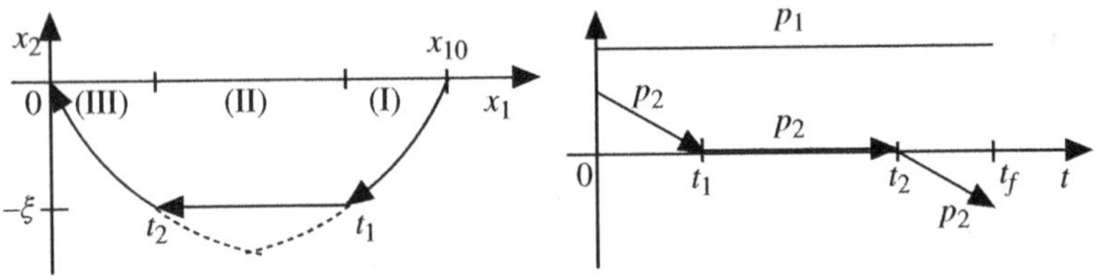


**Figure E.5** Optimal system trajectory in the phase plane and costate variables for Example E.2.

## References


**[Bry63]** A.E. Bryson Jr., W.F. Denham, and S.E. Dreyfus, "Optimal Programming Problems with Inequality Constrains I: Necessary Conditions for Extremal Conditions," *AIAA Journal* **1**, 2544 (1963).

**[Doy89]** J.C. Doyle, K. Glover, P.P. Khargonekar, and B.A. Francis, "State-Space Solution to Standard $H_2$ and $H_\infty$ Control Problems," *IEEE Trans. Autom. Control* **34**, 831 (1989).

**[Gol80]** H. Goldstein, *Classical Mechanics*, 1st ed. (1959), 2nd ed., Addison-Wesley (1980).

**[Jac71]** D.H. Jacobson, M.M. Lele, and J.L. Speyer, "New Necessary Condition of Optimality for Control Problems with State-variable Inequality Constraints," *J. Math. Anal. Appl.* **35**, 255 (1971).

**[Pon63]** L.S. Pontryagin, V.G. Boltyanskii, R.V. Gamkrelidze, and E.F. Mishchenko, *The Mathematical Theory of Optimal Processes*, Interscience Publishers (1963).

# Appendix F

# Kalman Filter Algorithm

This appendix presents a brief derivation of the Kalman filter, which plays an important role in various system and control modeling tasks, as discussed in the diagnostics example in Section 16.6. The Kalman filter is a minimum-variance parameter estimation algorithm that generates an optimal estimate of system state vector $x(t)$ given observation vector $y(t)$, duly accounting for modeling uncertainties for $x(t)$ and statistical fluctuations in $y(t)$. The optimal estimate $\hat{x}(t)$ is obtained so that the covariance of the system estimation is minimized. We begin with the development of the *linear Kalman filter* algorithm followed by an approach that could be taken to explicitly represent the nonlinearity of system models.

## F.1 Linear Kalman Filter

Consider a dynamical system represented by $x(t)$ subject to white Gaussian noise vector $w(t)$ with covariance $Q$,

$$\frac{dx(t)}{dt} = F(t)x(t) + w(t), \quad \langle w(t) \rangle = 0, \left\langle w(t)w^T(t') \right\rangle = Q\delta(t - t'), \tag{F.1}$$

where $x(t)$ is determined indirectly through observation $y(t)$ subject to white Gaussian noise vector $v(t)$ with covariance $R$,

$$y(t) = M(t)x(t) + v(t), \quad \langle v(t) \rangle = 0, \left\langle v(t)v^T(t') \right\rangle = R\delta(t - t'). \tag{F.2}$$

The optimal system estimate $\hat{x}(t)$ may be considered a statistical expectation of the true or exact system state $x(t)$ given observation $y(t)$,

$$\hat{x}(t) = \langle x(t) \,| y(t) \rangle, \tag{F.3}$$

such that the covariance matrix

$$P(t) = \left\langle \left[x(t) - \hat{x}(t)\right] \left[x(t) - \hat{x}(t)\right]^T \right\rangle \tag{F.4}$$

*Nuclear Reactor Physics and Engineering*, Second Edition. John C. Lee.
© 2025 John Wiley & Sons, Inc. Published 2025 by John Wiley & Sons, Inc.
Companion website: www.wiley.com/go/LeeNRPE2

is minimized. Although continuum formulations of the Kalman filter are possible [Jaz70], a discretized form of the filter is derived here with various practical applications in mind. For this purpose, the state transition matrix is defined over the time interval $[t_{k-1}, t_k]$,

$$\Phi(k|k-1) = \exp\left[\int_{t_{k-1}}^{t_k} F(t)\mathrm{d}t\right], \tag{F.5}$$

so that Eq. (F.1) may be written in a discretized form

$$x(k) = \Phi(k|k-1)x(k-1) + w(k) = \Phi x(k-1) + w \tag{F.6}$$

together with the measurement equation (F.2) similarly discretized,

$$y(k) = M(k)x(k) + v(k) = Mx(k) + v. \tag{F.7}$$

For notational convenience, the explicit time-step indices are suppressed in the last expression for each of the discretized equations (F.6) and (F.7). The covariance matrix of Eq. (F.4) may be written for time steps $k-1$ and $k$ as

$$P(k-1) = \left\langle \left[x(k-1) - \hat{x}(k-1)\right]\left[x(k-1) - \hat{x}(k-1)\right]^T \right\rangle, \tag{F.8}$$

$$P(k) = \left\langle \left[x(k) - \hat{x}(k)\right]\left[x(k) - \hat{x}(k)\right]^T \right\rangle. \tag{F.9}$$

The Kalman filter is formulated in a two-step recursive structure beginning with a *prior estimate* at time step $k$ before the measurement is taken:

$$\hat{x}^-(k) \equiv \hat{x}(k|k-1) = \Phi\hat{x}(k-1). \tag{F.10}$$

Here the superscripted estimate $\hat{x}^-(k)$ is used synonymously with the conditional estimate $\hat{x}(k\,|k-1)$ to indicate that the estimation is an initial prediction based on the optimal estimate $\hat{x}(k-1)$ of the previous time step $k-1$. Equations (F.6) and (F.10) yield a prior estimate of the covariance with $w = w(k)$,

$$\begin{aligned} P^-(k) &\equiv P(k|k-1) = \left\langle [x(k) - \hat{x}^-(k)][x(k) - \hat{x}^-(k)]^T \right\rangle \\ &= \left\langle \left\{\Phi\left[x(k-1) - \hat{x}(k-1)\right] + w\right\}\left\{\Phi\left[x(k-1) - \hat{x}(k-1)\right] + w\right\}^T \right\rangle \\ &= \Phi\left\langle \left[x(k-1) - \hat{x}(k-1)\right]\left[x(k-1) - \hat{x}(k-1)\right]^T \right\rangle \Phi^T + \left\langle ww^T \right\rangle. \end{aligned} \tag{F.11}$$

In the last estimation step, the term involving the cross product of the estimation error $\left[x(k-1) - \hat{x}(k-1)\right]$ in step $k-1$ and the modeling error $w = w(k)$ in step $k$ is dropped because the two errors are independent of one another. Equation (F.11) may be simplified by using Eq. (F.8) and a discretized form of covariance $Q$ in Eq. (F.1) to give

$$P^-(k) = \Phi P(k-1)\Phi^T + Q(k) \equiv \Phi P(k-1)\Phi^T + Q. \tag{F.12}$$

In the *correction step* of the Kalman filter, after a new measurement is taken at step $k$, the objective is to add to the prior estimate of Eq. (F.10) a term proportional to the measurement residual,

$$\xi(k) = y(k) - M\hat{x}^{-}(k), \tag{F.13}$$

so that the resulting *posterior estimate*

$$\hat{x}(k) \equiv \hat{x}^{+}(k) = \hat{x}^{-}(k) + K\left[y(k) - M\hat{x}^{-}(k)\right] \tag{F.14}$$

minimizes the estimation error

$$\varepsilon(k) = x(k) - \hat{x}(k) \tag{F.15}$$

or equivalently the covariance $P(k)$ of Eq. (F.9). Thus, the key remaining task is to derive an expression for the proportionality constant $K$ introduced in Eq. (F.14). Before inserting $\varepsilon(k)$ from Eq. (F.15) into Eq. (F.9), however, an alternate form of measurement residual $\xi(k)$ is obtained via Eqs. (F.6) and (F.7),

$$\xi(k) = Mx(k) + v(k) - M\hat{x}^{-}(k) = M\left[\Phi x(k-1) + w\right] + v(k) - M\hat{x}^{-}(k),$$

which can be rewritten with Eq. (F.15) at time step $k-1$ and Eq. (F.10):

$$\begin{aligned}\xi(k) &= M\left[\Phi\left\{\varepsilon(k-1) + \hat{x}(k-1)\right\} + w\right] + v - M\hat{x}^{-}(k)\\ &= M\left[\Phi\varepsilon(k-1) + w\right] + v.\end{aligned} \tag{F.16}$$

Substituting Eqs. (F.6), (F.14), (F.10), and (F.13) into Eq. (F.15) provides a more useful form:

$$\begin{aligned}\varepsilon(k) &= x(k) - \hat{x}(k) = \Phi\left[x(k-1) - \hat{x}(k-1)\right] + w - K\xi(k)\\ &= \Phi\varepsilon(k-1) + w - KM\left[\Phi\varepsilon(k-1) + w\right] - Kv\\ &= (I - KM)\left[\Phi\varepsilon(k-1) + w\right] - Kv.\end{aligned} \tag{F.17}$$

After substituting Eq. (F.17) into Eq. (F.9) and invoking the measurement covariance matrix $R$ of Eq. (F.2), the posterior estimate of the covariance matrix becomes

$$\begin{aligned}P(k) &= P^{+}(k) = \left\langle \varepsilon(k)\varepsilon^{T}(k)\right\rangle\\ &= (I - KM)\left[\Phi\left\langle \varepsilon(k-1)\varepsilon^{T}(k-1)\right\rangle\Phi^{T} + Q\right](I - KM)^{T} + KRK^{T},\end{aligned} \tag{F.18}$$

which can be simplified, via Eq. (F.12), to

$$P(k) = P^{+}(k) = (I - KM)P^{-}(k)(I - KM)^{T} + KRK^{T}. \tag{F.19}$$

Minimization of the posterior covariance matrix $P(k)$ may be accomplished by taking a derivative of the trace of $P(k)$ with respect to $K$ and setting it to zero,

$$-2(I - KM)P^{-}(k)M^{T} + 2KR = 0,$$

which may be rearranged as

$$P^{-}(k)M^{T} - KMP^{-}(k)M^{T} = KR$$

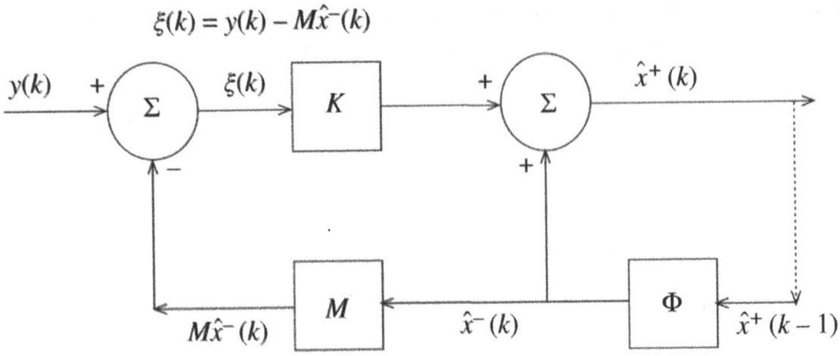


**Figure F.1** Flow of information for the Kalman filter.

and finally solved to give the *Kalman gain matrix* at time step $k$:

$$K(k) = P^{-}(k)M^{T}\left[MP^{-}(k)M^{T} + R\right]^{-1}. \tag{F.20}$$

Use of Eq. (F.20) also yields an alternate, simpler form of the posterior covariance matrix,

$$P(k) = P^{+}(k) = (I - KM)P^{-}(k). \tag{F.21}$$

In summary, the discretized linear Kalman filter algorithm can be recursively applied through the following steps:

(a) Obtain prior estimates before the measurement via Eqs. (F.10) and (F.12).
(b) Update the prior estimates into the posterior estimates via Eqs. (F.14) and (F.19) or (F.21), together with the Kalman gain matrix of Eq. (F.20).

The flow of information for the Kalman filter algorithm is illustrated in Figure F.1.

## F.2 Unscented Kalman Filter

When the system equation (F.1) or the measurement equation (F.2) is nonlinear, then the equations may be successively linearized as the system evolves in time. This approach is known as the *extended Kalman filter*. Alternately, Monte Carlo algorithms may be implemented to represent the nonlinear system evolutions explicitly in a *particle filter* algorithm [Aru02], which naturally will require significant computational resources. More recently, an *unscented Kalman filter* algorithm [Jul00, Jul04, Vos04] has been developed that allows for direct use of nonlinear equations, with only modest increases in computational costs. In this approach, a few system trajectories are evaluated judiciously along a set of characteristic points known as the *sigma points* to represent the nonlinear evolution of the systems so that the mean and variance of the system trajectory and associated covariance are estimated with sufficient accuracy.

With the objective in mind, the discretized system equation (F.6) and measurement equation (F.7) are rewritten to represent the nonlinearity explicitly:

$$x(k) = F[x(k-1), w(k-1)], \tag{F.22a}$$

$$y(k) = M[x(k), v(k)]. \tag{F.22b}$$

Given the optimal estimate $\hat{x}(k-1)$ and associated covariance matrix $P^+ \equiv P(k-1) \equiv P_{xx}(k-1)$ of dimension $n$ at time step $k-1$, choose $2n$ sigma points $x_i^*(k-1)$ and associated weights $W_i$

$$x_i^*(k-1) = \hat{x}(k-1) - (\sqrt{nP^+})_i, i = 1, \dots, n, \tag{F.23a}$$

$$x_{i+n}^*(k-1) = \hat{x}(k-1) + (\sqrt{nP^+})_i, i = 1, \dots, n, \tag{F.23b}$$

$$W_i = \frac{1}{2n}, i = 1, \dots, 2n, \tag{F.23c}$$

to construct $2n$ nonlinear estimates of the system state and observation for time step $k$:

$$x_i^-(k) = F[x_i^*(k-1)], \tag{F.24a}$$

$$y_i(k) = M[x_1^-(k)]. \tag{F.24b}$$

We obtain the consolidated estimates of the system state, observation, and covariance matrices for time step $k$:

$$\hat{x}^-(k) = \sum_{i=1}^{2n} W_i x_i^-(k), \tag{F.25a}$$

$$\hat{y}(k) = \sum_{i=1}^{2n} W_i y_i(k) = \sum_{i}^{2n} W_i M[x_1^-(k)], \tag{F.25b}$$

$$P_{xx}^-(k) = \sum_{i=1}^{2n} W_i [x_i^-(k) - \hat{x}^-(k)][x_i^-(k) - \hat{x}^-(k)]^T + Q(k), \tag{F.26a}$$

$$P_{xy}(k) = \sum_{i=1}^{2n} W_i [x_i^-(k) - \hat{x}^-(k)][y_i(k) - \hat{y}(k)]^T, \tag{F.26b}$$

$$P_{yy}(k) = \sum_{i=1}^{2n} W_i [y_i(k) - \hat{y}(k)][y_i(k) - \hat{y}(k)]^T + R(k). \tag{F.26c}$$

The updated estimates for $\hat{x}^-(k), \hat{y}(k)$ and three covariance matrices finally provide the equivalent of Eqs. (F.14), (F.20), and (F.21) in the linear Kalman filter formulation to advance the filtering to time step $k$:

$$\hat{x}(k) = \hat{x}^-(k) + K(k)\left[y(k) - \hat{y}(k)\right], \tag{F.27a}$$

$$K(k) = P_{xy}(k)P_{yy}^{-1}(k), \tag{F.27b}$$

$$P(k) = P_{xx}^{-}(k) - K(k)P_{xy}^{T}(k) \equiv P_{xx}(k). \tag{F.27c}$$

It can be shown [Jul00, Jul04, Sim06] that unscented Kalman filtering provides the system estimate $\hat{x}(k)$ and covariance $P(k)$ accurate up to the third order with the system trajectories evaluated along a few sigma points, retaining the general Kalman filter structure in Figure F.1 and without linearizing the nonlinear system and observation equations. Thus, with some additional effort exercising the system equations, optimal estimates of nonlinear system models may conveniently be obtained [Jul04, Vos04, Alp07] for system dynamics, control, and diagnostics. It should be noted that the choice of the sigma points in Eqs. (F.23) is not unique and should be considered as one convenient example.

## References

**[Alp07]** B. Alpay, J.P. Holloway, and J.C. Lee, "Degradation Monitoring in IRIS Steam Generators," *Proc. M&C + SNA, Am. Nucl. Soc.* (2007).

**[Aru02]** M.S. Arulampalam, S. Maskell, N. Gordon, and T. Clapp, "A Tutorial on Particle Filters for Online Nonlinear/Non-Gaussian Bayesian Tracking," *IEEE Trans. Signal. Proc.* **50**, 174 (2002).

**[Jaz70]** A.H. Jazwinski, *Stochastic Processes and Filtering Theory*, Academic Press (1970).

**[Jul04]** S.J. Julier and J.K. Uhlmann, "Unscented Filtering and Nonlinear Estimation," *Proc. IEEE* **92**, 401 (2004).

**[Jul00]** S. Julier, J. Uhlmann, and H.F. Durrant-Whyte, "A New Method for the Nonlinear Transformation of Means and Covariances in Filters and Estimators," *IEEE Trans. Autom. Control* **45**, 477 (2000).

**[Sim06]** D.J. Simon, *Optimal State Estimation*, Wiley (2006).

**[Vos04]** H.U. Voss, J. Timmer, and J. Kurths, "Nonlinear Dynamical System Identification from Uncertain and Indirect Measurements," *Int. J. Bifurcat. Chaos* **14**, 1904 (2004).

# Answers to Selected Problems

## Chapter 2

**2.2** (a) $2.30 \times 10^9$ neutron $\cdot$ s$^{-1}$ mg$^{-1}$ of $^{252}$Cf, (b) 0.116 neutron $\cdot$ Bq$^{-1}$ s$^{-1}$, (c) 2.645 yr.

**2.3** (a) $^{27}_{13}$Al, $^{26}_{12}$Mg, (b) 11.8 MeV.

**2.5** $\dfrac{E}{E_0} = \dfrac{A^2(1+\gamma)^2 + (1-A\gamma)^2 + 2A(1+\gamma)(1-A\gamma)\cos\theta_c}{(1+A)^2}, \gamma = \dfrac{V_\ell}{v_\ell},$
$E/E_0 \geq 1.0$ is possible for $A\gamma \geq 1.0$.

**2.6** (a) 0.68, (b) 1.935 MeV.

**2.9** (a) 223 Ci, (b) $^{12}_{6}$C, (c) $2.5 \times 10^8$ neutron $\cdot$ s$^{-1}$, 11.2 MeV.

**2.10** (c) $v'_c = v_0/2$, (d) $\theta_\ell = \pi/4$

**2.11** (b) $\lambda' - \lambda = (h/mc)(1 - \cos\theta)$.

**2.12** (a) $\dfrac{mc^2}{E'} = \dfrac{mc^2}{E} + 1 - \cos\theta$, (b) $\dfrac{32}{9} \times 0.511$ MeV.

**2.13** (b) $\overline{E} = E_0(1+\alpha)/2$, (c) 23.

**2.16** (a) $^{95}_{37}$Rb, (b) $^{141}_{59}$Pr and $^{95}_{42}$Mo, (c) 223 MeV, (d) $^{93}_{37}$Rb, 205 MeV.

**2.17** 0.33 mm.

**2.18** (a) After collision, $\mathbf{v}'_\ell = \mathbf{v}_{CM} + \mathbf{v}'_c = 0$, yielding $v'_c = v_{cM} = mv_0/(m+M)$.

*Nuclear Reactor Physics and Engineering*, Second Edition. John C. Lee.
© 2025 John Wiley & Sons, Inc. Published 2025 by John Wiley & Sons, Inc.
Companion website: www.wiley.com/go/LeeNRPE2

(b) Energy balance: $\frac{1}{2}mv_c^2 + \frac{1}{2}MV_c^2 = \frac{1}{2}mv_c^2\left(1+\frac{m}{M}\right) = E_0\frac{M}{M+m} = \frac{1}{2}m(v_c')^2\left(1+\frac{m}{M}\right) + Q, Q = E_0\left(1-\frac{m}{M}\right)$.
Momentum balance: $MV_\ell' = mv_0$, with $v_\ell' = 0, Q = \frac{1}{2}mv_0^2 - \frac{1}{2}M(V_\ell')^2 = E_0\left(1-\frac{m}{M}\right)$.

## Chapter 3

**3.1** (a) $1.59 \times 10^{13}$, (b) $\phi_0$, (c) 7.63 MW, (d) $\psi(\mathbf{r}, v, \mathbf{\Omega}) = mv\psi(\mathbf{r}, E, \mathbf{\Omega})$, $\phi(\mathbf{r}) = \Phi(\mathbf{r})$.

**3.2** $3.84 \times 10^{12}$.

**3.3** $\frac{R}{N} = I\sigma(|\mathbf{v}_0 - \mathbf{V}_0|)\frac{|\mathbf{v}_0 - \mathbf{V}_0|}{v_0}, |\mathbf{v}_0 - \mathbf{V}_0| = (v_0^2 + V_0^2 - 2v_0V_0\cos\theta)^{1/2}$.

**3.4** (a) $\frac{\partial I(E,x)}{\partial x} = -\Sigma(E)I(E,x), I(H) = \int_0^\infty dEI_0(E)\exp(-N\sigma_0E_0/E \cdot H)$
(b) With $N\sigma_0H = 73.06, I(H)/I_0 = 1.85 \times 10^{-6}$.

**3.6** $J_+(\mathbf{r}) = J_-(\mathbf{r}) = \phi(\mathbf{r})/4$.

**3.7** $R = \int_\mathbf{v} d\mathbf{v}\int_\mathbf{V} d\mathbf{V}n(\mathbf{v})|\mathbf{v}-\mathbf{V}|\sigma(|\mathbf{v}-\mathbf{V}|)N(\mathbf{V}) = N_0\int_\mathbf{v} d\mathbf{v}n(\mathbf{v})v\overline{\sigma}(v), \sigma(|\mathbf{v}-\mathbf{V}|) = \frac{\sigma(v_0)v_0}{|\mathbf{v}-\mathbf{V}|}$ with $v_0 = 2,200\ \mathrm{m\cdot s^{-1}}, N_0 = \int_\mathbf{V} d\mathbf{V}N(\mathbf{V})$.
(a) $\overline{\sigma}(v) = \frac{1}{vN_0}\int_\mathbf{V} d\mathbf{V}|\mathbf{v}-\mathbf{V}|\sigma(|\mathbf{v}-\mathbf{V}|)N(\mathbf{V}) = \sigma(v_0)\frac{v_0}{v}$.
(b) $\phi\left(\frac{E}{E_1}\right) = \phi_0\frac{E}{E_1}\exp\left(-\frac{E}{E_1}\right), \phi_0 = \text{total flux} = \int_0^\infty \phi\left(\frac{E}{E_1}\right)d\left(\frac{E}{E_1}\right)$,
$\overline{\sigma}(v) = \overline{\sigma}\left(\frac{E}{E_1}\right) = \sigma(v_0)\sqrt{\frac{E_0}{E}} = \sigma(v_1)\sqrt{\frac{E_1}{E}}, E = \frac{1}{2}mv^2, E_0 = kT_0 = \frac{1}{2}mv_0^2$,
$T_0 = 300\ \mathrm{K}, E_1 = kT_1, T_1 = 350\ \mathrm{K}$,
$\langle\sigma\rangle = \frac{1}{\phi_0}\int_0^\infty \overline{\sigma}\left(\frac{E}{E_1}\right)\phi\left(\frac{E}{E_1}\right)d\left(\frac{E}{E_1}\right) = \overline{\sigma}(v_0)\sqrt{\frac{E_0}{E_1}}\Gamma\left(\frac{3}{2}\right) = 1,641\ \mathrm{b}$.

## Chapter 4

**4.1** $S_0\frac{A}{4\pi(\sqrt{2}H)^2}\exp(-\sqrt{2}\Sigma_aH)\cos\frac{\pi}{4}$.

**4.3** (a) $p(\theta \to \theta') = \dfrac{\Sigma_s}{2\Sigma_t} \dfrac{\sin\theta'}{\cos\theta} \left[ \dfrac{1}{\cos\theta} + \dfrac{1}{\cos\theta'} \right]^{-1}$,
(b) $p(\theta) = \dfrac{\Sigma_s}{2\Sigma_t} \left( 1 + \mu \ln \dfrac{\mu}{1+\mu} \right)$, $\mu = \cos\theta$.

**4.5** With $E = 2.5$ MeV of energy released per decay and at radius $R = 1.0$ m from the source, dose rate in air $= \dfrac{SE}{4\pi R^2} \dfrac{\mu_e}{\rho} = 1.1\ \mathrm{rem \cdot hr^{-1}}$. Tissue dose $\simeq 1.1 \times$ air dose.

**4.6** With $S(\mu)d\mu = S_0 d\mu$, $R(x) = \Sigma S_0 \displaystyle\int_0^1 d\mu \frac{\exp(-\Sigma x/\mu)}{\mu} = \Sigma S_0 E_1(\Sigma x)$.

**4.7** $R(x) = \Sigma S_0 E_2(\Sigma x)$.

## Chapter 5

**5.1** $R(r)dr = \Sigma_a \phi(r) \cdot 4\pi r^2 dr$, $\phi(r) = S_0 \dfrac{\exp(-\kappa r)}{4\pi D r}$, $\langle r^2 \rangle = \dfrac{1}{S_0} \displaystyle\int_0^\infty r^2 R(r)dr =$ $6/\kappa^2 = 6L^2$.

**5.4** (a) $\phi(z) = \dfrac{S}{2\kappa D} \dfrac{\sinh \kappa(H-z) + 2\kappa D \cosh \kappa(H-z)}{\cosh \kappa H + 2\kappa D \sinh \kappa H}$,
(b) $\phi(z) = \dfrac{S}{2\kappa D} \dfrac{\sinh \kappa(H+2D-z)}{\cosh \kappa(H+2D)}$,
(c) $\phi(z) = \dfrac{S}{2\kappa D} \dfrac{\sinh \kappa(H-z)}{\cosh \kappa H}$, with $\Sigma_{tr} = \Sigma_a + \Sigma_s(1 - \overline{\mu}_0)$.

**5.5** $\phi(x) = \dfrac{Q}{\Sigma_a} \left( 1 - \dfrac{\cosh \kappa x}{\cosh \kappa H} \right)$.

**5.6** $\phi(x) = \dfrac{Q}{\Sigma_a} - \dfrac{\Sigma_a^{In} \phi(0) a}{2\kappa D} \exp(-\kappa x)$, $\phi(0) = \dfrac{Q}{\Sigma_a} \dfrac{2\kappa D}{\Sigma_a^{In} a + 2\kappa D}$, $F = \dfrac{\phi(0)}{\phi(\infty)} =$ $\dfrac{2\kappa D}{\Sigma_a^{In} a + 2\kappa D} = 0.83$, activation rate $= 1.02Q$.

**5.7** $\phi(x) = \displaystyle\int_{-\infty}^{\infty} S(x')\phi(x' \to x)dx'$
$= \displaystyle\int_{-\infty}^{\infty} \left[ Q - \Sigma_a^{In} \phi(0) a \delta(x') \right] \frac{\exp(-\kappa |x - x'|)}{2\kappa D} dx'$.

**5.8** (a) $\phi(r) = \dfrac{S}{4\pi D(1 + \kappa R)} \dfrac{\exp\{-\kappa(r-R)\}}{r}$.

**5.9** $1 - e^{-2\sigma NH}$.

**5.10** (a) $I(H) = I(0) \exp[-\{\Sigma_{a0} + \Sigma_{s0}\}H]$,
(b) $I(H) = I(0) \exp[-\{\Sigma_{a0} + \sqrt{2}\Sigma_{s0}\}H]$, $|\mathbf{V} - \mathbf{v}_0| = \sqrt{2} v_0$.

**5.12** $\phi_c(x) = \dfrac{S_0}{\Sigma_a} \sin Bx, B^2 = \left(\dfrac{\pi}{2H}\right)^2.$

**5.14** $\phi_1(x) = (Q/\Sigma_a)[1 - \exp(-\kappa H)\cosh \kappa x], \phi_2(x) = (Q/\Sigma_a)\sinh \kappa H \exp(-\kappa x).$

**5.15** $\phi_1(x) = \dfrac{Q}{2\kappa D}\left[\int_{-H}^{x} \exp\{-\kappa(x-x')\}dx' + \int_{x}^{H} \exp\{\kappa(x-x')\}dx'\right].$

**5.18** $B_g^2 = \dfrac{\pi}{(H_0 - H_1)^2} = \dfrac{\nu\Sigma_f/k - \Sigma_a}{D}, k = \dfrac{k_\infty}{1 + L^2B_g^2} = 0.9628.$

**5.20** $J(a)/\phi(a) = \kappa_M D_M \coth \kappa_M b, \kappa_= \sqrt{\Sigma_{aM}/D_M},$
$J(a) = (\nu\Sigma_f/k_{eff} - \Sigma_{aF})\phi(a)a \simeq J(0).$

**5.22** $\phi(r) = \dfrac{A}{r}\sinh \kappa r, \kappa^2 = \Sigma_a D$, leakage probability $= \dfrac{J^+(R)}{J^-(R)}.$

**5.25** (a) $BR_1 = \tan B(R_1 - R_0)$, (c) $\dfrac{\partial V_0}{\partial V_1} = \dfrac{B^2R_0^2}{1 + B^2R_1^2}, \lim_{R_1\to 0}\dfrac{\partial V_0}{\partial V_1} = \pi^2.$

**5.26** $\phi(x) = A\sin B\left(\dfrac{H}{2} - x\right)$, at midplane $\gamma = \dfrac{J_+(0+)}{J_+(0-)} = \dfrac{J_+(0+)}{J_-(0+)},$
$D\dfrac{d\ln\phi(0+)}{dx} = \dfrac{1}{2}\left(\dfrac{1-\gamma}{1+\gamma}\right) = -BD\cot\dfrac{BH}{2}.$

**5.27** $B\cot BH = -\kappa, B^2 = (\nu\Sigma_f - \Sigma_a)/D, \kappa^2 = \Sigma_a/D.$

## Chapter 6

**6.9** (a) $a_0 = \phi_0, a_1 = \phi_2 - \phi_1, a_2 = \phi_1 + \phi_2 - 2\phi_0,$
(b) $a_1 = -(h/2D)(J_2^+ - J_2^- + J_1^+ - J_1^-), a_2 = -(h/6D)(J_2^+ - J_2^- - J_1^+ + J_1^-),$
$\left(\dfrac{1}{2} + \dfrac{h}{12D}\right)J_1^- - \dfrac{h}{24D}J_2^+ = \dfrac{\phi_0}{4} + \left(\dfrac{h}{12D} - \dfrac{1}{2}\right)J_1^+ - \dfrac{h}{24D}J_2^-,$
$-\dfrac{h}{24D}J_1^- + \left(\dfrac{1}{2} + \dfrac{h}{12D}\right)J_2^+ = \dfrac{\phi_0}{4} - \dfrac{h}{24D}J_1^+ + \left(\dfrac{h}{12D} - \dfrac{1}{2}\right)J_2^-.$

**6.10** With the reflector represented by $\dfrac{d\ln\phi(x_1)}{dx} = \alpha = -\dfrac{1}{2D}\dfrac{1-\beta}{1+\beta}$ at the left-side boundary, the leakage term can be modified to $I_1 = -\langle D_1\rangle\phi_2 + \langle D_1\rangle\phi_1 + D_1\alpha\phi_1$ so that the tridiagonal matrix elements are modified to $b_1 = \langle D_1\rangle + \langle\Sigma_1^+\rangle + D_1\alpha, a_1 = 0, c_1 = -\langle D_1\rangle.$

## Chapter 7

**7.1** $\dfrac{\phi_2}{\phi_1} = \dfrac{\Sigma_{1\to 2}}{\Sigma_{a2} + \Sigma_{2\to 1}}.$

**7.4** $\phi_1(z) = A_1 \exp(-\kappa_1 z), A_1 = \dfrac{2S}{1 + 2\kappa_1 D_1}, \kappa_1^2 = \dfrac{\Sigma_{a1} + \Sigma_r}{D_1},$

$\phi_2(z) = A_2[\exp(-\kappa_1 z) - \exp(-\kappa_2 z)], A_2 = \dfrac{\Sigma_r A_1}{D_2(\kappa_2^2 - \kappa_1^2)}, \kappa_2^2 = \dfrac{\Sigma_{a2}}{D_2}.$

**7.5** (a) $\phi_1(z) = A_1 \sinh \kappa_1(H - z), A_1 = \dfrac{S}{2\kappa_1 D_1 \cosh \kappa_1 H}, \kappa_1^2 = \dfrac{\Sigma_{a1} + \Sigma_r}{D_1},$

$\phi_2(z) = A_2 \sinh \kappa_1(H - z) + C_2 \sinh \kappa_2(H - z), \kappa_2^2 = D_2/\Sigma_{a2},$

$\dfrac{A_2}{A_1} = \dfrac{\Sigma_r}{D_2(\kappa_2^2 - \kappa_1^2)}, \dfrac{C_2}{A_2} = -\dfrac{\kappa_1 \cosh \kappa_1 H}{\kappa_2 \cosh \kappa_2 H},$

(b) $\dfrac{J_2^+(H)}{S/2} = \dfrac{\Sigma_r}{2D_1(\kappa_2^2 - \kappa_1^2)} \left( \dfrac{1}{\cosh \kappa_1 H} - \dfrac{1}{\cosh \kappa_2 H} \right).$

**7.6** $\phi_2(z) = C_1 \exp(-\Sigma_r z) + C_2 \sinh \kappa z + C_3 \cosh \kappa z, \kappa^2 = \Sigma_2/D_2,$

$C_1 = -\dfrac{\Sigma_r S_0}{D_2(\Sigma_r^2 - \kappa^2)}, C_2 = \dfrac{\Sigma_r C_1}{\kappa}, C_3 = -\dfrac{C_1}{\cosh \kappa H} \left( e^{-\Sigma_r H} + \dfrac{\Sigma_r}{\kappa} \sinh \kappa H \right).$

**7.7** $k^* = k_\infty/\lambda, M^2 = D_1/(\Sigma_{a1} + \Sigma_r).$

**7.8** With the core-reflector interface at $x = 0, \phi_1(x) = C_1 \exp(-\kappa_1 x)$, $\kappa_1^2 = (\Sigma_{a1} + \Sigma_r)/D_1$, $J_1(0) = \kappa_1 D_1 \phi_1(0), \phi_2(x) = C_2 \exp(-\kappa_2 x) + C_3 \phi_1(x)$, $\kappa_2^2 = \Sigma_{a2}/D_2$, yielding

$$\begin{pmatrix} \alpha_{11} & \alpha_{12} \\ \alpha_{21} & \alpha_{22} \end{pmatrix} = \begin{pmatrix} \kappa_1 D_1 & 0 \\ -\Sigma_r/(\kappa_1 + \kappa_2) & \kappa_2 D_2 \end{pmatrix}.$$

For each group, each direction, illustrated by the current $J_1 = \alpha \phi_{1/2} = \dfrac{D_0(\phi_{1/2} - \phi_0)}{\Delta R_0/2}$, leading to $J_1 = \dfrac{\alpha D_0}{D_0 - \alpha \Delta R_0/2} \phi_0$.

**7.9** $\dfrac{\Delta k}{k} = \dfrac{k' - k}{k} = (k_1 + \eta p f') - (k_1 + \eta p f) = \dfrac{f' - f}{f} k_2 = \left( \dfrac{\Sigma_{a2}}{\Sigma_{a2} - \Delta \Sigma_b} - 1 \right) k_2,$

$\Sigma_{a2} = \Sigma_{a2}^F + \Sigma_{a2}^M + \Sigma_b.$

**7.10** With a reflecting boundary condition with 25 uniform meshes for the core and 10 meshes for the top and bottom reflectors with a $B_r^2 = 3.39 \times 10^{-4}\ \text{cm}^{-2}$, the ONED code provides $k = 0.9698$ and axial power peaking factor $F_z = 1.462$.

**7.11** $\phi(x) = \dfrac{I_0 \Sigma_r}{\Sigma_a - D\Sigma_r^2}(e^{-\Sigma_r x} - e^{-\kappa x}), x_m = \dfrac{1}{\Sigma_r - \kappa} \ln \dfrac{\Sigma_r}{\kappa}, \kappa^2 = \dfrac{\Sigma_a}{D}.$

**7.12** $J_2^-(R) = \beta[J_2^+(R) + J_1^+(R)], \phi_2(r) = A\dfrac{\sinh \kappa r}{r}, \kappa^2 = \dfrac{\Sigma_a}{D},$

$J_1^+(R) = \dfrac{1}{4\pi R^2} \displaystyle\int_0^R \nu \Sigma_f \phi_2(r) 4\pi r^2 dr.$

Criticality: $\tanh \kappa R = C_0 \kappa \left( \frac{1-\beta}{4} + \frac{C_0}{R} \right)^{-1}$, with

$C_0 = D \left( \frac{\beta \nu \Sigma_f}{\Sigma_a} - \frac{1+\beta}{2} \right)$.

## Chapter 8

**8.2** $sn(s) - n(0) = -\frac{n(s)}{\Lambda} + \lambda C(s) + S(s), S(s) = \frac{S_0}{s}(1 - e^{-\tau s}), sC(s) - C_0 = -\lambda C(s) + \frac{n(s)}{\Lambda}, n(s) = \frac{n_1}{s} + \frac{(s+\lambda)S(s)}{s(s+1/\Lambda)}, n(\infty) - n(0) \simeq S_0 \lambda \Lambda \tau.$

**8.3** $\left[ s + \frac{1}{\Lambda} - \frac{\lambda}{\Lambda(s+\lambda)} \right] n(s) = n(0) \left[ 1 + \frac{1}{\Lambda(s+\lambda)} \right] + S(s), n(s) - \frac{n(0)}{s} = \frac{(s+\lambda)S(s)}{s(s+\lambda+1/\Lambda)} = \frac{\Lambda S(s)}{s[\Lambda + 1/(s+\lambda)]}.$

**8.4** A realistic power-level transient is illustrated to contrast with the approximate one-group variation in Figure 8.7.

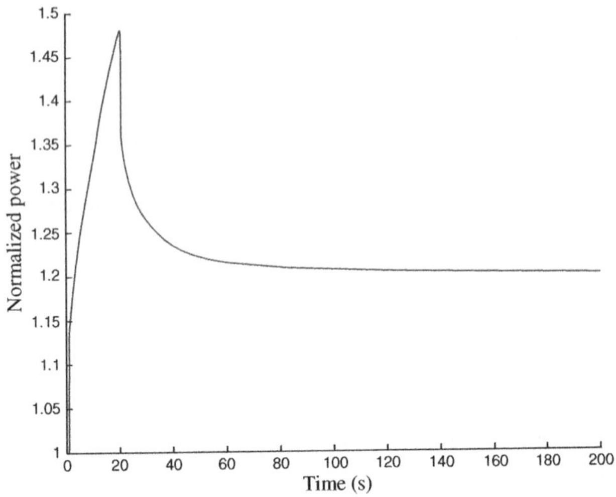


**8.8** (a) $K_0 = -S_0 \Lambda / n_0$,

(b) $\left[ \left( s - \frac{K_0 - 1}{\Lambda} \right)(s+\lambda) - \frac{\lambda}{\Lambda} \right] n(s)$

$= n_0 \left( \frac{1}{\Lambda} + s + \lambda \right) + \frac{s+\lambda}{s} [S_0 + S_1\{1 - \exp(-\tau s)\}], n(\infty) = n_0.$

**8.9** $C_1 = \frac{n_2(0)}{n_1(0)} + \frac{S_0 \Lambda}{n_1(0) K_1}, C_2 = -\frac{S_0 \Lambda}{K_1}.$

**8.11** For the rising portion: $n(t) = \left(\frac{Q_0}{\alpha_1} + n_0\right) e^{\alpha_1 t} - \frac{Q_0}{\alpha_1}, \alpha_1 = \frac{K_1 - 1}{\Lambda}$, $Q_0 = \frac{n_0}{\Lambda}(1 - K_0), Q = \frac{n_0 \Lambda (K_1 - K_0)}{(K_1 - 1)^2} \exp(\alpha_1 \tau)$.
For the falling portion: $n(t) = \frac{Q_0}{\alpha_0}[\exp\{\alpha_0(t - \tau)\} - 1]$ $+ n(\tau)\exp[\alpha_0(t - \tau)], \alpha_0 = \frac{K_0 - 1}{\Lambda} < 0$.

**8.12** (a) $n_{max} = 2.28$ GW, (b) $T_{max} = 2,622$ K, (c) $Q_{tot} = 6.82$ MJ, (d) $2.1 \times 10^{17}$ fissions, (e) FWHM = 2.63 ms.

**8.13**

$$G = G_2 G_1 = \left[\begin{array}{cc|c} A_2 & B_2 C_1 & B_2 D_1 \\ 0 & A_1 & B_1 \\ \hline C_2 & D_2 C_1 & D_2 D_1 \end{array}\right].$$

**8.14** $G = \frac{2s - 1}{(s - 2)(s + 1)}$.

**8.15** $n(\infty) = (1 + K_0 \lambda) n_0 = 1.1 n_0$.

**8.16** $\frac{dQ(t)}{dt} = \frac{K_0 - \alpha T}{\Lambda}, C\frac{dT(t)}{dt} \simeq n(0)Q(t), T = 2\pi\sqrt{\frac{\Lambda C}{\alpha n(0)}}$.

**8.18** $\mathrm{Re}(GK) = \frac{C(2 - \omega^2)}{(1 + \omega^2)(4 + \omega^2)}, \mathrm{Im}(GK) = \frac{-3C\omega}{(1 + \omega^2)(4 + \omega^2)}$, $\arg(GK) = \tan^{-1}\frac{3\omega}{\omega^2 - 2}$. System is stable with $C = 2$ and PM $= -\pi$.

## Chapter 9

**9.2** (a) With $\alpha E_1 \le E_2 < E_1 < \alpha^3 E_0 < \alpha E_0, q(E_2) = \int_{E_1}^{E_2/\alpha} dE' \int_{\alpha E'}^{E_2} dE'' \frac{F(E')}{E'(1 - \alpha)}$, $F(E) = \frac{Q}{\xi E}$ for $E > E_1$, $\frac{q(E_2)}{Q} = 1 - \frac{1}{(1 - \alpha)\xi}\left[1 - \frac{E_2}{E_1} - \alpha \ln \frac{E_1}{E_2}\right] = p(E_2)$.
NR approximation: $p(E_2) = \exp\left[-\int_{E_2}^{E_1} \frac{dE}{\xi E}\right] = \left(\frac{E_2}{E_1}\right)^{1/\xi}$.
(b) For $\alpha \ll 1, 1 - \frac{E_2}{E_1} = \epsilon \ll 1, p(E_2) = 1 - \frac{1}{\xi}\left(1 - \frac{E_2}{E_1}\right) \simeq (1 - \epsilon)^{1/\xi}$.

**9.3** (a) $\langle\sigma\rangle_a = \frac{\sigma_1 E_1}{E_2 - E_1} \ln \frac{E_2}{E_1}$, (b) $\langle\sigma\rangle_b = \frac{\sigma_1 E_1}{\ln E_2/E_1}\left(\frac{1}{E_1} - \frac{1}{E_2}\right)$, (c) $\langle\sigma\rangle_b > \langle\sigma\rangle_a$.

**9.4** $\Sigma_t(u)\phi(u) = \Sigma_{pM} + \Sigma_{sF}\phi(u)\lim_{\alpha_F \to 1}\int_{u-\Delta_f}^{u} \frac{e^{u'-u}du'}{1-\alpha_F}, \Delta_F \to 0.$

**9.5** $F(E) = \frac{Q}{(1-\alpha)E_0}\left(\frac{E_0}{E}\right)^{1/(1-\alpha)}, E_1 \le E \le E_0,$

$$F(E) = \frac{QE_0^{\alpha/(1-\alpha)}}{(1-\alpha)E_1^{\Sigma^*/[(1-\alpha)\Sigma_t]}} E^{-\Sigma_s/[(1-\alpha)\Sigma_t]}, \alpha E_0 \le E \le E_1.$$

## Chapter 10

**10.2** $\Delta\Sigma_a/\Sigma_a = -0.05 = \Delta \ln \rho = \Delta \ln \nu\Sigma_f = -\Delta \ln D, \delta\Sigma_a(x) = \Delta\Sigma_a \Delta x \delta(x),$ with $\Delta x = 0.05H, \delta\rho \simeq -2(0.05)^2(0.03)\Delta k/k = -15\,\text{pcm}.$

**10.3** (a) $q = B^2R^2/6$, (b) with $\delta \ln \Sigma_a = \delta\Sigma_a/\Sigma_a = \delta \ln \nu\Sigma_f = -\delta \ln D = \delta \ln \rho,$

$$w(r) = \frac{D[(\nabla\phi)^2 + B^2\phi^2]}{\beta\rho\int_V \nu\Sigma_f\phi^2 d\mathbf{r}} = w(0)\left(1 - \gamma\frac{r^2}{R^2}\right),$$

$$w(0) = \frac{6qD}{\beta\rho\nu\Sigma_f V(1 - 6q/5 + 3q^2/7)R^2},$$

$\gamma = 4q/3$, (c) $w(0) = 1.4 \times 10^{-5}/\rho, \delta K = \delta\rho w(0)(1 - 3\gamma/5)V = 2.18$ \$.

**10.4** $\delta\rho = -\frac{4\Sigma^*\pi^2}{\nu\Sigma_f\gamma H} \cdot \frac{1 - e^{-\gamma H}}{\gamma^2H^2 + 4\pi^2}.$

**10.5** $\frac{\delta k}{k} = \frac{(\Sigma_a - \nu\Sigma_f)\phi^2(x)\Delta x}{\nu\Sigma_f \int_0^H \phi^2(x)dx}.$

**10.6** With the normalized axial modes $\phi_0(x) = \sqrt{\frac{2}{H}}\cos\frac{\pi x}{H}$ and $\phi_1(x) = \sqrt{\frac{2}{H}}\sin\frac{2\pi x}{H}$, $a_1 = \frac{\langle\phi_1^\dagger, L\delta\phi_0\rangle}{1-\lambda_1} = \frac{8\Delta\Sigma_a}{3\pi\Sigma_a}\frac{\Sigma_a + 4DB^2}{3DB^2} \simeq 0.11$. This indicates that a small imbalance in the reaction rates between the top and bottom halves of a core could result in a large tilt in the axial power distribution or a large axial offset of power.

## Chapter 11

**11.1** $\Gamma_n = 0.08\,\text{meV}, \Gamma_\gamma = 39.82\,\text{meV}, \Gamma_f = 56.20\,\text{meV}, g = \frac{(2J+1)}{2(2I+1)} = \frac{3}{4}$ with $J = 1, \Gamma = 96.10\,\text{meV}, \sigma_0 = \frac{2.604 \times 10^6}{0.29562}\left(\frac{240}{239}\right)^2\frac{\Gamma_n}{\Gamma}g = 5.545\,\text{kb},$

$\sigma_t = \sigma_0 + 4\pi R^2 = 5.556\,\text{kb}$, $\sigma_\gamma = \sigma_0\Gamma_\gamma/\Gamma = 2.298\,\text{kb}$, $\sigma_f = \sigma_0\Gamma_f/\Gamma =$ $3.243\,\text{kb}$, $R = 9.46\,\text{fm}$.

**11.3** $\zeta_2 = \kappa_F a \coth \kappa_F a + \dfrac{a\Sigma_{aF}}{(b-a)\Sigma_{aM}}[b(1-\gamma)\kappa_M \coth \kappa_M b(1-\gamma) - 1]$, with $\gamma = a/b$ fixed, $\lim_{x\to 0} x \coth x = 1$.

**11.4** $\phi_F(x) = \dfrac{Q(b-a)}{\kappa_F D_F \sinh \kappa_F a} \cosh \kappa_F x, \phi_M(x) = \dfrac{Q}{2D_M}[(a-b)^2 - (x-b)^2] +$ $\dfrac{Q(b-a)}{\kappa_F D_F} \coth \kappa_F a, \zeta = \dfrac{\overline{\phi}_M}{\overline{\phi}_F} = a\Sigma_{aF}\left[\dfrac{\coth \kappa_F a}{\kappa_F D_F} + \dfrac{b-a}{3D_M}\right]$.

**11.5** $\phi_{NR}^{het}(u) = \dfrac{\Sigma_e + \Sigma_{pF}}{\Sigma_e + \Sigma_{pF} + \Sigma_{aF}} = \dfrac{1.221 + 0.418}{1.221 + 0.418 + 5.64} = 0.225, \phi_{NR}^{homog}(u) =$ $\dfrac{\langle\Sigma_s\rangle_{cell}}{\langle\Sigma_s\rangle_{cell} + \langle\Sigma_a\rangle_{cell}} = 0.298$.

**11.6** $\phi(x) = (Q/\Sigma_a)[1 - C\cosh \kappa(b-x)], C = [\cosh \kappa(b-a) +$ $2\kappa D \sinh \kappa(b-a)]^{-1}$,
absorption rate in fuel $= -J(a) = -J\_(a), f = \dfrac{-J(a)}{Q(b-a)} =$ $\dfrac{DC\kappa \sinh \kappa(b-a)}{\Sigma_a(b-a)}$.

**11.7** $\rho_{het}^{28} = \dfrac{0.4992 + 1/0.9382}{0.02347} = 66.6\,\text{b}, \rho_{homog}^{28} = \dfrac{0.7696}{0.02347 \cdot 0.3433} = 95.5\,\text{b}$.

**11.9** $I(\rho) = \dfrac{\langle\sigma(\rho)\rangle\Delta u}{1 + \langle\sigma(\rho)\rangle/\rho}, \rho = \Sigma_{sM}/N_F \neq f(u), I(\rho) = \displaystyle\int_{\Delta u} \sigma(u)\phi(u)du$,
$I(\infty) = \displaystyle\int_{\Delta u} \sigma(u)du = \langle\sigma(\infty)\rangle\Delta u, f_I = \dfrac{\langle\sigma(\rho)\rangle/\langle\sigma(\infty)\rangle}{1 + f_x\langle\sigma(\infty)\rangle/\rho} = \dfrac{f_x}{1 + f_x\langle\sigma(\infty)\rangle/\rho}$.

**11.10** $\langle\sigma\rangle = \dfrac{\sum_{i=1}^2 \beta_i f(\rho_i)\rho_i[\rho_i + \langle\sigma(\rho_i)\rangle]^{-1}\langle\sigma(\infty)\rangle}{1 - \sum_{i=1}^2 \beta_i f(\rho_i)[\rho_i + \langle\sigma(\rho_i)\rangle]^{-1}\langle\sigma(\infty)\rangle}$.

**11.11** $\Sigma_s^H = \int_{-1}^1 d\mu_0 \Sigma_s^H(\mu_0) = \Sigma_{s0}, \Sigma_{s1}^H = \Sigma_{s0}/6, \phi_1 = -\dfrac{iB\phi_0}{3\Sigma_{tr}}$,
with $\Sigma_{tr} = \gamma\Sigma_t - \dfrac{\Sigma_{s0}}{6}, \Sigma_a\phi_0 + \dfrac{B^2}{3\Sigma tr}\phi_0 = \dfrac{\nu\Sigma_f}{k_{eff}}\phi_0$, yielding
$k_{eff} = \nu\Sigma_f/(\Sigma_a + DB^2), D = 1/(3\Sigma_{tr}), B^2 = (\pi/H)^2$.

**11.12** (a) $S_0(B,u) \equiv S_0(u) = \delta(u), \Sigma_{sn}(u' \to u) = \Sigma e^{u'-u}P_n[e^{(u'-u)/2}]$, and
$\phi_n(u) = \phi_n^c(u) + K_n\delta(u), n = 0, 1$, yield
$\Sigma[\phi_n^c(u) + K_n\delta(u)] = A_{0n}\delta(u) + A_{0n}\rho_0(u) + 3A_{1n}\rho_1(u)$,
$\rho_n(u) = \rho_n^c(u) + A_{0n}e^{-u}P_n(e^{-u/2}), \rho_1^c(u) = 0$,

(b) $\dfrac{d\rho_0^c(u)}{du} = \Sigma\phi_0^c(u) - \Sigma\displaystyle\int_0^u e^{u'-u}\phi_0^c(u')du' = \Sigma\phi_0^c(u) - \rho_0^c(u)$, indicating $\rho_0^c(u) = q_0^H(u)$.

(c) $\dfrac{d\rho_0^c(u)}{du} = (A_{00} - 1)\rho_0^c(u) + A_{00}^2 e^{-u} + 3A_{10}^2 e^{-3u/2}$,

$\rho_0^c(u) = A_{00}[e^{(A_{00}-1)u} - e^{-u}] + \dfrac{6A_{10}^2}{1+2A_{00}}[e^{(A_{00}-1)u} - e^{-3u/2}]$.

(d) $\Sigma\phi_0^c(u) = \dfrac{d\rho_0^c(u)}{du} + \rho_0^c(u), \phi_0(u) = \phi_0^c(u) + \phi_0^u(u) = \phi_0^c(u) + A_{00}\delta(u)/\Sigma$,

$\phi_0^c(u) \simeq \dfrac{1}{\Sigma}\left[\left(1 - \dfrac{8}{9}y^2\right)e^{-y^2u/3} - \dfrac{1}{9}y^2 e^{-3u/2} + O(y^4)\right], y = B/\Sigma$,

indicating improvements over the Fermi age result $q(u) = \rho_0^c(u) = \Sigma\phi_0^c(u) = \exp(-DB^2u/\Sigma) = \exp(-y^2u/3) = \exp[-B^2\tau(u)]$.

**11.13** $D_1 = \dfrac{\Sigma_{tr,2} + \Sigma_{2\to1}\phi_2/\phi_1}{3(\Sigma_{tr,1}\Sigma_{tr,2} - \Sigma_{1\to2}\Sigma_{2\to1})}, D_2 = \dfrac{\Sigma_{tr,1} + \Sigma_{1\to2}\phi_1/\phi_2}{3(\Sigma_{tr,1}\Sigma_{tr,2} - \Sigma_{1\to2}\Sigma_{2\to1})}$.

## Chapter 12

**12.1** (a) With $T = 3$ yr, $N(T) = N(0)\exp(-\sigma_a\phi T), 1 - N(T)/N(0) = 0.882$,
(b) $\beta = \dfrac{\sigma_f}{\sigma_a}[1 - \exp(-\sigma_a\phi T)] = 0.61$.

**12.2** (a) $\Sigma_f\phi = 109.7\,\text{W}\cdot\text{cm}^{-3}/(200\,\text{MeV}\cdot\text{fission}^{-1})$
$= 3.43\times10^{12}$ fission $\cdot\,\text{cm}^{-3}\text{s}^{-1}$, $\phi = 5.28\times10^{13}$ neutron $\cdot\,\text{cm}^{-2}\text{s}^{-1}$,
$\sigma_X\phi = 7.92\times10^{-5}\,\text{s}^{-1}, \lambda^* = \lambda_X + \sigma_X\phi = 1.00\times10^{-4}\,\text{s}^{-1}, I_0^\infty = 7.60\times10^{-9}$ nucleus $\cdot\,\text{b}^{-1}\,\text{cm}^{-1}$, $X_0^\infty = 2.26\times10^{-9}$ nucleus $\cdot\,\text{b}^{-1}\,\text{cm}^{-1}$,
$X(t_m) = 3.28\times10^{-9}$ nucleus $\cdot\,\text{b}^{-1}\,\text{cm}^{-1}, \rho_X = \sigma_X X(t_m)/(\nu\Sigma_f) = -3.15\,\%\Delta k/k$.

**12.3** Fraction of initial $^{235}$U atoms fissioned $\beta = [1 - \exp(-\sigma_a^{25}\theta)]\sigma_f^{25}/\sigma_a^{25} = 0.7$, with $\theta = \phi T$, yields $\exp(-\sigma_a^{25}\theta) = 0.15$ and $\exp(-\sigma_a^{28}\theta) = 9.5\times10^{-3}$, $N^{49}(T)/N^{25}(0) \simeq \sigma_a^{28}\theta\beta/e = 0.16$, fifa = 0.78.

**12.5** (a) 23.6 s, (c) 7%.

**12.6** $P_d(t,T) = 13.3(P/Q)[t^{-0.2} - (t+T)^{-0.2}]\,\text{MeV}\cdot\text{s}^{-1}, Q = 200\,\text{MeV}\cdot\text{fission}^{-1}$.

**12.7** (b) $\xi = 0.026$.

**12.8** $\Psi = \dfrac{S+B}{F\cdot L} = 74$.

**12.9** For $^{14}C$ in the form of CO, $ALI = 2 \times 10^6 \mu Ci$, $DAC = 2 \times 10^6\ \mu Ci/2.4 \times 10^3\ m^3 = 700\ \mu Ci \cdot m^{-3}$, $Tx(^{14}C) = 1.4 \times 10^3\ m^3$. For $^{135}Cs$, $ALI = 10^3\ \mu Ci$, $DAC = 0.5\ \mu Ci \cdot m^{-3}$, $Tx(^{135}Cs) = 2 \times 10^6\ m^3$. Decreasing order of toxicity: $^{137}Cs$, $^{135}Cs$, $^3H$, $^{14}C$.

## Chapter 13

**13.2** Cross-sectional area of coolant channel $A = \pi(b^2 - a^2) = 89.1\ mm^2$, viscosity $\mu = 8.709 \times 10^{-5}\ kg \cdot m^{-1}\ s^{-1}$ at $\langle T_b \rangle = 576.4\ K$, $\rho = 0.721$ $Mg \cdot m^{-3}$, $W = 0.325\ kg \cdot s^{-1}$, $\langle v \rangle = 5.08\ m \cdot s^{-1}$, providing $Re = \rho \langle v \rangle D_h / \mu = 5.01 \times 10^5$, friction factor $f = 2.97 \times 10^{-3}$, $\Delta P_{fric} = 39.5\ kPa$, $\Delta P_{elev} = \rho g H = 30.1\ kPa$, $\Delta P_{local} = 139.3\ kPa$, for a total $\Delta P = 0.209\ MPa$, smaller than the DCD estimate of 0.271 MPa.

**13.3** (a) Core average enthalpy rise for $N = 41,448$ fuel rods $\Delta h = 252.4$ $kJ \cdot kg^{-1}$ and $\Delta T_b = 46.4\ K$, with heat capacity $C_p = 5.44\ kJ \cdot kg^{-1}\ K^{-1}$.
(b) Core average power density $P = 109.7\ MW \cdot m^{-3}$ of core volume and $270\ MW \cdot m^{-3}$ of fuel volume. Core average heat flux $q = PA/M = 642\ kW \cdot m^{-2}$, overall heat transfer coefficient $U_c = 1.12\ kW \cdot m^{-2}\ K^{-1}$, yielding fuel centerline temperature $T_c = T_b + q/U_c = 1,149\ K$.

**13.4** Power output for the reference coolant is proportional to $P_1 = WC_p(T_2 - T_1) = WC_p(T_m - T_1)/\theta_{fm} = WC_p(T_m - T_1)/2$ at the central channel. Power output with large heat capacity is proportional to $P_2 = 2MHU(T_m - T_1)/\pi$ and $P_2/P_1 = \sqrt{2}$.

**13.5** (a) Reference design: $q(z) = U[T_c(z) - T_b(z)]$, $T_c(z) - T_1 = \dfrac{q(z)}{U} + \dfrac{MHq_0}{\pi W C_p}\left(1 + \sin\dfrac{\pi z}{H}\right)$.

Alternate design: $q(z) = q_0 \dfrac{2}{\pi}$, $T_c(z) - T_1 = \dfrac{2q_0}{\pi u} + \dfrac{2Mq_0}{\pi W C_p}\left(z + \dfrac{H}{2}\right)$.

**13.8** $-\dfrac{1}{r}\dfrac{d}{dr}\left(rk\dfrac{dT}{dr}\right) = S$, $T(r) - T_1 = \dfrac{S}{4k}(R_2^2 - r^2)\left[1 - \dfrac{2}{(R_2^2 - r^2)/R_1^2}\ln\dfrac{R_2}{r}\right]$.
For the annular rod, $R_1/R_2 = 0.2$, $S = (P/L)/[\pi(R_2^2 - R_1^2)]$, $T_{max} = 1,724\ K$, to be compared with $T_{max} = T_s + (P/L)/(4\pi k) = 1,853\ K$.

**13.9** $-\frac{dp}{dz} - \frac{1}{r}\frac{d}{dr}(r\tau_{rz}) - \rho g = 0, \tau_{rz}(r) = \left(\frac{\Delta P}{\Delta z}\right)\frac{R}{2}\left(\frac{r}{R} - \lambda^2\frac{R}{r}\right) = -\mu\frac{dv_z(r)}{dr},$

$$\tau_{rz}(\lambda R) = 0, \tau_{rz}(r) = \left(\frac{\Delta P}{\Delta z}\right)\frac{R}{2}\left(\frac{r}{R} - \frac{1-\kappa^2}{2\ln\kappa}\frac{R}{r}\right),$$

$$v_z(r) = \left(\frac{\Delta P}{\Delta z}\right)\frac{R^2}{4\mu}\left(1 - \frac{r^2}{R^2} - \frac{1-\kappa^2}{\ln\kappa}\ln\frac{r}{R}\right),$$

$$W = \rho\langle v_z\rangle A = \left(\frac{\Delta P}{\Delta z}\right)\frac{\rho\pi R^2}{8\mu}\left[1 - \kappa^4 + \frac{(1-\kappa^2)^2}{\ln\kappa}\right].$$

**13.10** $\rho C_p v_z(r)\frac{\partial T(r,z)}{\partial z} = k\frac{1}{r}\frac{\partial}{\partial r}\left(r\frac{\partial T}{\partial r}\right) + Q,$

$$\frac{dT_b(z)}{dz} = \frac{AQ}{WC_p}\psi = \frac{\partial T(r,z)}{\partial z} = \text{constant}, \psi = \frac{\pi R^2 Q + 2\pi R q}{\pi R^2 Q}, q < 0.$$

$$T(r,z) - T(0,z) = \frac{Qr^2}{4k}\left[2\psi\left(1 - \frac{r^2}{4R^2}\right) - 1\right],$$

$$T_b(z) - T(0,z) = \frac{R^2 Q}{48k}(7\psi - 4), \theta(r) = \frac{6r^2}{5R^2}\left(4 - 3\frac{r^2}{R^2}\right),$$

$$\theta_{max} = \theta(\sqrt{2/3}R) = \frac{8}{5}.$$

**13.11** $\theta(z) = \frac{T_f(z) - T_1}{T_2 - T_1} = \frac{1}{2}\left(1 + \sin\frac{\pi z}{H}\right) + A\cos\frac{\pi z}{H}, A = \frac{\pi W C_p}{2MHU},$

$T_f(z)$ = fuel centerline temperature.

For the smooth surface, $\theta_{sm} = 1.5$, with $z_m = H/6, A_s = \sqrt{3}/2$.

For the roughened surface, $A_r = \frac{A_s}{3}, \theta_{rm} = \frac{1}{2}\left(1 + \frac{2}{\sqrt{3}}\right) = 1.077,$

$$\frac{P_r}{P_s} = \frac{T_{mp} - 1}{\theta_{rm}}\frac{\theta_{sm}}{T_m p - T_1} = \frac{3}{2}\frac{2}{1 + 2\sqrt{3}} = 1.39.$$

**13.12** $\Delta p = p_1 - p_2 = \frac{W^2}{2\rho}\left(\frac{1}{A_2^2} - \frac{1}{A_1^2}\right) < 0.$

**13.13** For the two-phase region, $W\frac{dh_s(z)}{dz} = MU(T_p - T_{sat}), h_s(H) - h_f = \gamma(T_p - T_{sat})(H - z_0) = x_e h_{fg}$, with $\gamma = MU/W$, providing $z_0$. For the single-phase region, $T_s(z) = T_p - (T_p - T_{in})\exp(-\gamma z/C_p), z \le z_0$.

**13.14** With heat flux $q(z) = \cos\frac{\pi z}{H} = h[T_s(z) - T_b(z)], \theta_s(z) = \frac{\pi W C_p}{hMH}\cos\frac{\pi z}{H} + \sin\frac{\pi z}{H}$ for Case B. The hot spots occur at $z_{m,B} = \frac{\pm H}{6}, \theta^B_{s,m} = \left[\sin\frac{\pi z_m}{H}\right]^{-1}$, with the power output $P_B = 2WC_p\frac{T^B_{s,m} - T_1}{\theta^B_{s,m}}, \frac{P_B}{P_A} = 1 + \sin\frac{\pi z_m}{H} = \frac{3}{2}$.

**13.15** $T_c(z) - T_1 = \dfrac{q(z)}{U_c} + \dfrac{M}{WC_p}\displaystyle\int_{-H/2}^{z} q(z')dz', q(z) = C_2 \exp\left(-\dfrac{MU_c}{WC_p}z\right)$
$= \dfrac{P(z)}{M}.$

**13.16** $-k\dfrac{1}{r^2}\dfrac{d}{dr}\left(r^2\dfrac{dT}{dr}\right) = S, T_m - T_s = \dfrac{SR^2}{6k}, P_{th} = \dfrac{4\pi R^3 S}{3} = 8\pi kR(T_m - T_s),$
$P_{th} = 26$ kW, $P_e = 1.3$ kW, $R = 0.11$ m.

**13.17** (a) $T_m - T_s = \dfrac{Q}{k\kappa^2}(\cosh\kappa a - 1)$, (b) $\dfrac{P}{A} = \dfrac{Q\sinh\kappa a}{\kappa} = q(a)$.

## Chapter 14

**14.1** $-35\,\text{pcm}\cdot\text{K}^{-1}$.

**14.2** 960 ppm.

**14.3** (a) For the initial design, $f = 0.76, f' = 0.7656, p' = 0.7945, \alpha_V = 8.0\,\text{pcm}\cdot(\%\text{void})^{-1}$. (b) For the alternate design, $f' = 0.7442, \alpha_V = -5.0\,\text{pcm}\cdot(\%\text{void})^{-1}$.

**14.4** $P(t) = 0.073$, when $F(t) = 0.05$.

**14.5** (a) $\delta\Sigma_a(z) = -2\left(\beta + \dfrac{1}{2\langle T_b\rangle}\right)\Sigma_{a,Li}\theta_b(z)\delta\langle T_b\rangle, \alpha_T = -\left(\beta + \dfrac{1}{2\langle T_b\rangle}\right)\dfrac{\Sigma_{a,Li}}{\nu\Sigma_f} =$
$-8.7\,\text{pcm}\cdot\text{K}^{-1}$, (b) $\alpha_p = \alpha_T(\langle T_b\rangle - T_1) = -0.013,$

## Chapter 16

**16.4** Objective function $J = B^2 = \left(\dfrac{\nu}{R}\right)^2 + \left(\dfrac{\pi}{H}\right)^2, \nu = 2.405$, subject to $V = \pi R^2 H =$ constant. Augment $J$ with $\lambda(\pi R^2 H - V)$ to form $J^*$, $H = 1.85R$.

**16.6** $\dfrac{dx_1}{dt} = x_2, \dfrac{dx_2}{dt} = u - g, 0 \le u \le u_m, H = C_1x_2 + (C_2 - C_1t)(u - g),$

minimal-time descent: $u = 0$, until $t_s = \left[\dfrac{2(u_m - g)h}{u_m g}\right]^{1/2}$ and switch to $u = u_m$.

**16.7** $H = p_0ux_2 + p_1x_2 + p_2(u - g)$, one switching when $p_0x_2 + p_2 = 0$.

**16.10** $x^+ = (\sigma_y^2 x^- + \sigma_x^2 y)/(\sigma_y^2 + \sigma_x^2)$.

**16.11** Initiating the control $K_c$ at $t_m$, where $n(t_m) = n_{max}$, requires $K(t) = 0$ or $K_c(t) = \alpha T(t) = \alpha T(t_m) + \alpha n_{max}(t - t_m)/C_p$.

**16.12** (a) $\dfrac{dx_1}{dt} = x_2, \dfrac{dx_2}{dt} = u - C_0 x_2, H = C_1 x_2 + p_2(t)(u - C_0 x_2)$,
$p_2(t) = (C_1/C_0 + C_2 e^{C_0 t})$. With $p_2(t)$ a monotonic function of time, there can be at most one switching.

## Chapter 17

**17.3** (i) Marshak condition satisfies the vacuum boundary condition in an integral sense: $\int_{-1}^{0} \mu\psi(0,\mu)d\mu = J\_(0) = 0$, representing the requirement that the return current from vacuum vanishes.
(ii) Mark condition satisfies the boundary condition for some values of the angle, e.g. $\mu = -1/\sqrt{3}$, corresponding to the negative root of $P_2(\mu) = (3\mu^2 - 1)/2 = 0$.

**17.4** $\phi(0) = \phi_0(0) = 2\phi_1(0) = 2J(0)$.

**17.5** (a) $\phi(x) = \displaystyle\int_0^1 \frac{S}{2\mu} \exp\left(-\frac{\Sigma_a x}{\mu}\right) d\mu, x > 0$, (b) $\dfrac{2}{3\Sigma_a^2}$.

**17.6** (a) $F = \dfrac{\phi_F(a)}{\overline{\phi}_F} = -\dfrac{\phi_F(a)}{J(a)}\dfrac{\Sigma_{tF} V_F}{A_F} = \dfrac{\Sigma_{tF} V_F}{\alpha A_F}$,

(b) $\alpha = -\dfrac{J(a)}{\phi(a)} = \dfrac{1}{2}\dfrac{1 - 2E_3(2\Sigma_{tF} a)}{1 + 3E_4(2\Sigma_{tF} a)}$.

**17.7** $\phi(a + b) = \dfrac{S_0}{2\Sigma_a}\dfrac{1 - e^{-2a\Sigma_0}}{2a\Sigma_0}$.

**17.8** $\phi(x_0 \to x) = \dfrac{1}{2} E_1(\Sigma |x - x_0|)$.

**17.9** $\beta_{FC}^* = (a/b)\beta_F^*$.

**17.10** With the transport kernel $\phi(x_0 \to x)$ from Problem 17.8, the collision probability $P_{i \to j} = \dfrac{1}{2h_i\Sigma}[E_3\{\Sigma(x_j - x_i - h_i)\} - E_3\{\Sigma(x_j - x_i)\}$
$- E_3\{\Sigma(x_j + h_j - x_i - h_i)\} + E_3\{\Sigma(x_j - x_i + h_j)\}]$.

**17.11** $\mu_j \dfrac{d\psi_j(z)}{dz} + \Sigma\psi_j(z) = \dfrac{\Sigma_s}{2}[\psi_1(z) + \psi_2(z)], j = 1, 2$. Combining the equations yields $\mu_1 \dfrac{d^2\psi^+(z)}{dz^2} - \dfrac{\Sigma\Sigma_a}{\mu_1}\psi^+(z) = 0, z_0 = \dfrac{\mu_1}{\sqrt{\Sigma\Sigma_a}} = \dfrac{1}{\sqrt{3\Sigma\Sigma_a}}$.

**17.12** (a) Integrating $\mu \dfrac{\partial \psi(x,\mu)}{\partial x} + \Sigma\psi(z,\mu) = q$, over $\Delta$ at $x_i$ yields the difference equation for $\mu_j > 0$, $\psi_{i+1/2} = \psi_{i-1/2} f_i + q_i(1-f_i)/\Sigma_i$, with $f_i = e^{-\epsilon_i}$ and $\epsilon_i = \Sigma_i \Delta/\mu_j$, $\overline{\psi}_i = \dfrac{q_i}{\Sigma_i} - \dfrac{1}{\epsilon_i}(\psi_{i+1/2} - \psi_{i-1/2})$, (b) for $\mu_j > 0$, $f_i < 1.0$, guaranteeing $\psi_{i+1/2}$ and hence $\overline{\psi}_i$ to be positive if $\psi_{i-1/2} > 0$.

**17.13** $(1 + C_i)\psi_{i+1/2} - (1 - C_i)\psi_{i-1/2} = q_i\Delta/\mu_j$, with $C_i = \Sigma_i\Delta/(2\mu_j)$.

17.12

# Index



*Nuclear Reactor Physics and Engineering*, Second Edition. John C. Lee.
© 2025 John Wiley & Sons, Inc. Published 2025 by John Wiley & Sons, Inc.
Companion website: www.wiley.com/go/LeeNRPE2

## *d*

## e

## g

## m

Printed and bound by CPI Group (UK) Ltd, Croydon, CR0 4YY
17/09/2026
14964333-0007